Introduction to Exploration Geochemistry

Introduction to
Exploration Geochemistry

A. A. Levinson
DEPARTMENT OF GEOLOGY
UNIVERSITY OF CALGARY
CALGARY, ALBERTA, CANADA

Applied Publishing Ltd.
Wilmette, Illinois, U.S.A.

Typeset in Canada by Alcraft Printing Co. Limited,
Calgary, Alberta.

Printed in the United States of America by Photopress
Incorporated, Broadview, Illinois.

Library of Congress Catalog Card Number: 73-92843

ISBN 0-915834-01-4

DEDICATED TO

E. Wm. Heinrich

The late C. B. Slawson

The late R. M. Denning

The University of Michigan

Preface

The methods employed in the search for ore deposits and hydrocarbon accumulations have changed considerably in the last few decades. Today, prospecting is an exact science and in most parts of the world is being carried out using the latest advances in geology, geophysics and geochemistry. The purpose of this book is to present an up-to-date introduction to exploration geochemistry sufficient to enable the reader to understand the literature and to discuss the subject intelligently with experienced individuals or consultants. The subject matter is presented in such a way that, apart from the usual undergraduate background in mineralogy, petrology and economic geology, only a one-year course in inorganic chemistry is considered essential for its understanding except for very specialized sections, such as those on statistics (Chapter 12) and the search for crude oil and natural gas (Chapter 13). The book is intended to be of value to (1) students, as an introductory course; (2) geologists, including those in administrative capacities, who wish to obtain a degree of proficiency in exploration geochemistry; (3) those in allied fields such as geophysics; and (4) serious and experienced prospectors. It is also possible that some economic geologists and practising geochemists will find the discussions of value as an up-to-date review, or appreciate a different approach, but the book is not directed towards the experienced exploration geochemist.

This book is a compromise between an exhaustive treatise and a generalized summary. In order to limit the book to a reasonable size, not all subjects are treated in as much detail as would perhaps have been desirable. In this field which embraces parts of several sciences, the literature is huge and the annual increment of relevant papers is increasing at a rapid rate. In view of this rapid increase in knowledge no single book could hope to approach full and detailed coverage, and thus one must be selective. In particular, weathering processes, soil formation, theories of ore genesis, pH and Eh, and other topics in chapters 2 and 3 deserve far greater space than it has been possible to give them. Nevertheless, it is hoped that the principles are presented in an adequate manner, and that the student will be stimulated to read some of the excellent new books cited on these subjects.

All teachers have their preferences in the matter of presentation of material. My own dictate that, at an early stage, students should be introduced to colorimetric tests for the common metals and have at least superficial exposure to the more important analytical methods used today in exploration geochemistry, such as atomic absorption spectrometry. This is because they are then better able to grasp the significance of trace elements in the parts per million range, element associations, problems of contamination, and will also develop a feeling for the historical development of the science. The value of this approach, which is reflected in the presentation of a simplified discussion of colorimetric and other analytical methods in Chapter 6 and Appendix B, has been borne out by the favorable reception it has received from students. At the present time, discussions of the analytical methods most commonly used in exploration geochemistry are appropriate as developments in analytical techniques seem to have reached a temporary plateau. Critical readers will notice that metric and English units are used interchangeably throughout this book, primarily because the illustrations selected use both systems. Students must quickly learn to correlate the two systems and, to assist them, conversion tables are presented in Appendix A.

Even though this is primarily an introductory textbook, a serious effort has been made to document statements and the source of illustrations. These references have not been screened for priority, and so there may be many instances in which credits for ideas and concepts have been inadvertently overlooked. Where two articles or references can be cited for the same data, or equivalent illustration, the more recent is generally given. Throughout the book an attempt has been made to present the material factually and in an unbiased manner but, in some cases, based on my own experiences, preferences and prejudices, a possibly less than objective stand has been taken, or at least implied. Also, wherever possible, North American (including Mexican) or Australian examples have been used. To most North American students with whom I have been in contact, Soviet and African geology and geography are complete unknowns.

The subject of exploration for hydrocarbons is very specialized and, accordingly, I am pleased that Dr. Brian Hitchon, Senior Research Officer at Alberta Research, Edmonton, Alberta, agreed to prepare a chapter on this topic. Similarly, Professor Richard B. McCammon, of the University of Illinois at Chicago Circle, has contributed the chapter on statistics as applied to exploration geochemistry. Both Dr. Hitchon and Professor McCammon have enjoyed complete freedom in the selection of their material, method of presentation, and interpretation of the present state of their specialties with respect to exploration geochemistry. I have complete confidence in their judgment and, accordingly, I accept full responsibility for all their statements.

Calgary, Alberta, Canada   *A. A. Levinson*
November, 1973

Acknowledgments

The preparation of this book has benefited immensely from the assistance, advice and cooperation of many individuals. At the outset it is important to acknowledge that Chapter 12 and Appendix C, on statistics and computers, were written by Prof. R. B. McCammon, and Chapter 13, on the application of geochemistry to the search for crude oil and natural gas, was written by Dr. Brian Hitchon. In both these subjects I lack expertise and am therefore extremely grateful to my two colleagues for their efforts. In addition, Dr. Hitchon thoroughly reviewed the first three chapters and Chapter 12.

In the way of critical reviews, I am deeply indebted to two individuals in particular. Dr. K. Bloomfield, Institute of Geological Sciences (Overseas Division), London, critically read the first 11 chapters in their entirety. His recommendations, careful attention to the manner of presentation, and advice on editorial matters, have been a great help in putting the manuscript into its present form. Further, the book has benefited from Dr. Bloomfield's comments based on almost 20 years experience with African exploration projects. Similarly, I am indebted to Dr. T. C. Mowatt, Division of Geological Survey, Department of Natural Resources, College, Alaska, who also read the first 11 chapters and offered his comments, criticism and advice. Dr. Mowatt's observations on geochemical exploration in northern latitudes, particularly in Alaska, have been incorporated into parts of Chapter 11.

Numerous colleagues at the University of Calgary have critically evaluated those parts of the book concerned with their specialties. I am grateful to Prof. P. Bayliss for his stimulating comments and efforts on many parts of chapters 2 and 3. Prof. J. W. Nicholls read parts of Chapter 2, and Prof. G. D. Osborn reviewed the glaciology section of Chapter 11. Prof. S. A. Harris spent many hours with me discussing aspects of soil geochemistry and limnology, much of which is incorporated within chapters 3 and 11. Prof. C. D. Bird assisted with the sections on bogs, muskeg, and limnology of Canadian lakes.

Dr. N. W. Rutter (Geological Survey of Canada) was extremely helpful with the preparation of the simplified soil map of Canada, with discussions on soils, and with his reading of the soil section of Chapter 2. Prof. P. L. Cloke (University of Michigan) reviewed the sections on pH – Eh in Chapter 3. Prof. J. A. C. Fortescue (Brock University) reviewed Chapter 10, and that part of Chapter 1 which discusses vegetation surveys. Dr. R. J. E. Brown (National Research Council of Canada) was of great assistance in the preparation of those sections concerned with permafrost.

Prof. McCammon wishes to acknowledge reviews of his chapter by Dr. E. Dahlberg, Prof. J. E. Klovan, and by Dr. Brian Hitchon. Dr. C. R. Evans, Dr. B. Tissot, Mr. W. M. Zarrella, Dr. M. H. Horn, Dr. R. Green and Prof. N. C. Wardlaw all reviewed Dr. Hitchon's chapter. Mrs. E. G. Pippard assisted with the preparation of Appendix B and the color plate. Mr. E. Okon graciously undertook the task of reading a large part of the proofs.

Drafting and photographic work was done by the Department of Communications Media, University of Calgary. I appreciate their cooperation on these aspects of the project. Very special thanks go to Mrs. Marilyn E. Croot who produced all the original drawings, and modified others taken from the literature. Typing was by Mss. P. R. O'Hara, L. M. Kind, A. Andre, M. D. King, B. E. Horne and G. E. Whyte, and I thank them, one and all, for their dedication to the task.

This book was begun during 1972 while I was on sabbatical leave. During that year I spent several months at the National University of Mexico, Mexico City, where the manuscript was started. I appreciate the cooperation of Professors D. A. Córdoba, L. de Pablo and A. Obregon, and Dr. J. H. Butler, and the hospitable and stimulating atmosphere they provided. I also appreciate the warm hospitality provided by various governmental agencies (particularly C.S.I.R.O.), universities and private laboratories in Australia which I also visited during 1972. Among the many people who assisted me in the collection of information on geochemical exploration in Australia I would especially like to thank E. Bettany, C. R. M. Butt, W. Ewers, A. J. Gaskin, W. R. Hesp, R. R. Keays, A. L. Mather, E. H. Nickel, K. Norrish, D. Sampy, D. J. Swaine, G. F. Taylor, S. R. Taylor, J. Wilmhurst and D. O. Zimmerman. And, last but not least, I sincerely appreciate the many scientific discussions I have had with Prof. Harold Bloom, Colorado School of Mines, over a 15 year period and, in fact, it was he who first introduced me to modern methods of geochemical exploration.

I wish to thank the Geological Survey of Canada and the U.S. Geological Survey for permission to use many illustrations which originated with their organizations. The Canadian Institute of Mining and Metallurgy, The Institution of Mining and Metallurgy (London), and the Colorado School of Mines Research Institute also generously permitted me to use numerous illustrations from their publications. One or a few illustrations were taken from various other journals and books, by kind permission of their editors and publishers.

Generous financial support for various aspects of the preparation of this book (e.g., field trips, laboratory work) came from the National Research Council of Canada. In addition, I sincerely appreciate support for secretarial and drafting services from the President's Fund, University of Calgary.

Contents

1 / *Introduction*

INTRODUCTION

Geochemistry, in its broadest sense, is the science concerned with the chemistry of the Earth and, like all sciences, it has many divisions which grade imperceptibly into each other. Exploration geochemistry is one of the major divisions; others include geochronology, stable isotope geochemistry, organic geochemistry, sedimentary geochemistry, and hydrogeochemistry, as well as some newly recognized fields such as lunar and environmental geochemistry. In recent years, regardless of their specialty, most geochemists have chosen studies which have one point in common, namely, an attempt to solve some geological problem by means of chemistry. Therefore, in a modern sense, the aim of all divisions of geochemistry is to solve geological (including lunar and environmental) problems.

Exploration geochemistry, also called geochemical prospecting, is the practical application of theoretical geochemical principles to mineral exploration. Its specific aim is to find new deposits of metals and non-metals, or accumulations of crude oil and natural gas, and to locate extensions of existing deposits, by employing chemical methods. The methods used involve the systematic measurement of one or more chemical elements or compounds, which usually occur in very small amounts. The measurements are made on any of several naturally occurring, easily sampled, substances such as rocks, stream sediments, soils, waters, vegetation, glacial debris or air.

For convenience, exploration geochemistry is usually divided into two broad categories: (1) the search for mineral deposits, both metallic and non-metallic, and (2) the search for crude oil and natural gas. Each of these categories is highly specialized and they are almost mutually exclusive at present, yet in both the geochemist is generally looking for concealed natural resources. In both cases the objective is the same, that is, to find some dispersion of elements or compounds sufficiently above normal to be called an *anomaly,* which, it is hoped, may indicate mineralization or hydrocarbon accumulations. However, in the search for mineral deposits and for hydrocarbon accumulations, different parameters are measured, significantly different instrumentation is used, and to a lesser extent,

1

emphasis is placed on different materials to be sampled. Accordingly, the discussion of exploration geochemistry as applied to hydrocarbons is treated separately in the last chapter of this book. It is recognized that this arbitrary separation may not be ideal because philosophically the exploration for any natural resource must be based on the same fundamental geological and geochemical principles (for example, the importance of structural geology, and the migration of fluids). Furthermore, when it is realized that a large percentage of all metals now mined are either in sedimentary rocks, or in rocks which were originally sedimentary but are now metamorphosed, then it is apparent that there should be more integration of exploration for both metals and hydrocarbons. With many of the large petroleum companies now firmly committed to diversification into non-hydrocarbon natural resources, and with both hydrocarbon and metal exploration groups sometimes under the same supervision, we may expect much closer liaison between them in the future, to the mutual advantage of both.

In the past, exploration was largely carried out by prospectors and geologists who tended to concentrate their efforts on outcrops, or related expressions of mineralization such as heavy minerals in stream sediments, especially in well-known mineralized belts of the world. This resulted in the discovery of many ore bodies, most of which could be recognized visually. Similar success also applied to some of the larger oil and gas fields because these too had visual expression, such as oil and gas seeps and occurrences of tar. However, the era of visual observation is essentially drawing to a close and most of the obvious mineral deposits of the world have probably been found. This fact is well illustrated by the statement of Ing. D. A. Córdoba (Director, Institute of Geology, National University of Mexico, Mexico City) that in Mexico about 90% of the metals mined today are from mines found by the Spaniards. It is also said that the locations of most of the mines operating today in those parts of Europe which were within the Roman Empire, were known to the Romans. Clearly, there is thus a need for a radical, imaginative new approach to mineral exploration, which geochemistry, along with other sciences such as geophysics, is attempting to fill. But as will be seen below, the application of chemistry to exploration is not entirely new — the newness simply involves the application of sensitive, sophisticated instrumentation in the "micro" chemical range, as opposed to the previous "mega" identification of visible mineral deposits. The so-called modern approach to what is really an old technique continues with such additional aids as airborne methods of sample collection and analysis, electronic data processing, and enlightened methods of interpretation. In all cases, the one underlying factor is the attempt to recognize some sort of chemical anomaly which will be indicative of mineralization of economic value.

As time progresses, it is apparent that mineral deposits of lower grade will become successively more economic. For example, about 25 years ago copper prospects were not economic unless the ore contained about 1.5% copper in sufficient tonnages. Today the large open-pit porphyry copper deposits in British Columbia and other favorable locations are economic at about 0.4% copper. From a practical point of view, these low-grade deposits are extremely difficult to locate visually especially if they are covered by soil, glacial debris or other types of overburden, and it is in just this type of situation that geochemistry works well. It is of interest that the economic success of these low-grade, high-tonnage porphyry copper deposits has essentially eliminated from further consideration any new vein-type discoveries of copper unless they are high grade and/or amenable to large volume, open-pit operations. (In this book 'porphyry copper' is used in the sense of a large low-grade copper deposit in which the copper minerals occur in disseminated grains and/or veinlets. The deposit may occur in any number of rock types, but quartz-bearing igneous rocks are always in close association).

Exploration geochemistry is ideally suited to the search for low grade deposits, especially those which are difficult to locate (e.g., porphyry copper) or impossible to recognize visually (e.g., the gold deposits at Carlin, Nevada). This is because geochemistry is a *direct* method which measures the actual element being sought, or an associated element, in order to detect an anomaly. There are, of course, many other exploration techniques, but at present all of them are indirect. Geophysics, for example, is generally based upon the measurement of some primary or induced secondary physical property of elements or minerals and, as far as can be determined at present, it will remain an indirect method. Geological methods, although of great value in selecting favorable areas and structures, are also indirect in that they cannot identify a concealed mineral deposit or chemical anomaly. Nevertheless, it must be stressed that geochemistry is just one method of exploration which should be used in conjunction with as many other techniques as possible. Geophysical methods, with which geochemical methods are very closely associated, have found and will continue to find geophysical anomalies, of which about 1 in 10,000 will become mines. With respect to actual prospects, Aho and Brock (1972, p. 27) state "It has been often said that one prospect in 1000 becomes a mine and perhaps one in 200 is explored intensively, some at costs of millions of dollars". Geochemistry will be able to assist in many cases (but not all) in the interpretation and evaluation of these geophysical anomalies, prospects and favorable geological areas. In practice, it is unusual for geochemistry to be the only technique used in an exploration project, and at some stage geophysics and geology are generally combined with geochemistry.

A recent development is the use of geochemical surveys in areas selected from study of remote-sensing data, particularly satellite photographs and related imagery. The Earth Resources Technology Satellite Program of the U.S. National Aeronautics and Space Administration has provided the opportunity to use data at an unprecedented scale, enabling new perspectives to be brought to bear in the continuing search for natural resources. Large-scale structural and lithological features are discernible, often suggestive of geological situations deserving of further, more detailed scrutiny by other means, including geochemical methods.

HISTORY

Several summaries of the history of geochemistry (Hawkes, 1957; Boyle, 1967; Boyle and Garrett, 1970) have pointed out that the beginnings of geochemical prospecting originated in antiquity. The prospector who panned for gold was, like the modern exploration geochemist, following dispersion patterns. Similarly, the ancients who searched for iron stains or gossans were, in fact, looking for clues to buried deposits. The main difference between the ancient prospector and the modern exploration geochemist is that the former was using mineralogical observations, whereas the latter uses chemical analyses.

There are frequent references in the classical and renaissance literature to the use of chemical analyses in prospecting, particularly in the analysis of natural waters and springs, and their residues. Hawkes (1957, p. 313) quotes a paragraph from Biringuccio published in Venice in 1540, and Boyle (1967) has selected a passage from Agricola's *"De Re Metallica"* (pp. 33-34) first published in 1546 in Basel, which states:

> "Now I will discuss that kind of minerals for which it is not
> necessary to dig, because the force of water carries them out of
> the veins. Of these there are two kinds, minerals — and their
> fragments — and juices. When there are springs at the outcrop
> of the veins from which, as I have already said, the above-
> mentioned products are emitted, the miner should consider
> these first, to see whether there are metals or gems mixed with
> the sand, or whether the waters discharged are filled with
> juices. In case metals or gems have settled in the pool of the
> spring, not only should the sand from it be washed, but also
> that from the streams which flow from these springs, and even
> from the river itself into which they again discharge. If the
> springs discharge water containing some juice, this also should
> be collected; the further such a stream has flowed from the
> source, the more it receives plain water and the more diluted

does it become, and so much the more deficient in strength. If the stream receives no water of another kind, or scarcely any, not only the rivers, but likewise the lakes which receive these waters, are of the same nature as the springs, and serve the same uses." And further, "The waters of springs taste according to the juice they contain, and they differ greatly in this respect. There are six kinds of these tastes which the worker especially observes and examines; there is the salty kind, which shows that salt may be obtained by evaporation; the nitrous, which indicates soda; the aluminous kind, which indicates alum; the vitrioline, which indicates vitriol; the sulphurous kind, which indicates sulphur; and as for the bituminous juice, out of which bitumen is melted down, the colour itself proclaims it to the worker who is evaporating it" (in, Hoover and Hoover, 1912).

As Boyle (1967) points out, these are remarkable statements about hydrogeochemical methods written in the middle of the 16th century. Other statements could be quoted from Agricola to show that the medieval men had knowledge of what we presently call the thermal effects caused by the oxidation of sulphide minerals over veins, and the use of biogeochemical methods in exploration. They also knew the significance of indicator plants, particularly the discoloration and physical changes in vegetation resulting from the toxic effects of excess trace elements in soils associated with mineralized zones. Botanical associations with mineralization have apparently been known for a very long time, at least since the 8th or 9th century A.D., according to a source quoted by Boyle (1967). The Chinese had observed that certain species of plants occur near deposits of silver, gold, copper and tin, and they knew that this fact could be used in exploration. They were apparently aware that plants contained metals, and they extracted mercury from certain species. Reports on the use of chemical and biogeochemical methods of prospecting become increasingly numerous in the literature from the 17th century onwards, becoming particularly abundant from the beginning of the 20th century. Clearly, present day methods of exploration geochemistry are not based entirely on new concepts, but rather on new, sensitive methods capable of extracting more information of high quality from materials which, in some cases, had been used by prospectors for more than 500 years.

Modern methods of exploration geochemistry were first used in the early 1930's in the U.S.S.R., and shortly thereafter in the Scandinavian countries, particularly Sweden. The first large-scale exploration programs were begun in 1932 by Soviet geologists who had perfected emission spectrographic analytical methods, as well as sampling procedures, for routine geochemical soil sampling surveys. Their first surveys were concerned with the search for tin, but these were quickly followed by other "metallometric" (soil and weathered rock) surveys for copper, lead, zinc, nickel, tungsten and other metals. Much of the analytical work was carried out in the field or in base-camp laboratories, using semi-quantitative emission spectrographic data. About the same time (middle to late 1930's), research was begun in the U.S.S.R., Sweden and Finland, into the use of vegetation for prospecting purposes, including both the analysis of plants and the recognition of indicator plants.

The geochemical prospecting initiated in the U.S.S.R. and Scandinavia was based on more than the ancient and medieval writings mentioned previously. From the beginning of the 20th century, strong schools of fundamental geochemistry were being built up in these countries led by such famous Soviet geochemists as V. I. Vernadsky, who did pioneering work in biogeochemistry, and his student A. E. Fersman, who was the first to stress the importance of primary and secondary halos associated with ore deposits. Other notable Soviet geochemists who contributed significantly in various ways included I. I. Ginzburg, A. P. Vinogradov and D. P. Malyuga. In Norway at the same time (1930's), V. M. Goldschmidt's classical work on the distribution of elements was attracting much attention. For the first time, not only were geochemists making quantitative estimates of the abundances of many important trace elements in all types of rocks, but Goldschmidt was formulating the laws governing the observed data which subsequently would be used for exploration purposes. Other studies were made by T. Vogt in Norway on the dispersion of weathering products in soils, vegetation and water from known sulfide deposits. In Finland, K. Rankama studied the nickel content of vegetation from several nickel deposits in the glaciated northern area of that country. Clearly, before 1940, the groundwork had been completed for a modern chemical approach to exploration by means of soil, vegetation and drainage basin surveys and, by the beginning of World War II, these were well under way in the U.S.S.R. and Scandinavia.

In most of the Western World, except for very isolated cases, exploration geochemistry did not attract much attention until after World War II. In 1945, Warren and his associates at the University of British Columbia undertook a research program on the use of the metal content of vegetation for exploration purposes. This first attempt at biogeochemical prospecting in Canada established the existence of copper and zinc anomalies over known mineral deposits (Warren, 1972). Later studies at the University of British Columbia involved the use of dithizone (a colorimetric chemical reagent) for rapid, simple chemical analysis of certain metals, which, according to Boyle and Smith (1968, p. 124) "probably more than any other single factor, gave impetus to the subsequent development of geochemical prospecting methods in Canada." By 1950, the investigation of soils, waters (and later stream sediments) was beginning to exceed biogeochemistry in importance. Boyle and Smith (1968) discussed other historical aspects of the development of the use of exploration geochemistry in Canada, including the many fundamental and practical contributions of the Geological Survey of Canada which began in the early 1950's. Although far too numerous to describe here, the importance of these contributions, many of which are associated with Dr. R. W. Boyle himself, can be gleaned from the references and illustrations selected for inclusion in this book. It is interesting and important to note that Boyle and Smith (1968) considered the introduction, in 1954, of chemical field kits for testing geochemical samples as a mixed blessing. Clearly, the use of kits by qualified persons for on-site testing is an advantage in many ways, and can permit decisions to be made immediately. Their use by untrained personnel, however, can and has resulted in many failures. Further, the initial confidence placed in the use of these kits by some of those in planning and administrative capacities was unjustified, based on fundamental facts which are now well known. As a consequence geochemistry suffered with a poor reputation which has only recently been repudiated following some unquestionable successes. In Canada, probably more than in any other country in the Western World, geochemistry must be used with careful attention to special problems, such as those caused by glacial cover and poor drainage in many areas. Wherever possible in this book, particular emphasis is placed on Canadian examples, although other areas with special environments, such as Australia with its arid, tropical terrain, are also discussed.

Exploration geochemistry began in a serious way in the United States at the U.S. Geological Survey about 1947. Workers in the first several years included Hawkes, Huff, Ward, Lakin, Lovering, Cannon and Bloom. The programs consisted of research on all types of experimental sampling surveys, under a variety of geologic and climatic conditions to determine characteristic dispersion patterns of the elements in rocks, soils, vegetation and natural waters. By the early 1950's geochemical exploration programs had been carried out by large mining companies in the southern Appalachians and the Pacific Northwest. From the very beginning, the U.S. Geological Survey's approach to the field analytical techniques was directed mainly towards colorimetric methods, in contrast to the Soviet reliance on the emission spectrograph. This approach, which resulted in several publications on colorimetric methods applicable to the detection of ore metals (e.g., Ward et al., 1963), was widely accepted, highly successful and applicable to situations in many other parts of the world. In addition, the U.S. Geological Survey also places significant emphasis on the emission spectrograph both in the laboratory and in their modern mobile field laboratories.

In the United Kingdom, the Applied Geochemistry Research Group at the Imperial College of Science and Technology, University of London, was established in 1954 under the direction of Professor J. S. Webb. Much research has been carried out there by Professor Webb, his colleagues and students, into analytical methods and basic principles, and this group has been responsible for a broad program of exploration in various parts of the British Commonwealth, particularly in Africa and the Far East. Graduates from this research center are now actively engaged in exploration geochemistry in many parts of the world, particularly in Canada, Africa and Australia, and they occupy important positions in government, industry and universities. In France, research related to exploration geochemistry began in 1955, and practical applications in that country and parts of French-speaking Africa were initiated shortly thereafter.

It is not practical to attempt to identify all the other countries in which exploration geochemistry is currently employed, both by government and the mining industry. It suffices to say that almost every nation with a geological survey, or the equivalent, now uses or considers the application of exploration geochemistry in the search for mineral resources, as well as in general mapping projects. The same applies to the geological surveys of some states of the United States (e.g., Alaska), and provinces of Canada (e.g., Quebec, Ontario). In those areas or countries which do not have geological surveys and where assistance is requested, the United Nations under its Development Program, has undertaken numerous geochemical prospecting programs, some of which have been notably successful, such as the discovery of porphyry copper in Panama (Lepeltier, 1971). It is clear that with the present unprecedented activity in mining, exploration geochemistry will be used with increasing frequency both by industry and by governments, especially as more successes are credited, in whole or in part, to the techniques.

Modern methods of geochemical prospecting owe their rapid development in the last 50 years to certain factors more than to others. Boyle and Smith (1968) have listed what they, in common with many other workers, believe to have been the main advances. Briefly summarized they are:

1. The recognition of the nature of primary and secondary dispersion halos and trains associated with mineral deposits. Although halos and trains were known for centuries to be associated with metal deposits, the work carried out in the U.S.S.R. and Scandinavia in the 1930's and early 1940's systematized the data and permitted it to be used in the most advantageous manner.

2. The development of accurate and rapid analytical methods using the emission spectrograph in the U.S.S.R. in the 1930's, and the use of specific, sensitive colorimetric reagents (particularly dithizone) by the U.S. Geological Survey in the late 1940's and early 1950's.

3. The development of polyethylene laboratory equipment which reduced the problems associated with contamination and permitted greater reliability in field analysis. The introduction of resins for the production of metal-free water for use in chemical procedures was also significant.
4. The development of atomic absorption spectrometry in the late 1950's permitted the rapid, accurate, sensitive and relatively interference-free analysis of many elements of interest in exploration geochemistry.
5. The development of gas chromatography which has had its major effect on the search for hydrocarbons. It is an extremely sensitive method and has replaced the earlier fractionation and condensation methods and equipment which were generally unsatisfactory.
6. The use of statistical and computer methods has aided the interpretation of analytical data from exploration geochemistry. These methods permit rapid evaluation of data by computing backgrounds, assisting in the recognition of anomalies and in graphical presentation.

All of these important developments are discussed in greater detail in appropriate sections of this book.

TYPES OF SURVEYS

In the preceding paragraphs several types of geochemical surveys were mentioned, but no indication was given of their relative importance. In Table 1-1, the most recent data bearing on this point are presented. Clearly, the data imply that there is no preference for methods used, for in Canada, soil samples were by far the most common type analyzed, whereas in the United States rock samples were more frequently used. The choice of

Table 1-1. Types of geochemical samples analyzed in Canada and the United States (in percent).

	Canada	United States
Soils	56.7	19.0
Rocks	14.6	44.0
Stream sediments	23.4	23.0
Water	2.7	8.3
Vegetation	1.9	4.3
Air	0.2	0.9
Other	0.4	0.5
	99.9%	100.0%
Number of samples analyzed:	812,456	337,370

Source: Report of the Analysis Committee of The Association Of Exploration Geochemists. Published in the July 31, 1971 *Newsletter* of the Association.
Period: June 1, 1970 - May 31, 1971.

which type of survey to employ is influenced by the particular project, and is governed by the geochemical nature of the element being sought, the type and availability of material for sampling, climatic conditions and weathering, and dispersion characteristics in the area. For example, in regions with generally low topographic relief, thick overburden, and little water such as in large parts of Australia, soil surveys would be expected to be far more common than rock, water, or other types of surveys. The relative importance of the various types of surveys (Table 1-1) also can change significantly from year to year depending on the objectives and locations of current exploration activity. As a generalization, in the more underdeveloped countries, stream sediment surveys of a reconnaissance nature are the most common at present.

The surveys represented in Table 1-1 are all classified according to the type of material being sampled and analyzed. However, geochemical surveys may also be classified as to whether they are *reconnaissance* or *detailed* in their scope and, to some extent, each of these two types uses different materials for sampling and analysis. The object of a reconnaissance survey is to evaluate a large area, from hundreds to tens of thousands of square miles, such as a large mineral concession, an area with specific geophysical or geological features which appear favorable, or a large number of claims, with the purpose of delineating mineralized areas for follow-up studies, and eliminating barren ground. Thus it is a method for area selection and not one to locate specific deposits. Only a relatively small sample density (perhaps one sample per sq mi or even one sample per 100 sq mi) may be required to determine the broad geochemical pattern, depending on the geological conditions and the element sought. Reconnaissance surveys are generally rapid, low in cost for the size of area sampled, and each sample is likely to represent a considerable area. Stream sediment (drainage basin) samples are particularly suitable for reconnaissance surveys. Detailed surveys, on the other hand, are carried out in a local, much smaller area, from a few square miles to tens of square miles, with the objective of locating as exactly as possible individual mineral deposits, or indications of structures (such as faults) where such deposits might occur. For the detailed survey, samples may be collected at an interval of 10 feet or even less, especially when veins and other small targets are being sought. In the case of detailed surveys, soil and vegetation are usually preferred, but rock, water and other types of samples may be collected depending on the particular situation. It must be remembered, however, that in differing circumstances every type of material may be used for either reconnaissance or detailed surveys. In the following sections an introduction is presented to the major categories of surveys based on the type of material sampled and analyzed. In Fig. 1-1 some methods for the presentation of analytical data obtained from stream sediment and soil surveys are illustrated.

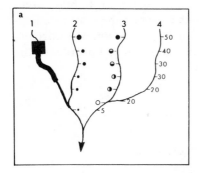
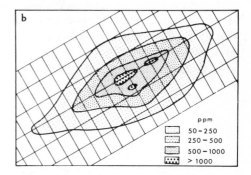

Fig. 1-1. Some commonly used methods for the presentation of analytical data. *Left* (a): For stream sediment (drainage basin) surveys. In 1 and 2, contents are proportional to the size of the symbols, whereas in 3, different symbols are used to represent different contents, and in 4, numbers (generally in ppm units) are used. *Right* (b): For soil surveys, a contour of equal metal contents (isograd) is used to depict element distribution on a geochemical map. From Andrews-Jones (1968).

Soil Surveys

Soil (pedogeochemical) surveys are widely used in exploration geochemistry (Table 1-1), and very successful results have been achieved, especially with surveys conducted in residual soils. Their success is based on the fact that during the process of weathering and leaching, anomalous concentrations of elements from underlying mineralization may become incorporated in the soil, as well as in other weathering products, and in groundwaters. Thus, the elements spread outward from the ore deposit, resulting in anomalous values (a secondary dispersion halo) which, in the overlying soil, is a considerably larger exploration target than the ore deposit itself (Fig. 1-2). Soils typically display layering, or a profile, and the individual horizons differ from each other in mineralogy and trace element composition, so that varying results can be obtained by sampling the different horizons. In general, wider anomalous zones will be obtained from both the A and B horizons (the true soils), by comparison with the C horizon which consists of bedrock fragments in various stages of degradation, with an anomalous zone similar to that of the parent bedrock. The contents of many metals in the B zone is normally higher than that in the A zone, and therefore, samples from the B zone are preferred in most soil surveys. However, because of extremely variable surface weathering conditions, as well as many other factors, there are situations in which the A or the C soil horizon gives more significant results. In areas of transported overburden (glacial deposits or wind-blown sand areas), or thick soil cover as in western Australia, experience has shown that deep drilling, and sampling near the bedrock is preferable, if not essential. In other areas it may be necessary to sample all soil horizons to obtain the most meaning-ful information. Ideally, in order to determine the optimum soil horizon, preliminary samples should be taken through a series of soil profiles in the area (see Orientation Surveys; Chapter 4).

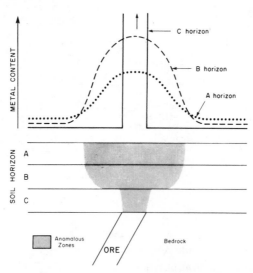

Fig. 1-2. Diagrammatic representation of the relative lateral dimensions of secondary dispersion halos, and associated metal contents, typically encountered in sampling and analyzing the three major soil horizons.

For routine geochemical surveys, it is standard practice to obtain the minus 80 mesh fraction of the soil (also stream sediments) and analyze it for the elements of interest. Experience has shown that, in general, this fraction is a good compromise between obtaining sufficient material, and providing maximum contrast between the background and anomalous sam-ples. However, there are exceptions to this generalization. For example, there are cases in which it is preferable to take the heavy (or resistate) minerals from soils, till or other weathered material, by means of heavy liquids, panning or magnetic methods, and analyze them either chemically or microscopically. Examples of the use of heavy mineral fractions would include the search for tin, niobium, titanium and other elements forming heavy or chemically resistant minerals. In these cases, a soil fraction coarser than 80 mesh would probably be used. In the laterite areas of Australia, fractions coarser than 80 mesh are generally used, as is discussed in more detail below.

Soil surveys may be used on a reconnaissance basis as in Australia and parts of Africa and Canada, or to screen geophysical anomalies, but in general they are preferably used on a detailed, local level to determine the exact location and dimension of an anomaly. Although residual soil anomalies are often related to an underlying mineralized zone, their element dispersion trains cover only a relatively small area by comparison with the dispersion trains found in stream sediment (drainage basin) reconnaissance surveys. Nevertheless, surveys in residual soils are preferred in many cases as they are considered highly reliable, and suffer from fewer variables and limitations than other types of surveys, even though many false (non-significant) anomalies can be encountered. In the case of immobile elements (e.g., chromium) which do not travel far except as detrital grains, soil surveys can be particularly useful for detailed exploration purposes. Whether in residual or transported material, soil samples should be collected on a relatively closely-spaced grid. Ideally, a soil survey should yield the highest values for the element(s) being sought directly over a deposit, with lower values fanning out from it. In actual fact, such simple situations are not generally encountered because of many factors such as soil movement and hydrologic effects, and therefore, careful interpretation is usually necessary to locate the source of the anomaly (Fig. 1-3).

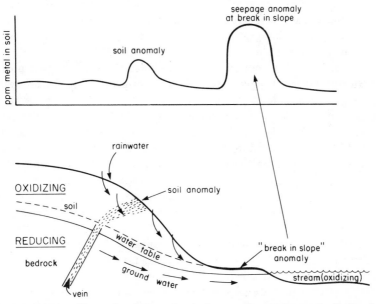

Fig. 1-3. Diagram illustrating that geochemical anomalies are frequently displaced, often considerable distances, from the source of mineralization. The soil anomaly is displaced by soil movement. Further, metals leached from the soil by water may be redeposited where the water comes to the surface (at the spring line); such an accumulation of metals is called a seepage or "break-in-slope" anomaly. Drilling directly over these displaced anomalies would miss the mineralization. After Bradshaw et al. (1972).

In Canada, soil surveys are generally useful in the Yukon, Northwest
Territories, British Columbia and the Maritimes, as well as in certain parts
of the other provinces. In the Canadian Shield, where essentially all the
overburden is transported, conventional soil techniques generally are not
applicable although research into methods of sampling overburden of all
types offers hope for the future once the methods of interpretation are perfected. A few examples have been reported in which soil sampling in parts
of the Canadian Shield have been successful but such instances are restricted to specific situations. The problems associated with geochemical exploration in Canada are discussed in Chapter 11.

Rock Surveys

Rock (also called lithogeochemical or bedrock) surveys are based
either on the analysis of a whole rock sample or the individual mineral(s)
within the rock. These materials may be obtained from outcrops or from
drilling. In most cases, unweathered rock materials are used. On a world-
wide basis, notwithstanding the figure for the United States in Table 1-1,
rock surveys have not been used as extensively as other types of surveys.
However, there has been a developing trend, especially at the research
level, toward greater integration of rock surveys into the overall context of
exploration geochemistry.

Rock surveys, on a regional reconnaissance basis, have great potential
for outlining favorable geochemical (metallogenic) provinces and for
recognizing favorable host rocks. The usual objective of the regional rock
survey is to determine some geochemically diagnostic feature that permits
the recognition of rocks which are likely to be associated with mineralization, and their distinction from barren rocks. It is well known, for
example, that certain granites have above average contents of tin, tungsten,
beryllium, copper or molybdenum, whereas others are high in uranium.
Therefore, an increasing number of studies deal with the general trace
element content of rocks from potentially favorable regions, as well as
trace elements within individual minerals. An example of the latter is the
study of the niobium, tin and chlorine contents of micas which may be
indicative of niobium, tin, and base-metal sulfide mineralization,
respectively, in the host rock.

Only a few large-scale regional rock surveys have been reported in the
Western literature, but examples can be cited for Australia, Saskatchewan,
Ontario, the Yukon, and Idaho. In the U.S.S.R., there are numerous
examples of rock surveys covering extensive areas, but details are lacking.
On a regional basis, the collection of rock samples representative of a large
area may be difficult, owing to variations in minerals, trace elements, and

texture often over very short distances, as well as a paucity of outcrops in many areas. In general, rock reconnaissance surveys are carried out on a uniform grid, across a geological terrain which may well include several rock types. Accordingly, interpretation is often very difficult. Great attention must be paid to the careful recognition of different rock types and their degree of alteration, as this information is needed for proper correlation with the analytical data obtained. Boyle and Garrett (1970) have pointed out that in several very carefully studied areas with important mineral deposits (e.g., Yellowknife, Northwest Territories; Keno Hill, Yukon; Bathurst-Newcastle district, New Brunswick), there are no definite elemental patterns in the country rocks to suggest the presence of metal-bearing deposits in these areas. Therefore, until the reason for this enigma is resolved, caution is necessary and it must not be expected that rock surveys will be able to accurately reflect mineralization in every area studied.

Detailed rock surveys have various objectives. In some instances they are used to determine the extension of known mineral deposits by sampling either underground or on the surface. A classic example is the east Tintic district of Utah, where rock geochemistry was used to locate previously unknown mineralization. It is becoming increasingly common for certain companies to send out crews with the specific purpose of collecting bedrock from all outcrops in areas where airborne geophysical surveys have indicated anomalies. The rocks are usually subsequently analyzed for nickel, copper and other base metals. Other examples are reported where prospecting crews will make only a qualitative field test for nickel on all rock outcrops (grab samples) within a specified area in the search for mineralization.

Most detailed rock surveys have different immediate objectives from the examples just cited, however. They may be follow-up surveys which have resulted from an interpretation of a drainage basin reconnaissance or some other type of geochemical or geophysical survey. Rock surveys undertaken under these circumstances, particularly during diamond drilling phases, are concerned with finding mineralization, or outlining a primary halo (dispersion pattern) of elements associated with mineralization, as there is often a reflection of mineralization beyond the actual margins of the deposit into the surrounding country rock. The region in which there is a decrease in the content of valuable elements to background values is called the *primary halo*. These halos are not large targets, often varying from the order of inches beyond the actual deposit to perhaps a few hundred feet as a practical limit, and in these situations the sampling grid will be closely spaced. However, some of the more volatile and mobile elements, such as mercury or zinc, may form a broader halo and, therefore, the sample spacing selected must take into account the dispersion characteristics of specific elements being sought, as well as other important factors, such as the probable size and shape of the potential ore body. The entire problem of primary halos, including their extent both vertically and horizontally, is receiving increasing attention, particularly with the object of discovering buried deposits. They are discussed in greater detail in chapters 2 and 7.

In their summary of geochemical prospecting methods, Boyle and Garrett (1970) noted that Soviet geochemists reported that some primary halos in rocks associated with massive pyrite deposits are quite extensive both laterally and vertically. The halos project above the ore bodies for distances of up to a kilometer in steeply dipping structures, and are hundreds of meters to several kilometers in lateral extent. No halos from concealed deposits approaching these dimensions have been reported in the Western literature possibly because methods of recognizing them by means of the analysis of groups of trace metals have not reached a sufficient degree of refinement. In the vicinity of an exposed sulfide deposit in Cyprus, however, Govett (1972) has shown that it is possible to recognize a halo 2 km from mineralization, by the use of statistical methods and multi-element variations in the interpretation of a rock geochemistry survey. Clearly, primary halos in rocks are likely to be of great importance in discovering deeply buried deposits, particularly those in favorable structural settings. The recognition of primary halos at great distances from mineralization is more feasible now than a decade ago because of the great advances in analytical methods capable of accurately detecting very small quantities of many elements, as well as the use of statistical methods to aid in interpretation. However, it must be kept in mind that some types of deposits, for example syngenetic sulfides in sediments, or platinum-bearing ultramafics, frequently seem to produce no primary halos, using present methods of detection and interpretation.

Stream Sediment Surveys

Stream sediment surveys are employed almost exclusively for reconnaissance studies in drainage basins, and if properly collected, the samples represent the best composite of materials from the catchment area upstream from the sampling site. When discussing drainage basin surveys, the usual implication is that stream sediments are being collected, but these surveys may refer to the collection of water samples or separation of heavy minerals (Fig. 1-4). Lake and swamp sediments may also be used for drainage basin surveys. The use of stream sediments for reconnaissance surveys became popular in the early 1960's after several studies, some of them in the Soviet Union, proved that this technique was effective and

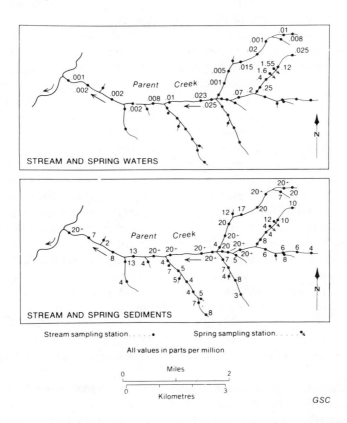

Fig. 1-4. Heavy metal content (mainly zinc) in waters from streams and springs *(upper)*, and from stream and spring sediments *(lower)*. Based on a study in the Keno Hill area, Yukon Territory (Geological Survey of Canada Maps 20-1964 and 21-1964). From Lang (1970).

economical, and could delineate targets for follow-up work. By sampling stream sediments, or heavy minerals separated from them, at perhaps an interval of 4 or 5 samples per mile, it is possible to recognize geochemical or mineralogical anomalies within a catchment area and trace them to their sources. As a generalization, the larger the stream being sampled, the larger the mineralization must be to have a significant effect on the trace element content of the stream sediments.

Reconnaissance sampling density varies greatly depending upon the objectives of the survey. For example, Webb (1971) reported a density of 1 sample per 100 sq mi for a stream sediment sampling program covering

80,000 sq mi in Zambia, but only 1 sample per sq mi in Northern Ireland. Only under very favorable circumstances, such as in the case of some surveys conducted in Zambia and Sierra Leone described in Chapter 9, should a stream sediment sample represent a catchment area of more than 10 sq mi. Usually only sediments in the process of being moved, and which are found in the active stream channel, are used; these are called *active sediments* by many geochemists. It is possible to make use of older sediments on terraces and flood plains dating from earlier erosional cycles in favorable situations. Stream sediments have been sampled extensively in the Soviet Union, in many parts of North and South America, East and West Africa, Australia and Ireland, and numerous deposits have been discovered by these programs. Stream sediment sampling has also proved valuable in outlining geochemical provinces and screening airborne geophysical anomalies. The technique is applicable under a wide variety of climatic conditions and, although areas of at least moderate rainfall are preferred, it is practicable to use sediments from water courses which are usually dry, such as are commonly found in Mexico and Australia. Generally, the less than 80 mesh size (clay and silt) fraction is collected, and this is then subjected to chemical tests in the field or laboratory. Various chemical digestions are used, ranging in severity from extraction by weak cold acids or buffers, to digestions using hot acids or fusions. The choice of digestion is dictated by the particular element being sought, by the weathering conditions in the area, and by a number of other factors discussed in detail in Chapter 6.

The underlying rationale behind stream sediment surveys is the fact that, in the weathering environment, many minerals, and particularly sulfide minerals, are unstable and will break down as a result of oxidation and other chemical reactions. This results in a dispersion of both ore and indicator elements in solution, in run-off and groundwater, sometimes for relatively long distances within the drainage basin. The mobility of the different elements varies considerably, as is discussed in more detail in Chapter 3, but it suffices to say here that many factors, such as adsorption and pH-Eh conditions, are important in controlling the relative mobility. Eventually, the fine-grained sediments in the streams within the drainage basin, particularly clays, reflect an increase in ore elements or indicator elements if there is mineralization within the basin. In the case of a porphyry copper deposit in Panama, anomalies in active stream sediments were detected 17 km from the deposit (Lepeltier, 1971). The streams transporting the mobile elements do not necessarily have to intersect the mineral deposit, as run-off or groundwater entering the stream can be the source of elements.

The discussion to this point has been based on the assumption that the metal content of stream sediments is introduced in solution. The metal content of stream sediments may also be derived by means of a second mechanism, that is, by clastic or mechanical movement. In those localities where physical (mechanical) weathering is predominant, such as in deserts or in regions of rugged topography, the ore minerals themselves may be mechanically dispersed in the form of discrete detrital mineral grains (clastics) within the stream sediments, and these can be detected by appropriate chemical analysis for selected metals or indicator elements. Larger amounts of the detrital grains, corresponding to increases in the abundance of metals or indicator elements, will be encountered as the source is approached. Clastic sediment anomalies may also be produced when a river intersects a mineralized zone. In addition to chemical analysis of the total sample or selected size fractions, the heavy minerals may be extracted for either mineralogical identification or separate analysis. The extraction of a heavy mineral fraction is analogous to the very successful historical practice of panning for gold, tin and other stable heavy minerals in stream sediments or gravels along a drainage basin in search for the lode deposit.

If we examine the metal values in an upstream direction, then the point at which the maximum metal values are found is called the *cut-off*. This usually corresponds to the position where metal-rich groundwater enters the drainage basin, or detrital ore minerals enter the river or, in the least probable case, where the river actually cuts the zone of mineralization. Ideally, upstream of the cut-off, metal values decrease rapidly, whereas downstream of the cut-off, the values will decrease gradually due to dilution. In actual practice, however, the recognition of a cut-off can be difficult, particularly when the anomalous samples have only slightly higher metal contents than the background samples or when there is contamination within the drainage basin from mining or other causes. Further, not all cut-offs represent mineralization, as many are related to any number of naturally-formed anomalies unrelated to mineralization. Some of these, termed false anomalies, are discussed in Chapter 4.

Once a cut-off is recognized, the next step is to locate the source of the anomaly. Visual examination is the first step, and the geochemist will seek mineralization, groundwater seepages, contamination, and similar phenomena. Assuming none are found, *bank sampling*, a form of follow-up survey, could be initiated. This consists of collecting soils and alluvium on both sides of the river, as well as materials from the base of slopes adjacent to the river valley, which may contain metals from groundwater seepages (Fig. 1-5), and should indicate from which side (or sides) of the river the anomaly originates. A bank sampling survey is a preliminary step to a detailed soil survey and can indicate the limits of the area to be covered by more detailed soil sampling. It should be noted that bank anomalies frequently are poorly developed or absent. Where they do occur, however, they can be particularly useful.

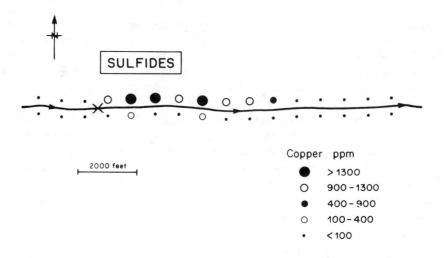

Fig. 1-5. Diagram illustrating bank sampling, based on an example from British Columbia. A stream sediment survey of a river flowing toward the east indicated a cut-off at point "X". Bank sampling on both sides of the river clearly shows that the anomalous values originate from the north side of the river, where sulfides were later found.

Water Surveys

From the point of view of exploration geochemistry, water (hydro-geochemical) surveys may be discussed on the basis of the source of the water, which is either (1) groundwater, or (2) surface water. These two types have distinctly different chemical and physical properties. Groundwater tends to have higher contents of dissolved components and, especially in the vicinity of oxidized sulfide deposits, it has a low pH (acid) and, therefore, a greater ability to dissolve and transport metals than does surface water. For this and other reasons, groundwater has great potential in exploration geochemistry, particularly if it is acid and shallow enough for sampling. Surface water, on the other hand, has less dissolving power, and one of its main functions in exploration geochemistry is to transport dissolved metals brought into the drainage basin by groundwater. As a result, stream sediments generally yield better, or a least more easily recognizable, anomalies than do associated surface waters, because the fine-grained sediments adsorb much of the metal carried by the water. Nevertheless, for some reconnaissance surveys of drainage basins, the collection and analysis of stream water and sediments are carried out simultaneously. Other types of water such as groundwater, lake, bog, and spring can be used on a reconnaissance basis, as well as for follow-up work. The chemical elements in surface water, and most lake and groundwater, are mobilized by the same mechanism described above for soils and sediments, that is, chemical weathering and oxidation, followed by leaching, in which some elements find their way into the drainage system. In the presence of oxidized sulfide deposits, shallow groundwater can dissolve appreciable quantities of metals and transport them to the surface drainage (Fig. 1-3). The dissolved elements form either dispersion trains (increasing amounts of elements upstream toward the source) in the case of streams, or halos (increasing amounts of elements in bodies of water toward the source) in the case of standing bodies such as lakes. The distinction is illustrated in Fig. 1-6.

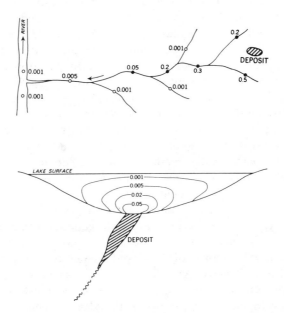

Fig. 1-6. Distinction between dispersion trains and halos in water. *Top:* Idealized dispersion train in water of a stream system draining an area containing a Zn-Cu deposit. *Bottom:* Idealized cross-section of a dispersion halo in lake water overlying a Zn-Cu deposit. All values in ppm Zn. From Boyle et al. (1971).

As indicated in the historical introduction, the use of hydro-geochemical methods in the form of the analysis of spring water and their residues has been known for centuries, yet the use of water in modern times has not been significant (Table 1-1). The reason is the many problems that have been encountered in attempts to use water. For example, the metal content of the water fluctuates with rainfall and season, making inter-pretation difficult; analysis is difficult in the low ranges (parts per billion in most cases) in which most metals occur; collecting and transporting the relatively large volumes of water needed for analysis in the past is inconvenient; and in many areas groundwater samples are contaminated. Recent advances in analytical methods now permit on-site determination by colorimetric or instrumental methods which require no more than 100 ml of water and these advances have made water surveys more feasible. In Canada, where many of the most favorable areas for mineralization are covered with lakes, muskeg and bogs, research has shown the utility of hydrogeochemical methods. Lake water in the vicinity of the copper deposits near Coppermine River, N.W.T. is demonstrably an effective sampling medium for regional geochemical exploration in lakes at least 1000 ft in length, although stream sediments are probably better (Allan and Hornbrook, 1971). Lake water in the Beaverlodge uranium district, Saskatchewan, has been used for reliable delineation of zones of pitchblende mineralization within the Canadian Shield (Macdonald, 1969). Similarly, ground and well water has been sampled extensively in the search for uranium in the Colorado sedimentary basin. In addition to the natural precipitates from springs, whose value in reflecting the presence of mineralization has long been known, manganese oxides, limonite and other precipitates from either surface or groundwater are also of great value in exploration geochemistry because of their ability to adsorb metal ions. These precipitates should be collected and analyzed separately from other samples. Hydrogeochemical methods have great potential in heavily forested, mountainous terrain, or other regions such as lake areas, in which it is difficult to sample sediments or soil.

A recent review by Boyle et al. (1971) on the application of hydrogeochemical methods to the Canadian Shield concludes that these methods provide a third dimension to prospecting because suitable waters result from leaching great volumes of rock, and frequently carry to the surface or into underground workings, clues to the presence of hidden mineral deposits. Boyle et al. (1971) rightly conclude that at present these methods are poorly developed either for use in the Canadian Shield or for other areas. Apart from general use in the Soviet Union, and the specific search for uranium in the Western World, hydrogeochemical methods have been largely ignored. Recent studies have attempted to relate the sulfate content, and a few other constituents, to sulfide occurrences but they have not been too encouraging (an exception is the work by Dall'Aglio and Tonani, 1973). Admittedly, the interpretation of groundwater anomalies requires the assistance of a hydrologist, especially in regional surveys and in more geologically and hydrologically complex areas. Nevertheless, the method appears to be attracting more attention, based on some recent encouraging studies not only for metals, but for non-metals as well, such as the use of fluoride in groundwater as an indicator of fluorite mineralization (e.g., Schwartz and Friedrich, 1973). New, extremely sensitive analytical methods, for use in the laboratory and by means of portable instruments capable of making accurate field determinations (Eh, fluorine, carbonate, etc.), have removed some of the earlier objections to the analysis of water. Further, there is a vast amount of trace element data now available in connection with environmental studies of surface and groundwater as well as associated sediments and this fundamental hydrogeochemical information can now be applied to exploration geochemistry. This is also true of the formation waters recovered in the search for crude oil and natural gas. Dissolved hydrocarbons (such as benzene), isotopic variations in sulfate, dissolved inorganic components (e.g., bromide, iodide, nickel, cobalt), and salinity of formation waters are now more easily determined than they were in the past. The use of some of these parameters in the search for hydrocarbon accumulations is discussed separately in Chapter 13.

Geochemical methods are also being applied to the search for mineral deposits under the sea. Near the coast of England some success has been reported in locating detrital heavy mineral and primary tin deposits. The geochemistry of submarine phosphate and manganese deposits has also been studied. The techniques employed appear to be similar to those that would be used on land, that is, sediments are obtained from the sea floor and then analyzed by conventional methods. The future will probably see the development of practical devices capable of detecting radioactivity on the ocean floor, as well as *in situ* radiometric (neutron activation) methods capable of at least qualitative analysis under water. The geochemistry of the Red Sea metalliferous hot brines, as well as of their associated sediments, have been studied with the intention of applying the data to the search for similar deposits elsewhere, but from an earlier geologic period (e.g., Holmes and Tooms, 1973). Similar studies are being made in the Gulf of California because of a comparable structural setting but no brines or significantly enriched sediments have been reported yet.

Vegetation Surveys

There are two general types of geochemical surveys which involve the use of vegetation. These are based on (1) geobotany, which consists of a *visual survey* of vegetation, and (2) biogeochemistry, which involves the collection and *chemical analysis* of whole plants, selected parts of plants, or humus (which in this book is discussed with soils). Often both types are classified under the general term "biogeochemistry", but this usage is incorrect and should be avoided. Brooks (1972) has pointed out that two-thirds of the world's land surface is covered with vegetation and, therefore, it is probable that much of the world's mineral resources are hidden beneath vegetation. Accordingly, vegetation surveys are likely to receive much more attention in the future.

Geobotanical studies include the recognition of (a) specific plant communities, or the presence or absence of specific plant varieties indicative of a particular element, and (b) malformed or oddly colored varieties, whose characteristics are the result of deleterious or toxic effects caused by an excess of certain trace elements on or near mineralization. The data presented in Table 1-1, which indicate that only about 3% of the samples analyzed in North America are vegetation, do not include any geobotanical statistics.

All vegetation methods are based on the fact that during chemical weathering, the mobilized elements are concentrated as dissolved ions in soil moisture, and, by complex mechanisms, become available to plants (Fig. 1-7). Elements migrate to various parts of the plant or tree and, by

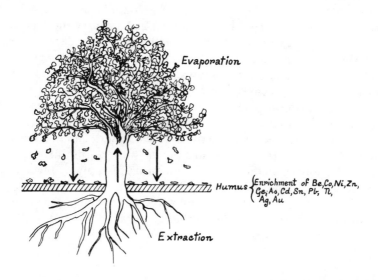

Fig. 1-7. Illustration showing that many elements, including metals, may be extracted from the soil by the roots of trees which act as sampling agents. The elements migrate through the various parts of the tree and finally to the leaves. When the leaves fall to the ground, they enrich the humus in metals. This mechanism also explains the enrichment of certain metals in coal. From Goldschmidt (1937).

judicious selection of the parts of the plant to be subjected to analysis (biogeochemistry), anomalies may be found which are possibly indicative of mineralization. In geobotanical surveys, discolorations, dwarfing or other physiological effects can be observed but usually only by someone specially trained to recognize this aspect of the flora of an area. Although indicator plants show no special physiological features, they grow preferentially in areas where there is an abundance of a specific trace element. Thus there are indicator plants for copper which have been used successfully in Central Africa, and others for zinc, selenium, sulphur, and many other elements. Again special, on site, training is often needed to recognize indicator plants, although some are rather distinctive in color or other properties.

Interest in the chemical analysis of vegetation, particularly trees, is based primarily on the fact that the roots of some species reach depths of 100 feet or more and often spread over considerable distances so that a large volume of soil is sampled by their root system. In such cases, analysis of appropriate parts of the tree (such as bark, twigs, needles, leaves) may reveal buried mineralization (Fig. 1-8). Simple as the concept appears, there are many problems associated with vegetation surveys, and the results may be difficult to interpret because of the many factors which enter into the uptake of elements by plants (e.g., plant metabolism, soil conditions and exchangeable ions in the soil). Critical factors in biogeochemical sur-

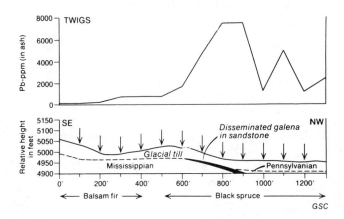

Fig. 1-8. Results of a biogeochemical survey in which the lead content of twigs indicate the occurrence of disseminated galena in sandstone buried beneath glacial till, Cape Breton, Nova Scotia. Based on a M.Sc. thesis by M. Carter, Department of Geology, Carleton University (1965). Figure from Lang (1970).

veys are the type of plant, the part selected, and the age of the particular part (usually two year-old growth is sampled). Numerous examples have been described in which the analysis of vegetation has indicated anomalous metal values but no mineralization has been found presumably because the roots have extracted metals from weathered silicates and other non-economic minerals. In other studies different results have been obtained from samples taken on different sides of a tree, possibly related to the direction of the prevailing wind. Clearly, the use of vegetation as a guide to mineralization is often considerably more complex than any of the other

methods discussed so far and may require more skill in execution and interpretation.

Notwithstanding the negative aspects noted, vegetation surveys have
proved useful in many places, particularly in unglaciated areas. In the
glaciated parts of Canada, the root systems of most plants are not very
deep, nor has there been sufficient time since the last glaciation for
geobotanical indicator plants to evolve in a particular habitat. In desert
areas of the western United States and northern Mexico methods employing
vegetation have been used successfully. Chaffee and Hessin (1971) have
described the use of chemical analyses of the ash of deep-rooted plants to
outline anomalous patterns related to concealed porphyry copper deposits
in Arizona. The biogeochemical anomalies extend as much as 800 ft
beyond the soil anomalies. However, in other places in which
biogeochemical methods have worked well, soil surveys have worked
equally well, if not better (Fig. 1-9). The reason for at least some of the
corresponding soil anomalies is related to the fact that dead leaves, and
eventually the tree or plant itself, fall to the ground and nourish the soil

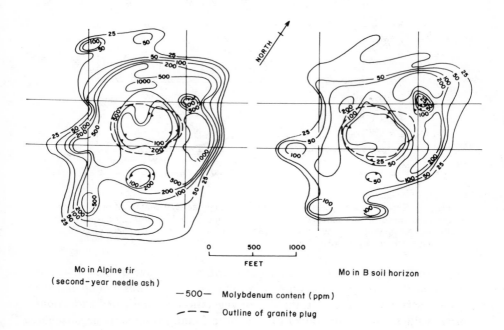

Mo in Alpine fir
(second-year needle ash)

Mo in B soil horizon

—500— Molybdenum content (ppm)

— — — Outline of granite plug

Fig. 1-9. *Left:* Distribution of molybdenum in Alpine fir (second year growth). *Right:*
Distribution of molybdenum in the B soil horizon. From the Lucky Ship molybdenum
deposit (Hornbrook, 1969). Although this biogeochemical survey clearly established
the effectiveness of plants (trees) in prospecting for this deposit, B soil horizon samples
outline the anomaly almost equally well.

with metals released upon decay, providing a more uniform sampling material (Fig. 1-7). One very practical use of biogeochemical surveys in Canada is based upon the fact that during the many months in which the country is covered with snow, this method is the easiest and least expensive way to investigate a new area or carry out assessment work. In general, in all environments, preliminary (orientation) surveys should always precede the use of biogeochemical methods in order to determine the applicability and the type of material to be collected during the biogeochemical survey, and to ensure that confidence can be placed in the results.

Vegetation methods are receiving increasing research attention in Canada (e.g., Fortescue and Hornbrook, 1967; Fortescue, 1970), in the United States (Cannon, 1971), and elsewhere (e.g., Cole, 1971). Both geobotany and biogeochemistry, as applied to mineral exploration, are the subject of an excellent book by Brooks (1972), and the prognosis is that these methods will be used with increasing frequency in the future, particularly in tropical countries. The subject is discussed in further detail in Chapter 10.

Vapor Surveys

Vapor surveys, also called gas surveys, are among the most recent developments in geochemical exploration even though certain types have been in use in the Soviet Union for at least 20 years, particularly the sampling of mercury vapor. Vapor surveys aid in the location of buried deposits through the detection of halos of sulfur dioxide, hydrogen sulfide, mercury, iodine, radon, or other gases or volatile elements and compounds, often at considerable distances from the source of mineralization. Vapors which can be sampled, and the type of deposits from which they emanate, are listed in Table 1-2. Some, such as the organometallics, are still in the very early stages of development. Included in the general category of vapor surveys are (1) air (atmogeochemical) surveys, and (2) soil gas surveys. Some vapors (e.g., He, Rn, Hg) may be detected in groundwater. The subject of vapor surveys has been reviewed by Bristow and Jonasson (1972), McCarthy (1972) and Ovchinnikov et al. (1973). Experience has shown that the results from most vapor surveys are very difficult to repeat. Also, vapors present in the soil are not necessarily present in the air, and vice versa. Further, it must be realized that volatile elements may be released to the atmosphere (or soil) from three major sources: (1) through oxidation of ore deposits; (2) through radioactive decay (e.g., He and Rn from uranium); and (3) during volcanism (e.g., H and He).

In the Soviet Union, mercury was suggested as a pathfinder for locating sulfide deposits as long ago as 1946 but only since about 1964 has it

Table 1-2. Vapor indicators of ore deposits. Modified from Bristow
and Jonasson (1972).

Vapor	Type of Deposit
Mercury (Hg)	Ag-Pb-Zn sulfides; Zn-Cu sulfides; U ores; Au ores; Sn-Mo ores; polymetallic (Hg, As, Sb, Bi, Cu) ores; pyrites
Sulfur dioxide (SO_2)	all sulfide deposits
Hydrogen sulfide (H_2S)	all sulfide deposits
Carbon dioxide and oxygen (CO_2, O_2)	all sulfide ores; gold ores
Halogens and halides (F, Br, I)	Pb-Zn sulfides; porphyry copper deposits
Noble gases (He, Ne, Ar, Kr, Xe, Rn)	U-Ra ores; Hg sulfides; potash deposits
Organometallics such as $(CH_3)_2HgAsH_3$, and compounds of Pb, Cu, Ag, Ni, Co, etc.	possibly all sulfides; Au-As deposits
Nitrogen oxides (N_2O, NO_2)	nitrate deposits

attracted much attention in the western countries. Now, because of the ease of detection of mercury in soils (and waters) by atomic absorption and other methods, this element is determined in many surveys, both reconnaissance and detailed. Vapor surveys for mercury and other elements and compounds which require special systems of collection and analysis, have been investigated by the U.S. Geological Survey with encouraging results. Some consulting-service companies (e.g., Barringer Research Ltd., Rexdale, Ontario) offer airborne mercury surveys utilizing advanced types of instrumentation such as correlation spectroscopy, usually as an adjunct to geophysical methods. Most airborne surveys are conducted from airplanes flying between 200-400 feet above the ground.

Sulfur dioxide, hydrogen sulfide and other gases liberated during the oxidation of sulfide deposits can be determined in soils, waters, or in air but at present very little has been published concerning either actual surveys or their results. Contamination associated with industrial and urban environments will be difficult to isolate and evaluate in many instances. Radon (a gas), has been measured successfully in soils (and waters) in the search for uranium, and although there are problems related to the interpretation of some of the data, contamination is not one of them. Iodine is associated with porphyry copper and other types of metal deposits, and soil gases above these deposits are potentially valuable sampling media. The analysis of gases in the search for hydrocarbons has also been used extensively and offers great potential for the future especially with the availability of gas chromatography and other analytical methods capable of producing the necessary analyses at extremely low levels of abundance. These are discussed in Chapter 13.

At present, soil vapors appear to offer the greatest potential because the vapors of interest are more concentrated in them than in air, perhaps by a factor of 1000. Despite this, no new mineral deposit has yet been found by soil gas surveys. Atmospheric sampling has special problems which result from diverse factors such as wind, rain, and even exhaust fumes from airplanes. Nevertheless, atmospheric sampling has been proved technically successful, particularly in desert areas, for example by the detection of mercury over the Mt. Isa ore deposits in Australia, and in this circumstance the sulfides, particularly sphalerite, have high contents of mercury (Table 2-5). On the other hand, mercury does not appear to be a useful indicator element for most porphyry copper deposits as it occurs only at background levels. Air surveys appear to have value for reconnaissance purposes augmenting data from airborne geophysical techniques, photogrammetry, and remote sensing methods.

Dr. P. M. D. Bradshaw (personal communication) has stated that recent airborne surveys by Barringer Research Ltd. have been successful in delineating known mineralized areas by determining organometallic compounds of Hg, Cu, Pb, Zn, Ag, Ni and Co by means of correlation spectroscopy (Table 1-2). This extremely sensitive technique determines molecules, thus organometallics, as opposed to atoms. Atomic mercury apparently does not exist more than a few feet up into the atmosphere, according to Dr. Bradshaw. Correlation spectroscopy for the detection of minute quantities of organometallic compounds during airborne surveys is an exciting development which offers great future potential. As organometallics are probably formed in the A soil horizon, the determination of these compounds in that horizon may be of value in the interpretation of data obtained in soil surveys.

In the case of soil gas, the ratio of $CO_2:O_2$ (Table 1-2) seems to offer particular potential for the future. This is because during the oxidation of a sulfide ore, the natural gas mixture will become depleted in O_2 and richer in CO_2. In addition, if the sulfides are in carbonate rocks, the sulfuric acid formed during the oxidation of the sulfide minerals will react with the carbonates to yield additional CO_2, thus the $CO_2:O_2$ ratio will be increased further. Studies in the Soviet Union have shown that CO_2 anomalies are as much as 3 to 6 times background over polymetallic deposits covered by 70 - 80 feet of overburden. It appears, therefore, that the measurement of $CO_2:O_2$ ratios in soil gases may be a useful prospecting technique.

The analysis of many of the vapors listed in Table 1-2 is particularly difficult, even with the availability of gas chromatographs and mass spectrometers capable of great sensitivity. In the field, instruments may not be necessary, because dogs have been trained to help in the search for sulfide deposits by sniffing out boulders of ore occurring in dispersion trains and fans from sulfide deposits. The use of dogs to detect vapors of SO_2 and possibly other gases, from oxidizing sulfide minerals is a relatively new innovation. The method was pioneered by Dr. A. Kahma of the Geological Survey of Finland, and was based upon the assumption that the keen sense of smell of dogs should enable them to detect the vapors from weathering sulfides, even when the sulfides are buried up to a few meters. Subsequent experiments have shown that some dogs can be trained to discern sulfide-bearing rocks not normally found by conventional or geochemical prospecting. Dogs are likely to be particularly valuable in boulder-tracing in glaciated areas. However, dogs failed to sense porphyry copper mineralization in the Highland Valley, British Columbia (Coope, 1973).

The most complete report on the use of dogs as an aid in exploration for sulfides has been presented by Brock (1972). After a brief review of the history of the use of dogs in Scandinavia, he described the results of several years of experiments with two Alsatian dogs in Canada, particularly in British Columbia. Among his numerous observations and conclusions are:

1. The dogs should commence training as puppies. They are not able to work steadily until the age of two, but from three years of age, they work quite well. Ideally, the handler should be a prospector so that he can interpret any float discoveries.

2. The training of the dog may be quite long and requires that the handler learn to interpret the dog's reaction to subtle situations. Eventually, the Geological Survey of Finland anticipates that the dogs will work eight hours a day for eight years.

3. Weather conditions play an important part in the dog's ability to detect SO_2 vapors. Ground temperatures below 50°F tend to reduce the rate of oxidation and hence the release of vapors. When relative humidity is too high or too low, evaporation and vapor release is affected, and these affect the success rate in finding sulfide-bearing boulders. As a result, dogs are not particularly effective in the coastal areas of British Columbia and Alaska. Also rugged areas, and those which have heavy underbush, do not appear favorable for dogs.

4. In Finland, dogs are reported to have indicated the presence of sulfide-bearing boulders up to two feet below the surface, and in one case a dog indicated an area where trenching yielded sulfide boulders at a depth of 10 feet. In both of these cases the discoveries were associated with good soil anomalies which may have aided the dogs in their searches through relations yet unknown.

In summary, from the reviews of Bristow and Jonasson (1972) and McCarthy (1972) it is concluded that the use of some vapors, particularly mercury, is successful and reasonably well understood. But problems are encountered with surveys made above the surface, whether from vehicles or aircraft. The collection and analysis of vapors from soils, or in the air just above ground surface, offer excellent possibilities for the detection of deeply buried deposits, but more research is needed to outline the exploration parameters.

Other Types of Surveys

There are numerous other types of surveys (Table 1-1) which are conducted from time to time, but they are not significant with respect to the number of samples analyzed. However, some may well be more important in the future. They include:

1. Heavy minerals: These surveys are geochemical, by definition, only if the heavy minerals are analyzed by chemical methods. Some surveys require the separation of magnetic and non-magnetic fractions prior to analysis.
2. Bog materials: These materials are discussed in Chapter 11.
3. Fish and other fauna: The copper, zinc and lead contents of trout livers have been used as pathfinders for mineralization (Warren et al., 1971). These surveys may be classed as biological. Watson (1972) analyzed termite mounds for gold.
4. Isotopic surveys: These involve the determination of the isotopic abundances of carbon, sulfur, oxygen, silver or lead, some of which have been used on a reconnaissance basis, and others on a detailed basis. These are discussed in Chapter 7.
5. Overburden surveys: This non-specific term includes any survey in areas which are covered by glacial debris, laterites, or any other transported or residual materials (including soils). Overburden, by definition, includes all material above bedrock.

PRESENT STATUS

The technique of exploration geochemistry is held in high regard and, as successes continue to be recorded, the use of geochemical methods will increase further. A survey conducted in 1971 by The Association Of Exploration Geochemists showed that most mining companies use geochemical methods, and that about 12% of their exploration budgets are alloted for this purpose. In addition, many mining companies anticipate future needs for exploration geochemists on their staffs. Presumably other resource-oriented companies engaged in the search for metals, such as the petroleum companies, have similar budgets and needs for exploration geochemists. In the Soviet Union, where modern exploration geochemistry was first employed, this field continues to enjoy unprecedented growth as indicated by the number of scientific papers published and the increasing number of institutes devoted to the study of geochemical prospecting.

It is difficult to determine how many geochemical samples are collected each year; such information would provide an indication of activity in the field. It is especially difficult to determine the number of chemical analyses performed on each sample. A survey conducted by the Analysis Committee of The Association Of Exploration Geochemists reported that for the one year period from June 1, 1970 to May 31, 1971, 143 laboratories, representative of industry, government and universities throughout the world (except the Soviet Union and other Communist countries), analyzed about 3,800,000 samples which were arranged geographically in the following order of decreasing number of samples analyzed: Australia-New Zealand, Canada, United States, Europe, South Africa, Mexico-Central and South America, Japan, and Southeast Asia. Of this group, Australia-New Zealand and Canada accounted for 55% of the total number of analyzed samples. If all laboratories in the Western World which analyze geochemical samples were included in the compilation, the total would probably have been at least 5,000,000 and possibly 7,000,000, and these figures do not include tests made in the field. Possibly, the total for the non-Communist world is less than that for the Soviet Union alone, but exact figures do not appear to be available.

Although many universities in North America offer courses in exploration geochemistry, none have undergraduate degrees or specialization in this field, though several do at the graduate level. Among those in the profession, it is generally agreed that the most effective method of training exploration geochemists is at the graduate level, after students receive a thorough training in both geology and chemistry. This is necessary so that they may understand and interpret such fundamental geochemical principles as geochemical dispersion, and have some proficiency in analytical methods. Interest by students has grown steadily in recent years with many finding summer employment in exploration geochemistry.

In Canada, exploration geochemistry has overcome all of the poor publicity it received during early attempts to use it in the Canadian Shield. This geological province has some very special problems and limitations to the use of geochemistry, which are discussed in greater detail in Chapter 11. After a slow start, geochemistry is now being applied to exploration within the Canadian Shield with moderate success, and the future should see more momentum in this direction. Much of the original suspicion of the use of geochemical methods in the Canadian Shield was the result of incorrect methods and, fortunately, fundamental studies by the Geological Survey of Canada and other government agencies, as well as by industry, are leading the way to a better understanding of the correct methods of sampling, analysis and interpretation. Similar problems were experienced in the late 1960's in the thick lateritic, arid areas of Australia, and these too are being overcome by the combined efforts of industry and government.

At the present time, as in the past, geochemical methods are often used in conjunction with other techniques, such as geophysics and geology. In such joint surveys, geochemistry is used to detect anomalies or favorable structures found by other methods. In some circumstances, such as in areas of deep overburden as in Australia or tropical parts of Central and South America, or in mountainous areas such as British Columbia where geochemical drainage basin surveys are particularly valuable, geochemistry is increasingly being used as the sole exploration method, at least at the reconnaissance level. This tendency to rely more fully on exploration geochemistry is a result of greater confidence in the methods of sample collection, analysis and interpretation, as well as some unquestionable successes.

During the Third International Geochemical Exploration Symposium in Toronto, Canada (April 16-18, 1970), The Association Of Exploration Geochemists was organized (Coope, 1971). The Association now has more than 500 members from over twenty-five countries, attesting to its international status. Its aims include fostering exploration geochemistry by maintaining high professional standards, and working closely with universities, government and industry. Several very useful committees were established, one of which has been referred to above (the Analysis Committee). Other important and active committees include Education, Case History and Computer Applications. The Publications Committee (recently disbanded) started the Association's official publication, *The Journal Of Geochemical Exploration*, published by Elsevier Publishing Company, which made its appearance in 1972 and promises to be an important journal.

GEOCHEMICAL SUCCESSES

The success of geochemistry in finding mineral deposits is difficult to evaluate because in many cases more than one method has been employed to locate a particular discovery, and it is not always possible to assign credit to a single method. Also, the techniques employed in discoveries are not always announced by the successful companies. Nevertheless, numerous discoveries can be credited to exploration geochemistry. The Casino, Yukon, porphyry copper deposit is considered a geochemical success (Archer and Main, 1971), and geochemistry played a significant role in the discovery of a zinc deposit in Newfoundland, copper deposits in British Columbia and several other deposits of various types in Canada, Australia, Ireland, the Soviet Union and elsewhere. Coope (1973) has noted five instances (one each in Bougainville, Mexico and Wales, and two in British Columbia) where geochemical prospecting in areas of previously known showings has led to the discovery of much more extensive mineralization. But as pointed out by Derry (1971), the record of geochemistry has not been outstanding as a percentage of the total number of discoveries in comparison with other methods such as geophysics. Modern geochemical methods have only recently reached the point of refinement at which better work can be carried out, and only recently have geochemists with the proper training and experience become available in increasing numbers. Further, the percentage of exploration budgets allocated to geochemistry has been much lower than that allocated to geophysics and other methods.

Another point to be considered is that an ore deposit in one part of the world may not be an ore deposit in another. Geochemistry is credited with the discovery of several very small ore bodies in Mexico, which are only defined as ore because they can be worked profitably by one or two men on a part-time basis; in many other countries these very small ore deposits would be of interest only to mineral collectors. In non-capitalist countries such as the Soviet Union, items like the purchase of property, royalties, labor costs, taxes and many others do not enter significantly into the decision as to whether a deposit is economic or not, regardless of the method by which it was found. Therefore, the attribution of an economic or technical success to a geochemical exploration program becomes equivocal without adequate knowledge of all the circumstances concerning the discovery.

RESEARCH IN THE IMMEDIATE FUTURE

Study of current trends in exploration geochemistry makes it possible to suggest the course that new advances and research will take for a short time into the future (perhaps 5 years). Some of these include:

1. The continued search for methods effective in finding buried ore bodies. Soviet geochemists have claimed that they have had success in discovering buried bodies as deep as "some hundreds of meters" (e.g., Tauson et al., 1971) by using suites of elements (inter-element relationships) in soils, overburden and stream sediments as path-

finders. This intriguing possibility will likely be the subject of many studies in the future. Suggested geochemical approaches to this problem include study of (a) geochemical behavior of ore and trace metals in fault zones; (b) geochemical criteria for determining the ore mineralization potential of intrusives; (c) primary zoning and attempting to extrapolate these to depth; (d) new pathfinder elements, including such non-metals as the halogens which may be involved in transport of ore metals; and (e) geochemistry of rare elements in the altered zones around ore deposits.

2. Airborne detection methods for mercury, iodine, and other volatile materials (Table 1-2) will be improved and interpretation simplified. Gamma-ray spectrometry will probably be used with increasing frequency, not only for the detection of U, Th, and K, but also to provide useful geological and structural information. Various other "remote sensing" methods will be improved, applied to exploration, and integrated with present geochemical exploration programs.

3. Developments in instrumentation for chemical analysis will continue to be made with lower limits of detection obtained for many elements. Further, practical multi-element capabilities will be sought for many instruments (such as atomic absorption spectrometers), and improved in others (such as emission spectrographs). Research into the development of portable field instruments using radioactive sources will continue.

4. Electronic data processing of all types will attract much attention, particularly as the ability to analyze samples rapidly for many elements increases the need for interpretation of the data.

5. Biogeochemistry, and associated environmental geochemistry, will receive more attention as the ability to apply the data in these fields to exploration problems advances.

6. Much effort will be expended in developing geochemical methods to discover large, low-grade deposits, such as porphyry copper ores and nickel-bearing laterites, together with advances in technology for making these low-grade deposits more economic. Exploration geochemistry, assisted by statistical treatment of the data, is particularly well adapted to the discovery of these generally broad, extensive targets.

LITERATURE

The literature of exploration geochemistry is widely scattered among many scientific journals, some trade journals, an increasing number of textbooks, as well as the proceedings of the bi-annual International

Geochemical Exploration Symposium. The more important literature may be summarized as follows (of the Russian books, only those which have been translated are cited):

1. General summaries in textbook or phamphlet form: Hawkes (1957), and Hawkes and Webb (1962) based mainly on American and African experiences; Ginzburg (1960) based on Soviet experiences; Granier (1973) based on French experiences.
2. General summaries within scientific journals which are short and tend to stress recent advances: Boyle (1967), Andrews-Jones (1968), Boyle and Smith (1968), and Boyle and Garrett (1970).
3. Textbooks on biogeochemistry and geobotany: Malyuga (1964) and Brooks (1972).
4. Symposia with particular reference to geochemical prospecting in the tropics: United Nations (1963, 1970).
5. Compilations of papers concerned with geochemical prospecting in Fennoscandia and other glacial areas (which have considerable application to Canadian problems): Kvalheim (1967) and Jones (1973).
6. Symposia covering a wide range of topics on applied geochemistry from the following International Geological Congresses: XX in Mexico City (Lovering et al., 1958); XXI in Copenhagen (Marmo and Puranen, 1960); XXIII in Prague (Tugarinov and Grigorian, 1968).
7. Proceedings of the following International Geochemical Exploration Symposia; these volumes present a combined total of more than 150 papers and more than 50 abstracts, cover a wide range of subjects, represent the most advanced state of the science at the time they were released, and are indispensible to the serious student of this subject:
   a. First Symposium, Ottawa; Cameron (1967).
   b. Second Symposium, Denver; Canney (1969).
   c. Third Symposium, Toronto; Boyle (1971).
   d. Fourth Symposium, London; Jones (1973a).
8. Exploration Geochemistry Bibliography (from 1965-1971): lists about 1300 titles as well as an extremely useful cross-index of subjects and localities. Available from the Secretary, The Association of Exploration Geochemists (as Special Volume No. 1). Supplements will be published annually in *Journal of Geochemical Exploration*.
9. Journal devoted exclusively to exploration geochemistry: *Journal of Geochemical Exploration,* official journal of The Association of Exploration Geochemists. Publication began in 1972.
10. Major journals and other publications which often include articles of interest to exploration geochemists (a partial list only): *Economic Geology; Canadian Institute of Mining and Metallurgy Bulletin; Transactions of the Institution of Mining and Metallurgy, London* (Section B, Applied Earth Science); *Geological Survey of Canada Bulletin;*

Geological Survey of Canada Paper; U.S. Geological Survey Bulletin; U.S. Geological Survey Circular; Geokhimiya; Geochemistry International; and *Geochimica et Cosmochimica Acta.* Trade journals: *Northern Miner; Western Miner; Canadian Mining Journal.*

11 Abstracting journals: *Chemical Abstracts; Mineralogical Abstracts.*

12. Miscellaneous: Lang (1970) has published an unusually valuable book presenting a discussion of many phases of exploration in a simplified form, including a chapter on geochemical prospecting; Stanton (1966) has published a book on colorimetric, rapid methods of trace analysis; and Bradshaw et al. (1972) have presented an outstanding seven-part series of articles on exploration geochemistry with particular application to Canadian problems.

2/ The Primary Environment

INTRODUCTION

Geochemical methods of prospecting are based, to a large extent, upon a systematic study of the *dispersion* of chemical elements in natural materials surrounding, or associated with, ore bodies. Dispersion may be defined as a process causing the distribution or re-distribution of elements by physical and chemical agencies. Hawkes (1957), as well as Hawkes and Webb (1962), have related dispersion processes to the geochemical environment which they divided into two parts: (1) the *primary environment* which embraces those areas extending downward from the lower levels of circulating meteoric water to those of the deep-seated processes of igneous differentiation and metamorphism, and (2) the *secondary environment* which comprises the surficial processes of weathering, soil formation, and sedimentation at the surface of the Earth. The subdivision adopted by Hawkes (1957) and Hawkes and Webb (1962) is followed here as their classification into primary and secondary, with the use of additional descriptive terms such as primary and secondary dispersion patterns, is a practical one. The primary environment is generally one of relatively high temperature and pressure with little free oxygen, and limited movement of fluids. The secondary environment, on the other hand, is characterized by lower temperatures and pressures, abundant free oxygen and other gases, particularly carbon dioxide, and the relatively free flow of fluids.

Movement of materials between the primary and secondary environments may be presented graphically in the form of a simplified closed system known as a geochemical cycle (Fig. 2-1). The geochemical cycle is defined as the sequence of stages in the migration of elements during geologic changes. It can be thought of as beginning with the primary environment, in which there is formation of magma and crystallization of igneous rocks. Rocks formed in the primary environment may reach the secondary environment and, by a variety of geologic processes, of which the most important are weathering, erosion, sedimentation, diagenesis and biological action, many changes can result. Minerals formed under the primary conditions are often unstable in the secondary environment and will weather, with the result that elements contained within them may be released, transported and re-distributed. It is during this process of trans-

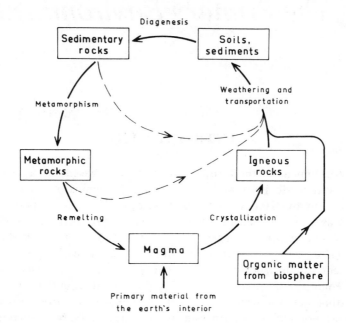

Fig. 2-1. The geochemical cycle. From Saxby (1969).

portation and re-distribution that extremely sensitive geochemical methods are particularly applicable, and may be used either to find the primary source of the elements being weathered, or new ore deposits resulting from the re-distribution.

In this chapter, basic geochemical principles relating to the primary environment will be considered. These are followed in Chapter 3 by those principles characteristic of the secondary environment. Any attempt to discuss subjects as complex as element associations, weathering or soil formation within the confines of two chapters must, of necessity, be little more than reviews whose main purpose is to present the reader with highlights essential for an understanding of what is to follow. Details can be found in the literature in the form of articles covering specific items, and some excellent books which summarize these segments of geochemistry. The following are recommended:

1. Goldschmidt (1954), and Rankama and Sahama (1950) cover similar material, and discuss the general principles of element distribution, as well as details of the geochemical behavior of each element. Although these volumes are essentially classics, they still contain much valuable information that has stood the test of time. Mason (1966) may be regarded as a simplified version of the above two volumes in which the data have been further synthesized.

2. Garrels and Christ (1965), and Krauskopf (1967) cover topics of importance to exploration geochemistry, such as oxidation-reduction, weathering, and mobility of elements in solution, in a more rigorous chemical manner. Krauskopf (1967) also includes excellent discussions on many other topics such as crystallization of magmas, ore-forming solutions and metamorphism.

3. Recent reviews and books on weathering, soil formation, and the chemistry of soils include publications by Mitchell (1964; 1972), Bear (1964), Loughnan (1969), and Hunt (1972).

DISTRIBUTION OF ELEMENTS IN IGNEOUS ROCKS AND MINERALS

Bowen's Reaction Series

Although the final words have certainly not been written on the origin of igneous rocks, nor on the mechanisms by which elements are distributed in them, for the purposes of this book, we can assume that the classical view of differentiation is correct. An acceptance of differentiation ignores such questions as the possible origin of granite by "granitization", but has the advantage that it presents a framework within which to unify the discussion. By accepting the hypothesis that most igneous rocks formed by a differentiation process, it is then possible to write of "early formed minerals" and "late differentiates". This is the general procedure used in most of the geochemical literature and it will be continued here.

Differentiation starts with a parent magma often of basaltic composition and, as cooling proceeds, early crystallizing minerals settle out to form cumulate rocks. The remaining melt is, therefore, changed in composition because it has lost those elements that have been incorporated into the early-formed minerals. By this process the magma eventually may pass through a differentiation series such as gabbro-diorite-granodiorite-granite-pegmatite, and at each point in the series, rocks with distinctive mineralogy and chemical composition crystallize. Accompanying the crystallization and separation are important changes in the abundance of the major (>1%) and minor (0.1-1.0%) elements. For example, in the differentiation sequence there is usually a gradual overall decrease in the content of such elements as iron, magnesium, calcium and titanium with a corresponding enrichment in silicon, aluminum, sodium and potassium in the residual liquid.

The differentiation sequence can also be discussed mineralogically in terms of Bowen's reaction series (Fig. 2-2) again starting with a magma of basaltic composition. This actually consists of two lines-of-descent called

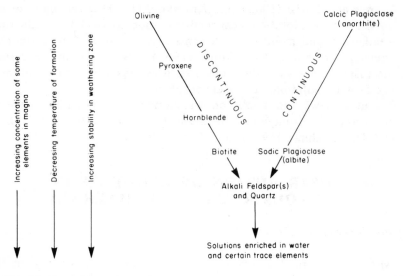

Fig. 2-2. Bowen's reaction series.

the *discontinuous series* and the *continuous series*. The former, illustrated by the sequence of minerals from olivine to biotite on the left side of Fig. 2-2, shows that as the basaltic magma cools, olivine, the first mineral to separate, reacts with the melt to form pyroxene, the next mineral in line, from which it is different in both structure and composition. In other words, the growth of pyroxene uses up olivine. This reaction, and all others in Bowen's series, take place at a definite temperature or over restricted temperature intervals. If no fractionation (separation) of minerals occur, then the melt with pyroxene and associated calcic plagioclase from the continuous series (see below), will eventually crystallize as a basalt or gabbro. If, however, there is fractionation of olivine from the system, then the bulk composition of the residual magma will be changed, the reaction process may continue, and the remaining melt may react with the pyroxene to form hornblende. Meanwhile, in the continuous series, which is only concerned with the plagioclase feldspars, similar reactions between crystals and melt are proceeding simultaneously with those in the discontinuous series, and the composition of the plagioclase changes from calcic to sodic. The more fractionation of minerals from the system, the more extensive the process. The early crystallizing minerals are generally SiO_2-poor, contain little or no K_2O and have a high MgO/FeO ratio. Therefore, the final liquid will be enriched in silica, water and other volatiles, as well as certain elements such as rubidium and cesium which are not incorporated into the common early-forming minerals in significant quantities.

Distribution of Trace Elements

As the major and minor element composition of a magma changes with differentiation, so too does the trace element composition. A comparison of the minor and trace (<0.1%) element composition of important igneous rock types is presented in Table 2-1, which lists the average abundance of 65 elements of particular interest in exploration geochemistry.

Table 2-1. Average abundance (or range) of selected minor and trace elements in the Earth's crust, various rocks, soil and river water. (All values in ppm, except those for river water which are ppb).

Element	Earth's crust	Ultra-mafic	Basalt	Grano-diorite	Granite	Shale	Lime-stone	Soil	River water
Ag	0.07	0.06	0.1	0.07	0.04	0.05	1	0.1	0.3
As	1.8	1	2	2	1.5	15	2.5	1-50	2
Au	0.004	0.005	0.004	0.004	0.004	0.004	0.005	—	0.002
B	10	5	5	20	15	100	10	2-100	10
Ba	425	2	250	500	600	700	100	100-3000	10
Be	2.8	—	0.5	2	5	3	1	6	—
Bi	0.17	0.02	0.15	—	0.1	0.18	—	—	—
Br	2.5	1	3.6	—	2.9	4	6.2	—	20
Cd	0.2	—	0.2	0.2	0.2	0.2	0.1	1	—
Ce	60	8	35	40	46	50	10	—	0.06
Cl	130	85	60	—	165	180	150	—	7800
Co	25	150	50	10	1	20	4	1-40	0.2
Cr	100	2000	200	20	4	100	10	5-1000	1
Cs	3	—	1	2	5	5	—	6	0.02
Cu	55	10	100	30	10	50	15	2-100	7
Dy	3	0.59	3	3.2	0.5	5	0.4	—	0.05
Er	2.8	0.36	1.69	4.8	0.2	2	0.5	—	0.05
Eu	1.2	0.16	1.27	1.2	—	1	—	—	0.07
F	625	100	400	—	735	740	330	—	100
Ga	15	1	12	18	18	20	0.06	15	0.09
Gd	5.4	0.65	4.7	7.4	2	6	0.6	—	0.04
Ge	1.5	1	1.5	1	1.5	1.5	0.1	1	—
Hf	3	0.5	2	2	4	3	0.5	—	—
Hg	0.08	—	0.08	0.08	0.08	0.5	0.05	0.03	0.007
Ho	1.2	0.14	0.64	1.6	0.07	1	0.1	—	0.01
I	0.5	0.5	0.5	—	0.5	2.2	1.2	—	7
In	0.1	0.01	0.1	0.1	0.1	0.1	0.02	—	—
Ir	0.0004	—	—	—	—	—	—	—	—
La	30	3.3	10.5	36	25	20	6	—	0.2
Li	20	—	10	25	30	60	20	5-200	3
Lu	0.50	0.064	0.20	—	0.01	0.5	—	—	0.008
Mn	950	1300	2200	1200	500	850	1100	850	7
Mo	1.5	0.3	1	1	2	3	1	2	1
Nb	20	15	20	20	20	20	—	—	—
Nd	28	3.4	17.8	26	18	24	3	—	0.2
Ni	75	2000	150	20	0.5	70	12	5-500	0.3
Os	0.0004	—	—	—	—	—	—	—	—
Pb	12.5	0.1	5	15	20	20	8	2-200	3
Pd	0.004	0.02	0.02	—	0.002	—	—	—	—
Pr	8.2	1.02	3.9	8.5	4.6	6	1	—	0.03
Pt	0.002	0.02	0.02	—	0.008	—	—	—	—
Rb	90	—	30	120	150	140	5	20-500	1
Re	0.0005	—	0.0005	—	0.0005	—	—	—	—

Table 2-1. (Continued)

Element	Earth's crust	Ultra-mafic	Basalt	Grano-diorite	Granite	Shale	Lime-stone	Soil	River water
Rh	0.0004	—	—	—	—	—	—	—	—
Ru	0.0004	—	—	—	—	—	—	—	—
Sb	0.2	0.1	0.2	0.2	0.2	1	—	5	1
Sc	16	10	38	10	5	15	5	—	0.004
Se	0.05	—	0.05	—	0.05	0.6	0.08	0.2	0.2
Sm	6	0.57	4.2	6.8	3	6	0.8	—	0.03
Sn	2	0.5	1	2	3	4	4	10	—
Sr	375	1	465	450	285	300	500	50-1000	50
Ta	2	1	0.5	2	3.5	2	—	—	—
Tb	0.9	0.088	0.63	1.3	0.05	1	—	—	0.008
Te	0.001	0.001	0.001	0.001	0.001	0.01	—	—	—
Th	10	0.003	2.2	10.	17	12	2	13	0.1
Ti	5700	3000	9000	8000	2300	4600	400	5000	3
Tl	0.45	0.05	0.1	0.5	0.75	0.3	—	0.1	—
Tm	0.48	0.053	0.21	0.5	—	0.2	0.1	—	0.009
U	2.7	0.001	0.6	3	4.8	4	2	1	0.4
V	135	50	250	100	20	130	15	20-500	0.9
W	1.5	0.5	1	2	2	2	0.5	—	0.03
Y	30	—	25	30	40	25	15	—	0.7
Yb	3	0.43	1.11	3.6	0.06	3	0.1	—	0.05
Zn	70	50	100	60	40	100	25	10-300	20
Zr	165	50	150	140	180	160	20	300	—

Notes:
1. Dashes (—) indicate no data are available.
2. Earth's crust: All data from Taylor (1964, 1966) except for Mn, Ti and Se (Saxby, 1969); Re, Ir, Os, Pd, Pt, Rh and Ru (appropriate chapters in Wedepohl, 1969); Te (Parker, 1967).
3. Igneous rocks, shales and limestones: All data from Taylor (1964, 1966, 1969) except for Mn, Ti and Se (Andrews-Jones, 1968; Saxby, 1969); Re, Ir, Os, Pd, Pt, Rh and Ru (appropriate chapters in Wedepohl, 1969); Te (Parker, 1967); F, Cl, Br and I (Turekian and Wedepohl, 1961).
4. Soil: All data from Taylor (1966) except for Ag, Cs, Hg, Hf, Mn, Sb, Ti, Tl and Zr (Saxby, 1969, who also reports total rare earth elements in soils as approximately 100 ppm); Th (Andrews-Jones, 1968).
5. River water data from Turekian (1969) except for chlorine (Livingstone, 1963) and uranium (Reeder et al., 1972). All rare earth element data (except La) are estimates by Turekian.

The list also includes average abundances of these trace elements in the crust (which is itself considered equivalent to the average value for all igneous rocks), shale, limestone, soil and river water, since these values will be discussed later. The compilation of "average" amounts is beset with many difficulties, particularly as the Earth's crust is not uniform and we can sample only to relatively shallow depths. Thus, although reliable sources (chiefly Taylor, 1966) were used in compiling Table 2-1, the values are subject to change as new data are obtained. Further, as will be noted in the discussion of geochemical provinces, the trace and minor element content of rocks with the same name (e.g., granite) can vary significantly. This is particularly true for the rarer elements in rocks generally, and especially in soils. Many recent environmental studies of soils are now enabling geochemists to compile more reliable trace element averages based on significantly more data. Because soils are derived from so many rock types with varying composition, ranges were used by Taylor (1966), instead of averages, and this procedure is followed in part in Table 2-1.

The data in Table 2-1 show that the abundances of most minor and trace elements vary from ultramafic to granitic igneous rocks. They also show how remarkably widespread are small amounts of most elements including those of interest in exploration geochemistry. It has been mentioned previously that both the minerals formed during Bowen's reaction series, and the minor and major element abundances, vary with the degree of differentiation and hence rock type, and therefore it is reasonable to expect the trace element content of these rocks and their associated minerals to vary. Table 2-2 lists most of the common igneous minerals specifically mentioned in Bowen's reaction series, as well as others, along with the approximate content of a large number of minor and trace elements which may be found in them. The trace elements within these common rock-forming minerals are the origin of the background values found in residual soils formed on igneous or metamorphic rocks, as well as the background values for the rocks themselves. With the exception of zircon which is a source of Hf, none of the minerals listed in Table 2-2 is an economic source of any minor or trace element. These common rock-forming minerals are mostly silicates from which it would be extremely expensive to extract the elements contained within their structures. The two non-silicates are apatite, a phosphate ore which also has potential as a source of uranium and possibly rare earths, and magnetite, which may yield commercial by-product vanadium and tin, but both minerals can only be considered as future sources for rare elements. In geochemical surveys, therefore, it is important to determine what proportion of the total amount of an element of interest is in a form amenable to mining, milling and metallurgical extraction and what quantity is in an economically unrecoverable form, primarily in silicates or other minerals such as magnetite (Table 2-2). The chemical analysis of pure mineral separates for trace elements, or the differential chemical treatment of bulk samples prior to analysis, are the two methods usually used to determine the proportion of elements in the various minerals, although electron probe studies are now being used with increasing frequency.

The controls on the occurrence of trace elements in rock-forming minerals (Table 2-2) are fairly well understood, in general terms. They are generally explained by a combination of three factors: ionic size, valency, and type of chemical bond, all of which will be discussed in further detail. However, at this point it is sufficient to note that certain elements prefer not to enter rock-forming silicates during crystallization of the magma.

Table 2-2. Selected elements found in small amounts in common rock-forming minerals of igneous rocks, and relative stability of the minerals.

Mineral	X%	0.X%	0.0X%	0.00X% and less	Stability
Olivine	—	Ni, Mn	Ca, Al, Cr, Ti, P, Co	Zn, V, Cu, Sc	
Amphibole	—	Ti, F, K, Mn, Cl, Rb	Zn, Cr, V, Sr, Ni	Ba, Cu, P, Co, Ga, Pb, Li, B	
Pyroxene	—	Ti, Na, Mn, K	Cr, V, Ni, Cl, Sr	P, Cu, Co, Zn, Li, Rb, Ba	Easily Weathered
Biotite	Ti, F	Ca, Na, Ba, Mn, Rb	Cl, Zn, V, Cr, Li, Ni	Cu, Sn, Sr, Co, P, Pb, Ga	
Plagioclase	K	Sr	Ba, Rb, Ti, Mn	P, Ga, V, Zn, Ni, Pb, Cu, Li	
Epidote	RE	Mn, Ti	Th, Sn	V, Nb, Zn, Be, U	
Sphene	—	RE, Nb, Sn, Sr	Mn, Ta, V, Cr	Ba	
Apatite	—	Sr, RE, Mn	U, Pb	As, Cr, V	Moderately Stable
Garnet	Mn, Cr	Ti, RE	Ga	—	
Feldspar (potash)	Na	Ca, Ba, Sr	Rb, Ti	Pb, Ga, V, Zn, Ni, Cu, Li	
Muscovite	—	Ti, Na, Fe, Ba, Rb, Li	Cr, Mn, V, Cs, Ga	Zn, Sn, Cu, B, Nb	
Tourmaline	—	Ti, Li, Mn	Cr, Ga, Sn, Cu, V	Rb	
Magnetite	Ti, Al, Cr	Mn, V	Zn, Cu, Sn, Ni	Co, Pb, Mo	
Zircon	Hf	RE, Th	Ti, Mn, P	Be, U, Sn, Nb	Stable
Quartz	—	—	—	Fe, Mg, Al, Ti, Na, B, Ga, Ge, Mn, Zn	

Note: The above groupings are generalizations; for example, only the allanite variety of epidote contains abundant rare earth (RE) elements, and only some varieties of muscovite will have Rb and Li in the quantities indicated.

These elements include Li, Be, Nb, Ta, Sn, U, Th, W, Zr, and the rare earths, and they tend to be preferentially concentrated in the residual fluids high in water, together with other components such as HF, HCl and CO_2 which facilitate increased mobility within the residual silicate fluids.

These elements are characteristic of pegmatites, although some are found in small quantities in rock-forming minerals (Table 2-2). Their eventual concentration in the final residual stage when the water-rich, increasingly fluid magma cools to below 600°C, is an example of *primary mobility,* as they were continually mobile throughout the crystallization history of the magma. The residual fluids with their associated late-stage elements may also give rise to hydrothermal solutions, however, the mechanism for deriving ore-forming fluids containing Sn and U, for example, from the pegmatite stage of crystallization is obscure.

From this discussion it is clear that not only is Bowen's reaction series one of changing chemical and mineral composition, but it is also one of decreasing temperature, and increasing concentration of certain elements and volatile compounds. It is also one of apparently increasing mineral stability in the weathering environment because the last formed minerals, such as quartz, are much more stable than the first formed minerals, such as olivine. These features are shown in Figure 2-2 and Table 2-2, and indicate which trace elements are separated in the igneous environment, and which are the first to be available for remobilization in the weathering environment.

Post-magmatic (Hydrothermal) Ore Deposits

Numerous studies have attempted to confirm the classical hypothesis for the derivation of post-magmatic hydrothermal ore deposits. Holland (1972) has discussed the problem and noted that the hypothesis is based on three propositions: (1) that magmas contain between 1-5 wt. % water, much of which is released during crystallization as hydrothermal solutions; (2) that hydrothermal solutions produced in this manner contain a sufficient concentration of metals to give rise to the observed base metal deposits; and (3) that it is the decrease in temperature and pressure, together with the reactions of these solutions with wall rocks which leads to the precipitation of ore minerals. Holland (1972) has demonstrated the validity of the three propositions and the classical hypothesis for the origin of post-magmatic hydrothermal ore deposits, by means of experimental and geological data, certainly with respect to zinc, manganese, and probably lead ore deposits (which were the three elements he considered). Further, he suggested that water, chloride, base metals, and sulfur are the most important ingredients in the production of such ore deposits. It is especially important to note that he considers that the criteria for assessing the availability of these components in particular plutons may turn out to be of value in prospecting. Chloride, in particular, is apparently a critical component, presumably because it forms complexes with the metals and thus assists in their transport. Stollery et al. (1971) have shown that biotites in a granodiorite stock which is closely related to the base metal deposits at Providencia, Zacatecas, Mexico, contain 0.3 to 0.5 weight percent chlorine, whereas the average chlorine content of biotites from granitic rocks, in general, is apparently near 0.1 percent.

A magmatic source for the metallic and gangue elements in hydrothermal deposits is not favored by all geologists, at least for all deposits. For example, Boyle (1968) presents very convincing arguments as to why he believes that the elements in hydrothermal deposits came from the enclosing country rocks. From their sites in the country rocks, the elements can be mobilized during metamorphism, or collected by deep-seated circulating groundwater, and subsequently concentrated into ore deposits. Boyle and Lynch (1968) have speculated on the source of zinc, lead and other elements in Mississippi Valley type lead-zinc deposits in carbonate rocks. They visualize an initial concentration of the elements from sea water by marine organisms, followed by biochemical and bacterial processes, and finally a concentration of the elements by brines in favorable porous zones or structural sites. Derry (1971) has presented evidence that supergene processes, specifically those involving groundwater, are extremely important in the formation of some uranium, lead-zinc, copper and other metal deposits. Numerous alternative sources for metals in assumed hydrothermal deposits have been proposed over the past century. These include acid and basic magmatism, metamorphic fluids, brines and groundwater. For a source of sulfide, "biogenic" action, which means that the sulfide was derived by the bacterial reduction of sulfate, cannot be overlooked as an important factor in many sedimentary environments (Trudinger et al., 1972).

The important point to be gained from the above discussion is that highly competent geochemists and economic geologists can visualize, and convincingly explain, alternative mechanisms for similar ore deposits. Those who are prepared to accept concepts, models and mechanisms which represent multiple views of ore genesis are likely to be in a better position to interpret, and benefit from geochemical data. Many of these aspects are covered in more detail in textbooks on geochemistry and economic geology, such as those of Krauskopf (1967), Park and MacDiarmid (1964), and Stanton (1972), as well as in many journal articles.

Complex Formation

Many elements are neither mobile nor soluble in ionic form, to any appreciable extent, either in magmas, late-stage magmatic fluids, or in ground and surface waters. Rather than moving as ions, these elements are believed to form complexes with other elements or radicals, with the resulting complexes having different chemical characteristics from ions. Metal complexes with the halogens, particularly chlorine and fluorine, are believed to be very important within magmas and the hydrothermal solutions derived from them. This belief is based on laboratory and theoretical studies (e.g., Helgeson, 1964), geological observations such as in fluid inclusions (Roedder, 1967, 1972), and the analysis of modern metalliferous brines, such as those from the Salton Sea and Red Sea. Krauskopf (1967) mentioned that chloride complexes such as $PbCl^+$ and $AuCl_4^-$, and sulfide complexes like AsS_2^- and $HgS_2^=$, are particularly important in this context. In ground and surface waters, uranium and molybdenum may form complexes with oxygen (for example, the uranyl complex, UO_2^{++}), which explains their mobilities in these environments. Dunham (1972) has observed that 16 elements (Mo, W, U, Cu, Ag, Au, Zn, Cd, Hg, Tl, Ge, Sn, Pb, As, Sb, and Bi), all of which are of interest in exploration geochemistry, form a compact group in the center of the periodic chart. However, they have no other obvious common characteristics, such as bond type, ionic size and valency. Nevertheless, Dunham (1972) noted that these elements must have some common chemical property which enables them to be concentrated in ore deposits by an essentially similar process of solution, transport and eventual deposition, as evidenced by the fact that they are concentrated in a wide variety of combinations. Dunham (1972) also pointed out that these elements all have the common property of forming complexes with chloride, sulfide and bicarbonate below 350°C.

The subject of complex formation in geochemical systems is in its infancy, and many aspects are not well understood. Still to be investigated are complexes involving metals with chloride, bicarbonate, sulfide and polysulfides, while complexes with Sb, As and Te have yet to be confirmed. The study of metal complexes is of importance not only with respect to element mobility, but also in relation to pathfinder elements (see below). This is because halogens and the other complex-forming elements may possibly migrate further than their associated metals, once the complex is broken and the metal precipitated.

PRIMARY HALOS AND PRIMARY DISPERSION

In igneous rocks, the distribution of major elements is controlled by the stability of individual minerals, which in turn is governed by conditions of temperature, pressure and the availability of elements in the original magma. It was mentioned previously that certain trace elements (metals) are able to enter the structures of rock-forming minerals (Table 2-2). If this occurs, these elements are then removed from the magma and are thus eliminated from any further possibility of concentration into ore deposits, at least in the primary environment. Other trace elements, such as those that commonly occur in pegmatites and certain hydrothermal deposits, remain mobile until they reach an environment in which they are able to crystallize as stable minerals, occasionally in economically significant quantities. As the content of elements which are not incorporated into the lattices of the common rock-forming minerals increases during differentiation, so the mobility of the residual fluid also increases owing to the increasing abundance of water and volatiles, which reduce the viscosity in the silica-rich end stages of the magma. With decreasing temperature and pressure, the composition of the residual fluids may approach the composition of dilute aqueous solutions. Eventually the elements within these residual fluids will crystallize and it is believed that most hydrothermal vein deposits precipitated from such late-stage material.

Many mineral deposits of igneous or hydrothermal origin are characterized by a central zone, such as a vein, in which the valuable elements (or minerals) may be concentrated in economic percentages. The degree of concentration may be in the high percent range as in the case of barite, fluorite and manganese ores, in the low-percent range in the case of lead and zinc ores, or in the parts per million(ppm) range as in the case of gold and platinum. In most cases there is a progressive decrease in the amount of the valuable element surrounding the deposit until the content reaches that of the enclosing rock, and is classified as background. The zone within which the content of valuable elements diminishes to background values is called the *primary halo* (Fig. 2-3). Primary halos (also called primary aureoles) represent the distribution patterns of elements which formed as a

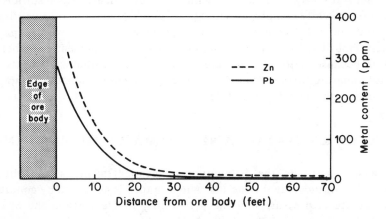

Fig. 2-3. Primary halo, as defined by zinc and lead contents, developed in limestone. Generalized from data obtained from the Vieille Montagne mine, Nenthead, England (Finlayson, 1910).

result of *primary dispersion,* that is, the distribution or re-distribution of elements in the primary environment. The word "primary" implies first, that the dispersion originated with processes from within the Earth, and second, that the halo was formed at the same time, or nearly so, as the central core. As will be seen later (Chapter 7), the halo may vary considerably in size and shape from that of the ore body; in fact, there is almost an infinite variety of forms it may assume owing to the variables of fluid movement in rocks.

Many halos, especially those in which elements have been introduced into massive (unfractured) wall rock, characteristically show a logarithmic decay away from the mineralized zone. Some halos can be detected for distances of hundreds of feet, whereas others are no more than a few inches in width. Further, because of the different mobility characteristics of the many elements within crystallizing fluids, each element may have its own halo characteristic of a particular deposit. Clearly, primary halos are complex. Some of the major factors determining their size and shape are: mobility characteristics of the elements in solution; microfractures in the rocks; porosity and permeability of the host rock; tendency of fluids to react with host rocks such as limestones; volatility of the elements; physical factors such as the viscosity and pressure of the magma; and the tendency to form stable minerals.

Studies of primary halos have resulted in the recognition of several types based on both the time of emplacement, and on the geometry. With respect to time of emplacement, two types of halos are recognized which are referred to as syngenetic and epigenetic. *Primary syngenetic halos* are those formed essentially contemporaneously with the enclosing rock, for example, halos associated with pegmatites or ultramafic segregations. *Primary epigenetic halos* are formed after the rock has crystallized, and result from the introduction of mineralizing solutions along fractures or faults, with a resultant halo being produced either upwards and/or outwards into the surrounding rock (Fig. 2-3). These terms are not to be confused with the terms endogenic and exogenic, which apply to the origin or place of formation of ore deposits. *Endogenic* ore bodies are those of deep-seated origin (primary environment), whereas *exogenic* ore bodies are of near-surface origin (secondary environment). It is also possible to speak of endogenic and exogenic halos.

Based on geometrical characteristics, Hawkes (1957) classified primary halos as: (a) areal patterns, resulting from widespread impregnation (up to several miles) of a large mass of rock by hydrothermal solutions or other fluids emanating from depth, (b) leakage patterns, with well-defined systems of solution channels, and (c) wall rock patterns, where the rock adjacent to the solution channels has been modified by hydrothermal activity. A leakage pattern generally implies that the disper-

sion (or migration) of the element(s) is along channels, faults, microfractures and other paths leading from the ore body, and a *leakage halo* is
developed in the overlying rocks during (or shortly after) emplacement of
the ore (Fig. 2-4). Leakage halos, like other forms of primary dispersion,

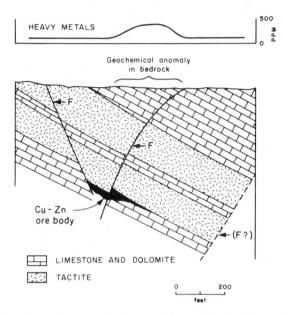

Fig. 2-4. Leakage halo defined by the heavy metals content of rocks exposed at the surface about 400 feet above the mineralization. Rocks above the ore body are faulted and
fractured. Generalized from data at the Moore ore body, Johnson, Arizona, described
by Cooper and Huff (1951).

vary widely in size and shape. They are commonly narrow over vertically
dipping structures but may be broad and oval or circular over highly fractured stocks. They are especially important because they offer much hope
in the search for some types of buried ore deposits. Leakage halos have
been identified in fault and fracture systems at more than 500 feet from
mineral deposits and they have been detected in most types of geochemical
samples, particularly soil, rock and water.

    Gaseous leakage halos offer even more hope for the discovery of buried
deposits. This is because certain elements such as mercury, emanating from
mineralizing solutions move through the pore spaces of rocks and soils as
gases (Table 1-2). As a result of their extreme volatility and mobility they
move farther than other elements, resulting in broader halos which even
extend to the atmosphere. Other elements, such as iodine and bromine, are
also gases at the temperatures and pressures likely to be encountered in
hydrothermal fluids and offer similar possibilities although no gaseous

leakage halos based on these elements have yet been described. Radiogenic gases, such as radon and helium, which are produced by the decay of uranium and thorium, form gaseous leakage halos which may be considered secondary or epigenetic. Mercury and radon have been analyzed successfully from soils, waters of all types, and rocks. Airborne surveys for these and other gaseous elements are being conducted with increasing frequency and the future will no doubt see more effort to perfect these methods (McCarthy, 1972).

The recognition and interpretation of primary halos is one of the main objectives of geochemical rock surveys. Once a primary halo is outlined, it is usually not difficult to narrow down the source of the element or mineral accumulation because of the limited extent of the halo (generally less than 500 ft from the deposit). Boyle et al. (1969) and Dass et al. (1973) have shown conclusively that the native silver veins at Cobalt, Ontario can be located by detecting the primary halos. Analysis of rocks from Cobalt for As, Sb, Co, Ni, Ag, Cu, Pb, Zn and Hg indicates that the contents of these elements increase as the silver veins are approached. Soviet geochemists claim to have detected primary halos associated with massive pyritic deposits up to a kilometer vertically above steeply dipping structures, and hundreds of meters laterally. It is this type of halo that is attracting most interest in the West, as evidenced by the increasing attention given to them in the recent literature (e.g., Boyle, 1971). In view of the great importance of primary halos and primary dispersion, they are covered in greater detail in Chapter 7.

PATHFINDERS

From the foregoing discussion of primary halos, it is clear that many elements are found in hydrothermal solutions. It has also been stressed that certain elements are more mobile than others either because of the physical-chemical conditions of the solutions in which they are found, or because of their physical state (e.g., gaseous). It is this mobility that permits the development of extensive primary halos. The element constituting the most extensive or furthest halo is generally not the same element that comprises the main mineral deposit; however, it is one that is closely related geochemically. The fact that one or more elements (or minerals) can be closely associated and can constitute a halo, assists in finding sought-after mineral deposits, and has led to the concept of a *pathfinder,* a term originally proposed by Warren and Delavault (1953; 1956). Pathfinders (often called indicator elements) are defined as relatively mobile elements (or gases) occurring in close association with the element being sought, but which can be more easily found either because they form a broader halo, or because they can be detected more easily by present analytical methods. Pathfinders are particularly useful in the search for buried ore deposits because they generally form larger halos. They are used in both the primary and secondary environments.

Representative pathfinders are listed in Table 2-3, including some unusual elements, for example, the use of gold as a pathfinder for porphyry copper deposits (Learned and Boissen, 1973). In some cases the pathfinder element may be in the gangue, whereas in others it may be substituting in the structure of the ore mineral. In the case of polymetallic ores, even one of the ore elements may be the pathfinder for the deposit. Some geochemists restrict the word *indicator* to the situation in which one of the major elements of a polymetallic deposit is determined, for example zinc in a Pb-Ag-Zn deposit, but such terminology is not followed here.

There are two basic reasons for the successful use of pathfinders: (1) they are more mobile than the element being sought and thus form a broader, more extensive halo; for example, the use of mercury or arsenic as a pathfinder for gold, and (2) the analytical methods used for the pathfinder are simpler, less expensive, and more sensitive than the methods for the element in the ore body; the use of easily analyzed copper, nickel and

Table 2-3. Examples of pathfinder elements used to detect mineralization.

Pathfinder Element(s)	Type of Deposit
As	Au, Ag; vein-type
As	Au-Ag-Cu-Co-Zn; complex sulfide ores
B	W-Be-Zn-Mo-Cu-Pb; skarns
B	Sn-W-Be; veins or greisens
Hg	Pb-Zn-Ag; complex sulfide deposits
Mo	W-Sn; contact metamorphic deposits
Mn	Ba-Ag; vein deposits; porphyry copper
Se, V, Mo	U; sandstone-type
Cu, Bi, As, Co, Mo, Ni	U; vein-type
Mo, Te, Au	porphyry copper
Pd, Cr, Cu, Ni, Co	platinum in ultramafic rocks
Zn	Ag-Pb-Zn; sulfide deposits in general
Zn, Cu	Cu-Pb-Zn; sulfide deposits in general
Rn	U; all types of occurrences
SO_4	sulfide deposits of all types

Note: In most cases, several types of material (e.g., rock, soil, sediment, water and vegetation) can be sampled. In some cases, such as radon, only water and soil gas are practical. In the case of sulfate, only water is practical.

chromium as pathfinders for platinum is a good example. Some potential pathfinder elements have yet to be exploited primarily because present analytical methods are not sufficiently sensitive to detect them at the extremely low levels in which they occur. Rhenium is such a potential pathfinder for porphyry copper deposits (Coope, 1973). It occurs in small amounts (generally less than 2000 ppm) within molybdenite which is associated with porphyry copper deposits, and is very soluble in water, thus permitting it to form extensive, but so far undetected, halos in the secondary (weathering) environment.

The selection of a pathfinder requires that the element(s) used occur in the primary environment with the element being sought, or else be derived from it by radioactive decay, such as the use of radon as a pathfinder for uranium. It is also essential that there be a direct and interpretable relation between the geochemical distribution of the pathfinder and the mineralization. Interestingly, recent Soviet studies have shown that, consideration must be given to shape and position of the ore body, and particularly whether a vertical or lateral halo is expected when choosing pathfinders. This is because the mobility of pathfinder elements varies considerably, both vertically and laterally (Ovchinnikov and Grigoryan, 1971).

Although this discussion has been concerned mainly with the nature of pathfinders in rocks (primary environment), there is no reason why pathfinders cannot be used extensively in the secondary environment. Historically, in fact, pathfinders have been used chiefly in the secondary environment such as in soils, waters, and sediments. Only recently has interest been changing to rock geochemistry and pathfinders in the primary environment. Further discussion of the use of pathfinders in the secondary environment (secondary dispersion halos) is presented in Chapter 8.

GEOCHEMICAL PROVINCES

Geochemical provinces are relatively large, well-defined, areas of the Earth's crust that have a distinctive chemical composition. This definition does not indicate how large these areas are because size is less significant than chemical composition. Bradshaw et al. (1972) suggested "up to tens or sometimes hundreds of miles", which is certainly acceptable for purposes of discussion. They also pointed out that geochemical provinces are the largest example of primary halos. Although most people think of geochemical provinces in terms of the accumulation of elements, there can just as easily be provinces of element depletion.

Numerous examples of geochemical provinces could be cited. Among the better known are the copper-producing areas of Peru and Chile, the porphyry copper belt extending from northern Mexico and the south-western United States through British Columbia to the Yukon, the car-bonatite area (rare earths, Sr, etc.) of East Africa, the uranium province of the Colorado Front Range, and the gold fields of South Africa. Recently there has been a tendency to include reference to association of rock types with the specific mineralization, instead of only the geographic area. For example, certain granitic rocks have higher than normal accumulations of copper and molybdenum (e.g., porphyry-copper in altered "granites" of the southwestern United States), whereas others are higher in tin (e.g., tin-bearing granites of Bolivia). Mafic rocks may also occur in groups or clusters which are high in the associated elements nickel-platinum-chromium as in South Africa, and these likewise constitute a geochemical province. Geochemical provinces may also be based on the composition of sedimentary rocks as in the case of the Mississippi Valley carbonates con-taining lead and zinc in larger than average amounts, or the copper-enriched shales of Germany and Zambia. The most recently recognized geochemical province is on the Moon where the Apollo 11 basalts at Mare Tranquillitatis contain at least five times as much titanium (approximately 12% TiO_2) as a typical terrestrial basalt, and several times as much as the basalts from other maria (Levinson and Taylor, 1971). The great variety of terrestrial rocks and ores which give rise to areas large enough to be classed as geochemical provinces suggest origins due to many geological processes. Some are probably the immediate result of element variations in that part of the crust or mantle from which the magmas were separated. Sedimentary provinces are the result of favorable source areas and sedimentological environment, whereas in metamorphic rocks favorable conditions of remobilization are necessary to produce a geochemical province. It is important to remember that the relatively large areas for-ming a geochemical province rarely comprise one rock type, but are com-posed of suites of rocks, in which either all or most of the rocks, often of different ages, exhibit similar geochemical characteristics of enrichment or depletion.

Geochemical provinces are useful in exploration primarily because they outline broad areas (primary dispersions) in which there are proven occurrences of a metal. They are, therefore, ideal places to start ex-ploration programs. On the basis of our current understanding of the prin-ciples of economic geology, the presence of one type of occurrence suggests that similar mineralization is likely within the province. No better exam-ples can be offered than the discovery within this past decade of at least four major porphyry copper deposits within a relatively small area of southern British Columbia, and the large number of nickel discoveries in the "nickel-belt" near Kalgoorlie, Western Australia. Bradshaw et al. (1972) also pointed out that in the future the day may come when entire rocks (as opposed to ores, which actually are rare concentrations), will have to be mined to supply world needs, especially for the rarer metals. Certainly it is advisable to consider the mining of rocks from geochemical provinces which are enriched, albeit only slightly, in the element sought. Examples of such rocks are the Conway granite in New Hampshire, and the Chattanooga shale, which are relatively enriched in thorium and uranium, respectively. As most geochemical provinces have been recognized in either igneous or metamorphic terrains, it is likely that there will be more interest in lithogeochemical techniques in the future in order to outline these provinces, although under favorable circumstances they can be detected by means of reconnaissance stream sediment, or water drainage basin surveys. Based on our present understanding of the geochemistry of trace and minor elements in igneous rocks and minerals, it seems that geochemical provinces may be detected, or at least suspected, following the analysis of relatively few well-selected specimens.

The term "metallogenic province" has been used to characterize an unusual spatial abundance of a particular metallic element or association of elements. The examples cited for geochemical provinces also illustrate metallogenic provinces, such as the copper province of the southwestern United States, and the gold province of South Africa. Boyle (1967), however, made an excellent point when he noted that the term metallogenic province is a misnomer, because a province is not only indicated by one (ore) element, but by characteristic variations in several other elements which need not necessarily be ore elements. Hence the term geochemical province is preferred. He pointed out that associated with the ore element there is usually a concentration of other metals and non-metals, including gangue material. For example, the great gold province of the Canadian Shield is really concentrations of silica in the form of quartz, or of calcium, magnesium, iron, and carbon dioxide in the form of carbonates. Gold, from which the "gold province" obtains its name, is only an incidental element, generally in amounts less than 15 ppm (0.5 oz/ton). Other examples can be cited of base metal provinces which are really sulfur provinces. This is an important and fundamental point because the more we learn about element occurrences and associations, and particularly the relations between major elements and ore elements, the better we will be able to delineate favorable geochemical provinces. By including the non-ore (gangue) elements in a geochemical province, we can more easily recognize pathfinders, such as fluorine and boron in geochemical provinces characterized by abundant tin, or manganese in barium and silver geochemical provinces. Throughout the remainder of the book only the term geochemical province will be used, although most probably the term metallogenic province is too deeply rooted in the literature of economic geology to be changed by the decision made here.

A word of caution with respect to the use of geochemical provinces was mentioned in the discussion of rock surveys (Chapter 1) and it is worth repeating. Based on a study of the trace element content of certain Canadian veins and massive sulfide deposits and their enclosing rocks, Boyle and Garrett (1970) concluded that there are localities in which it may not be possible to recognize definite elemental patterns in the country rocks to indicate the presence of metal-bearing deposits. In the areas they studied the metal content of the country rocks did not reflect the deposits present, being neither above nor below normal by comparison with non-productive areas. This implies that important metal deposits may occur in host rocks which give no indication of their presence. Boyle and Garrett (1970) are prompt to point out that the lack of correlation may be the result of only very small variations between barren and productive areas, and that more positive results would possibly be obtained with sophisticated statistical treatment of the analytical data. A complicating factor in some of the areas studied is that there are black shales and schists, presumably enriched in metals, and that these rocks, rather than the enclosing igneous and metamorphic rocks, could have been the source of the ore elements. Nevertheless, their conclusions do present an enigma, the solution of which requires more geochemical and statistical work.

Finally, "biogeochemical" provinces have been recognized by Vinogradov (1963). These are based on chemical changes in organisms resulting from variations in the local geochemical environment, and two types are described. The first is related to a deficiency of certain elements in soils, such as the deficiency of Co, Cu, I, and Ca in podzols and brown forest soils of the temperate zone; this type is closely related to climatic zones. The second is characterized by excesses of certain elements in the soils, and is more directly related to geochemical provinces than to climate. The latter is of obvious interest in exploration geochemistry. Biogeochemical provinces are discussed in more detail by Malyuga (1964) and Brooks (1972).

GEOCHEMICAL ASSOCIATIONS

General

In the discussions of the differentiation of igneous rocks, Bowen's reaction series, primary halos and pathfinders, the fact was frequently mentioned that certain elements are commonly associated. One of the most practical applications of this knowledge has been in the use of pathfinders. However, there are some associations in which none of the elements are apparently suitable for use as pathfinders, usually because of their lack of mobility in either the primary or secondary environment. But these associations still have important geochemical significance, as the presence of one element can commonly suggest, or even require, the presence of other elements. For example, in those pegmatites which contain easily recognizable Li, Be, Ta and Sn minerals such as spodumene, lepidolite, beryl, tantalite and cassiterite, the occurrence of pollucite (cesium silicate) is a distinct possibility. In addition, hafnium is always found within zirconium minerals, the "light" rare earth elements La-Ce-Pr-Nd (and usually Sm) are always associated and, porphyry copper deposits commonly contain some Mo and Au. Table 2-4 lists some of the more important element associations likely to be encountered in geochemical exploration. In some cases the associations will be the same as those included in the selected list of pathfinders (Table 2-3) because a pathfinder element must also be an associated element. A point to be kept in mind is that an exploration geochemist must be alert to all clues to mineralization and therefore to all element associations, and thus the reported occurrence of a mineral should be recognized immediately for its significance. Further, element associations can be of great value in interpreting data obtained from soil and other types of geochemical surveys.

The associations in Table 2-4 include those from both the primary and secondary environments. Some associations, such as Pb-Zn-Cu, may be found in either environment, but others are restricted. This is because of the differing mobilities of elements under extremely different environmental conditions. For example, some elements such as Li, Be, Rb, Cs, Nb, and Ta remain in association throughout the entire igneous environment and are concentrated in the pegmatite phase, but they are disassociated during the supergene cycle of weathering and sedimentation. In order to understand the distribution of elements in the primary igneous environment, it is first necessary to consider some fundamental geochemical observations and relations, many of which were formulated in the 1930's by Goldschmidt.

In the igneous environment, there are only eight major elements (O, Si, Al, Fe, Ca, Na, K, Mg) which are present in the Earth's crust in amounts greater than 1%, and these comprise nearly 99% of the crust. Oxygen is unquestionably dominant, making up about one-half the weight and over 90% of the volume. Silicon and aluminum are also abundant, and, consequently silicates and aluminosilicates of other cations (Fe, Ca, Na, K, Mg) are the most common minerals in the crust. Four elements (Ti, H, P and Mn) are present in amounts between 0.1 - 1%, but all the remaining elements together make up less than 0.5% of the crust, and these include almost all the important ore elements of concern in exploration geochemistry. Inspection of Table 2-1 shows that for the 65 elements listed, the highest crustal value is 5700 ppm (0.57%) for titanium. The table also reveals that many familiar elements which are important to man and industry are extremely rare, whereas other elements are unexpectedly relatively abundant. For example Hg, Mo, Sb, Bi and Au are rarer than Ce, Dy, Hf, Sc and Ga.

Table 2-4. Selected geochemical associations of minor and trace elements. Modified from Andrews-Jones (1968) and other sources.

Rock Type or Occurrence	Association
1. *Plutonic Associations*	
Ultramafic rocks	Cr-Co-Ni-Cu
Mafic rocks	Ti-V-Sc
Alkaline rocks	Ti-Nb-Ta-Zr-RE-F-P
Carbonatites	RE-Ti-Nb-Ta-P-F
Granitic rocks	Ba-Li-W-Mo-Sn-Zr-Hf-U-Th-Ti
Pegmatites	Li-Rb-Cs-Be-RE-Nb-Ta-U-Th-Zr-Hf-Sc
2. *Hydrothermal Sulfide Ores*	
General Associations	Cu-Pb-Zn-Mo-Au-Ag-As-Hg-Sb-Se-Te-Co-Ni-U-V-Bi-Cd
Porphyry copper deposits	Cu-Mo-Re
Complex sulfides	Hg-As-Sb-Se-Ag-Zn-Cd-Pb
Low temperature sulfides	Bi-Sb-As
Base metal deposits	Pb-Zn-Cd-Ba
Precious metals	Au-Ag-Cu-Co-As
Precious metals	Au-Ag-Te-Hg
Associated with mafic rocks	Ni-Cu-Pt-Co
3. *Contact Metamorphic Rocks*	
Scheelite-cassiterite deposits	W-Sn-Mo
Fluorite-helvite deposits	Be-F-B
4. *Sedimentary Associations*	
Black shales	U-Cu-Pb-Zn-Cd-Ag-Au-V-Mo-Ni-As-Bi-Sb
Phosphorites	U-V-Mo-Ni-Ag-Pb-F-RE
Evaporites	Li-Rb-Cs-Sr-Br-I-B
Laterites	Ni-Cr-V
Manganese oxides	Co-Ni-Mo-Zn-W-As-Ba-V
Placers and sands	Au-Pt-Sn-Nb-Ta-Zr-Hf-Th-RE
Red beds, continental	U-V-Se-As-Mo-Pb-Cu
Red beds, volcanic origin	Cu-Pb-Zn-Ag-V-Se
Bauxites	Nb-Ti-Ga-Be
5. *Miscellaneous*	
	K-Rb; Rb-Cs; Al-Ga; Si-Ge; Zr-Hf; Nb-Ta; RE; S-Se; Br-I; Zn-Cd; Rb-Tl; Pt-Pd-Rh-Ru-Os-Ir

(Note: RE = rare earth elements)

The reason why one group of elements is familiar and the other is not, is a reflection of the fact that the rarer elements mentioned above all form easily recognizable minerals and, under favorable conditions, they are concentrated into veins or other deposits which can be mined. Ga and Hf never form individual minerals but instead are *dispersed* within the crystal structures of aluminum and zirconium minerals, respectively. Even though Ga is relatively abundant in the Earth's crust, its ionic size, valency and chemical characteristics are so close to those of aluminum that it is essentially always found within the structure of aluminum silicates. For the same reasons, Hf substitutes for Zr, Rb substitutes for K, and Re generally substitutes for Mo. In the case of Sc, this element is mostly found substituting for Mg or Fe^{+3}, although rarely, relatively small amounts of individual scandium minerals form, particularly in pegmatites as the rare mineral thortveitite ($Sc_2Si_2O_7$). The rare earth elements (RE), such as Ce and Dy mentioned above, are likewise found substituting for Ca within the structures of many minerals (e.g., apatite, garnet, epidote; Table 2-2) which crystallize throughout the magmatic sequence, although during the late stage phases they sometimes form their own minerals such as monazite, xenotime and allanite. On occasion, rare earth minerals are found in deposits sufficiently large to be considered rare earth ores (e.g., carbonatites and beach sands). The fact remains, however, that regardless of crustal abundance, it is those elements forming the easily recognizable minerals and concentrated by geological processes into deposits amenable to mining, that are the basis of our mineral economy.

It may be asked why certain elements, even though their crustal abundance is very low, form mineable amounts of minerals, whereas other more abundant elements rarely, or never, form their own minerals. The answer has been alluded to previously by referring to ionic radius, valency and type of chemical bonding (which comprise part of the field of crystal chemistry). However, a more complete explanation is necessary. It will then become apparent that the recognition of element and mineral associations, together with an understanding of the simple principles of crystal chemistry are the first steps in the recognition and interpretation of geochemical processes and environments, which data can in turn be used for exploration purposes.

Goldschmidt's Geochemical Classification

V. M. Goldschmidt devoted his outstanding career in geochemistry to the study of the distribution of elements in the Earth, and to an attempt to formulate a set of laws by which the observed element distributions could be explained. When Goldschmidt began his work in the 1920's the basic facts relating to the regularities inherent in the periodic chart (Fig. 2-5) were well-known: arrangement of electrons in atomic structures; active metals at the left of the chart, with non-metals at the right; within each vertical column the valence is the same, but metallic activity increases in a downward direction; etc. From these facts it was possible to predict accurately the similarity in geochemical distribution between Zr-Hf and S-Se-Te, but at the same time the prediction of other associations, such as Nb-V and Mo-Cr, was incorrect.

Goldschmidt's approach to the problem of element distribution was based on the assumption that the primary (initial) distribution of elements probably took place during, or shortly after, the time of formation of the primitive Earth. In its very early history the entire Earth was molten, and on cooling three phases separated: a metal phase, a sulfide phase, and a silicate phase. Goldschmidt based his concept on the composition of meteorites which can be divided into three major groups: iron-nickel, troilite (FeS), and the silicates. He conceived the idea of studying trace element distributions in these meteorite phases, as well as the analysis of metal, matte (sulfide) and slag (silicate) formed during metallurgical processes. In addition, he presumed that the composition of silicate rocks and sulfide ores on Earth would also yield valuable data on the distribution of the elements.

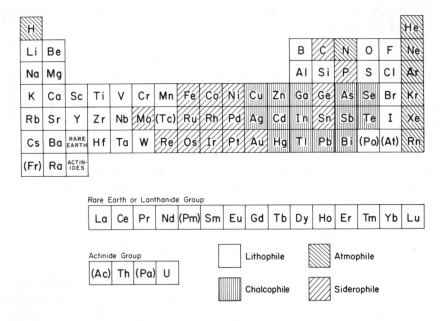

Fig. 2-5. Goldschmidt's geochemical classification of the elements.

The results of these studies showed that the distribution of elements in meteorites, smelter products, natural rocks, sulfide ores (and rare terrestrial native iron) all agreed reasonably well, and from this Goldschmidt (1937; 1954) concluded that the elements can be classified on the basis of geochemical affinity:

1. *Siderophile*: with affinity for iron; concentrated in the Earth's core
2. *Chalcophile:* with affinity for sulfur; concentrated in sulfides
3. *Lithophile:* with affinity for silicates; concentrated in the Earth's crust
4. *Atmophile:* as gases in the atmosphere.

Figure 2-5 schematically illustrates Goldschmidt's grouping of the elements, but it is only generalized for he realized that some elements have characteristics common to two groups. For example, Au is primarily siderophile, but it often occurs with sulfides (chalcophile). Chromium is strongly lithophile in the Earth's crust, but, if oxygen is deficient, as in some meteorites, it is chalcophile and may form an iron-chromium sulfide mineral (daubréelite, $FeCr_2S_4$). Nevertheless, the groupings are an approximate qualitative indication of natural associations and offer partial explanations for phenomena such as the scarcity of the Pt-group metals and Au in crustal rocks (Table 2-1). In this instance the explanation is that these elements are siderophile and thus concentrated in the Earth's core, a fact indirectly confirmed by studies of the lunar rocks. From Fig. 2-5 it is clear that the geochemical character of an element, using Goldschmidt's terminology, is closely related to its position within the periodic chart, and that this in turn is governed by electronic configurations and other periodic characteristics. The geochemical character of elements can also be correlated in a general way with electrode potential and free energy of the oxides of the metals, but all attempts to render the classification quantitative failed. Nevertheless, the terms siderophile, chalcophile, lithophile and atmophile, as originally defined by Goldschmidt, are still valuable and are widely used in the geochemical literature.

The geochemical character of an element and its position within the periodic chart can also be correlated with the type of bonding it prefers. Ionic, metallic and covalent bonding are clearly related, at least in a general way, to the lithophile, siderophile and chalcophile groups. Lithophile elements ionize readily and tend to form, or be associated with, silicate minerals in which ionic bonding (the transfer of electrons as in sodium chloride) is found. Chalcophile elements form covalent bonds (the sharing of electrons as in the case of sphalerite or diamond) with sulfur, or with selenium or tellurium if these are present. Siderophile elements normally prefer the metallic bond characteristic of metals and do not tend to form compounds with oxygen or sulfur, thus explaining why gold and the platinum metals commonly occur as native elements. Goldschmidt (1937) also recognized a fifth type of chemical affinity. This he called biophile and in it he included those elements commonly concentrated in organisms, primarily C, H, O, N, P, S, Cl and I, as well as certain trace metals (V, Mn, Cu, Fe and B).

At the time Goldschmidt proposed his system of classification, X-ray diffraction studies of crystals were just permitting scientists to determine the atomic structure of minerals and to understand the significance of the various types of interatomic bonding (ionic, metallic, covalent, and Van der Walls'). Goldschmidt was very quick to realize the importance of crystal structure as a major controlling factor in the distribution of elements. The oxygen ion is extremely large in comparison with most other elements, so that mineral structures can be considered a framework of oxygen ions with the cations filling the interstices. In such a situation, the size of the cation is a major factor in determining whether it can fit into a particular structure. Fig. 2-6 illustrates the ionic radii of the elements, which are a periodic feature of the chart in that they change in a systematic manner, for example gradually increasing in the direction Li-Cs in the first group.

Bowen's reaction series was proposed in 1928 at about the time that Goldschmidt was doing his fundamental work. Integrating information on the affinity of elements for particular environments (siderophile, chalcophile, lithophile, atmophile), the facts of crystal structure (ionic, radii, valency, type of bonding), and the known sequence of mineral crystallization (Bowen's series), Goldschmidt was able to determine the general distribution of minor elements during the crystallization of igneous rocks, assuming slow crystallization of an orderly sequence of minerals from a differentiating melt. Chalcophile elements, and some lithophile elements such as chromium, settle out early possibly as sulfide or oxide minerals. Many of the other trace elements are incorporated into the common rock-forming minerals, mainly on the basis of ionic size. As long as the ionic radii of minor (or trace) and major elements do not differ by more than 15%, substitution is possible. Ionic charge is another important factor, and on a practical basis it is necessary that the charges between the two elements do not vary by more than one, and also that electrical neutrality be maintained by other substitutions. For example, Nb^{+5} may substitute for Ti^{+4} in rutile, provided neutrality is maintained by another substitution such as Fe^{+3} for Ti^{+4}; but Cu^{+2} is extremely unlikely to substitute for Sn^{+4} because the valency differs by more than one. On the basis of what has been discussed so far it is not surprising that there are many element substitutions and associations (Tables 2-2 and 2-4). Thus crystals are three-dimensional networks, in which any ion may replace another of similar radius and charge, at least, as a starting hypothesis.

The nature of the structure (or the lattice sites available) takes into account the nature of the chemical bond, and is also very important in deter-

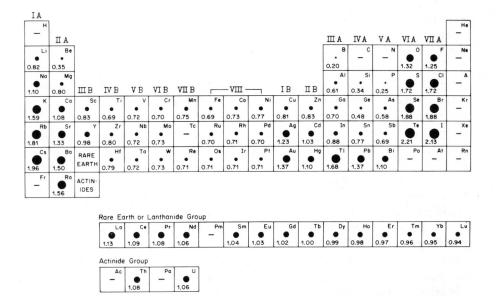

Fig. 2-6. Diagrammatic representation of the ionic radii of the elements (in Ångstrom units). Based on data from Whittaker and Muntus (1970), except for Te and Au (Pauling, 1960).

Notes:

1. Coordination numbers for all elements are VI, except for Be, B, Si, Ge and P which are IV; K, Tl, Ba, Sr, Pb and Ra which are VIII; and Rb and Cs which are XII.

2. The following valence states were used where there is a reasonable choice: 1+ for Tl; 2+ for Mn, Fe, Co, Ni, Cu, Hg and Pb; 3+ for V, Cr and B; 4+ for Sn, Mo, W and the platinum group; 5+ for Nb, Ta, P, As and Sb.

3. Data for unstable elements, noble gases, H, C and N, not included.

mining the degree of possible element substitution. Some structures, such as the micas, apatite and amphiboles, exhibit extensive substitution, whereas the quartz structure permits very little substitution, owing in large part to a lack of ions of small size and high charge which can substitute for Si^{+4} without distortion. Ge^{+4} is the most likely, but because it occurs in extremely small quantities in terrestrial rocks (Table 2-1), very little substitution is observed. The term *diadochy* is used to describe the ability of different elements to occupy the same lattice position in a crystal. By way of example, Rb, Cs and K are diadochic in the mica structure, particularly well seen in lepidolite in which each occupies the large interlayer positions. Lithium, on the other hand, is considerably smaller in ionic radius, and, although present in large amounts in the same lepidolite structure, it is not diadochic with Rb, Cs and K. Instead, it substitutes in other positions (octahedral) and is diadochic with smaller ions, particularly Mg. This example stresses that diadochy (referred to in the older literature as "isomorphous substitution"), in its strict sense, requires substitution in the same position and in the same structure.

These explanations for substitution could be discussed at greater length and it would be found that they work extremely well for the first three columns of the periodic chart (Fig. 2-6), but for the other elements there are many exceptions. For example, the Cd^{+2} ion (ionic radius 1.03Å) is very similar to Ca^{+2} (1.08Å), but Cd is not found in Ca minerals. The nature of the chemical bond in minerals, particularly the degree of ionic character, is extremely important in determining the degree of association between pairs of elements. Goldschmidt's generalizations, although a very useful guide to the distribution of elements during crystallization, are predicated on a purely ionic basis. However, minerals are not purely ionic and many studies have attempted to resolve the lack of universality of the generalizations, primarily by studying the significance and effect of the chemical bond in controlling the geochemical distribution of the elements in the igneous environment (summarized by Taylor, 1966; Mason 1966; Krauskopf, 1967). The use of the concept of electronegativity (a measure of the tendency to form covalent bonds) as a measure of bond type has met with some success, and it was found that those elements with lower electro-negatives are preferentially incorporated into silicate structures (because lower electronegativity implies higher ionic bonding). However, electronegativity values are unable to explain all element associations, nor indeed are other proposed methods. The problem of a quantitative measure of bond type and element distribution is not yet solved. The most recent approaches involve the study of partition coefficients, and the application of crystal field theory to the problem. Crystal field theory, which explains why the transition elements have preference for certain crystallographic positions, is beyond the scope of this book but is well explained by Burns (1970).

Applications

For purposes of exploration geochemistry, it is sufficient to know that Goldschmidt's methods for explaining the distribution of many trace and minor elements in igneous minerals and rocks, although far from per-fect, are good enough as a first approximation, especially for the lithophile elements. It is particularly important to realize that some lithophile trace elements (e.g., Cs, Rb) are much larger in ionic size than the abundant mineral-forming elements, and that others have much higher charges (e.g., Nb^{+5}, Ta^{+5}, Zr^{+4}), and therefore, it is possible to explain why these elements are not accommodated in the structures of the major rock forming minerals to any appreciable extent. As a result, these elements are concentrated in the residual fluids, where they eventually crystallize, generally in the pegmatite stage. Other elements, particularly certain chalcophile and siderophile elements which are of particular interest to exploration geochemists, have significantly different bonding characteristics from the silicate-forming elements, and they eventually crystallize in sulfide veins, or as native elements, in various late-stage environments.

Table 2-5. Summary of selected trace and minor element contents of galena, sphalerite, chalcopyrite, pyrite, pyrrhotite and arsenopyrite, arranged in decreasing order of relative frequency of occurrence within each mineral. All values in ppm except where indicated.

Element	Maximum Content	Most Common Range (when present)	Total No. of Samples Considered	Relative Frequency of Occurrence (%)
Galena (PbS)				
Cu	3000	10-200	51	96
Ag	3%	500-5000	233	94
Sb	3%	200-5000	224	84
Bi	5%	200-5000	327	62
Fe	5000	10-50	89	43
Mn	2000	10-50	90	41
Ni	100	10-50	40	38
Tl	1000	<10-50	148	36
Sn	1300	<10-50	338	24
As	1%	200-5000	229	22
Sphalerite (ZnS)				
Cd	4.4%	1000-5000	921	100
Se	900	<10	41	100
Mn	5.4%	1000-5000	652	87
Ag	1%	10-100	448	84
Cu	5%	1000-5000	297	80
Ga	3000	10-200	962	66
In	1%	10-50	938	52
Co	3000	10-100	413	49
Ge	5000	50-200	959	48
Sn	1%	100-200	585	40
Hg	1%	10-50	225	37
Ni	300	10-50	211	33
As	1%	200-500	235	25
Tl	5000	10-50	310	25
Sb	3%	10-50	197	24
Bi	1000	10-50	186	22

Table 2-5. (Continued)

Element	Maximum Content	Most Common Range (when present)	Total No. of Samples Considered	Relative Frequency of Occurrence (%)
Chalcopyrite ($CuFeS_2$)				
Se	2100	10-50	43	100
Ag	2300	10-1000	8	100
Sn	770	10-200	10	90
In	1000	<10-100	33	60
Ni	2000	10-50	85	54
Co	2000	10-50	88	38
Mn	2%	10-50	36	17
Pyrite (FeS_2)				
Se	300	10-50	115	97
Ni	~2.5%	10-500	1055	89
Cu	~6%	10-1%	785	87
Co	>2.5%	200-5000	1094	86
V	~1000	10-50	17	83
Pb	5000	200-500	24	79
As	~5%	500-1000	99	67
Ti	600	200-500	21	67
Mn	1%	10-50	927	54
Ag	200	<10	73	47
Sn	400	10-50	18	39
Zn	~4.5%	1000-5000	722	36
Tl	100	50-100	17	35
Bi	100	10-50	17	35
Sb	700	100-200	35	23
Pyrrhotite ($Fe_{1-x}S$)				
Cu	7000	100-200	26	100
Se	63	10-50	20	100
Ni	7.47%	50-500	244	96
Mn	3000	200-500	155	88
Ag	100	<10	20	70
Co	8500	200-500	252	65
Arsenopyrite (FeAsS)				
Mn	3000	10-50	40	98
Co	3.36%	1000-5000	54	87
Ni	3000	200-500	54	85

Notes:
1. Based on data in the literature compiled and interpreted by Fleischer (1955). For details of the significance, interpretation and uncertainties in the data, see Fleischer.
2. The "most common range" has been arbitrarily selected. In some cases it is quite limited (e.g., 10-50 ppm); in other cases it is quite broad (e.g., 200-5000 ppm).

3. In some cases, for example the Sn and Ag data in chalcopyrite, there are insufficient data for statistically meaningful interpretations.
4. Some data, such as that for Hg and Se in sphalerite, are likely superseded by more recent information.
5. In all cases, the detection limit for each element was at least 100 ppm, and in many cases was 10 ppm or less.
6. The above tabulation of trace and minor elements in the six minerals assumes that the elements substitute within the mineral structures. However, recent electron probe studies of chalcopyrite, for example, have shown that those samples with high Ag (Sb and Bi) often contain inclusions of galena or tetrahedrite, and those with high Sn may contain stannite, although some of these elements are definitely present substituting within the structures of the various minerals. Even though mineral inclusions may account for some of the elements reported above, the data are useful because they suggest possible pathfinders and by-products, and the abundances in which they are likely to be found.

Table 2-5 is a list of trace and minor elements found in several important sulfide minerals and is based on an extensive literature review by Fleischer (1955). The elements found within the structure of these sulfide minerals are predominantly chalcophile, but as would be expected from the previous discussion, some of the diadochic substitutions are difficult to explain on the basis of Goldschmidt's work alone. Nevertheless, the data are important for several reasons: (1) they indicate which elements are possible pathfinders for that particular sulfide mineral; (2) they suggest that caution is warranted in interpreting anomalous concentrations as being economically recoverable (high contents of Cu or Co within the pyrite structure, for example, are non-economic); (3) they point out which trace elements should be considered for by-product metallurgical recovery (such as Cd and Ge from sphalerite). Because by-products are often important, the trace elements which may possibly occur in each of the sulfide minerals listed in Table 2-5 are tabulated in decreasing order of their relative frequency of occurrence. For example, cadmium has been detected in every sphalerite and, in fact, sphalerite is the chief commercial source of cadmium. Similarly, silver should always be considered as a by-product in the smelting of galena. On the other hand, silver in pyrite is not recoverable economically. Similarly, up to one-half of the total gold content of some porphyry copper deposits in British Columbia is reported to be within the pyrite structure, and therefore, it is in an unrecoverable form. (Fleischer, 1955, did not include gold in his compilation which is the basis of Table 2-5).

Boyle (1969) has compiled the most complete report on element associations in mineral deposits and indicator (pathfinder) elements of interest in geochemical prospecting. He pointed out that concentrations of only one element are rare; usually, a suite of elements is concentrated in any particular deposit. Because of the many interacting processes, it is not generally possible to predict *a priori* which elements will be concentrated together. However, most elemental associations are known from empirical data which have been gathered during the long history of geochemistry. This data is of fundamental importance and can be used in a practical way by exploration geochemists such as to predict the probable occurrence of elements that may be unsuspected in deposits.

3/ *The Secondary Environment*

INTRODUCTION

The secondary environment is extremely important in exploration geochemistry and historically, most emphasis has been placed on the analysis of materials from it. Minerals and rocks that are stable in the primary environment are often unstable in the secondary environment. By various processes, primarily weathering, elements are released to soils and waters, and volatiles are released to the atmosphere resulting in secondary dispersion halos. The processes by which elements move in the secondary environment are distinctly different from those in the primary environment, and secondary halos, which are usually much larger in areal extent than primary halos, are formed and dispersed by entirely different mechanisms.

Geologists generally recognize two main types of weathering: chemical and physical. Chemical weathering involves the breakdown, by chemical means, of rocks and minerals and dispersion of the released elements, generally by water and sometimes for considerable distances from the source. Abundant water, carbon dioxide and oxygen are usually essential in this type of weathering. Physical weathering includes all those processes of rock disintegration which do not employ chemical reactions or mineralogical changes. It is generally most effective in those environments which either lack water, such as the Australian desert, or have relatively cold climates like the polar regions, but it is also important in all areas with steep and rugged topography. Both types of weathering, often assisted by biological agents, may be operative simultaneously and on the same rock. Chemical weathering is generally capable of causing much greater changes in the constituents of rocks than is physical weathering.

Chemical weathering and related topics (including biological materials such as bacteria) will be discussed first, followed by physical weathering. The aim of the discussions will be to present some of the essentials of the weathering processes for a later discussion of element mobility and dispersion, soil formation and other features of the secondary environment. Particular emphasis will be placed on how an understanding of these features is essential for efficient operation of a geochemical survey and interpretation of the results. It should be stressed that the secondary environment is of the utmost importance in exploration geochemistry, as most sampling and analysis is carried out on the materials of this environment (soil, sediments, water, etc.), and secondary halos are significantly larger

71

than those in the primary environment. For more details of chemical weathering and the oxidation of sulfides, Krauskopf (1967, particularly chapters 4 and 18) and Park and MacDiarmid (1964, particularly chapters 18 and 19) are recommended.

CHEMICAL WEATHERING

Introduction

Chemical weathering can be defined as the chemical reactions between rocks and minerals and the constituents of air and water at or near the Earth's surface. In this environment oxygen, carbon dioxide and water are abundant, and the temperatures and pressures are low in comparison with the conditions of the deep-seated primary environment discussed in the previous chapter. Many minerals, including both silicates and sulfides, which were stable in the primary environment, are unstable in the secondary environment and consequently chemical changes take place during weathering in an attempt to reach equilibrium. As a result, elements which may have been associated during the entire crystallization history of magmatic rocks, such as Li, Rb, Cs, Be, Nb, Ta and Sn, may become separated in the secondary environment, because their behavior is now governed by an entirely different set of parameters which result in different dispersion characteristics for each element. The order of resistance to chemical weathering of rock-forming minerals is generally:

oxides > silicates > carbonates and sulfides.

Weathering reactions on the Earth's surface are progressing where they can be studied and, in many cases, it is possible to reproduce in the laboratory the conditions of low temperature and pressure, and an abundance of oxygen, carbon dioxide and water, in order to confirm by experimentation what has been observed in the field. Krauskopf (1967) has observed that all chemical reactions related to chemical weathering involve four relatively simple processes: ionization, addition of water and carbon dioxide, hydrolysis and oxidation. One might therefore expect that weathering reactions are among the best understood of geochemical processes, and although to a large extent this is true, there are numerous complications. These include the fact that chemical weathering also involves physical and biological processes, such as expansion of water upon freezing, and swelling of minerals upon hydration, which affect the chemical processes. An additional problem is the extreme slowness of chemical weathering which is difficult to simulate in the laboratory. This problem is especially true for low temperature reactions in contrast with those taking place at high temperatures. The net effect is to deny us the details of some of the reactions, and an understanding of certain complexities but the overall processes are sufficiently well understood for the purposes of this book.

Some Chemical Weathering Reactions

Hydrolysis. As mentioned previously, and as indicated in Fig. 2-2 and Table 2-2, there is a sequence to mineral alteration in the secondary environment. Field observations and microscopic and laboratory studies have established that mafic minerals generally alter much more rapidly than felsic minerals and this sequence follows very closely the order of crystallization in Bowen's reaction series. In other words, minerals formed at the highest temperatures generally weather more rapidly than those formed at lower temperatures. Thus the elements present in mafic minerals are available for mobilization much earlier than those associated with alkali feldspars and micas. The weathering of silicates is primarily a process of *hydrolysis*, whereby the ionic species H^+ and OH^-, become incorporated into the structure of minerals; more specifically, there is a reaction between water and the ion of a weak acid or a weak base. Hydrolysis can be illustrated by the following equation for the alteration of iron-rich olivine (fayalite) by water at pH of 7:

$$Fe_2SiO_4 + 4H_2O \longrightarrow 2Fe^{+2} + 4OH^- + H_4SiO_4 \qquad (1)$$

Further, rainwater can result in a lower pH for waters in the zone of weathering. Rainwater when it is in equilibrium with CO_2 from the atmosphere, has a pH of 5.7 (containing carbonic acid). Analyses of rainwater, in fact, with pH values as low as 3.0, are found in industrial areas where SO_2, H_2SO_4, HCl and other gases or acids are discharged into the atmosphere. Natural contamination with these and other acidic components takes place in areas with active volcanoes (e.g., Iceland). Therefore, water, especially if reinforced with additional H^+ from rain and other sources (e.g., organic acids in soils), can readily attack silicates. In the vicinity of sulfide-bearing veins, the acids can become very strong (see below). The main exception to the above generalization is found in those waters draining limestone areas which have pH values above 7. Another important exception can be found in the interior of certain continents, such as in Canada, where most of the waters, even those draining igneous rocks, may be neutral or alkaline with pH values commonly ranging from 7.0 – 8.4 (Reeder et al., 1972).

Feth et al. (1964) in a study of the effects of weathering on granitic rocks in the Sierra Nevada of California and Nevada, found that about half of the solute content of groundwater in this terrain is acquired in the first few hours to weeks of contact of snow melt water with surface soil. They concluded that the aggressive attack of the water on the rock minerals resulted from carbonic acid in the water derived from solution of carbon dioxide in the soil. An analysis of rainwater is presented in Table 3-1, but it must be realized that precipitation may be very variable in composition and, in general, the total dissolved solids is 10 ppm or less, and that contents of components decrease rapidly towards the interior of continents. An exception is snow in central Canada which may contain appreciably more dissolved salts (up to 100 ppm) even allowing for contamination from wind-blown soil (Professor S. A. Harris, personal communication).

Table 3-1. Major components of selected river water, groundwater, geothermal water, and precipitation (in ppm or mg/l).

Component	1. River	Groundwater 2. Limestone	3. Shale	4. Rhyolite	5. Geothermal	6. Rain
Na	6.3	8.1	362	62	352	9.4
K	2.3	5.7	14	2.0	24	0.0
Mg	4.1	28	143	1.0	0.0	1.2
Ca	15	79	416	8.0	0.8	0.8
HCO_3	58.4	267	104	131	—	4
SO_4	11.2	51	2107	22	23	7.6
Cl	7.8	29	38	16	405	17
NO_3	1	28	0.2	6.7	1.8	0.0
SiO_2	13.1	8.4	26	52	363	0.3
Others	0.67	0.07	64.8	0.92	42.06	0.02
Total	120	504	3300	302	1310	38
pH	—	7.3	6.3	7.9	9.6	5.5
Temperature (°C)	—	—	6.1	15.6	94	—

Source:

1. Livingstone (1963, Table 81); mean composition of rivers of the world.
2. White et al. (1963, Anal. 10, Table 6); groundwater from well in Bayport dolomitic limestone, Grand Rapids, Michigan.
3. White et al. (1963, Anal. 16, Table 5); groundwater from well in Pierre shale, Langdon, North Dakota.
4. White et al. (1963, Anal. 5, Table 1); groundwater from spring in rhyolite, Beatty, Nevada.
5. Hem (1970, Anal. 3, Table 12); spring from Upper Geyser Basin, Yellowstone National Park, Wyoming.
6. Hem (1970, Anal. 4, Table 6); rain at Menlo Park, California, January 10, 1958. Rain taken at the same location a few hours earlier had significantly less total dissolved solids (8.2 ppm).

Other processes involving hydrolysis of silicates are more complicated, such as that for orthoclase, because two reactions must be considered, one for the potassium and one for the aluminum. In the following equation orthoclase is weathered to kaolinite, soluble KOH and SiO_2:

$$2KAlSi_3O_8 + 3H_2O \longrightarrow Al_2Si_2O_5(OH)_4 + 4SiO_2 + 2KOH \qquad (2)$$

Hydrolysis reactions involving carbonates also occur:

$$MgCO_3 + H_2O \longrightarrow Mg^{+2} + OH^- + HCO_3^- \qquad (3)$$

An important point with respect to equations (1) and (2) is the statement of Krauskopf (1967, p. 115) that "Any solution in contact with silicate minerals cannot long remain appreciably acid, and if contact is continued, the solution must eventually become alkaline". The alkalinity does not increase much above a pH of 9 because of reactions of OH^- with various forms of silica. In the case of equation (3), the solution will also become more alkaline.

A second important point illustrated in the above equations is that not only are iron, potassium and magnesium set free, but so also are all of the trace elements present in the mineral structures. Thus the trace and minor elements in the common rock forming minerals (Table 2-2) are liberated. Some remain in the soils, whereas others become mobile and enter the groundwater and eventually the drainage system (Table 3-1). These comprise the background values in soil, sediments and water samples collected as part of an exploration program.

Oxidation. Notwithstanding the importance of hydrolysis in the weathering of silicates and carbonates, *oxidation* is the dominant weathering process of interest in exploration geochemistry. This is because oxidation produces gossans, iron and manganese oxides, and secondary dispersion halos from sulfide minerals, all of which are of great importance. Therefore, an understanding of oxidation is essential. In discussing oxidation in exploration geochemistry, or in any aspect of weathering for that matter, it is sufficient to illustrate the effects on only three elements: iron, manganese and sulfur. These are the only abundant elements in rock-forming minerals and ores for which oxidation is an important factor in chemical weathering. Although other elements present in trace and minor amounts are oxidized, for example, uranium, antimony and vanadium, these can be more conveniently discussed elsewhere.

In igneous and metamorphic rocks, most iron is present in the ferrous (Fe^{+2}) state as long as the rocks and minerals remain in a reducing environment. But as soon as they are exposed to weathering, the ferrous iron will tend to oxidize to the ferric state (Fe^{+3}). Compounds containing ferrous iron are usually green in color, whereas those containing the trivalent form of iron are brown or red. Ferric oxides are found in several mineral forms: anhydrous Fe_2O_3 (hematite); and two hydrated forms of $FeO(OH)$, lepidocrocite and goethite. Goethite and lepidocrocite along with clay and other impurities, are the main constituents of limonite (which is actually a rock). These ferric oxides are extremely stable and, in fact, other ferric compounds in the zone of oxidation are uncommon.

When an iron-bearing silicate, such as olivine comes into contact with air or oxygenated water, oxidation results, as in the following equation:

$$2Fe_2SiO_4 + O_2 + 4H_2O \longrightarrow 2Fe_2O_3 + 2H_4SiO_4 \qquad (4)$$

This equation is similar to equation (1) in which hydrolysis was demonstrated, but, in addition, it shows that the ferrous iron is oxidized. In nature, oxidation is usually accompanied by hydrolysis and often by hydration and carbonation, because CO_2 is almost always present in the weathering environment. Under natural conditions in the zone of weathering (except, for example, in highly reducing environments, such as peat bogs), the ferrous iron in equation (1) would rapidly oxidize to ferric. In fact, on exposure to air any ferrous compound will be oxidized relatively rapidly, because the formation of Fe_2O_3 represents an exothermic energy change (the newly formed products possess less energy than the original ones, i.e., there is a decrease in free energy). The equation for the oxidation of siderite is:

$$4FeCO_3 + O_2 + 4H_2O \longrightarrow 2Fe_2O_3 + 4H_2CO_3 \qquad (5)$$

Even with the more complex silicates, the process of oxidation of iron operates in the same way, that is, insoluble iron oxides are formed. The decomposition of biotite, in which carbonic acid is included among the reactants is:

$$4KMg_2FeAlSi_3O_{10}(OH)_2 + O_2 + 20H_2CO_3 + nH_2O \longrightarrow$$
$$4KHCO_3 + 8Mg(HCO_3)_2 + 2Fe_2O_3.nH_2O + 2Al_2Si_4O_{10}(OH)_2$$
$$+ 4SiO_2 + 10H_2O \qquad (6)$$

And as in hydrolysis, other elements are released and ionized, and become mobile in waters. Thus trace metals in the biotite are released and as long as they do not form insoluble compounds they will remain mobile, at least initially.

Manganese is oxidized in a manner similar to that of iron. Thus Mn_2SiO_4 may be substituted for Fe_2SiO_4 in equation (4), and $MnCO_3$ (rhodochrosite) for siderite in equation (5). MnO_2 is the insoluble product of both reactions. In rocks in the primary environment, manganese is present in the manganous state (Mn^{+2}), which is readily oxidized to the manganic state (Mn^{+4}). This oxidation is a little more complicated than that of iron as manganese may also occur in a + 3 state and, in fact, some manganese minerals contain manganese in two valence states (e.g., braunite, $3Mn_2O_3 \cdot MnSiO_3$). The eventual product of oxidation in air is MnO_2, commonly pyrolusite, but other polymorphs of this compound are now known to be reasonably abundant (e.g., birnessite). All forms of MnO_2 are black and are difficult for exploration geochemists to recognize and classify because they are likely to be found only as thin coatings or films on rock surfaces and fractures, and on stream sediments.

However, it is the oxidation of sulfur that is of most interest to geochemists and economic geologists. In the primary environment this element occurs in the reduced form (S^{-2}) as sulfides. Like manganese there are several higher valence states, from S^0 (elemental sulfur), through to the highest, and most stable state of +6 in sulfate $(SO_4)^{-2}$. Pyrite and marcasite (FeS_2) are among the most abundant sulfides, and they are very easily weathered. The reaction is best written in steps.

$$2FeS_2 + 7O_2 + 2H_2O \longrightarrow 2FeSO_4 + 2H_2SO_4 \qquad (7)$$

The ferrous sulfate is then oxidized:

$$4FeSO_4 + O_2 + 10H_2O \longrightarrow 4Fe(OH)_3 + 4H_2SO_4 \qquad (8)$$

The ferric hydroxide is finally transformed to goethite or lepidocrocite:

$$4Fe(OH)_3 \longrightarrow 4FeO(OH) + 4H_2O \qquad (9)$$

By combining equations (7), (8) and (9) it is seen that the weathering of iron sulfides involves the oxidation of both iron and sulfur:

$$4FeS_2 + 15O_2 + 10H_2O \longrightarrow 4FeO(OH) + 8H_2SO_4 \qquad (10)$$

Once sulfuric acid has been formed, it may react with more pyrite and marcasite:

$$FeS_2 + H_2SO_4 \longrightarrow FeSO_4 + H_2S + S \qquad (11)$$

This equation explains why small amounts of native sulfur are sometimes found on outcrops and why H_2S is a common gas around coal mines and mine dumps. Sulfur dioxide may be generated by subsequent reactions.

The ferrous sulfate produced in equation (11) is able to react and produce more sulfuric acid as outlined in equation (8). Clearly, the ferrous sulfate and sulfuric acid generated by the oxidation of iron sulfides are powerful agents in the decomposition of other sulfide minerals, for example those of copper, zinc, arsenic and silver. Once sulfuric acid is formed, the pH is lowered drastically, perhaps from 7 to 2.5 or even lower. In this acid environment many other sulfide minerals (e.g., sphalerite, chalcopyrite) or native elements (e.g., silver) are attacked, and it is this attack which permits the initial mobilization of major metals from their respective minerals, and also those elements present in trace and minor amounts (Table 2-5). For example, when acid waters formed by the oxidation of pyrite percolate through an ore deposit, elements such as copper, zinc, silver, cobalt, selenium and nickel may be leached from other sulfide minerals, in addition to which any elements present within the pyrite itself will also be mobilized.

Reactions similar to those in equations (7), (8), (9) and (10) can be written for the oxidation of many other sulfide minerals, such as for chalcopyrite ($CuFeS_2$) and bornite (Cu_5FeS_4). Those simple sulfides, such as sphalerite and galena, that do not contain iron, may be oxidized directly, or may be dissolved by ferrous sulfate or sulfuric acid. Regardless of the mechanism, the oxidation of any sulfide mineral leads to the formation of acid solutions, the strength of which depends on a number of factors, the most important being the extent of the hydrolysis of the particular metal, and/or the insolubility of its hydroxide. Ferric oxide and hydroxide are extremely insoluble and therefore solutions resulting from the oxidation of iron sulfides are the most acid. The sulfuric acid generated during the oxidation of any metal may react not only with sulfides, but also with carbonate rocks, such as limestone, resulting in the formation of gypsum:

$$CaCO_3 + H_2SO_4 + 2H_2O \longrightarrow CaSO_4 \cdot 2H_2O + H_2CO_3 \qquad (12)$$

In addition, sulfuric acid may encounter other rocks and minerals with which it may react, such as aluminous clays, in which case aluminum sulfate may be formed.

In the discussion of hydrolysis, mention was made of the presence of carbonic acid in rainwater and groundwater. This acid may also contribute to the oxidation of sulfur in sulfide minerals, for example in the case of galena:

$$PbS + H_2CO_3 + 2O_2 \longrightarrow PbCO_3 + H_2SO_4 \qquad (13)$$

In this case the insoluble lead carbonate (cerussite) is precipitated. Anglesite (lead sulfate) is formed by the reaction of galena with sulfuric acid and is also insoluble, commonly forming a coating on galena, thus inhibiting further oxidation. These examples illustrate common reactions involving galena, and show that the mobility of elements is dependent, in large part, upon the products of each reaction. If insoluble compounds are formed, as in the case of lead sulfate, hydrogeochemical dispersion is not likely to be an important factor. However, the sulfates of zinc and copper are very soluble in water, and in these cases elements tend to form significant secondary halos.

Carbonic acid can also oxidize iron in silicate minerals, for example in the weathering of iron-rich olivine:

$$Fe_2SiO_4 + 4H_2CO_3 \longrightarrow 2Fe^{+2} + 4HCO_3^- + H_4SiO_4 \qquad (14)$$

This is followed by the oxidation of iron:

$$4Fe^{+2} + 8HCO_3^- + O_2 + 4H_2O \longrightarrow 2Fe_2O_3 + 8H_2CO_3 \qquad (15)$$

And carbonic acid can also dissolve limestone:

$$CaCO_3 + H_2CO_3 \longrightarrow Ca(HCO_3)_2 \qquad (16)$$

The importance of carbonic acid cannot be over-emphasized in the chemical weathering processes involving all types of rocks. This is because most rocks contain a limited amount of sulfides from which sulfuric acid can be generated, and therefore most chemical weathering involving acids utilizes carbonic acid.

Many of the equations presented in the preceding discussions are over-simplifications of what are actually complex processes, and some of the products may not be found in Nature. Krauskopf (1967) has summed up the net results of the chemical weathering of sulfide minerals by observing that this process (1) gets the metal ions into solution or into stable insoluble compounds under surface conditions; (2) converts sulfide to the sulfate ion; and (3) produces relatively acid solutions. From the point of view of the exploration geochemist, chemical weathering of sulfide deposits causes a fractionation of ore-forming and associated elements.

It is extremely important to realize that chemical weathering of sulfide minerals may completely remove any evidence of mineralization from the surface. One very good example is found at the Blind River, Ontario, uraninite deposit (Heinrich, 1958). Pyrite at this deposit generally ranges from 2-8% in the unweathered ore, but at the surface the pyrite has been oxidized, sulfuric acid formed, and thus surface expression of uranium is low because it has been leached away by the acids. However, radium sulfate is insoluble and has remained on the surface (sometimes it is found within barite) and it is this sulfate that is the source of the radioactivity that originally attracted attention to the locality, although assays for uranium proved disappointing. Credit for opening the district belongs to F. R.

Joubin, who determined through drilling that, although surface values were low, large tonnages of mineable rock existed below the leached outcrops.

Bacteria

Microorganisms play a significant role in the geochemical cycle of many elements. Certain organisms, particularly iron-sulfur (oxidizing) and sulfur-reducing bacteria, are of significance in the search for mineral deposits because of their capacity to oxidize or reduce certain elements and thus affect the mobility of these elements. One such group of bacteria, to which the generic name *Ferrobacillus* has been given, has the ability to oxidize ferrous iron, while another group, known as *Thiobacillus,* is able to oxidize sulfide and other reduced inorganic sulfur compounds, to sulfate. Some species, such as *Thiobacillus ferrooxidans* found in acid mine waters, are able to oxidize both iron and sulfur compounds, whereas *Thiobacillus thiooxidans* found in soils, oxidizes sulfur but not iron compounds. All these bacteria grow in an acid environment (pH $= 2 - 3$). The iron-sulfur bacteria inhabit acid mine waters and oxidize ferrous sulfides, such as pyrite and chalcopyrite, with the production of sulfuric acid, soluble iron and, in the case of chalcopyrite, iron and copper. The oxidation of non-ferrous sulfides is more complex. In the case of the bacterial release of zinc from sphalerite, or of molybdenum from molybdenite, such release is increased about ten-fold in the presence of pyrite (Trudinger, 1971). Reactions such as:

$$4FeS_2 + 15O_2 + 2H_2O \longrightarrow 2Fe_2(SO_4)_3 + 2H_2SO_4 \qquad (17)$$

can be written to describe the bacterial oxidation of pyrite and this reaction is similar to equation (7). Approximately 10% of the primary copper now produced in the United States is from leaching of dumps using bacteria which are able to dissolve copper sulfides (Roman and Benner, 1973). Recent research on methods to increase the rate of bacterial attack on sulfide concentrates has shown that some of the variables are temperature, speed of agitation, particle size, acidity, nutrients, additives and the supply of oxygen. These and related factors are also likely to be important in soils in which *Thiobacillus thiooxidans* is present. These bacteria may thus augment the inorganic oxidation processes. Several species of bacteria oxidize Mn^{+2} to Mn^{+4} resulting in the precipitation of manganese oxides. The action of bacteria may account for the partial separation of manganese from iron in Nature, because although both elements can be oxidized by bacteria, the iron oxide would be more readily precipitated (Krauskopf, 1967).

Bacteria also play a part in reducing reactions. The most important are those belonging to the genera *Desulfovibrio* and *Desulfotomaculum*. They are anaerobic and reduce sulfate (from sulfate in water, for example) to H_2S (Trudinger, 1971):

$$SO_4^{-2} + 8e + 10H^+ \longrightarrow H_2S + 4H_2O \tag{18}$$

Any metals present in solution would form metal sulfides by reaction with the H_2S. The action of sulfur-reducing bacteria has been used to explain the formation of many stratiform base metal deposits. For further discussion on bacteria in oxidation-reduction reactions, and the significance of bacteria in ore-forming processes, Park and MacDiarmid (1964), Krauskopf (1967), Trudinger (1971), Trudinger et al. (1972), Roman and Benner (1973), or other books or articles in the vast literature, including those in microbiology and bacteriology (e.g., Alexander, 1961; 1971), should be consulted. It will become evident that bacteria are important agents in soil formation, and in changes which occur in the chemical composition of ore minerals at or near the surface of the Earth.

Products of Chemical Weathering

Three types of products result from the chemical weathering of rocks and minerals: (1) soluble constituents, (2) insoluble minerals formed in the secondary environment, and (3) insoluble residual primary minerals.

Soluble Constituents. The formation of soluble products resulting from the breakdown of silicates, sulfides and carbonates was discussed above. These soluble elements move downwards with meteoric water which percolates through soil and rock (e.g., Fig. 1-3). Most of this water will eventually emerge as seepages or springs or will mix with river and lake water. At any place along the flow path, the dissolved constituents may leave evidence of the original mineralization. The evidence may be in the form of metals in the stream sediments, adsorbed onto precipitated manganese or iron oxides, or anomalous values in the waters themselves. In Table 3-2 two acid groundwaters from Cobalt, Ontario, with pH of 3.0 and 6.4, respectively, illustrate that anomalous quantities of many metals may be carried in groundwater draining oxidizing ore deposits and that the more acid water carries the greater amount of the total metals. The high sulfate content of the spring at Fault 1 (Table 3-2) in conjunction with a low pH and trace metals in the parts per million range, is a clear indication of possible oxidizing sulfides in the area. Near-surface waters draining non-mineralized sulfate rocks do not usually contain more than 1000 ppm SO_4 and river water averages only 11.2 ppm SO_4 (Table 3-1). The dissolved

Table 3-2. Partial analyses of acid groundwaters draining native silver veins, Christopher Mine, Cobalt, Ontario (in ppm). From Boyle and Dass (1971).

Fault	T°C	pH	Pb	Zn	Cd	Cu	As	Ag	Ni	Co	Mn	Fe	SO₄	Cl
1	7	3.0	1.36	29	1.14	0.84	5.91	0.05	75.0	445	100	780	9,370	2.9
2	7	6.4	0.84	885	0.02	178	0.05	0.02	1.5	8	6.4	39	790	<0.4

Notes:

1. Dr. R. W. Boyle (personal communication) suggests that the pH of the water from Fault 2 may be too high. Later *in situ* determinations on this water showed the pH to be closer to 5.

2. The waters in Fault 2 apparently have access to copper and zinc minerals (generally uncommon in the immediate area), which accounts for the high values.

metals in acid groundwater are often several hundred times as abundant as in the average river water in which they usually occur in the parts per billion (ppb) range (Table 2-1).

In the case of non-mineralized rocks it is possible to determine the parent rock material by examination of the soluble constituents of waters draining each rock type. For example, water in contact with calcareous parent rocks (Table 3-1) is characteristically high in Ca, Mg and HCO_3^-, and has a high pH (about 8.2), whereas water associated with felsic igneous rocks is high in silica and has a lower pH (about 6.6). (The pH values are generalizations and are not always applicable). White et al. (1963) have presented data on the soluble constituents in sub-surface water, and this type of information sometimes has been applied to exploration projects with success, such as in the search for halite and sylvite. However, there are numerous difficulties. For example, low contents of aluminum and iron inhibit the use of these elements (they are not very mobile) for determining igneous rock types.

Insoluble Minerals. A variety of insoluble minerals are formed in the secondary environment, some of which have been discussed previously. These include iron and manganese oxides, clay minerals, sulfates, carbonates, chlorides (e.g., cerargyrite), silicates (e.g., chrysocolla), and native metals (e.g., mercury, silver). The minerals which form in any specific case depend upon the elements present, and the solubilities (or insolubilities) of the metal compounds formed. Of special interest are the secondary iron oxides, particularly the hydrous forms, because they are abundant, easy to recognize, and have the ability to adsorb, or co-precipitate with, trace metals. In Table 3-3, a list of the content of elements in a limonite from the Cobalt area, Ontario is presented. Contents of 1000 ppm or greater for Ni, Co and As are clearly anomalous and illustrate the value of sampling and analyzing such material. Abundances of up to several percent have been reported for some metals in hydrous iron oxides.

Normally, hydrous iron oxides are found as coatings, cement, or small masses in the area of oxidized iron-containing sulfides. A *gossan* is the

Table 3-3. Trace element content of oxidized vein material (limonite-clay complex) from Wood's vein, Keeley Mine, South Lorraine, Cobalt area, Ontario. From Boyle and Dass (1971).

Element	Content (ppm)	Element	Content (ppm)	Element	Content (ppm)
Ba	175	Cu	750	W	<2
Sr	40	Pb	450	Sn	<10
Zr	250	As	4,135	Cd	0.7
B	<10	Sb	50	Zn	500
Ni	1,000	Bi	540	Hg	1.7
Co	3,185	V	180	U	<0.3
Ag	350	Mo	6		

name given to a large mass of residual limonitic material (usually with some quartz and clay) that remains after the removal of the soluble products of the weathering of a sulfide-bearing deposit. The mechanism of formation is the same as that described above for goethite (e.g., equation 10), but it is important to realize that many different types of iron-containing minerals may weather to form gossans. These include sulfides (e.g., pyrite, marcasite, pyrrhotite, arsenopyrite, chalcopyrite) and carbonates (e.g., siderite and ankerite). The mineral composition of the gossan depends on the degree of oxidation and leaching. If the processes of oxidation and leaching are well advanced there will be no evidence of sulfide minerals at the surface, and the ore metals may be entirely removed. Boyle and Dass (1971) estimate that at least 75% of the arsenic and antimony and one half of the sulfur, as well as major amounts of other elements originally present at the Keeley mine, Cobalt, Ontario, have been removed during intensive oxidation. At this particular mine a great many uncommon secondary minerals have been formed (e.g., erythrite, annabergite) due to the large number of elements present in the native silver veins, as shown by the re-mobilized trace elements now adsorbed onto hydrous iron oxides (Table 3-3). In addition, relatively large quantities of some elements have dispersed in the ground-water and ultimately into the till, soil or surface waters of the area.

Fig. 3-1 is a diagrammatic sketch illustrating the zones of alteration associated with a gossan overlying a copper sulfide deposit. In the zone of oxidation and leaching. If the processes of oxidation and leaching are well Soluble elements are removed and if carbonate is present, either in the form of dissolved carbon dioxide in water or as carbonate rocks, malachite or azurite may form. In this zone of oxide enrichment, native copper and chrysocolla may also be found. If zinc were present in the ore, smithsonite or hemimorphite might precipitate and, if lead were present, anglesite and cerussite might be found. As the metal-containing solutions are carried

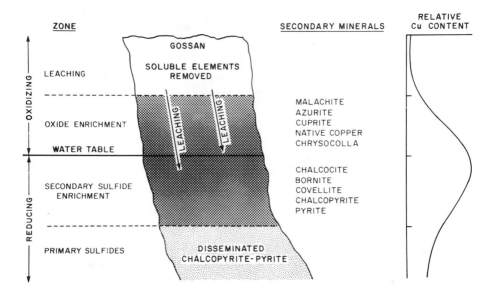

Fig. 3-1. Sketch of a gossan and zones of alteration after weathering of a copper sulfide vein. (Not shown is the thin layer of soil which is usually found over gossans except where they have been exposed by soil movement on slopes, glaciation, or other mechanisms.)

below the water table, they move from an oxidizing to a reducing environment. The zone of secondary sulfide enrichment is in the reducing environment, and in the example shown in Fig. 3-1, the copper-containing solutions have reacted with pre-existing chalcopyrite and pyrite to form secondary copper sulfides such as chalcocite, bornite, and covellite which have higher copper contents than the original chalcopyrite. This process is referred to as *supergene enrichment*. Copper is the one element which exhibits extensive supergene enrichment because it fulfills two requirements: (1) high solubility of certain compounds (e.g., sulfate) in the zone of oxidation, and (2) very low sulfide solubility. Silver is also found in supergene enrichment zones, to some extent, but other elements such as Zn, Pb, Ni, Co and Hg, lack at least one of the basic requirements, and so are never found in this environment.

A thorough knowledge, understanding and ability to interpret gossans is of great value to economic geologists and geochemists. Even though gossans represent extreme weathering conditions, they may still retain distinctive characteristics which permit an indication of the nature of the original mineralization. It is important to know, if possible without drilling, whether a gossan originated from economic sulfides, or whether the enrichment is from some non-economic sulfide (pyrite) or carbonate (e.g., siderite). Several very comprehensive studies have been made of gossans with the object of determining characteristics which will enable one to use them as indicators of economic mineralization. The criteria studied include color, texture, shape of cavities, quantity and type of associated clay minerals, and the presence of insoluble oxidation products (e.g., anglesite). Pisolitic structures, for example, normally imply a laterite source for the iron oxides. Most emphasis is usually placed on the "boxwork" (replica) textures, which are the porous parts of the gossan material remaining after the original sulfide minerals have been leached out. Blanchard (1968) has made an exhaustive study of boxwork textures and is able to interpret these characteristic textures in terms of pre-existing minerals, although great care and experience are essential for correct evaluation and a detailed knowledge of local geology and environmental conditions is often necessary.

From a geochemical point of view, the analysis of gossans (Table 3-3) will often yield conclusive evidence regarding the presence or absence of mineralization. In general, gossans resulting from economic mineralization are characterized by anomalous geochemistry in comparison with hydrous iron oxides from non-economic sources, or with laterites. Recent terminology, particularly in Australia, limits the word gossan to those hydrous iron oxides which overlie economic mineralization, whereas the term "false gossan" is used for any iron oxide, whether from sulfide veins, laterites or any other source, which does not represent economic mineralization. Historically, false gossan has also been used to describe a displaced iron oxide zone, such as at a break-in-slope geochemical anomaly (Fig. 1-3). Cases are known, for example in Australia, where extreme leaching causes the removal of all geochemical characteristics indicative of economic mineralization, and the result is a gossan indistinguishable from one originating from a non-economic source. Conversely, anomalous metal enrichment can be found in gossans from non-economic mineralization, for several reasons, such as the scavenging action of the oxides of iron and associated manganese. Therefore, great care is required in investigations of gossans and usually other criteria must be employed. These include mineralographic (polished section) examination and mineral identification, often by X-ray diffraction methods. Even in very ancient gossans, such as those of Australia, many of the textures have been preserved by subsequent silicification.

Electron probe analyses have proved of great value, for example, in determining whether goethite is related to pre-existing sulfides. As a generalization, it has been found that locally high contents of certain metals (e.g., Cu, Ni) within the goethite indicate buried mineralization, whereas a uniform content does not. Manganese oxides within the gossan can result in false anomalies because of their high adsorptive capacity for metals. If manganese oxides are detected by means of the electron probe, and the trace metals are associated with them, any anomaly would be downgraded. It must be stressed that the electron probe is only an accessory to the whole problem of gossan geochemistry. Clema and Stevens-Hoare (1973) have recently described methods by which nickel gossans in Australia are distinguished from those that do not represent mineralization. Finally, it is worth noting that some very productive ore deposits, such as some silver-galena veins in Mexico, do not produce gossans. This is because they do not contain iron minerals.

Residual Primary Minerals. These are often encountered in the zone of weathering because they are not affected by the chemical reactions around them. Table 2-2 and Fig. 2-2 list some of the stable minerals of igneous and metamorphic origin. The resistant minerals include oxides (e.g., cassiterite, rutile, magnetite, chromite, ilmenite), native elements (e.g., gold, platinum, diamond), some silicates (e.g., beryl, zircon), and a few other chemical types such as phosphates (e.g., monazite). Not only are these minerals resistant to chemical attack, but they are generally hard (or else ductile and malleable), lack cleavage, or have other physical properties which permit them to be transported and deposited as placers. Other minerals which are also chemically resistant in the zone of weathering, such as wolframite and barite, are either too soft or friable to permit them to be transported any great distance. Geochemical methods of detecting the ore minerals in this last category almost always require a total analysis, because these minerals do not yield ions into solution which can be detected by dilute acid or cold-extractable techniques (discussed in Chapter 6). Because minerals in this category are generally heavy, preconcentration often enhances their detection.

PHYSICAL WEATHERING

Physical weathering, which causes rocks to be fragmented with little or no chemical change, was mentioned briefly in the preceding paragraphs. Overall, it plays a minor role in weathering processes on the Earth, but it is important in very cold and/or very dry climates where chemical weathering is inhibited, and in areas with steep and rugged topography. Where chemical weathering is predominant, physical weathering plays a supplementary role.

Examples of purely physical weathering of interest in exploration geochemistry include glacial plucking, with the subsequent reduction of the plucked rock debris into smaller and smaller fragments, and their eventual deposition in till or other glacial deposits. The products of frost heaving, and talus formation, are other examples. The cracks formed when water alternately freezes and thaws will eventually be enlarged, thus permitting water and other agents of chemical weathering to enter and begin to decompose rocks. In desert areas, because of the lack of water, physical weathering by wind is the main agent of both weathering and erosion. In all the cases cited, physical destruction of the rocks is important primarily because it "prepares" rocks for subsequent, and more potent, chemical weathering (except in the extremely dry areas) primarily by increasing the surface area susceptible to chemical attack. In general, physical weathering such as found in glacial regions, does not result in a separation of ore elements as is so prevalent in chemical weathering.

ENVIRONMENTAL FACTORS AFFECTING WEATHERING

In previous sections it was established that chemical weathering proceeds through hydrolysis, oxidation, and a number of other processes, which in turn are followed by leaching, precipitation, and so forth. Weathering requires air for oxygen and carbon dioxide, as well as water to participate in the reactions and transport the various soluble products. There are also many environmental factors which influence chemical weathering of which the more important are: (1) climate, (2) biological activity, (3) parent material, (4) topography, and (5) time (Jenny, 1941). The first four are discussed at this point, and during the discussion of soils (see below), all of the factors are considered in connection with their influence on soil formation.

Climate

Climate, which includes precipitation and temperature changes, is a major factor in chemical weathering. Precipitation controls the amount of water which is so essential for the oxidation reactions to proceed and for the removal of the soluble products. In fact, *without water chemical weathering is essentially non-existent.* Temperature is also extremely important, because for each 10°C rise, the rate of chemical reactions increases by a factor of 2 or 3. Temperature also affects the rate of evaporation and freezing thus it is an important factor in physical weathering. The combination of precipitation and temperature controls the amount of vegetation. Certain plants have very high acidity at the tips of their roots and this is a very powerful force in the overall process of chemical weathering. Roots also exert a significant mechanical effect and they contribute to the breakdown of rocks by opening channels for the flow of water and movement of burrowing animals. Plants sometimes preferentially remove certain elements from rock materials, and at a later stage in their biological cycle, these elements may re-appear in forms more susceptible to chemical weathering. The influence of climate is best seen by comparing the products of chemical weathering in a desert and in a tropical rain forest. Suffice it to say that weathering is minimal in the desert area, whereas it is at a maximum in the tropical area. Weathering can be observed at depths of 500 ft and it is by no means strictly a surface phenomenon.

Biological Activity

The ability of some bacteria to oxidize iron and sulfur, and others to reduce sulfur, has been mentioned previously. These are examples of what is sometimes called biological weathering. Numerous other instances in which biological activity is important in the weathering processes can be cited. For example, organic acids are generated around the root hairs of living plants, and bacteria decompose vegetation. The products formed by these biologic processes interact with rock materials and weaken their surfaces, thus making them more susceptible to both chemical and physical weathering. Biologic weathering is generally restricted to the near-surface soil zones. The main bacterial biomass is confined to the top 6 inches of soil and has been estimated at anywhere from 1 to 2 tons per acre (Bear, 1964, p. 243).

Parent Material

The nature of the parent material is more a geological factor than an environmental one. Such features as texture, porosity, permeability and the nature of the rock itself, all affect the rate of chemical weathering. For example, porous rocks may be more readily attacked chemically than dense, non-permeable rocks because water is able to penetrate more easily. In addition, fine-grained minerals are more susceptible to chemical attack than are coarse-grained varieties.

Topography

Topography strongly influences chemical weathering in several ways, such as by controlling (a) the rate of surface run-off and hence the amount of moisture available for chemical reactions, (b) the rate of groundwater movement and therefore the rate of removal of the soluble products, and (c) the rate of erosion of the weathered products and consequently the rate of exposure of fresh surfaces.

In mountainous areas, especially those with high rainfall as on the west coast of British Columbia and in southeastern Alaska, physical erosion is extremely rapid, and the debris is removed faster than it can be weathered. In such places geochemical anomalies are best detected in the clastic debris in the valleys and in the deltas where the streams enter fjords or the ocean. In this type of environment (high rainfall and steep slopes), wind, running water, landslides, soil creep and other processes are active, and physical disintegration is likely a major factor and chemical weathering a minor one. The other extreme is a flat lying area with abundant rainfall. Here, little run-off would occur and infiltration of rainwater would be at a maximum. But, because the sub-surface drainage is very slow, soluble products are not freely moved and chemical weathering is inhibited.

The ideal topographic conditions favoring chemical weathering are found in areas of moderate relief where surface run-off is not excessive, and groundwater is able to circulate freely. Even so, local variations in topography result in different degrees of weathering.

SOIL

Introduction

Rock weathering processes lead to the formation of soils, and in fact, soil formation is an integral part of weathering. Soils can be defined simply as the products of weathering that remain *in situ* above weathered parent rock. This definition does not emphasize the influence of climate, biological activity, parent material, topographic relief, and time which are extremely important in any study of soil, whether it be for exploration geochemistry or for other purposes. Buckman and Brady (1969) integrate these factors into another definition of soil: "The collection of natural bodies occupying parts of the earth's surface that support plants and that have properties due to the integrated effect of climate and living matter acting upon parent material, as conditioned by relief, over periods of time." Even from this expanded definition such important features as soil horizons, and the nature of the clay minerals formed during the weathering processes, are not immediately evident. For example, kaolinitic soils which are common in temperate climates with moderate to heavy rainfall, are significantly different in certain properties, such as the ability to adsorb metals, in comparison with the montmorillonitic and illitic clays which are most abundant in the soils of semi-arid regions. It is obvious that such a complex subject as soils (pedology), which integrates aspects of geology, chemistry, microbiology and other sciences, is far too detailed to discuss here in more than a cursory manner. Fortunately, several extremely good books on soils are available, such as Buckman and Brady (1969), and from a geological point of view, Hunt (1972) is highly recommended. Many aspects of the chemistry of soil are well covered in Bear (1964), and Legget (1967) has presented an interesting summary of soil from the point of view of geology and engineering. Books with emphasis on soils of specific countries and geographic regions are also available, such as Leeper (1964), who stresses Australian examples. The field of pedology has a voluminous literature, and the above references are merely suggestions.

From the point of view of exploration geochemistry, it can be stated safely that it is impossible to know too much about soils and soil forming processes. This is because soils are an important sampling medium, and the dispersion (or lack of dispersion) of metals in soils forms the basis of many exploration programs. Legget (1967) has pointed out that at least 72% of the land surface of the globe, excluding the areas covered by ice, permafrost and fresh water, is covered with soil. Yet it is surprising how little most geologists know about soils. The topics selected for discussion which are of most general application to exploration geochemistry are soil profiles, soil classification, factors affecting soil formation, laterites, and trace elements in soil profiles.

Soil Profiles

One of the characteristic features of soils is the layering evident in vertical sections. The sequential layers are called horizons (Fig. 3-2). Each horizon differs in composition, color, texture, and/or structure, and the boundaries between them are often very sharp. From the surface downwards the horizons are designated as A, B, C, and D, and they often have defineable sub-horizons such as A_0 or B_1. When a complete succession is found, it is referred to as a *soil profile,* regardless of the thickness of each layer. Some definitions require only two layers to be present (e.g., U.S. Dept. of Agriculture, 1957). Soil profiles may develop on parent material which can either be residual, or transported as in the case of glacial materials, alluvium or colluvium. The A and B horizons constitute the solum, or "true soil" above the parent material, and consist of organic matter, a leached layer, and a layer of deposition. Changes occur below the B horizon but they are generally minimal. The profile grades downward into the weathered and then the fresh parent material referred to as the C and D horizons, respectively. The C horizon of residual soils consists of variously weathered fragments of parent material, with any anomalous metal values similar to those in the parent bedrock (Fig. 1-2). The C horizon may also comprise old soils from which new soils are formed.

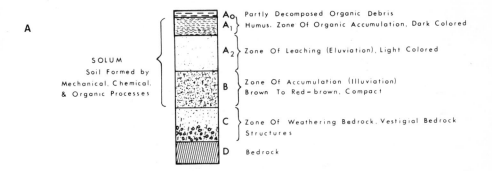

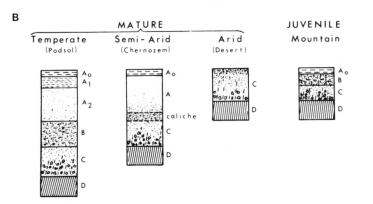

Fig. 3-2. Diagrammatic representation of soil profiles. (A) A typical mature podzol soil profile in a temperate zone showing all the important horizons of interest in exploration geochemistry. (B) Variations in soil profile in four climatic environments. Note that all horizons are not present in every profile, but that every profile has some horizons. From Andrews-Jones (1968).

    A simplified description of selected horizons in a soil profile is presented in Table 3-4. Included in the table is the system of conventions used by most geochemists (column 1) in which such designations as A_0 and A_1 are found. A second system of conventions (column 2) which is now being used with increasing frequency in Canada is also presented (also see Fig. 3-5); it is described in detail by the Canada Department of Agriculture (1970) and for more precise definitions readers should consult this publication. The symbols listed in Table 3-4 are the ones most likely to be encountered in exploration geochemistry. Descriptions of the main soil horizons follow.

Table 3-4.   Description of selected horizons in a hypothetical soil profile comparing general geological usage and Canada Department of Agriculture (1970) terminology

General Geological Usage	Canada Department Of Agriculture (1970)	Description
A_{00}	L	Loose leaves and organic debris, largely undecomposed
A_0	F	Organic debris partially decomposed or matted
A_0	H	Organic debris fully destroyed
A_1	A_h	A dark-colored horizon with a high content of organic matter mixed with mineral matter
A_2	A_e	A light colored horizon of maximum eluviation of clay, iron oxide and/or organic matter. Prominent in podzolic soils; faintly developed or absent in chernozems
$\left. \begin{array}{c} A_3 \\ B_1 \end{array} \right\}$	AB	Transitional between A and B
B_2	$\left\{ \begin{array}{c} B_t \\ B_f \\ B_h \\ B_n \\ B_m \end{array} \right.$	Maximum accumlation of silicate clay minerals, or of iron and organic matter; maximum development of blocky or prismatic structure, or both
B_3	BC	Transitional to C
$\left. \begin{array}{c} C \\ C_{ca} \\ C_{cs} \\ C_{sa} \end{array} \right\}$	C	Horizons C_{ca}, C_{cs} and C_{sa} are layers of accumulation of calcium carbonate, calcium sulfate, and soluble salts found in some soils. C horizon is comparatively unaffected by soil-forming processes
D	D	Bedrock

Notes:

1. This table is very brief and oversimplified, but, it is sufficient for most needs in exploration geochemistry.

2. General geological usage is based on the U.S. Department of Agriculture system.

3. Prepared with the assistance of Dr. N. W. Rutter, Geological Survey of Canada.

A Horizon. The uppermost layer of the A horizon consists of organic debris lodged on the soil and is the zone of maximum biological activity. The designation A_{00} is given to loose leaves and organic litter which is largely undecomposed (not shown in Fig. 3-2). A_0 is the designation for the partially decomposed, or matted organic debris commonly called *humus* and the A_1 horizon is characterized by humus mixed with some mineral matter. A_{00} and A_0 layers are usually absent on soils developed from grasses. The A_0 horizon closely reflects the chemistry of the vegetation from which it is derived and analysis of material from this horizon has much in common with the biogeochemical prospecting (Fig. 1-7). In the Soviet Union soil sampling, particularly of the humic horizons, is considered to be a part of biogeochemical prospecting in view of the interdependence of soils and vegetation. Bog and peat materials are analogous to A_0 horizon material. Numerous geochemical studies have been made using humus rather than biogeochemical methods (*sensu stricto*) and the results have indicated that strong anomalies can be found over mineralization. Even gold has been found enriched in humus (Curtin et al., 1968). The same enrichment generally applies to A_1 horizons which also have a high humus content. However, interpretation of this type of data is often complicated by the fact that the organic matter may contain anomalous values unrelated to mineralization due to adsorption and other mechanisms.

As water filters through the decomposed organic matter, carbonic and various organic acids are formed, with the result that the pH may be 4 or less. Although the acids are weak, they are continually being replenished by the decomposition of humus. They move downward to lower levels where they react with other mineral matter and, as a result, the soluble products released by weathering processes, as well as some colloidal and mineral matter, are also continually moved downward in solution or suspension by circulating waters. This depletion is characteristic of the entire A horizon, and is called *eluviation* (leaching). Where leaching is intense an A_2 zone will be found. This horizon of maximum eluviation (A_2) is prominent in podzolic soils, and is devoid of most of the organic matter, trace elements and much of the clays, and as a result, is light in color, as opposed to the dark colors (black and brown) characteristic of higher horizons. The A_2 horizon is chiefly a mixture of sand and silt, with some clay (loam), and is a very poor sampling medium for geochemical purposes because of the intense leaching. Even in highly mineralized areas, negative responses can be obtained and therefore every effort must be made to avoid sampling this particular horizon. Its white (or light) color is quite characteristic, and its thickness is generally less than 1 foot, so that avoiding this horizon should present no difficulties to properly trained samplers. Both the A_1 and A_2 horizons can be recognized in many uneroded soil profiles in humid regions although they do not often co-exist in soil profiles in dry regions (Fig. 3-2, B). The A_3 horizon is transitional to the B horizon but it is often very thin or absent.

B Horizon. The B horizon lies immediately beneath the A horizon and it has some characteristics of both the A and C horizons. Living organisms are less abundant in the B horizon than in the A, but are more abundant than in the C. It is generally brown to orange-brown in color, harder when dry (and stickier when wet), by comparison with the horizons above and below, and will often have a blocky or prismatic structure. These properties are a reflection of the high concentrations of iron and/or aluminum oxides, usually in combination with manganese oxides and organic matter. Whereas the A horizon is one of eluviation, the B horizon is one of *illuviation* (accumulation), and part of the leached material from the A horizon is deposited here. The most soluble elements (e.g., the alkalis and alkaline earths as well as some metals) may be carried further downward, enter the groundwater system, and possibly be removed from the area by drainage, perhaps forming anomalies a considerable distance from the source. Where removal is complete as in a well-drained soil in a humid area, the profile is referred to as an "open" chemical system and soils in this environment are likely to be *acid soils. Alkaline soils,* on the other hand, are characteristically found in dry or semi-arid regions, where water is insufficient to drain completely through to the water table — hence a "closed" chemical system. The lack of water in such areas, plus sparse vegetation from which organic acids can be derived, results not only in correspondingly less leaching, but also in the precipitation of a calcium carbonate layer, called *caliche.* It is found at the general level of the B horizon (Fig. 3-2, B), the exact position being determined by the depth to which the water can penetrate before it evaporates. Caliche is the most striking result of illuviation in semi-arid regions. It often occurs as nodules, generally pea-sized although, on occasion, they may be up to a foot across. In dry areas the caliche will be near (or on) the surface but, if it is present in more humid areas (e.g., 25 inches of rainfall), it is likely to be encountered (if at all) at depths of up to 20 ft. The caliche may form impervious layers, sometimes of considerable thickness (several tens of feet) which prevent the upward migration of solutions or vapors from forming dispersion halos (at a later period of time). In places it is possible to distinguish a B_1 from a B_2 horizon, the former being relatively enriched in iron and aluminum oxides and trace elements.

Because the B horizon is one of accumulation of elements, and because the clay minerals and iron and manganese oxides which are found there have the capacity to adsorb metals to varying degrees, this zone is the one normally sampled during geochemical exploration soil surveys. It is generally easy to identify by virtue of its brownish color and clayey texture. As a rule, the contrast between anomalous and background values will be higher in the B (preferably B_1) horizon than in the A (Fig. 1-2), although the C horizon is likely to give a sharper, but narrower anomaly. There are many exceptions, however to the general "rule" that the B horizon is the best soil sampling horizon. For example, Boyle and Dass (1967) have shown (Fig. 3-3) that the A_0 horizon gives better results in many parts of Canada than does the B horizon, although a preliminary (orientation) survey is necessary to ascertain this in each region. As these authors pointed out, if the humic horizon is well developed it represents the sum total of many biogeochemical and geochemical events that in many cases lead to enrichment of metals, especially in the vicinity of veins and other types of deposits. Thus a sample from the humic horizon is likely to be more representative of the total environment than other horizons. (By using only the minus 80 mesh soil fraction Boyle and Dass were able to avoid roots and undecomposed organic matter, which often requires ashing for proper analysis). If it is not possible to conduct a preliminary survey to determine the optimum soil horizon for exploration purposes, and only one horizon can be sampled, the B horizon would normally be preferred. However, it must be realized that the explanation so far presented (Fig. 3-2) for the accumulation of trace metals is most applicable to podzols, and that other types of soil are likely to show variations in their trace element behavior. This point is expanded upon below. The preceding discussion has also assumed that the soil profile is on residual parent material but, if transported glacial material is involved, or if the profile is developed on

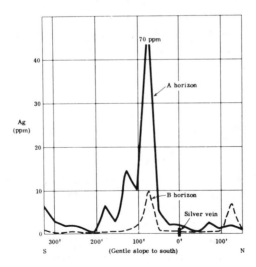

Fig. 3-3. Comparison of silver content obtained from A and B soil horizons over O'Brien No. 6 silver vein, Cobalt, Ontario. In this case, a greater anomaly was obtained by sampling the A horizon, in comparison with the more usual B horizon. Similar relations were observed with arsenic and other elements. From Boyle and Dass (1967). Reproduced from *Economic Geology*, 1967, v. 62. p. 275.

alluvium, colluvium, or on top of an ash layer, then special precautions are necessary and the above generalizations are not applicable.

C Horizon. The upper part of the C horizon consists of loose and partly decayed rock which is the parent material for the bulk of the soil, and except in the case of transported soils, it grades downward into unaltered parent material (Fig. 3-2). Organic matter is very low in the C horizon; illuviation is at a minimum and generally restricted to the upper part, and the color of the rock material is usually lighter than in the B horizon. The C horizon is only sampled in a limited number of situations such as where the A and B horizons are missing, or where anomalous values are expected to be found there (e.g., in some laterites).

The soil horizons are commonly more complex than has been described above. Multiple sub-horizons may exist, parts of the profile may be missing because of erosion or very poorly developed, and complexities may be introduced by multiple and vastly different cycles of weathering which may be related to climatic changes. In geologically relatively short periods of time, surface weathering conditions at any place, such as in northern Canada, may have changed from physical to chemical, with a major effect on the soil forming processes and secondary dispersion patterns.

The thicknesses of the individual soil horizons vary considerably, depending on climate, biological activity, parent material, and the length of time the profile has been developing. As a generalization, the A horizon varies from 1 in to 4 ft, and the B horizon from a few inches to about 6 ft. The C horizon is generally thicker than the other horizons, and in well-leached tropical soils in which weathering has been a long uninterrupted process (e.g., parts of Africa and Australia) it may extend to 300 or more ft.

Clay Minerals

In the discussion of weathering, and in the description of the B horizon, mention was made of clay minerals. Clay minerals are the insoluble products of chemical weathering of silicates, and they are often concentrated in the B soil horizon. There are many types of clay minerals formed under varying conditions and, although most are capable of adsorbing metals with which they may come into contact, this ability varies greatly. All are hydrated silicates, generally with a sheet (layer) structure.

There are five common groups of clay minerals, and a sixth, the mixed-layer clay minerals, that includes more than one type (Loughnan, 1969). The most common types of clay minerals are listed below together with their *cation exchange capacity,* sometimes called base exchange capacity, which is a measure of their ability to adsorb cations (expressed in milliequivalents of the adsorbed cation per 100 grams of the clay). The cation exchange capacity of clay minerals arises from the fact that between the plates and along the exposed edges of their plate-like structures they contain unsatisfied bonds, and the overall charge of the plate is negative. In addition, some clays "exchange" hydrogen for other cations.

Group	Minerals	Cation Exchange Capacity (meq/100gm)
kaolinite (kandite)	(halloysite, dickite, etc.)	3 - 15
mica	(illite, glauconite, etc.)	10 - 40
montmorillonite (smectite)	(beidellite, montmorillonite, etc.)	80 - 150
chlorite	(vermiculite, etc.)	10 - 150
fibrous clays	(palygorskite, sepiolite, etc.)	low
mixed-layer	(montmorillonite-illite, etc.)	variable

Numerous other minerals, such as talc and pyrophyllite, are usually included in detailed clay mineral classifications, but they are of minor importance to exploration geochemistry. Further discussion of clay mineralogy, with particular application to soils, is found in Bear (1964), Loughnan (1969), Buckman and Brady (1969), and Hunt (1972). The identification of clay minerals is a specialized field, and is almost exclusively accomplished by means of X-ray diffraction techniques which are described by Brown (1961). Considering the wide scope of clay mineralogy, it is possible to discuss only those outstanding features of clay minerals which are of significance in exploration geochemistry.

The cation exchange capacity of clay minerals enables large amounts of certain elements, including some metals, to be adsorbed. As a result of its high cation exchange capacity, montmorillonite can adsorb 5-50 times as many cations as kaolinite. In most cases, the cation exchange capacity increases with pH and thus at low pH values little adsorbed metal will be found on clay minerals. For example, studies of montmorillonite have shown that only above a pH of 6.0 does it adsorb and retain cations. (Humus, on the other hand, is able to adsorb cations at very low pH values; Buckman and Brady, 1969, p. 97). Particle size, surface area, amount of soil moisture, degree of crystallinity, and other factors all affect the ability of clay minerals to adsorb metals. Not all elements in soils are adsorbed to the same extent by clay minerals. The univalent cations, for example rubidium and cesium, are less strongly adsorbed than are the divalent alkaline earth elements. The general order of adsorption of some cations on clay minerals (in decreasing order is): $Ba > Sr > Ca > Mg > Cs > Rb > K > Na > Li$, although this will vary depending on the particular mineral species. The common metals, which are not listed in the above compilation, are also adsorbed to some degree. There is convincing evidence, however, to indicate that it is manganese and iron oxides commonly found in soils and sediments, which are the controlling factors in the fixation of the important heavy metals (Zn, Cu, Ni, Co), rather than the clay minerals (Jenny 1968). Regardless of the mechanism, the B soil horizon, which often has abundant clay minerals, is generally the most reliable for the detection of secondary dispersion patterns. This is illustrated in Fig. 3-4, in which the variations of selected physical and chemical properties in a soil profile are compared. Again, this figure is based on podzols or podzolic soils.

Clays (argillaceous rocks) generally have low permeability, which affects the ability of metal-containing fluids, or vapors, to penetrate them. Montmorillonitic clays are especially impermeable, whereas illitic and kaolinitic are less so. Particle size of the clay minerals, and the thickness of the rock unit are among the critical factors in this respect and, if the clays are mixed with coarser particles, such as silt and sand grains, they may be more permeable. Bentonites (composed chiefly of montmorillonite) and clays deposited in a glacial lake (e.g., the glacial lake deposits of the Clay Belt of Ontario), can effectively mask an expression of buried mineralization.

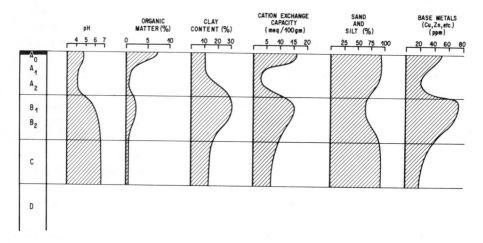

Fig. 3-4. Variations in selected physical and chemical properties of different horizons of a soil profile. Generalized for a podzol formed in a humid climate.

The nature of the clay mineral formed is probably more dependent on the environment of weathering than it is on parent material. For example, the formation of kaolinite is favored in areas of temperate climate, with high rainfall, thorough leaching of alkali and alkaline earth elements, and good drainage which results in a slightly acid environment. Montmorillonite, on the other hand, is formed in drier areas with incomplete leaching and poor drainage which result in a neutral to alkaline environment. Illite is typically formed in an environment similar to that of montmorillonite but, in addition, it requires some K, hence it is commonly an alteration product of feldspar-rich rocks. The most striking example of the influence of environment is the formation of the conspicuous red soils of the humid tropics. The extreme end-products of soil formation in this environment are laterites and bauxites in which everything except alumina, ferric oxide, and a few trace elements, are leached away including silica.

Classification of Soils

Attempts to classify soils have been made by many scientists in the past 100 years, for a multitude of purposes such as agriculture, engineering and geology. Yet today there is still no universally accepted terminology and, in fact, we are probably in the midst of one of the most significant transitions ever to affect soil classification and terminology. Reference was made to the fact that new designations for the soil horizons have recently been introduced in Canada (Table 3-4) but what is of more importance is that an entirely new terminology for soils is being used by soil scientists in the United States, which is slightly different from the one in use in Canada, and both of which differ from that in use in parts of the rest of the world.

Soils can be classified in many ways. In North America, the system generally in use from about 1938 until the 1960's (and still employed by many pedologists), was based on soil genesis. In this context, climate is the most important factor to be considered, with less weight given to the nature of the parent rock, topography and so forth. This resulted in the terms *zonal* — which indicates a soil characteristic of a large area where climate and vegetation are the dominant factors controlling soil type; thus a chernozem in Russia, is the same as one in Canada; *intrazonal* — soils whose characteristics reflect the dominating influence of some local factor such as relief, parent material or age, over the normal effects of climate and vegetation; and *azonal* — soils without distinct genetic horizons, that is, their profiles are poorly developed or absent; the term is particularly applicable to youthful surface deposits, such as soils developing on alluvium. In this classification, which generally is called the "zonal classification", the terms zonal, intrazonal and azonal are called "orders" and they are further subdivided into "suborders" and "great soil groups". Each great soil group is again subdivided into numerous soil series and soil types.

The classification recently introduced in the United States, at least on an evaluation basis, was developed by the Soil Survey Staff of the U.S. Department of Agriculture (1960). During the period of its development (from about 1951), a similar classification was being developed in Canada. The Canadian classification, however, differs in terminology from that in the United States and it is somewhat simpler. Even so, it contains 8 orders, 22 great groups, 138 subgroups, 800 to 1000 families, about 3000 series, and about 4000 types (Canada Dept. of Agriculture 1970, p. 12). The United States and Canadian systems are based on the properties of soils as found in the field today — properties which can be measured. This eliminates the dependence on genesis, which has strong geological and climatic influences and is subject to controversy in some cases. These new systems of classification, and specifically that used in the United States, have been developed and evaluated through several stages, or "approximations". The Seventh Approximation is now being used in the United States, and both it and the Canadian classification are likely to be changed from time to time as experience dictates.

Table 3-5. The names of the soil orders in the Seventh Approximation (1960) and their *approximate* equivalents in the Canadian and other systems.

U.S. Department of Agriculture (1960)	Canada Department of Agriculture (1970)	General Geological Usage (approximate)
—	Tundra	Tundra
—	Sub-Arctic	Arctic Brown, etc.
Entisol	Regosol	Azonal
Vertisol	—	Gumusols
Inceptisol	Brunisol	Brown Forest, Gray-Brown Forest, etc.
Ardisol	Solonetz	Desert, Reddish Desert, Solonetz, etc.
Mollisol (Borolls)	Chernozem	Chestnut, Chernozem, Prairie, etc.
Spodosol	Podzol	Podzol, Brown Podzolic, etc.
Alfisol	Luvisol	Gray Brown Podzolic, Gray Wooded Soils, Degraded Chernozem, etc.
Ultisol	—	Red-yellow Podzolic, Reddish Brown Lateritic, etc.
Oxisol	—	Lateritic soils, Latosol
Histosol	Organic	Bog soils

Notes:
1. In those cases where there are no representative soils in the United States or Canada, no names have been assigned in the respective classification. For example, there are neither tundra nor sub-arctic soils in the United States (except in Alaska), nor ultisols or oxisols in Canada.
2. Under General Geological Usage, only a few of the more common soil names are given.

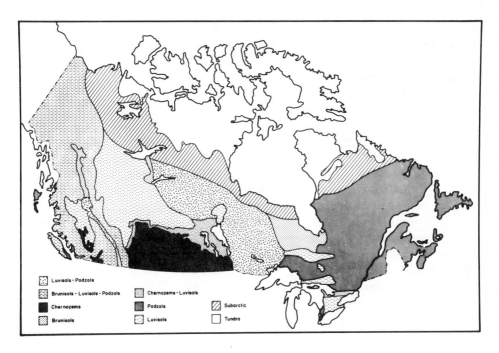

Fig. 3-5. Generalized soil map of Canada using the new terminology (Canada Department of Agriculture, 1970) (excluding terminology for the Subarctic and Tundra).

Most geologists and geochemists do not use the new systems at present They are likely to find them very confusing, and are often appalled by the terminology, even though the names are supposed to give a definite connotation of the major characteristics of the soils in question (Latin and Greek root words are the basis of the names). The degree of frustration of many is probably no better summed up than by the statement of Hunt (1972, p. 180), "America is first in technology; the Seventh Approximation would keep America first in terminology!" To illustrate his point, he quotes several passages from the Seventh Approximation such as (p. 105) "The Entisols are those soils, exclusive of Vertisols, that have a plaggen horizon or that have a diagnostic horizon other than an ochric or anthropic epipedon, and albic horizon, and agric horizon, or, if the N value exceeds 0.5 in all horizons between 20 and 50 cm. (8 and 20 inches), a histic epipedon".

It is difficult for a geochemist or a geologist, whose interests in soils are primarily related to process, genesis and environmental factors, to disagree with Hunt's evaluation. However, the fact remains that soil scientists are using the new nomenclature, and if there is to be a cross-fertilization of ideas and data among disciplines, it is necessary to know their terminology. Further, exploration geochemists are now watching more closely than ever developments in the soil sciences, the release of soil maps in potentially interesting areas, and other phases of pedology for clues to mineralization. In exploration geochemistry it is necessary to know that a transition in soil nomenclature is taking place, and the approximate equivalents of the major "orders" in the various systems. These are presented in Table 3-5. Other proposed soil classifications, such as the "International System", are not included. In Fig. 3-5, a generalized soil map of Canada is presented with the nomenclature based on the new Canadian classification. Some approximate equivalents in the zonal classification are also given. Good simplified descriptions of the Seventh Approximation can be found in recent textbooks on soils, such as Buckman and Brady (1969) who present a complete discussion of the old classification (zonal, intrazonal, azonal) as well. Inasmuch as the soil terminology used in exploration geochemistry has been based exclusively on the zonal concept, at least until the present, this will now be discussed in more detail.

Zonal Classification

In the zonal classification of soils, climate, vegetation and soil groups are related: this is illustrated in Fig. 3-6. Here it can be seen for example, that *podzols* generally form in temperate, humid areas with a forest cover, where conditions favor the surface accumulation of organic matter. In this case, similar to the formation of certain soil profiles in humid, vegetation-covered regions (Fig. 3-2), the soils are acid and leaching of the primary minerals of the rocks can take place. Podzols are characterized particularly by a highly-leached, light colored A_2 horizon. Such soils are often called *pedalfers,* which is a name used to stress the fact that aluminum (Al) and iron (Fe) accumulate as weathering products from the decomposition of primary minerals. If the process of podzol formation (podzolization) continues for an extended period under hot, humid, forested conditions, the relative abundance of iron and aluminum oxides increases as the silica is removed, with the result that *laterites* form; laterite would appear in the low right-hand corner of the soil diagram in Fig. 3-6. In arid or semi-arid regions with only grass cover, alkaline soils occur in a closed system, with the result that carbonates such as caliche are reprecipitated, commonly in the B horizon (Fig. 3-2). Soils in which calcium carbonate has been re-distributed and accumulated in specific zones are called *pedocals*. Montmorillonite is often found in pedocals because it tends to form in an alkaline environment. Under certain conditions it is possible for a soil to form in a mixed (open and closed) system. *Gley* is such

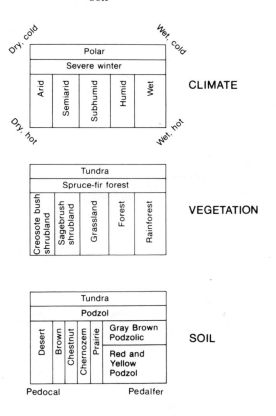

Fig. 3-6. Schematic diagram showing the relations of climate, vegetation and soil groups, based on the zonal (climatic) classification as applied to North America. Also indicated are the relations of pedocals and pedalfers. Modified from Hunt (1972).

a case as it forms under humid conditions, but with poor drainage as in bogs, with a resulting mixture of organic matter and iron largely in the reduced form (Fe^{+2}).

Fig. 3-7 illustrates the variations found in modern soils and shows how profiles reflect differences in climate, moisture and vegetation from the Hudson Bay region of Canada to the southwestern United States. This is essentially a pictorial representation of the data in Fig. 3-6. In this northeast-southwest direction, the climate changes from cold and wet to warm and dry, the soils from acid to alkaline, and the vegetation from coniferous forest to desert shrubs. Also illustrated in Fig. 3-7 are calcium carbonate (caliche) accumulations, open and closed systems, and other characteristics of soil profiles discussed above. The boundary between a dry and wet climate in most of North America can be taken as an average precipitation of 25 inches per year.

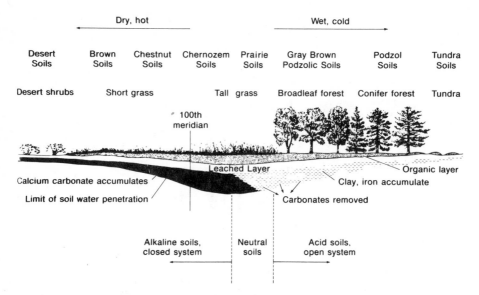

Fig. 3-7. Diagram illustrating the changes in soil profiles and soil type that accompany changes in climate and vegetation between the tundra of northern Canada and the deserts of the southwestern United States. At the 100th meridian the annual precipitation is about 20 inches; there and to the west the soils are alkaline. The easternmost grasslands are about neutral; farther east the soils are acid. From GEOLOGY OF SOILS by Charles B. Hunt. W. H. Freeman and Company. Copyright © 1972.

The concept of zonal soils has been described above with some examples, and it is illustrated in Fig. 3-6. The most important zonal soils will now be discussed briefly with particular reference to North America. The first to be discussed are the zonal soils of humid regions (most tundra and podzols, podzolic and prairie) and then the soils of the semi-arid and arid regions (chernozem, chestnut, brown and desert). The relations between the soils can be seen in Figs. 3-6 and 3-7. The concept of zonal soils as applied to North America, bears a close relation to climate and geography — tundra soils occur in northern Canada, red podzolic soils are found in the southeastern United States, desert soils in the western states, and so forth.

Tundra soils. These occur in arctic and subarctic regions where the vegetation consists primarily of mosses, lichens and shrubs (Fig. 3-5), and where the weather is too cold for trees or grass. Tundra soils are found under high moisture efficiency conditions (Fig. 3-6). Approximately 40% of the Soviet Union has this type of soil. Tundra soils may contain a large amount of organic matter because in a cold climate decay is slower than accumulation, as biological activity is low. Generally, tundra soils overlie permafrost, and as a result the soil is poorly drained and boggy. There is abundant peat, and this overlies a bluish-gray, water-logged compact sub-soil. The pH of the soil varies from mildly alkaline to strongly acid. Where developed, soil profiles are generally very thin.

Podzol Soils. Podzols (and podzolization) have been described above where it was shown that they are leached soils with good profile development. Three podzolic regions are generally recognized ,(Fig. 3-6) but in some classifications, six types are recognized. Much of the podzol shown in Fig. 3-6 is formed in a cool-temperate humid climate (such as northern Minnesota, northern New England, and large parts of Ontario), under a coniferous or mixed coniferous and deciduous forest, and it is characterized particularly by a highly-leached, whitish-gray A_2 horizon (see Fig. 3-2). The B zone has an accumulation of humus and of precipitated iron, manganese and aluminum oxides. Because of the relative abundance of organic matter these soils are similar to tundra soils, are very acid (pH values as low as 3.5 in the humus), but are well-drained in comparison with the tundra soils.

Gray-brown Podzolic Soils. The gray-brown podzolic soils are found in North America south of the zone of the podzols, where the climate is warmer, and the vegetation and parent material differ from those to the north. The profile is somewhat different from that of the true podzol, as it shows a thin, moderately dark A_1 horizon (from organic matter) and a grayish-brown (as opposed to whiteish) A_2 horizon. It is the presence of the thin A_1 horizon, and the darker color of the A_2 horizon in the podzolic soil that distinguish it from a podzol. In effect, the podzolic soil has not been as severely leached as the podzol (and is less acid), but in general, both groups are formed by the same process. The gray-brown podzolic soil also has an appreciable quantity of illuviated clay in the B horizon. It is commonly developed on glacial deposits, and is found in great abundance in the Middle Atlantic states of the United States (north of Tennessee), and in southern Ontario.

Red and Yellow Podzolic Soils. These podzolic soils are found in warm, humid regions, under deciduous or coniferous forest vegetation and usually under conditions of good drainage. They cover most of the Southern States of the United States. The distinction between the red and yellow podzolic types is based on the color of the B horizon. The iron oxides in the yellow podzolics are slightly more hydrated (limonitic) than those in the red podzolics, as they are found in slightly more humid areas. Laterites and latosols are special cases of this group, and are discussed separately below.

Prairie Soils. Prairie soils are formed in temperate, or cool-temperate, humid regions under tall grass vegetation. The climate is much the same as that responsible for the gray-brown podzolic soils but, because the vegetation is tall grass in contrast to forest, the soils are different. The abundance of grass cover results in a deep, rich dark-brown A horizon (sometimes reddish-brown), with much microbiological activity. Prairie soils are found in abundance in the central United States (e.g., Iowa, Illinois), Russia, Argentina and Romania.

Chernozem Soils. Chernozem soils, also called black earths, occur in areas of fertile grassland but with less rainfall than the prairie soils. Chernozems are particularly well developed in the Ukraine but are also abundant in the central part of North America from Saskatchewan to Texas. The climate is generally temperate, and semi-arid (12-25 inches of precipitation per year) and, as a result, acid leaching does not occur to any great extent. The precipitation is insufficient to remove all the calcium and other divalent elements from the soil, and they accumulate in the profile. The A horizon is thick (1-4 feet) and black, due to the decay of organic matter and the lack of sufficient precipitation for effective leaching. The A horizon is also high in exchangeable calcium and magnesium. It is because of the high organic matter and the unleached cations that chernozems are among the best agricultural soils. The A zone is underlain by a lighter colored transitional zone and a B horizon and with a C_{ca} horizon consisting of calcium carbonate accumulations (see Fig. 3-7). The depth to the upper limit of the calcareous layer (caliche) is variable and is dependent on the amount of precipitation. The top of the profile with the high organic content is slightly acid (pH about 6), but with increasing depth it becomes alkaline near the bottom of the A horizon. As many trace metals have reduced solubility under alkaline conditions, their mobility is reduced in this environment. In addition, the tendency for montmorillonite to occur in chernozem soils is also detrimental to the formation of dispersive halos because of the high cation exchange capacity of this clay mineral.

Chestnut Soils. Chestnut soils, like chernozems, are developed on grassland (or shrubland) but in areas with even less precipitation (10-15 inches annually), such as the Great Plains. Because of the low precipitation, vegetation is scant, there is less organic matter than in the chernozems, and the carbonate layer is closer to the surface (Fig. 3-7). The A zone is moderately thick, and has a dark brown color. It overlies a lighter colored horizon that is above the zone of calcium carbonate accumulation. It may be noted that in all arid-zone soils, calcite is the most abundant calcium mineral, but locally gypsum will be found.

Brown Soils. Brown soils are found in areas that are even more arid than the chestnut types, in a temperate to cool climate. They develop under short grasses (sagebrush) and have a brown surface and a light-colored transitional subsurface horizon over calcium carbonate accumulation. Profile development is less marked than in chestnut soils, and the organic matter content is even lower. Under warmer conditions (but with the same low precipitation) the brown color gives way to reddish colors as iron is oxidized, and there is a reduction in the organic content on the surface.

This effect is seen in some parts of the Great Plains.

Desert Soils. These are the soils of arid regions (less than 10 inches of precipitation per year) and sparse shrub vegetation. They are thin, with a light-colored surface which may be vesicular and is ordinarily underlain by calcareous material near the surface. Soil profiles are poorly developed (Fig. 3-2), and organic matter is sparse or absent, which explains the light color. As is shown in Fig. 3-2, there is a high proportion of relatively fresh material (C horizon) in the profile. It is in this environment that deep-rooted vegetation is likely to be the most effective agent of secondary metal dispersion. Several types of desert soils are recognized: (1) sierozem (gray desert); (2) desert; and (3) red desert. Sierozem soils develop under slightly higher moisture conditions (as in the intermountain regions of the Rocky Mountains) than the desert soils, whereas the red desert soils are found in hot arid climates (as in the southwestern United States).

Intrazonal Soils

Intrazonal soils are those which have well developed characteristics that reflect the dominating influence of some local factor such as relief, parent material, or age, rather than the normal effects of climate and vegetation. Three principal types are recognized.

Hydromorphic soils. These soils are characterized by an excess of water due to poor drainage, either because of profile characteristics which prevent drainage, or because the soil is located in a depression or low-lying area. Marshes, swamps, seep areas, flats, and bogs are environments where these soils will form. A high water table close to the surface influences the soil profile. Such soils can be found in the glaciated regions of Canada and the northern United States, and also along the Atlantic and Gulf coasts. Soils known as bog, half-bog, gley, and groundwater podzols are examples of this type of intrazonal soil. They are also found in Africa, where they are called dambos, and are very important in exploration geochemistry (Hawkes and Webb, 1962). In all cases, precipitation exceeds evaporation and drainage. Organic matter accumulates in hydromorphic soils, and therefore most of them have a dark, peaty zone on the surface. Conditions in this environment are generally reducing and much of the iron will be in the ferrous state, which imparts a bluish-gray color to the horizon underlying the dark surface zone. Alternate periods of dry and moist conditions result in red to brown streaks and a characteristic mottled appearance as some of the iron and manganese are oxidized. The pH may be as low as 4 in some bogs, but in others it can be neutral to slightly alkaline if bases are present. Tundra soils, although usually considered zonal because their characteristics are determined by climate, may also be considered as intrazonal hydromorphic soils because of impeded drainage from the underlying permafrost.

Halomorphic soils. Halomorphic soils are formed under imperfect drainage in arid regions, and are characterized by high salt concentrations in the upper horizons. They are usually found in association with chernozems, chestnut, and other soils of arid areas, and occur around springs in deserts such as in the Basin and Range Province of the western United States. The salts may be chlorides, carbonates and sulfates of sodium, calcium, magnesium and potassium, most of which are very soluble in water. Their source may be weathering of the parent rock, but usually it is from dried up lake beds, as in Australia. By means of capillary action, the soluble salts are carried to the surface where they are easily recognized, especially in one particular variety, called *solonchak* or alkaline soils, which commonly show a white efflorescence of sodium chloride, generally with some calcium and magnesium salts, during the dry periods. Another variety, called *solonetz,* has a high concentration of sodium carbonate.

Calcimorphic soils. These soils are characterized by their high lime content, which originates in the parent material. Two groups are recognized: (a) the *brown forest* soils which form in humid regions where the calcium-rich parent material (e.g., marl or shaley limestone) is easily weathered under deciduous forests, and (b) *rendzina* which develops from soft, highly calcareous parent material under grass (or mixed grass and forest) vegetation, generally in semi-arid climates.

Azonal Soils

The characteristics of these soils are determined by the nature of their parent material, rather than by any soil forming process or climatic factors. Usually they have been subjected to the various soil-forming processes for only a relatively short period, and so have indistinct, or only slightly developed, horizons. Three groups are generally recognized: regosols (developed on deep, unconsolidated deposits such as sand dunes, loess and steeply sloping glacial drift), alluvial (developed on recently deposited alluvium), and lithosol (developed on steep slopes, and they consist of fresh and imperfectly weathered rock fragments). Several examples of azonal soils are shown in Fig. 3-9 (particularly profiles 3, 4, 8 and 9).

Factors Affecting Soil Formation

In the previous discussions it was noted that there are many different kinds of soil, each of which has a distinctive profile. There are five major factors affecting the formation of soil and they are now briefly discussed. They are (1) climate; (2) biological activity; (3) parent material; (4) topography; and (5) time. Four of these were discussed previously as environmental factors affecting weathering. The additional comments presented below are essentially extensions of the earlier discussions.

Climate. Climate, as has been stressed already, is the most influential factor in chemical weathering and in the development of most soils. Temperature and precipitation strongly influence the amount and type of vegetation, the rate of decay of this vegetation, the rate of weathering and leaching, the type of soil formed, and the position of calcium carbonate accumulations. Climate also exerts an influence on the type and abundance of living organisms such as bacteria which assist in soil formation processes, especially near the surface. It is also a major factor in determining the type and abundance of humus in soil horizons and this in turn affects the pH. Humus generates acid conditions, as in the top layers of podzols and bogs, which increase the rate of mineral decomposition and the migration of elements to lower horizons, such as the B horizon of podzols. Those areas low in humus, such as arid regions, often have soils with high pH values, a slow rate of soil formation, and low mobility of elements. The pH, and the availability of certain elements, particularly the divalent cations, influence the type of clay minerals which are likely to form. Clearly, climate has a strong influence on many factors in soil formation, all of which are interrelated to some extent. However, it must be remembered that climatic conditions can change drastically in geologically very short periods. About 15,000 years ago most of Canada was covered with glaciers, and most soil present today has formed since that time. All these factors have important implications for exploration geochemistry because they determine the optimum horizons for sampling, the extent of secondary dispersion, and in turn, the best analytical methods to be used.

Biological Activity. Vegetation is the main biologic factor in soil formation, although microorganisms do influence weathering and soil processes, perhaps more so than has been previously realized. Plants supply most of the organic material to the soil, both as dry matter added to the surface and as roots in the subsurface. The amount of dry matter added annually to the surface will vary with climatic conditions, for example from about 0.7 tons/acre in semi-arid regions, through 2 tons/acre in pine forests in more humid areas, to as much as 90 tons/acre in tropical rain forests. In some grasslands, roots comprise about 2 tons/acre of the top 4 inches of soil. In general, there is less vegetation in the drier climates and hence less organic matter in soils; with higher temperatures, organic matter tends to be more rapidly decomposed. Most of the organic matter (including humus) is found within the top 5 feet of any soil profile.

As soon as plants gain a foothold on a rock surface, soil formation begins. With decomposition, vegetation is converted to carbon dioxide, water and various organic acids, and elements in the rock are then mobilized. Some of these elements will come directly from weathered rock, whereas others will be released from the decaying vegetation which originally obtained the elements through their roots (Fig. 1-7). Residual plant material (humus) tends to accumulate in some soils, and is mixed with mineral matter forming the A soil horizon, and thus a profile begins to develop. The accumulation of humus is largely controlled by climatic conditions, which in turn determine the degree of leaching. Under oxidizing conditions, either in arid or extremely humid environments, organic matter will be sparse or absent.

Humus developed from the decay of organic matter is very complex and varied material and its composition will depend on many factors, including the nature of the original decaying vegetation. Most humus give an acid reaction but some, such as that developed from certain grasses with high calcium contents, may give a basic reaction. Two general types of humus are recognized, humic acid and fulvic acid. Each differs greatly in many properties, such as its mobility, stability, and relative abundance in different soil groups. Both humic and fulvic acids have the ability to chelate (to form a type of chemical compound) with iron and aluminum, and this partially explains the mechanism by which these elements are moved from the surface to lower horizons. Other elements may be similarly mobilized. Some reports indicate that fulvic acid and not humic acid, is the chelating agent, and that an increase in humic acid, in chernozems for example, will result in less leaching of metals (Brooks, 1972). Bear (1964) discussed various aspects of soil organic matter, which is considered one of the most complex materials existing in nature.

Baker (1973) has shown that humic acids extracted from several types of vegetation growing on podzolic soils in Tasmania exhibit strong solvent activity towards a number of minerals. Fine-grained galena, sphalerite, chalcocite, chalcopyrite, and other primary and secondary minerals of metalliferous deposits, were all at least partially decomposed, and their metals mobilized, after being placed in dilute humic acid solutions for 24 hours. Hematite and pyrolusite were quite vulnerable to attack by humic acid and Baker (1973) suggested that this may be one reason for the scarcity of extensive gossans over mineralization in parts of Tasmania. Smaller amounts of gold and silver were released by humic acid after a longer period of time (6 weeks). Although different humic acids show variable effects on minerals, there appears to be little doubt from these and previous experiments, that under favorable climatic conditions (e.g., cool temperature and high precipitation), they are powerful solvents in the weathering environment. This conclusion probably applies also to other acids derived from organic materials. It is possible that where leaching by humic acids (and inorganic acids as well) is strong, secondary dispersion patterns may be weakened or obliterated, and geochemical exploration would have to be based upon the search for primary halos in the country rock. This effect, as well as the role of humic and other organic acids in exploration geochemistry, needs further research.

Other aspects of the effect of biologic activity include the stabilization of soil by vegetation, and consequent protection from erosion and creep, and decrease in the rate of evaporation, all of which help in the development of soil profiles. By contrast, worms, burrowing animals, and root growth, tend to destroy the distinct horizons.

Parent Material. The nature of the parent material even though under-emphasized in favor of climate in the discussions so far, is an important factor in determining the nature of soils. In the simplest case, a clean quartz sandstone can yield only silica upon disintegration, whereas in the same environment a granite with feldspars, mica and quartz, may release many elements upon weathering and probably will develop a different profile. In the case of a limestone in a humid area, the soils developed on it will have a tendency to be alkaline, thereby counteracting the acidity of decaying organic matter. The nature of the parent material will also influence the nature of vegetation which grows upon it. Trees which prefer calcium and magnesium grow on carbonate rocks, and as a result these elements are continually being supplied to the surface, thereby inhibiting acidification, soil profile development, and the mobility of metals.

The physical nature of the parent material also affects soil-forming processes. A porous, permeable parent rock assists leaching and the movement of fluids, and increases the rate of soil development. Sandstones and loose friable sands are particularly susceptible to leaching above the water table as rainwater infiltrates easily. On an overall basis, the effect of parent material is mainly that of influencing the rate of weathering and soil formation, except in localized situations. The fact remains that climate, rather than rock type (lithology) is the major factor in soil formation, and this has been used as the basis of the zonal classification.

Topography. Topography may either inhibit or hasten soil formation. Gradient of slope and elevation are two of the topographic factors which influence run-off and drainage conditions, and indirectly soil formation. In mountainous areas, temperature decreases and precipitation increases with elevation. In Fig. 3-8 the relation between topography, precipitation and vegetation is shown. The effect of relief is to produce a series of zonal soils in the mountains, similar to those across much of the North American continent (Fig. 3-7), but on a much more localized scale. Other processes are also illustrated on Fig. 3-8. For example, the higher elevations characteristically have acid soils (because of effective leaching), whereas soils of the lower slopes have a higher pH because of the deposition of calcium car-

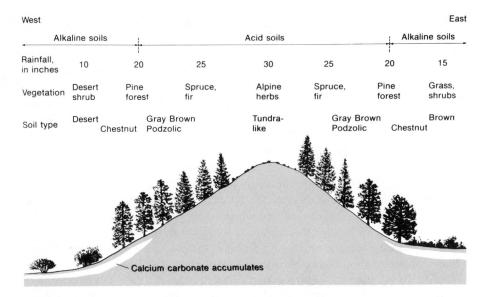

West   East

Alkaline soils   Acid soils   Alkaline soils

Rainfall, in inches	10	20	25	30	25	20	15
Vegetation	Desert shrub	Pine forest	Spruce, fir	Alpine herbs	Spruce, fir	Pine forest	Grass, shrubs
Soil type	Desert	Chestnut	Gray Brown Podzolic	Tundra-like	Gray Brown Podzolic	Chestnut	Brown

Calcium carbonate accumulates

Fig. 3-8. Generalized relations between topography, rainfall, vegetation and soils on mountains in the western United States (also applicable to parts of British Columbia). At the summits the growing season is less than 60 days (and may be underlain by permafrost in British Columbia; see Fig. 11-1); at the base of the mountains the growing season may be 4 to 5 months. From GEOLOGY OF SOILS by Charles B. Hunt. W. H. Freeman and Company. Copyright © 1972.

bonate. On very steep slopes most of the precipitation is lost by direct run-off, but in these situations physical weathering is more active. In low lying humid areas drainage is likely to be poor, with the water table near the surface, and a higher content of organic matter may ultimately result in peat. Therefore, even when conditions of precipitation and leaching are favorable, soil horizons will not form if it is not possible to remove the leached material.

In summary, topography exerts an influence primarily by controlling (a) the rate of surface run-off of precipitation, and hence the rate of moisture intake by the parent rock; (b) the rate of subsurface drainage, and therefore the rate of leaching of the soluble constituents, and (c) the rate of erosion of the weathered products and thereby the rate of exposure of fresh rock and mineral surfaces (Loughnan, 1969).

Topography and related effects (e.g., slumping, valley fill) may result in very significant variations in soil profiles in relatively short distances. Although the concept of zonal soils, the regional variations in the soil map of Canada (Fig. 3-5), and the cross sections shown in Figs. 3-7 and 3-8 imply very orderly and gradual changes over large distance, this is not always the case. Fig. 3-9, illustrates 9 soil profiles along a 3 mile traverse

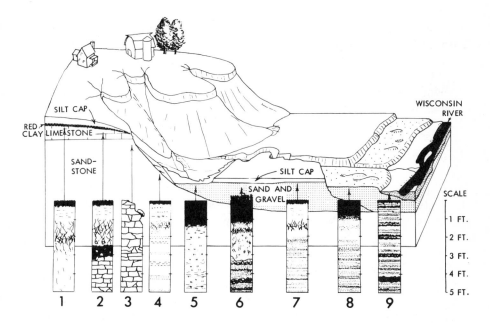

Fig. 3-9. Illustration of the variations in nine soil (including some sediment) profiles obtained in the hilly country of southwestern Wisconsin (Richland County). The horizontal distance across the drawing is 3 miles; the top of the bluff is 500 feet above the Wisconsin River. The dark portions of each profile represent organic matter. Variations due to topography, parent material and drainage are represented. Courtesy of Professor F. D. Hole, University of Wisconsin.

in southwestern Wisconsin. These were obtained from soils developed on several rock types, including a sand dune (No. 8) and alluvial soils (No. 9). The steep stony outcrop (No. 3) does not have a soil horizon. The differences between profiles are pronounced, and for purposes of exploration geochemistry, such possible variations must be considered in any interpretation. Such rapid variations are not unusual, as inspection of soil survey maps will confirm.

Time. Time is an important factor in all geological processes, and soil development is no exception. Under favorable conditions of a hot, humid climate, soils will form more quickly than in cold, dry conditions. Studies have been made in glaciated areas, regions of recent volcanism (e.g., Hawaii, Krakatoa) and other localities in which the time of soil development can be measured by radiocarbon methods, or is known historically. Predictably, the results show a wide range of values. Soil was developed to a depth of 14 inches after 45 years on Krakatoa, which is located in a favorable, tropical climate. It has been estimated that 500-1000 years were necessary for a well-developed profile to form on a moraine near Juneau,

Alaska, in a cool temperate area with considerable rainfall. As an extreme case, soil has still not developed in many places in the Canadian Arctic, after a period of perhaps 10,000 years since glaciation. Soils are sometimes classified as *juvenile* or *mature*, depending on the state of development of their horizons. In part this is dependent on time, but other factors such as climate and topography are obviously important.

Laterites

Laterites are of special interest in exploration geochemistry because they are abundant in many parts of the world, such as Australia, South America and Africa, where mineral exploration has great potential but where prospecting always presents special difficulties. Mineralized outcrops are often obscured by thick soils, or by a featureless blanket of hard lateritic material, as well as by vegetation in some areas.

Many meanings and interpretations have been applied to the term laterite, and this has led to much confusion. The term was originally proposed to describe the earthy, vesicular ferruginous crusts which are found in parts of India. Although the term laterite is still used for ferruginous crusts, the definition has been expanded to include the entire deeply-weathered profile of tropical soils, consisting of earthy, granular or concretionary masses, composed chiefly of iron and aluminum oxides and hydroxides. If iron predominates, they are called ferruginous laterites, and if aluminum predominates, they are called *bauxites.* Mineralogically ferruginous laterites are mixtures of hematite and goethite, and bauxite is a mixture of gibbsite $(Al(OH)_3)$, boehmite $(AlO(OH))$, and more rarely diaspore. Bauxite commonly develops on aluminum-rich rocks, such as syenite, or residual aluminum-rich clays weathered out from limestones, but laterites are found over a much wider variety of rocks. Other terms, such as lateritic soil, are sometimes applied to residual deposits, for example, those in the southeastern United States, but these are not true laterites according to the generally accepted definition because most of the iron and aluminum are in clay minerals (silicates), whereas oxides or hydroxides of iron and aluminum dominate in true laterites. Latosol is another term which is similar to lateritic soil.

Formation of laterites and bauxites. The optimum conditions for the formation of laterite appear to be high rainfall, high temperature, intense leaching, strongly oxidizing environment, subdued topography, long duration of weathering and a chemically unstable parent rock. These conditions are met in the tropics, particularly where the drainage is good, but it must not be construed that all soils in the tropics are lateritic. Under conditions favorable for the formation of laterite, organic matter accumulation is inhibited, and intensive leaching removes all the alkalis, alkaline earths, silica and other mobile elements. This results in the enrichment of iron, aluminum, titanium, a few trace elements (e.g., Nb, Ga, Cr), and some elements such as nickel, which are sufficiently concentrated to be recovered economically in New Caledonia and elsewhere. The process is essentially that of podzolization, but more intensive, and is called lateritization. However, there are significant differences in the mechanisms by which these two processes form their final products. In Fig. 3-6, laterite occurs in the lower right-hand corner of the climate diagram, in the field of the red and yellow podzols.

The details of the mechanism of the formation of bauxite and laterite have long been argued by soil scientists and geologists. Excellent reviews of all aspects of laterite (morphology, composition, distribution, classification, origin, etc.) have been prepared by Sivarajasingham et al. (1962), Maignien (1966) and Valeton (1972). Gordon et al. (1958) have studied bauxite in Arkansas, and Bleackley (1964) has described the bauxite and laterite of Guyana. These references are just a small part of the voluminous literature and their perusal will reveal that several systems are used to describe the soil horizons found in laterite and bauxite, and it will not always be clear to the reader how to integrate these systems.

Bayliss (1972) has correlated some of the terminology, which is shown in Fig. 3-10. The profile is divided into four horizons, A, B, C and D. These are equivalent to the terms soil, concretionary, leached and parent material of some soil scientists (Column 2 of Fig. 3-10). Comparison with Fig. 3-2 shows that the letter designations for the podzol and laterite profiles are not equivalent. Bayliss (1972) has also summarized the generally accepted interpretation of the formation of bauxite and laterite, starting with the C horizon.

The C horizon is divided into two parts, the lower pallid zone, and the upper mottled zone. The boundary between them is the lower limit of the fluctuating water table (Fig. 3-10), whereas the boundary between the C and B horizon is its upper limit. In other words, at some time or other, possibly yearly, the entire C horizon is below the water table. The pallid zone is composed largely (up to 80%) of a white kaolinite (hence the name clay zone; column 4, Fig. 3-10), but because the leaching of this material is in a saturated water environment (always below the water table), only the more mobile elements are removed, and the texture of the parent material is retained.

The upper part of the leached horizon, that is the mottled zone, is entirely within the zone of water table fluctuation, and during the wet season this zone is completely submerged. The vertical movement of the water table is essential to active lateritization and is responsible for leaching and moving certain elements not only downward, but also

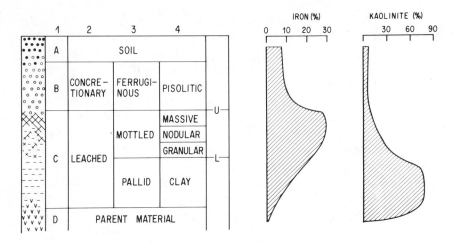

Fig. 3-10. Correlation of four different classifications used for bauxite and laterite. Column 1 is the classification often used by geologists. Columns 2, 3 and 4 are various classifications used by soil scientists. (However, the reader is cautioned there are other classifications). The upper limit (U), and the lower limit (L) of the fluctuating water table are also shown. Generalized distributions for iron and kaolinite are indicated. Approximate depth of this idealized profile is from 30 – 50 feet. Modified from Bayliss (1972).

upward. In other words, lateritization most readily takes place in the zone of water table fluctuation, between the annual high and low positions. Leaching of the soluble and partially soluble elements in the mottled zone is more effective than in the pallid zone because, in it, water is more mobile. Iron, after its release from a mineral, migrates *upwards* in the ferrous state, either with the rising water table or by capillary action, where it is later oxidized to an insoluble ferric oxide, initially goethite but later this may change to hematite. As a result of the movement, iron becomes separated from aluminum. This upward movement, or movement within or associated with a fluctuating water table, is a notable difference from podzolization which can be thought of as due to downward movement only. It is likely that other sparingly soluble elements, such as cobalt and arsenic, are also moved upward by the same mechanism, but others are likely to move downward (the alkalis, for example). The largest concentrations of iron are found in the massive zone (column 4, Fig. 3-10), which is at the highest level reached by the water table. The distribution of iron is graphically illustrated in Fig. 3-10. In the granular zone, iron minerals form the cement between isolated rock fragments, and the proportion of cement increases in the nodular zone, until in the massive zone they cement the whole rock. The mottled color is formed by periodic oxidation during fluctuations of the water table.

The overlying B horizon (ferruginous, etc.) forms above the water table through direct alteration of the parent material (D horizon) by an intense process directly comparable to podzolization. After a surface layer of silicate minerals decomposes during podzolization, the soluble and partially soluble elements are completely leached and removed by water. The insoluble elements, particularly iron and aluminum, are re-deposited with a concretionary or pisolitic texture (hence the names in Fig. 3-10). Therefore, not only is the iron in the B zone morphologically different from that in the C zone (pisolitic versus massive), but also the process of formation is different. Quartz appears to be relatively stable in the B horizon because it is protected by coatings of iron or aluminum oxides. Gibbsite is the first aluminum oxide mineral to form, but it slowly converts to boehmite, and eventually even diaspore may form.

Large areas of lateritic manganese concretions are common in some places, such as India (where they are called gondite) and Cuba, and they form by processes similar to the pisolitic laterites, except that they overlie manganese-rich rocks. Laterites are subject to further changes with time, which may include erosion, a change in groundwater chemistry, and deposition of a wind-blown cover, all of which result in variations from the ideal situation described above.

Distribution of trace elements in laterite and bauxite. From an exploration point of view it is necessary to consider the distribution of the trace elements in individual horizons of the laterite profile. Regrettably, there is little systematic information on this subject from which generalizations can be made. Mazzucchelli and James (1966) have observed that arsenic is concentrated in amounts up to about 250 ppm in the coarse lateritic fragments in the area of gold mineralization near Kalgoorlie, Western Australia (Fig. 3-11). Analysis of these fragments for arsenic enabled broad geochemical patterns to be defined which are of use in exploration. Although insufficient data precluded any definite conclusions, Mazzucchelli and James (1966) suggested that considerable lateral migration of arsenic took place both during and after the formation of the laterite. The important points here are: (1) arsenic was not leached with the soluble elements, but was concentrated by some unknown mechanism with iron in the nodules; and (2) the *coarse* fraction gave better indications of an underlying source of arsenic (and possibly gold) than did the minus 80 mesh fraction which is conventionally taken in many surveys. More recent studies in Australia (e.g., Zeissink, 1971) have confirmed the results of Mazzucchelli and James (1966) and have even extended it to the point where the coarse fraction (pisolitic) is analyzed for Ni, and the pathfinder elements As, Pt and Te in the search for nickel in Western Australia.

Study of the nickel laterites of New Caledonia yielded information on the distribution of nickel, cobalt and manganese, in a different type of

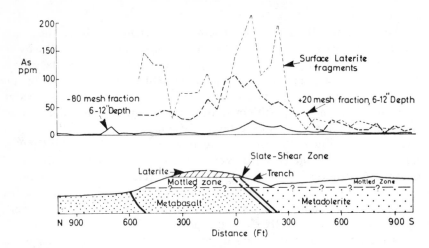

Fig. 3-11. Distribution of arsenic in lateritic materials, Kalgoorlie area, Western Australia. Comparisons of three types of material are shown: (coarse) surface laterite fragments; +20 mesh fraction; -80 mesh fraction. Note the larger anomaly in the coarse surface fragments. From Mazzucchelli and James (1966).

laterite. Here, laterite is developed over ultramafic rocks which originally contained about 0.25% combined nickel and cobalt. These elements were present within the structure of certain minerals, particularly olivine. Many of the ultramafic rocks have been altered and weathered, the main process being lateritization (Fig. 3-12), although serpentinization has also taken place. Low-grade nickeliferous laterites (about 1% Ni) thus formed are economically recoverable. The most valuable nickel ores occur below the laterite near the top of the altered peridotite. However, cobalt and manganese are concentrated above the nickel at the base of the porous laterite (Fig. 3-12). This illustrates the separation of iron from cobalt and manganese by the process of lateritization, and shows that the trace element profile for each of these elements is different. In other respects the nickeliferous laterites of New Caledonia are similar to those from elsewhere in the world. All alkalis, alkaline earth elements and much of the silica have been leached, the upper ferruginous layer is pisolitic, and the best nickel concentrations are where topographic conditions aided leaching. Garnierite, a nickeliferous silicate (serpentine, in part), is the principal ore mineral. In places in New Caledonia and elsewhere quite pure deposits of this mineral occur, but the low-grade nickeliferous laterite is more common in tropical environments throughout the world.

The study by Gordon et al. (1958) gives an excellent account of the soil profiles developed in the Arkansas bauxite area as well as the relative concentration or depletion of minor elements compared to the original nepheline syenite source rock. Suffice it to say that certain elements are

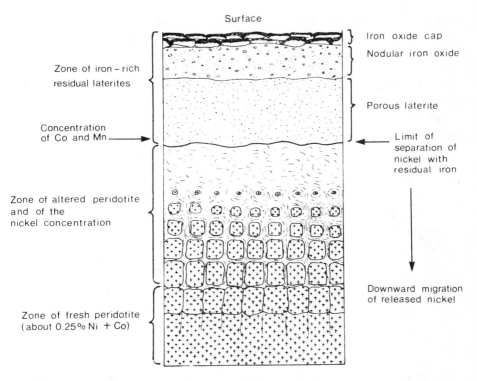

Fig. 3-12. Typical section through nickeliferous laterite in New Caledonia showing how economic nickel deposits may be formed by lateritization. Modified from Chêtelat (1947).

concentrated even more than the very insoluble aluminum under the conditions which formed the bauxite. These include Cr, Cu, Ga, Nb and Mo. Some elements (Zn, Ti, Sc, V, Be, Mn, Y and Pb) are concentrated to some extent, but less than aluminum, whereas others (Sr, La, Ba, Ca and Mg) are essentially completely leached.

The conclusion to be drawn from these examples is that great differences exist in the trace element profiles in laterites for individual elements and for different soil types, reflecting the nature and extent of the various weathering processes (e.g., podzolization, lateritization) which are operative. In addition, the trace elements may be distributed in various size fractions.

It would be inappropriate to conclude this discussion without stressing that the mechanisms invoked to explain the formation of laterite and bauxite have been based largely on visual and empirical observations. In the past few years several important theoretical and laboratory studies have been made to better define the conditions of formation of these materials.

For example, Norton (1973) applied Eh-pH diagrams (described below) to the problem. Using these diagrams he was able to explain more accurately the limiting conditions for the formation of laterite and bauxite, and the accompanying enrichment of Mn, Co and Ni. Finally, it is important to note that some of the principles invoked in connection with the formation of laterite and bauxite have application elsewhere. For example, fluctuating water tables in many localities can explain the upward movement and the dispersion of elements.

Trace Elements In Soil Profiles

Sampling the particular soil horizon which will yield the most significant anomaly is essential for realizing the greatest potential from a soil geochemical survey. In the preceding section it was shown that trace element contents vary in different parts of the profile (Figs. 3-3, 3-4, 3-12). Further, the distribution of trace elements also varies with the size fraction analyzed (Fig. 3-11). In view of the importance of soil sampling in exploration geochemistry, further comments are necessary in order to stress that trace elements vary with depth, soil type, and size fraction, and that this is to be considered a normal phenomenon.

Relatively little information has been published in the geochemical literature which has direct bearing on trace element variation in soil profiles. Most of the work which has been reported is from the Soviet Union (for example, Vinogradov, 1959), but there many soil types, such as laterites, which are rare in the U.S.S.R. A few studies have been mentioned from the western literature, such as those by Mazzucchelli and James (1966) and Zeissink (1969; 1971) on trace element behavior in laterite profiles, and by Gordon et al. (1958) on the distribution of elements in Arkansas bauxite profiles but, in general, these are exceptions rather than the rule. It is because of the paucity of data in this field that Boyle and Dass (1967) and Mazzucchelli and James (1966) saw fit to stress the points, respectively, that the A horizon gave better geochemical anomalies than the conventional B horizon, and that the coarse fractions of laterites (in Australia, at least) gave stronger anomalies than did the conventional minus 80 mesh fractions.

Most of the information which is available on the distribution of trace elements in soil profiles can be found in the soil science literature. For example, Turton et al. (1962) observed that some laterites in Western Australia had higher contents of some elements (Mo, Ga, V, P) in the gravel of the top 30 inches of soil profiles (probably equivalent to the pisolites of the B horizon in this chapter), in comparison with the finer fractions, and that the abundance of these elements decreased with depth.

In many cases the objectives of these studies have been concerned with the micronutrients (Fe, Mn, Zn, Cu, B, Mo, Cl) essential for the growth of plants and animals, or with certain elements (e.g., Se) known to be toxic to animals. Nevertheless, they do give data on the variations which have been observed, and insight into the factors that control the trace element distribution. Excellent summaries on trace elements in soils have been published by Mitchell (1964; 1972).

Based on data obtained from many different types of soils over a period of many years, soil scientists have been able to determine the range of trace and minor elements in soils which can be expected from the more common parent materials. These ranges are illustrated on a logarithmic scale in Fig. 3-13, which values are similar to those given in Table 2-1. Extreme values reported in the literature have a much wider range, and very high values are usually found in the vicinity of ore deposits, neither of which are indicated on Fig. 3-13. The data in Fig. 3-13 do not indicate from which soil horizons the figures were obtained, but it can be assumed that most of the ranges are based on analyses of the A horizon. Fig. 3-13 is a useful indicator of the background values that might be expected in areas not containing mineralization.

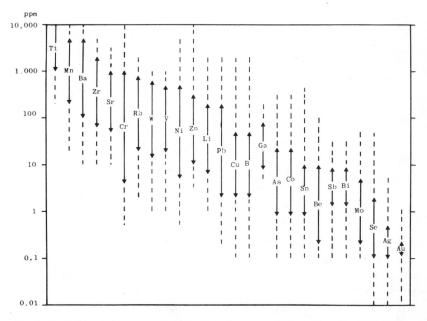

Fig. 3-13. Range of abundance of trace elements commonly found in soils. Dashed lines indicate the more unusual values. From Andrews-Jones (1968).

The distribution of elements in soil profiles is dependent upon two fundamental factors: (1) the abundance of the elements in the parent rock, and (2) the nature of the weathering and soil-forming processes operating on these rocks. Both of these points have been discussed in detail above, and on the basis of these discussions, variations of trace elements in soils and soil profiles should be expected. It would be very desirable to be able to make generalizations about the distribution of trace elements in soil profiles, now that it is realized that their abundance is not constant. Unfortunately this is extremely difficult because of the great number of variables involved. Nevertheless, an attempt must be made so that the degree of variation may be emphasized. Fig. 3-14 is the result of such an attempt for one element, much of which is based on soil studies in the Soviet Union.

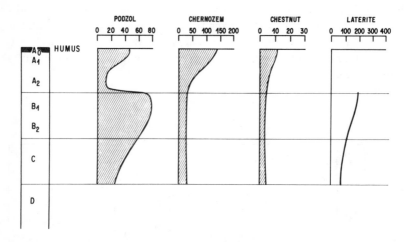

Fig. 3-14. Generalized representation of the distribution of copper in four different types of soil. Distributions for podzol, chernozem and chestnut soils are based on data from the Soviet Union (Vinogradov, 1959). The laterite distribution is based on data from Zeissink (1971). In the laterite profile, insufficient information is available for the A horizon to permit generalizations. All values in ppm.

Vinogradov (1959) found that over vast areas of the Russian plain, the trace element content of the zonal soils varied within narrow limits, primarily because of the uniformity of the parent material. Thus the trace element variations in the soil profiles from the Soviet Union have been selected for the generalized profiles in Fig. 3-14. As data were not reported for laterite by Vinogradov (1959), that of Zeissink (1971) has been used. In all cases, the profiles are generalizations for copper. Profiles for some other elements such as zinc, would be similar, but chromium which is much less mobile and usually found in the relatively stable mineral chromite, is distinctly different. The profiles of many other elements, such as lead and cobalt, would be expected to lie between the extremes.

In podzols and forest soils, copper accumulates in the upper humic layer and also in the illuvial B horizon (Figs. 3-4, 3-14). In chernozems there is a marked accumulation of copper in the humic layer, but in the B and C horizon the copper contents are similar. In chestnut soils the copper content is directly related to the amount of organic matter. In general, in the less humid regions there is an accumulation of trace metals in the organic-rich portions of the soils, and no marked increase in other horizons.

The reason for the shape of the trace element profile in chernozems has been explained by Mitchell (1964). As chernozems have a high content of organic matter and clay, the trace elements tend to remain *in situ* and not to be leached out, except those taken up by plants. Chernozems and related soils, therefore, often have a higher trace element content than other soils derived from the same parent material, and display enrichment in the organic surface horizon. Average values for several trace elements in Russian chernozems are: Mn, 830 ppm; Ba, 400 ppm; Cr, 400 ppm; Sr, 400 ppm; V, 290 ppm; Ni, 49 ppm; Zn, 72 ppm; Cu, 30 ppm; Co. 6.1 ppm (from Mitchell, 1964, p. 357).

In Australian laterites, Zeissink (1971) noted that Zn, Mn, Ni and Co were concentrated in the upper ferruginous zone (B horizon of Fig. 3-10) of one profile, but in a second, the greatest concentrations of these four elements occurred in the intermediate montmorillonite zone (pallid zone of Fig. 3-10). Many other elements, for example, Cu, Cr, Ga, Ge, V, Ni, Mn, Ag, Pt and Pd, were concentrated in the upper ferruginous horizon. Thus the profile in Fig. 3-14 is similar, not only for copper but for many other elements. Zeissink (1971) attributed the difference in the behavior of Zn, Mn, Ni, and Co to the fact that these elements are more mobile, and that probably one of the profiles (the one in which they are concentrated in the C horizon), is more mature, enabling them to move farther downward with increasing time. Many African laterites appear to have the highest concentration of trace elements in the C horizon, whereas the B horizon is generally depleted in these elements.

A survey of the literature reveals a great variation in trace element distributions for the same element in soil profiles developed in soils called by the same name. From this it can be concluded that, for purposes of exploration geochemistry, a preliminary (orientation) study should be run on the soils in every new area in order to determine their trace element characteristics and to ensure that the optimum horizon is sampled. This will usually require that samples be collected through a series of vertical sections, obtained either by digging a pit or augering. Profiles should also be obtained away from any mineralization so that background values can be determined.

It has now been established that during the process of soil formation there is usually a partition of the major and trace elements between the various horizons, and that certain horizons preferentially concentrate different elements which are of interest for exploration purposes. Therefore, it is *essential* that samples should always be taken from the same horizon so that meaningful comparisons can be made and significant anomalies recognized. If this is not possible, then any information obtained from the analysis of samples from different horizons must be interpreted with extreme caution. In some cases it is not always possible to recognize that samples have been collected from different horizons, and in these situations false anomalies may result. Collecting soil samples from a uniform depth is generally not a satisfactory substitute for determining the best soil horizon.

APPLICATION OF pH AND Eh

Introduction

In previous discussions of soils, waters, and weathering processes, frequent mention was made of the acidic or basic (alkaline) nature (pH) of these materials, and that this fundamental property has an important bearing on mobility of elements. Among the many possible examples, it is sufficient to recall that during the oxidation of sulfide minerals, low pH (high acidity) values are generated, and in this environment many metals are mobile, but mobility is decreased as the pH becomes higher (more alkaline). Similarly, elements are more mobile in well-drained soils which are generally acidic, whereas poorly drained soils, or soils in dry areas, are more likely to have high pH (alkaline) values, and as a result, mobility is generally reduced, at lease for many of the base metals. Thus pH is one major controlling factor in the mobility of elements in many environments.

Another fact mentioned on several occasions concerned the precipitation of iron and manganese oxides. When these elements are in a reduced state, that is as Fe^{+2} and Mn^{+2}, they are mobile in many natural environments. However, as soon as they are oxidized to the $+3$ and $+4$ states, respectively, their mobilities in the secondary environment are greatly reduced and they are likely to form sols or precipitates of iron and manganese oxides. In other words, the valence state determines in large part whether these elements remain in solution or precipitate. This is an example of the importance of oxidation-reduction processes. A measure of the oxidizing and reducing tendency of a solution is known by the synonymous terms Eh, redox potential, oxidation potential, and oxidation-reduction potential.

The oxidation potential (Eh) is actually a measure of the voltage involved in the addition or removal of electrons during the change of the valence state of various elements (e.g., Fe, Mn, Cu), and the energy required is modified by pH. Therefore, both Eh and pH determine the concentration of an ion that will remain in solution, and a change in either may be sufficient to cause precipitation. In view of this interdependence, Eh-pH diagrams, which permit an evaluation of both parameters, have found widespread use in certain branches of geochemistry, such as in explaining certain low temperature phenomena of the secondary environment (e.g., the formation of sedimentary iron deposits), and in the understanding of water chemistry. Reviews of Eh-pH as applied to exploration geochemistry have been published by Cloke (1966), Hansuld (1967) and Krauskopf (1967), and Hem (1970) stressed the application to the interpretation of the chemistry of natural waters. Garrels and Christ (1965) have written the classical and fundamental book on the subject, which contains many examples, as applied to low temperature geochemistry and mineral deposits.

In the next section pH is discussed, and this will be followed by Eh. Emphasis is on the importance of these concepts in the secondary environment. Eh and pH are of importance in the magmas and hydrothermal fluids of the primary environment, but this is of minor concern in exploration geochemistry.

pH

pH is a numerical expression of the relative acidity or alkalinity of an aqueous system. It refers to the concentration of H^+ and OH^-, and is defined as the negative logarithm (to the base 10) of the hydrogen ion activity. (The use of "activities" of substances, instead of concentration, facilitates exact computation of chemical relations; activity may be defined as an "ideal" concentration). From this is derived the pH scale, which ranges from 0 – 14, where the neutral point is defined as the condition when the activity of the hydrogen ion equals that of the hydroxyl ion. For dilute aqueous solutions at 25°C and 1 atmosphere, this means a pH of 7.0. Pure water under these conditions is neutral, that is, it has equal concentrations of both hydrogen and hydroxyl ions. The pH of the neutral point changes slightly with temperature, pressure and salinity. It must be remembered that pH is a logarithmic function, and therefore, a change of one pH unit represents a ten-fold change of ion concentration. pH values lower than 7 represent acid conditions, whereas those above 7 represent alkaline conditions.

Measurement of pH is a simple procedure and should be standard practice for all water samples at the time of collection. It is easily accomplished electronically with a pH meter which measures the voltage developed between a specially designed glass electrode and a reference electrode, when both are immersed in the same solution (e.g., water). This may be done in the field by means of portable, battery operated pH meters (Fig. 6-11). pH may also be measured by means of colored indicator reagents, such as indicator paper or the recently developed plastic indicator strips which are generally sufficiently sensitive and accurate for field purposes.

The pH of soil is defined as the pH of an aqueous mud produced from it. Soil chemists (Bear, 1964, p. 500) determine the pH of soil by placing 50 gm of a less than 2 mm grain size air-dry soil sample in a suitable container (e.g., a 100 ml beaker, or paper cup) adding 50 ml of water, and stirring occasionally for 1 hour. Soils with high organic matter content may require more water to become thoroughly wet. While the solution is being stirred the pH is determined, usually with a pH meter, although in the field pH indicator paper is acceptable. In addition, various dyes may be used to determine pH, particularly under field conditions (Buckman and Brady, 1969), and although this method is simpler, it is less accurate. It is based on the fact that many dyes change color with an increase or decrease in pH, making it possible to estimate the approximate pH of a soil after it has been saturated with the dye.

Certain types of pH meters can also be used to measure Eh and this requires a special pair of electrodes. In addition, some instruments, such as that manufactured by Orion Research Inc., may also be used as specific ion meters (Fig. 6-11).

As stated previously, the pH of pure water at 25°C is 7.0. Most groundwater found in North America has pH values from around 6.0 - 8.5, but water having a lower pH is not uncommon in thermal springs and springs draining oxidizing sulfide minerals (and from various types of pollution). Water with a pH much over 9.0 is unusual, but values as high as 11.7 have been observed in certain springs, usually in desert areas. In areas not influenced by pollution, river water has a pH between 6.5 and 8.5, the higher values generally being from carbonate areas. Where photosynthesis by aquatic organisms takes up dissolved carbon dioxide during the daylight hours, a diurnal fluctuation of pH may occur and the maximum values may sometimes reach as high as 9.0. One example where the pH exceeded 12 during diurnal fluctuations is reported by Livingstone (1963, p. 9). As a generalization, most waters and soils have a pH from about 4.0 to 9.0, and thus most chemical reactions in the weathering cycle take place within this range.

Natural water whose pH is below 4.5 commonly involves oxidation of some form of sulfur, usually sulfides, and should be considered a very definite indication of probable sulfide mineralization, but caution is always warranted. For example, Hem (1970, p. 95) gives data which show that the oxidation of ferrous iron can also result in low pH values. The data below represent successive determinations of ferrous iron and pH on a sample of water collected from the overflow of a coal mine in Pennsylvania where the ferrous iron most likely was originally present as pyrite (see equations (7) to (10)).

Date (1963)	Time after sampling (days)	$Fe^{+2}(ppm)$	pH
July 16	0	—	4.98
July 30	14	135	3.98
Aug. 20	35	87	3.05
Sept. 24	70	41	2.81
Oct. 29	105	2.2	2.69

The sample contained a very high content of iron, and its pH decreased by more than 2 units as the ferrous iron was oxidized and precipitated as ferric hydroxide. Many other examples of low pH values which do not represent mineralization can be cited from present and past mining and refining activities, and pollution of other types.

The solubility of most elements and the stability of their compounds is extremely sensitive to the pH of the aqueous environment. Only a few elements, for example, the alkali metals (such as Na, K and Rb), alkaline earths (e.g., Ca, Mg and Sr), and some elements that form acid radicals (e.g., nitrogen and chlorine) are normally soluble throughout the entire pH range. Most metallic elements are soluble only in acid solutions, and tend to precipitate as hydroxides (or basic salts) with increasing pH. The pH at which these elements precipitate as hydroxides is known as the *pH of hydrolysis*. In Table 3-6 several elements are presented in the order of increasing pH of hydrolysis. The pH of hydrolysis depends on the metal ion concentration but for the purposes of this discussion low concentrations are assumed. Thus, for example, when the pH very near an oxidizing pyrite deposit is extremely acid (perhaps pH of 1), ferric iron will be in solution. However, as soon as the pH reaches 2, ferric iron is likely to precipitate. Other elements do not precipitate until much higher pH values, such as copper at a pH of 5.3, and zinc about pH 7.0. Clearly, pH is a major factor in element mobility in solution, and explains among other things, why elements are much less mobile in alkaline environments (e.g., limestone areas) by comparison with some acidic environments. It also explains one mechanism by which trace elements accumulate in stream sediments. As acid groundwater or springs are diluted with waters of high pH, or as the acid waters react with rocks over which they are passing, the pH of these waters will tend to become neutral (higher pH), and trace metals will

Table 3-6. pH of hydrolysis (hydroxide precipitation) of some elements from dilute
            solutions (from Britton, 1955, vol. 2, p. 102)

Element	pH	Element	pH	Element	pH	Element	pH
Fe^{+3}	2.0	Al^{+3}	4.1	Cd^{+2}	6.7	Pr^{+3}	7.1
Zr^{+4}	2.0	U^{+6}	4.2	Ni^{+2}	6.7	Hg^{+2}	7.3
Sn^{+2}	2.0	Cr^{+3}	5.3	Co^{+2}	6.8	Ce^{+3}	7.4
Ce^{+4}	2.7	Cu^{+2}	5.3	Y^{+3}	6.8	La^{+3}	8.4
Hg^{+1}	3.0	Fe^{+2}	5.5	Sm^{+3}	6.8	Ag^{+1}	7.5-8.0
In^{+3}	3.4	Be^{+2}	5.7	Zn^{+2}	7.0	Mn^{+2}	8.5-8.8
Th^{+4}	3.5	Pb^{+2}	6.0	Nd^{+3}	7.0	Mg^{+2}	10.5

precipitate out as their pH of hydrolysis is reached. Finally, in the way of examples, pH also explains the separation of certain elements (e.g., Fe) from others (e.g., Zn) which may be found in the same deposit, but have extremely different dispersion halos in the secondary environment.

In addition to its effect on the solubility of elements in solution, pH also plays an important part in determining the adsorption capacity of clays. As mentioned previously, montmorillonite in an acid environment will have little if any adsorbed metals, by comparison with an alkaline environment. This is because the hydrogen ion is preferentially adsorbed (or exchanged) in place of metals. And finally, among the many variables which can adversely affect the reliability of biogeochemical methods, the pH factor dominates (Brooks, 1972, p. 103). pH affects elemental uptake by plants, the availability of the elements in the soil (whether the elements are in a soluble or insoluble form), and whether and how strongly, the elements are adsorbed onto clays.

Although extremely valuable, the data presented in Table 3-6 must be considered as no more than a first approximation of actual situations found in nature. Mobility based on the pH of hydrolysis does not take into account such factors as the formation of organic and inorganic complexing agents, for example, the formation of uranyl carbonate and sulfate complexes which extend the solubility of uranium well into the alkaline pH range from the pH of 4.2 shown in Table 3-6. Also, the presence of certain elements and radicals can cause almost immediate precipitation of certain other elements considerably before their pH of hydrolysis is reached. An example of this is the precipitation of insoluble anglesite when sulfate occurs in waters associated with an oxidizing lead deposit.

Eh

Eh of a solution, as defined previously, is a measure of the oxidizing or reducing tendency or, more specifically, the potential of a system. Because oxidation or reduction involves a transfer of electrons, it is fundamentally an electrical property which can be measured in volts or millivolts. All measurements are referred to the standard hydrogen electrode the potential of which is arbitrarily taken as zero, at pH = 0 and a pressure of 1 atmosphere. Therefore, the oxidation potential is a relative figure, the reference standard being the reaction

$$2H^+ + 2e = H_2 \qquad E^0 = 0.00 \text{ volts} \qquad (19)$$

At equilibrium, chemical reactions taking place in water are limited to those redox potentials (at pH = 0) which range between 0.00 volts (at lower, i.e. negative values, hydrogen would be liberated) and 1.23 volts, which, if exceeded would result in the liberation of oxygen from water by the reaction:

$$H_2O = \tfrac{1}{2}O_2 + 2H^+ + 2e \qquad E^0 = + 1.23 \text{ volts} \qquad (20)$$

Oxidation potentials are indicated by the symbol E^0 when the relevant reactions take place under the standard conditions of unit activities of the reactants and products(e.g., $Fe^{+2} = Fe^{+3} + e$; $E^0 = 0.77$ volts)and by Eh when compared with, or measured against, the standard hydrogen half cell (equation 19). Hence the term Eh means that reactions occur in reference to the hydrogen scale. The Eh scale of oxidation and reduction potentials extends on either side of zero, the arbitrarily fixed value for reaction (19) under standard conditions. The more positive the potential, the more oxidizing it is relative to the hydrogen half cell reaction. However, a negative value does not necessarily mean a reducing potential, only that it is reducing relative to the standard half cell.

Both reactions (19) and (20) are strongly affected by pH. The E^0 values given for these reactions are based on a pH of 0, and the potentials decrease 0.06 volt for each unit increase in pH (at 25°C). This accounts for the slope of lines on the Eh-pH diagram in Fig. 3-15. As mentioned previously, the pH of soils and waters generally lies between 4 and 9, which boundaries are also indicated on Fig. 3-15. For a pH of 7, the potential for equation (19) is − 0.41 volt, and for equation (20) it is + 0.82 volt. These figures indicate that potentials in natural environments at pH of 7 should lie between − 0.41 and + 0.82 volt, the former value representing extreme reducing environments as in marine bottom sediments rich in organic matter, and the latter extreme oxidizing conditions, such as may

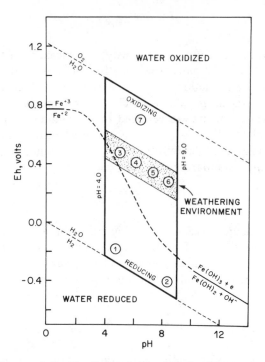

Fig. 3-15. Framework for Eh-pH diagram in which the most usual limits of Eh and pH in the near-surface environments are outlined. The upper limit of the weathering environment is in direct contact with air, whereas the lower limit is considered to be the water table. Numbers indicate other environments: (1) bogs and waterlogged soils; (2) reducing marine sediments; (3) acid mine waters; (4) rain; (5) river water; (6) ocean water; (7) oxidizing lead sulfide deposits. The oxidation-reduction potential for the simple ions and hydroxides of iron are indicated (at 25°C and 1M). The horizontal dashed portion of the line (top) represents the reaction of $Fe^{+2} = Fe^{+3} + e$, and the solid portion (bottom) $Fe(OH)_2 + OH = Fe(OH)_3 + e$, and the high slope dashed portion $Fe^{+2} + 3H_2O = Fe(OH)_3 + 3H^+ + e$.

possibly be found in certain oxidizing metal deposits. However, most reactions in the zone of weathering take place within the shaded area on Fig. 3-15.

The oxidation potential of an aqueous system can be determined by measuring the voltage between an inert electrode such as platinum, and a reference standard hydrogen electrode. In actual practice a secondary standard half cell, such as a calomel electrode is used as a reference electrode in place of the hydrogen electrode, and as a result a correction or conversion is necessary to obtain Eh. Hansuld (1967) reports that this correction, which in the case of a calomel electrode requires that 0.241 volt be added to the measurement, is not often made, and so confusing and misleading results are often reported. Numerous instruments (especially pH meters) are available on which Eh can be measured, including the portable, battery operated Orion instrument mentioned previously. Care must be taken to ensure that these instruments read in Eh units, and if not, the necessary conversions must be made. Recently Bolviken et al. (1973) have described an instrument for *in situ* measurements of the pH and Eh of groundwater in drill holes to depths of at least 350 meters. Results from a Norwegian drill hole showed pH variations of as much as one pH unit within a vertical distance of 1 foot. Information obtained from such an instrument may lead to a better understanding of the chemical processes in groundwater. Noteworthy is the observation of Boyle and Garrett (1970, p. 59) that Eh, as well as pH, measurements are influenced by many factors that are generally unknown without a large amount of investigative work. As a result, interpretation of the measurements is not always satisfactory or even possible.

Eh – pH Diagrams

Many chemical reactions of interest in geochemistry involve both pH and Eh, and a graphical representation of the relations as shown in Fig. 3-15, is known as an Eh-pH diagram. These are extremely valuable for describing and explaining characteristics of various elements under a given set of conditions. The pH range outlined is from 4 to 9, which is the range in which most reactions in the weathering environment take place.

The relation between ferrous and ferric iron under differing pH-Eh conditions is illustrated in Fig. 3-15, and this is typical of the behavior of other elements (e.g., Mn, V, U, Cu) which occur in two or more valence states. Of particular importance in exploration geochemistry is the fact that the properties of mobility and solubility vary considerably from one state to another. As far as mobility is concerned, ferrous iron has no more similarity to ferric iron than it does to magnesium, vanadium or other elements. The curve in Fig. 3-15 represents the oxidation-reduction potential for ferrous and ferric iron in solution (dashed portion), and the hydroxides for these valence states (solid portion). The portion with the greatest (negative) slope represents iron in solution (Fe^{+2}) in equilibrium with iron as a solid (Fe_2O_3, $FeO(OH)_3$). At a pH of 5.5, the ferrous iron in solution begins to precipitate as ferrous hydroxide (Table 3-6) but, as may be seen from the diagram, ferrous hydroxide is only stable under very reducing conditions. Thus the redox-potential, as represented approximately by the dashed-solid line, is the minimum potential at which the ferrous ion or hydroxide will convert to the ferric form. The stability relations show that the potential decreases with increasing pH, and from experimentation (not shown in Fig. 3-15) it is known that oxidation proceeds more rapidly in an alkaline environment. The potential for the oxidation of ferrous hydroxide, at a pH of 5 or higher, lies far below the potential available in the weathering environment and it is easily oxidized. When ferrous ions are oxidized to ferric, ferric hydroxide is immediately precipitated because the pH of hydrolysis of the latter is only 2 (Table 3-6). In acid solutions, from a pH of about 4.2 or lower, the oxidation potential is more than 0.6 volt. As these voltages are not normally encountered in the weathering environment, ferrous iron is stable in solution.

From the above discussion and the interpretation of Fig. 3-15, it is possible to explain the factors which govern the mobility or deposition of iron compounds. They can be summarized as: (1) oxidizing conditions promote the precipitation of iron, whereas reducing conditions promote the solution of iron; and (2) acid solutions generally promote the solution of iron, whereas alkaline solutions promote precipitation of iron. From the point of view of exploration geochemistry these factors explain the precipitation of ferric iron under conditions of increasing pH as in many soil horizons, or in the weathering residues of rocks which have a distinct alkaline reaction (e.g., carbonates). In addition, where groundwater enters a drainage basin as from springs, or at a break-in-slope (Fig. 1-3), the change from a reducing to an oxidizing environment usually results in the precipitation of those elements, particularly iron and manganese, whose mobility is significantly less in such an environment. On the other hand, solution of iron also takes place in numerous environments when reducing and acid conditions prevail. Examples are the leaching of iron from humic soils, especially under waterlogged conditions, and from rock surfaces under peat. Although ferric hydroxide is very abundant in the secondary environment, ferrous carbonate and sulfide (siderite and pyrite), and many other iron compounds (e.g., phosphates) are also found, but a discussion of these is beyond the scope of this book. Garrels and Christ (1965) and Hem (1972) are additional sources for further information on these and other chemical systems.

One more example will suffice to indicate the utility of Eh-pH diagrams. Fig. 3-16 illustrates the oxidation potential of some simple sulfides. From this diagram it can be seen that sphalerite and galena are easily oxidized in most weathering environments by comparison with chalcocite, covellite, molybdenite, and especially argentite. That is, the weathering environment possesses an oxidation potential much greater than that needed to oxidize sphalerite and galena, but only in some parts is the potential high enough to oxidize the other minerals. In fact, for argentite in particular, the potential necessary is so high, that dispersion of silver from this mineral would most likely be by mechanical processes. Zinc and lead would be chemically mobilized more easily than copper, molybdenum and certainly silver from a deposit containing the minerals mentioned above, as

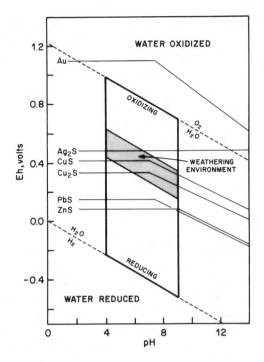

Fig. 3-16. Oxidation-reduction potentials for some simple sulfides. (Temperature, 25°C; Pressure, 1 atmosphere; metal ion activity; 10^{-6}M). Modified from Hansuld (1967).

oxidation of sulfides is a prerequisite to the chemical dispersion of the metals they contain. The sequence of oxidation is sometimes seen by visual inspection of outcrops, in which certain minerals are oxidized, whereas others are not. This is completely compatible and explicable using the data in Fig. 3-16. Again, the illustration shows that the potentials are pH dependent. The potential for native gold is shown on the top of the diagram for comparative purposes and its position explains why gold is not oxidized in nature.

On a practical basis, it is necessary to remember that in addition to abundant oxygen, free drainage is required for oxidation to be a significant process. Outcrops will show little oxidation if the rock is not permeable to water and air. Also, where organic activity (e.g., bogs) consumes all or most of the oxygen, oxidation will be inhibited and the environment will tend to be reducing, even on the surface.

Where organic matter is abundant, especially if the drainage is poor, as in bogs and waterlogged soils, conditions of low pH and Eh may result in higher solubility of iron, manganese and certain other elements. These elements will be leached from rocks and soils and transported in the groundwater, to be precipitated later should the conditions of pH and Eh change. The degree to which groundwater and soil water retain their trace metal content is dependent to a large measure on the pH and Eh conditions. Usually the pH and Eh conditions are such that metals are more soluble in groundwater than in surface water, but nevertheless, the use of surface water for exploration purposes is still justified, provided that the mobility characteristics of the different elements are kept in mind (for example, the pH of hydrolysis; Table 3-6).

ADSORPTION

Hydrous Iron And Manganese Oxides

In the preceding section on pH and Eh, the point was made that Fe^{+2} and Mn^{+2} remain in solution only under certain conditions in the secondary environment. From Fig. 3-15 the conditions for iron to remain in solution (manganese is similar) are shown to be either strongly acid or strongly reducing. As these circumstances are not normally met in the weathering environment, iron and manganese oxides and hydrous oxides are commonly precipitated under surface weathering conditions, which are generally oxidizing. These compounds of iron and manganese, referred to as hydrous iron and manganese oxides, may be found at the break-in-slope (Fig. 1-3), as coatings around sediments in stream beds, in soils, on the bottom of lakes, or anywhere where metal-containing waters are oxidized. This is generally where groundwater (under reducing conditions) is exposed to the surface environment (oxidizing conditions). There is also a tendency for the hydrous iron and manganese oxides to occur suspended in water as finely divided particles (Hem, 1972).

The hydrous iron and manganese oxides are of particular concern in exploration geochemistry for two reasons. First, during their precipitation they may incorporate, by coprecipitation, other elements which would normally be unaffected by changes in pH and Eh. Second, once formed they have the tendency to adsorb (scavenge) elements with which they come in contact. (In this book, the term adsorption includes sorption, and is defined as the adhesion of ions or molecules to the surfaces of solid bodies with which they are in contact). The net result of both processes is that many metals, such as zinc and copper, which are normally very mobile, at least in slightly acidic solutions (Table 3-6), tend to become concentrated in the iron and manganese precipitates. Normally this is detrimental to exploration geochemistry because dispersion halos of mobile elements in the secondary environment can be greatly reduced, and also because non-significant (false) anomalies may result from the accumulation of normal background amounts of trace elements. An added complication is caused by the fact that it is often very difficult to recognize the presence of these hydrous oxides, especially manganese, in soils and stream sediments. In soils, for example, the manganese oxides rarely comprise more than 1% of the total weight, and usually considerably less, perhaps $0.1 - 0.3\%$. It is for this reason that many geochemists determine the manganese (and iron) chemically in such materials (they are soluble in several acids, but this treatment also will dissolve soluble salts and some sulfide minerals). They then compare the manganese (and iron) abundance with that of the base metals either by plotting both on a map, or by means of ratios. On the other hand, where base metals are only sparingly oxidized, and background values are low, the collection and analysis of pure hydrous oxides of iron and manganese (obtained by scraping them off rocks and sediments) may be desirable because of their ability to scavenge trace metals. This technique is used primarily in drainage basin surveys. Without question, the common occurrence of these materials in soils and sediments, allows them to exert an influence on geochemical surveys far out of proportion to their total amounts.

Adsorption of metals by the hydrous oxides of iron and manganese is of great importance to soil scientists and soil chemists because an understanding of the controls on the concentration of certain metals in soils and natural waters is of significance in maintaining and improving the fertility of soils, and in understanding the incidence of certain diseases in animals and man. Therefore, as in the case of the distribution of trace elements in soils, much information on adsorption by hydrous oxides of manganese and iron can be found in the soil and related literature. Literally hundreds of papers can be found, many of which report on the pH, Eh, and other conditions under which these minerals form, and also on their trace element contents. In the way of examples, in a series of papers on Australian soil (e.g., Taylor and McKenzie, 1966; McKenzie, 1967; McKenzie and Taylor, 1968; Taylor, 1968) it has been established that where there are alternating wet and dry conditions and restricted drainage; (1) manganese nodules and stains are common constituents of the soils; (2) the manganese minerals are most commonly birnessite (a complicated manganese oxide mineral with alkalis and alkaline earth elements; an alternate name is $\delta - MnO_2$), and lithiophorite (a hydrated lithium, aluminum manganese oxide), but hollandite (a barium, manganese mineral) and numerous other phases can form; (3) most of the manganese minerals contain varying amounts of Ni, Co, Mg, Ba, Al, K, Na, Fe, Cr, Mo, Cu, Li, Ca and other elements; (4) most (about 80%) of the cobalt (up to 200 ppm) in the soil is contained within the manganese minerals, but much smaller proportions of the total Ni, Mo, Ba, Zn, V, Cr and Pb in the soils are associated with the manganese; (5) manganese nodules from soil samples may adsorb very large amounts of certain elements, such as Cu, Co and Ni (perhaps up to 10% of the weight of the nodules), but this varies with surface area of the nodule; and (6) the initial adsorption sequence is Cu > Co > Ni but, upon ageing, cobalt is most strongly held. Co, Ni, Cu and many of the other elements mentioned are constituents of manganese nodules on the ocean floor, but their mechanism of formation is still subject to debate.

Jenny (1968), in reviewing the available information on Mn, Fe, Co, Ni, Cu and Zn in soils, sediments and fresh water, concluded that the hydrous oxides of Mn and Fe generally furnish the principal controls on the fixation of heavy metals in soils and fresh water sediments. Adsorption (or desorption) of these heavy metals occurs in response to the following factors: (1) aqueous content of the metals in question; (2) aqueous content of other heavy metals; (3) pH; and (4) amount and strength of organic chelates and inorganic complex ion formers present in the solution. But other possible controls on the content of the heavy metals in soils and fresh waters, such as organic matter, or clays (cation exchange capacity) are not significant in the case of the heavy metals. Studies under closely controlled conditions (24°C and pH of 4) by Loganathan and Burau (1973) have shown that the adsorption process by manganese oxide is very complicated and, in the case of Co and Zn, during adsorption Mn is released to the aqueous phase. The process is very rapid, equilibrium being reached in 1 or 2 days. Despite its importance, our knowledge of adsorption by hydrous manganese and iron oxide minerals is still largely empirical.

With respect to exploration geochemistry, the true significance and importance of the scavenging (and coprecipitation) ability of the hydrous manganese and iron oxides only became apparent within the past decade, although it had been mentioned on numerous occasions previously. Boyle et al. (1966) stressed that the strong adsorptive capacity of hydrous manganese oxides and ferric hydroxide had a noticeable effect on the metal values obtained from stream and spring sediments in the Bathurst-Jacquet River district of New Brunswick. They also pointed out the necessity of assessing the true significance of metal anomalies when these hydrous oxides are present in stream sediments.

Nichol et al. (1967) observed that during a stream sediment regional reconnaissance in England and Wales, freshly precipitated iron and manganese oxides adsorbed various trace metals, and therefore, considerable caution had to be exercised in the interpretation of the results owing to the ubiquitous occurrence of these materials. Values of up to 850 ppm Co, and 90 ppm As were determined in stream sediments from poorly drained moorlands. In agreement with what has been said previously, they concluded that the Mn and Fe passed into solution and migrated with circulating groundwater in acid, waterlogged soils, but were precipitated on entering the drainage channel due to an increase in pH and Eh. Nichol et al. (1967) devised a method of metal extraction capable of discriminating between Mn weathered in the soil and that which may have been freshly precipitated in or near the stream channel. This involved the determination of exchangeable, reducible and total manganese (based on the methods used by soil scientists; Sherman et al., 1942). In one part of their study area they found that there was a high content of total Mn, accompanied by a high ratio of reducible to total Mn. This is characteristic of those situations in which Mn is enriched by precipitation in or near the stream channel. Although the studies were preliminary, it appears that, in some circumstances at least, the proportion of the total Mn in the sediment which is exchangeable or reducible may provide a method of determining the true significance and origin of the metal contents of stream sediments.

Dyck (1971) investigated the adsorption of silver on freshly precipitated hydrous iron and manganese oxides in the pH range of 4-8. He showed that the adsorption is essentially complete in 5 minutes, and is strongly dependent on pH and the manner in which the precipitate was prepared. Adsorption is independent of temperature in the range of 0 – 50°C, of moderate concentrations (1%) of such compounds as sodium nitrate, and of traces (1 ppm) of sodium chloride. The amounts of silver coprecipitated or adsorbed were identical under equilibrium conditions and similar chemical environments. In natural materials it is not usually possible to determine whether coprecipitation or adsorption (or both) is the mechanism by which the trace metals are associated with the iron and manganese oxides. The methods which can be employed include differential chemical treatment and possibly electron probe techniques as in the case of gossans described above, although in many cases the results are equivocal.

Horsnail and Elliott (1971) studied the scavenging effects on heavy metals of iron and manganese oxides in drainage channels and soils from several localities in British Columbia. In the drainage channels they noted that Mo and As are frequently found to accumulate in the iron-rich seepage precipitates, whereas the manganese-rich precipitates have high cobalt contents. (The high cobalt content of manganese oxides in Australian soils was mentioned above.) The British Columbia case is illustrated in Fig. 3-17. The locality is on Vancouver Island, where the precipitation exceeds 50 inches per year, and acid, freely-drained podzols predominate. The depressions are poorly drained and waterlogged. Attention was first attracted to the area because molybdenum values of up to 20 ppm were found in the stream sediments. Fig. 3-17 shows the results obtained from the determination of Mn, Fe, Co and pH in soils on the hillside, in the swamp sediment, and in iron and manganese precipitates in the stream sediments. The contrast is particularly evident for manganese which reaches 1.7% in the sediments, by comparison with 300 – 450 ppm in the hillside soils.

Similarly, 200 ppm cobalt is found in the sediment adsorbed on the hydrous iron and manganese oxides, whereas its abundance in the soil does not exceed 10 ppm. Samples of the ferric oxide-cemented bank material (see Fig. 3-17) show enrichment of Mn (0.4 – 1.3%), Fe (9.4 – 13.0%), Mo (6 – 18 ppm), Cu (200 – 300 ppm) and Co (100 – 150 ppm). Notwithstanding the high trace element contents for Co, Cu and Mo, the area contains no commercial deposits, but it does illustrate the non-significant anomalies which can result from the scavenging effects of hydrous iron and manganese oxides.

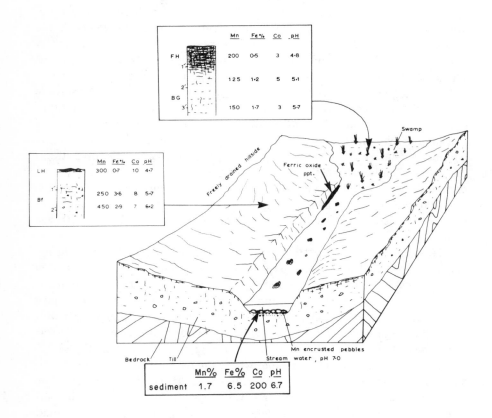

Fig. 3-17. Illustration of the high cobalt values which can result from the scavenging (adsorption) ability of hydrous iron and manganese oxides commonly found in stream sediments. Profiles for a hillside soil (well-drained, acid, podzol) and a swamp (poorly-drained, acid) are also shown. The Canadian notations for the soil horizon are used (see Table 3-4). From Horsnail and Elliott (1971).

Organic Matter

Organic matter is also an important factor in the adsorption of certain elements, although diverse opinions exist as to its importance in comparison with the hydrous iron and manganese oxides. As mentioned previously, Jenne (1968), believes that adsorption by organic matter is not an important control for Co, Ni, Cu and Zn. Other authors note that coprecipitation is the main mechanism by which the hydrous oxides incorporate trace metals, whereas organic matter only adsorbs. For our purposes, we can assume that both the hydrous oxides and organic matter adsorb, as long as it is realized that the matter is unresolved.

That the organic (humus) layers of soils accumulate certain metals, cannot be denied (Figs. 1-7; 3-4; 3-14). In fact, the exchange capacity of humus can be as high as 500 meq/100 g, whereas clays rarely exceed 150 meq/100 g. Organic matter in swamps and bogs is also likely to show high metal values. However, it cannot be assumed that all metals are adsorbed equally by organic matter, because field observations have shown that this may not be the case, even at the same locality. For example, Horsnail and Elliott (1971) found that in some swamps in British Columbia, copper and molybdenum were markedly enriched in organic swamp deposits, but Fe, Mn, Co, Ni and Zn were not. This is illustrated in Fig. 3-18 (Co, Ni and Zn not shown), where the copper content in the organic matter is from 8000 -

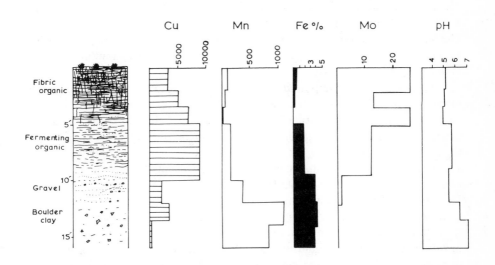

Fig. 3-18. Variation of Cu, Mn, Fe, Mo and pH with organic matter and other materials in a swamp profile from southern British Columbia. From Horsnail and Elliott (1971).

10,000 ppm (0.8 - 1.0%). That this copper is loosely-held (adsorbed) was confirmed by the fact that most of it was extracted by shaking a sample with a cold, 2.5% EDTA solution (a weak acid). Molybdenum is concentrated in the fibrous organic layer, again attesting to the association of organic matter with this element, at least in this locality.

The mechanism by which organic matter adsorbs, or combines with, metals is very complex, and is part of the field of metal-organic chemistry. Saxby (1969) has reviewed metal-organic chemistry as it applies to geochemistry, with particular emphasis on the transition metals. He concluded that during weathering, sedimentation and diagenesis, metal-organic compounds form, but the only well-characterized compounds are the metal porphyrin complexes isolated from sediments. Two major practical difficulties faced by those working in this field are: (1) the extraction of metal-organic compounds from geological specimens without decomposition; and (2) the separation and identification of the species which are almost always present in small quantities in complex mixtures. Therefore, it is not surprising that from the point of view of exploration geochemistry, little is known about the properties of metal-organic compounds in soils, waters and sediments. Some of the metal-organic compounds are soluble in water, whereas others are not, but even the soluble compounds have chemical characteristics different from those of dissociated ions in aqueous solutions. Differences in the properties of the metal-organic complexes in the area illustrated in Fig. 3-18 probably account, at least in part, for the fact that whereas Cu and Mo are concentrated in the organic material, other elements are not. In addition, pH, Eh, and the nature of the decaying material are probably important factors. The degree of enrichment of elements in the humus layer of soils depends very largely on the type of vegetation, since plants vary greatly in their capacity to accumulate various elements.

Metal-organic compounds have been employed to explain the mobilization of uranium, lead and other elements. However, until more studies of these metal-organic compounds in relation to exploration geochemistry have been made, little more can be said apart from stressing that geochemists should be aware of their presence and probable significance. The role of organic substances in the migration and concentration of chemical elements has been discussed by Manskaya and Drozdova (1968) as one phase of a review of the entire area of the geochemistry of organic substances. From an exploration point of view, they have presented enlightening discussions on the concentration of uranium, germanium, vanadium, molybdenum, copper and manganese by organic constituents in sedimentary rocks, bogs, peat and muds. A more recent review of humic substances in the environment by Schnitzer and Khan (1972) also has application to geochemical exploration.

Other Examples of Adsorption

Adsorption can be accomplished by means of clay minerals, the mechanism being primarily cation exchange described previously. Usually the alkaline earth elements (e.g., Mg, Ca, Ba) are preferentially adsorbed by clays in comparison with the base metals. However, one of the fundamental premises of stream sediment drainage basin surveys is that hydromorphically dispersed metals will be adsorbed by fine-grained sediments. Clays are fine-grained and likely sites for adsorbed metals in both sediments and soils, but in addition, adsorption may take place at many other solid (mineral) – liquid (water) interfaces. Although the mechanisms are not well understood, it can be assumed that ions (or molecules) are taken up by the mineral phase and held by various physical or chemical forces, usually on the surfaces of the solid phase. Finely divided solids with a large surface area per unit volume have a considerable capacity for adsorption.

Hem (1970) has discussed some aspects of adsorption (including such variations as chemisorption and ion exchange). He pointed out that the methods of collection, preservation and treatment of water and particulate matter have a great influence on the results and interpretation. For example, if a water sample containing clay particles is acidified (the usual procedure to preserve trace metals in a water sample), the adsorbed metal ions, as well as the hydrous iron and manganese oxides, will be brought into solution and thus the chemical significance of the results may be difficult to establish. Also, coarser suspended material of varying mineralogy (including rock-forming silicates) may carry adsorbed ions. Some metals may be transported as suspended colloidal oxide particles, or these colloids may be in the sediments. Clearly, there are many sources for real or apparent adsorption, both in the suspended and sediment load of a river, and using present techniques of analysis and interpretation, recognition of these may be an impossible task.

Mason (1966) has summarized some of the general principles governing adsorption as follows:

1. The amount of adsorption increases as the grain size of the adsorbent decreases and hence its surface area increases.
2. Adsorption is favored if the adsorbate forms a compound of low solubility with the adsorbent (an example is the adsorption of phosphate ions by ferric hydroxide).
3. The amount of a substance adsorbed from solution increases with its concentration in that solution.
4. Highly charged ions are adsorbed more readily than lower charged ions.

MOBILITY IN THE SECONDARY ENVIRONMENT

An indirect objective of the foregoing discussions of trace elements in soil profiles, adsorption, pH and Eh, has been to set the foundations for a consideration of element mobility (that is, chemical mobility, as opposed to physical (clastic) mobility, which will be discussed in Chapter 8). Mobility is defined as the ease with which an element can move within a specific environment. Mobility of elements, particularly the metals, in the secondary environment is the basis of the dispersion halos used in both regional and detailed surveys. The previous discussions have shown that many factors affect mobility although only a few of the more obvious have been mentioned so far.

Element mobility is a very complex matter. In addition to those mentioned above, the following variables must also be considered: (1) the nature of the medium, as mobility in groundwater will differ from that in surface water and water in soils and sediments; (2) the mechanism of transport, since mobility of clastic grains is an entirely different process from mobility in solution; (3) changing rock type, as for example, when certain heavy metals from acid water precipitate when this water is neutralized by contact with carbonate; (4) the presence and effects of microorganisms which can reduce sulfate to sulfide, resulting in the precipitation of insoluble metal sulfides; (5) the presence of microorganisms such as certain fresh water algae which can incorporate trace elements into their body structures and may contain several tenths percent of Zn, Cu and Pb; (6) the solubility of salts (compounds) which a metal may form with anions in the same solution, e.g., the low solubility of anglesite and cerussite which severely limits the mobility of lead in the presence of sufficient sulfate and carbonate ions; (7) the formation of complex ions which may either reduce molybdenum mobility as in the case of $HMoO_4^-$, or increase it as in the case of MoO_4^{-2} (this is discussed further below); (8) membrane effects, such as the high concentrations of metals and all other constituents (including Br, I, Cl, Na, etc.) developed in association with certain shales and clays, particularly at depth, in waters generally classed as brines; (9) the presence of dissolved gases, such as CO_2, which control the solubility of certain components; and (10) such mechanical factors as permeability, porosity, grain size, fluid viscosity and fluid velocity.

Many attempts have been made to determine the relative mobility of elements in different metalliferous areas. These involve determining the trace element abundances in water, stream sediments, soils, wall rock and more recently, in the atmosphere, at varying distances from each deposit. Another approach particularly applicable in non-mineralized areas, involves the comparison of the contents of elements in natural water, with their content in the rocks drained by these waters. The results will generally differ from one area to another, particularly with respect to the trace elements, which is not surprising considering the great number of variables involved. With full awareness of the problems, Andrews-Jones (1968) has empirically estimated the relative mobilities of elements in the secondary environment (Fig. 3-19). His classification is based upon Eh and pH, as he considers oxidizing, acid, neutral to alkaline and reducing as the

RELATIVE MOBILITIES	ENVIRONMENTAL CONDITIONS			
	Oxidizing	Acid	Neutral to Alkaline	Reducing
VERY HIGH	Cl, I, Br S, B	Cl, I, Br S, B	Cl, I, Br S, B Mo, V, U, Se, Re	Cl, I, Br
HIGH	Mo, V, U, Se, Re Ca, Na, Mg, F, Sr, Ra Zn	Mo, V, U, Se, Re Ca, Na, Mg, F, Sr, Ra Zn Cu, Co, Ni, Hg, Ag, Au	Ca, Na, Mg, F, Sr, Ra	Ca, Na, Mg, F, Sr, Ra
MEDIUM	Cu, Co, Ni, Hg, Ag, Au As, Cd	As, Cd	As, Cd	
LOW	Si, P, K Pb, Li, Rb, Ba, Be Bi, Sb, Ge, Cs, Tl	Si, P, K Pb, Li, Rb, Ba, Be Bi, Sb, Ge, Cs, Tl Fe, Mn	Si, P, K Pb, Li, Rb, Ba, Be Bi, Sb, Ge, Cs, Tl Fe, Mn	Si, P, K Fe, Mn
VERY LOW TO IMMOBILE	Fe, Mn Al, Ti, Sn, Te, W Nb, Ta, Pt, Cr, Zr Th, Rare Earths	Al, Ti, Sn, Te, W Nb, Ta, Pt, Cr, Zr Th, Rare Earths	Al, Ti, Sn, Te, W Nb, Ta, Pt, Cr, Zr Th, Rare Earths Zn Cu, Co, Ni, Hg, Ag, Au	Al, Ti, Sn, Te, W Nb, Ta, Pt, Cr, Zr Th, Rare Earths S, B Mo, V, U, Se, Re Zn Co, Cu, Ni, Hg, Ag, Au As, Cd Pb, Li, Rb, Ba, Be Bi, Sb, Ge, Cs, Tl

Fig. 3-19. Relative mobilities of the elements in the secondary environment. From Andrews-Jones (1968).

four main environmental conditions controlling mobility of aqueous phases of the elements. This classification (Fig. 3-19) is extremely useful, and is as accurate as any classification could be considering the great number of variables.

Inspection of Fig. 3-19 reveals many interesting facts, and stresses that mobilities vary according to specific environmental conditions. For example, the group Cu, Co, Ni, Hg, Ag and Au, which has medium mobility under oxidizing conditions and high mobility under acid conditions, becomes immobile under neutral to alkaline conditions. Eh and pH of the water are the main controlling factors. Similarly, molybdenum and copper have either medium or high mobilities in oxidizing and acid environments but in the neutral to alkaline environments, molybdenum is very mobile, whereas copper is immobile (its pH of hydrolysis has been exceeded; see Table 3-6). This fact has great practical significance in geochemical exploration for those porphyry copper deposits which contain molybdenum, and most porphyry copper deposits in the western hemisphere are of this type. Because molybdenum is so much more mobile than copper under neutral to alkaline conditions, which are the conditions of most surface water, molybdenum is used as the pathfinder for such deposits. Either stream water or sediments may be analyzed for molybdenum as it forms a larger halo than does copper. In the reducing environment both elements have limited mobilities and produce small halos, but again, the molybdenum halo would be expected to be relatively larger.

The above discussion of the significance of Fig. 3-19 stresses the importance of environmental conditions in the secondary environment. Andrews-Jones (1968) has considered what are probably the main conditions, but at least ten others have been listed at the beginning of this section although it is usually very difficult, if not impossible, to evaluate them all. Returning to molybdenum again as an example, it is not obvious from Fig. 3-19 that even where molybdenum is supposed to be soluble, the presence of lead in solution may well precipitate it as wulfenite ($PbMoO_4$). Fortunately, lead is a rare constituent of porphyry copper deposits, so in this case it is not a problem. Also, in the presence of Fe, molybdenum has reduced mobility either because it is adsorbed on hydrous ferric oxide, or precipitated as ferrimolybdite, $Fe_2(MoO_4)_3 \cdot 8H_2O$. Further, in the presence of chloride, silver forms an insoluble chloride (cerargyrite, AgCl). These examples are by no means isolated cases, as the formation of many insoluble compounds (minerals) are described in the soil and mineralogical literature — again illustrating the wealth of information available in related sciences. For example, members of the plumbogummite group, $XAl_3(PO_4)_2(OH)_5 \cdot H_2O$ have been identified in soils (Campbell et al., 1972). These alumino-phosphates are believed to be common in soils, and represent an end product in the weathering sequence of certain rocks. The various end members include plumbogummite (where $X = Pb$), crandallite (Ca), goyazite (Sr), gorceixite (Ba) and florencite (Ce and other rare earths). The significance of plumbogummite in exploration geochemistry is not the presence of the mineral itself, but the fact that it shows that insoluble minerals and precipitates may form in soils indicating that the mobility of many other important elements may be severely reduced. Even if specific minerals (e.g., wulfenite) of the elements of interest are not formed, the elements may be incorporated into the structures of other minerals, such as those of the plumbogummite group.

The ability of an ion to remain in solution in an aqueous system is limited by the stability of the minerals it forms by reaction with other components in solution. In view of their abundance in natural water (Table 3-1) it is worthwhile to list the generalized solubilities of carbonates, chlorides and sulfates of certain elements at temperatures, pressures and concentrations commonly found in fresh water of the surface environment; complexes are not considered. Only the following elements are considered: Sb, Ba, Bi, Cd, Ca, Cr, Co, Cu, Au^{+1}, Fe^{+2}, Fe^{+3}, Pb, Mn^{+2}, Hg^{+1}, Ni, K, Ag, Na, Sn^{+2}, Sr, Zn and Pt (source: Solubility Chart, *Handbook of Chemistry and Physics*):

1. Carbonates: only the carbonates of Cr, K and Na are significantly soluble in water; the remainder are either insoluble or only sparingly soluble.
2. Chlorides: the chlorides of Sb, Ba, Cd, Ca, Co, Cu, Fe^{+2}, Fe^{+3}, Pb, Mg, Mn^{+2}, Ni, K, Na, Sn^{+2}, Sr, Zn and Pt are all soluble in water; the chlorides of Bi, Cr, Au^{+1}, Hg^{+1} and Ag are insoluble.
3. Sulfates: the sulfates of Cd, Cr, Co, Cu, Fe^{+2}, Mg, Mn^{+2}, Ni, K, Na, Sn^{+2}, Zn and Pt are soluble in water; the remainder are either insoluble or only sparingly soluble.

The above generalizations as to element mobility based on the insolubilities of their carbonates, chlorides and sulfates are not particularly applicable to the more concentrated natural waters, such as those in some soils and brines. Further, these generalizations must only be considered as approximations of what actually happens in dilute waters as there are numerous complications and variables, many of which are still incompletely known. A case in point would be the implied immobility of zinc in the presence of carbonate because zinc carbonate is insoluble, or sparingly soluble. But as Hem (1972) pointed out, the solubility of zinc carbonate is sufficiently high, especially at or below a pH of 7.0, to give zinc values of 1000 ppb in surface and groundwater, although the actual value is influenced by the abundance of dissolved carbon dioxide and other factors. Hem (1972) suggested that silica, also a constituent of river water (Table 3-1), is perhaps a more effective inorganic solubility control over zinc contents (willemite, zinc silicate, is very insoluble). Biological factors and adsorption (sorption) by stream sediments may also be important controls operating simultaneously with inorganic precipitation.

When considering the application of mobility to practical geochemical problems it is first necessary to determine the extent to which the two basic types of movement, physical and chemical, are involved. If the dispersion is physical, the problem is simplified to the extent that only the various erosional processes in the area need be considered, especially as they are generally subject to fewer unknowns than are chemical ones. Ideally, the most advantageous use of chemical dispersion requires a study of (1) the oxidation of the primary sulfides in an area, and (2) the mobility of the various metal ions which may be released (Hansuld, 1967).

Hansuld (1967) has made a classic study of the relative mobility of copper and molybdenum with specific application to geochemical exploration for porphyry copper in the southwestern United States, western Canada and other areas in the western hemisphere. The relative mobility of each element is shown on Eh-pH diagrams in Fig. 3-20. These diagrams portray the fundamental relations in terms of Eh and pH, demonstrating why molybdenum is a pathfinder for copper in those porphyry deposits which contain both elements.

The following two paragraphs are a direct quotation from Hansuld (1967, p. 184), except for the figure numbers which have been changed for consistency here.

Fig. 3-20 (left) "shows some stability relations in a simple chalcocite-water system together with plots of soil and stream water environments typical of the porphyry copper deposits of the southwestern United States. The solubility of cupric ion decreases with increasing pH and to a lesser extent decreasing Eh. For example, at a pH of 6 the solubility of copper is less than 3 ppb. The decreasing solubility is depicted schematically by the arrow at the top of the diagram. Note that the copper formed in the oxidation of chalcocite is soluble only under strongly acid and highly oxidizing conditions. At higher pH's the oxidized copper is precipitated *in situ* as the metastable copper hydroxide. The acid character of the soils overlying mineralized bedrock implies that it should be leached of most of its oxidized copper. This explains the low copper values commonly found in soils directly over known mineralization in the southwestern United States. Similarly, the alkaline surface drainage waters, with very limited copper solubility, account for the negligible amount of copper usually

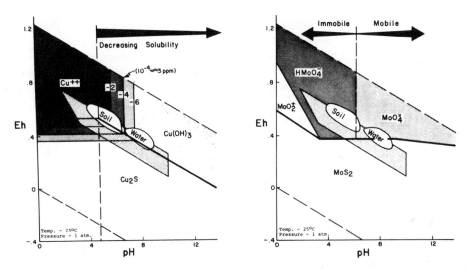

Fig. 3-20. Application of Eh-pH diagrams to an understanding of the relative mobilities of copper and molybdenum in the search for porphyry copper deposits. *Left.* Some stability relations in the chalcocite-water system. *Right.* Some stability relations in the molybdenite-water system. From Hansuld (1967). Hansuld considers the weathering environment to cover a larger pH-Eh range than that indicated on Fig. 3-15.

found in the main drainage systems. Most of the oxidized copper has been carried vertically downward in solution by acid mine waters to eventually form supergene copper sulphides."

Fig. 3-20 (right) "shows some stability relations in a molybdenite-water system. Of particular interest in this diagram is the boundary at pH 6 between acid molybdate and molybdate anions. The acid molybdate anion is relatively insoluble in contrast to the molybdate anion. The relative mobilities are depicted schematically by the arrows at the top of the diagram. Looking again at the soil and stream environments of the southwest we see that in the soil environment oxidized molybdenum occurs predominantly as the immobile acid molybdate anion and in the stream environment as the relatively mobile molybdate anion".

The mobility of metal ions in water of the secondary environment is strongly influenced by the composition of the country rocks. The most important example from the point of view of exploration geochemistry is the reduction of mobility of many heavy metals in acidic (and slightly acidic) water when subjected to neutralization by reaction with carbonates. In its simplest terms, the mechanism can be thought of as one in which the pH of hydrolysis is exceeded, and the metals precipitate as hydroxides. The pH can also be raised by dilution with water of higher pH, but more particularly those draining carbonate areas. Therefore, surface water surveys for heavy metals in carbonate areas have a limited chance of success for the simple reason that these elements are more likely to be found precipitated in the sediments. Although this type of reaction of water with carbonates has been known qualitatively for many years, little is known of the quantitative aspects. Carbonates neutralization has been applied to acid coal mine waters, and although much is known about this process, it is mainly concerned with methods of reducing the pH and removing dissolved iron. Research along these lines of mine water pollution control in the United States has also been concerned with such items as mixing ratios, residence times, and improving the settling characteristics of the resulting mixed iron-hydroxide and calcium-sulfate sludge (van Everdingen and Banner, 1971).

A very interesting study on natural acid spring water with extremely high heavy-metal contents, and their precipitation, has been described by van Everdingen (1970) from the Paint Pots (the name of a group of mineral springs), in southeastern British Columbia. The waters of these springs have a pH of from $2.5 - 3.5$, with a high content of Fe, Mn, Zn, Pb and SO_4^{-2} (representative values: 100, 0.3, 50, 0.2 and 1000 ppm respectively) which are similar in many respects to acid drainage from mines. The components are derived from sulfide mineralization in the Cambrian rocks of the area. Oxidation of the Fe^{+2} in the water and hydrolysis of the resulting Fe^{+3} produce a still lower pH, a high redox potential, and thus a deposit of ferric hydroxide builds up around the springs (hence the name "Paint Pot"). Less than one half of the available iron, approximately three-quarters of the Mn and Zn and most of the lead are discharged into the fast-flowing Vermilion River, about a mile away. The gravels in and along the river channel below the confluence are heavily rust-stained, indicating that the high pH, resulting from both dilution and reaction of the acid water with bicarbonate in the river water, has caused precipitation, at least of the iron. In a companion study, van Everdingen and Banner (1971) circulated synthetic acid water (of a composition similar to that at the Paint Pots) through crushed limestone. This resulted in neutralization of the acidity and removal of over 99% of the dissolved Fe, Cu and Pb, 97% of the Zn and about 40% of the Mn. Minimum contents of the metals remaining in solution were (in ppm): Fe^{+2} (0.05), Cu (0.006), Pb (0.002), Zn (3.7) and Mn (0.22), indicating the process is very effective in removing trace metals.

Theobald et al. (1963) described a case in Colorado where the oxidation of disseminated pyrite in schists and gneisses provided abundant dissolved iron sulfate and sulfuric acid to groundwater and surface water, so that pH values of about 3.5 – 4.4 were generated. When this surface water joined with that of another river with a pH of about 8, dilution changed the pH to about 6 – 7 below the confluence, and precipitation of aluminum oxides, as well as those of iron and manganese, took place. These oxides contained over 100 ppm of Pb, Cu, Zn, Ni and Co, as well as lesser amounts of Mo, Ag, V, Be and several other elements. This is another example of the scavenging ability of various oxides (including aluminum, not previously mentioned as it is apparently not as common), and also shows that non-significant anomalies are common. It emphasizes the value of pH measurements during the collection of either water or sediment samples. Low pH values should suggest that below the junction of such a river with one of higher pH, trace elements might be concentrated in precipitated hydrous iron and manganese oxides in the stream sediments.

To conclude the discussion of chemical mobility, it is worthwhile to refer to the observation of Goldschmidt (1937) that the behavior, including mobility, of various ions in weathering reactions can be related to their ionic potentials. Ionic potential is expressed numerally as the ratio of the charge in valence units (Z) to the ionic radius (r) in Ångstrom units (Fig. 3-21). Although ionic potential has no physical or chemical entity in nature, it is useful as a concept to explain certain chemical reactions.

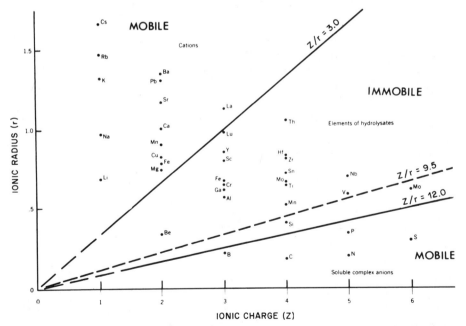

Fig. 3-21. Grouping of certain elements according to their ionic potentials (Z/r). From Gordon et al. (1958).

Many elements are plotted on the graph (Fig. 3-21) to show the relation of ionic radius to ionic charge. Three general groups of elements are recognized: (1) those with low ionic potential $(Z/r<3.0)$, such as Na, K, Ca, Mg, Sr, Fe^{+2} and Mn^{+2} which go into true solution during weathering and transportation process and are mobile; (2) those with intermediate ionic potentials (Z/r between 3.0 - 12.0, such as Al, Fe^{+3}, Ti and Mn^{+4} which are precipitated by hydrolysis — hence the name "elements of hydrolysates", and become concentrated in weathering residues such as laterites and bauxites; and (3) those with high ionic potential $(Z/r>12)$ which form very soluble complex anions containing oxygen during weathering, and are very mobile. The broken line representing the ionic potential of 9.5 represents the upper limit of elements concentrated in bauxites, according to Gordon et al. (1958). In general, ionic potential is a useful guide to the behavior of various ions, such as iron and manganese which are mobile when divalent, but in the $+3$ and $+4$ forms they are immobile. However, pH, Eh, complex formation and other factors previously mentioned are not taken into account. Consequently, molybdenum is shown as an immobile hydrolysate, when in fact, it may be extremely mobile when it forms an appropriate complex (Fig. 3-20).

WATER — GENERAL COMMENTS

The importance of water in the secondary environment has been stressed in the discussions of weathering, soil formation and the mobility of elements. From Table 1-1 it can be seen that water is fourth in the list of materials used for exploration purposes. Fig. 3-22 shows the inter-relations between rock, soil, stream sediment and water (part of the hydrologic cycle) in a very simplified form and, from this, the fundamental role of water in the use of soils and sediments for exploration purposes becomes more obvious. For example, after oxidation has resulted in the release of elements in the secondary environment, it is water which assists in the transport of these elements to the soils, streams and finally, lakes. Water also transports particulate matter, such as clastic sulfide grains or metal colloids, which may themselves produce anomalous values in stream sediments. Therefore water is much more important in exploration geochemistry than its relative position in Table 1-1 would indicate.

As explained previously, all water used in exploration geochemistry can be classed as either (1) groundwater or (2) surface water, and the use of any of these waters for exploration purposes is known as a hydrogeochemical survey. As used here, groundwater includes all water in the zone of saturation beneath the Earth's surface (except that water chemically combined in minerals). Surface water, as the name implies, con-

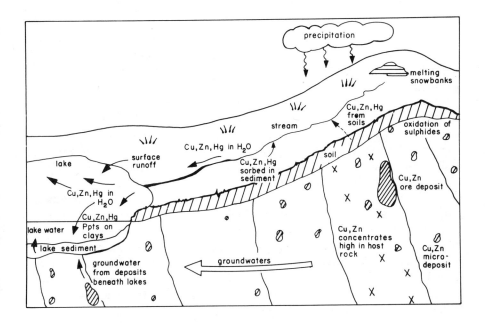

Fig. 3-22. Dispersion of metals from sulfide deposits by groundwater and streams into lakes and lake sediments. From Allan et al. (1973a).

sists of all water found on the surface of the Earth, such as stream and lake water. Vadose (or suspended) water is found in the zone of aeration between the top of the water table and the surface. Vadose water is unquestionably important because it includes water in the soil zone and, as such, it supports plant life, assists in the transport of elements and is itself transported to the water table by the process of recharge. However, details of the process by which vadose water descends to the water table and its actual composition are very poorly known (Fairbridge, 1972), and therefore, it is not considered further.

Groundwater will be considered next, followed by a discussion of the most abundant types of surface waters: stream water and lake water. They are treated separately in view of their different importance in exploration geochemistry. Other types of water, such as connate, juvenile, hydrothermal, geothermal, brines and meteoric, are only mentioned in connection with other topics. Snow, also a hydrogeochemical sampling material, is discussed in Chapter 11 in conjunction with permafrost. Water as used for exploration purposes in Canada is also discussed in Chapter 11.

GROUNDWATER

Occurrence, Movement and Fluctuation

The water table is the upper limit of completely saturated rock material and for practical purposes it may be determined by the level of standing water in open holes or wells penetrating into the zone of saturation. The water table may be at the surface, as along the banks of rivers and lakes, or it may be at depths of hundreds of feet in arid areas. Groundwater extends at depth until eventually the pore spaces of rocks are reduced essentially to zero porosity. Deep wells drilled to explore for oil and gas show evidence of water (usually highly saline) at 30,000 feet and more, but groundwater at these depths is of little practical value in exploration for mineral deposits since mining of metals from great depths is not possible with present or foreseeable technology. However, waters from deep environments are potentially important as sources of metalliferous brines, easily recoverable halogens (Br, I), or in the search for hydrocarbons (Chapter 13). Interestingly, by extrapolating the Earth's average temperature gradients in a large number of areas in the United States, it has been calculated that the normal boiling point of water is reached at a depth of less than 7000 feet in one third of the cases. However, other factors such as pressure permit water to exist at greater depths.

For all practical purposes, the actual source of groundwater in the near surface environment of interest in the search for mineral deposits is precipitation, commonly called meteoric water (Fig. 3-22). Fig. 3-22 also shows that storage in the zone of saturation is temporary, the groundwater moving toward discharge areas at the land surface. Precipitation is very irregular both in time and location, and this results in variations in the level of the water table, the rate of flow, water composition, and many other factors which are of great concern in hydrology, the science concerned with the character, source and mode of occurrence of groundwater. In Fig. 1-3 seasonal fluctuations in the water table are indicated, but in that example it is always above the bottom of the stream bed. In such circumstances, groundwater can supply metals from any oxidizing metal deposits to form a break-in-slope (seepage), river sediment or river water anomaly. In dry areas the water table may be below the stream most of the time, and therefore it is not often possible for groundwater to supply elements to the stream bed or to the river water. It must be remembered that conditions today do not necessarily reflect conditions in the immediate geological past. For example, there is historical evidence that many desert areas were once covered by vegetation and supported considerable animal life (e.g., the deserts of Israel) and as a result the analysis of ancient stream beds in such areas may reveal significant anomalies.

Fluctuations in the level of the water table may also be caused by natural events, such as air pressure changes and earthquakes, but these are short term variations and are measured in inches or fractions of an inch. Longer term fluctuations are due to supply of recharge water and are more often seasonal. Such fluctuations can vary greatly in magnitude, and in most areas of Canada range from 5 to 20 feet (Brown, 1967). Marked changes in the water table level can greatly affect the groundwater supply to streams, lakes or springs, as inspection of Figs. 1-3 and 3-22 will reveal. Variations in composition also occur. Fluctuations caused by man's interference with the natural flow system by pumping, irrigation, altering the course of rivers or lakes, building dams, changing or removing the natural cover, and so forth, may be significant, and are difficult to evaluate from the point of view of exploration geochemistry. A number of methods may possibly be used to interpret natural and artificially-effected groundwater flow systems, such as the measurement of hydraulic gradient or the use of tracers (dyes, chemicals, or radioactive tracers), but these require the assistance of an expert hydrologist. The importance of groundwater fluctuations in the development of laterites has been discussed previously (e.g., Fig. 3-10).

The rate of movement and fluctuation of groundwater are considerably slower and more uniform than those of either surface or atmospheric water. The rate of groundwater movement depends on such factors as hydraulic gradient and permeability of the rocks and is only a fraction of that of most surface water. Rates of movement vary from zero to several hundred feet per year and herein lies one of the reasons why groundwater is a good geochemical sampling medium. Because of the slow rate of movement, such water remains in contact with mineral deposits long enough for solution of detectable amounts of metals from the ore or gangue. The slow moving waters then form dispersion patterns (or halos) in the groundwater systems which may be traced back to the source. However, tracing dispersion patterns in groundwater systems is not as simple as might be supposed. It is desirable, if not essential, to know the hydrologic gradient, direction of groundwater flow, topography and geologic structure. Data from a large number of wells can greatly assist in the interpretation. A complicating factor is that groundwater at any particular location may be moving through several aquifers, each of which may have its own characteristic flow rate and chemical composition.

Fig. 3-23 shows a geochemical map with contents of molybdenum in groundwater, copper in basal alluvium, and molybdenum in the ash of mesquite leaves. All anomalies result directly or indirectly from the dispersion of the ore metals in groundwater. The mesquite leaves obtained their metal contents because their roots sampled molybdenum-containing groundwater, whereas the copper in the basal alluvium was transported by

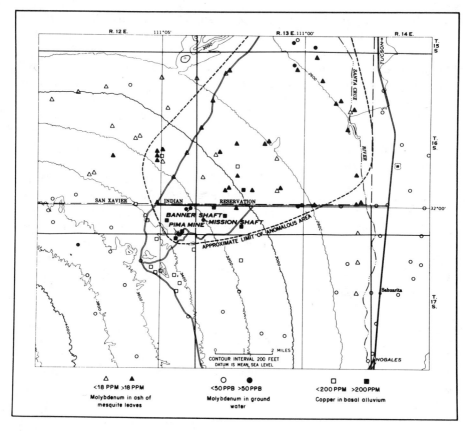

Fig. 3-23. Geochemical map showing anomalies resulting from the dispersion of metals from a porphyry copper deposit by groundwater, Pima district, Arizona. From Huff and Marranzino (1961).

groundwater. A more recent study in the same area (Huff, 1970) confirmed the earlier work on which Fig. 3-23 was based, and further concluded that the analysis of groundwater for molybdenum appears to be the most promising prospecting technique in that particular Arizona copper district. Newer analyses of waters obtained from wells showed anomalous molybdenum contents eight miles northeast of the Pima mine, but copper values were not anomalous. As discussed previously (e.g., Fig. 3-20), the relative mobilities of molybdenum and copper can be explained on the basis of Eh and pH conditions. The report by Huff and Marranzino (1961) illustrated in Fig. 3-23, and the more recent study by Huff (1970), confirm this. Groundwater in the Pima district are alkaline with the result that the mobility of copper is much more restricted than that of molybdenum. From this the fact emerges that the mobility of elements in groundwater, and like-wise in surface water, is governed by the same fundamental principles (e.g., Eh, pH, solubility of compounds) discussed previously for soils.

However, it must not be assumed that molybdenum will always form halos in waters draining molybdenum-bearing deposits. For example, Rostad (1967) found that molybdenum was generally absent in water from a creek draining a molybdenite prospect in Idaho. He suggested that this may be because groundwater did not usually percolate directly through the molybdenum-containing rock. Barakso and Bradshaw (1970) emphasized that bacteria may affect the relative mobilities of copper and molybdenum in British Columbia porphyry copper deposits. Also the adsorption ("fixing") of molybdenum by iron-bearing oxides or hydroxides may result in copper being more extensively dispersed than molybdenum in some water and soil. Clearly, the mobility of elements in groundwater cannot always be predicted or explained solely on the basis of pH and Eh (Fig. 3-19), as other natural phenomena are important, particularly on a local level.

Groundwater Aquifers and Element Transport

For the purpose of groundwater studies, geological materials may be divided into three broad groups: surficial materials, sedimentary rocks and crystalline rocks (Brown, 1967).

Surficial materials such as sand and gravel contain many of the best local aquifers. They comprise all the unconsolidated material lying on top of the bedrock, including the broad range of glacial and alluvial materials, some of which are extremely fine-grained and essentially impervious, like the materials laid down in glacial lakes. Glacial deposits are discussed in Chapter 11.

Hydrogeochemical anomalies are typically fan-shaped when they are developed in surficial materials that are uniform and gently sloping. Fan-shaped anomalies may also develop in sedimentary and crystalline rocks provided that fracturing does not alter the path of metal-containing groundwater.

Sedimentary rocks provide the largest aquifers, and water passing through such rocks may well sample mineralization at great depths. Water circulating through sedimentary rocks may dissolve, and thus destroy, a metal deposit whose elements may be re-deposited at some distant location, or may be completely dispersed. An example of the ability of groundwater to transport elements and subsequently precipitate them in the form of a mineral deposit may be found in the occurrence of the "roll-front" (sand-stone) type of uranium deposits in the Shirley Basin, Wyoming (Fig. 3-24).

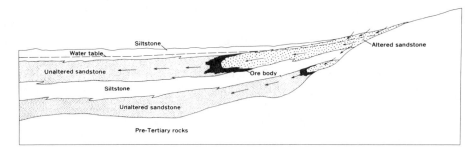

Fig. 3-24. Possible circulation paths of uranium-bearing ore solutions in the Shirley Basin area, Wyoming. Arrows indicate direction of groundwater flow. From Harshman (1972).

Uranium in these deposits is believed to have been leached from tuffs possibly during devitrification. These tuffs probably only contained average abundances, or slightly higher, of uranium (perhaps $10 - 20$ ppm), but such rocks are easily leachable by meteoric water. Once leached, the uranium-bearing waters (also alkaline and probably high in carbonate) moved from an oxygen-rich surface environment, in which uranium is in the hexavalent (U^{+6}) state (occurring as the uranyl ion UO_2^{+2}, or complexed with carbonate) into the sub-surface (groundwater) environment. In Fig. 3-24 the sandstone aquifer is bounded by impervious siltstone. The dilute, uranium-bearing groundwater migrated down-dip where alteration of the sandstone occurred by processes such as oxidation of organic matter and pyrite. In the process, oxygen in the groundwater was depleted, and when the Eh conditions changed from oxidizing to reducing, uranium was reduced to the tetravalent state (U^{+4}) and precipitated as uraninite (UO_2). Several elements, particularly molybdenum, selenium, vanadium and copper are often associated with uranium in the roll-front type of deposit and obviously were carried by the groundwater. These elements precipitate at different conditions of Eh and pH with the result that some elements migrated farther. For example, molybdenum may sometimes be detected about one mile beyond the uranium ore, whereas selenium is found up-dip of the ore body, because it has lower mobility than either molybdenum or uranium. These facts can be used for exploration purposes. Thus the presence of selenium in the sandstone aquifer indicates the relative position of the uranium (down-dip). Other characteristics, such as the degree of alteration, can also indicate the relative position of the ore zone. This process of groundwater transport of uranium and related elements may well be in progress in the Shirley Basin today.

The example just cited is but one of many which could be used to illustrate the importance of groundwater (including metal-rich formation

waters and brines) in the transport of metals which eventually form
economic mineral deposits.

Groundwater anomalies in sedimentary rocks can take many forms
depending on the geometry and permeability of the aquifers through which
they pass. They may be fan-shaped as in surficial deposits, in which case
the analytical data usually can be contoured, but in highly fractured rock
the groundwater pattern is more likely to be irregular and not amenable to
contouring. Groundwater may also follow high permeability channels in
the bedrock in which case anomalies tend to be linear.

Crystalline rocks (igneous and metamorphic) are, with few exceptions,
the poorest aquifers because of their low porosity and permeability.
However, they often contain interconnected joints and fractures, as evidenced by the large volumes of water which must be pumped from some mines
in crystalline rock. Precipitation may enter such fracture systems and it can
be used for exploration purposes by sampling wells and springs. The
analysis of groundwater and materials which have precipitated from it may
also be used to locate extensions of ore in mines, or in the evaluation of
drilling results in the vicinity of mines, or in unexplored areas (Fig. 3-25).
Drill holes frequently miss large ore bodies, but they may tap water which
has access to such bodies. By means of the analysis of the water, evidence
for nearby mineralization may be obtained.

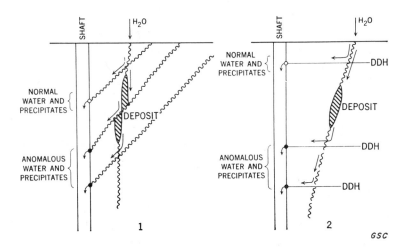

Fig. 3-25. Sketch illustrating the use of groundwater and precipitates in mine exploration. (1) Analyses of water and precipitates along fault zones and other planar
structures. (2) Analyses of water and precipitates from diamond drill holes (DDH).
From Boyle et al. (1971).

Chemistry of Groundwater

Although absolutely pure water can be considered chemically inert, such water does not exist in nature. In Table 3-1 rainwater, the major source of near-surface groundwater, is generally considered the "purest" of natural waters, and at one locality contained 38 ppm total dissolved solids with a pH of 5.5. The total dissolved solids in the three groundwater analyses presented in Table 3-1 range from 302 – 3300 ppm. Obviously, groundwater may carry significant amounts of elements in solution, obtained from soils and bedrock with which it comes into contact.

Precipitation dissolves oxygen and carbon dioxide from the atmosphere and on contact with soil or exposed rock surfaces it is both acid and oxidizing, both of which are conditions favoring chemical attack (weathering). Initially, such water is very active chemically, and changes take place rapidly until chemical equilibrium is reached with its surroundings, after which the water reacts only slowly. However, if groundwater crosses into another formation, it may become chemically active again as the new formation may have totally different chemical characteristics which require that the groundwater approach a new equilibrium. Clearly, the chemistry of groundwater is closely related to enclosing rock materials. Another possibility to be considered is that groundwater of one composition may encounter another of a different composition. In this case the resulting groundwater may be a mixture of the two, or precipitation of one or several components may take place. An example of the latter would be the precipitation of barite when a barium-containing groundwater high in chloride encounters one high in sulfate. Examples of this reaction are known from several localities, such as in England, although the original source of the barium-containing chloride groundwater is a matter of speculation.

Carbonate rocks, such as limestones and dolomites, dissolve easily in acid water with the result that waters in contact with these rocks are high in calcium, magnesium and bicarbonate. Waters in contact with halite will be high in sodium and chloride, whereas those in contact with gypsum or anhydrite will be high in calcium and sulfate. Shales and some crystalline rocks may contain abundant pyrite and marcasite which can result in high iron (much of which is quickly precipitated) and sulfate, but most sulfate in groundwater and surface water originates from gypsum or anhydrite, and the groundwater is not necessarily acid. As geological materials may be mixed in almost any possible combination, groundwater can have great variations in the proportions of the elements dissolved in it, as well as in the total amount of dissolved solids. The range of the latter may vary from perhaps 30 ppm for groundwater in crystalline rocks to several hundred thousand ppm for groundwater dissolving bedded halite (sea water contains about 35,000 ppm dissolved solids).

Notwithstanding the fact that the chemical characteristics of groundwater vary greatly as a result of the diversity of rock materials, most groundwater contains only eight major (more than 1 ppm) dissolved constituents: Ca, Mg, Na, K, HCO_3, SO_4, Cl and SiO_2 (Table 3-1). (Nitrate found in groundwater originates from the soil, biological materials and the atmosphere, if it is not from contamination.) This small number of major ions may appear surprising in view of the many abundant elements in rocks, but the fact remains that elements such as Al, Fe, Mn and Ti are often not detected in groundwater because of their low solubilities, tendency to hydrolyze (Table 3-6), or other factors. Among the eight most abundant constituents Ca, Mg, Na and K tend to be adsorbed or base exchanged onto clay minerals, which may change their abundances in groundwater. Actually all minor and trace elements which occur as cations are similarly affected to some extent. Base exchange reactions are, however, reversible under certain circumstances. Low pH values will tend to reduce the ability of clay minerals to adsorb and base exchange. Groundwater with very high total dissolved solids (perhaps 100,000 ppm or more), such as brines associated with oil fields, will often have high (more than 1 ppm) contents of Br, I, Li, Rb, Cs, B and other highly soluble elements, several of which are recovered commercially. Such groundwaters are usually encountered at depths of at least several thousand feet. The more normal abundance for these elements in near-surface groundwater, as well as the metals, is in the parts per billion (ppb) range. Small amounts of gases, for example radon and helium, may also be found dissolved in near-surface groundwater. Clarke and Kugler (1973) have shown that the helium content of such waters may be used as a possible method for uranium and thorium prospecting.

It would be desirable to be able to list the average abundances of trace elements in groundwater in Table 2-1 but regrettably, information on this subject is insufficient. In general, it can only be expected that the trace element content will be greater than that of river water (Table 2-1). Very little information is available on the theoretical maximum contents of metals in groundwater, which is not surprising in view of the hazards involved in attempting to correlate such calculations with observed abundances. Nevertheless, Cadek et al. (1968) suggested maximum theoretical contents for certain elements in moderately mineralized (containing about 1000 ppm sodium) subsurface waters as (in ppb): Be, 150; Ge, 140; U, 90; Pb, 560; Mo, 60; and (in ppm) Cu, 2.5; Zn, 2.85 and F, 23. These values may be useful as guides but they certainly do not apply to acid waters from oxidizing sulfide deposits (compare Pb and Zn values with those in Table 3-2). Further discussion of the theoretical content of trace and minor elements in waters may be found in Hem (1970) where lower maximum contents of metals are suggested, based on solubility restrictions imposed by certain compounds (e.g., lead carbonate). Clearly, the recognition of anomalous metal values in groundwater is not a simple process, but rather one which requires careful attention to all the local and geochemical parameters and, ideally, access to a considerable amount of reliable data.

In general, groundwater above the water table is oxidizing (Fig. 1-3) and slightly acidic (owing to the presence of humic acids and precipitation, both of which are acid), whereas that below the water table is reducing (low Eh) and tends to be alkaline. However, in groundwater systems in recharge regions with high permeabilities, oxidizing conditions may persist well below the water table, such a condition being necessary, at least in part, to account for the movement of uranium into the subsurface (Fig. 3-24). Well-defined wet and dry seasons may result in significant variations, not only in the composition of groundwater near the water table, but also in a slightly higher Eh and lower pH in the upper part of the zone of saturation. In tundra areas, especially those where marshes and swamps are plentiful and the drainage is poor, the water table is close to the ground surface. Subsurface waters in these areas are generally acid and have a low Eh because the slowly decaying organic matter produces humic acids and a reducing environment. These conditions are favorable for the transport of some metals, particularly iron and manganese, as has been discussed previously, and may result in bog or lake iron-manganese deposits. Because of the changes which are possible in the Eh and pH of a groundwater after it is collected these parameters should be measured in the field at the time of collection, and elements likely to precipitate should be analyzed at the sampling site or as soon thereafter as possible. Metals and silicates may well be lost from solution from hot spring and other thermal waters upon cooling.

White et al. (1963) have prepared the most complete compilation on the chemical composition of subsurface waters. Among the approximately 300 analyses of subsurface water (many obtained from springs), examples can be found for those from all the common rock types, as well as analyses of various thermal, connate, juvenile, magmatic, metamorphic, soil waters and brines. White et al. (1963) discussed the composition of the various waters, with special emphasis on median values of ratios of the various constituents, as these ratios can be used as a guide to the recognition of different genetic types of water. In Table 3-7, characteristic ratios for several genetic types of water are presented, based on the data from White et al. (1963, Table 29). The advantage of ratios, as opposed to absolute values, is that the effects of dilution, concentration or even mixing of these two types of subsurface waters are minimized and identification of the origin of such water is more likely to be possible. In exploration geo-

Table 3-7. Approximate median ratios (by weight), total dissolved solids, and pH of selected subsurface waters, compared to ocean water. Modified from White et al. (1963, Table 29).

Water	Ca/Na	Mg/Ca	K/Na	Li/Na	HCO₃/Cl	SO₄/Cl	Total (ppm)	pH
Ocean water	0.038	3.2	0.036	0.00001	0.0074	0.14	34,500	8.0
Oilfield brine (NaCl type)	.04	.4	.015	.0003	.02	.0005	30,000	7.0
Na–Ca–Cl type spring	.2	.1	.05	.0005	.03	.005	20,000	7.1
Geyser water	.03	.06	.10	.006	.1	.1	2,000	8.4
Springs: acid SO₄–Cl type	.8	.3	.2	.01	.00	.7	9,000	2.2
Springs: acid SO₄ type	1.5	.4	.4	.00	.00	400.	2,000	1.9
Springs: acid HCO₃–SO₄ type	1.	.2	.4	.005	50.	10.	500	7.0
Springs: mercury deposits	.04	.5	.03	.001	2.	.4	3,000	7.0
Springs: depositing travertine	1.	.3	.15	.001	.2	1.4	2,000	6.5
Heated meteoric water	.2	.2	.1	–	15.	4.	200	9.2

Notes:

1. Additional ratios presented by White et al. (1963) are F/Cl, Br/Cl, I/Cl, B/Cl, as well as the average contents for SiO_2 and NH_4.

2. Similar data for five other types of subsurface waters also are presented by White et al. (1963).

chemistry the use of ratios is well established as, for example, the ratio of a metal to the chloride content, one metal to another metal, or uranium to sulfate, as such ratios are often diagnostic.

Williams (1970) discussed the applicability of mathematical models of groundwater flow systems to hydrogeochemical exploration, and Barnes and Hem (1973) have reviewed the chemistry of subsurface waters. These studies, as well as those by White et al. (1963) and Hem (1970), are of importance to those seriously interested in the composition and use of groundwater in exploration geochemistry. Brown (1967) is recommended for those interested in the use of groundwater for exploration purposes in Canada.

RIVER WATER

Sources and Fluctuations

The water carried in streams consists of two components as illustrated in Fig. 3-22: (1) direct runoff which enters the drainage system during or soon after precipitation, and (2) the "base-flow", or groundwater, fraction which infiltrates into the channel. The direct runoff has only a brief contact with soil or vegetation and no residence time in the groundwater system, hence it is likely to be low in dissolved solids. Even so, reaction with the soil is often sufficiently rapid and extensive so that a higher dissolved solids content is usually found by comparison with the original rain or snow. Nevertheless, generally the base-flow has a greater dissolved solids content than the runoff. During periods of heavy precipitation, the effects of runoff on the chemical composition of a river greatly exceed those of the base-flow, so that the total dissolved solids content of a river tends to be inversely related to discharge, although the relation is seldom as simple as illustrated in Fig. 3-26. At very high flow rates, the water may have very few dissolved solids.

Fig. 3-26 illustrates the variations in flow rate of the Saline River, Kansas, for a portion of 1946 and 1947. The great variations in flow rate of this river can be correlated with heavy precipitation and lack of vegetation cover within its basin. Runoff will result in dilution of any metal-containing groundwater (base-flow is about 100 cfs) and it will also modify certain properties such as pH and Eh. In the case of a continually flowing stream such as the Saline River, geochemical sampling of the water should be done during periods of low flow, that is, when groundwater is a major component of the river water. An ideal period would be during the winter months when a river is covered with ice, in which case groundwater can be considered the sole source of water for the river. If a sampling program is

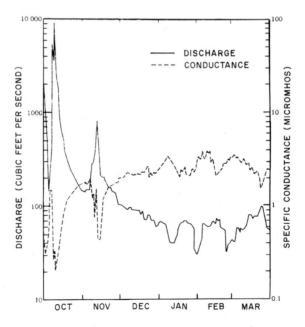

Fig. 3-26. Relation of discharge (flow rate) to specific conductance (an electrical property which is proportional to the dissolved solids content) of the Saline River, near Russell, Kansas, during part of 1946 and 1947. From Livingstone (1963).

conducted during periods of high flow, the dilution effects are likely to make the recognition of base-flow anomalous values extremely difficult, if not impossible. In general, the contribution of groundwater to a river tends to be relatively stable, but the contribution of surface water tends to be variable.

It is not usually possible to separate and evaluate the contribution of runoff and base-flow to the overall composition of a river with any degree of certainty. There are many reasons for this, among them the fact that runoff can come from different sub-basins within the river system each of which may have different rock types and hence different water composition, and because the quantity (and to a lesser extent the quality) of the groundwater can change with time. The result is a complex fluctuation of the content of dissolved solids, especially for some medium to larger sized basins in areas with regional variations in climate and diverse rock types. In addition to the mixing of groundwater and runoff, there are many other important natural factors that influence stream composition. These include reactions of water with solids in the stream bed and in suspension, reactions among solutes, losses of water by evaporation and by transpiration from plants growing in and near the stream, and effects of organic matter and water-dwelling biota (Hem, 1970). This latter set of natural factors result in fluctuations in composition that bear little relation to discharge rate or to the composition of the components of river water. Superimposed on all these factors are the influences of man (pollution, flow diversions, etc.).

In those areas where there has been a long interval between heavy precipitation as in some arid and semi-arid regions of the world, a sudden rainfall may flush elements released by weathering since the previous precipitation. This may result in anomalously high values at approximately the period of maximum flow which need not necessarily reflect economic mineralization. But because the amount of metal released during the period of weathering is small by comparison with the amount of water available, the sharp increase in metals is usually detectable for only a short interval. There is no specific relation between the peak flow and the increase in the dissolved solids (including trace metal) content of a river due to flushing. Each case is different, depending on the local conditions of weathering and climate which take into account the effectiveness of the weathering process, the intensity and interval between precipitation, the position of any metal deposit with respect to the stream, the amount of precipitation which enters the soil before entering the stream, and so forth. If metals are flushed from a soil horizon 10 feet below the surface and 1 mile from the nearest stream, it may well be several weeks after a heavy precipitation before anomalous metal values are detectable in the stream water. This would probably not correspond with the increased content of the common constituents (e.g.,

Ca, Mg, HCO_3) as these are flushed from the more accessible surrounding surface rocks and soils in the area. Similarly, in snow covered areas, the initial increase in flow rates at the time of the spring thaw is not likely to correspond with an increase in dissolved constituents. This is because snow melts from the surface downward, and the high metal values are found near the snow-soil interface (Fig. 11-7).

Dilution must also be considered in the discussion of fluctuation in the composition of river water. Precipitation and runoff are very dilute (Table 3-1) and once the initial flushing is completed the influence of groundwater may be overshadowed by the diluting effects of the high flow rate from these waters. In the case of the Amazon River, Gibbs (1972) noted that the minimum salinity of this river which supplies about 1/5 of the world's total river water into the oceans, occurs approximately two months after maximum discharge (but nevertheless still a period of high flow).

Another factor of importance in exploration geochemistry is the cross-section homogeneity of a stream at any particular locality. In other words, it is important to know if a sample collected near the bank is equivalent to one collected near the bottom at mid-channel. Again, generalization is difficult owing to numerous physical factors such as turbulence, proximity of inflows, and size of the river. Usually small creeks, brooks and streams, as well as larger streams if they are turbulent, can be considered to be well-mixed, and any sample collected in a convenient place will be representative. Streams tend to be homogeneous primarily because of turbulence, which in turn, is controlled by the physical nature of the stream bed and by the relief. Most of the larger streams will exhibit poor lateral and vertical mixing, at least locally. Inhomogeneity will usually be found where turbulence is low or absent, in streams which are both shallow and wide, and especially where waters of great differing physical properties join.

Outstanding examples of the lack of mixing across a stream are afforded by the Mackenzie River at two locations. One is where it is joined by the Liard River (Mackay, 1970) and the other at the confluence with the Great Bear River (Mackay, 1972). In both cases, several hundred kilometers are required for complete mixing (500 km in the latter case). These are, admittedly, rather exceptional examples, and are caused by great differences in the temperatures of waters of the respective rivers (the Great Bear River, which receives its water from Great Bear Lake, is extremely cold; see Fig. 11-1 for location). Mackay (1972) was successful in showing that water temperature, at least in this locality, may serve as a very sensitive tracer in river mixing studies (temperatures should be read to ±0.01°C or better). Electrical conductance, or conductivity, a physical property of water which is easily measured in the field by means of a relatively inexpensive instrument, is another method of determining the degree of lateral and vertical mixing of a river, and is applicable in many areas provided a stream and its tributary have different conductivities. The use of temperature and conductivity together should ensure a reasonably accurate, inexpensive and rapid indication of mixing in many circumstances. Where available, infrared imagery is also very effective for this purpose, as demonstrated for the Mackenzie River (Mackay, 1972). Because the degree of turbulence is often an indication of the effectiveness of mixing, this characteristic should be noted at the time water samples are collected. Simple methods (e.g., a Secchi disk) are available for measuring turbidity in the field.

From the discussion to this point it is clear that great variations are possible in the flow and mixing of rivers and hence in the composition of river water. Only rarely will a river have a uniform flow rate and/or composition throughout the year and, when this occurs, it is the result of some special set of conditions. A case in point is the Great Bear River (Fig. 11-1) which shows almost no seasonal variation in its discharge into the Mackenzie River (always about 20,000 cfs) whether it flows under the winter ice or in the open summer season. The reason for this is that this river is fed from Great Bear Lake, which acts as a storage basin and supplies the river with a uniform volume of water at all times.

Because of the great variation often encountered in the quantity and composition of river water, any one sample from a stream can only be considered as equivalent to a grab sample (unless there is definite evidence to believe otherwise). It is for this reason that variable, misleading and sometimes contradictory results can be obtained from river water surveys taken at different times. For example, Warren et al. (1967) noted that one stream in British Columbia they initially sampled with negative results, yielded anomalous values for copper several months later. This suggests that the source of the water comprising the stream flow had changed drastically in the interim.

At present there are no efficient, inexpensive and effective methods of sampling a stream over an extended period of time and, until such methods are devised, the use of river water data for exploration purposes will be subject to some degree of uncertainty. Although automated, continuous monitoring instruments are available for both water flow measurements and the analysis of selected elements or components of river water, these installations are expensive and not yet in wide use. However, they are becoming increasingly common especially in connection with environmental studies and controls. The future holds forth the likelihood that resins with selectivity for certain elements, or groups of elements, will be developed to the point where they can be placed in a stream for weeks or months. Among the practical problems to be overcome before this method is generally applicable is the requirement that the resins be selective for the base metals, and that they reject Ca, Mg, Na and other elements of limited interest.

Notwithstanding the difficulties in its use, river water should be considered for exploration purposes where it is available. In fact, streams suitable for such purposes are likely to be more abundant than is commonly realized. New York state (area 49,576 sq mi), for example, has 70,000 miles of streams that are sufficiently large and permanent to have been seriously considered for stocking with trout (Frey, 1963, p. 197). This figure, which almost certainly does not include the extremely large or heavily polluted rivers, can be enlarged for geochemical purposes if the intermittent streams are included, so that perhaps there are two miles of streams for each square mile of surface area. Yet, there does not appear to be a single reference in the literature to the use of river water for exploration purposes in this state. Similar, although not as inclusive, statements can be made for the other states in the Appalachian region of the eastern United States.

In summation, success with hydrogeochemical surveys based on the use of river water is dependent, to a large extent, on climatic conditions. In regions with abundant precipitation the best results are likely to be obtained by sampling small and medium-sized streams at low flow (dry season) when the groundwater component is the most significant. In this case, the geochemist should look for (a) an increase in metal content, (b) an increase in sulfate and especially in the SO_4/Cl ratio, and (c) a low pH. By extrapolation of hydrologic records it is generally possible to predict the periods when sampling will be most effective, as well as to recognize anomalous values for a particular area. Hem (1970) has discussed the methods by which hydrologic data can be extrapolated. Although detailed studies are lacking, it is possible that in desert areas the best period for sampling may be when the groundwater level is at its highest.

Mechanisms Of Trace Element Transport In Rivers

Whereas trace elements in groundwater are transported almost exclusively in solution (although a small amount may travel in colloidal form) the mechanisms of transport in rivers are much more varied and complicated. A river may carry material in any of three ways: (1) as bed load, (2) as suspended load, and (3) in solution.

The *bed load* consists of solid material too coarse or too heavy to be carried in suspension and which is rolled or pushed along the bottom by currents. Clastic sulfide grains released by the process of physical weathering, as well as heavy and insoluble accessory minerals and rock and mineral fragments, will be found in the bed load. The bed load normally moves downstream at an uneven pace, depending on the turbulence and on fluctuations in the flow rate. During periods of flood, particles normally in the bed load may be found temporarily in suspension, thus they can be moved great distances in relatively short periods of time. It is also during these periods that rivers acquire much material from bank erosion, soil creep, sheet wash and by other mechanisms associated with downslope movement. Conversely, during periods of very low flow, material normally in suspension, such as clays with adsorbed metals, metal-containing colloids or iron-manganese oxides, may form part of the bed load. Particle size and shape, rate of river flow, and turbulence are the factors determining the location of a particle in a stream at any specific time.

In fast flowing rivers, or those which have a high suspended load (e.g., muddy rivers), as a first approximation it is reasonable to assume that the bed load materials will yield much less loosely-held (adsorbed) metals than the suspended load. This assumption is based on the fact that in these cases the bed load will have only a small amount of metal-adsorbing clay and other fine-grained minerals. Accordingly, total analyses, as opposed to partial analyses (discussed in Chapter 6), should be performed on bed load materials. However, such an assumption should always be confirmed for each particular area, keeping in mind the fact that changes in the flow rate and turbulence can rapidly affect the position of the fine-grained mineral components in a river.

The *suspended load* consists of clay-size and finer (colloidal) particles, some of which are so small that they can remain in suspension even when a river is essentially stagnant. However, when a river loses its velocity, or suspension-laden waters become isolated in low turbidity pools, much of the suspended load will be deposited into the bed load. Turbulence is the main factor controlling the amount and type (particle size) of the materials in the suspended load and this, in turn, is influenced by such factors as water flow, and the shape and characteristics of the stream channel. Because the suspended load is made up of fine-grained material, particularly clays which have a large surface area and high cation exchange capacity, it will tend to have a large amount of loosely-held metal ions by comparison with the bed load.

Colloids, which can be considered as inorganic or organic particles that range in diameter from about 5 millimicrons (50 Ångstroms) up to about 0.2 micron (Hem, 1970), are very abundant in some river waters. Because of their extremely small size they are usually in suspension and are likely to carry a significant portion of the trace element load of a river. For example, Perhac and Whelan (1972) showed that the colloidal fractions of the suspended load of a river draining a mineralized area in the zinc district of northeastern Tennessee had markedly anomalous amounts of zinc and lead by comparison with bottom sediments and water in the same stream.

As Hem (1970, p. 87) pointed out, the suspended material in a river poses something of a dilemma for the water chemist and, by analogy, to the exploration geochemist attempting to use hydrogeochemical methods. In the usual water chemistry or hydrogeochemical survey, part or all of the suspended solids are removed before analysis, either by filtration or by settling. The suspended solids are then discarded, and the clear solution analyzed. However, some kinds of suspended materials may contain important amounts of metals either adsorbed onto the colloids or clays, or as an integral component of colloidal material. A case in point might be those metals from reducing and acid groundwater which are precipitated as colloidal hydroxides or oxides when they enter the oxidizing, more alkaline river environment (Table 3-6). As a result, significant anomalies can be missed if suspended matter is overlooked. Numerous studies have shown that an important part of the total ion load in a river is carried on the suspended sediment particles, and particularly so for the heavy metals. Although very few hydrogeochemical surveys have analyzed the suspended load as well as the filtrate, there is every reason to believe that such a practice would be very rewarding.

When water samples are filtered before analysis, the common practice is to use a plastic membrane filter with pores 0.45 micron in diameter. Thus particles nearly 100 times as large as the lower limit of the colloidal particle size would pass this filter. It is not valid to assume that only dissolved species are present in filtered water samples. Filtration provides a rather inexact means of separating suspended from dissolved constituents, but nothing better seems to be available. Centrifugal techniques are slow, costly, and do not lend themselves to practical and routine geochemical prospecting (Perhac and Whelan, 1972).

Water samples are usually acidified upon collection (discussed in Chapter 5) and this procedure has a tendency to release into true solution metal ions adsorbed onto clays (possibly by replacing the metal ions with H^+ from the acid) and to dissolve some metal-containing colloids and soluble minerals. As a result, some elements which might have been lost or discarded after filtration are available in true solution for analysis. Since so little is actually known of the details of adsorption reactions on suspended sediments (including colloids), the subject cannot be usefully considered further here. Suffice it to say that the behavior of many minor constituents in river water cannot be fully explained without consideration of the suspended fraction.

Rivers carry a large amount of material in *solution*. The dissolved load comes chiefly from inflowing groundwater, with minor amounts from runoff which has been in contact briefly with soils, and a very small amount directly from precipitation (Table 3-1). Relatively little is dissolved from the river banks except in the case of rivers flowing on limestone, halite and other readily soluble rocks. The elements in true solution may occur in several forms: (1) cations (e.g., Na^+, Cu^{+2}, Zn^{+2}); (2) anions (e.g., SO_4^{-2}, Cl^-); (3) complex ions (e.g., UO_2^{+2}, MoO_4^{-2}); (4) undissociated inorganic solutes (e.g., H_4SiO_4); and (5) soluble organometallic complexes. Depending upon environmental conditions, changes in the relative abundances of suspended and bed load material of the river system, these mobile phases may be removed from true solution by any one of the various processes mentioned above (e.g., adsorption), or by floating organisms, plants rooted near the bank, as well as by fish and other aquatic biota. Photosynthesis of plant species which are rooted in the stream (or lake) bottom, as well as floating species, produces oxygen and consumes carbon dioxide, whereas respiration and decay consume oxygen and produce carbon dioxide. These processes result in well-defined diurnal cyclic fluctuations of pH which can be observed in rivers and lakes (Livingstone, 1963). Such variations can affect the location of trace metals at any particular time. Seasonal decay in vegetation, particularly where it is isolated from the atmosphere by ice cover, can result in a depletion of oxygen and a change in Eh. In turn, variations in Eh can result in the precipitation of some elements, and the mobility of others.

Gibbs (1973) determined the methods of transport of the soluble and suspended Fe, Mn, Cu, Cr, Co and Mn by the Amazon and Yukon rivers, and found the distributions to be similar for these widely-separated rivers. Copper and chromium are transported mainly in crystalline solids and manganese in mineral coatings, whereas iron, nickel and cobalt are distributed equally between precipitated metallic coatings and crystalline solids. For these particular elements, in these particular rivers, transport in true solution (including soluble organic complexes) is minor, never exceeding 17%, and generally averaging less than 5%. Clearly the distribution of elements transported by river water in its various fractions (suspended, solution, etc.) varies considerably, and is influenced by the local conditions of weathering, mineralogical nature of the bed and suspended load, availability of organic compounds, and turbidity, among many other factors. Of particular significance in exploration geochemistry is the high proportion of important elements transported as metal coatings (iron and manganese oxide stains with or without other elements). For reasons discussed above, this fraction would normally be overlooked because it would probably be discarded with the suspended matter following filtration.

Although the preceding discussions may give the impression that the use of river water for geochemical purposes is of limited value, in actual fact, if it is used with a full appreciation of the problems involved (e.g., the partition of the elements among the various phases present in a river, fluctuations in composition which may result from dilution or flushing), useful information can be obtained. Historically, river water has been analyzed chiefly for U, Zn, Mo, Cu and SO_4, but atomic absorption and other analytical methods (Chapter 6) are now sufficiently sensitive and accurate for many other elements (e.g., Co, Ni, B, Li, Hg, F) that they can be analyzed routinely in the parts per billion (ppb) range. Gases, for example radon, can also be determined in river water.

Chemistry of River Water

Livingstone (1963) has published the most complete compilation of analyses of river waters of the world (Table 3-1). He determined the average dissolved solids to be 120 ppm (0.012%). Using more recent data, primarily reflecting more accurate data on the Amazon, Gibbs (1972) has modified this figure slightly to 117 ppm. From the thousands of analyses now available on river waters from various parts of the world, several generalizations can be made: (1) few rivers have more than 1000 ppm (0.1%) total dissolved solids; (2) most surface waters (rivers and lakes) are characterized by a predominance of bicarbonate and sulfate ions; (3) the trace metals of interest in exploration geochemistry usually occur in amounts of 20 ppb or less (Table 2-1), and (4) the relative proportions of the major constituents are greatly influenced by the composition of the rocks over which the rivers flow.

The factors which control the geochemistry of a river are discussed in detail by Hem (1970) and Fairbridge (1972). Briefly they can be grouped into two categories: (1) those that are variable with time, and (2) those that are relatively unchanged over a seasonal (one year) period. Some of the factors which do not vary over a period of one year are type of rock over which a river is flowing, and vegetation, and they do not pose any sampling or geochemical problem. Environmental conditions, on the other hand, vary with time, and result in sampling problems. These include climatic factors (e.g., temperature, precipitation, evaporation) which cause variations in flow rate, flushing, dilution and so forth. All the factors which vary with time will result in variations in the total dissolved solids (composition) of a river.

According to Fairbridge (1972) the rivers of the world can be classified into three groups according to the chemical composition of their major cations and anions: (1) those rivers high in $Ca–HCO_3$ (a dilute-type is also recognized), (2) those high in $Ca–SO_4$, and (3) those high in Na-Cl. Compositions of rivers representative of the three types are listed in Table 3-8. By comparing the average composition of rivers of the world (Table 3-

1) with the compositions of the three types listed in Table 3-8, it will be seen that most rivers are the Ca–HCO$_3$ type. Further, the Ca–SO$_4$ and Na-Cl types have higher total dissolved solids than the more abundant Ca–HCO$_3$ type.

World averages for the components and total dissolved solids of rivers, like average abundances for elements in rocks (Table 2-1), must be considered only as guides to the interpretation of hydrogeochemical data. The average compositions of rivers varies within wide limits on both a regional and intercontinental basis. The rivers of South America and Australia average only about 60 ppm total dissolved solids. Their dilute nature is the result of a combination of factors, such as heavy rainfall in the case of South America, and in both continents by the presence of lateritic

Table 3-8. Composition of the major types of rivers (chemical basis). From Fairbridge (1972). All values in ppm.

Type	HCO$_3$	SO$_4$	Cl	Na	Mg	Ca	SiO$_2$	Total
1. Dilute Ca–HCO$_3$ river	17	3	1.7	1.8	1.1	4.3	7.0	36
2. Ca–HCO$_3$ river	101	41	15	11	7.6	34	5.9	221
3. Ca–SO$_4$ river	183	289	113	124	30	94	14	853
4. Na–Cl river	238	174	473	253	71	80	–	1310

1. Amazon River

2. Mississippi River near Baton Rouge, Louisiana

3. Colorado River at Yuma, Arizona

4. Jordan River at Jericho

soils from which much of the alkali and alkaline earth elements already have been leached. African, Asian and North American rivers have between about 120 and 140 ppm total dissolved solids, whereas European rivers average 182 ppm (Livingstone, 1963; Gibbs, 1972). In general, the content of dissolved matter of river water tends to increase from source to mouth.

Extent of Geochemical Anomalies In River Water

The environmental and physical conditions of groundwater, particularly Eh and pH, can be considered to be relatively constant over long distances and for relatively long periods of time. Therefore, once elements are in solution in groundwater, they are likely to remain in that state, and geochemical anomalies can be detected, in favorable circumstances, at great distances from the source (Fig. 3-23). River water, on the other hand, is subject to more changes in environmental and chemical conditions and, as a result, in many cases, the length (or persistence) of a hydrogeochemical anomaly is relatively limited.

The factors which control the length of geochemical anomalies in river water are mainly: (1) contrast at the source; (2) dilution effects; (3) precipitation of the elements; and (4) inherent solubility of the elements, or complexes of such elements, in water. Eventually all anomalies in river water are reduced in intensity to the point where they are indistinguishable from background values (Figs. 1-4, 1-6).

Contrast is the difference between anomalous and background values. At the source it is determined by the degree and rate of chemical weathering, abundance of the minerals being weathered, and climatic factors such as amount of precipitation. The latter, if great, will tend to dilute metal-containing solutions. Thus, the differences between background and anomalous values at the source are strongly controlled by local conditions. In general, areas with moderate relief and precipitation will be favorable locations for high contrast. Physical properties of the host rock, such as degree of fracturing, influence the accessibility of water to oxidizing metal deposits and these, likewise, can be considered local conditions. High contrast at the source is a condition favorable to the development of long anomalies.

Dilution effects can occur anywhere from the immediate source of metals, to many miles down-river. Such effects will usually be more rapid and drastic during periods of heavy precipitation, as well as in the circumstance where anomalous water enters into a larger stream in close proximity to the source of an anomaly. Where dilution is gradual, as is likely to occur during periods of low flow when the main decrease in contrast is caused by the inflow of groundwater, anomalies in river water often can be traced to their source, or to a cut-off.

Precipitation of the elements in solution in river water takes place primarily when changes occur in Eh and pH. Assuming that water emanating from an oxidizing metal deposit is acidic, upon mixing with surface water of a different composition, or neutralization by reaction with limestone or other reactive rocks, the resultant change in pH will cause many metals to precipitate (see Table 3-6).

Numerous detailed studies could be cited which show that geochemical anomalies in river water tend to decay much faster than would be expected from simple dilution alone. These observations can be explained by assuming that the sources of the anomalies are groundwater or springs which emerge from a reducing into an oxidizing environment, where certain elements are no longer soluble, and then precipitate as oxides or hydroxides. These precipitates will either be found in the bed load, or as suspended matter (or colloids), depending on the flow rate, turbidity and the other physical factors discussed above, which control the distribution of materials in rivers. It is partly for this reason that many exploration geochemists prefer stream sediment surveys to river water surveys although, as has been mentioned above, there is every likelihood that some of the anomalous values are dispersed as colloids or adsorbed onto suspended matter.

It is entirely possible that the environmental conditions at the locality in which precipitation has taken place will change over a period of time (e.g., from oxidizing to reducing). Therefore, re-solution of certain precipitates can occur, although this reaction may be difficult to recognize and interpret as such. Tooms and Webb (1961) were able to recognize a secondary release (re-solution) of copper from soils into groundwater in the Zambian Copperbelt. Initially this copper was concentrated during soil-forming processes on underlying copper-bearing sulfides which are now deeply buried. Re-solution of precipitated metals is an important process in certain lake sediments (see below).

Elements and radicals which are *soluble* under a wide range of Eh and pH conditions will form the longest anomalies in river water. These include all the alkali elements, Ca, Mg, Sr, all the halogens, sulfate, nitrate, boron, and a few other elements (Fig. 3-19). Uranium and molybdenum are soluble in most surface waters when they form complexes. These dispersions can be very extensive. For example, sodium and chloride anomalies from salt springs dissolving buried halite beds can be detected over 100 miles from their source in the Mackenzie River drainage basin (Hitchon et al., 1969). Admittedly such examples are rare as most base metal anomalies in river water are detectable from perhaps 100 ft to 1 mile, as a practical range. Adsorption by clay and silt-sized particles, and by organic matter, is one mechanism by which some soluble elements are removed from solution in river water. Most metal anomalies, however, have limited ranges because of dilution, or precipitation as oxides or hydroxides.

It must be emphasized that this discussion on the extent of geochemical anomalies in water has been concerned primarily with anomalies resulting from elements which are in solution. Mechanically dispersed anomalies either in the bed load or in suspension, are not detected by hydrogeochemical methods (except possibly in the case of some colloids).

LAKE WATER AND SEDIMENTS

Introduction

The Earth's land areas are dotted with hundreds of thousands of lakes. The greatest number of lakes lie in regions that were once covered by glaciers. Glaciers scour and deepen valleys as they travel, and glacial deposits often restrict drainage. As a result, glacial water, and subsequently precipitation and groundwater, tend to fill these depressions and poorly drained areas, and lakes form. Although the action of glaciers and their deposits explain the great abundance of most of the lakes in Canada, the northern part of the United States, Alaska and parts of Scandinavia, there are many other processes by which lakes can form. Many lakes are found in limestone areas because this rock is relatively soluble in groundwater (Lake County, Florida, has 1400 lakes formed in this manner). Lake Tanganyika is an example of a lake associated with a rift valley (graben). Crater Lake occupies an extinct volcano, but this type, like others found in meteorite craters, are relatively rare and of little importance in exploration geochemistry. Because of the nature of their origins (e.g., glacial), lakes often occur in definite geographical and geological districts and may be related in composition and age.

Lakes disappear in relatively short periods of time, geologically speaking. Some of the more important processes involved are: climatic changes which may cause rivers and springs supplying a lake to dry up; changes in the course of a river; infilling by sediment and/or organic matter; and headward stream erosion.

Lakes are often referred to as large or small, and deep or shallow. Both of these terms are relative and require quantifying. For the purpose of this book, large refers to any lake which contains more than five cubic miles of water. Deep lakes are defined as those in which the maximum depth exceeds 30 ft (lakes in temperate regions deeper than this usually develop thermal stratification, described below). Lakes deeper than 400 ft are uncommon. Because of the absence of any relationship between volume and depth, it is possible to have some lakes with essentially identical surface area, but with distinctly different volumes of water. Most lakes of interest in exploration geochemistry are fresh water lakes and, therefore, discussions here will be concerned with these types. Saline lakes are equivalent in magnitude to fresh water lakes on a world wide basis although their distribution is quite different. The Caspian Sea alone has about 75% of the total world saline lake volume, and most of the remainder is also in Asia. North America's shallow Great Salt Lake is comparatively insignificant, with a volume of seven cubic miles.

Lakes may have many shapes, such as circular, subcircular, elliptical, dendritic, triangular and irregular. The shape is most satisfactorily expressed by means of a bathymetric map (contour map of the depth) which also assists in selecting the most desirable sampling locations for lake sediments. For example, a bathymetric map may assist in the location of deltaic sediments moved by currents.

Limnology is the science of lakes, and it has a voluminous literature.

The most outstanding source of information on the geological, geographical, physical and chemical aspects of limnology is Hutchinson (1957). From the point of view of exploration, the compilation on the chemical composition of lake water by Livingstone (1963) is also important. In recent years interest in the pollution of lakes has resulted in a large amount of new data, particularly on the chemical and biological characteristics. But before any of this can be incorporated into the framework of exploration geochemistry, it is necessary to present a brief introduction to some classical limnological theory, particularly stratification and overturn, lake terminology, and the influence of oxidizing and reducing conditions upon the distribution of some elements in lake sediments. Simplified reviews of various topics in limnology may be found in Fairbridge (1972) and Hutchinson (1973).

Stratification, Overturn And Classification

Every lake in a temperate region that is deep enough (in most cases over 10 meters) develops what is called *thermal stratification* (Fig. 3-27). A curve (profile) representing the temperature in a deep, cold lake in typical summer conditions shows an upper layer of warmer and lighter water called an *epilimnion,* which is primarily the result of heating by the sun. This layer is well-mixed by wind, and the water circulates freely. The zone of rapid change in temperature is called the *thermocline* (or metalimnion). It is often observed at depths of 8 – 13 meters below the surface, and has a thermal gradient in excess of 1°C/meter. The lowest layer of water, which is cold and dense, is called the *hypolimnion.* The temperature of the water at the bottom of a typical deep lake is about 39.2°F (4°C), which is the temperature of maximum density of fresh water. The hypolimnion is colder and darker than the epilimnion and thermocline, and protected from wind action by the density barrier. Thus, the lake is divided into two distinct parts by the thermocline: the well-mixed epilimnion and the cold, dense and stagnant hypolimnion. A temperature profile in a warm lake will be similar except that the temperature gradient in the thermocline will be less steep and the temperature in the hypolimnion will be perhaps 55°F. Clearly, mineralized spring water or groundwater entering the bottom of a stratified lake during the summer will not be reflected in a water sample taken from the epilimnion or the thermocline.

Lakes go through seasonal warming and cooling which profoundly affects the distribution of water in them. In Fig. 3-28, the seasonal patterns are illustrated. In the spring, there is a period in which the water has the same temperature (isothermal water) from top to bottom (about 39.2°F). At this time the density of the water is the same throughout and, under the

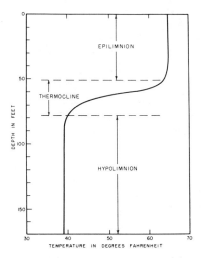

Fig. 3-27. Vertical temperature profile of a cold, deep lake in a temperate climate. A typical summer temperature distribution is illustrated, with warm water in the upper zone (epilimnion) and cold water in the lower zone (hypolimnion). The thermocline is the zone of rapidly changing temperature. Similar profiles are found in most lakes 10 meters or more in depth (in a temperate climate). From Hough (1958).

influence of wind and currents, the water circulates and mixes (Fig. 3-28A). This phenomenon is known as *spring overturn,* and at this period, the lake is vertically homogeneous. During the warmer months the surface water gains heat from the sun and atmosphere. As the temperature rises the water becomes lighter and can no longer sink, and thus a warm layer is formed which floats on the denser water below (Fig. 3-28B). Only the upper layer circulates, by means of the action of wind, and the colder, heavier bottom water stagnates throughout the summer. During this period of stagnation, metal-containing groundwater seeping into the lake will tend to increase the metal values in the hypolimnion.

During the fall, there is a net loss of heat from the water to the atmosphere, and the temperature of the surface water gradually lowers to the point where it is equal to that of the bottom water. As an isothermal condition exists, similar to that in the spring, complete circulation occurs and a *fall overturn* results (Fig. 3-28C). Metal contents which might have been building up in the hypolimnion during the summer are now redistributed throughout the lake. With further loss of heat the temperature of the surface water drops below that of the deep water but, because of the unique property of water whereby its greatest density is at 39.2°F (4°C), the water on the surface is both lighter and colder than the deep water (Fig. 3-28D). If the lake does not freeze over entirely, the surface water will circulate due to wind action while the warmer, denser water will stagnate during the winter, or the period of *winter stagnation*. Thus, in the winter, a lake will show an inverse temperature gradient with the warmest water at its greatest depth. If a lake is covered with ice, a similar effect will be observed. Just below the ice the water is colder (hence lighter than the deep water) and, although the lake is protected from wind, there are gentle currents beneath the ice. In warmer (tropical) climates, as long as a density difference occurs between the epilimnion and hypolimnion (Fig. 3-27), especially at higher elevations where seasonal effects are somewhat more

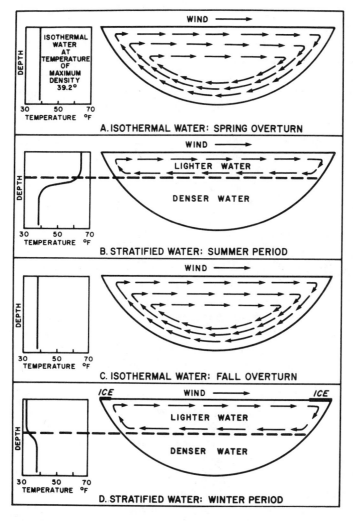

Fig. 3-28. Illustration of the annual temperature cycle, seasonal circulation patterns, and water overturn in typical deep lakes of the temperate zone. From Hough (1958).

pronounced, stratification and overturn will take place. Therefore, in all parts of the world, geochemists must be aware of the fact that any given water sample taken in a lake may not be representative or homogeneous. The obvious exception is at the time of the spring and fall overturns, which, for the sake of discussion, may last a few weeks but actually are variable in duration from place to place.

The seasonal temperature phenomena described above are characteristic of *dimictic* lakes, that is, lakes that turn over twice a year, in spring and fall, and during the winter and summer their deep waters are stagnant. However, there are many modifications to the ordinary type of thermal stratification. Some lakes are *monomictic,* that is, they turn over only once a year. Two general types of monomictic lakes are recognized. The more common type is found in sub-tropical and warm temperate zones where such lakes circulate freely in the winter and stratification is found only in the summer. In the polar regions, on the other hand, monomictic lakes are found which circulate freely only in the summer; these lakes though open in the summer, always have ice in contact with water (Hutchinson, 1957). *Oligomictic* lakes, characteristic of the humid tropics at low altitudes, circulate only rarely and at irregular periods. *Polymictic* lakes, characteristic of high mountains in equatorial latitudes in which there is little seasonal change in air temperature but always enough heat loss to prevent development of stable stratification, continually circulate. Some lakes are monomictic some years and dimictic in others. All of the lakes described above are *holomictic,* that is, their entire water mass circulates at periods of overturn. However, some lakes are *meromictic,* which is a term applied to those lakes whose bottom waters are permanently stagnant, with circulation limited to the upper layers. Such conditions prevail in the case of certain very deep lakes where the bottom topography prevents circulation, or where dense salty water from springs is introduced into the bottom of a fresh water lake, as well as in other rare situations.

Although the characteristics of lakes vary greatly, from the point of view of exploration geochemistry it suffices to recognize that the thermal history of a lake depends primarily on its geographical location and depth. Because most lakes, and correspondingly, interest in lake geochemistry are centered in the colder, temperate latitudes, it is reasonable to be guided by the concept of a dimictic lake, provided depths of more than ten meters are encountered. In shallow lakes, if there is sufficient wind, uniform mixing can be assumed. The wind is able to work effectively on a lake several miles long and mix it to a depth of some tens of feet, although at any specific time water from the epilimnion or an incoming stream may be "piled up" by local strong winds from one direction, and other special conditions may prevail (e.g., the hypolimnion may be exposed on rare occasions). Very small lakes, and farm ponds sheltered by thick woods may be mixed only to a depth of a few inches. For further details on these topics, Hutchinson (1957), other limnology textbooks (e.g., Ruttner, 1963), or the extensive literature on stratification effects should be consulted.

Influence Of Oxidizing And Reducing Conditions

The chemical differences that develop between the surface and deep water in a stratified lake are of interest to both biologists and geochemists. Biologists are concerned with photosynthesis, availability of oxygen for living organisms, the occurrence of fish, and other topics of limited interest in some areas of geology. From the point of view of exploration geochemistry, the exchange of metals between the mud and water in lakes is of extreme importance in determining the proper method and time for conducting lake geochemistry surveys, as well as in the interpretation of the results.

As a consequence of stratification, there is always an abundance of oxygen in the epilimnion of a stratified lake, but conditions in the deep water are quite different. Oxygen is used up and carbon dioxide produced during the decay of organic matter by bacteria and by other living members of the plant and animal community. This may result in a reducing environment in the hypolimnion, especially in those lakes where there is a small hypolimnion. There is no possibility of oxygen replenishment until the next overturn, as this area is effectively sealed from the atmosphere by the epilimnion and the thermocline (Fig. 3-28). The change from oxidizing to reducing conditions results in a change in the solubility of many metals introduced into the lake sediments from groundwater seepage, clastically, or absorbed into clay-sized particles (Fig. 3-22). An accumulation of H_2S may occur in the hypolimnion, and this may quantitatively precipitate certain metals (e.g., zinc and iron) as sulfides.

A classic study of the exchange between mud (sediment) and water was carried out in the English Lake District by Mortimer (1941-42), and it has general application elsewhere. Some of his results, which were concerned primarily with major constituents and biological aspects, are shown in Fig. 3-29 (from Esthwaite Water; depth about 16 meters). The change from oxidizing to reducing conditions in the summer months leads to the appearance of much ferrous iron, manganese (divalent), phosphorus, ammonia and some silica (not shown) in the water. These components are released from the sediments in the lake, which stratifies very strongly during the summer, and to some extent during the winter (dimictic). The reverse situation also exists. During the change from reducing to oxidizing conditions in the fall and winter the dissolved constituents return to the

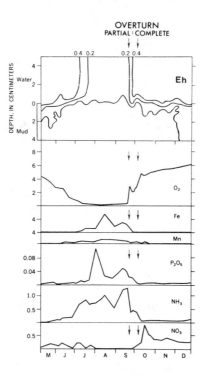

Fig. 3-29. Selected results from the study of Mortimer (1941-42) on exchange between dissolved substances at the interface between water and mud, English Lake District. The abundances of the dissolved components in the water are shown (all values in ppm). Oxidation-reduction potentials (top) shown are 0.2 and 0.4; less than 0.2 is reducing, and greater than 0.2 is oxidizing (note the oxygen content). MJJASOND are months of the year. From Livingstone (1963).

sediments (mud). Mortimer (1941-42) did not analyze for trace metals in his study, nor does there appear to have been any subsequent similar study which shows the temporal variations which can be expected for trace metals. However, it is evident that any lake showing as many changes in chemical content will be inadequately represented by the chemical analysis of a single water sample, whether it is taken at the surface or at depth. Limnologists are aware of this problem, but exploration geochemists and geologists tend to sample lake water and sediment as if they were temporally and spatially homogeneous. An exception is the study by Arnold (1970) discussed in connection with Fig. 11-12, which does consider limnological aspects.

    Gorham and Swaine (1965) expanded on some aspects of Mortimer's work by analyzing oxidate crusts, oxidized surface muds, reduced surface muds, and glacial clays collected from the same lakes in the English Lake

District. They found a distinct difference in the occurrence of many elements of interest in exploration geochemistry among the four types of solid samples. Details of their results are discussed below (Fig. 3-32).

In Canada, Alaska, Scandinavia, the USSR and other localities in which lakes are covered with ice for a large part of the winter, oxygen may be used up and such lakes tend to become reducing. This applies equally well to both shallow lakes which are not stratified, and to the deep, stratified types. Those which contain organic matter are particularly likely to become reducing. Canadian limnologists are well aware that as far as winter conditions are concerned, the most important factor is ice cover. *In waters that are cut off from contact with the atmosphere, the entire chemistry is governed chiefly by the extent of depletion of dissolved oxygen, by means of the oxidation-reduction potential.* It would appear that emphasis should be placed on whether the elements of interest are more mobile under oxidizing or reducing conditions (Fig. 3-19) before selecting waters or sediments for sampling purposes at any particular time of year. Clastic sulfides in lake sediments are less likely to be affected by changes in oxidation-reduction conditions by comparison with those metals introduced by groundwater or surface water.

Lakes are also classified as *oligotrophic* or *eutrophic* (Greek. *oligos* small; *eu* well; *trophein* to nourish), terms which have been defined in several ways. In general, oligotrophic lakes can be thought of as youthful, deep and containing clear, blue water, whereas eutrophic lakes are mature, shallow and their water is yellow or green (Fig. 3-30). The latter has much more organic matter and nutrients, which result in a high plankton population, and hence the characteristic color. An oligotrophic lake has a

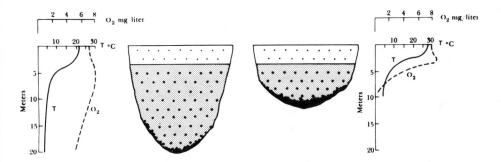

Fig. 3-30. Illustration of stratified (late summer) oligotrophic (left) and eutrophic (right) lakes with moderate amounts of nutrients and organic matter (dots). In the lake on the left, dead organic detritus falls through a large volume of water, but the oxygen content is not used up and the lake is oxidizing. In the lake on the right, the decaying organic detritus has used up most of the oxygen from the small hypolimnion, and the lake bottom is reducing. Modified from Hutchinson (1973).

high oxygen and low nutrient content, and if it is sufficiently deep, any organic detritus produced in the surface layers in the course of a season will fall through a large volume of oxygenated water in the hypolimnion of a stratified lake, which is sufficient to decompose the organic matter. A shallow lake in the same locality will have a small hypolimnion, which is insufficient to supply enough oxygen to decompose all falling organic matter. As a result, all lakes in the same area may be stratified in the summer, but the deeper ones may be oxidizing and the shallow ones reducing. Because of different oxygen contents, each lake may have a different partition of elements between the water and sediment phases. The term *mesotrophic* is applied to lakes whose properties are intermediate between oligotrophic and eutrophic.

Water Movement In Lakes

Water movement in lakes is affected and modified chiefly by currents, turbulence and waves, in addition to stratification. The surface of the water (air-water interface) is particularly important because it is here that wind and radiant energy exert their greatest influence. Mention was made above to the fact that strong winds can "pile-up" water of the epilimnion on one side of a lake so that the hypolimnion may be exposed, but this is just one of many features which must be taken into account when sampling lake water. When the strong wind referred to above dies down, a *seiche* may result. This is a periodic oscillation occurring when strong winds cease after having blown water to one side of a lake. The water flows back to the former (windward) side, and then again to the leeward shore. This oscillation, or rocking, results from the fact that the current does not loose its energy immediately. These currents, and their periods, are dependent on the shape, length and depth of the basin. Seiches are particularly obvious in large lakes, especially those with gentle marginal slopes. Similar phenomena may be caused by other mechanisms, such as lunar pull. Thus hydrogeochemical anomalies, whether introduced at the surface from inflowing streams, or at the margins by groundwater or springs, can be displaced by water movements of many types.

That winds and currents may not be totally successful in mixing either surface or deep waters of lakes is illustrated in a vertical profile of zinc in Lake Ontario (Fig. 3-31). This lake at the time of sampling was stratified, but considering its size, and the winds and currents commonly found on the lake, mixing of the surface water, at least, would be expected. However, according to Chawla (1971) such profiles, and horizontal variations on the surface water (not shown), are typical of this lake, as well as Lake Erie. It is interesting to note that high values for zinc are indicated near the bottom

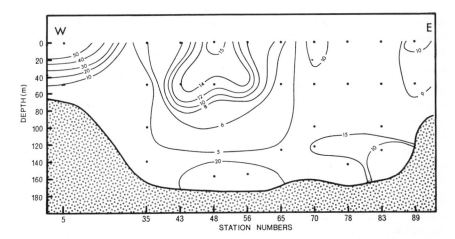

Fig. 3-31. Distribution of zinc along an east-west section of Lake Ontario, September 4-10, 1969. Note the high zinc values on the bottom at stations 48 and 56 (see text for further details). From Chawla (1971).

at stations 48 and 56 in the central part of the lake. Chawla (1971) reported that iron and copper also consistently give high values at the same stations "suggesting the possibility of a rich mineral deposit in the central section of Lake Ontario" (Chawla, 1971, p. 63). Whether the cause of this anomaly is a rich mineral deposit or groundwater seepage is academic at this time, but Fig. 3-31 does show that a carefully organized hydro-geochemical survey can detect anomalies in waters which are undeniably heavily polluted. The Hamilton-Toronto area (west-side of Fig. 3-31) is the origin of high iron and other trace metals in the upper water levels, but apparently they can be distinguished from the anomalous concentrations at the bottom in the central region. It might be suggested that the high concentrations "slide down" to their present position by means of bottom currents along the western slope of the lake, but presumably this and other mechanisms of surface water movement were considered by Chawla.

The rate at which water moves through a lake (flow-through) is an important point to be considered in exploration geochemistry, usually in conjunction with other factors such as stratification, volume of water in the lake, groundwater contribution, and water chemistry. Another very important factor is the depth at which the river water travels as it moves through a lake, and this is determined primarily by the relative density of the water masses. During the summer the inflowing water may be colder and heavier than the surface water of a lake and, therefore, it may sink or even travel along the bottom.

If swift-moving river water enters a shallow, non-stratified lake and has only a short residence time in the lake, the lake can be considered as a wide area in the river in which some sediment is deposited. On the other hand, if the flow-through is very slow, as in the case of many lakes on the Canadian Shield where precipitation is low and rivers move sluggishly, the lakes can develop their own chemical characteristics and, in these cases, groundwater influence will be much more significant. In many such lakes the water may be clear near the inlets but darker, and more organic in nature towards the center. Sediments may change, correspondingly, being more clastic near the inlets and more organic in the deeper parts.

Although quantitatively minor, some currents may be developed in a situation where warm groundwater tends to rise as it enters a cold lake. This process may result in some mixing in the normally stagnant hypolimnion of a stratified lake. Possibly the distribution of zinc in the bottom waters of Lake Ontario may be explained, at least in part, by this mechanism (Fig. 3-31).

Chemistry Of Lake Water

A lake which has an outlet represents a storage and mixing basin for river water and groundwater (Fig. 3-22). Thus, the composition of most lakes can be described in terms of their river and groundwater sources (Tables 3-7; 3-8), since the contribution of dissolved solids by direct precipitation or surface runoff is minor. The composition of most lakes tends to approach that of river water (Table 3-1), although some lakes are supplied entirely by groundwater. There are many exceptions, however, to these observations, particularly for the lakes from closed basins in arid regions where evaporation leads to the precipitation of certain constituents (e.g., $CaCO_3$) and an enrichment of the remainder (e.g., Na, Cl).

The chemistry of lake water, like that of river water, reflects climatic conditions, source material and relief of the basin. Examples are known of lakes at both high and low elevations in the same area, with identical lithologies in the surrounding rocks (e.g., some mountainous areas of British Columbia), but with different contents of total dissolved solids. This is explained by two factors: (1) a greater evaporation rate in the lower, warmer lakes, and (2) greater precipitation, and thus dilution, in the higher lakes. The use of ratios of the constituents, as opposed to the total dissolved solids content, will generally show the similarity of such waters. In the oxygen-poor environment of some eutrophic (Fig. 3-30) and meromictic lakes, sulfate is reduced to H_2S by sulfate-reducing bacteria, and this can result in the precipitation of small amounts of base metals as sulfides.

As indicated above, the composition of any lake, and especially a stratified lake, cannot be evaluated properly without taking into account the inherent chemical differences in the various layers and their degree of mixing, the organic matter content, and the influence of oxidizing and reducing conditions in or near the lake sediments.

Biological Aspects

Many species of plants and animals may be found in a lake and these may have a profound effect on the distribution of elements. Biologists and chemical limnologists are greatly concerned with this subject, but geologists and geochemists have generally tended to overlook it. Photosynthesis, which is a biological process, results in the removal of carbon dioxide from the water, and the addition of oxygen. However, this takes place only in the upper zone where sunlight is plentiful. Conditions in the deep water are different and photosynthesis is greatly reduced. In these waters, a host of reducing organisms, particularly bacteria, thrive and obtain their energy for life by breaking down organic substances that settle from the epilimnion.

Of more direct concern to exploration geochemists is the fact that living cells have the ability to scavenge elements from the water in which they live. Although it is most obvious and well-studied for marine organisms, especially algae, similar processes take place in fresh water lakes and rivers, saline lakes, acid mine and other types of water.

Bowen (1966) has summarized the voluminous literature available on the uptake and excretion of elements by organisms in marine and inland waters, as well as in soils. Among the many points made by Bowen (1966), and others, is the fact that organisms have different affinities for different elements. Generally speaking, the order of affinity for cations among living matter is: tetravalent and trivalent elements > divalent transition metals > divalent metals > univalent metals. Differences can also be found between organisms in their affinities for metals. With divalent metals, for example:

Plankton: $Zn > Pb > Cu > Mn > Co > Ni > Cd$

Brown algae: $Pb > Mn > Zn > Cu > Cd > Co > Ni$

The order of affinity of living matter for anions is: nitrate > trivalent anions > divalent anions > univalent anions. Although the above observations are based mainly on the study of marine organisms, comparable concentrations and orders of affinity probably take place in fresh water.

Bowen (1966) quotes a Soviet study on the uptake of 19 elements by 32 species of plants (algae, bryophytes and vascular hydrophytes). The results of this study are shown in Table 3-9, where they are presented as "concentration factors" (a comparison of the concentration of an element in the dry plant species compared to the concentration of that element in the solution in which it lives, i.e., fresh water in this case). A large con-

Table 3-9. Mean concentration factors for 19 elements by 32 freshwater plant species. From Bowen (1966).

Element	Concentration Factor	Element	Concentration Factor	Element	Concentration Factor
Ca	265	Ge	305	Ru	1700
Cd	1620	Hg	5915	S	165
Ce	7100	I	370	Sr	475
Co	4425	Nb	7640	Y	6880
Cr	695	P	5480	Zn	4600
Cs	480	Rb	1230	Zr	6230
Fe	4935				

Note: "Concentration Factor" is a comparison of the concentration of an element in the dry plant species compared to the concentration of that element in the solution in which it lives.

centration factor was found for all elements studied, and for the divalent cations the order of concentration is: Hg > Zn > Co > Cd > Sr > Ca. Regrettably no data is available for such important elements as Cu and Mo, but it can be assumed that they, likewise, are scavenged by at least some organisms. The fact that copper is scavenged by plankton and other organisms has been demonstrated by numerous studies, some of which have shown that upon the death of large numbers of these organisms, the copper content of lake water will rise significantly.

During periods of algal blooms so commonly occurring in fresh waters, some limnologists feel that the entire supply of certain metals are scavenged (the blooms themselves are related to the abundance of nitrogen and to other factors). Thus, when sampling water from the epilimnion for trace metals, negative results can be obtained even in normally anomalous waters. Some limnologists state that the living matter in the epilimnion is a major control on the distribution of trace elements. Assuming this is true, it need not necessarily eliminate lake water as a sampling medium, because in some areas, e.g., in the Canadian Arctic, the living matter is low or negligible, or it has a very limited growing season. However, in tropical and temperate lakes, the effect of organisms on the trace element content of water must be considered a potentially serious problem in geochemistry at all times. Collecting (by means of nets) and analyzing plankton and other appropriate organisms for geochemical purposes may be worthwhile in some cases, but there do not appear to have been any studies reported employing this concept.

At the termination of an algal bloom, or upon the death of any of the scavenging organisms, such organisms will settle, and after a slight delay at the thermocline (due to the density change), they will tend to reach the bottom. During their descent and on the bottom they will decay and their contained elements will be released into the hypolimnion. If the lake water is freely circulating, the released elements will move up towards the top again. If, however, iron and manganese oxides are present in the lake sediments, the metals may be scavenged by these inorganic compounds. In a eutrophic lake the decay may be incomplete and some of the elements may be permanently removed from the lake system by incorporation into the sediments.

It is obviously impossible to summarize all the information available on the effect of living matter on the chemical aspects of lakes. The important point to realize is that organisms must be considered in the design and interpretation of any fresh water geochemical survey. Organisms are likely to be most important in tropical (as opposed to Arctic) lakes, and during certain seasons (e.g., at the time of blooms) they may deplete the epilimnion of all base metals. Arctic fauna and flora would be expected to be significantly different from those in the more temperate climates in their ability to scavenge, and equally important, is the fact that the growing season in the Arctic is very limited.

The concepts presented in this section also apply to river water and soils, although in modified form. The most complete discussion of the nature of fresh water biota in lakes, their seasonal variations, vertical distribution, and migration characteristics has been written by Hutchinson (1967), as a companion volume to his work on the geography, physics and chemistry of lakes (Hutchinson, 1957).

Lake Sediments

The use of lake sediments as an aid to mineral exploration relies on two basic concepts (Fig. 3-22). The first is that the detrital or clastic portion of a fine-grained lake sediment is a good composite sample of the mineralization in an area. The sediments may receive clastic sulfide grains from areas of physical weathering, or fine-grained clays with adsorbed metals from those areas in which chemical weathering is predominant, as well as metal-containing colloids large enough to settle. The second concept is that the fine-grained particles in the sediments are an excellent medium for the adsorption of metal ions released during the weathering of a sulfide deposit, or similar mineralization, and subsequently transported by groundwater. In theory and practice these concepts are acceptable, but most success with lake sediment surveys will be achieved when the following are taken into account: the type of weathering in an area; the most probable methods of transport of the metals (groundwater or surface

water); the most suitable sampling site on a lake; the mineralogical nature
and particle size of the sediments; the depth from which the sample should
be collected; the effects of oxidation and reduction on the mobility of
elements in the sediment and adjacent water; the organic content of the
sediment; the sedimentation rate; and many other factors.

Mention was made above to the importance of oxidation and reduction
in lake water, and now it is appropriate to discuss this with reference to
lake sediments. The study by Gorham and Swaine (1965) on the sediments
of lakes in the English Lake District is one of several published in the limnological literature in recent years which stresses the influence of these
conditions. They analyzed four fractions of the sediments: white glacial
clay, reduced subsurface mud, oxidized surface mud, and some iron-
manganese crusts, and they determined the distribution of elements among
them. Some of their results are illustrated in Fig. 3-32. From this figure the
following facts emerge: (1) Zn, Co and Mo are most abundant in the iron-
manganese crusts, but significant abundances are found in both types of
mud; (2) Pb is concentrated in the crusts, with very little in other fractions;
(3) Cu is about equally abundant in both types of mud, as is Ni, but the latter reaches its highest levels in the reducing subsurface muds; (4) P is low
in the clay, but about equally abundant in the other fractions. Gorham and
Swaine (1965) reported on the distribution of 18 other elements, and
similar variations were observed, depending on their affinities for the
various fractions. Clearly, it is not possible to generalize on the
distribution of elements in lake sediments; each element is likely to have its
own characteristic distribution, and this will be likely to vary from one
lake to another.

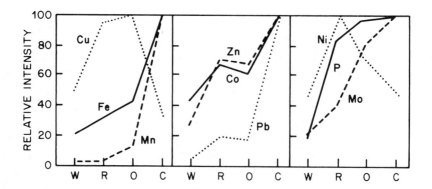

Fig. 3-32. Distribution of elements in various fractions of a lake sediment, English Lake
District. W = white clay; R = reducing subsurface mud; O = oxidizing mud; C =
iron-manganese crusts. Modified from Gorham and Swaine (1965).

Iron-manganese crusts and nodules from North American lakes have been described and analyzed on several occasions. Cronan and Thomas (1970) have studied some from Lake Ontario, and also summarized the results from previous studies. Table 3-10 lists the abundance for Ni, Co, Cu and Zn from such materials from several locations. The Lake Ontario materials have the highest trace element contents and this may be explained by their lower rate of accumulation (Cronan and Thomas, 1970).

Table 3-10. Average composition of iron-manganese oxide concretions from various environments. From Cronan and Thomas (1970).

	Mn(%)	Fe(%)	Ni(ppm)	Co(ppm)	Cu(ppm)	Zn(ppm)
1	17.0	20.6	2385	643	363	1996
2	20.5	20.0	725	305	90	460
3	30.5	19.1	220	200	10	1240
4	29.0	3.8	75	200	17	—
5	14.0	22.5	750	160	48	80
6	16.0	12.6	5450	3110	2930	3500

1. Average composition of 33 iron-manganese coatings, Lake Ontario.
2. Average composition of nodules from one locality, Lake Ontario.
3. Average composition of concretions from three other Canadian lakes.
4. Average composition of two nodules from Scotland.
5. Average composition of Baltic Sea concretions.
6. Average composition of deep-sea nodules.

Some geochemists have recognized the importance of oxidation-reduction on the mobility of elements and, to overcome the problem, they sample about 8 – 12 inches below the lake sediment surface. This environment presumably will always be reducing and, thereby, at least consistent. In general this may be true, but such an assumption should be confirmed. The water in a sandy, low-organic lake bottom could well be oxygenated at this depth. Further, there is always the possibility that the trace element composition will vary significantly in short distances below the sediment surface. This has certainly been established in the case of many interstitial waters from oceanic environments. In Canadian lakes, for example, manganese, which tends to migrate upward from the sediments toward the lake water, will commonly be 2000 - 3000 ppm at the top of the sediment layer, but only 200 ppm at a depth of one meter. Similar trends have been observed in some lakes in arctic Alaska. In surveys which collect lake sediments at depth, the objective would likely be a groundwater transported anomaly, although a clastic anomaly may be preserved under certain environmental conditions. One distinct advantage of this approach is that sediments at such depths would probably pre-date any human contamination.

The sedimentation rate in lakes is a very important, but generally overlooked, factor. Intuitively, it should be obvious that if one lake receives twice as much sediment as another, and if trace metal values from any source are the same, sediment from the first lake will yield values only half of those found in the second lake. It is also possible to have non-significant anomalies from groundwater with normal trace elements appearing in lake sediments where the sedimentation rate is very slow or, alternatively, true anomalies may be masked by a high sedimentation rate. Determination of the sediment rate is not a simple matter, as it requires such techniques as radiocarbon dating, pollen studies, and so forth. The measurement of the sediment input from rivers entering a lake represents only the sedimentation rate at the time of measurement, and cannot be projected with any degree of confidence. As a general rule, the higher the sedimentation rate, whether it be clastic or organic, the lower the trace element content of the lake sediments. In addition, it is important to recognize that sedimentation rates are affected by many factors which, although often disregarded by geologists, are well known to limnologists. A forest fire, for example, can result in increased erosion, which will be reflected by a higher sedimentation rate in nearby lakes.

Certain heavy metals, particularly those that are transported in solution such as zinc, copper and uranium, may be enriched in sediment samples containing organic matter, in a manner similar to the process described previously for soils. For this reason the organic matter content, as well as that of the iron-manganese oxides, is often determined along with the heavy metals. Should high values for the heavy metals coincide with those of the well-known adsorbers, the significance of the heavy metal values is downgraded. Lynch et al. (1973) have described a rapid method for the estimation of organic carbon in lake sediments which is applicable for this purpose.

Other important processes operative in lakes include reactions of metals with phosphates, arsenates and other anions in the water which result in the precipitation of insoluble mineral phases (e.g., vivianite) in the sediments, analogous to their occurrence in soils. In addition, clays adsorb (scavenge) certain elements. Bacteria may be a major factor in some reduction processes. In several localities, for example in British Columbia, native copper is found in reducing muds. The copper was probably transported by groundwater, and the reduction from the ionized state (probably Cu^{+2}) to the elemental state has been attributed to bacteria, although it is difficult to separate the effects of bacteria from those of other types of organic matter as reducing agents in such an environment.

Summary

Lake water and lake sediments are potentially important in exploration geochemistry in certain parts of the world, particularly in those areas which have been recently glaciated. At this time, the use of these sampling media is receiving great attention, especially in Canada, for reconnaissance and mapping purposes (see Chapter 11). The best results with lake surveys will require a thorough understanding of the limnology in the area being investigated as groundwater, for example, may enter either a reducing or oxidizing environment depending on the condition (e.g., stratification) of a lake. Care must be taken to determine the nature of the sediments, as they may be a mixture in any proportion of mud, organic matter, or clastic grains, with or without iron-manganese oxides, each of which has characteristic adsorption properties.

The selection of the most appropriate sampling site in a lake will depend upon mixing rates and other limnological factors in the case of water, and many geological variables (e.g., type of weathering, movement of deltaic sediments by currents, rate of water flow-through, particle size and sorting, sedimentation rate) in the case of sediments.

Further discussions on the use of lake water and sediment surveys in Canada will be found in Chapter 11.

4 / *Some Basic Principles*

This chapter is concerned with four topics which are common to exploration in both the primary and secondary environments. These include: (1) contamination; (2) orientation surveys; (3) false anomalies; and (4) interpretation of geochemical data.

CONTAMINATION

Contamination is an ever present risk in exploration geochemistry and its possible presence, particularly in sediments, water and soil, must constantly be in the minds of those collecting, analyzing and interpreting samples and data. In some cases it can be suspected immediately, for example, when data does not fit the general geochemical pattern. In other cases, contamination which gives the appearance of an anomaly is recognized only after considerable effort and expense have been expended.

There are many sources of contamination but, in this section, only those related to industrial and human causes are discussed. Contamination in the laboratory, which may occur during sample preparation and analysis, is discussed in the following chapter.

Mining Contamination

Contamination due to mining activity is a major problem in some of those areas in which exploration geochemistry, theoretically at least, could be most useful. Previous or present mining activity in these areas usually results in contamination from waste dumps, mine workings, fragments and dust from ore and rock piles, and smelter operations. The sulfides (usually pyrite) in mine dumps are especially susceptible to oxidation and produce acid waters which are then able to leach the small amount of ore minerals not completely recovered during benefication processes. The acid waters, now containing trace metals, can then make their way into the drainage system, and may result in metal trains and halos great distances from an ore body. This makes new geochemical surveys very difficult, if not impossible, in some localities. Further, dispersion of solid particles can occur by mechanical movement, by means of wind or water. Mechanically transported particles will usually result in contamination of the surface only, perhaps to a depth of only a few inches. In the Coeur d'Alene mining district of Idaho, Gott and Botbol (1973) found that soil samples collected from a depth greater than 6 inches were generally free from contamination. However, in some circumstances these metal-bearing particles will weather, and contamination of the soil to a depth of 6 ft or more can be found near ancient workings. Vegetation may absorb those metals which reach the root zone and thus give a false biogeochemical anomaly which in many cases is especially difficult to interpret.

In Mexico, where mining has been a major industry for at least 400 years, the effects of contamination can be especially well illustrated with the following examples (Mr. J. Pantoja and Dr. J. H. Butler, personal communication):

1. River sediments in the Nacozari district, Sonora, are contaminated for 75 km from the Mina Pilares; stream sediment surveys would lead directly to the mine.
2. The paths followed by mule trains carrying ore from a mine worked in Sonora during the period of Spanish occupation, but now long-abandoned, is clearly definable today by geochemical analysis of soil along paths. The area is one of low rainfall and chemical weathering is a relatively slow process.
3. Water analyzed for copper in the Pichucalco area of Chiapas (southern Mexico; high rainfall) leads directly to another abandoned mine (Mina Santa Fe), some 40 km distant. Presumably the sediments, if analyzed, would also show copper contamination.
4. Acid mine waters from an abandoned mine in the Charcas district, San Luis Potosi, resulted in high metal values along a drainage channel which was also the location of a major fault, giving the impression of a large mineralized zone in this structural setting.

These and many more examples which could be cited, may be of great importance to anthropologists and historians interested in older civilizations, but they clearly present great difficulties to the exploration geochemist interested in finding new mines, as opposed to old ones, unless of course, modern technology will permit these mines to be reopened or the area to be re-evaluated. This is not to say that the exploration problems are insurmountable, because, in fact, they can usually be overcome or at least compensated for. As mentioned above, one characteristic feature of contamination in soils is that it is usually superficial, another is that the ratio of the cold extractable to the total heavy metal content (discussed in Chapter 6) in both soil and sediment is usually higher than would be obtained from similar but uncontaminated samples from the same area. Also, the geometry of an anomaly caused by contamination will not usually correspond with geological evidence, such as rock type and structure. Regrettably, in many cases contamination can only be recognized after the samples have been collected and analyzed, and an ancient mine, for example, has been re-discovered. However, careful preliminary investigation into the mining history of an area before a survey is undertaken, should certainly indicate possible pitfalls of this nature.

Industrial Contamination

Until recently, it was commonly thought that only copper and zinc were the industrial contaminants of concern in exploration geochemistry. Now, however, many more elements can be added to the list from the ever-increasing number of smelters, refineries, chemical plants, coal-burning power plants, and industries of all types (Fig. 4-1). The contaminants are

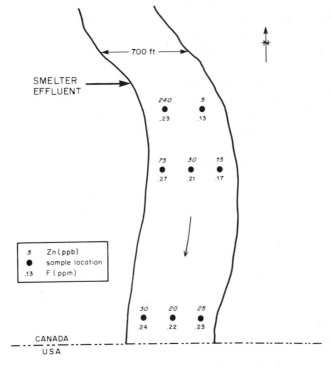

Fig. 4-1. Effect of smelter effluent discharge on the zinc and fluorine contents of the Columbia River, Trail, British Columbia, October 19 and 20, 1967. River flow was about 44,000 cfs. Distance from the smelter to the U.S. border is approximately 11 miles. Background values for the river are 5 ppb Zn and 0.13 ppm F (station near east bank, one-half mile below effluent outlet). Complete mixing is indicated at all three stations about 1 mile from the U.S. border with the zinc content raised to at least 20 ppb, and F to 0.22 ppm, on the day sampled. Data selected from Reeder et al. (1971); no data are available on associated sediments. Water pollution from the smelter has since been eliminated.

introduced primarily by means of fumes (either as solid particles or after condensation), or by water. Some of the more notable examples of industrial pollution in recent years (now stopped) range from mercury, derived from chlorine-alkali chemical plants which have polluted certain parts of the Great Lakes, to sulfur emitted from natural gas plants in western Canada whose effects could be detected many miles downwind. These contaminants, like others described above, are usually restricted to the top few inches of soils or stream beds, and exhibit geometric patterns unrelated to geology.

Agricultural Contamination

The use of insect sprays, fertilizers and other materials to increase agricultural yields has a great effect on geochemical exploration activity in some areas. The use of mercury for the past decade is well-known, and although it now appears to have been stopped, in most cases by legislation, the effects on lakes, rivers and sediments may be an annoyance for a long time to come as background values will have been raised. But other metals and elements have also been and continue to be introduced, such as copper-containing solutions as spray against potato blight, zinc additives to increase the rate of growth of potatoes, uranium and fluorine in phosphate fertilizers, and probably rubidium in potash (KCl) fertilizers. All are soluble under most conditions, and will eventually find their way into the drainage system. On the other hand, there is some evidence that certain fertilizers, particularly phosphates, may tend to form insoluble compounds with elements such as copper and zinc, and this fixation may result in reduced values for these elements in solution. The use of lime for agricultural purposes may result in lead or zinc contamination, if it is produced from metal-containing limestones. This list is by no means exhaustive, and new compounds are continually being introduced for agricultural purposes.

Constructional Contamination

Contamination from constructional activities includes a multitude of possibilities, ranging from metals originating from dam sites, railroad trestles, bridges and culverts of all types, to other highway features containing valley and road fill which could well be mine waste, or from abandoned machinery (Fig. 4-2). These effects are most prominent in water and stream sediment surveys, and can be avoided by collecting upstream of building activity and construction sites.

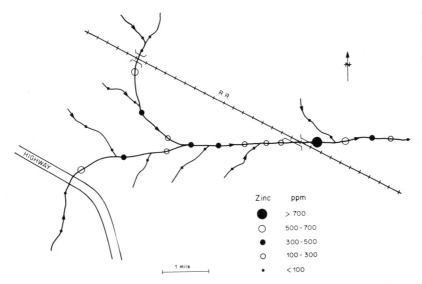

Fig. 4-2. Diagram illustrating the effects of contamination from constructional activity on the zinc values of stream sediments. Metals dissolved from structural parts of rail-road bridges, and from highway culverts, can increase the zinc values of stream sediments below these sources of contamination by hundreds, or thousands, of ppm. Based on an example from British Columbia.

Domestic, Municipal, Human and Animal Contamination

Domestic contamination contributes a great number of elements, first to the drainage system and then into stream sediments. The effect is essentially that of pollution of rivers and lakes with all its unpleasant ramifications. Included are contaminants from household products (phosphates and boron in laundry detergents and related products), raw sewage in some areas (copper and other elements as organo-metallic compounds in garbage; metals in urine and excreta), pipes (usually copper, lead or zinc, but nickel from stainless steel sinks has recently appeared in the waters from some Canadian cities), well casings in rural areas (usually zinc), fertilizers and fungicides for lawns, and so forth. Municipalities that treat sewage and garbage reduce immediate metal contamination, but the recent tendency to use this treated material as land fill, will eventually yield metals to the drainage basin. Frequently rivers below habited places are "dead" (low oxygen content) and, have modified pH and Eh conditions which affect the solubility of elements in natural water systems. Many municipalities in northern climates use large quantities of salt (both NaCl and CaCl) for snow and ice control, all of which eventually enters the drainage system and may effect the mobility of elements. Lead from gasoline additives has been determined to be present in vegetation and soils in relatively high quantities (100-1000 ppm), in some cases up to hundreds of feet away from the sides of major highways.

Even in more remote areas, contamination from vacationers, hunters and others can be observed. Exploration parties are known to have followed lead anomalies in water to find the source to be a lead storage battery. There is one story told of an exploration party which followed a geochemical stream sediment anomaly to a small airplane which had crashed in the mountains of British Columbia but, unfortunately, they were several years too late to be of any help to the occupants. In Mexico, one exploration party followed a lead anomaly in stream sediments to a rifle range. In another situation, an exploration party followed a zinc anomaly to a pasture frequented by moose; apparently moose excrement in northern British Columbia is high in this element.

Discussion

From the above case histories it is clear that contamination can appear from many sources, not all of which have been mentioned by any means. The analytical methods of exploration geochemistry are extremely sensitive and permit the detection of contamination even in low amounts. When planning to use geochemical methods in old mining districts, areas of industrial activity and intensive farming, preliminary or orientation surveys should be run to determine the degree of contamination (if any) and what soil horizon can be safely sampled without fear of contamination. In general, contamination of soil samples is less severe than it is in either stream sediments or water.

Relatively few studies have been reported on contamination, and this is perhaps understandable in view of the tendency to only report successes. An excellent paper by Hosking (1971) not only discusses contamination problems in relation to exploration in Cornwall, but offers encouragement in that it points to circumstances in which contamination in old mining districts may actually be of help, such as in locating parent lodes from ancient mining operations. Also on the positive side, is the large amount of data now available from pollution control agencies, such as the Inland Waters Branch (Environment Canada). This agency, like its counterparts throughout the world, has a vast amount of data obtained from water surveys, particularly for the past few years, including trace metal data. Those willing to obtain and digest this data, may not need to collect and analyze water samples, and they may be able to recognize favorable areas or some significant anomaly. Similar data on lakes is often available from various agencies concerned with fisheries and limnology and, in some cases, trace element data on associated sediments is also available.

It is sometimes possible to check on the likelihood of contamination by considering element associations. For example, cadmium is closely associated with zinc, and cobalt with nickel, in many geological oc- currences. If cadmium or cobalt correspond with a zinc or nickel hydromorphic anomaly, respectively, such anomaly would be less likely to be caused by contamination, as it is not probable that the associated element (e.g., cadmium) would be introduced as a contaminant in direct proportion to the other element (zinc, in this case). An actual case of how Zn/Cd ratios may be used to recognize contamination is illustrated by an example from the Coeur d'Alene district, Idaho. Anomalously low Zn/Cd ratios are found in the soils in the area directly east of Kellogg, Idaho, for a distance of about 6 mi. The high abundance of cadmium is derived from smelter fumes; apparently there is a fractionation of zinc and cadmium at the elevated temperatures of the smelting process. On the other hand, were the contamination to originate with mechanically (wind) derived grains, fractionation of zinc and cadmium would not take place, and Zn/Cd ratios could not be used to detect contamination.

ORIENTATION SURVEYS

In the discussions in chapters 2 and 3, it was shown that there are many variables in both the primary and secondary environments that have profound effects on the dispersions of elements, the detection of which is the basis of exploration geochemistry. As a result, every area in which ex- ploration geochemistry is to be used is likely to be different from a previously studied area in some way. Therefore, a preliminary study, called an *orientation survey,* should be conducted in every new area. The objective of an orientation survey is to determine the optimum field, analytical and interpretive parameters which can distinguish an anomaly from background. Some of the important parameters include: (1) the type of geochemical dispersion that exists in the area; (2) the best sampling medium; (3) the optimum sampling interval; (4) the soil zone and depth from which samples should be taken; (5) the size fraction to be analyzed; (6) the element or group of elements which should be analyzed, and by what analytical technique; (7) the effects of topography, hydrology, drainage, glacial history, climate, rainfall, vegetation, organic matter and Fe-Mn oxides on the dispersion of metals; (8) the upper limit of background values (threshold) in rocks, soils and waters; (9) the most ef- ficient manner of sample collection and analysis; (10) whether or not geochemical methods are feasible; and (11) if contamination is likely to be a factor in the area. Clearly the list of factors to be considered is as large as the subject of exploration geochemistry itself, and there is no point in reiterating it all here. Suffice to say that all factors so far discussed, and those that follow, must be considered during an orientation survey. The time to ponder all the variables, and to test them out, is *before* the actual detailed exploration begins, not after the samples have been collected. However, there may, of course, be some circumstances in which the area is so well known that orientation surveys are not necessary.

Most discussions of orientation surveys in the literature appear to assume that they are run on soil samples generally in the vicinity of known mineralization. There is no doubt that such orientation surveys based on soils are very important, but there are also stream sediment, vegetation and other types of orientation surveys as well. Before starting any orientation survey, the first factor to be considered is whether a reconnaissance or detailed survey is in mind, and the second is whether or not the area to be investigated is known to be mineralized.

Let us assume that an area is being considered for reconnaissance purposes, and that evidence of copper mineralization has been reported from time to time in reputable government publications. Immediately, a great number of possible variables come to mind such as: (1) the type of rock which might be host to the copper — if porphyry copper is the likely geological target, then molybdenum immediately comes to mind as a pathfinder in either soils or stream sediments, but if it is a vein system or massive sulfide deposit, then zinc and other elements would be more appropriate; (2) the copper content in the silicate minerals as opposed to that in mineable sulfides; (3) the climatic conditions — if the area is in northern Canada, hot extractable methods on sediments or soils would be preferred if field tests are contemplated, but in a tropical climate cold extractable methods could be used; (4) if there are lakes in the area, as in many parts of Canada, these waters could be sampled, but this would require a large sampling program to obtain background values for the lakes, as well as a knowledge of the variations in rock types in the area, again for background interpretation; (5) if the area is sufficiently remote from cities and previous or present mining activities, and is reasonably flat, consideration can be given to an airborne survey (mercury, for example), and so forth. The above discussion could be extended, but the point to be stressed is that there are many variables in a reconnaissance-type survey which only an orientation survey can help elucidate.

In the case of a follow-up, or detailed survey, the procedure is simplified somewhat because, in these cases, the objective is very much more finite. For example, strong zinc-lead stream sediment anomalies may have been narrowed down to an area between two drainage basins, about twelve sq mi in size, in a region with known zinc-lead-silver mineralization, and an average of 8 ft of residual overburden. The orientation survey would now set out to determine the geochemical characteristics of the known mineralization in the area which, it is stressed, need not be economic. Immediately, traverses would be conducted across the known mineralization, making sure that the sample spacing was appropriate to detect the mineralization, and that the traverse was of sufficient length to determine the true background metal values. Then a vertical soil profile would be obtained to bedrock, by trenching or coring if necessary, from both the known (anomalous) mineralized area as well as from unmineralized ground. The soil types would be described and samples representative of each zone would be taken, but if a certain soil zone was several feet in thickness, several samples, one representative of each foot would be collected. The samples would then be separated into various size fractions (e.g., − 80 and − 150 mesh), and analyzed for numerous elements by various extraction techniques (in this case a multi-element emission spectrographic analysis should be considered). Thus for this situation, the important variables of depth of sample (soil horizons), size fraction of the soil, and the most appropriate element (or group of elements) to be analyzed would be known, and this information would permit the best samples to be collected and analyzed so as to obtain the maximum contrast against background.

It must be remembered that, in some areas, owing to rapidly changing rock types, variations in topography and drainage, and other factors, soil profiles can show considerable lateral differences over small distances (Fig. 3-9). Accordingly, a sufficient number of samples must be taken during the orientation survey and studied sufficiently well to ensure that this possibility, if it exists, is understood. It is also necessary to collect enough background samples to give a reliable estimate of background and threshold values. Biogeochemistry, from what can be learned from many sources, should never be attempted without first running vegetation and soil orientation surveys. Orientation surveys in glacial soils introduce special problems, and it is important to understand the nature of the glacial material, its source, and the glacial history of the area if possible (Chapter 11).

In some exploration companies, it is difficult to convince management that the expense involved in an orientation survey is justified. However, if properly executed, such a survey can be extremely valuable in that it can increase the efficiency and permit more confidence in the interpretation of the data. Further, in the long run, it can result in a considerable saving of time and money by determining the optimum amount of analytical data required and ensuring proper sampling the first time. Actually, it is generally not necessary to complete an entire orientation survey at once, but each phase should be completed before each successive part of the overall project is undertaken.

Many basic studies and case histories reported in government publications, as well as in numerous journal articles, are research or preliminary in nature, and in this respect they are, in fact, orientation surveys. With this in mind, it is possible to find many excellent orientation surveys covering a wide variety of situations, which are worth further scrutiny. Some may be called by slightly different names, such as preliminary studies, pilot studies, or feasibility studies, but the result is the same. By way of examples, the following may be cited: Allan and Hornbrook (1971) who made a feasibility study of the value of geochemical exploration methods in an area of continuous permafrost in the Coppermine River area, N.W.T.; Bloomfield et al. (1971) who conducted orientation surveys over carbonatite complexes in Uganda, preliminary to major surveys; Chaffee and Hessin (1971) who evaluated biogeochemical methods in the search for concealed porphyry copper-molybdenum deposits in

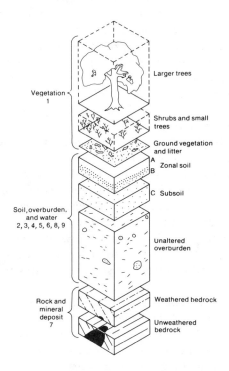

Fig. 4-3. The concept of landscape geochemistry as proposed by Fortescue (e.g., Fortescue and Hornbrook, 1967). The diagram illustrates some of the various environments and sampling materials of importance in exploration geochemistry. The various types of geochemistry represented are: 1. plant; 2. soil; 3. bog; 4. water; 5. stream sediment; 6. overburden; 7. rock; 8. pathfinder minerals; 9. boulder tracing, etc. This illustration may also be considered as a simplified representation of one part of the geochemical cycle. From Lang (1970).

Arizona; Reed and Miller (1971) who discussed an orientation survey in Alaska; and Plant (1971) who described in great detail a very thorough orientation study on stream sediments which was undertaken to define the regional geochemistry in northern Scotland.

Closely associated with orientation surveys is the concept of *landscape geochemistry*. This very useful term has been defined as "the study of the principles and patterns governing the circulation of chemical elements at, or near, the daylight surface of the Earth" (Fortescue, 1973, p. 10), and its utility has been demonstrated in many situations (e.g., Fortescue and Hornbrook, 1967). It is illustrated diagrammatically in Fig. 4-3, in which the various components of the landscape are shown in the form of a prism. The importance of the concept is that it stresses, like orientation surveys, the need to study systematically the geochemistry of the numerous sampling media in the vicinity of a mineral deposit, in order to fully interpret the results of geochemical prospecting methods. In addition, it illustrates that biogeochemical studies involve complex inter-relationships between geology, soil science, botany and ecology.

A well-planned geochemical exploration program is carried out in a logical, step-wise sequence, such that each step is more detailed and serves to locate mineralization more closely than the previous one. In this manner, the maximum area is covered at a minimum cost and most effort can then be expended on the most favorable areas, as the less favorable regions are eliminated at an early stage. Bradshaw et al. (1972), with full realization that local conditions may cause variations, and that geochemistry should always be used in conjunction with, and modified by, all available geological and geophysical information, have outlined the normal order of a reasonably complete geochemical exploration program:

1. Orientation survey (discussed above).
2. Reconnaissance stream sediment survey for rapid, low cost coverage of large areas.
3. Preliminary follow-up stream sediment survey to define more closely the target areas, and to provide further confirmation on regions whose importance was not clear from the reconnaissance results.
4. Detailed follow-up stream sediment survey to define as closely as possible the location of the source of an anomaly with respect to drainage channels.
5. Bank sampling which, if applicable, may reduce the area to be covered in the following (soil) survey by one half (Fig. 1-5).
6. Soil sampling, which is frequently carried out in several stages using finer and finer grids, to define the source of an anomaly as closely as possible.
7. Profile soil sampling to give more information as to the exact location of the anomaly.

8. Bedrock sampling to locate the deposit or primary halos associated with the deposit.

FALSE (NON-SIGNIFICANT) ANOMALIES

Under normal conditions, anomalously high concentrations of an element, or association of elements, indicate that economic mineralization may occur in a particular area. However, experience has shown that in some circumstances high concentrations of an element do not necessarily mean that mineral deposits of economic value will be found. In other words, high metal content cannot always be used as a guide to ore. High geochemical values which are related to ore are called *significant* anomalies, whereas those unrelated to ore are termed *false* or *non-significant* anomalies. Excluding those anomalies caused by any of the many types of contamination discussed above, or the erroneous anomalies which might result from analytical errors described in the next chapter, there are many examples of false or non-significant anomalies that have resulted from natural processes. For example:

1. High copper values (over 3000 ppm) have been found in soils developed on sedimentary rocks in south-central British Columbia having a background value of from 80-100 ppm, and yet there is no economic mineralization.
2. High zinc values (over 4000 ppm) occur in soils overlying limestones in north-central British Columbia but, after drilling several feet into the soils, the values rarely exceeded 100 ppm, and no mineralization has been found.
3. Anomalously high values for uranium and molybdenum have been reported in soils which overlie sedimentary rocks in different parts of the Colorado Basin, but no economic deposits have been found in the particular areas.
4. Anomalously high lead-zinc values are found in some soils of the mid-continent region (Kansas-Oklahoma) which overlie carbonate rocks but, in some localities, there is no economic mineralization.
5. Anomalously high gold values (0.4 oz. per ton) occur in the top one foot of the organic layer in soil overlying a copper deposit at the Mina Santa Fe, Pichucalco, Chiapas, Mexico, but, within the copper ore itself, gold occurs in much smaller quantities.

Additional examples could be cited from Australia and other localities, but the point is sufficiently well illustrated. They are all instances of false or non-significant anomalies. All these examples have one point in common, which is that they are developed only to a depth of a few feet in soils, but there are probably several mechanisms to account for this. Such mechanisms seem to be complex, and are related to pH-Eh conditions, groundwater effects, adsorption, formation of organo-metallic complexes, rock type, and climatic conditions, which may operate individually or together.

Some false anomalies in soils could be due to certain species of plants which preferentially concentrate a metal, and produce, upon death and decay, anomalous concentrations in the organic layer, certainly after several generations (Fig. 1-7). In other cases, reactions may occur between certain types of organic matter (including saps from specific types of trees) and selected metals such as gold, thereby reducing their mobility and causing anomalously high values. Clay-metal reactions are known to occur during the weathering process which can also decrease mobility and increase background values considerably. Mn-Fe oxides in certain types of soils may, for instance, adsorb anomalously large amounts of metals. Mobility may also be decreased by the alkaline nature (high pH) of limestone environments. Finally, in the reducing environment of carbonaceous sediments of all types, sulfate can be reduced to sulfide which precipitates out as metal sulfides. (In this connection, there are some who believe that many high organic soils with, for example, more than 500 ppm zinc, will be found to contain sphalerite upon detailed mineralogical study.)

It should be noted that the above possibilities are merely suggestions. Very little detailed work, such as differential leaching experiments, detailed mineralogical identifications, and isolation of any insoluble metal compound, has been done on the origin of the many false anomalies in soils. However, some observations of anomalously high metal values unrelated to mineralization have been reported from time to time. For example, Dyck (1971) has described the adsorption and coprecipitation of silver on hydrous oxides of iron and manganese in soils. Cannon (1955) observed abnormal concentrations of zinc in peaty deposits overlying the zinc-bearing Lockport dolomite in New York. Some reports of false anomalies are explained by the behavior of trace metals in bog or swamp environments in which the trace metals are associated with the organic matter (e.g., Gleeson and Coope, 1967; Usik, 1969). False anomalies have also been found in soils over pyritic bodies, where the pyrite itself not uncommonly contains as much as 1% copper, 0.2% lead, 0.2% zinc, as well as smaller amounts of cobalt, nickel and other metals, substituting within the mineral structure (Table 2-5).

Another type of false anomaly results from the sampling and analysis of high-background source rocks. This reflects geology, and changes in rock type can result in anomalies up to an order of magnitude greater than background. A well-known example of this is the high nickel, and to a lesser extent chromium, cobalt and copper, content of ultramafic rocks. As mentioned previously, the average nickel content of an ultramafic rock is 2000 ppm (Table 2-1) and, although soils derived from these rocks are relatively high in nickel, no economic mineralization occurs, as the nickel is mostly contained within silicate structures. Other examples of rocks (Table 2-4) with high background values (or residual concentrations) of metals include black shales (many metals, such as U, V, Cu), coal (Ge), and phosphorites (U, V, Mo, rare earths). Some of these false anomalies can be recognized by element associations, such as Ni-Co-Cr in the ultramafics, which tend not to be released and separated during the weathering process except after prolonged laterite formation, as in Western Australia. The easiest way to recognize these anomalies for what they are, is to super-impose a geological map on the geochemical map, and in this manner the effect of rock type becomes immediately evident. If geological maps are not available, as is often the case, an indication that the anomaly is not due to mineralization can be gleaned from the fact that values often form a "plateau" of relatively high and uniform values which may cover a con-siderable area.

Although the examples given above have described anomalies in rocks, or soil derived from them, false anomalies caused by changes in rock type can also be encountered in stream sediments. Clearly, the upper limit of background values (threshold) for the materials under discussion is significantly different from that for the area in general, and a "local thres-hold" (see below) must be determined before any significance can be at-tached to values in the high background area. There are also rocks with low metal background values that give rise to negative anomalies which, if they are not the result of natural acid leaching, normally do not have much utility in exploration geochemistry except to outline the occurrence of specific rock types.

Two interesting examples of false anomalies can be cited from Australia, where the thick laterite cover on a very flat surface makes it especially difficult to correlate soil (laterite) anomalies with underlying bedrock. In the first (Fig. 4-4), the result of sampling a high background (non-economic) source rock is shown. Here a false copper anomaly is illustrated which may be interpreted as a mafic copper-bearing intrusive, but actually originates from an intercalated black shale. The false anomaly can be recognized as such by analyzing for zinc, and as zinc is not normally high in mafic copper-rich intrusives, the anomaly is suspect. The low nickel content confirms the black shale source. This example not only illustrates a false anomaly, but also the importance of element associations discussed above (e.g., Table 2-4). In the second example (Fig. 4-5), a false anomaly related to the present topographic expression is illustrated. On the left side of Fig. 4-5, the trace element distribution of nickel in laterites is shown,

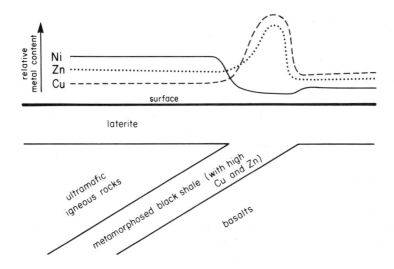

Fig. 4-4. False copper anomaly caused by sampling soil developed on a high background source rock (black shale). See text for details on how to recognize this as a false anomaly. Based on example from Western Australia.

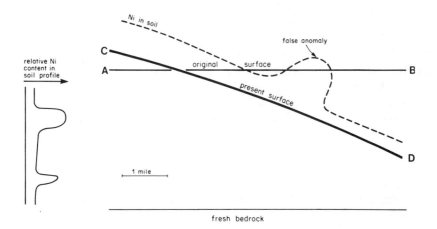

Fig. 4-5. Diagrammatic representation of a false anomaly related to topography, based on a common problem in Western Australia. A-B represents the original surface formed after the development of a thick lateritic soil. C-D is the present topographic expression (slope is greatly exaggerated). The relative Ni content in the soil profile is illustrated on the left. Nickel is usually found in high concentrations (up to 2%) in the upper part of the soil profile, and in smaller concentrations above the bedrock. A soil survey run parallel to the present surface (C-D) would yield a false anomaly, as the nickel would be in the form of silicates or oxides, or within iron oxides, but not as economic sulfides. This example would be analogous to sampling different soil horizons.

especially those being formed over an extremely long period of time. A soil survey conducted along the present topographic surface (C-D) would, in effect, be sampling several soil horizons. As a result, a false anomaly would be obtained. Although these examples are from Australia, similar situations can and do exist in many other localities.

Normally, false anomalies are found in soils or in the waters which drain them, but as mentioned above, they may also occur in stream sediments and any other sampling medium, even snow (Fig. 11-7). The presence of Fe-Mn precipitates in stream sediments which are able to adsorb (scavenge) metals such as cobalt, copper, arsenic, nickel and zinc has been discussed previously. If these precipitates containing adsorbed metals were to be included in the analysis of stream sediment samples, anomalously high values would be obtained for certain elements. Such situations may be classed as false anomalies caused by sampling errors and it is for this reason that Mn-Fe coatings should be carefully avoided during drainage basin surveys, or collected and analyzed separately (they can often be concentrated by ultrasonic techniques in the laboratory). Several studies have now been published on the false anomalies which can result from sampling errors due to naturally enriched Mn-Fe oxides. These include the study by Horsnail et al. (1969) on sediments from the United Kingdom, as well as one specifically related to molybdenum and copper in British Columbia (Horsnail and Elliott, 1971). Several metals are similarly enriched in bog manganese precipitates from drainage basin sediments in eastern Canada (Boyle et al., 1966). Perhaps the most spectacular is the occurrence in a seepage area called the "copper swamp" in Dorchester County, New Brunswick, where the swamp material runs as high as 5% copper, but the rock immediately below is unmineralized.

Regardless of the origin of false anomalies, they must be distinguished from significant anomalies. This also applies to the equally difficult problem of recognizing false anomalies formed only in part by the mechanisms mentioned above. False anomalies originating as a result of soil-forming processes are the most common. They can often be recognized, or suspected, by the fact that they cover extensive areas by comparison with what might be expected for a reasonable ore deposit. In other words, when a geochemical soil anomaly extends for several miles in each direction, caution is always warranted. It is worth noting that many of the false anomalies in soils enumerated above have been found in areas with known mineralization, and that many of the anomalies appear to be developed on or near rocks of higher than normal (world average) metal content. It might be logical, therefore, to concentrate exploration in these areas on the assumption that somewhere in the area a significant anomaly can be found.

Up to this point consideration has been given primarily to false anomalies developed in soils in direct or close association with the rocks from which the metals originated. But, as has been mentioned previously, geochemical anomalies are commonly displaced (Fig. 1-3) either by mechanical or by hydromorphic means, and this also applies to false anomalies (Fig. 4-6). Groundwater plays a major role in the migration of metal-containing fluids and, as a result, displaced false anomalies may be formed many miles distant from the source of the metals. Some secondary dispersions, for example break-in-slope accumulations, are extremely difficult to evaluate, in some cases. Certainly one of the most important factors is a knowledge of the direction of groundwater movement. It is necessary, nevertheless, for the exploration geochemist to determine whether such accumulations are significant or false anomalies, which often means finding the source of the metals.

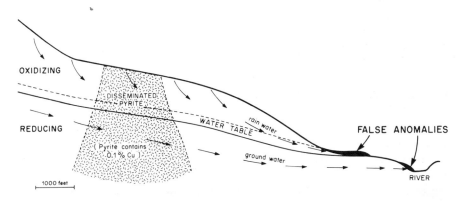

Fig. 4-6. Diagram illustrating a displaced false anomaly at the break-in-slope, and in river sediments, resulting from the leaching of copper-bearing pyrite (0.1% Cu) by meteoric and groundwater. The high values of copper are encountered where these waters come to the surface, but because the copper values originate from an uneconomic source, these are called displaced false anomalies. See also Fig. 1-3.

Finally, there exist many examples, in Canada and elsewhere, where soil and stream sediments over known mineralized deposits do not yield geochemical anomalies. This is referred to as masking (sometimes as "blank out"), and is, in essence, the opposite of a false anomaly. At this point it need only be mentioned that masking effects present serious problems which must be recognized for their significance. Some of the many proven explanations include the fact that the mineralization is covered by glacial debris or alluvium. In other cases, bentonite beds above the mineralization are effectively impervious to secondary (hydromorphic) dispersions, as are caliche zones within the soil profile. In all the instances mentioned, soils which have the appearance of residual soils may be developed on the debris or alluvium, or above the impervious layers. In situations such as these, a thorough understanding of the Pleistocene and Recent history of the area is extremely valuable.

INTERPRETATION OF GEOCHEMICAL DATA

Introduction

From the previous discussions it is clear that exploration geochemists do not search for ore directly, but rather for indicative anomalies. As soon as the analytical results are obtained, whether they be from field determinations or from a distant laboratory, this information should be screened visually and then plotted on a map (e.g., Figs. 1-1, 1-4, 1-9). These simple methods probably will disclose any obvious anomalies but even so those interpreting data as anomalous must be constantly aware of possible pitfalls. Three general possibilities exist with regard to any anomaly: (1) it is genetically related to a mineral deposit or concentration of hydrocarbons; (2) it is genetically related to sub-economic accumulations of minerals or hydrocarbons such as a large low-grade deposit; (3) it is due to a concentration of elements resulting from any one or a combination of factors (false anomalies, sampling or analytical errors, contamination, etc.) which do not represent mineralization. On the other hand, the absence of an anomaly does not necessarily mean that an economic metal deposit does not exist in an area being explored. Such absence may mean: a low rate of weathering of the mineral deposit is taking place; the metallic deposit occurs either too deep, or below an impervious layer that prevents the transportation of metal ions by groundwater; or, in the case of stream sediments and water, dilution or precipitation of metal ions is taking place somewhere above the sample site.

The decision as to the true meaning of geochemical data is probably the most difficult part of exploration geochemistry. Even assuming that a geochemical sampling program has produced anomalous values genetically related to a mineral deposit (possibility 1, above), the question still arises as to how to recognize the data as anomalous when, more often than not, the anomalies themselves are not obvious. This is especially true in the case of stream sediment or soil reconnaissance surveys, and to a lesser extent for detailed surveys, attempting to find large, low-grade deposits. A case in point would be the difficulty in recognizing true anomalies from porphyry copper deposits containing about 0.5% copper in the Highland Valley of British Columbia. Here some of the first companies in the area are reported to have dropped certain properties which are now mines because of their inability to distinguish anomalous from background values obtained from soil surveys. Another very common problem concerns the interpretation of one or a few high values from stream sediments obtained during regional reconnaissance drainage basin surveys. One of the questions often asked of geochemists is, "What do these two high values mean?" Unfortunately, there is no simple answer to the question, nor is there a guaranteed way to correctly, and unequivocally, interpret the results of a geochemical survey.

The correct interpretation of geochemical data requires that every available piece of information be brought to bear on the problem, together with a thorough understanding of the fundamental principles of exploration geochemistry as applied to the specific area. These include the geology of the area, trace element abundances in different rock types, primary dispersion, groundwater and surface water movement, climatic conditions, type of weathering, soil formation, glacial movement, type and distribution of vegetation and humus deposits, Eh and pH conditions, mobility of the elements, characteristics of the sampling medium (e.g., soils, sediments, waters), sampling methods employed, and the analytical methods used to obtain the data. If one factor were to be singled out for special consideration, it would be the geology (in its broadest sense). A thorough knowledge of the geology would include such items as rock types, nature of weathering, soil formation, mineralization, structural geology and glacial history of an area; such information is the foundation for reliable interpretations. Regrettably, the geology of many areas in which exploration geochemistry is used is poorly known, and this adversely affects attempts at interpretation. Of course, if geophysical or other types of data are available, they should be incorporated in the interpretation process.

At this point it is necessary to discuss some fundamental concepts which are essential to the interpretation of geochemical data. These concepts include: maps and diagrams, background values, threshold and anomalies. They are discussed here in only the most general terms. The more advanced statistical techniques and computer approaches to interpretation are discussed in Chapter 12. The presentation follows the normal sequence used in everyday interpretation, that is, the preparation of maps and diagrams, the determination of background and threshold values, and the recognition of anomalies. Both histograms and frequency distribution curves are normally prepared as well.

Maps and Diagrams

The data obtained from geochemical surveys are commonly presented as maps or diagrams. These methods greatly assist geochemists in the interpretation of the data and, at the same time, provide a means of permanently recording and storing the data in a concise form.

Geochemical maps, usually of potentially interesting areas, are now published by numerous governmental agencies, and these can be used as examples of proper data presentation. Generally, the results obtained for each element are plotted on a separate sheet, but should it be possible to include the results for several elements on the same map by means of appropriate symbolization, this is a desirable convenience. Alternatively, overlays for each element, as well as overlays for geological and other features, can be prepared. Normally the analytical results from each sampling medium (e.g., soils or stream sediments), and for each soil horizon, are placed on separate maps, but occasionally as in Fig. 1-4, it is possible to incorporate the results from two sampling media without causing undue difficulties in comprehension. One of the main advantages of plotting data on a map is that geological, topographic and other types of information (such as possible sources of contamination) also may be included to facilitate interpretation.

Geochemical data obtained from stream sediment surveys are plotted on maps of the stream courses, at each sampling point, by any one of several methods to indicate element concentration (Fig. 1-1). Data from soil and vegetation surveys, and glacial dispersion trains, are also plotted at each sampling point, but the results of these surveys generally are also amenable to contouring, color shading, and other visual aids to assist in interpretation (e.g., Figs. 1-1, 1-9). Contours of equal metal content are called isograds. Hawkes and Webb (1962) state that it is usual to select the contour interval so that the threshold value (see below) is taken to be the first contour, and above this, contours are simple multiples of the threshold value. Thus, if the threshold is 200 ppm, contours above this would be 400 ppm, 600 ppm and so forth. Then all background values are blanketed below the threshold value. There are, however, many instances where it will not be possible to contour geochemical data, such as when the sample spacing is too great, multiple soil horizons are sampled, or when the data is discontinuous. The best visual effect in these situations is a series of dots, each representing a range of values (see Figs. 1-1, 1-5).

Hydrogeochemical data may be plotted in any one of several ways. Data from stream and spring waters are shown in the same manner as stream sediments (Fig. 1-4) and the direction of stream flow is always indicated. Lake waters (Fig. 1-6) can be contoured either in a vertical profile or in planar configuration, although both methods require more data than are usually obtained. Data from groundwater are also contoured if sufficient information is available. Although the metal content of the sampling medium is usually plotted, ratios of metals (e.g., Cu:Ni), or the ratio of metals to a major constituent (e.g., $Cu:SO_4$ in waters), may be used for all sampling media, if this is more meaningful or significant.

Experience has shown that in addition to geology, for convenience, topographic contours superimposed on a geochemical map can be of great value in interpretation. Other features of importance are the surface drainage patterns, areas of groundwater seepages, and the location of lakes, springs, swamps and flood plains. These assist in the interpretation of hydromorphic anomalies (e.g., Fig. 1-3). In glacial areas the direction of glacial movement should be indicated. In lake surveys the depth of water (bathymetric maps) and the flow of currents should be shown, as this may assist in interpreting the source of metal-containing sediments or groundwater seepages.

Data may also be plotted as profiles where it is advantageous to indicate the distribution of elements along a single traverse, or where the sample spacing is too far apart to permit contouring. Examples of profiles are illustrated in Figs. 1-2, 1-3, 1-8, etc. Although the metal contents are usually drawn on arithmetic scales, logarithmic and semi-logarithmic scales are commonly used where there is a large difference between the anomalous and background (high and low) values.

Background

Background is defined as the normal range of concentrations for an element, or elements, in an area (excluding mineralized samples). Obviously, before anomalous conditions can be recognized, it is necessary to establish background values against which they can be compared. Background values must be determined for each element, for each area, and for each type of rock, soil, sediment, and water. From the data presented in Table 2-1, it can be seen that the background values for each element will vary significantly between rock types. And even within rocks called by the same name, the abundance of any element can vary significantly. As mentioned in the discussion of geochemical provinces (Chapter 2), certain granites have higher than normal accumulations of copper and molybdenum, others are high in tin, whereas granites are also known which appear to be depleted in these elements. It is for these reasons that a knowledge of the geology in an area, and especially the rock types, is of such extreme importance in the interpretation of geochemical data.

In order to determine the background values in any given area, a relatively large number of samples of the materials to be used in the geochemical survey should be analyzed. These may be soils, stream sediments, rocks, waters, and so forth, but obviously mineralized materials should be excluded, or at least considered separately. The values obtained will show a large scatter, or range, rather than a fixed value, as only rarely will element distribution be uniform in geological materials, even those of the same type. This process must be repeated for each new area to be investigated.

Although the range of values obtained by analyzing a large number of samples will be large, the most frequently occurring values tend to be found within a relatively restricted range. This distribution of values is discussed more fully in Chapter 12. Suffice to say at this point, that the restricted range (or modal value) is generally considered to be the normal abundance, or background value for that particular element, in that specific rock type (or stream sediment, soil, water, etc.), in that specific area. Table 2-1 can be considered as a guide to normal background values for rocks, soils and waters and Fig. 3-13 presents the range for some trace elements commonly found in soils. It must, however, be stressed that economic mineralization can be represented by values which appear to be "normal" based upon these data. This is possible, among other reasons, when the background values in the particular area are much lower than the so-called normal background value, or when other influences mask the true background value.

As a first approximation, it is sometimes possible to determine the background values for many elements in rocks by analyzing the residual soil or glacial till which cover them. However, the nature of the soil forming processes, particularly leaching, must be kept in mind when attempting to extrapolate trace elements in soils to background values in bedrock. In the case of glacial tills, a few studies (e.g., Bayrock and Pawluk, 1967; Pawluk and Bayrock, 1969) have shown that they are often locally derived, at least for the purposes of reconnaissance surveys. However, orientation surveys should confirm the validity of such an approach to the determination of background values.

Background values for vegetation are extremely variable from one locality to another and orientation surveys must be run, or large numbers of samples collected during a survey, so that background values can be determined for the specific area of interest.

Threshold

Threshold is defined as the upper limit of normal background values. Threshold values, like background values, will vary for each element, in each rock type, and in each area. Values higher than the threshold are considered anomalous and worthy of careful scrutiny. A statistical analysis of the sampled data quite often allows a threshold to be more precisely defined.

In mineralized areas there are likely to be two background values and, therefore, two threshold values. These are called (1) *regional threshold* which is based on the normal (regional) background discussed above, and (2) the *local threshold* which is related to a local background. The local background gives higher values (negative local backgrounds are generally of no concern in exploration geochemistry) and generally is in the vicinity of a primary dispersion halo of an ore. The concept is illustrated in Fig. 4-7. From this figure it can be seen that the regional threshold can be considered a "plain", the local threshold a "plateau", and the anomalies are represented as peaks. The terms *contrast*, or *geochemical relief*, refer to the degree of variation between the local threshold and the peaks. In the absence of a local threshold, the contrast would be between the regional threshold and the peaks. The definition of local thresholds, and the distinction between local and regional thresholds, are of great importance in the planning and interpretation of geochemical surveys. Orientation surveys are generally able to assist in the determination of these values.

An actual case in which a regional threshold may be higher than an anomaly is illustrated in Fig. 4-8. In the example shown, the uranium content of surface waters draining Paleozoic sedimentary rocks (mainly carbonates and shales) averages about 2 ppb, whereas those surface waters (mainly lakes) over Precambrian rocks average about 0.4 ppb. Values for lake waters in the Precambrian areas above 1.2 ppb are considered anomalous. Such values are significantly lower than the regional background to the west.

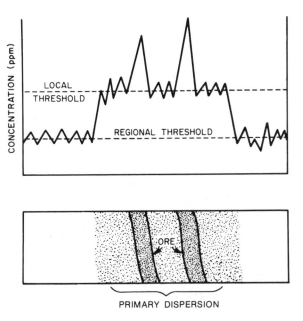

Fig. 4-7. Diagram illustrating regional and local threshold.

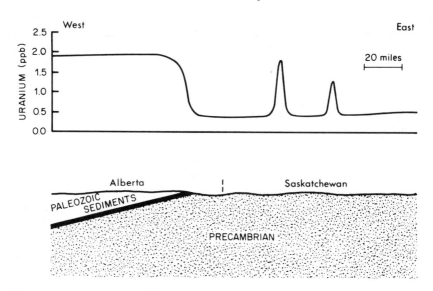

Fig. 4-8. Diagram illustrating a change in the regional background of uranium in surface waters on different rock types. Based on lake and river waters in northern Saskatchewan and northern Alberta, Canada, at about 60° latitude. Two significant anomalies are indicated over the Precambrian rocks.

Anomaly

An anomaly is defined as a deviation or a departure from the norm. From the point of view of exploration geochemistry, Hawkes (1957) defined an anomaly as an area where the chemical properties of a naturally occurring material indicate the presence of a mineral deposit in the vicinity. Other definitions will stress that an anomaly is an unusual measurement (or value), as opposed to an area, which is believed to indicate ore. Because most anomalies are determined by means of the interpretation of analytical data, the concept of an anomaly as a measurement, or an abundance which deviates from the norm, is more in line with current usage in exploration geochemistry.

Once the background has been determined for a specific element in an area, an attempt can then be made to recognize the anomalous values (or area). Several rules of thumb have been proposed. Boyle (1971a) suggests that samples which contain amounts of elements twice background, or more, are generally considered to be anomalous. Others determine the threshold empirically by means of an orientation survey, or by statistical methods, and then assume all values greater than this are anomalous. Hawkes and Webb (1962), for a single population of values (for example, values for one element from one type of rock), suggest that the threshold for that material may be conveniently considered to be the mean plus twice the standard deviation, which is equivalent to saying only 1 in 40 of the samples is likely to exceed the threshold value. Alternatively, if there are insufficient data to compute the mean, or if the statistical distribution is irregular, Hawkes and Webb (1962) suggest the best approximation is to take the median value as background, and to consider only the highest $2^{1}/_{2}$ percent as anomalous (or above the threshold value).

The discussion to this point has considered only single populations (and eliminated false anomalies from consideration). In practice, geochemical surveys are conducted in an environment of multiple populations, such as is usually found in a regional drainage basin survey. Where the data are made up of two or more populations, great care must be taken to recognize and isolate each population. Statistical methods, such as cumulative-frequency plots, can assist in this process.

A classic, and to an extent extremely simple, case of multiple populations which, if not recognized as such, can result in an erroneous interpretation, is illustrated in Fig. 4-9. In this figure the frequency distribution of nickel in soils from an area underlain by shale, sandstone, and an ultramafic rock are shown. The average nickel contents of the rocks are, respectively, about 50, 200 and 2000 ppm. The average values for the shale and the ultramafic rock are close to the average abundances to be expected for these rocks (Table 2-1) whereas the average value for the sandstone is anomalous. Thus, the anomalous population is not the one with the highest values, but rather the one from the sandstone, which is possibly sulfide-impregnated. Multiple populations are often encountered in exploration geochemistry because of the great variability in rock types. Large areas which are underlain by homogeneous rocks are relatively rare.

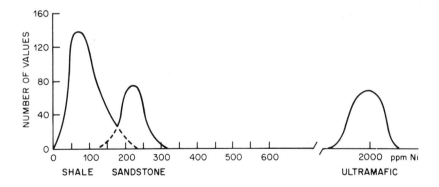

Fig. 4-9. Hypothetical frequency distribution of Ni in soils from three populations. The values from the shale and ultramafic rock are normal for these rock types, whereas those from the sandstone are anomalous.

When discussing the distribution of elements, such as those plotted in Fig. 4-9, it was assumed that the arithmetic values (in ppm) would give normal (bell-shaped) distributions, from which an "average" abundance for each element would be obtained. In actual fact, when histograms of the data from many (but not all) geochemical programs are plotted arithmetically (with linear coordinates, as in Fig. 4-9), a skewed distribution will often result, as in Fig. 4-10 (left). Such a distribution is of little value in establishing background, threshold, and in the recognition of an anomaly. When the logarithms of the same data are plotted, as in Fig. 4-10 (right), an entirely different distribution will be found. The log distribution has a shape which is expected in rocks, and is useful in the interpretation of geochemical data, that is, the symmetrical bell-shaped curve.

In the case of the log plot illustrated, the median of the data corresponds approximately to the geometric mean (average), and the population is said to be lognormally distributed. Instead of plotting histograms, it is possible to calculate whether the data are normally or lognormally distributed. If there is agreement between the median and geometric mean (Fig. 4-10, right), the distribution is lognormal, whereas it is normally distributed if there is a correspondence between the median and arithmetic mean. In the case cited, the median was 55 ppm, and the arithmetic and geometric means were 380 ppm and 64 ppm respectively, hence the data are more closely lognormally distributed. (The geometric mean

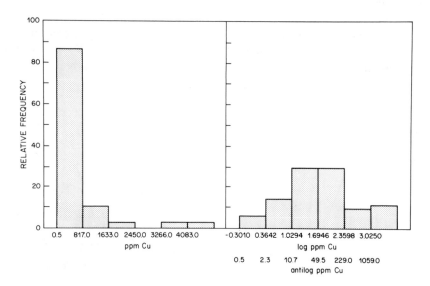

Fig. 4-10. Histograms illustrating the differences obtained by plotting geochemical data arithmetically (left) and by means of the log (base 10) of the same data (right). Based on 57 copper analyses from Canada. The class limits were set by a computer program.

was obtained from antilogs of the log ppm data.) Again, it must be assumed that just one population was involved.

The problem of whether the abundances of chemical elements are normally or lognormally distributed in rocks has been debated for about 20 years, since Ahrens (1954) first suggested that the concentrations of most elements in igneous rocks are lognormally distributed. Ahrens' paper stimulated many discussions in the literature, among them is the article by Shaw (1961) which has attempted to reconcile some of the different points of view on this subject which, to this day, is still a matter of controversy. From the point of view of exploration geochemistry it is sufficient to know that the problem of normal versus lognormal distribution exists, and that in order to get a normal (bell-shaped) distribution, the logarithms of the data may have to be plotted.

Multiple Populations

Mixed populations are very frequently encountered in geochemical exploration. These may result from any number of factors. For example, different rock (soil, vegetation, etc.) types may be included in a set of samples collected; a hydromorphic dispersion halo may be superimposed on a physically dispersed anomaly; or samples representing mineralization may be included with samples containing only normal background values. It is

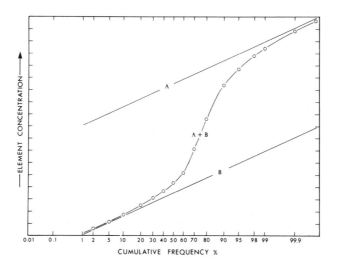

Fig. 4-11. Hypothetical cumulative frequency diagram of a multiple population. A is the anomalous population; B is the background population. A + B is the mixed population. From Andrews-Jones (1968).

sometimes possible to recognize these mixed populations by plotting cumulative frequencies on logarithmic probability paper. This is illustrated in Fig. 4-11. Tennant and White (1959) showed that if the cumulative frequency is plotted on probability paper, each population (A and B of Fig. 4-11) will be represented by a straight line or a curve. The intersection of the lines or curves represents a point which in some circumstances, may be taken arbitrarily as being the threshold between background and anomalous values (A + B in Fig. 4-11). The populations will appear either as straight lines or curves depending upon whether their distributions (normal or lognormal) are the same as the type of scale (linear or logarithmic) used as the ordinate of the probability paper. Fig. 4-12 (left) shows that the distribution of lead is from one population, whereas that of zinc (Fig. 4-12, right), determined on the identical samples, has two populations. A possible explanation for the two zinc populations is that about 20% of the zinc is groundwater transported, whereas all of the lead, copper and the remainder of the zinc is physically transported, as clastic grains. The threshold in this case is at the 80th percentile.

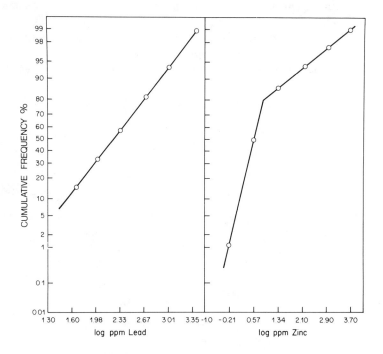

Fig. 4-12. Distribution of lead (left) and zinc (right) determined on the same 57 samples on which copper was determined (Fig. 4-10). A cumulative frequency diagram for copper (not shown) indicates it has one population. See text for further details.

Computers

Because large numbers of samples are often collected in geochemical surveys, and many elements are often determined on each sample, statistics and computers have been employed extensively in recent years. These techniques (Chapter 12) are well suited for processing large amounts of data, presenting the data in graphical and map form, and otherwise assisting in the interpretation. Many of the simpler calculations discussed above can be programmed on small desk computers.

Despite the notable advances made in the application of statistics and computers to exploration geochemistry, there is still no substitute for sound geological judgment. Statistical and computer techniques should be regarded as aids to the geochemist to assist him in making more informed decisions. It must be remembered that geochemical anomalies, and numerous mines, were found long before computers came into being, and since then as well, by virtue of the proper interpretation of available geochemical data.

The value of statistics and computers should not be over-emphasized by believing, for instance, that more data (numbers) for the computer to work with will automatically ensure better decisions, or that the computer will always be able to recognize an anomaly. This point is illustrated by the 100,000 Cu, Pb and Zn analyses recently obtained by a government survey in a South American country for which no geological or other supporting information was available. Because of the lack of additional information, very little could be done with the data by computer techniques that hadn't already been done by the simple methods described above.

Practising geochemists vary in their opinion as to the role of statistics and computers in exploration geochemistry. None will argue about the advantage of speed in calculations, the value of visual display of the data, or in checking the precision of analyses. There are those who say that once the data have been plotted on maps, histograms and frequency distribution curves have been prepared, and the mean and standard deviations have been calculated, most (perhaps 80%) of the value of statistical and computer techniques has been realized. Others, in basic agreement with the above, will say that for reconnaissance surveys computers are of value only for mathematical assistance, and that computers are mainly valuable for the analysis of masses of data at the follow-up stage, or perhaps in finding extensions of known deposits on the same structure.

The truth of these statements is difficult to ascertain. There can be no doubt that with the *proper* kind of data, statistical and computer techniques can be of great assistance in interpretation. Very often there is considerable redundancy in the data and with these techniques the redundancy can be removed. However, the crucial criteria as to the usefulness of computers depend on what kind of data are available. If only Cu, Pb and Zn values are available, maps, histograms and frequency distribution plots, and the determination of the mean and standard deviations, probably will yield as much information as is reasonable to obtain. Even with the six elements commonly determined in Canada (Cu, Pb, Zn, Ni, Ag and Mn) experience has shown that advanced statistical and computer techniques usually add little to the simple interpretation methods discussed above. However, if it is possible to relate the analytical results to geology, interpretations can often be extended significantly.

Several practising geochemists with experience in computer techniques were recently asked (by the author), "What percentage of the typical geochemical surveys presently being run can realistically and effectively profit from the more sophisticated statistical and computer techniques?". Their answers left no doubt that, in their experiences, to date very few surveys were candidates for the advanced techniques. The purpose of this discussion has not been to downgrade the role of statistics or computers in exploration geochemistry but rather to emphasize that it has a definite place in the overall geochemical program. The success of statistical analysis depends on how critical the variables are to the phenomena, and how well they are measured.

5 / Field Methods

The three fundamental parts of exploration geochemistry are: (1) sample collection; (2) sample analysis; and (3) interpretation of results. Certainly all three aspects are essential to the success of a geochemical exploration program, but should there be errors in analysis or interpretation, these can be re-checked or re-interpreted at a fraction of the cost of re-sampling an area. In fact, cost analysis will reveal that sample collection is by far the most expensive part of any geochemical program.

In this chapter general comments are presented on field procedures, particularly as they relate to the collection of samples. It is almost impossible to discuss all the ramifications and details of sample collection, as specific environments and situations will require modifications of any generalized method. Some of the more important items always to be considered are: (1) the best sampling material for the element being sought; (2) the optimum sampling pattern; and (3) the sample spacing which will reveal the presence of an ore body of the desired size. These points are most conveniently discussed in terms of the type of material being sampled. Only the major types (stream sediments, soil, bedrock, water) are considered in detail. Methods for collecting unusual materials, or materials requiring special equipment (e.g., air), are not discussed. The requirements for the collection of vegetation are considered briefly in Chapter 10, and glacial and bog materials are discussed in Chapter 11.

STREAM SEDIMENTS

Stream sediments, specifically silts and clays, are the basis of most drainage basin surveys, and as such sediments are considered to be representative of the whole (or part) of a catchment area, collection of the correct material is essential. Usually 50 grams of the material is sufficient for analytical purposes. If it is necessary to include coarser material because of the lack of fines, a larger amount may be necessary, but under no condition should pebbles or twigs be included. To ensure sufficient material of the correct size, it is advisable to sieve the wet sediments as they are collected. A series of screens may be necessary if there is much material of different sizes. The minus 80 mesh screen size is usually preferred, but an orientation survey should confirm this. If the heavy mineral content of a stream sediment is of interest, a special effort must be made to collect material as close to the bedrock as possible, as this is where the heavy minerals will be most abundant. If the bedrock is buried under much alluvium, augering or digging a pit may be necessary.

Stream sediments are usually collected by hand, or with a plastic or aluminum scoop (kitchen utensils or garden tools are satisfactory), just below the top of the sediment layer, and placed wet in small, high wet-strength, metal-free Kraft (brown) paper envelopes (these are often supplied free by commercial geochemical laboratories). Alternatively, brown Kraft bags available from grocery stores are acceptable provided they do not have trace metals either within the paper itself, or in the glues used to seal them. Envelopes with metal clips, even if covered with paper should not be used, if possible. Cloth sample bags which are preferred for rocks (because the sharp corners can cut plastic or paper), if previously used, are potential sources of contamination. Some types of cloth bags are impregnated with potentially contaminating substances including talc, pyrophyllite and kaolin minerals, and their use should be avoided. In all cases the samples may be dried without removal from the envelope, usually being hung out or placed in the sun at a base camp, but in the laboratory they may be dried in an oven. (Samples, generally soils or rocks, intended for the analysis of mercury or other volatile elements, must not be placed in the sun or in an oven). The use of polyethylene bags, and plastic or glass jars, is discouraged as drying is difficult, and under the oxygen-poor environment, various types of changes related to bacterial activity may occur. This may result in changes to the original metal-silt bonding, and make some types of interpretations, for example, cold extractable metal values, more difficult. The same restriction on plastic bags and glass jars applies to soil samples (see below). Many geochemists choose to pre-number sample bags before leaving for the field, or else they use pre-numbered tape which can be torn off as samples are collected. These procedures tend to ensure sequential numbering.

Stream sediments should be collected from as near the middle of the stream as possible, so that the sample will be representative of the entire drainage area, and the location documented. Initially the location need only be accurate enough to permit revisiting anomalous areas but it is sometimes desirable to mark the sample location with (biodegradable) tape, although this is a certain indication to others of exploration activity in the area. Every effort should be made to avoid collapsed bank material as this is not representative of the drainage basin (Fig. 5-1). Some geochemical parties, realizing the significance of a drainage basin sediment sample, will collect sediments from a dozen sites within a 10-50 foot radius of each sample location. They will collect a few grams each from stagnant pools, under and behind boulders, and in the main stream bed, to obtain a representative 50 gram sample. This care is to be commended. Organic

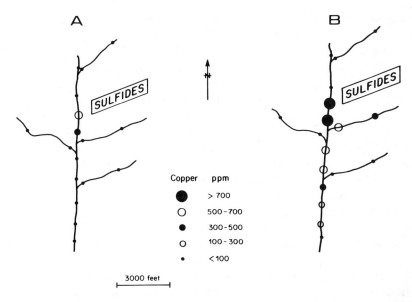

Fig. 5-1. Diagram illustrating the result of incorrect (A) and correct (B) stream sediment sampling. A. Data obtained from samples collected close to the bank of a fast moving stream. Although the samples contained sufficient fine-grained (—80 mesh) sediment, it was mostly collapsed bank material. The anomalous values extend only a limited distance from the mineralized area, which might well have been detected visually. B. Data obtained from correctly sampled, active sediments, in the middle of the stream, without included bank material. The anomalous values can be recognized at least a mile from the mineralization. Based on an example from British Columbia.

matter should be carefully avoided because it often has high, erratic metal concentrations related to adsorption, and also because it interferes with some analytical methods, particularly the colorimetric techniques. Yet in some situations organic matter, and manganese and iron oxide coatings, may be scraped off rocks in the stream bed, and analyzed separately because, owing to their scavenging effects, they can yield indications of metals not otherwise recognized (discussed in Chapter 3). However, interpretation of results may be difficult, as experience has shown that many apparently anomalous concentrations of metals in these materials are not related to mineralization (false anomalies).

In mountainous areas where stream flow is rapid, as in certain parts of British Columbia and Alaska for example, streams may contain much suspended matter but there is little easily recoverable sediment in the stream bed. In such situations it is possible to collect the sediment-laden water in appropriate vessels (e.g., milk-shake containers), set them aside to settle, and, upon return along the same route to a base camp later in the day, collect the sediment which should have settled to the bottom of the container, by merely decanting off the top layer of water.

Generally during reconnaissance drainage basin surveys in North America between four and six samples are taken per mile (although the sampling density can vary considerably) and, in addition, every confluence is sampled. Samples are always taken immediately upstream of confluences as illustrated in Fig. 1-4. During follow-up studies of a drainage basin, the sampling density will be increased and the general rule of thumb is that at least two contiguous anomalous samples are necessary for mineralization to be considered significant. In those areas with streams that flow only intermittently, such as the arid regions of northern Mexico, the western U.S. and parts of Australia, the fine-grained stream sediments may be sampled either during the wet or dry season. Also in any area where streams transport clastic sulfide grains, whether it be in a desert or in the Yukon, sediments may be collected at any time. However, where metals are transported only in groundwater, it is necessary for the water table to intersect the stream, at least occasionally, and if this occurs, then the sediment may be collected whether the stream is wet or dry. Lake sediments, swamp sediments, sediments behind beaver dams and other natural obstructions such as at outlets of cirques, in which either clays with adsorbed metals, or detrital sulfide grains are trapped, can be profitably collected as part of a drainage basin survey. Lee (1965) has sampled eskers, which, as part of a glacier's drainage, may perhaps be considered as a special type of drainage basin sample. Occasionally it may be desirable to separate the heavy minerals (e.g., detrital sulfide grains) from drainage basin sediments to detect or enhance anomalies, in the way described by Gleeson and Cormier (1971) and Garrett (1971) in various parts of Canada.

In recognition of the high cost of sampling by comparison with the analytical and interpretation phases of a drainage basin reconnaissance survey, as for example in some of the more remote parts of Canada, there has been a recent tendency to increase the productivity of sampling parties, and the size of the area covered by means of helicopters (e.g., Dyck, 1973; Allan et al., 1973). Several such surveys have been carried out in which as many as 16 stream sediment samples per hour (about one every 4 minutes) are reported to have been collected for 6 hours per day for several days, at an interval of about four samples per mile. In situations such as this, the exploration party must be experienced, and have all aspects extremely well-coordinated. Details such as pre-numbered sample bags, and pre-selected sample locations from study of topographic and geologic maps, must be completed beforehand. Nevertheless, the careful and thoughtful selection of sediment samples should not be sacrificed for mere speed with resultant false economy. In more rugged topography, where helicopters cannot land, there have been requests to at least one western Canadian university where there are students seeking summer employment in exploration geochemistry, for men willing to be lowered on specially designed buckets to collect stream sediment samples while the helicopter hovered overhead. Fortunately, there were no acceptances. Pontoon-equipped helicopters and airplanes have been successfully, and safely, used in sampling sediments in Canadian lakes. Soils, sediments, rock fragments and vegetation samples have been collected on the run from fixed wing, light aircraft in some areas, such as on the Seward Peninsula, Alaska (Sainsbury et al., 1973).

SOILS

Soil sampling differs in several very important respects from stream sediment sampling. It is generally employed in follow-up surveys where a drainage basin reconnaissance survey has isolated an anomaly, or narrowed an anomaly down to a particular watershed, or where geophysical, radiometric and other methods have indicated the possibility of mineralization. However, soil surveys have also been successfully used as a reconnaissance technique in flat, arid areas where other methods are not applicable, for example in parts of Australia. Soil surveys are almost always conducted on a grid system, preferably square but usually rectangular, with sample locations between 50-200 feet apart for the follow-up survey and from 1000-5000 feet apart for the reconnaissance survey. The line and sample spacings, especially for the follow-up survey, must be selected so that they will cross the strike of an ore body, and outline it. Therefore, the probable size and shape of the body will be a factor in determining the grid system, but it is obvious that the more detailed the sampling, the more confidence one can place in any interpretation. In the initial phase of soil sampling, the grid should be selected so that at least two adjacent lines will intersect any ore body, based on the probable minimum economic length in that area. Sample points along these lines should be spaced at intervals such that at least two points fall on each line within the anomaly. In places where the veins are discontinous, or the ore body not homogeneous, a closer sampling interval will be necessary. Base maps should be prepared, indicating such information as geology, soil types and topography, so that the analytical data can be properly interpreted. Very accurate locations of the sample sites is only necessary in the more detailed stages of exploration. Initially, pace and compass traverses are usually sufficient.

The importance of the correct sampling interval is illustrated by data from Warren et al. (1967) presented in Table 5-1. These data were obtained from soils directly over, and 100 ft at right angles to, the fault at the Sullivan mine, British Columbia, one of the world's most important sources of lead and zinc. Obviously, there would be little difficulty in suspecting the presence of mineralization if soils over the fault were sampled.

Table 5-1. Trace element content of soils (in ppm) and thickness of soil horizons (in inches) over a fault and 100 feet at right angles to the fault, Sullivan Mine, British Columbia. From Warren et al. (1967).

Horizon	Over fault			Thickness (inches)	100 feet at right angles to fault			Thickness (inches)
	Cu	Zn	Pb		Cu	Zn	Pb	
L-H	160	7900	2000	3	6	74	500	1
A_h or A_e	32	2800	2400	2	6	450	trace	1
B	20	1400	1400	10	6	400	2	8

However, as Warren et al. (1967, p. 255) state, "The point we wish to make is that a mere hundred feet from the fault it is possible to sample after removing the forest litter, a normal procedure, and have this sample report lead, a trace!" In that area glaciation probably removed any previous dispersion patterns. Interestingly, an anomalous value (500 ppm) was obtained from the forest litter (L-H) at the sample location 100 ft from the vein.

Where bedrock has been mineralized, some kind of chemical evidence will be found in any overlying residual soil. In those cases where no chemical expressions of the mineralization have been found in a residual soil, the problem has usually been traced to incorrect sampling or analytical procedures or, occasionally, to incorrect requests of the analyst. Without question, residual soil surveys are the most reliable and the most straightforward method of exploration geochemistry providing sampling, analysis and interpretation are carried out correctly. Fortunately, large parts of the world are amenable to the residual soil sampling technique, but in other parts of the world, soils are transported (e.g., in the glaciated areas of Canada, or in areas of high topographic relief, especially those accompanied by high rainfall) and conventional residual soil sampling methods are subject to many problems and often are not applicable. The use of transported overburden in lieu of soil, especially in the glaciated areas of Canada, is receiving further study and is discussed more fully below (chapters 8 and 11). At this point it suffices to say that transported overburden, and the soils developed on it, can be used for exploration purposes, but interpretation is likely to be difficult.

Different soil horizons will yield different amounts of trace elements (discussed in Chapter 3), so it is essential that the type of soil collected be consistent. Any deviation from the selected horizon must be duly noted since random selection of soil from different horizons may be difficult, if not impossible, to interpret. Further, it is essential that fine-grained clay or silty-clay be sampled, and sands, which have little adsorptive capacity for metals, should be taken only as a last resort (Fig. 5-2). If a soil survey is being conducted as a follow-up to a drainage basin or promising geophysical survey, it is often advisable to collect soil samples from several accessible horizons, considering the relatively minor cost of analyses by comparison with collection. If time and facilities (field laboratory) permit, it is also advisable to determine the optimum sampling horizon at the time of collection, if it was not done previously during an orientation survey. Drilling or digging to expose a vertical cross section through the soil profile may be necessary, and at the same time, it is advisable to determine the general level of trace element content in soils some distance from the area of interest (background levels). Caution should always be maintained to avoid areas of slumping, and other evidences of local soil transport. Any change in the type of material being sampled, even within the same soil horizon (e.g., color change, or change from chiefly clay to a high proportion of sand) should be noted during sampling, as this information may be of help in interpretation at a later date. Soil samples are collected in the same high wet-strength Kraft envelopes described above for stream sediments, and they are also dried without being removed from the envelope. Fifty grams is usually sufficient.

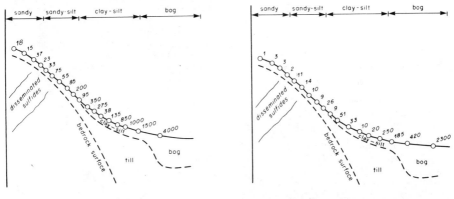

A. 20% HNO₃ EXTRACTION   B. COLD EXTRACTION

Fig. 5-2. Diagram showing variations in copper content with change of soil type down-slope from mineralization. Soils above the mineralization are sandy, well-drained, and have low metal values because of the low adsorptive capacity of sands. There is a progressive increase in fine-grained clays and silt down slope, which have higher adsorptive capacity and resultant higher copper values. The highest values are found in the poorly drained bog. The same relationship is found in both the hot extraction (A) and cold extraction (B) analyses. (A) can be considered a total analysis and (B) a partial analysis in this case. Drilling at the point of the highest copper values (within the bog) would not encounter mineralization. This diagram also illustrates a displaced anomaly. Based on an example from British Columbia. Modified from Bradshaw et al. (1972).

The type of equipment used to collect soil samples will vary with the depth selected for sampling, and the nature of the overburden. If the soil horizon is near the surface (e.g., less than 6 inches) as it is in many mountainous areas, then the chisel-end of a geological hammer will be satisfactory. With deeper horizons, up to about 2 feet, shovels or hoes are usually used. In stony soils, picks and shovels are usually best, but if the soils are primarily clays, soil augers have been found to be more satisfactory. For greater depths, various types and sizes of augers, post hole diggers and drills are available or it may even be necessary to dig a pit (Fig. 5-3). If convenient, some exploration groups will use a bulldozer to investigate promising locations, in an attempt to get below the soil to any bedrock mineralization. In many areas of Canada the glacial debris often contains sufficient coarse-grained rock fragments to make the use of soil augers impractical and experience has shown that digging is the most satisfactory method of sample collection. However, new types of augers are always being designed such as that by Curtin and King (1972) which has been successfully used for geochemical purposes where soils contain as

Fig. 5-3. *Left:* Combination percussion drill and diamond drill developed by the Outokumpu Company, Finland, for use in regions with thick overburden (from Wennervirta et al., 1971). *Right:* Overburden drilling, under winter conditions in Canada, with gasoline-operated Pionjar hammer drill. The drill weighs about 60 pounds, and with accessories (jack, sampler, crowbars, etc.) it weighs about 350 pounds. It can be transported by Ski-doo in the winter and carried in a pack-frame in the summer (from Gleeson and Cormier, 1971).

much as 60% pebbles and cobbles. Similarly, Dyck and Meilleur (1972) have designed a soil gas sampler for difficult overburden and this probe has performed well in glacial till containing boulders.

ROCKS

The proper sampling of rocks is often more difficult than the sampling of either sediments or soils. This is because of the uneven distribution patterns of elements and minerals throughout the rock mass, the uneven occurrence of some minerals in segregations or in fracture fillings, and because of textural variations within some rock types. In other words, there is no averaging effect such as exists in the secondary environment, and rock samples are often representative only of the material actually collected. The solution to this problem cannot be generalized, and each rock mass must be considered individually. If the rock is fine-grained and relatively homogeneous, then a single grab sample might suffice. If the rock is fine- to coarse-grained with spotty or erratic distribution of minerals, then channel or chip samples taken along several traverses will be necessary. To obtain a representative sample of cores, the entire length should be split, and the entire split ground.

When sampling rocks to determine primary dispersion patterns, consideration must be given to the probable dimensions of the halo. Wallrock alteration patterns are generally narrow whereas the patterns of a geochemical province naturally cover large areas (the dimensions of primary dispersion patterns are discussed further in Chapter 7). The areal extent of leakage anomalies will vary with the shape and inclination of the ore body, as well as with the amount of fracturing and other factors discussed previously. Leakage halos are best detected by sampling rocks and gouge material above, within, and in the vicinity of faults, fractures, veins, alteration zones and underground workings. Special care should be taken to collect those rocks with even the minutest evidence of fractures.

Regardless of the type of dispersion pattern being sampled, it is essential to include rocks from background areas, and traverses must be sufficiently long to ensure that these rocks are obtained. In some cases, channel or composite samples are collected in background areas, but these should be composed of just one rock type. It is important to document the rock types, as well as the degree and nature of alteration and weathering in both mineralized and background areas. Hand specimen and petrographic studies, paying particular attention to wallrock alteration patterns, are often necessary before proper interpretations can be made of the analytical results. Many of these details can be attended to during an orientation survey, and such surveys are extremely valuable in connection with any rock geochemistry program. If at all possible, orientation surveys should include the study of known ore deposits in the area.

Experience has shown that it is generally not possible to apply litho-geochemical techniques perfected for one mineralized district elsewhere because the primary dispersion patterns as well as the mineralogical nature of the ore are likely to be different. Further, because of the two- and three-dimensional variations in the size and shape of primary dispersion halos of different elements in the same deposit (see Fig. 7-5), caution is necessary in any interpretation.

Drill cores and rock chips from underground workings are materials which can yield valuable geochemical data. It is essential that the relationship of these materials to a particular rock type, and to structural features such as faults, be known. It may be necessary to sample cores and chips at very close intervals if there are great lithological variations. The sampling interval for rocks or drill cores can only be determined by careful consideration of the geological variables in the area being studied, as well as the aims of the project.

The amount of sample collected must be governed by the homogeneity of the mineral distributions and textures mentioned above. It must be remembered that from one ground and quartered sample, generally weighing from 1-5 g, an analysis and interpretation will be made which is supposed to be representative of perhaps a million times this weight. If a rock is very coarse-grained and minerals occur in specialized zones, as in the case of certain rare-element pegmatites, a truly representative sample may be almost impossible to obtain. An indication of the ideal amounts required, supposing the rock body to be homogenous and the grains approximately equi-dimensional, is given in the following tabulation from Wager and Brown (1960):

Grain size	Textural term	Sample weight (g)
3 cm	pegmatitic	5,000
1-3 cm	coarse-grained	2,000
1-10 mm	medium-grained	1,000
0-1 mm	fine-grained	500

Clearly, the tabulation shows that for coarse-grained rocks larger samples are required. Wager and Brown (1960) specifically recommended that if a rock is uneven in grain size, e.g., porphyritic, then it would be safest to consider the size of the largest grains, in using the above tabulation. Should the rock be heterogenous, then much larger samples must be used to obtain an average material.

It has been suggested that in the case of primary dispersions, the soils may be more representative of the true element content than any rock materials collected, and the soil is certainly easier to collect. This is unquestionably true in many cases, particularly in residual soils, because of the averaging effect of weathering and soil formation with the tendency toward homogenization. However, care must be taken to ensure that certain elements are not selectively and preferentially leached and, if they are, account taken of this. Should an orientation survey show that soil anomalies do correspond with primary rock anomalies, then a soil sampling program can be undertaken with confidence. There are numerous studies published in which soils do represent unexposed deposits. These include a wide range of examples, ranging from residual soils over porphyry copper deposits to those over carbonatite deposits in Africa (the latter described by Bloomfield et al., 1971).

In those cases in which only qualitative chemical analyses are to be made on the rocks to determine, for example, the presence or absence of nickel in response to a geophysical anomaly, the problem of sampling is greatly simplified because it is less critical. But the trend in rock geochemistry is toward more quantitative data, with attempts to outline primary halos, determine geochemical provinces, and for other uses which require careful sampling. In recent years numerous statistical studies have been made which specify the requirements for, and the evaluation of, many aspects of rock sampling procedures (see Chapter 12). In some cases the sampling plan is designed for the purpose of reducing the cost of geochemical surveys while keeping the thoroughness of coverage at an acceptable level.

WATERS AND LAKE SEDIMENTS

Water

Sampling of waters is extremely simple, but at the same time there are certain potential complications. Hem (1970) has stated that incorrect sampling is perhaps the major source of error in the whole process of obtaining water quality information and, by analogy, this probably applies to exploration geochemistry as well. The first point to keep in mind is the absolute necessity for the sampling bottles to be exceptionally clean, as befits a situation in which element variations in the parts per billion (ppb) range may be extremely significant. Very stringent rules for the preparation or cleaning of bottles before sample collection, and the prevention of contamination, have been formulated in recent years by hydrogeochemists and particularly chemists concerned with water quality. For water collection, all new bottles, as well as those used previously for any other purpose, should be cleaned with a strong metal-free acid before being taken to the field, and in the field rinsed three times with the water being sampled, before the sample to be sent for analysis is taken. Bottles which may have had high metal contents from any source, should never be used if at all possible. Soft polyethylene bottles, especially the common colored varieties, should be avoided and "hard" polyethylene should be used. Bottles made of the "hard" variety of polyethylene but with thin walls (which are satisfactory) are now readily available and inexpensive. Also, owing to the extreme sensitivity of the analytical methods used, exceptional caution must be taken to avoid man-made sources of contamination by, for example, always sampling above bridges and culverts and recording the presence of pipes in wells. Hem (1970) presented a discussion of some of the points mentioned above, the problems of obtaining a representative sample, and an excellent summary of the interpretation of chemical analyses performed on waters.

As mentioned previously, for exploration purposes waters can be grouped into (a) groundwater and (b) surface water. The geochemical significance, content of dissolved matter, and other physical-chemical characteristics of these two classes of water are distinctly different. Groundwater includes not only conventional fresh and saline subsurface water, but also oil-field brines, some of which have high contents of metal, hydrothermal brines (e.g., Salton Sea), and spring water of various types. These groundwaters are all potentially very important in exploration geochemistry because they have the ability to dissolve as well as transport metals. In the vicinity of oxidized or oxidizing sulfide deposits, as in Ireland, for example, groundwater has been used successfully for exploration purposes. However, it has been mentioned above that when this groundwater is brought to, or reaches the surface, changes in the physical and chemical conditions result, such as a change from a reducing to an oxidizing environment, and for this reason, it is preferable to carry out some analytical tests at the sampling site. The more common tests done on the spot are pH, Eh, conductivity, and the sulfate, carbonate and bicarbonate content. The first three can be carried out with portable meters, and the last three by inexpensive water analysis kits available from many chemical supply houses. Further, if possible, it is advisable to extract the metals from these waters with an organic solvent (MIBK is often used) for later analysis in the laboratory. One reason given for the metal extraction is that metals in ionic form in the groundwater may form complexes when they reach the surface and, as such, be indistinguishable from metals present on suspended matter, in suspension, or as non-ionic solutes.

The chemistry of surface water, which includes river, stream, lake and bog water, is not significantly affected by collection. Their contents of sulfate, carbonate and bicarbonate are generally much lower than those of groundwater, and no attempt is made to analyze them on the spot for anything other than pH and temperature (Eh is usually difficult to interpret, therefore, the measurement is not usually made). In the special case of uranium exploration, radon in water should be determined at the site by means of a portable radon detector. In the past when large volumes of water were required for analysis of metals in the ppb range, some geochemists would pass several liters of water through ion exchange resins in the field to adsorb the metals, thus making it unnecessary to bring large amounts of water to the laboratory. Other forms of pre-concentration by solvent extraction, or by filtering after co-precipitation, have also been used in certain situations. Because new analytical methods require only small volumes of water for analyses, even in the ppb range, these techniques are no longer necessary.

There is no uniformity of opinion as to the best method of sample collection, preservation and analysis of either groundwater or surface water. Water quality chemists concerned with surface water seem to feel that only pH and temperature should be measured in the field, and beyond this the most essential item is to get the samples to a laboratory for analysis as soon as possible. Some analyses, such as nitrate and phosphate, which are rarely of interest to exploration geochemists, must be completed within about 3 or 4 days of collection or else the sample will have deteriorated to an extent that meaningful determinations cannot be made. Other elements and radicals are not as time-critical. It is recommended that those contemplating using water for exploration purposes should consult the analyst who will make the analyses, so that he will be prepared to start the analyses as soon as the samples are received. Further, he will advise the collectors as to the quantity of water he will need, which will usually be 100 ml and will rarely be more than 1 liter, but will depend upon the number of elements to be determined and the techniques he will use. For the fluorometric analysis of uranium in water, 5 ml is commonly used, although usually 100 ml is collected.

It is common practice among exploration geochemists and water quality chemists to acidify water samples with a few drops of acid (usually HNO_3) to a pH of between 1 and 2 upon collection to prevent the "loss" or "plating out" of the metals, especially in dilute surface waters. The various reasons suggested for the loss of trace metals from solution include adsorption into the walls of the polyethylene bottles, and co-precipitation with hydrous iron oxides during hydrolysis. Care must be taken to ensure that the nitric acid added to the water is metal-free. Some nitric acid has been reported to contain zinc, uranium and other elements in sufficient amounts to cause difficulties.

From the discussion of river water and lake water in Chapter 3, it is obvious that any one water sample may not be representative of water in other parts of the river or lake (Figs. 3-28 and 3-31). Therefore, to assist in the interpretation of data obtained from such waters, a special set of field observations should be made. These include: (1) weather conditions at the time of sampling and, if possible, in the immediate past; (2) amount of suspended matter and turbidity; (3) water color; (4) relief in the immediate vicinity; (5) presence or absence of iron-manganese oxides; (6) sample depth; (7) biological aspects, such as the abundance of algae; (8) the presence of vegetation and organic matter; and (9) possible contamination. For river water the mixing characteristics, particularly with respect to any tributaries, should also be noted. In the case of lakes, additional observations should include surface conditions, such as wind strength, and the sediment type and color (when wet), if it is collected.

As the transport of metals is particularly dependent upon pH, this measurement should be made on all water samples. It is easy to determine pH, and this simple measurement may, for example, show values of 7 to 7.5 in an area in which copper was not detected in the water. This should not be interpreted negatively because copper would have precipitated out at a lower pH (Table 3-6). Eh measurements are obtained less commonly because of the difficulty in interpreting the results, especially in river water.

In the case of lake water, because of the problems inherent in stratification, the thermocline should be located. (This is accomplished by lowering a special thermometer into the lake while an operator on the surface records the temperature). If only one sample is to be taken, it probably should be obtained below the thermocline where conditions are more reducing, groundwater is likely to influence the water chemistry, algae and other organisms are relatively low in abundance, and the effects of fresh water dilution are likely to be absent. Some geochemists will suggest the opposite, that is, that if only one sample is to be taken it should be a near-surface sample. Arnold (1970), based on a study of lakes in the Canadian Shield (see Fig. 11-12), believes that a near-surface sample gives a closer estimate of normal metal content under both stratified and mixed water conditions, since a near-bottom sample can be highly enriched in metals mobilized from the sediments during periods of stratification. Such high values can give misleading results.

In the ideal lake water sampling program a series of samples (profiles) will be collected as illustrated in Figs. 3-31 and 11-12. The samples should be from one foot below the surface, in the middle of both the epilimnion and the thermocline, and at the top, middle and bottom of the hypolimnion. Canadian limnologists have found that a continuous profile of oxygen and temperature is extremely valuable for interpreting the lake environment, and geochemists will also find this data extremely valuable (Fig. 11-12).

The oxygen profile is obtained by means of an oxygen-sensitive electrode which is lowered into the water and then read by an operator in a boat. Turbidity is another very valuable measurement which can be determined on-site by means of a Secchi disk. This simple device is based upon the determination of the depth of visibility of a white plate (or light) lowered on a calibrated line until it just disappears. The transparency in some lakes can be used as an indication of their oligotrophic or eutrophic character.

Rivers are not stratified, but the surface water will often contain floating debris of various kinds (e.g., leaves, plants) and, therefore, a sample should be taken about a foot below the surface. Ideally an integrated water sample should be collected, that is, one representative of the total depth at the sampling location (which is rarely more than 10 feet in most small and medium sized rivers). This is accomplished by lowering a weighted bottle which fills as it falls to the bottom.

Lake Sediments

Lake sediments may be conveniently considered here. These materials have received much attention in recent years in Canada, and are discussed more fully in Chapter 11. Suffice it to say that there are numerous opinions as to the best sampling site, size of lakes to be sampled, type of materials to be collected, and so forth.

Some geochemists avoid lake sediments near the river inflow and prefer samples near the outflow as they reason that such a sample represents a homogenizing effect of the entire drainage. Others prefer a sample near the inflow, especially sediments in shallow bays. Still others prefer an outlet sample for small lakes but for larger lakes they will take several samples at river inlets, around the perimeter, and in the middle. This approach is especially useful where more detailed studies are in progress. Most recommend that samples be taken 15-20 ft from the shore so as to avoid the influence and effects of vegetation, contamination and coarser particle size. Some geochemists who have had considerable experience with lakes on the southern and western parts of the Canadian Shield prefer a sediment sample collected in three feet of water. It is assumed that samples at this depth are always in an oxidizing environment and trace metals of interest will be found adsorbed or co-precipitated with iron-manganese oxides. Clearly, there is no unanimity of opinion, and rightly so, because each lake sytem, and perhaps each lake itself, must be treated individually. Further, the objectives of exploration programs will differ as will the methods of metal dispersion. On a practical basis, as with other types of geochemical surveys, there are times when samples have to be collected wherever they can be obtained.

The size of a lake may be important in determining its potential value for exploration purposes. Very small lakes (<0.1 sq mi) have generally not given very good results, based on an orientation survey using lake sediments in the Yellowknife area of northern Canada, unless a major river passes through them (Nickerson, 1973). The smallest lake sampled by Allan et al. (1973) in their regional reconnaissance in the northwestern part of the Canadian Shield was 1000 feet long. On the other hand, Arnold (1970) found that sediments in small lakes (about 30 acres) were very satisfactory for locating adjacent mineralization. Presumably he was detecting detrital sulfide grains.

To obtain the most information from lake sediments, as with other types of sampling media, the various size fractions should be separated and analyzed. This is especially so in the case of sediments whose contained metals have multiple origins (e.g., detrital sulfides, chemical precipitates, groundwater). Some lakes have sandy bottoms, whereas others will have organic bottoms, and the trace element content will differ in each. Clay bottoms are preferred for most exploration purposes as the results will likely be easier to interpret. Allan et al. (1972) analyzed the -80 mesh fraction in their study in the Coppermine River area of the Canadian Shield, but used the -250 mesh fraction in their reconnaissance survey in the Bear and Slave structural provinces (Allan et al., 1973). Results of the two studies cannot be directly comparable.

Because of the chemical variations which are possible in lake sediments at or near the sediment-water interface, some geochemists attempt to obtain samples from 6 inches to 1 foot below the top of such sediments. This may be accomplished by means of various coring and augering devices, analogous to those used for soils. However, relatively simple devices may be used, such as a 2 inch diameter polyvinylchloride pipe fitted with a rubber bung at one end (Nickerson, 1973). When sampling in shallow water, and a surface sediment sample is considered satisfactory, this may be collected by hand.

DUTIES AND RESPONSIBILITIES
OF SAMPLE COLLECTORS

From the above discussions it should be clear that proper sample collection is a very important part of a geochemical survey and it requires training and experience. Experienced geochemists invariably do the job properly as they are well aware of the pitfalls which can result from improper sampling. However, in the case of inexperienced samplers, for example students hired for the summer, it is essential that they be thoroughly trained by an experienced geochemist, and be especially aware of the importance of their phase of the operation.

A sample collector should do more than merely collect samples. He should make notes of anything which may be of value in the interpretation phase of the survey. The observations he should record will depend to a great extent upon the type of survey. For example, in the case of a soil survey, the collector should record: the color of the soil and especially any changes in color; any change in the constitution of the soil, such as a change from clay to a sandy-clay; possible soil slumping; the presence of organic matter; possible contamination from any source; the soil zone being sampled and its thickness; changes in the zone being sampled and the reason why; the relationship of the soil to geologic formations, should he be working from geologic maps; and the presence of any mineralization. Collectors should also note the presence of any transported material in an area, such as alluvium, moraine, or fluvio-glacial materials. This is a particular problem in some soil surveys, as anomalies can be obliterated or lost as a result of the introduction of transported material. In the case of a rock survey, many of the above points are also applicable, but in addition, the collector should make observations of rock textures, degree of alteration, and rock types being sampled. Above all, regardless of the type of survey being conducted, collectors must be constantly alert to the fact that changing geologic and environmental conditions may require immediate modifications of procedures.

Many employees of geochemical survey parties are temporary and have little or no previous experience in this type of work. Most of them can be readily trained by an experienced collector or geochemist, and can be motivated for as long as the field season or project lasts. As a result, the samples collected are meaningful. A more serious problem is the employee who appears to have little interest in his work. If this person cannot be motivated in a short time, experience has shown that he should be moved to a non-critical phase of the survey (or discharged), as there is nothing worse than trying to interpret data from samples collected by a person in whom there is no confidence.

To aid the collector in his work, some organizations have adopted punch-card techniques. The Institute of Geological Sciences (London) has prepared field data cards which have been used successfully in the field in the Far East and South America (Fig. 5-4). All the variables mentioned above (e.g., geology, pH) are on the card together with a grid reference. A trained collector can enter the data in about a minute. When this information is used in conjunction with similar punch-cards from the chemical analyst, the whole survey becomes very easy to computerize. Similar systems have been developed and adopted by other organizations. Several field data cards used by the Geological Survey of Canada are illustrated in reports by Boyle et al. (1966) and Allan et al. (1973). Other advantages of the system are that no observations are omitted or forgotten, and that notetaking time is considerably reduced over conventional methods.

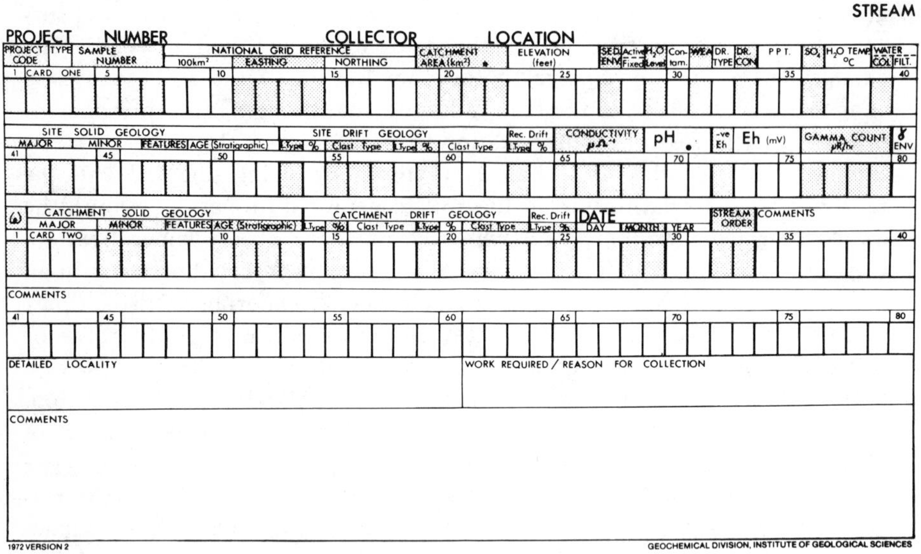

Fig. 5-4. Field data sheet for stream sediment survey. The top sheet (illustrated) has a carbon-type backing, so that a copy of the data is automatically recorded on an attached second sheet which contains the same format as the top sheet. The top sheet is sent for data processing, whereas the second sheet is retained by the field party. Actual size is 17 x 9.4 cm. Courtesy of the Geochemical Division, Institute of Geological Sciences, London.

6/ *Analytical Methods*

INTRODUCTION

After a sample has been collected, it is then prepared (dried, sieved, ground, etc.), and later analyzed. In some cases, as in the analysis of water and air, several steps in the preparation process may be eliminated or simplified. It cannot be over-emphasized, however, that an understanding of the correct preparation procedures, the various methods of sample decomposition (digestion), the capabilities and limitations of the important analytical methods, the correct chemical analysis for a particular situation, and other basic facts about trace element analyses are critical to the realization of the full potential of exploration geochemistry. The geologist engaged in mineral exploration is normally conversant with the field and interpretation phases of an exploration program, but is often unfamiliar with, and confused by, the analytical aspects. Part of the problem the geologist faces is the result of a barrage of well-intentioned, but confusing, literature received from commercial laboratories and instrument companies.

The purpose of this chapter is to set forth the basic concepts of trace element analysis, and to present brief descriptions of the analytical methods most commonly used in exploration geochemistry today, in all cases stressing the practical aspects. This requires that the methods be sufficiently: (1) accurate and precise for the particular project; (2) sensitive to detect background levels; and (3) rapid and inexpensive. It is hoped that the discussions in this chapter will remove some of the confusion and difficulty so prevalent among practicing geologists, and at the same time, those entering the field will obtain an understanding of the present status of elemental analysis as applied to exploration geochemistry. It should be constantly kept in mind that analytical instrumentation is a dynamic and vigorous field. Products are being improved and new ones are being introduced all the time. Research to increase the capabilities (e.g., lower detection limits for certain elements) of existing equipment is constantly under way. Geochemists should be aware of such advances for obvious reasons, not the least of which is to maintain a competitive advantage.

Before discussing the main substance of this chapter a few general principles will be presented.

241

GENERAL PRINCIPLES

Accuracy Versus Precision

In exploration geochemistry, precision (which is the ability to reproduce and repeat the same result) is usually more important than accuracy (which is the approach to the true content), at least in the initial stages of a project. The distinction is illustrated in Fig. 6-1. Thus if a laboratory reports 200 ppm Zn, and on repeating the analysis values of between 180 and 220 ppm are always obtained, the precision is plus or minus 10%. The true answer (accuracy) may in fact be 250 ppm, but this inaccuracy of 20% is not normally a problem as long as all samples from the same survey are analyzed in the same manner, and the precision and accuracy are within the above limits. It is then possible to compare samples relative to each other in looking for anomalies. In general, however, it is not possible to compare analytical data between different surveys or areas because of the differences in geological and geochemical factors, such as rock type and amount of secondary dispersion; this is the reason why accuracy is less important than precision. Of course, there are practical limits as to what deviations from accuracy and precision can be tolerated. There will be times when it will be necessary to check results with other laboratories, or by other methods within the same laboratory, and in these circumstances great deviations in accuracy may become apparent. Because of the variations in accuracy and precision which are possible, it is

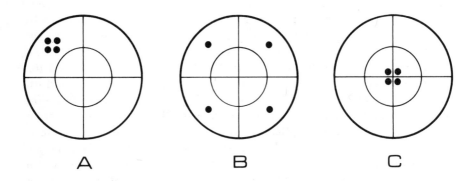

Fig. 6-1. Distinction between precision and accuracy. These figures represent targets, and the four dots in each represent bullet holes. "A" is very precise but has poor accuracy. "B" is imprecise but highly accurate, as the average of the four bullets corresponds with the bull's eye. "C" is both highly precise and accurate.

desirable to have all samples from a particular sampling program analyzed in the same laboratory, by the same method, and even by the same analyst. Should the results appear geologically or geochemically improbable, erroneous or suspicious, they should be checked in another laboratory using the identical samples. (This is one reason why most commercial laboratories store samples for at least six months before discarding them). In any event if significant sums of money are to be invested in any property, or in any drilling project, it is highly recommended that at least some representative samples be checked by at least one other analytical laboratory.

Limit of Detection

The limit of detection (commonly called sensitivity) is the minimum content of an element that can be measured by a specific analytical method. This value will vary greatly from one method to another, and from one matrix (host rock material) to another. For example, uranium can be routinely detected in rocks down to about 1 ppm by fluorimetry, 50 ppm by X-ray fluorescence, and 500 ppm by emission spectrography. There is, therefore, a great variation in the limit of detection by these three well-established methods. In addition, when the matrix changes significantly, such as for example from a basalt to a granite, there are further changes in the limit of detection. In using all three above methods, owing to the effect of variations in associated elements, the limit of detection will probably be lower (more sensitive) for uranium in the granite than in the basalt.

Instrument manufacturers will always quote limits of detection as low as is ethically and legally possible, but in all cases these low limits of detection are only practicable under ideal conditions. For example, a manufacturer of a flame spectrophotometer might state a lower limit of detection for cesium in water as 1 ppm. Upon investigation of this claim, the geochemist may find that such is indeed true, provided that the water is free of other elements, but should there be as little as 5 ppm sodium present, the limit of detection for cesium might be raised to 10 ppm. The above example is not designed to single out any particular manufacturer or piece of equipment, but to alert geochemists to the general problem. Fortunately, most analysts do not over-state their limits of detection, but, when receiving reports which indicate an element was not detected (n.d.), it would not be inappropriate for the geologist to ask for the limit of detection and details of the method used. In older literature and reports, the word "nil" was often used, but fortunately it is not common at the present time. The term implied that the element was not detected and, if present, is below the limit of detection (which is generally not known exactly), for the particular method employed. Regardless of which analytical technique is being discussed, at the lower limits of detection, accuracy and precision tend to fall off.

Wet Analyses

This term is used as a general expression for all methods which require that the sample be put into solution. Once in solution, the metal or element to be determined is separated from the rest of the sample by various operations of chemical analysis. The four general types of wet analyses are: colorimetric (comparison of colors, either visually or with a spectrophotometer), gravimetric (precipitating the metal and weighing it), volumetric (reacting the metal with some other reagent which is more easily measured), and electrochemical (the element is electroplated onto an electrode and weighed). In exploration geochemistry only colorimetric procedures are commonly used today, whereas the others are used for special cases (for example, it is possible to determine platinum by electrochemical analysis, but other methods are preferred by most analysts). Gravimetric and volumetric methods are used for the determination of high abundances (>5%) of many elements, including copper, lead and zinc, because atomic absorption methods lose precision and accuracy in the higher ranges. Atomic absorption, which generally requires the samples to be digested into an acidic solution before being aspirated into the instrument, can be considered as a wet chemical procedure, in part, although it is more appropriately considered an instrumental method. Wet analyses require "classical" analysts with great skill and experience, especially for rock materials, as possible complications are numerous and subtle.

Instrumental Analyses

Instrumental analyses determine the quantity of an element in a sample by measuring some physical property of that element. For example, in emission spectrography the wave-lengths of light emitted when a sample is vaporized are detected and their intensities measured. In X-ray fluorescence, the characteristic X-ray wave lengths of elements are detected and measured. These techniques require many standards, as the elements are not measured directly, but rather their physical characteristics, and it is for this reason that they are sometimes referred to as "relative" analyses. In general, the results obtained by instrumental methods are subject to significant errors because of variations in sample matrix, and these variations must be considered and corrected for, in accurate analytical determinations. Nevertheless, instrumental techniques are important, and with atomic absorption included, about 75% of all geochemical samples are presently analyzed by such methods.

Partial and Total Analyses

The methods of sample digestion (or decomposition) used in exploration geochemistry are extremely varied. As a result, the amount of a trace element which is determined in a sample can be a very small portion (perhaps 15% or less) of the total amount actually present, hence it is a "partial (differential) analysis", as in the case of the cold acid extraction techniques (Fig. 5-2). On the other hand, 100% of a trace element can be determined, using emission spectrography, X-ray fluorescence, or after complete digestion by fusion or with a strong acid treatment prior to colorimetric or atomic absorption analysis; this is a "total analysis".

At first glance the great variability in analytical results which are possible because of the different methods of digestion may be considered as a factor limiting the ability to interpret geochemical data. Actually, the opposite is true because, as is shown in examples below, if the digestion procedures are wisely selected, the interpretation of the data may be simplified, and the results quite significant.

In a soil, sediment or rock sample, an element may occur in a number of forms. For example, in one particular deposit copper may be present in rock-forming silicates (Table 2-2), as a sulfide, or loosely adsorbed onto clays or other weathered particles. By means of different types of digestion (partial and total analyses), the amount of copper in each of the three forms can be determined. The loosely adsorbed copper on clay particles can be removed and determined by means of a dilute, cold acid leach. The sulfide copper can be put into solution and determined by means of a hot acid leach after a digestion of an hour or so, whereas the silicate-bound copper can be liberated only by very strong attack, such as by an alkali (Na_2CO_3) fusion, or a hydrofluoric acid leach, after which it can be determined by atomic absorption or colorimetric methods. Certain instrumental techniques, such as X-ray fluorescence and emission spectrography, would yield *total* copper if a sample in its original state were analyzed, because these techniques are inherently unable to distinguish between the various associations enumerated above.

It is evident from the above discussion that the total metal content of a rock sample, no matter how geochemically enticing, may not always be recoverable by any practical metallurgical process. A well-known example of this would be the nickel in unaltered ultramafic rocks, normally about 2000 ppm (Table 2-1), most of which is usually within the olivine (a silicate) structure. This nickel is not present as sulfides, and the only way it could be recovered at present is by costly chemical fusions, which are completely uneconomic. If cold or hot acid methods of digestion and analysis were used on these samples, they would probably yield little or no nickel, as olivine and other primary silicates are not normally soluble in common acids. Only alkali fusions, hydrofluoric acid, or total methods of digestion or analysis (e.g., emission spectrography) would indicate the presence of nickel in the 2000 ppm range. Thus with a knowledge of the different methods of chemical treatment and the capabilities of analytical instruments, criteria for geochemical interpretations are available which can often discriminate between significant and non-significant values for metals. A recent example of a nickel occurrence in which nickel is present in four phases (silicate, sulfide, oxide and nickel-iron alloy) has been described from Quebec by Eckstrand (1971). If all the nickel were present as sulfides this deposit would probably be workable, but at least part of it is present within serpentine and other minerals which cannot be processed economically. Differential solution or extractions, and different combinations of partial and total analyses, followed by detailed mineralogical studies should always be employed in situations such as this to interpret the geochemical significance of apparently anomalous values.

Partial analysis in geochemical exploration had its origin in cold extraction techniques developed for use in both the field and laboratory. Loosely-bonded metals in stream sediments have traditionally been sought in an effort to detect metal concentrations which may be related to oxidizing sulfide deposits. The technique has been particularly successful under tropical weathering conditions where relatively long anomalous dispersion trains have formed. Many factors are involved in determining what proportion of the metal content will be released by any particular extraction method. These include the nature and strength of the extractant, time of digestion and temperature at which the reaction takes place. Several of these variables are illustrated in Fig. 6-2. Only experimentation will reveal the best extraction technique for each survey.

All laboratories have the potential to do cold extractions and other partial analyses, and of course, some determinations can easily be done colorimetrically in the field. Yet the tendency of most exploration geochemists and geologists is to request "total" values for an element, on the assumption that they do not want to miss anything of economic value. Generally speaking this is the correct approach, at least initially, because any mechanically derived (clastic) sulfide will be detected, whereas it could be missed by weak (partial) extraction methods. Therefore, in commercial laboratories, unless specifically requested otherwise, analyses are usually performed either by a fusion method or by a perchloric-nitric acid digestion, both of which get all loosely-held metals and all sulfides into solution, as well as metals in some of the less stable silicates (e.g., micas).

This is certainly the best compromise when those sending in samples are not aware of alternative or supplementary methods of analysis. However, for a small extra charge in comparison with the overall cost of the project, significantly more information can be obtained by the deliberate use of various partial analyses which are particularly useful in recognizing transported seepages in soils, and other loosely held (hydromorphically transported) metals in stream sediment samples which often reflect the presence of oxidizing sulfide minerals. Bradshaw et al. (1972) have shown the utility of various analytical extractions in interpreting geochemical anomalies (including false anomalies) in various parts of Canada and elsewhere. Some geochemists recommend that at least one partial and one total extraction be used on samples collected in any new area.

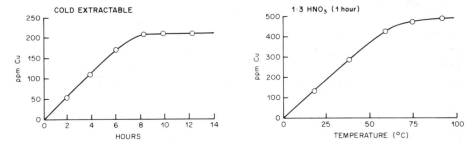

Fig. 6-2. Diagrams illustrating several of the variables involved in partial analytical techniques. *Left:* Eight hours were required before all the cold extractable (dilute acid) copper was removed from a sample. *Right:* A temperature of about 75°C was necessary for all the copper to be extracted from a sample placed in 1:3 HNO_3 for 1 hour.

Sample Attack (Decomposition) Commonly Used

In the above section various degrees of sample attack, or methods of decomposition (digestion), were mentioned. The most common of these are now discussed in order of increasing severity. In reading the descriptions of these methods, it should become clear that the results obtained from different types of sample attack cannot be compared directly and, for this reason, as well as others, it is essential to standardize the sample digestion procedures for a particular survey. It should also become clear that a knowledge of metal bonding in a sample will help to determine the optimum analytical and extraction procedure for the particular project.

1. **Cold Extraction (including weak acid attack).** Cold extractions commonly use (a) special solutions called buffers, whose function it is to maintain pH within specific limits, and as a result, extractions can be either selective or restrictive (e.g., for copper, zinc, total heavy metals), and (b) a

variety of dilute acids (e.g., nitric, hydrochloric, acetic, EDTA). The tests are often colorimetric and particularly useful in the field because of their simplicity, but they can also be performed in the laboratory by colorimetric or atomic absorption methods. Normally, only a small percentage (less than 15%) of the total amount of an element in a sample, the loosely bonded type, is determined but, as mentioned above, interpretation is sometimes easier than with results obtained by other methods of digestion. Fig. 6-3 illustrates the type of results which may be obtained.

Cold extraction field methods, using both buffers and dilute acids, have been particularly successful in tropical areas where loosely held metals from oxidizing deposits are adsorbed onto sediments and soil. However, these methods generally are not applicable in glaciated northern areas where oxidation and the release of metals have been slow since the last glacial period. The dilute acids will usually release somewhat more adsorbed metal than the buffers. In all cases the time of digestion, the strength of buffers or acid, the amount of sample used, and all other variables must be carefully controlled and uniformity maintained if meaningful results are to be obtained (Fig. 6-2). When a buffer is used, the process is essentially an exchange reaction with NH_4^+ from the buffer, exchanged for the metal.

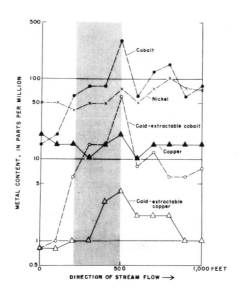

Fig. 6-3. Distribution of cold-extractable and total cobalt and copper, and total nickel, in stream sediments across mineralized gabbro-norite, Somerset Co., Maine. Stippled area indicates approximate position of mineralized zone. The cold-extractable and total content of the metals reveal the anomalous conditions equally well, even though the magnitude of the respective anomalies vary almost by a factor of ten. From Canney and Wing (1966). Reproduced from *Economic Geology,* 1966, v. 61, p. 199.

Cold extractable metals are sometimes called "readily-extractable". Both are generally abbreviated as "cx", or cxCu in the case of cold extractable copper.

2. **Hot Extractions.** In hot extractions, samples are treated with an individual acid (e.g., HNO_3), or a mixture of acids (e.g., aqua regia; nitric-perchloric), usually at 100°C or more for approximately an hour. The acids most commonly used, in varying strengths from dilute to concentrated, are perchloric ($HClO_4$), nitric, hydrochloric and sulphuric. In the analytical procedures presented in Appendix B, 25% nitric is preferred because it gives consistent results, and is sufficiently strong to get all loosely adsorbed metals and all common sulfides, except molybdenite, into solution. Hydrochloric acid is preferred if the analyses are to be done by atomic absorption because nitric acid causes some analytical interferences. Neither nitric nor hydrochloric acid decompose the more stable rock-forming silicates of no commercial value.

A mixture of nitric-perchloric is the acid commonly used for sample digestion in most commercial geochemistry laboratories for a number of reasons, the most important ones being: (a) the mixture can be kept at a higher temperature (180°C) than other acids, thus being more rigorous in its attack on minerals; (b) the decomposition is faster than certain fusion methods (e.g., potassium bisulfate) which are otherwise equally effective in decomposing minerals; and (c) greater productivity, hence lower costs, can be realized. The mixture has the advantage of decomposing certain copper and nickel silicates (such as chrysocolla and garnierite) which are important in exploration geochemistry in Australia and other areas, and are not decomposed by nitric acid alone. Some primary silicates, such as the micas and feldspars are attacked but the degree of attack is variable and unpredictable. Other primary silicates such as pyroxenes and amphiboles, and oxides such as magnetite and chromite, are attacked to a lesser and equally unpredictable degree. Most reports indicate that molybdenite is decomposed by perchloric acid, but Bradshaw et al. (1972) state that the attack is not very efficient (perhaps 70%). They find they get better, more reproducible results, using a carbonate fusion (colorimetric) method for molybdenum similar to that described in Appendix B. At this time, the efficiency of perchloric acid in decomposing molybdenite must be considered an open question, but there is no doubt the acid is 100% efficient in digesting secondary molybdenum minerals.

A mixture of nitric and perchloric or perchloric acid alone, is *potentially explosive* and for this reason is not recommended except for well-trained analytical chemists. Actually, the perchloric acid is the effective reagent, the nitric acid being added to react with readily oxidizable material which might cause a violent reaction with the perchloric acid. As pointed out by Stanton (1966), for the full exploitation of this mixture, the attack must be continued to dryness followed by leaching of the residue with 1 molar hydrochloric acid. Although the resulting solution will be suitable for the determination of cobalt, copper, iron, lead, manganese, nickel, zinc, and many other elements of geochemical interest, it is not suited for antimony, arsenic and selenium. In spite of these limitations, a nitric-perchloric mixture will yield about a 95% extraction of the metals in sulfides and weathered products, which is almost equivalent to a total extraction.

Aqua regia (1:3 mixture of nitric and hydrochloric acids) is used in some laboratories for sample attack. This acid will digest all the sulfides and the noble metals as well. About 90% extraction is usually obtained on a practical basis.

Hydrofluoric acid is used in those cases where it is desired to decompose silicates, and it is successful with all but the most refractory minerals. Alone, it has limited application in exploration geochemistry. It is usually used in combination with other acids, such as a perchloric-nitric-hydrofluoric mixture, when a total analysis is desired but even with this mixture, certain accessory minerals such as sphene, magnetite, zircon, chromite and ilmenite, are not completely decomposed. Hydrofluoric acid must be used with great care as it is potentially *dangerous* because of its reaction with the skin and is especially dangerous if it gets under the fingernails.

Foster (1971) has reported on some preliminary investigations carried out on the efficiencies of various acids in decomposing a number of rock-forming silicates and limonite (but not sulfides). One example of his results is presented in Fig. 6-4 which demonstrates a considerable variation in the proportion of the total metal extracted. In a subsequent paper, Foster (1973) showed that a similar situation applied to the analyses of a number of common rock types. In addition, grain size variation has an effect on the amount extracted by certain digestions. The most consistent results, and greatest amount of extraction, will be realized when samples are ground extremely fine.

3. **Fusions.** There are two common types of fusion (fluxes), acid and alkaline. Both require temperatures of 800°C or more to be entirely effective. Acid fusions involve the use of either potassium pyrosulfate ($K_2S_2O_7$), or potassium bisulfate ($KHSO_4$). In both cases, a fused powder is used, and it is sulfur trioxide (SO_3), which is formed during the decomposition of both compounds, that is actually the attacking agent. These compounds are effective against sulfides, but the degree of decomposition depends on the temperature of the fusion. At 350°C the attack will be slow, whereas at 800°C the attack will be rapid and even molybdenite is decomposed. However, for samples very rich in sulfides or organic matter, the attack is not totally effective as the flux does not incorporate an oxidizing

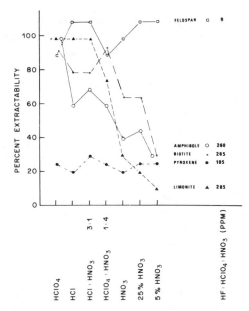

Fig. 6-4. Variation in the extraction of zinc from various rock-forming silicates and limonite following different acid attacks. All samples weighed 0.25 gm., were ground to − 80 mesh, and were leached with 10 ml of 1M HCl prior to analysis. The data from the HF-HClO₄-HNO₃ attack can be considered as the total metal content. From Foster (1971).

agent and oxidation must, therefore, be accomplished atmospherically. After the melt has cooled, it is generally leached with dilute hydrochloric acid, although under special conditions dilute sulfuric or nitric acids can be used, and the leachate then analyzed either colorimetrically or by atomic absorption. The fusion, generally performed on soil or sediment samples, can be carried out in borosilicate test tubes, each of which can be used about a dozen times.

Harden and Tooms (1964) have discussed the efficiency of the potassium bisulfate fusion in extracting metals from rocks, soils and stream sediment samples from Zambia. They found that copper, nickel, cobalt and zinc were removed from feldspars and micas, but only copper was removed from pyroxenes and amphiboles. They concluded that the efficiency of the bisulfate attack on rocks, soils and stream sediments is a reflection of the mineralogy and the degree of weathering and alteration. The potassium bisulfate-extractable metal is not equivalent to the total metal in all cases, yet this incomplete sample attack is sometimes preferable, because that portion of the total metal within rock-forming minerals is not of economic significance, and if included within a total geochemical analysis, could be misleading.

For alkaline fusions, a number of compounds, either individually or in combination, can be used. Sodium carbonate, sometimes mixed with potassium nitrate, is the most commonly used, and is extremely effective in decomposing chromates, vanadates, molybdates, tungstates, as well as silicates and aluminates. Because of the high temperatures attained, and the almost total decomposition, crucibles of either platinum, zirconium or nickel are required. After the melt has cooled, the residues can be taken up in dilute acids, or in water in the case of water-soluble molybdate, tungstate and a few other compounds, for colorimetric or other types of analyses. Other combinations of reagents occasionally used for alkaline fusions include sodium peroxide and sodium hydroxide, and for uranium analyses a flux mixture of potassium and sodium carbonates and sodium fluoride. In special circumstances, ammonium salts (ammonium fluoride, chloride or iodide) have been used to decompose samples.

4. **Others.** Two other types of sample decomposition can be briefly mentioned. The first, volatilization, breaks down a sample by heating, either with an electric discharge (as in emission spectrography) or with the heat of a flame (as in flame spectrometry). Mercury, which is distilled when heated, would be an example of an element made available by volatilization. The second, oxidation, is used to liberate or remove organic constituents of soils, or the organic constituents of plant material so that only the ash remains. As discussed below, oxidation can either be "dry" by ignition, or "wet" in which case oxidizing acids are used.

Reporting Units Used In Exploration Geochemistry

Most materials collected during the course of geochemical exploration contain only traces (<0.1%) or minor (0.1 – 1.0%) amounts of the elements of interest. It is common practice in all branches of geochemistry and economic geology to express the trace element content of soils, sediments, rocks and vegetation as parts per million (ppm) where possible, in preference to percent. The ppm notation permits the use of convenient, small whole numbers, instead of cumbersome percentages, as for example:

$$
\begin{aligned}
1000 \text{ ppm} &= 0.1\% \\
100 \text{ ppm} &= 0.01\% \\
10 \text{ ppm} &= 0.001\% \\
1 \text{ ppm} &= 0.0001\% = 1000 \text{ ppb} \\
.1 \text{ ppm} &= 0.00001\% = 100 \text{ ppb} \\
.01 \text{ ppm} &= 0.000001\% = 10 \text{ ppb} \\
.001 \text{ ppm} &= 0.0000001\% = 1 \text{ ppb}
\end{aligned}
$$

As the ppm designation becomes more cumbersome with lower concentrations, usually less than 0.1 ppm, it is then more convenient to use the part per billion (ppb) designation which is included in the above tabulation (some geochemists will use ppb for quantities less than 1 ppm). In general, for concentrations above 1000 ppm, the percentage is reported (e.g., 0.18% in preference to 1800 ppm). Wherever possible in this book, the ppm notation is used from 0.1 – 1000 ppm, but there are exceptions, such as in Table 2-1 where, for convenience, values greater than 1000 ppm are listed.

Although the above systems of ppm and ppb are well established, they are by no means universal. Other designations which have equivalent meaning are preferred by some geochemists, and even required by the editorial policies of some leading scientific journals. Some of these modifications are based on the unit known as the microgram, by definition a millionth of a gram, which is abbreviated as "μg" or sometimes designated by the Greek letter "γ" (gamma). Therefore, 1 μg/gm is equivalent to 1 ppm, and 1 μg/kg is equivalent to 1 ppb. In the case of water samples, many analysts will consider that one ml (1cc) weighs one gram, and will report weight per volume units (e.g., mg/l) rather than weight per weight (e.g., ppm) units (the former, strictly speaking, is preferable in the case of waters). Thus, the zinc content of a river may be reported as 6 μg/l, which identical in meaning to 6 ppb. This, of course, is true only for water very low in salt content. Brines, for example, contain enough salts for 1 ml to weigh significantly more than 1 gm, with the result that their densities can be 1.1 – 1.2, or even more.

The use of multiple systems of reporting units is confusing to exploration geochemists, especially those who may want to read the literature in theoretical geochemistry, water resources, environmental and other geochemical disciplines, but there is no alternative to being aware of all systems of unit notation which may be encountered. Below, as additional examples, is a list of equivalents for 1 ppm, and in Appendix A, further examples are found: 0.0001% = 1 ppm = 1 μg/gm = 1 γ/gm = 1 μg/ml = 1 mg/l = 1000 ppb = 1000 μg/l = 1000 ng/gm = 1000 ng/ml = 0.029 troy oz/ton $\approx$ 1 gm/ton.

Most geochemical determinations are now reported in the ppm range, but certain analyses, such as those for uranium and mercury, are more likely to be in the ppb range, as are almost all trace elements determined in waters. In the analysis of air samples, for mercury for example, the nanogram (ng = a billionth of a gram) is the most commonly used unit, and in such cases the number of nanograms found is based on a specific volume of air (e.g., Hg = 8 ng per cubic meter of air). Should it ever be necessary to detect smaller amounts of an element, a unit is available (parts per trillion, ppt) but except in certain very specialized circumstances, as for example in neutron activation analysis, the analytical capabilities are presently lacking. Furthermore, the utility of the ppt values obtained will be minimal as most natural substances used in exploration geochemistry would normally contain most elements at these levels of concentration (Table 2-1). Contamination will also always be a problem and, in fact, it is common in concentrations below 10 ppm. The only possible use for the ppt range in the forseeable future is the analysis of water and vapor samples.

PREPARATION OF SAMPLES FOR ANALYSIS

There are many variables which enter into the preparation of a geochemical sample for analysis, not the least of which is the nature of the sample itself. In the simplest of cases, water samples brought to the laboratory are often directly aspirated into an atomic absorption unit without any preparation, but simple preconcentration by means of evaporation or organic extractions is performed if the metal content of the water is extremely low. In the case of air samples, analyses are performed on-site, usually within a low flying airplane, which has equipment for filtering undesired fine particles before the air containing mercury, sulfur dioxide, radon or other vapors enters the appropriate instrument. However, most geochemical samples are soils, sediments, rocks or vegetation which require some preparation, and the following brief discussion is restricted to these materials.

The main processes to be considered in preparing a sample for chemical analyses are: drying, crushing, sieving, quartering and grinding. No matter how carefully a sample may be collected in the field, the data obtained from it will be worthless if it is not prepared correctly. It must constantly be kept in mind that the small amount of sample which is actually analyzed, anything from 0.1 gram for a reconnaissance sediment sample, to perhaps 100 grams in the case of a platinum fire assay, represents an extremely small part of any potential area or of an ore body. Most analysts try to use a representative 1 to 5 gram sample for digestion or fusion. Reproducibility and accuracy on a sample of 0.1 gram is likely to be low, and 100 gram samples are generally too cumbersome for most geochemical analyses.

1. **Drying.** Drying is necessary because damp samples cannot be properly sieved. If samples are collected and placed in special geochemical paper envelopes, they can be hung to dry by the hundreds from hooks on specially prepared boards, sometimes covering an entire wall, as the envelopes have holes designed for this purpose. Otherwise the samples must be treated differently, such as being laid out in pre-numbered watch glasses, or evaporating dishes which can be placed in low temperature

ovens (about 105°C) to speed the drying process. If time is not important, the samples may be left to dry in their bags or on plastic sheeting. Samples intended for the analysis of mercury and other volatile elements must not be dried in an oven.

2. **Crushing.** Crushing is necessary to reduce the size of lumps in the case of stream sediments and soils, and to reduce rock samples so they may pass through an appropriate screen. In the case of rock samples, in order to obtain a truly homogeneous and representative sample, they should be pulverized to pass through a 200 mesh screen.

3. **Screening.** Sediment and soil samples, after crushing, are generally passed through an 80 mesh screen unless an orientation survey indicates another fraction (e.g., a − 120 mesh fraction in the case of some residual African soils), is preferable. The screen should be made of nylon or aluminum to prevent contamination from brass or stainless steel. In commercial laboratories, banks of up to 100 screens are loaded with samples, and mechanically agitated for periods of up to an hour. All twigs, stones and other undesirable materials are thus separated. Hawkes (1957) has suggested an economical non-contaminating sieve for field use that can be made by knocking out the paper disc from the cap of a half pint ice-cream container, and using the remaining band to clamp a piece of sieving cloth to the bottom section of the container. Alternatively, sieves can be made up simply from nylon bolting cloth and wide-mouthed polyethylene beakers the bottoms of which have been sawn off.

It is occasionally necessary to deviate from the − 80 mesh screen size. As mentioned previously, rock sample data is desirable at − 200 mesh. In the case of some coarse pegmatite minerals, such as beryl, high and more significant values from the point of view of milling may be obtained from coarser fractions, such as between − 10 to +60 mesh. In general, it is advisable to restrict the sieve mesh range to a small interval, because different grain sizes of the same sample may have different metal contents. But as each project is likely to be different, an orientation survey should determine the optimum sieve range or size.

A most important example of soils in which a coarser than 80 mesh fraction is essential for correct interpretation of the geochemical data, may be cited from the laterite areas of Western Australia. Mazzucchelli and James (1966) showed that arsenic, which is used as a pathfinder for gold mineralization, is found concentrated to some degree within the ferruginous fragments coarser than 80 mesh. Their study preceded the great nickel searches of the late 1960's in the same area but, regretably for many companies, orientation studies were not made to determine the best sieve size for the detection of nickel and associated pathfinders. Had this been done, it would have been found that, in agreement with the earlier study by Mazzucchelli and James (1966), the coarser lateritic soil fractions also had higher nickel which could yield significant anomalies. In fact, those who used the conventional – 80 mesh fraction so successful in North America and East Africa, would have missed many of the significant nickel deposits in Western Australia. In a more recent study on secondary dispersion patterns associated with nickel deposits at Kambalda, Western Australia, Mazzucchelli (1972) confirmed that nickel is preferentially concentrated in the coarse fraction of the soils, although both Ni and Cu display strong anomalies in the minus 80-mesh fraction in the immediate vicinity of sub-outcropping mineralization at that locality. Clearly, this example illustrates the necessity, in some cases, of changing the standard (minus 80-mesh) procedure to detect an anomaly. Other examples could be shown where the analysis of different mesh sizes has enabled anomalies to be considerably enhanced.

4. **Quartering.** Mixing and quartering are particularly necessary to obtain a representative portion of a large sample, such as a rock, or drill core submitted for assay purposes. In the case of a well-ground 50 gram soil or stream sediment sample, it is usually possible to omit this step.

5. **Grinding.** Small portions (10-15 gms) of the rock, sediment, or soil which have been quartered should be ground in a ceramic or agate mortar by hand, or by means of a non-contaminating mill, to ensure a representative homogeneous sample. Special precautions must be taken with samples containing minerals of greatly differing physical properties, such as metallic gold in certain soils.

For the preparation of vegetation for analysis, several preliminary steps are necessary, the most important and difficult being to destroy the organic constituents of the plant material, so that the metal content of the ash can be determined. The vegetable material is "ashed", which is a term that includes all methods of eliminating the organic matter. This results in a 20- to 50-fold increase in the concentration of the metals and facilitates the recognition and interpretation of anomalies. Dry ashing involves only the use of heat (450 – 550°C for about 12 hours) to destroy much of the organic material, but under these conditions, a large proportion of the more volatile elements (e.g., Hg, Sb, As, Se, Te, Pb) may be lost. (Some workers report that zinc is lost when some vegetation is heated above 450°C, but confirmation is lacking). Should these elements be of interest, "wet" ashing is necessary. In this procedure, the dried vegetable matter is gently digested in a boiling mixture of sulfuric and nitric acids, with small amounts of nitric acid continually being added until the liquid is pale yellow, and finally a few milliliters of hydrogen peroxide is added to remove the last traces of organic matter. The wet ash is then dried, and it can be analyzed like the dry ash, in the same manner as soil, sediment or rock samples. Further details on ashing have been presented by Ward et al., (1963) and Brooks (1972).

From the above brief description it should be clear that sample preparation is an important facet in the laboratory aspect of exploration geochemistry. In fact, Ingamells et al. (1972) have shown that the effects of improper grinding, splitting and other laboratory operations can result in errors which are devastating, sometimes reaching 100% or more. Garrett (1969) has examined the whole question of the determination of sampling and analytical errors in exploration geochemistry, and discussed methods whereby these parameters can be calculated. Others who have discussed the problems of sample collecting and/or analysis include Miesch (1967), Howarth and Lowenstein (1971) and Griffiths (1971), whereas Nicholls (1971) has considered the problems of sampling and preparation in the laboratory, in an attempt to isolate the various factors that contribute to imprecision of geochemical analyses. If done correctly, sample preparation is time and space consuming. In large-scale commercial operations the preparation of samples is more likely to be a source of delay in the reporting of results than any other part of the analytical operation. In recognition of these facts, most commercial laboratories impose an additional charge on samples which are bulky, difficult to dry or sieve, or otherwise present unusual problems in preparation.

ANALYTICAL METHODS — GENERAL COMMENTS

Analytical methods used in exploration geochemistry have changed significantly over the past 30 years. The emission spectrograph was the principal instrument used in the U.S.S.R., Europe, and, later, in parts of Africa, from the 1930's until the 1960's and it still is widely used in all these, as well as other localities. Colorimetric methods were the chief analytical techniques used in the 1950's and early 1960's in the Americas and some parts of Africa. Since the introduction and acceptance of atomic absorption for exploration geochemistry in the early 1960's, this particular method has overtaken, by far, all the other techniques, at least in North America and Australia. Table 6-1 gives a summary of analytical methods used in North America during 1970-71, and indicates that atomic absorption is used for the analysis of about 70% of all geochemical samples with colorimetry, emission spectrography and X-ray fluorescence in decreasing order of importance accounting for most of the remainder. However, the figures presented in Table 6-1 must be kept in perspective because they represent only a limited number of laboratories (even though together they analyzed over a million samples), do not include colorimetric or other tests done in the field, and do not include the U.S.S.R. which still relies heavily, although not exclusively, on emission spectrography. Also, owing to rapid changes in exploration objectives, analytical methods which

Table 6-1. Summary of analytical methods used in exploration
geochemistry in Canada and the United States.

Analytical Method	Canada		United States	
	Percent of total number of samples	Number of elements by each method	Percent of total number of samples	Number of elements by each method
Atomic absorption	72.6	38	61.7	40
Colorimetry	12.5	35	16.1	38
Cold extraction colorimetry	3.0	7	3.8	12
X-ray fluorescence	4.0	41	0.1	7
Emission spectrography	3.8	34	13.3	70
Paper chromatography	0.6	7	0.2	1
Selective ion electrodes	0.4	5	1.0	9
All others	3.1	33	3.8	35
Total	100.0		100.0	

Notes:

1. Source of above data: *Newsletter* of The Association of Exploration Geochemists (July 31, 1971). Percentages recalculated to sum to 100%.
2. Based on data received from 27 Canadian and 16 United States laboratories, which include some government, university, company, and commercial laboratories. These selected laboratories analyzed 812,456 samples in Canada, and 337,370 samples in the United States, during the period June 1, 1970, to May 31, 1971 (see Table 1-1).

are employed can change from year to year. Nevertheless, the data in Table 6-1 does form the basis for the following discussions and descriptions of the four main analytical techniques which are used for the greater number of all geochemical samples. These are followed by brief mention of other techniques currently used, generally for special situations. General descriptions of the methods used in geochemistry have been given by Smales and Wager (1960) whose discussions are still valid, and, more recently, by Wainerdi and Uken (1971) who cover much of the same ground but also include some additional modern methods such as atomic absorption.

ATOMIC ABSORPTION SPECTROMETRY

Atomic absorption spectrometry, commonly referred to as "A.A." (or A.A.S.), is a relatively new technique, having been suggested as an analytical method as recently as 1955 by Walsh, in Australia. It has had profound effects not only in exploration geochemistry, but also in metallurgy, water quality studies, biology, agronomy, environmental control and all the other fields in which inexpensive, accurate, precise and rapid methods for the determination of a large number of elements, at low levels of detection (high sensitivity) are necessary. Every serious geochemical laboratory has at least one instrument, and some mobile field laboratories are now equipped with atomic absorption. The method is especially applicable to the determination of trace or minor quantities of an element in geochemical samples, and the precision and accuracy are comparable to most other methods. One special appeal of atomic absorption in exploration geochemistry is the fact that, from one sample digestion, many elements can be determined, thus resulting in a considerable saving in costs while at the same time providing a wealth of information. About 40 elements can be determined at sensitivities and costs competitive with other methods, and perhaps 70 elements are theoretically possible.

There are at least half a dozen major manufacturers of equipment, some of the more prominent names in North America being Perkin-Elmer, Jarrell-Ash, and Techtron (Varian). Several books have been published on atomic absorption (e.g., Angino and Billings, 1972; Slavin, 1968), as well as thousands of journal articles describing special analytical procedures and results. Atomic absorption methods used at the U.S. Geological Survey for determining 15 elements (Ag, Bi, Cd, Cu, Pb, Zn, Co, Ni, Ca, Li, Na, K, Au, Te and Hg) are described by Ward et al. (1969), and Abbey (1968) has described the methods used by the Geological Survey of Canada. The field is dynamic in the sense that new equipment is constantly being introduced, and modifications of existing methods permit ever lower limits of detection to be attained. "Atomic Absorption Newsletter", published by Perkin-Elmer (Norwalk, Connecticut, U.S.A.), permits analysts to keep up to date with the latest developments, and is considered a very valuable source of information. A schematic diagram illustrating the optics and essential components (lamp, burner, monochromator (grating) and detector (photomultiplier)) of a typical instrument is shown in Fig. 6-5.

As the name implies, atomic absorption involves the absorption of light-type energy by atoms. For this to be achieved, the sample has to be brought into solution (with the exceptions noted below), which usually is very simple. The solution, containing the element(s) to be analyzed, is aspirated (dispersed or vaporized) by means of a flame (generally fueled by a mixture of air and acetylene) which has a temperature sufficiently high, perhaps 2000°C, to reduce most of the droplets into an atomic form. The result is that most of the elements in the vapor exist as neutral, unbound atoms in what is called the "ground" state. If the temperature is very high (about 8000°C), as in the case of emission spectrography, the atoms may be

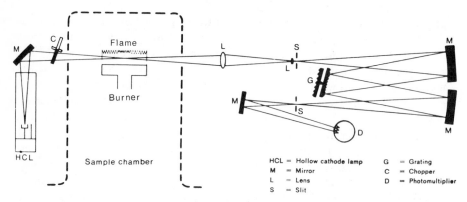

Fig. 6-5. Schematic diagram illustrating the optics of an atomic absorption spectrometer (Perkin-Elmer model 300). Courtesy of The Perkin-Elmer Corporation, Norwalk, Connecticut, U.S.A.

raised to the "excited" state, which, while good for emission spectrographic analysis, is detrimental to atomic absorption work.

The vapor containing the element to be analyzed, is illuminated by a light source, usually a hollow cathode lamp of the same element as that being determined, as shown in Fig. 6-5. Thus, for example, in the analysis of zinc, a zinc hollow cathode lamp radiates a light with a characteristic radiation. It is an inherent characteristic of atoms in the ground state (in the vapor) that they will absorb, at discrete frequencies, the incident energy radiating from the hollow cathode lamp; in the example above, zinc atoms in the vapor will absorb radiation from the zinc lamp. Zinc radiation from the hollow cathode lamp is then isolated by means of a monochromator, and its intensity, now reduced because of absorption in the flame, measured on a detector. The amount of this absorption is proportional to the concentration of the element in the vapor and, by measuring the decrease in energy being received by the detector, a quantitative determination of the amount of the element in the vapor (sample) can be made. One reason for the great sensitivity of the atomic absorption method is that about 95% of the atoms are in the atomic state in the vapor, and thus available for absorbing radiation from the hollow-cathode tube. By somewhat analogous methods, particularly emission spectrography, in which the ionized or excited state is required, only about 1 – 5% of the atoms are excited and thus capable of supplying the necessary radiation for analytical purposes.

As a generalization, it could be said that the atomic absorption method is free from interferences because only the wave length emitted by the hollow cathode source is measured, and no element other than that being analyzed in the vapor will absorb light. However, there are actually interferences of various types. One of the better known in the analysis of geo-chemical samples is concerned with a matrix effect (variations in the bulk composition of the sample) in lead determinations which affects accuracy, but not precision. Although this effect is minor in high concentrations of lead, at low concentrations the error can be considerable, giving results that are approximately 30 ppm too high, and corrections must be made (Fletcher, 1970). Other elements, such as Ag, Mo, Cd, Zn, Ni and Co are similarly affected, particularly at low concentrations (about 1 ppm), and this can be important when looking for subtle variations in the element distribution near threshold levels, especially in rocks (Govett and Whitehead, 1973). Various types of interference, referred to as chemical interferences, are caused when refractory compounds are formed within the vaporizing flame. Spectral interferences can also result when the mono-chromator is unable to resolve the closely spaced resonance frequencies at which the absorption is measured. Changes which can occur within the flame are extremely complex and are influenced by many factors which in-clude its fuel to oxygen ratio, the rate of aspiration, the size of the aspirated droplets, and the specific nature of the flame (several types of flames are used, e.g., air-acetylene for most analyses, but nitrous oxide or other gases are employed when a hotter or special flame is needed). From the above brief discussion of interferences and some possible problems, it should be clear that, although atomic absorption analysis may be relatively simple and straightforward in comparison with other methods, there has been a serious tendency to oversimplify it. Most technical persons can be trained to operate the equipment and produce good results, but it requires care, experience, and a large number of standards.

The instrumentation required for atomic absorption consists basically of a hollow cathode lamp, a burner, a monochromator, and a photodetector (Fig. 6-5). The simplest of instruments can be purchased for about $4,000, but all types of optional equipment are available, which, while of great value to the analysts from the point of view of increasing precision, ac-curacy, sensitivity and rate of sample analysis, also have the effect of ex-tending the cost to perhaps as much as $20,000. However, this is still con-sidered "relatively" inexpensive in the field of laboratory instrumentation. The optional equipment includes a hollow cathode lamp for each element to be determined (average cost of each lamp is approximately $150; multiple-element lamps using mixed metal cathodes should not be used, ex-cept for Ca-Mg, as spectra resolution is difficult), different types of burner heads, digital or strip-chart recorder readouts, carbon-rod (graphite) fur-naces, small computers with visual display of the metal content of an element in ppm, and so forth. Most commercial laboratories appear to prefer several medium-priced instruments (about $8,000), in preference to one of the more sophisticated, computer-connected types most commonly

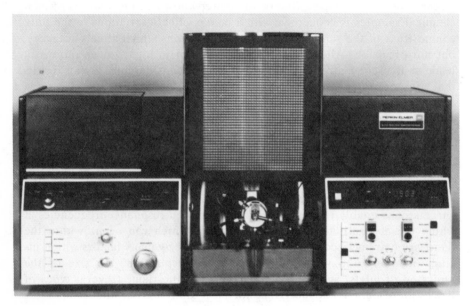

Fig. 6-6. The Perkin-Elmer model 503 atomic absorption spectrometer. Courtesy of
The Perkin-Elmer Corporation, Norwalk, Connecticut, U.S.A.

found at universities or research establishments. The newest Perkin-Elmer
atomic absorption instrument to be released is illustrated in Fig. 6-6. This
model costs in the vicinity of $15,000 when purchased with the accessories
normally advised for geochemical purposes.

Most of the time required for the atomic absorption analysis of soil,
sediment, rock and vegetable material is taken up with sample preparation
and the digestion process but waters, unless they contain significant quantities of dissolved salts such as in brines, can generally be directly aspirated.
Commercial laboratories have automated the sample preparation and
digestion processes, so much so that a well organized laboratory, with only
one instrument, can make about a quarter of a million determinations per
year. The acids and fluxes used for the decomposition of geochemical samples have been discussed above, but in all cases there is analyst preference.
As mentioned previously, most commercial laboratories use a nitric-
perchloric acid leach, but other institutions, such as the U.S. Geological
Survey, prefer a nitric acid leach for much of their work (Ward et al.,
1969). In some cases, after digestion, it is possible to use various organic
extraction reagents (most commonly APDC-MIBK) to concentrate the
elements and then directly aspirate the organic phase containing the metals
into the flame, which has the effect of increasing sensitivity. A list of such

extraction reagents for different elements is given in Wainerdi and Uken (1971). Another organic extraction process used specifically for mercury (Pyrih and Bisque, 1969) employs MIBK to extract a metal dithizonate for direct injection into the atomic absorption flame. In the case of very low concentrations of trace metals (e.g., Zn, Cu, Ni in the 1-10 ppb range) in river or lake water, preconcentration of the metals by solvent extraction may be necessary (Reeder et al., 1972).

Special procedures are required for several elements before they can be analyzed satisfactorily by atomic absorption. Gold and tellurium must be dissolved by combinations of hydrobromic acid and bromine, and this technique has been so successful as far as precision, accuracy and lower limits of detection (5 ppb) for gold are concerned (Ward et al., 1969), that relatively costly fire assays are no longer needed for exploration purposes. However, they are, of course, still needed for samples requiring the highest accuracy. Whereas most elements are determined from a solution, several are not. Mercury in solids, for example, is introduced directly into the atomic absorption instrument by the application of heat. This is possible because of the high volatility and high vapor pressure of elemental mercury and several of its compounds, the procedure being referred to as "flameless" atomic absorption. The ability of the elemental mercury vapor to absorb discrete energy frequencies has been used for almost a century, at least qualitatively, and, historically, this is the oldest example of atomic absorption. For the determination of arsenic and selenium at very low levels (ppb range), samples are chemically treated with hydrogen, forming gaseous arsenic and selenium hydrides, which are then introduced into the atomic absorption instrument for analysis by means of special apparatus. However, in higher quantities (ppm range), both arsenic and selenium can be determined by conventional atomic absorption methods from solutions.

The above example of the use of gaseous compounds of arsenic and selenium, in the place of the usual solutions, points out the various modifications used to extend the capabilities of atomic absorption spectrometry. The future holds forth the hope of converting solid samples directly into an atomic cloud for analysis and the possibility of perfecting multi-element systems also exists. A system has been described (MacLiver et al., 1969) which uses resonance detectors for each element of interest, around a common atomizing flame. This multi-element system does not employ a spectrometer, the most expensive part of present instruments. Jarrell-Ash markets at least one atomic absorption instrument, employing a spectrometer, on which two elements can be determined simultaneously. The future will see more effort to perfect multi-element capabilities, as well as to increase sensitivities and to analyze all types of samples without putting them in solution. The recently introduced carbon rod (graphite) furnace has achieved the last two objectives.

COLORIMETRY

General Comments

Colorimetry was the first method commonly used for the analysis of geochemical samples in the Western World, and was especially important during the period from the 1950's until the mid-1960's. From the data in Table 6-1 it is clear that colorimetry (including cold extraction colorimetry) is still a very important technique as about 16% of the samples analyzed in the 43 laboratories represents a very significant proportion of the total. On a world-wide scale, the 16% figure is probably low, because smaller laboratories and field colorimetric determinations are not included (Fig. 6-7). Therefore, although the use of colorimetric methods has declined in recent years, most probably the low point has been reached and the future will see continued reliance on some of these very important techniques which have both field and laboratory advantages.

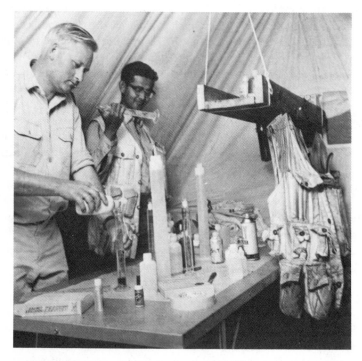

Fig. 6-7. Field (tent) geochemical laboratory in Canada in which colorimetric methods are used. From Lang (1970).

The methods are very versatile, and they can produce a large amount of moderately precise data (generally accepted as plus or minus 15 to 25% of the amount present) simply and very inexpensively. In fact, there are almost no capital costs associated with colorimetry as used in exploration geochemistry, although some laboratories do have spectrophotometers. In addition, technicians and summer field parties can be trained in relatively short periods of time to produce reliable analyses. For these reasons, plus the fact that it is extremely good experience for students to be exposed to colorimetric procedures from the point of view of realizing the low quantities which need to be measured, such methods are discussed in more detail than might be expected in a book of this type.

A decade ago, before atomic absorption equipment became readily available, most exploration groups had at least one person experienced in a few of the simplest of the colorimetric field and laboratory methods. Today many organizations, at least on the North American continent, lack the expertise to arrange for, or understand the full value of, these analytical methods. This is regrettable, as the simplest of field methods have very great applicability in many circumstances. For example:

1. Rapid, on-site decision-making during staking rushes.
2. On-site decision-making as to the direction exploration parties should follow during drainage basin (reconnaissance) programs. This is particularly applicable to those parts of the world with short exploration seasons (such as northern Canada) where an anomaly, if found after analysis in a distant laboratory, cannot be investigated until the following year. In most situations of this type it is quite costly to return to the same location. Similar reasoning would also apply to anomalies found in remote locations with longer field seasons.
3. On-site decision-making during the evaluation of follow-up studies of recognized anomalies, or for the direction of drilling programs. The author has personally witnessed one exploration group in northern British Columbia, without analytical capabilities, waiting a week for inclement flying weather to clear so that a helicopter could pick up their samples which were then to be flown to Vancouver for commercial analysis. In this case, the exploration party could not make a decision until the analytical results were obtained and these required an additional week of the relatively short field season.
4. It has been suggested by some exploration supervisors that field parties are likely to be more motivated if they actually analyzed the samples they collect, either in base camp laboratories or on-site.
5. The cost of colorimetric determinations in the field, or in the exploration geochemists' own laboratory, are minimal. Savings in exploration budgets can be achieved by judicious screening of samples by the appropriate colorimetric method(s).

The above discussion of the advantages of the colorimetric field or laboratory tests should not be misconstrued to mean that the central laboratories for the large organizations, or commercial laboratories, are not needed. Quite the contrary. Certainly well-equipped laboratories with modern equipment capable of determining a wide range of elements accurately are essential. Even if colorimetric methods are used in the field in the cases mentioned above, a certain percentage of the samples should be sent for verification, and all samples collected and analyzed on-site or in base camps should be saved until it is certain that they have no further value.

In all fairness it is necessary to point out some of the objections which have been voiced in connection with colorimetric analyses, particularly those made in the field. These include the fact that they are generally less quantitative than those performed in the laboratory, and examples are known where some field parties have spent a considerable amount of time investigating the first anomaly they encountered whereas stronger and more significant anomalies would have been found if the regular sampling program has been followed.

The fundamental theory of colorimetry is simple. The determinations involve the formation of colored solutions by appropriate chemical treatment of the samples. The color can then either be measured visually as is done in most geochemical applications, or, provided a mono-colored solution is being determined, with a spectrophotometer as is common in other fields of chemistry, or when greater accuracy and precision are required. Colorimetric methods require a series of standards for comparison. These series are of two types, one being the mono-colored referred to above, and the other being a mixed color. The mono-color series consists of increasing intensities of one color (e.g., pink in the case of copper determined by biquinoline) corresponding to an increase in the metal content, whereas the mixed color series results from a mixture of unreacted reagent (e.g., dithizone in the zinc determination), in varying proportions, with a colored metal-containing solution (see color plate in Appendix B). All measurements are done in solution, which makes it relatively easy to prepare synthetic standards with which to calibrate the procedures. The average person is reported to be able to discriminate, in favorable ranges of concentration, between differences of 5% (Wilson, 1966, p. 222); only about 3-5% of the male population is more or less color blind, and unable to match colors.

In attempting to analyze geochemical samples colorimetrically, whether in a permanent well-equipped laboratory or in the field, there are certain fundamental procedures and precautions which must be adopted, but the manner in which these are followed will vary greatly. It is beyond the scope of this book to describe in detail the various modifications and, therefore, only some general comments will be presented. More detailed discussions can be found in Ward et al. (1963) and Stanton (1966).

Apparatus

Most colorimetric methods are designed primarily to produce large amounts of moderately precise data rapidly, simply, and inexpensively. The type of equipment used to achieve these objectives need not be elaborate. For example, in the procedure described in Appendix B for the determination of zinc, a hot water bath is needed in which the samples simmer in test tubes for ½ hour. The normal tendency is to use commercial water baths warmed on hot plates, but equivalent results can be achieved inexpensively from any type of vessel more than about 3 or 4 inches deep (e.g., baking utensils or a common kitchen roaster), on top of which is placed some type of cover with holes of a size appropriate to keep test tubes in an upright position (even chicken wire is satisfactory). Alternatively, a test tube rack may be placed directly in the water bath. Similarly, for a test tube holder, a wire mesh of appropriate dimensions will suffice. Heat for the water bath can be obtained from stoves in cities, or from wood or charcoal fires or propane burners in more remote locations.

Pre-weighing of reagents into small bottles or vials eliminates the need for an analytical balance in the field. Reasonably accurate measurement of sample weight can be achieved by means of scoops, with depressions calibrated to be roughly equivalent to 0.1 gm (Fig. 6-8). These scoops are generally made of plastic or aluminum bars about three inches long, 0.5 inch wide and 0.5 inch thick, and have a hole about 1/8 inch deep and 1/4 inch in diameter, which is placed near one end of the bars (the weight of powder contained in this hole should be calibrated). Inexpensive automatic pipets or calibrated test tubes, the latter made by scratching with a diamond point, eliminate the need for burettes which are especially difficult to transport. In other words, the principal limiting factor in determining the amount of apparatus taken into the field, or purchased for an

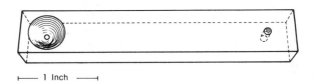

⊢——— 1 Inch ———⊣

Fig. 6-8. Plastic scoop with concial shaped hole at one end calibrated to accept 0.1 gram of soil or sediment. Thin hole at the opposite end is for a cord or chain to prevent loss.

established laboratory, is the ingenuity of the geochemist. Yet there are certain types of equipment which are almost always essential to most field operations: test tubes, polyethylene test tube caps (not cork), separatory funnels, polyethylene bottles of various sizes and shapes, pipets, precalibrated scoops, graduated cylinders, nylon or aluminum screening, and crucibles for special analyses such as for molybdenum and tungsten. The required number of each item will depend on many factors, such as the size of the operation, the number of different tests being made, and the number of analysts and their assistants.

The amounts of chemicals needed are predictable in the case of buffers, acids and most other reagents. For example, for most of the colorimetric tests described in Appendix B, one liter of buffer is good for 200 tests as each determination requires 5 ml. The amount of solvents required is less predictable as variable quantities are used in the purification of water. Water from streams and other local sources is usually of sufficient purity to be used for preparing reagents, but it can be easily purified with dithizone solutions if necessary. There are, however, special circumstances in which water may have to be brought to field laboratories, for instance if water from any source is in short supply, since relatively large quantities are needed for such purposes as cleaning test tubes and other apparatus. The discarding of test tubes, rather than cleaning them with hard-to-obtain water, may be a more practical solution in some cases, but is to be discouraged in this pollution-conscious era. In some analytical procedures described in the literature, particularly those of the Geological Survey of Canada, the exact amounts of each reagent and the pieces of equipment required for a specified number of determinations (usually 1000) are presented, as for example in Gilbert (1959), Holman (1963) and Smith (1964).

There are several companies in North America which offer "kits" for field parties and these contain all the essential analytical apparatus, premixed chemicals, instructions, and a price list for the replacement of chemicals and solvents, for the most common field tests (usually coldextractable total heavy metals with dithizone, and copper tests with either biquinoline or dithizone). Experience has shown that these are generally acceptable analytically, but they are usually quite costly in comparison with what can be obtained by following the instructions in the Geological Survey of Canada publications mentioned above, or compiling what is necessary from the instructions in Ward et al. (1963), Stanton (1966) or the methods as outlined in Appendix B. In some cases, however, there are no individuals capable of preparing the necessary solutions or assembling the equipment, and then the commercially available kits are the answer, especially if time is of the essence. Yet with most kits, the field geochemist is not certain as to exactly what method he is using, the chemicals and solvents being labelled as solution "A" or "B", or referred to by trade names. In these cases it is advisable to obtain kits from those suppliers who are the most informative with respect to the nature of their product so that interpretation of the results will be more meaningful. One very convenient feature in at least one commercial kit is pre-weighed dithizone in capsule form. Notwithstanding the simplicity of the kits, given an understanding of the essential facts relating to colorimetric methods presented here, a geochemist or geologist should not only be able to prepare his own equivalent, but in doing so he will be in a better position to interpret the significance of the findings. Most commercial geochemistry laboratories, upon request, will prepare solutions and supply the necessary chemicals and equipment for field testing (e.g., the various tests in Appendix B), and when ordered in this manner, costs should be lower in comparison with pre-assembled kits.

General Comments On Reagents

Any discussion of reagents in exploration geochemistry must include two topics: purity and safety precautions. In the case of purity, every effort must be made to obtain reagents which have as low a metal content as is possible. In some cases, this means purchasing the more expensive reagent grades, but, even so, one cannot be absolutely certain that the reagents are metal-free. It is not unusual, for example, to find several ppm Pb in analytical grade nitric acid, and aqueous ammonia often contains zinc. In many cases the reagents used in preparing the buffers contain significant amounts of heavy metals, which are still within the accepted limits specified for these reagents for general analytical purposes. The geochemist is thus faced with two choices: (a) searching for a source of metal-free acids, solvents and reagents from among several suppliers, or (b) accepting what is available and following the "cleaning" procedures, especially for reagent chemicals, given in the individual analytical procedures in Appendix B, and also in Ward et al. (1963), Stanton (1966), and in the Geological Survey of Canada publications. Although, in fact, within North America generally little trouble is experienced with high purity reagents from prominent manufacturers, most practicing geochemists choose the latter alternative. With respect to the organic solvents, there are conflicting reports in the literature as to their likelihood of being contaminated, but most workers have had little trouble in recent years. Methods of purification of these solvents are described by Stanton (1966). The general tendency among most geochemists is to find one source of solvents whose purity has been confirmed, and to continue with this supplier. In field situations this may not be feasible, and awareness of possible contamination must be uppermost in the analyst's mind and checks run frequently, especially when new reagents are used.

No discussion of purity reagents and chemicals can be complete without a special word of caution with regard to water. Whenever water is mentioned in analytical procedures, it is automatically assumed that it is at least metal-free, and preferably deionized. The average dissolved solid content of rivers of the world is about 120 ppm, most of which is Ca, Mg, Na, K, HCO_3, SO_4, SiO_2, and Cl, and these will have no effect on the heavy metal colorimetric tests. This is the reason why waters of many streams can be used directly for analytical purposes in exploration geochemistry. Should the stream contain as much as 0.1 ppm zinc or other metals combined, as is common, for example, in certain parts of Colorado and other mineralized areas, and in the tap water supplies of some cities, it results in high blanks in colorimetric or other instrumental tests, as large amounts of water are used in preparing solutions in the dilutions required in the procedures. However, the heavy metals can be easily removed from water by "cleaning" with dithizone solutions (the technique is described in the colorimetric method for zinc; Appendix B), or by means of passing the water through small, easily available, demineralizers containing mixed-bed type ion-exchange resins which also remove other cations and anions. Any contamination from water can be readily recognized by running blanks along with samples being analyzed, assuming the other reagents are pure. In geochemical laboratories equipped with all-glass distillation apparatus, the deionized water is trouble-free, but in those laboratories which receive their deionized water from a large, central distillation installation, as is common in many universities, metal contamination is likely to be introduced somewhere in the system and a content of 10 ppb total heavy metals is not at all unusual.

With regard to safety precautions, it must be remembered that most reagents and chemicals used in analytical chemistry are poisonous. Fortunately, the chemicals used in the colorimetric techniques described in Appendix B are less dangerous than some, but nevertheless great caution is advised, especially for students. In the case of those organic solvents with high volatility (e.g., chloroform, toluene, carbon tetrachloride), even small amounts inhaled over an extended period can be dangerous to the lungs and liver, and therefore, excellent ventilation is required. Bloom (1963) has discussed the toxic properties of several organic solvents. These same organic reagents may have serious effects on the skin of some individuals and it is recommended that contact be minimized by using plastic gloves. The geochemist must also be aware of other potential hazards such as inflammability, explosions and burns from acids for which the best protection is good laboratory procedures and a knowledge of the characteristics of reagents and solvents. Dangerous materials are always noted on the labels for North American and many other suppliers. In order to eliminate unnecessary hazards, techniques which require especially dangerous chemicals such as ethers, cyanide solutions, perchloric acid or hydrofluoric acid, have been omitted from the colorimetric tests described and recommended in this book.

Colorimetric Reagents

Of the various colorimetric reagents which are of importance in exploration geochemistry, by far the most important is dithizone, which is used in three of the five important colorimetric methods described here. 2,2′-biquinoline required for the determination of copper, and dithiol (as zinc dithiol) required for the determination of molybdenum and tungsten, are also important. Among the many other colorimetric reagents, there has been recent interest in α-furildioxime for the determination of nickel both in the laboratory (Stanton, 1966) and in the field (Nowlan, 1970), and for palladium and platinum (Grimaldi and Schnepfe, 1968).

Dithizone, or diphenylthiocarbazone, was first prepared by Emil Fisher in 1878 who noted that when it reacted with heavy metals, brilliantly colored products resulted. However, this observation did not attract much attention until about 1925 when H. Fisher was able to show the great value of dithizone in many trace element analyses (for a detailed discussion on dithizone, and many references to earlier papers, see Sandell, 1959). Dithizone began to be used as a colorimetric reagent in exploration geochemistry in the late 1940's and early 1950's, and for many years thereafter, formed the analytical basis of many exploration programs. It is commonly abbreviated, or spoken of, as "Dz". Although in recent years the use of colorimetry has declined, dithizone colorimetry is, nevertheless, still an important factor in exploration geochemistry and it will continue to enjoy its well-deserved status in the field and in many of the developing countries which lack atomic absorption equipment. Dithizone is inexpensive and exhibits the most desired characteristics of colorimetric reagents: high sensitivity, specificity (or selectivity), and good reproducibility under proper analytical conditions. Its sensitivity exceeds that of spectrography in many cases, and equals it in others.

Dithizone is a violet-black solid which dissolves to different degrees in many organic solvents, the most popular being chloroform, toluene, and carbon tetrachloride, but benzene, hexane, xylene, and other petroleum distillates such as "Shellsol", "Varsol", "White Spirit" (James, 1964), and even un-leaded (white) gasoline (Anthony, 1967) have been used. The resulting solution is green in all solvents, but its stability depends on the stability and purity of the solvent used. Usually hydrocarbons, such as benzene and toluene, yield more stable solutions than the chlorinated solvents such as chloroform or carbon tetrachloride. Under any conditions, whether in solution or in the solid form, dithizone is very sensitive to strong light, particularly sunlight and other ultra-violet sources, and to high temperatures. To prevent decomposition under these conditions, the dithizone should be kept in a dark, cool place, and solutions stored in tightly-sealed thermos bottles, preferably in a refrigerator. The more dilute the solution, the more susceptible it is to the effects of heat and light, and for this reason, dithizone solutions are always transported in the field in dispensing bottles wrapped in aluminum. The problem is especially serious in the tropics, but, with proper precautions, difficulty can be avoided. Dithizone stock solutions are generally stored as 0.1% or 0.01% solutions, and are stable for several weeks under refrigeration. Test solutions are generally 0.001% and must be prepared fresh daily, and perhaps twice daily in warm climates.

Dithizone reacts with the following 19 metals, in appropriate valence states, to form colored dithizonates: Mn, Fe, Co, Ni, Cu, Zn, Pd, Ag, Cd, In, Sn, Te, Pt, Au, Hg, Tl, Pb, Bi and Po. These are commonly abbreviated to, e.g., ZnDz, in the case of the zinc dithizonate. The reaction can be made specific by means of (a) controlling the pH of the solution to be extracted, and/or (b) adding complex-forming (masking) agents which tie up other reacting metals. Details of the chemistry of the reactions are given by Sandell (1959), who also describes the method of purification of dithizone. However, in recent years, dithizone purchased from first-class suppliers has rarely, if ever, required purification. Hawkes (1963) has discussed the steps in the reaction between dithizone and metal in non-chemical terms, which can be summarized as:

1. extraction of metal from the sample into an aqueous solution of either acid or buffer;
2. formation of metal dithizonate, which occurs when the sample is shaken after the addition of dithizone, followed by;
3. estimation of the metal content of the original sample by comparing the color of the organic layer with a set of standards.

The most common type of reaction between metals and dithizone results in a "mixed-color" series, in which unreacted green dithizone is mixed in varying proportions with a colored metal dithizonate such as the red zinc dithizonate (see color plate). This results in a series of varying colors consisting of green, blue-green, blue, purple and red, depending on the relative proportions of the unreacted dithizone and the colored dithizonates. However, in some determinations involving the use of dithizone, mono-colored series are produced (e.g., varying degrees of orange in the case of mercury; Ward et al., 1963).

Although most users of dithizone have experienced little difficulty with the results, or the chemical itself, geochemists should be aware of possible difficulties which may arise. Hawkes (1963) has summarized the likely analytical problems.

1. Yellowing of dithizone: probably caused by reactive components in the organic solvent; the dithizone solution in this case would be spoiled and must not be used again. A different source of the solvent should be sought.
2. Fading of dithizone: the loss of color is most commonly caused by the presence of organic matter in the sample and, therefore, can be avoided by eliminating this when the samples are collected, or by igniting the sample for a few minutes (to 450–550°C) prior to analysis (a propane torch is satisfactory in the field).

3. Poor reproducibility of data:this is generally related either to inhomogeneity in the samples or to the lack of standardization of the analytical method. These problems can generally be corrected by restricting samples to the same grain size and same organic matter content, fine-grinding, and by standardizing the shaking time (e.g., exactly 30 seconds if the procedure calls for that) during the extractions, as the amount of metal extracted is often time-dependent.
4. Contamination: this problem has been discussed above, and is especially serious when dithizone is being used as a colorimetric reagent because of its great sensitivity, and also because it is able to react with so many of the more common heavy metals.

Biquinoline and dithiol are the two other colorimetric reagents most commonly used for exploration purposes and both form mono-colored series. They present no special problems other than those mentioned in Appendix B in the description of the copper method, and the molybdenum and tungsten methods, and hence they are not discussed further here. Additional information on the chemistry of these reactions may be found in Sandell (1959) and Stanton (1966).

Standard Solutions And Matching Standards

Closely associated with the calculation of trace element contents are two other items which merit brief discussions: (1) standard solutions and (2) matching standards.

(1). Standard solutions are prepared for all colorimetric tests and contain a certain amount (weight) of the element under consideration in a specific volume of solution. Usually, standard solutions are expressed as micrograms per milliliter ($\mu g/ml$), but occasionally the essentially equivalent ppm, is used. In the zinc colorimetric procedure presented in Appendix B, the first standard solution prepared contains 100 μg of zinc per·ml of solution (100 $\mu g/ml$), which is reasonably stable and may be kept for several months. For purposes of making a standard series to be used for matching colors (see color plate), a more dilute standard solution is necessary and, in the case of zinc, the 100 $\mu g/ml$ standard solution is diluted with metal-free water so that a $10\mu g/ml$ solution is obtained. The more dilute the resulting solution, the more likely it is to deteriorate for any number of reasons, such as the metals becoming adsorbed onto the walls of the container. Although there are no definite rules as to when a dilute standard solution should be discarded, it probably should be replaced every few weeks, and considered as a likely source of error at all times. Directions for preparing standard solutions are given with each colorimetric procedure. Standard solutions for many elements are now available from most chemical supply houses, usually as "atomic absorption standard solutions". These are commonly offered in concentrations of either 500 ppm or 1000 ppm, and are likely to remain stable for at least a year. As might be expected, a considerable premium must be paid for the convenience of obtaining these pre-mixed standard solutions.

(2). Matching color standards are necessary for all colorimetric procedures in which visual comparisons are to be made (the color plate illustrates four color standard series). Even for those mono-colored series which are determined instrumentally by means of a spectrophotometer, a standard series of some type is necessary. In all cases, the standard series are expressed as micrograms (μg or "γ") of the element under consideration. In the case of the mixed-color zinc series, for example, the standards are 0, 1, 2, 3 and 4 micrograms of zinc. In the mono-color tungsten method, the standard series ranges from 0-3 micrograms of tungsten, but there are also several standards in the range of from 0-1 micrograms, as these are often necessary because of the low content of tungsten in many soils, sediments and rocks. It is possible to modify the range of a standard series for some procedures by, for example, changing the strength of the colorimetric reagent, but the ranges presented in Appendix B, and by Ward et al. (1963), and Stanton (1966) are applicable to most geochemical exploration analyses likely to be encountered.

Other than the obvious color changes which will result from deterioration of dithizone or other colorimetric reagents with time, the color obtained when preparing standard series can vary when new chemicals are used. Therefore, standard color series used for matching purposes must be prepared frequently. Of even greater concern in this regard is the requirement that all standard series, and samples being analyzed, be compared under identical lighting conditions. The ideal conditions call for a white fluorescent color-comparator (available from all chemical supply companies) but as a poor alternative, a sheet of white paper may be placed behind the test tubes being compared in as white a light as can be obtained. Ward et al. (1963) discuss methods of preparing plastic suitable for color standards by means of adding dyes and pigments to certain materials. Also, suppliers of geochemical kits will often supply plastic colored standards, but it must be remembered that because of deterioration of chemicals and changes in operating conditions, these can only be considered as approximations.

The color plate included in this book has been prepared in such a manner that it is as representative of the true colors as is possible. Those following the procedures outlined in Appendix B may well obtain colors which vary slightly from those on the color plate. This is to be expected considering the slight variations in chemicals, operating conditions, and the difficulties in preparing the color plate.

Calculation of Trace Element Concentrations

Most analytical procedures used in exploration geochemistry are based on the gram (a weight) for solid samples. In the case of water samples, analyses are based on a volume, either a milliliter (ml) or liter (l). (Assays for precious metals are reported on larger units, such as ounces per ton, or pounds per ton in the case of uranium and other metals). If, in an analytical determination, a one gram sample was digested and the total metal content was found to be, for example, two micrograms (μg), the calculated metal content would simply be 2 $\mu g/gm$, or 2 ppm. Or, if instead of an exact one gram sample, 0.2 gm was scooped for a geochemical analysis of zinc, and the same 2 micrograms were found, the zinc content would be 2 $\mu g/0.2$ gm, or 10 ppm. In other words, if the entire sample is digested and its entire metal content is weighed, or determined by comparison with known colorimetric standards, a simplified equation to express the content of the sample would be:

$$\text{ppm (of metal)} = \frac{\text{micrograms } (\mu g) \text{ of metal found (by comparison with standards)}}{\text{weight of sample (gm)}}$$

Remembering that $\mu g/gm$ is equivalent to ppm, in the latter example for zinc this would reduce to:

$$\text{ppm (Zn)} = \frac{2}{.2} = 10$$

However, in actual practice it is not usually possible, or desirable, to use an entire sample, and two other factors then enter into the calculation of metal concentration: (a) the volume of the sample solution (in ml); almost all procedures call for dilutions of varying amounts, and (b) the aliquot of the diluted solution from which the metal content is finally determined. Adding these two additional factors to the above equation (and shortening the above so all can be printed on one line) we have:

$$\text{ppm (metal)} = \frac{\text{micrograms } (\mu g) \text{ of metal found}}{\text{weight of sample (gm)}} \times \frac{\text{volume of sample solution (ml)}}{\text{aliquot (ml)}}$$

In the descriptions of the colorimetric methods for zinc, copper, molybdenum and tungsten, examples of the use of this formula are presented (Appendix B).

There is a great advantage in having the two added components of the formula, because it permits extension of the range of values which can be determined for the sample to be analyzed. For example, if it was found that with a 2 ml aliquot a red-colored zinc dithizonate was obtained in the organic layer, and this exceeded the highest (4 μg) zinc standard (see color plate), it is only necessary to repeat the final part of the analysis using a smaller aliquot (perhaps 0.5 ml) in order to extend the analytical range by a factor of four. By careful examination of the above formula, it can be seen that it is also possible to vary the volume of the sample solution (dilutions), and should these factors be insufficient for any specific purpose, then another sample using either a larger or smaller weight (gm) can be analyzed to extend the analytical range lower, or higher, respectively. Thorough understanding of the above formula is essential for those expecting to obtain the maximum from the analytical methods of trace analysis. Although the optimum range of applicability of the colorimetric methods described here is generally less than 0.1%, the methods can easily be extended to about 0.5%, and by very accurate weighings and dilutions in the laboratory, it is possible to extend some to the percent range, although accuracy is likely to suffer.

Erroneous Analytical Data

Mention is made above to possible impure chemicals which may produce erroneous results, and contamination in exploration geochemistry in general is discussed in Chapter 4. However, one of the most important sites for contamination and erroneous data is the chemical laboratory itself and particularly when using colorimetric methods. In most instances this source of error can be traced either to careless laboratory procedures or to the use of personnel unfamiliar with the pitfalls possible when analyzing for trace elements. Below is a discussion of twelve common sources of contamination and erroneous data and/or methods for detecting or avoiding them. It must, however, constantly be borne in mind that sources of erroneous data other than those mentioned below are also possible.

1. At least one blank with a known metal content should be run with each group of unknowns, whether the analysis is performed in a permanent or field laboratory. It is even preferable to run several standard samples covering a suitable range of values. If the values obtained on the standards are not acceptable, the source of error must be found before proceeding further.

2. Analyses which yield unusually high results and which are likely to be of special interest, should be repeated before being reported. This will ensure that time, effort, and money will not be wasted in the investigation of apparently anomalous areas which are entirely the result of faulty analyses. The same advice is applicable to results obtained from commercial laboratories which are not immune to reporting errors from any number of causes, not the least of which are the mislabelling of samples, or errors introduced during the typing of reports.

3. When any new source of chemical reagents, solvents, or water is used, this should be noted and the possibility of contamination considered. Even shipments from normal suppliers should be recorded, as supply houses occasionally change their sources.

4. Metal-free water must be used in all tests. Changes in the metal content of distilled water from central installations in universities over a period of time have been reported frequently.

5. The use of polyethylene or other similar plastics, and pyrex or other hard boro-silicate-type glassware, is essential. Soft glass is a source of heavy metal contamination.

6. All glassware and polyethylene, including test tubes, must be thoroughly cleaned before

use, and cleaned frequently during the course of analytical work. Test tubes must be thoroughly cleaned and rinsed with metal-free water after each use. Reagent bottles, whether of polyethylene or glass, should be labelled and used only for one specific purpose. If other uses are necessary, thorough washing with strong acid is essential.

7. All rubber items such as stoppers, tubing, and gloves must be avoided as rubber is a source of zinc contamination. Dithizone solutions in rubber-sealed thermos bottles have been known to be contaminated because certain solvents dissolve the rubber, thus releasing the zinc. Cork test tube caps should also be avoided as they are porous and satisfactory cleaning is difficult; plastic caps are now readily available. Stopcock grease and filter paper are also possible sources of contamination, as is any metal part of equipment, or coins in the analyst's pockets.

8. All plastic items such as test-tube caps, pipets and funnels should never be placed directly on laboratory bench tops as contamination may result.

9. Analysts should maintain careful working habits, ensuring that their hands are always clean, and their fingers never touch those parts of the test tube caps or samples which can result in contamination. Personnel to work in laboratories should be chosen with care.

10. The possibility of contamination during sample preparation and grinding is always a problem. Sample grinding and sieving should always be carried out in an area which is completely separate from the laboratory, and dust, which is a source of contamination, must be eliminated. If grinding equipment is not thoroughly cleaned, contamination from one high-grade sample can be reflected in erroneous values after several additional grindings. Steel grinding plates will often be alloyed with W, Co and Mn, among other elements and, therefore, ceramic grinding plates are recommended even though they are considerably more expensive and do not last as long. In shaker-type grinders, ceramic balls should be used, if possible, instead of the more common steel or tungsten carbide types which usually have other elements alloyed.

11. Screening and sieving should be done with nylon or aluminum screens, and stainless steel screens should be used only as a last resort. Even though some studies have reported that contamination from brass or copper screens is minimal under controlled conditions, these should not be used. Aluminum tea strainers (or other kitchen utensils) are satisfactory for screening, especially in the field.

12. Polyethylene tends to adsorb elements from solutions, particularly the inexpensive, "soft" polyethylene bottles so common as containers for household products. This is particularly a problem with very dilute solutions, such as 1 μg/l standards.

Selected Colorimetric Analyses

There are about 30 elements of interest in exploration geochemistry which can be analyzed by relatively simple colorimetric methods and which do not require sophisticated or expensive equipment. In some cases there are several different colorimetric methods for the same element, and also specific methods for the same element in different environments (for example, in soils and water). Therefore, those interested in doing their own analyses in the field or in the laboratory have available a wide choice of possible methods. For exploration geochemistry purposes, the analytical procedures presented by Ward et al. (1963), and Stanton (1966) should be consulted first, as the methods they have published are all proven to be satisfactory under the conditions they outline. The serious student of colorimetric methods of trace metal analysis should consult Sandell (1959), clearly the most detailed and fundamental study on the subject. In addition, the International Union of Pure and Applied Chemistry (IUPAC) (1964) report on reagents and reactions for inorganic analysis contains recommendations and comments on colorimetric reagents of geochemical interest with particular application to qualitative spot tests. Feigl (1958) is also an excellent source for data on spot tests.

From the large, and sometimes bewildering, choice of available colorimetric methods, five basic ones have been selected for inclusion here because of (1) their wide application to the most general type of exploration geochemistry program; (2) their relative simplicity with regards to techniques and equipment; (3) their use of chemicals which are non-explosive or not exceptionally dangerous; and (4) in the case of tungsten and molybdenum, to illustrate that there are still certain practical situations in exploration geochemistry where modern instrumental methods of analysis simply cannot fulfill all the requirements. A list of 25 additional elements of interest in exploration geochemistry which can be analyzed by colorimetric methods, and appropriate references, are listed in Table 6-2.

The five colorimetric methods described below and in Appendix B illustrate a broad spectrum of: geochemical methods of sample digestion (e.g., acid digestion, fusion); materials which may be analyzed (soil, sediments, rocks and water); colorimetric reagents (dithizone, biquinoline and dithiol); a specific element reaction (e.g., copper-biquinoline); and a non-specific reaction (cold extractable "total heavy metals"). For each of the five general methods, a brief description is presented below which attempts to summarize its particular usefulness and limitations, but only essential facts as to the nature of the chemical reactions are included. For convenience, details of the procedures and the color plate are combined in Appendix B.

It must be stressed that those anticipating using these methods should be *thoroughly* familiar with them before leaving for the field, or setting up analytical facilities in a base camp. If a base camp is to have only a minimum amount of equipment, as is usually the case, then it is preferable to bring to the field prepared buffers, solvents whose purities have been confirmed, as well as chemicals pre-weighed into appropriate receptacles in case it proves necessary to prepare additional reagents. This same advice is also applicable to some large cities in supposedly advanced countries where, in reality, appropriate supplies are often difficult to obtain at short notice. It is rarely necessary to bring water to the field, as has been mentioned previously, as the water of most streams in exploration areas is usually sufficiently pure for direct use (exceptions would be found in areas contaminated by present or past mining operations). In those cases where the metal content of a stream or public water supply is found to be unacceptable after testing with a dilute (0.001%) dithizone solution, it is a simple matter to extract the metals from the water with a 0.1% dithizone solution, a separatory funnel, and other very simple apparatus which can be easily obtained (the technique is described in Appendix B). Experience has shown that it is wiser to concentrate on one or two of the elements for field analysis rather than to attempt to have many readily available for any one project. Careful thought is clearly required for the proper selection of the method(s) most suitable for any specific project.

The colors which should be obtained for each of the colorimetric tests are illustrated on the color plate. The colored areas on the plate are the actual size of the colored organic layers for the zinc and copper reactions, if the procedures outlined in Appendix B are followed exactly. In the case of molybdenum and tungsten the colored organic layers have been doubled in size to permit better reproduction.

Table 6-2. Additional colorimetric tests of interest in exploration geochemistry.

Element	Material	Reference	Element	Material	Reference
Antimony	S,R	1,2	Nickel	S,R	1,2,5
Arsenic	S,R	1,2	Niobium	S,R	1,2
Barium	S,R	1,2	Palladium	S,R	6
Beryllium	S,R	2	Phosphorus	S,R	1,2,11
Boron	S,R	3	Platinum	S,R	6
Bismuth	S,R	1,2,9	Selenium	S,R	1,2
Chromium	S,R	1,2	Silver	S,R	7
Cobalt	S,R,W	1,2,4	Thorium	S,R	8
Germanium	S,R,C	1,2	Tin	S,R	1,2
Iron	S,R	1,2	Titanium	S,R	1,2
Lead	S,R,W	1,2	Uranium	S,R,W	1,2
Manganese	S,R	1,2	Vanadium	S,R	1,2,10
Mercury	S,R	1,2			

Notes:
1. Material: Plant material can be analyzed in most cases by the methods used for soils and sediments after it has been dry ashed.
   S = soil and sediments; R = rocks; W = water; C = coal
2. References: 1. Ward et al. (1963); 2. Stanton (1966); 3. Stanton and McDonald (1966); 4. Canney and Nowlan (1964); 5. Nowlan (1970); 6. Grimaldi and Schnepfe (1968); 7. Nakagawa and Lakin (1965); 8. Stanton (1971a); 9. Stanton (1971b); 10. Stanton and Hardwick (1971); 11. Peachey et al. (1973).
3. In all the above methods the desired metal is extracted from an aqueous solution into an immiscible solvent, and the color can be determined visually except for arsenic (a confined spot test), uranium (a paper chromatographic separation), and palladium and platinum (a spectrophotometer is necessary.)

Zinc In Soils, Sediments And Rocks. Zinc is the element most commonly determined in exploration geochemistry programs because (1) it is sought specifically in the exploration for sphalerite-galena deposits, and (2) it is a valuable pathfinder element for other types of metalliferous deposits. Both a hot acid extractable method (for the detection of zinc in sphalerite and within other sulfide minerals), and a cold extractable method (for loosely-bonded, readily soluble, zinc in stream sediments and soils) are presented in Appendix B. As discussed previously, the method of sample attack used will greatly affect the proportion of the total metal extracted. It is not always necessary to determine the total amount of zinc or other metals in a sample by a hot acid or fusion method. There are many instances in which values obtained for the readily soluble zinc or other metals provide greater contrast for the recognition of anomalies than do the total values (Fig. 6-3). Although hydrochloric acid is used in the hot acid digestion method described here, sulfuric acid or perchloric acid are often employed, as well as pyrosulfate fusions for the total metal values. For cold extractable zinc, a dilute acid can be added to the procedure presented (followed by the zinc buffer) should the situation warrant it; most likely the results will indicate higher readily soluble zinc with the cold acid leach in comparison with the zinc buffer alone. For areas containing zinc deposits in which chemical weathering is important, and in which there are clay-type minerals available to adsorb zinc ions from solution (soils or active sediments), the zinc buffer alone should suffice.

The zinc buffer maintains the pH of the solution being analyzed at about 5.5. The presence of the thiosulfate is designed to suppress the reaction of Cu, Pb, Ni, Co and other possible interfering elements with the dithizone by forming complexes, so that effectively only zinc reacts to form colored dithizonates. The thiosulfate is certainly effective for Cu, Ni and Pb (in the case of Pb, 6000 micrograms are required to produce the same red dithizonate as 4 micrograms of Zn). For cobalt, however, it appears that 8 micrograms will produce a dithizonate comparable to the 2 microgram zinc color, and with increasing cobalt, purple and red shades are formed. This will cause few problems because cobalt is encountered only rarely. Ward et al. (1963) report that a positive error is obtained for samples containing 10 or more times as much cadmium, cobalt, nickel or copper, as zinc.

The zinc standard series illustrated on the color plate shows a range of green (0 μg), blue-green (1 μg), blue (2 μg), purple (3 μg) and red (4 μg). This is a mixed color series, from unreacted green dithizone to the red zinc dithizonate. By variations in the weight of sample (gm) and aliquot (ml) used, the method is quite practical for the range of 10-800 ppm. There is rarely any need to detect less than 10 ppm zinc (except in waters) and more than 800 ppm can be determined by using smaller samples or aliquots than is mentioned in the procedure (a 0.1 ml aliquot will extend the concentration range to 4000 ppm). The method is extremely rapid and at least 100 samples per man-day can be analyzed.

Ward et al. (1963) use a potassium pyrosulfate ($K_2S_2O_7$) fusion for the determination of zinc with dithizone, and from the same acidified fusion they also determine copper by biquinoline (see below), and lead by dithizone. Because the lead method requires the use of potentially *dangerous* potassium cyanide (as lethal hydrogen cyanide might accidentally be generated), it is not discussed further here. Gilbert (1959) described in considerable detail the colorimetric determination of zinc, copper and lead, all based on dithizone reactions.

Copper In Soils, Sediments And Rocks. Copper is generally determined colorimetrically by its reaction with 2,2'-biquinoline, which requires it to be in the cuprous state, and yields a pink colored complex soluble in isoamyl alcohol (see color plate). The reaction is specific for copper, which is a major advantage. Another advantage of this method is that the (monocolor) pink complex is very stable, and can retain its color for months. The best pH range for the complex formation and extraction is 4-6 and this is achieved with the buffer. Hydroxylamine hydrochloride is added to reduce cupric to cuprous copper, although Stanton (1966) prefers ascorbic acid for this purpose. If organic matter is present in appreciable quantities, a yellow color will develop which can mask the pink copper-biquinoline complex. Samples with high organic content, therefore, should be ignited for a few minutes prior to analysis (to $450-550°C$) to prevent this interference.

The recommended standard series ranges from 1-6 micrograms of copper, and the range of copper which can be determined by the hot acid extractable method is from 50-300 ppm using the procedure as described. The method can easily be extended to 10-6000 ppm with variations in the aliquots used, and even further by variations in the weight of the sample. About 100 samples can be determined per man-day. An alternative method using dithizone to determine copper is described by Gilbert (1959). Kauranne and Nurmi (1967) have described a method for the determination of copper using neocuproine which they found applicable for glacial till samples in Finland.

For the cold acid extractable copper, biquinoline is also recommended and the range using the described procedure will be from 10-60 ppm copper. A well-established alternative method for readily-soluble copper in soil and sediments is described by Holman (1963). The method, commonly called the "Holman test", is based on the reaction of copper with dithizone and has been widely used in Africa and other tropical areas as a field test, but under laboratory conditions, the biquinoline method is preferred. A slightly modified Holman method is also recommended by Hawkes (1963).

Cold Extractable "Total Heavy Metals" (THM) In Soils And Sediments (Bloom Test). This test is one of the simplest, most rapid and most commonly used in exploration geochemistry, and is particularly suited to drainage basin reconnaissance surveys. It is based upon the reaction of dithizone at a pH of 8.5 with a large group of non-specific "heavy metals", which includes Cu, Pb, Zn, Ni, Co, Ag and Sn. In actual practice the test is most sensitive to zinc, followed by copper and lead; nickel and silver give practically no reaction. Zinc is what is usually determined, as even small amounts will mask copper and particularly lead. In other words, all metals do not react to the same degree with dithizone under the experimental conditions. The test is most valuable in the analysis of fine-grained soils and sediments, and is based on the premise that trace metals loosely adsorbed onto clays and other mineral grains, or in organic matter, will be removed easily by ammonium citrate, and these trace metals are then available to form colored metal dithizonates. The method was first described by Bloom (1955) and has since been commonly referred to as the "Bloom Test". Other synonyms, such as "ammonium citrate soluble heavy metals test", "cxHM" (cold extractable heavy metal), "cxME" (cold extractable metal), or "THM" (total heavy metals), are also used.

The method is designed for simplicity and, therefore, only a few items of equipment are needed. The test requires less than a minute to complete. In actual practice a quantitative determination is not made, but rather only a record of the volume of dithizone required to reach a uniform reference point (a blue-colored metal-dithizonate), although it is possible to extend the test by means of calibration curves to obtain a quantitative estimate of the exchangeable-metal content in ppm (zinc equivalent). In some areas the exchangeable-metal content of the samples may be high, and then the strength of the dithizone test solution can be increased to 0.003% (from 0.001%).

An interesting innovation has been proposed by Brown (1964). Rather than determine the volume of dithizone required to reach a uniform endpoint, he found it advantageous to titrate the ammonium citrate extract (i.e., buffer solution with contained cold extractable metals) against a fixed volume (2 ml) of dithizone. The endpoint in this procedure was taken to be the first show of color change from green towards red. The extract was added dropwise in amounts which ranged from one to several hundred, and the number of drops required to cause a color change was recorded. In this case, a small number would indicate a high metal content. By a judicious choice of dilutions and aliquots, it is possible to make the test versatile. Although this procedure is probably less accurate and precise than the more commonly used method described here, it does have very practical advantages, the main one being that the requirements for solvent, buffer and reagents which ordinarily must be carried to the field, can be reduced by a factor of 40 in some cases.

The cold extractable total heavy metal (Bloom test) procedure generally determines only a small fraction, perhaps 5-20%, of the heavy metals present in the soil or sediment. In some areas, such as recently glaciated parts of Canada where only detrital sulfides are in the soils or sediments, the test may give negative results as unweathered sulfides do not yield readily extractable metals (only acid digestion or fusion will yield metals from such sulfides). On the other hand, where oxidation of sulfides is extensive or where metal-rich waters have influenced the metal content of soils and sediments, the percentage of the total metal content detected by this method can be much higher, as the metals will be loosely held and, therefore, readily exchangeable by ammonium in the ammonium citrate. At the break-in-slope, cold extractable results may be especially high, and anomalies greatly accentuated in comparison with those obtained by "total" methods. After a soil or sediment sample has been dried, a somewhat smaller amount of cold extractables will be recorded because some of the metals will have become "fixed" through oxidation and other mechanisms.

Hawkes (1963) discussed this test in great detail, and gave a good summary of common operational problems likely to be encountered. Smith (1964) also described this test and noted that in samples high in copper, a brownish color may appear in the organic layer rather than the normal red. This should be taken as a positive result, and more dithizone solution added. Even though this cold extractable method generally measures only a small fraction of the total heavy metal content in a qualitative manner, geochemical dispersion patterns related to base-metal deposits are sometimes more clearly revealed by it than by methods that measure the total metal content.

Heavy Metals In Water. The heavy metals in water test is another simple and rapid field test, and is the most sensitive of all. It takes just over a minute to complete, yet sensitivities of 10 ppb are easily achieved using the method as described in Appendix B and these can be extended to as low as 2 ppb should the need arise (see procedure in Ward et al., 1963, p. 31). A complete description of the method and its application is given by Huff (1948) who reported detection limits as low as 2 ppb for zinc, and 10 ppb for lead and copper. From the tabulation included in the description of the procedure (Appendix B), and Huff's (1948) estimates of sensitivities, it is clear that zinc is the most likely metal to react with dithizone to produce colored dithizonates. Organic matter in the water may be a problem, as it tends to cause stable emulsions to form between the organic solvent (i.e., toluene) and water. Because of the great sensitivity of the method, extreme care must be taken to ensure that contamination, either from the chemical reagents or within the waters themselves, is eliminated or taken into account.

The method is particularly applicable to those areas with a large number of small streams, lakes or springs, and permits the geochemist to follow water anomalies almost instantly. It should be used in conjunction with the cold extractable heavy metals test (cxHM; Bloom test; THM), because with changes in pH-Eh conditions, metals are sometimes in the water and at other times are adsorbed onto the fine-grained sediments. By using both methods together, it is highly unlikely that a geochemical anomaly will be missed in areas where chemical weathering is an important process. Normally the water samples analyzed during the survey are not saved, but where anomalous values are obtained, a minimum of 100 ml of the water should be collected in polyethylene bottles, acidified, and sent for a detailed analysis, probably by atomic absorption.

Molybdenum And Tungsten In Soils, Sediments And Rocks. The zinc dithiol method for the colorimetric determination of molybdenum and tungsten in soils, sediments and rocks has been selected for inclusion in this book because of its exceptional importance in exploration geochemistry. In commercial laboratories, molybdenum is often determined by atomic absorption using a perchloric acid digestion, but according to Stanton (1970 a,b) this method is sometimes not suitable because of its inadequate sensitivity. Others (e.g., Bradshaw, 1972) have found perchloric acid attack on molybdenite not entirely efficient, and they prefer a carbonate fusion, zinc dithiol method (Appendix B). In the case of tungsten, atomic absorption analysis is unsatisfactory, and X-ray fluorescence and emission spectrography are not sufficiently sensitive for most exploration purposes (X-ray fluorescence is satisfactory only when more than about 100 ppm tungsten is present). The zinc dithiol colorimetric method for both molybdenum and tungsten fills an important gap, as it is sufficiently sensitive and can be made specific for each of these elements. Further, both elements can be determined from the same fusion. The analysis can be accomplished in a well-equipped base camp.

There are several published and unpublished variations of the zinc dithiol technique involving, for example, different digestion procedures. The method presented here (Appendix B) incorporates the better features of the several variations based upon the needs and experiences of the author's laboratory, as well as those of some commercial laboratories, particularly in the case of tungsten. The molybdenum procedure is very similar to that published by Stanton (1966), whereas the tungsten method has been principally modified from unpublished procedures of the Geological Survey of Canada.

The method for both elements is based upon the reaction of molybdenum and tungsten with dithiol (obtained in the form of zinc dithiol) after fusion with an alkaline flux in nickel crucibles. The molybdenum and tungsten in the sample form water-soluble sodium molybdate and sodium tungstate, but one of the more difficult parts of the procedure is the removal of this leach from the crucibles within strict volume constraints (see step 5 of the molybdenum procedure). Stanton (1970a) has published a dithiol method for molybdenum in which the samples are digested in test tubes. Although this eliminates the nickel crucibles, the fact that perchloric acid is used to decompose the sample makes the method less than desirable for student purposes even though up to 150 determinations per man may be made daily, in comparison with about 60 determinations for the method suggested here. In the case of tungsten, Stanton (1970b) uses potassium hydrogen sulfate ($KHSO_4$) as the flux in test tubes and can analyze about 100 samples per man per day, again opposed to 60 for the method proposed here. Nevertheless, the fact that both molybdenum and tungsten can be determined from the same alkaline fusion outlined in Appendix B, reduces some of the advantages of Stanton's (1970a,b) recent modifications as his methods require separate fusions.

In the procedure presented here, dithiol forms a yellow-green molybdenum complex in a cool solution, which eliminates interference from tungsten, as the tungsten-dithiol reaction is favored at high temperatures. Tungsten forms a blue (or blue-green) complex with dithiol in a hot solution containing stannous chloride. The stannous chloride reduces any molybdenum present, which eliminates molybdenum intereferences. Once formed, the complexes of both molybdenum and tungsten are very stable and the colors will last for a long time. Interferences will occur when either element is present in amounts of the order of 10 – 100 times the other. Arsenic, if present in large amounts (>2000 ppm), is likely to interfere with the molybdenum determination, but all other elements known to react with dithiol (e.g., Sn, Bi, Sb, Co, Cu) are effectively eliminated if the procedures are followed exactly. The sensitivities of the method as described are 2 ppm for molybdenum and 4 ppm for tungsten. In the case of sulfide minerals, it has been suggested that small (50 mg) samples be used to ensure complete fusion and to avoid attack on the nickel crucibles. Several alternative methods of analysis for molybdenum and tungsten have been proposed, for example, those of Ward et al. (1963). Stanton et al. (1973) recommend that organic-rich soils be evaporated with a mixture of nitric and perchloric acids prior to the colorimetric determination of molybdenum.

EMISSION SPECTROGRAPHY (SPECTROSCOPY)

The emission spectrograph was the first instrument widely used for geochemical exploration. This stemmed from the work of Goldschmidt in the period 1925 – 1940 which showed that the spectrographic technique had great utility in geochemistry, especially for the determination of trace amounts of many elements. The early geochemical efforts in the Soviet Union were based on the emission spectrograph, and it is still employed for the analysis of 13.3% of all geochemical samples in the United States and 3.8% of the samples in Canada based on the data in Table 6-1. It is also employed extensively in the United Kingdom and parts of Africa. Although these figures may appear small by comparison with atomic absorption and colorimetric methods, there is no doubt that the emission spectrograph will continue to occupy its relative position for years to come. In fact, with the trend toward multiple-element analysis of samples, which is one of the main advantages of the emission spectrograph, a resurgence of interest in this instrument is likely especially for the more expensive, direct-reading types. Emission spectrograph detection limits are often in the low ppm range, but regrettably the sensitivity for some important elements (e.g., Zn, U) is not as good as would be desired (Table 6-3). Still the technique is particularly valuable in regional and orientation surveys and is described in several publications such as Ahrens and Taylor (1960), Smales and Wager (1960), and Wainerdi and Uken (1971).

The emission spectrograph is basically a simple instrument, consisting of three main components: (1) the excitation source, usually a direct current (dc) arc although other sources, such as "ac" spark and laser beams, are also used; (2) the dispersing unit which, in newer instruments, is usually a high-quality grating, although prisms were used extensively in the past; and (3) the detection equipment which can be either visual, by photographic means, or electronic using photomultiplier tubes. As a generalization, the cost of the spectrograph, which can range from $2,000 – $150,000, can be correlated with the type of detection equipment, as is discussed below. A schematic diagram of a prism and a grating-type instrument is shown in Fig. 6-9.

The excitation of the sample by most methods takes place between graphite electrodes, and is essentially the same for all instruments. This excitation, at temperatures of up to 8000°C, changes an atom from the "ground" state to the "excited" state. After a very short period of time the atom, following appropriate electronic transitions, returns to a lower energy state, with the resultant energy difference emitted in the form of electromagnetic radiation. Because a considerable number of transitions are possible for most elements, a large number of electromagnetic radiations, each with a characteristic wave length recorded as a line, may be detected after dispersion through a grating or a prism. Because a large number of lines are possible (about 5000 are known for iron alone), great care must be taken to ensure that only interference-free lines are identified. Usually the strongest and most characteristic lines are detected for

Table 6-3. Approximate lower detection limits (sensitivity) for 61 elements by atomic absorption, emission spectrography and colorimetry (all values in ppm).

Element	Atomic Absorption	Emission Spectrography	Colorimetry	Element	Atomic Absorption	Emission Spectrography	Colorimetry
Ag	1.	0.5	.01	Ni	4.	5.	1.
As	2.	1000.	10.	Os	1.	50.	—
Au	.005	20.	.05	Pb	1.	10.	10.
B	40.	20.	10.	Pd	.2	2.	—
Ba	.3	2.	500.	Pr	12.	100.	—
Be	.03	2.	1.	Pt	2.5	50.	—
Bi	10.	10.	20.	Rb	.05	—	—
Cd	.2	50.	—	Re	12.	50.	—
Ce	—	200.	—	Rh	.5	2.	—
Co	4.	5.	1.	Ru	.5	10.	—
Cr	.1	1.	100.	Sb	.6	200.	1.
Cs	.3	—	—	Sc	.65	5.	—
Cu	.1	1.	5.	Se	2.	—	1.
Dy	.8	50.	—	Sm	7.	100.	—
Er	.9	50.	—	Sn	3.	10.	10.
Eu	.6	100.	—	Sr	.15	5.	—
Ga	2.5	5.	—	Ta	40.	500.	—
Gd	16.	50.	—	Tb	8.	300.	—
Ge	2.5	10.	4.	Te	.2	2000.	—
Hf	15.	100.	—	Th	—	200.	4.
Hg	.001	—	2.	Ti	2.	2.	100.
Ho	1.5	20.	—	Tl	.5	50.	—
In	.9	10.	—	Tm	.7	20.	—
Ir	12.	50.	—	U	50.	500.	4.
La	30.	50.	—	V	1.5	7.	4.
Li	.5	100.	—	W	20.	100.	4.
Lu	6.	30.	—	Y	2.	10.	—
Mn	.05	10.	50.	Yb	.2	1.	—
Mo	.6	3.	2.	Zn	.5	300.	5.
Nb	20.	10.	50.	Zr	6.	10.	—
Nd	13.	70.	—				

Notes:

1. Atomic absorption data for Ag, Au, Bi, Cd, Co, Cu, Hg, Li, Ni, Pb, Te and Zn are from Ward et al. (1969) and represent detection limits obtained by the U.S.Geological Survey. All other data are from the literature of the Perkin-Elmer Corporation (dated March, 1971), in which the limit of detection is based on an instrument reading of 1% absorption; in many cases special burning conditions are required to obtain the listed detection limits. Many of the Perkin-Elmer detection limits appear to have been obtained under ideal conditions, and are not normally attainable on geochemical samples. Dashes (—) for Ce and Th indicate that no lamps are available for these elements.

2. Emission spectrography data from Myers et al. (1961) and more recent U.S. Geological Survey data; also some data from Wainerdi and Uken (1971) are included. Dashes (—) for Cs, Hg, Rb and Se indicate these elements are not normally determined for exploration purposes by this method, primarily because of their volatility.

3. References and comments for colorimetry are the same as those given in the footnote to Table 6-2. Only field or laboratory tests applicable to exploration geochemistry are included. Dashes (—) indicate that no colorimetric geochemical tests are applicable.

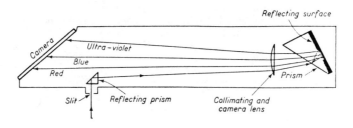

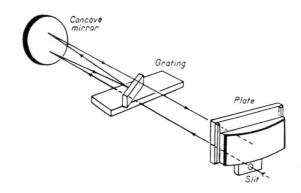

Fig. 6-9. A schematic diagram illustrating a prism and a grating type of emission spectrograph. From Ahrens and Taylor (1960).

qualitative identification, and their intensity measured for semiquantitative or quantitative determinations. Actually, under normal operating conditions, only about 5% of the atoms are in the "excited" state, so that the technique is relatively inefficient in comparison to atomic absorption, where about 95% of the atoms are in the atomic (ground) state and capable of contributing to the analysis. Generally about 10 mg of sample is used for each analysis and this small amount makes it essential that it should be truly representative.

The method of detection of the spectral lines varies widely. In the less expensive ($2000 – $8000) equipment, detection is usually visual, the differences in the price of the instruments being related to the quality of the grating or prism and other variables in construction. Although some of the visual detection instruments can be equipped with cameras for recording the spectra on film for eventual "quantitative" analysis, the fact remains that instruments of this type are basically only suitable for qualitative work, at least for geological samples. They are generally good for major or minor elements, but only rarely for elements in lower concentrations. One inexpensive visual-detection instrument has been found to be unexpectedly sensitive for low abundances of yttrium which, of course, has application in the search for rare-earth minerals, some types of carbonatite deposits, and possibly certain types of fluorite deposits.

Photographic means of detection are used extensively by several large and experienced organizations. The U.S. Geological Survey relies heavily on this method not only in permanent laboratories, but also in the many mobile laboratories equipped primarily with emission spectrographs. This accounts for a large portion of the emission spectrograph usage in the U.S. indicated in Table 6-1. In addition, in Africa photographic-type emission spectrographs, using mainly the method described by Nichol and Henderson-Hamilton (1965), are in operation in several countries where their value for orientation and regional surveys is particularly well-established (e.g., Bloomfield et al., 1971). In the Soviet Union most of the emission spectrographs use photographic methods of detection.

The data obtained after a film or plate is developed, and interpreted with a densitometer, can be either semi-quantitative or quantitative, depending on the needs of a specific program. The U.S. Geological Survey semi-quantitative laboratory method in which 68 elements are looked for is described by Myers et al. (1961). For the mobile field units, 14 elements are determined (Grimes and Marranzino, 1968). Instruments based upon photographic detection methods cost about $20,000 and all require very well-trained personnel. Productivity with the photographic-type emission spectrographs is not great, as one experienced spectroscopist can probably only prepare, analyze and interpret quantitatively about two geological samples per hour. Qualitative and semi-quantitative analyses are significantly faster and direct-reading (electronic detection) instruments give the most rapid results.

Instruments which detect the radiation electronically are often referred to as direct reading (D.R.) spectrographs. Such instruments, generally constructed to detect at least 15 elements, require a considerable capital outlay, in the range of $60,000 – $150,000, but include an associated computer for making matrix and other corrections. These are necessary for accurate results as the matrix variations in geological samples greatly affect the results. Actually, direct-reading spectrographs have been widely used for at least 20 years in, for example, steel and magnesium metal plants where perhaps a dozen trace metals have to be determined constantly for quality control. However, in all such cases the matrix is essentially constant which greatly simplifies quantitative analysis. As the problems of matrix variability can now be handled practically by computers, direct-reading spectrographs have recently begun to attract the attention of exploration organizations requiring a multiple-element capability. Several manufacturers now offer a direct-reader with 60 channels, about 10 of which must be programmed for major elements so matrix corrections and inter-element compensations can be made. An analysis for perhaps 40 elements, including the necessary computer matrix corrections, can be made in less than two minutes. It appears that, with this type of instrument, the main problem will be the collection of meaningful samples and sample preparation. For those organizations anticipating a large amount of work requiring multiple-element capabilities over an extended (perhaps 5-year) period, the large capital outlay for the direct-reader is justified as the price per sample will be extremely low on an averaged basis.

Sample preparation in emission spectroscopy has been, and continues to be, very crucial because rarely is more than 10 mg used, and hence it must be representative. Further, for quantitative work the sample must be uniformly mixed with a buffer (e.g., lithium borate) to present a uniform mixture to the arc, graphite to stabilize the arc, as well as with a specific amount of an appropriate internal standard. The internal standard, which may be Ge, Pt, In or some other metal, must be selected so that it has similar volatilization characteristics to the element(s) being sought, and of course, it must not be present in the sample as originally received. Standards of similar composition to the unknown should be run frequently. Accuracy of 5-10% can usually be obtained and, while this is quite satisfactory for trace and minor elements, it is unsatisfactory for major constituents. Among the many special problems encountered with any emission spectrograph are those which result from differing volatilities of elements. Thus the operating conditions for volatile elements such as Hg, Te and Tl are significantly different from those for non-volatile elements such as Th, Ce and Be. Therefore, although the emission spectrograph is ideally suited for multi-element analysis, poor precision and accuracy may result when a large number of elements are determined at one time on a routine basis unless special precautions are taken such as firing at both low and high amperages, or obtaining two plates, each with different burning times. For this reason, as well as other pitfalls, emission spectroscopy requires a well-trained, experienced analyst to obtain the maximum information.

"Semi-quantitative" emission spectrographic analyses have long been used in exploration geochemistry in the form of a "general scan" for all metals of possible interest. The so-called "30 element spectrographic analyses" which presently cost between $15 – $20 per sample, have serious shortcomings, but nevertheless usually do satisfy the curiosity of certain individuals, the needs of certain projects, and can act as confirmatory tests for some analytical methods. A good emission spectroscopist, by careful examination of a photographic film, can classify the elements present (arced without internal standard) in the following ranges of: (1) present, less than 0.01%; (2) between 0.01% and 1.0%; (3) between 1% and 10%; which roughly correspond with the accepted usage of trace, minor and major. This is often sufficient for many geochemical purposes. By means of the addition of an internal standard and buffers, somewhat better accuracy can be obtained, but it must be remembered that by generally accepted convention, "semi-quantitative" usually means that the results are within one-third to three times the amount actually present (Wainerdi and Uken, 1971, p. 179). In other words, if the true amount is 100 ppm, the "semi-quantitative" analyses could reasonably report anything between 33 – 300 ppm. However, experienced analysts can usually do much better than this.

The main advantage of the emission spectrograph is its multiple-element capability, although this is limited somewhat by problems such as volatility mentioned above. The photographic-type instruments are sturdy, moderately priced and well suited for mobile laboratories. The disadvantages of the emission spectrograph are that the spectra are complex and, to obtain good results with the method, experienced personnel are essential. Except for the direct-reader, production rate for data is slow. The fact that the method destroys the sample so that other analyses cannot be performed on identical material is partly offset by the fact that only a very small amount of sample is required, but sample preparation is a very important and exacting aspect. Notwithstanding these disadvantages, barring some major unexpected breakthrough in the multiple-element capabilities of other techniques, the use of direct-reading emission spectrographs will probably increase in the near future especially in the larger governmental, as well as private, organizations.

X-RAY FLUORESCENCE

Laboratory Analysis

X-ray fluorescence analysis (also called X-ray spectrometry, X-ray emission, or XRF) is the fourth most important method of analysis used in exploration geochemistry (Table 6-1). Its use in Canada and Australia (Fisher, 1971) is significantly greater than appears to be the case in the United States. Usage in Canada has increased recently since several commercial laboratories have purchased instruments. The equipment generally costs from about $30,000 to perhaps $60,000, if computer attachments are included, but computers are not normally used in analyses for exploration geochemistry. X-ray fluorescence equipment is found in many geology departments in North American and European universities where it is commonly used for the analysis of major and minor components of rocks, as well as adjuncts to geochemistry and economic geology courses. The basic requirements of the equipment are a generator capable of supplying a stabilized high voltage, a tube from which X-rays are emitted, an analyzing crystal and goniometer, appropriate detectors to record the spectra, a strip-chart recorder to display the spectra, and reliable counting equipment for quantitative analysis. Clearly the apparatus is complex.

For X-ray fluorescence analysis, in its simplest form, a sample is irradiated by a beam of X-rays which are generated from a sealed tube with either a tungsten or molybdenum cathode (other tubes, such as chromium or gold, are used only for special applications). This beam of X-rays displaces orbital electrons in the atoms of the sample being irradiated, and the vacant positions are immediately filled by other electrons which jump in from a higher energy level. The process is illustrated in Fig. 6-10 in which it can be seen that, when an electron in the K (inner) shell of the hypothetical atom is displaced, its position is filled by electrons from one of the outer shells. When the electron from the outer shell moves into the lower energy K shell, a quantum of energy is released which is charac-teristic for each element. It is this quantum of energy, in the X-ray wave-length region which is referred to as the X-ray fluorescence. In actual fact, there are several orbital levels (K,L,M) which can lose electrons when bombarded with X-rays, and there are several other different orbital levels from which electrons can jump to fill vacant sites. As a result, there are a number of different wave-lengths (lines) which can be produced for certain elements, with designations such as K_α, $K_{\beta1}$, L_α, and $L_{\beta2}$, depending on the energy level from which the replacing electron came (Fig. 6-10). On a prac-tical basis, however, the spectra obtained by X-ray fluorescence are simple in comparison with those from the emission spectrograph. In a typical situation, in which the content of an element is 0.1% or less, only two lines will usually be observed. In cases where the element is a major constituent, rarely more than ten lines will ever be seen, of which several will be of low intensity.

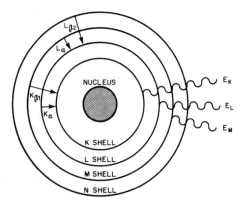

Fig. 6-10. Diagram illustrating how electron transitions give rise to characteristic X-ray fluorescence radiations.

The simplicity of the spectra obtained by X-ray fluorescence is one of the major advantages of the method. Although there are numerous examples of line interferences, such as Pb-As, Zr-Hf, and Rb-U, in most cases these problems can be overcome by selection of a different line for analysis, use of pulse-height discrimination, or changes in operating condition (voltage). There are numerous review articles and books covering theoretical and practical aspects of X-ray fluorescence in detail (e.g., Smales and Wager, 1960; Wilson, 1966; Adler, 1966; Jenkins and DeVries, 1967; Wainerdi and Uken, 1971). Perusal of any of these articles or books will bring out many practical aspects of the method, such as matrix effects, absorption edges, the importance of sample preparation, choice of tubes, detectors and crystals, and other topics beyond the scope of this book.

Because of its comparatively simple spectra, and the ease of analysis of elements heavier than Ca, X-ray fluorescence has great applicability in exploration geochemistry. The detection limit of all elements from Ca to U, assuming no unusual matrix effects, is about 100 ppm under routine operating conditions, and in the more favorable range of the spectrum (e.g., the transition metals), 50 ppm is realistic. If there is special interest in a particular element, lower limits of detection can be achieved with the proper set of operating conditions, and detection limits of 20 ppm or lower are possible, but the difficulty of the analysis generally increases rapidly as lower concentrations are sought. A qualitative scan of a sample for all elements from calcium to uranium, with detection limits in most matrices of 200 ppm, can be achieved in about an hour. Qualitative checks for even the most unusual of elements in exploration geochemistry at the 200 ppm level (e.g., Re, Sc, Ce, Se) can be achieved in about a minute. Samples may be either solids (soils, sediments, rocks) or liquids (bromine in brine, for example) and, in all cases, the method is non-destructive. Elements below calcium and as low as boron can be detected, but these require different operating conditions (Cr tube, different analyzing crystals, vacuum) and detection limits fall off considerably with decreasing atomic number.

The major problem facing analysts using this technique for quantitative determinations is the matrix effect, similar to the difficulties found in emission spectrography. For qualitative determinations there are few difficulties and an experienced analyst can estimate the abundance of an element about as well as with an emission spectrograph. For quantitative work in order to overcome the variations in matrices, a dilution technique in which the sample is fused (usually with lithium tetraborate) is often employed. Thereby however, some sensitivity is lost and, of course, the method is destructive except that the fused button is available for other analyses. In order to overcome the matrix problem, and at the same time increase productivity, several "rapid methods" have been proposed, all based on some simplifying assumptions concerning matrix variations. One which is especially attractive has been developed at C.S.I.R.O. in Australia by Hudson (reported in Fisher, 1971). Hudson developed a procedure in which 17 elements can be determined on a sample, without fusing, in quantities down to 10 ppm, with an accuracy of about 20% or better. Analyses for all 17 elements can be completed within 20 minutes. The method is based upon a single background measurement, which permits a simultaneous matrix correction for all 17 elements.

Although most X-ray fluorescence instruments (in the $30,000 range) are of relatively simple design, more expensive and sophisticated instruments are available, all the way up to multi-element direct-readers with automatic sample changers capable of holding perhaps 40 samples and running unattended for long periods of time. As with the emission spectrograph, most of the multi-element instruments are presently used for fixed-matrix industrial applications, such as quality control or research in metallurgy and plastics.

The main use of X-ray fluorescence in exploration geochemistry at present is for the analysis of elements which simply cannot be analyzed economically, or practically, by any other method. This reduces to the following: Ta, Nb, W, Th, Zr, and rare earths, and possibly Sn, V, Ba and Ti in certain cases. The list includes refractory elements which cannot be analyzed at all by atomic absorption, or elements for which the emission spectrograph is insensitive. In most cases geochemists are satisfied with a limit of detection of 100 ppm for these elements. In the case of tungsten, a good colorimetric method is available, and described in detail in Appendix B, but it is most effective in the range below 100 ppm. This fact brings out the point that atomic absorption, emission spectroscopy and colorimetric methods, although ideal for many trace determinations, lose accuracy and precision in samples containing higher quantities of most elements. This is because cumulative and serious errors are likely to be introduced during any type of dilution, whether it be in solution (atomic absorption, colorimetric) or in a solid form (emission spectrograph). It is inherent in X-ray fluorescence, with modern counting circuits, that high abundances of all elements can be accurately determined, and, in general, accuracy and precision increase with higher abundances, provided that adequate standards are available. As a result, X-ray fluorescence is often used preferentially for barium, uranium, molybdenum and many other elements in high concentrations, if not as the only method of analysis, then as a check on other techniques.

Recent increased interest in rock geochemistry may well result in much greater use of X-ray fluorescence techniques, especially for the analysis of the major constituents of rocks. For example, the distinction between several types of basaltic rocks in the volcano-sedimentary belt of the

Canadian Shield can be made on the basis of their contents of K_2O, Na_2O and SiO_2 (discussed in Chapter 7). These components, as well as Al_2O_3, MgO, CaO, Fe_2O_3 and other major constituents of rocks, can be determined satisfactorily by X-ray fluorescence. Commercial laboratories are now performing these analyses at very reasonable costs (11 elements for less than $40).

Field Analysis (Portable Equipment)

From a description of the laboratory equipment used for X-ray fluorescence analysis it is clear that the equipment is far too bulky and complex for field application. However, there is portable field equipment presently available which, under certain circumstances, can have great value for exploration purposes. These instruments are basically inexpensive (approximately $5000) and lightweight (from 15 to 75 pounds) but, in exchange for these advantages, they are not nearly as sensitive as the laboratory instruments, nor are they as versatile. The first instrument designed for geological use was developed in England jointly by the Institute of Geological Sciences and the U.K. Atomic Energy Authority. Early models are described by Bowie et al. (1965) and Gallagher (1967).

In laboratory equipment, X-rays from a high-voltage tube are used to generate secondary (fluorescent) X-rays from a sample. In addition, fluorescent X-rays can also be generated by γ-radiations from radioactive isotopes thus eliminating the need for a generator and an X-ray tube. Further, to eliminate the bulky goniometer with its analyzing crystal, filters or pulse height discriminators can be employed in the portable units which permit only selected radiations (wave-lengths) to be detected by a counting device. All portable X-ray fluorescence spectrographs incorporate the above two simplifying features. The result is that geochemists can walk through an area with an instrument, much as prospectors do with a gamma-ray spectrometer, looking for indications of the mineralization for which their instrument is programmed (a typical analysis will take from one-half to two minutes).

Several isotope sources can be used for excitation. Fe-55 can excite vanadium, titanium, or chromium, but, for lead, Co-57 would be used, and, for molybdenum and barium, Cd-109 would be necessary. All radioactive sources have half-lives, in the case of Fe-55 it is 2.6 years. After this period the source will yield only half the amount of radiation that it originally emitted, and thus the intensities of the fluorescent X-rays from the material being analyzed will be reduced by about one-half. Throughout the period of use, portable instruments must be calibrated periodically to compensate for the gradual decay of the radioactive source and subsequent loss of counts (sensitivity), and eventually the source must be replaced. Of course, constant attention must always be paid to possible radiation hazards, and a licence to possess these radioactive sources is required in most countries.

Portable isotope-source X-ray fluorescence units have found acceptance in many manufacturing industries where quality control on a routine basis is necessary. These range from the analysis of alloys, ceramic materials, and solutions (e.g., iron in lubricating oils), to the measurement of coatings (e.g., titanium dioxide on paper), in which sensitivities as low as 0.05% or less are sometimes quoted. However, for exploration purposes, where rock surfaces are variable in shape, as opposed to the uniform surface configurations in all the above-cited examples, and variable matrices of rocks are the rule, such low sensitivities are not normally attainable. In a recent study of one such instrument, Wollenberg et al. (1971), using two different radiation sources, found that the detection limits were 0.44% for Zr, 0.15% for Nb, and 0.17% for combined La + Ce, for their instrument. This leaves much room for improvement as far as sensitivities are concerned, but does show that these instruments have applications in the proper situations. In a more recent study of isotope-source X-ray fluorescence apparatus, Kunzendorf (1973) has been able to attain somewhat better sensitivities than those indicated above. He also pointed out that the portable instruments are especially valuable in the analysis of outcrops in the search for disseminated ores.

Although at first glance it appears that successful construction and operation of this equipment should not be very difficult considering the technological advances of the last decade, the fact remains that at least two prominent pioneering manufacturers of portable X-ray spectrographs have dropped this line for one reason or another. This is not to say that others will not overcome some of the practical problems, and if they do, they will no doubt find a ready market among exploration groups, especially if the price remains in the $5000 range. The main problems at present, from the point of view of exploration, seem to be a need for greater sensitivity, and greater versatility in the equipment so that radioactive sources and filters need not be changed whenever a different element is sought. Other portable types of equipment using the principles of neutron activation are discussed below.

Somewhat analogous to the portable X-ray units, is the method for determining beryllium both by means of portable field equipment, and comparably small laboratory units. These instruments depend upon nuclear reactions induced in beryllium by a radioactive source yielding gamma rays or alpha particles, the radioactive product so formed decaying with the emission of neutrons. The rate at which neutrons are emitted is counted and is a measure of the beryllium content of the rock. As the conditions of analysis are specific for this element the instrument has been called a "beryllometer", and although some models weigh about 100 pounds with the necessary shielding, it has been used successfully by two-man parties. On finely ground powders, the small laboratory models have sensitivities as low as 20 ppm but, owing to the practical field problems (solid rock, irregular surfaces), the field models are not nearly as sensitive.

OTHER METHODS

The data in Table 6-1 indicates that several methods additional to those described above are used in the analysis of geochemical samples. As the importance of these methods is significantly less, they are only briefly explained and discussed. Polarography, flame photometry, turbidimetry, mass spectrometry, and a few other methods mentioned only occasionally in the literature, are not included because of their very limited use at present.

Paper Chromatography

This is one type of colorimetric procedure in which an element to be analyzed is separated from others by reaction with a suitable compound to produce, on a special type of paper, a colored band whose width and intensity of color is proportional to the amount of the element present. The main factor in separating the various elements in an appropriate solution is the rate of movement of the various ions by a process similar to capillary action. Ward et al. (1963) mention chromatographic methods for uranium, cobalt, copper and nickel. Ritchie (1969) has reviewed the more recent advances in chromatographic analysis of geologic material. Research has shown that 21 metals, including elements such as Be, Nb, Ta, Ti, U, Th and Ge, that are normally difficult to determine, can be analyzed by paper chromatography, and the platinum-group can be quantitatively analyzed by a modification, thin layer chromatography. Although chromatographic methods are widely used in analytical chemistry, they are not normally used in exploration geochemistry, and the figures of 0.6% and 0.2% (in Table 6-1) of all geochemical samples analyzed by this method in Canada and the United States, respectively, are surprisingly high to this writer. Ritchie (1969) pointed out the simplicity of some of the methods, and their applicability to base camp operations. He made a suggestion (p. 434) for a more thorough assessment of the place of the various chromatographic methods in exploration geochemistry, and the data in Table 6-1 seem to suggest that some analysts in Canada have found it a worthwhile technique, but details as to which specific elements are analyzed are lacking.

Chromatographic methods are reportedly widely used in France for the determination of U and Th, and possibly also in Norway. Many articles and books on chromatography have been published, and ion-exchange chromatography is reviewed in Wainerdi and Uken (1971).

Specific Ion Electrodes

Certain electrodes have been developed which are sensitive and selective to specific ions in solution. The fluoride specific ion electrode is the one most widely used in geochemistry, for the analysis of small quantities of fluorine in natural water (Schwartz and Friedrich, 1973) and rocks (Ficklin, 1970; Kesler et al., 1973). The rock must be placed in solution by suitable digestion. This is the method of analysis now most preferred for fluorine. Specific ion electrodes are also available for other halogens (e.g., chlorine, bromine), sulfide, and some metals (e.g., Ca, Ag) but care must be taken for possible interferences and other analytical problems. (Technically, ion electrodes do not actually respond specifically to the ion of interest, but rather measure ion activity, and as a result, corrections are

Fig. 6-11. A portable combination specific ion, pH and Eh meter, with several electrodes immersed in a solution. Courtesy of Orion Research Inc., Cambridge, Massachusetts, U.S.A.

sometimes necessary). Although, superficially, an electrode which is specific for sulfide in waters may appear extremely valuable for detecting sulfide deposits, the fact is that in natural conditions (lakes, rivers, groundwater) sulfide would precipitate immediately in the presence of lead and many other metals. The specific ion equipment is suitable for field operations and, in fact, several field studies on the fluorine content of natural waters have been reported. Specific ion electrodes cost from about \$200 – \$400, and are used in conjunction with portable pH-Eh meters such as that manufactured by Orion Research Inc., Cambridge, Massachusetts (Fig. 6-11).

Gravimetric, Volumetric And Fire Assay

In gravimetric analysis the element to be determined is isolated by various chemical reactions from other constituents in a sample, and weighed either as the element itself (e.g., gold) or as a compound (e.g., barium as barium sulfate). In most cases, the separation is a precipitation from a solution and special chemical precautions are taken to ensure that it is quantitative and selective. The technique, one of the oldest in analytical chemistry, does not rely on standards or the measurement of some associated physical property, both inherent in all of the instrumental methods. Unfortunately, the lower limits of detection of most gravimetric methods are poor, and generally at least several grams of sample are required for each determination. Because of the complexities and interferences in many of the gravimetric methods, especially in the low abundance range, skilled "classical" analysts are necessary. Gravimetric methods are best employed in the higher ranges of element abundances, such as in the determination of silicon, aluminum and magnesium in rocks, and because the methods for many metals are extremely well established, they have found great use in those instances requiring analyses of high precision and accuracy, such as in ore samples.

Volumetric (titrimetric) analysis, another classical method, consists of the determination of the volume of a standard solution or reagent of known concentration, which is required to react quantitatively with the element being determined. Usually a sharp change of color is an indication that the reaction is completed. As with gravimetric analysis, no standards are required, although it is always wise to run a sample with a known quantity of the element being sought to ensure that the determination is not marred by unsuspected errors. Volumetric methods can work well with low quantities of the element being sought, such as the determination of iodine and bromine in the low ppm range in brines. In addition, some well-established methods are available for ore elements (e.g., Cu, Pb, Zn), and these methods are preferred by many analysts when the content of these elements exceed 5%, particularly for assay purposes. In general, however, volumetric analysis is not widely used in exploration geochemistry.

Fire assay, a special type of gravimetric analysis, occupies an important place in exploration geochemistry because this relatively ancient method is still the best for the determination of the platinum metals, since it ensures that all are isolated and included in the analysis (some of the newer methods extract only a portion of the platinum metals from a rock sample). It is also excellent for assaying ores containing gold and silver, although the new atomic absorption methods for these elements are satisfactory for exploration purposes. The procedure involves heating the ore with a flux, which results in the separation of the noble metals and they then can be weighed. Fire assays are exceedingly sensitive with less than 100 ppb easily obtainable.

Usually, from 50-150 grams of an ore are mixed with a flux consisting of sodium carbonate, lead oxide (litharge), borax or sodium fluoride, and charcoal, and heated in a clay crucible at red heat. The precious metals alloy with lead (reduced from the lead oxide), whereas the silica, alumina and other elements form a slag. The molten mass is poured into a steel mold, whereupon the lead containing precious metals forms a "button" (pril) on the bottom of the mold. After cooling, the button can be broken free of the slag. The lead-precious metal button is then placed on a cupel (a flat roasting dish) of magnesium oxide and other materials and heated under oxidizing conditions. The lead is partly volatilized, oxidized and adsorbed by the cupel, leaving a bead of only precious metals which, after cooling, can be weighed and converted back to a content for the original sample. Depending on the elemental content of the ore, any one or all of the noble metals can be within the button. To determine the exact proportions of these elements any number of alternatives are possible, including chemical separation by selective solution (silver can be removed by HCl), X-ray fluorescence analysis, emission spectrographic analysis or atomic absorption.

One very important aspect to be considered in the analysis of the precious metals is their general non-uniform distribution in the host rock. This, accompanied by the fact that only very low quantities of these elements (say 5-10 ppm) in a rock can make it an ore, require that very large samples be collected for analysis. In the case of platinum, 600 pounds is not an unusual sample to be collected, ground, split and a portion analyzed by the fire assay method. The fact that the fire assay is amenable to samples in the 100 gram range, simplifies the preparation of a representative sample. Fire assays range in price from $10 – $25, or more, depending on the number of samples submitted at one time, and whether gold, silver and/or the platinum-group as a whole are to be determined (charges for the preparation of large samples are additional). Identification of the individual elements of the platinum-group is considerably more costly.

Fluorometry

Fluorometry (also fluorimetry) is the most important method used for the analysis of uranium. It is extremely sensitive, for example, in waters as little as 0.2 ppb uranium can be detected, and in rocks 1 ppm is easily attainable. The procedure is based upon the fact that when a uranium-bearing sample is fused with the proper flux, and then cooled, it will emit a fluorescence in the visible range when excited by ultra-violet radiation (analogous to the fluorescence of some uranium minerals under a "black light"). This is quantitatively recorded by a photoelectric cell. Carefully controlled conditions are necessary, chemicals must be very pure, and standards must be constantly run with the unknowns.

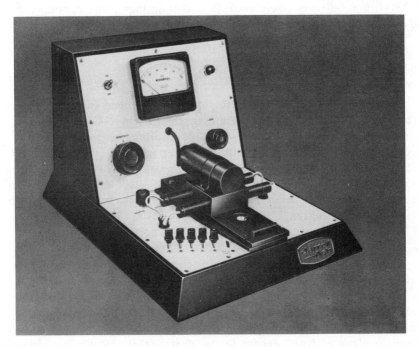

Fig. 6-12. A fluorometer, the most commonly used instrument for the determination of small quantities of uranium in water and solid samples. Courtesy of Jarrell-Ash Co., Waltham, Massachusetts, U.S.A.

Certain problems are, however, encountered in fluorometry, the most important of which is "quenching". This phenomenon is caused by certain impurities (manganese is commonly mentioned) which have the effect of reducing, or "quenching" the fluorescence. Because of this, it is sometimes desirable to selectively extract the uranium with ethyl acetate from a water or digested solid sample (rock, soil sediment or ash sample), before the fusion. Notwithstanding the quenching problem, in exploration geochemistry the fluorometric method is used exclusively for the determination of uranium in natural waters, and in most cases for the determination of uranium in solid samples containing less than 100 ppm, although it is satisfactory up to perhaps 0.1%. If specimens are known beforehand to have more than 100 ppm uranium, X-ray fluorescence would be preferable. For the analysis of water, 5 ml samples are needed with a fluorometer capable of handling this size sample, if the lowest limits of detection are to be attained. The Jarrel-Ash Company manufactures the most widely used instrument (Fig. 6-12) which costs about $2000, but small fusion-evaporation dishes of gold or platinum are also required. In commercial laboratories, uranium determinations by fluorometry cost about $3.00 each. Fluorometry is a possible method for the determination of beryllium, but it is rarely used. It is also an important technique in the identification of certain organic compounds.

Radon Counter

Historically the exploration for uranium has been carried out almost exclusively using gamma-ray detecting instruments (Geiger counter or gamma-ray spectrometer). An alternative approach is to measure the α-decay of radon (Rn^{222}), one of the daughter products of uranium. In the decay scheme for U^{238}, there are several unstable isotopes as follows: $U^{238} \rightarrow U^{234} \rightarrow Th^{230} \rightarrow Ra^{226} \rightarrow Rn^{222} \rightarrow Po^{218} \rightarrow Pb^{214} \rightarrow Bi^{214} \rightarrow Po^{214} \rightarrow Pb^{210} \rightarrow Pb^{206}$. In addition to Rn^{222}, several other isotopes are also α-emitters but their half-lives are in the range of minutes or seconds as opposed to the 3.8 days of Rn^{222}, and so interferences are negligible. The same applies to Rn^{220}, one of the daughter products of the decay of Th^{232}.

To determine radon in soils, it is necessary to flush (force or suck) it out, along with air, through a perforated or open-ended pipe inserted in the soil to a depth of 3 – 6 feet (Fig. 6-13). The air-radon mixture is then forced into the special activated zinc sulfide chamber of a radon detector from which all light is excluded, but which is open to a very sensitive photocell. When an atom of Rn^{222} decays in the chamber, it releases an α-particle which causes a fluorescence of the zinc sulfide and this is then detected by the photoelectric cell. In the case of water samples, after one liter has been

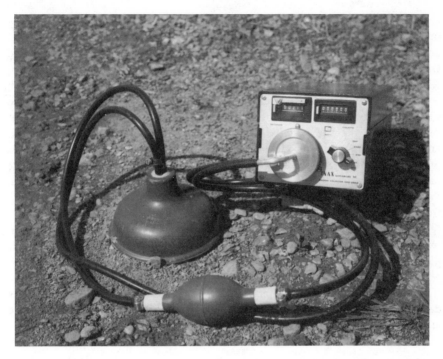

Fig. 6-13. A portable radon counter in operation with its collecting apparatus inserted into the ground. Courtesy of INAX Instruments Ltd., Ottawa, Ontario, Canada.

collected in a polyethylene bottle, air is circulated from the counting chamber to the water bottle and back to the counting chamber, bringing with it any radon dissolved in the water. As with radon collected from soil, decay of the radon results in a fluorescence of the zinc sulfide, and the light is detected by the photocell. The instrument shown in Fig. 6-13, including all accessories, is lightweight (less than 10 lb) and suitable for the field. In fact because of the short half-life of Rn^{222} measurements should not only be made in the field, but preferably on the spot. The instrument costs about $2000, and has been used widely in the lake waters of northern Saskatchewan and other parts of Canada, and in soils in the United States and Australia.

Radiometric Techniques

Radiometric techniques constitute a large and complex field, all stemming from the natural radioactivity of certain elements within the

Earth. For exploration purposes, apart from radon, discussed above, only the gamma radiations emanating naturally from uranium, thorium and potassium are important. The entire field is reviewed in Wainerdi and Uken (1971). It is shown that not only are the individual elements (U, Th, K) determined for exploration purposes, but that the ratios between them (e.g., Th/U) can be used to delineate favorable areas for exploration. The surveys can be airborne, carborne, or on foot, utilizing detection systems ranging from the inexpensive Geiger counter, to extremely sophisticated and expensive airborne spectrometers with thallium-activated NaI crystals of considerable size. In the laboratory, gamma-ray spectrometry can be used for the simultaneous and quantitative determination of uranium, thorium, and potassium, but on a practical basis, fluorimetry and X-ray fluorescence are preferred. Other types of natural radiations, and the results of radiation damage from the natural decay of elements (e.g., fission track studies), have been used, but only on a very limited scale, in exploration geochemistry. The application of fission-track methods to geochemical exploration for uranium in Scotland has been discussed by Bowie et al. (1973) and Wollenberg (1973) has shown how these methods can be used to determine the location and abundance of radioactive minerals in rocks.

Neutron Activation Analysis

Neutron activation analysis is a method based on the detection and interpretation of nuclear transmutations, after a sample has been suitably irradiated — it does not measure a naturally-occurring phenomenon as do the radiometric techniques described above. Samples are usually irradiated in a nuclear reactor, particle accelerator or cyclotron by neutrons, protons or any one of several other types of particles thus converting some of the atoms present into radioactive isotopes with known nuclear characteristics. The radiations emitted from the sample are identified and measured and, under favorable circumstances, very accurate analyses can be made, and very low sensitivities obtained. The techniques are extremely complex, require a very large capital outlay and, as a result, analyses are very costly by comparison with other methods commonly used in exploration geochemistry. The analysis of one element in a sample may cost $50, but inasmuch as this amount includes the irradiation costs, the determination of the second and subsequent elements in the same sample may cost only $25 each. Several commercial laboratories specialize in neutron activation analysis, such as Gulf Atomics (Los Angeles, California). Reviews of the neutron activation analysis methods have been published in Smales and Wager (1960), Wainerdi and Uken (1971), and in many other places.

Notwithstanding the power of this technique, it is not normally used in exploration geochemistry, for other reasons apart from cost. Although very low sensitivities are obtainable, they may be affected by interferences just as are other methods. For example, the analysis of uranium in water is affected by the radiation from a sodium isotope, which can result in apparent high uranium values. The platinum-group elements in basic igneous rocks may require a prior chemical separation and preconcentration before they can be determined. Nevertheless, neutron activation analysis is very important and valuable in certain cases, such as the analysis of the lunar rocks when neither time nor money are significant, and the utmost in accuracy, precision and sensitivity are the only governing factors. For exploration purposes, samples are considered for neutron activation analysis in special circumstances, such as confirmatory analyses on extremely important ore samples, or the analysis of elements not easily possible by other methods. Plant and Coleman (1973) found it advantageous to determine gold in placer concentrates by neutron activation analysis because large samples (up to 500 g) could be analyzed, hence more representative results could be obtained.

Neutron activation may in the future make its greatest contribution to exploration geochemistry in the area of portable equipment for field analysis. The situation is analogous to the field X-ray fluorescence instruments discussed above, but instead of X-rays, neutrons are used to irradiate samples from which the induced radiations are detected and interpreted. Field activation methods have been described for uranium in low grade ores (Philbin and Senftle, 1971), 30 elements in soils and rocks (Senftle and Hoyte, 1966), the exploration for manganese nodules on the ocean floor (Senftle et al., 1969), as well as others. Most, if not all, of the field instruments are prototypes, and not commercially available at this time, but they do point to the possible future utility of this powerful analytical method.

Mercury Detector

The importance of mercury as a pathfinder for both base and precious metals has been discussed in chapters 1 and 2 (see also Chapter 7). Accordingly, mercury detectors have been introduced from time to time. Recently a new portable mercury spectrometer, developed for the measurement of mercury vapor in soils, rocks and in the atmosphere, became commercially available (Fig. 6-14). The instrument illustrated (Scintrex model HGG-3) is adaptable to field operations and costs about $7500. It is a dual-beam dual-wavelength spectrometer utilizing the Zeeman effect, and reportedly eliminates some of the undesirable features

Fig. 6-14. A portable mercury detector in operation. This instrument is adaptable to studies of mercury vapor in soils, rocks and in the atmosphere. Courtesy of Scintrex Ltd., Concord, Ontario, Canada.

of single beam mercury detectors which are susceptible to interference problems. The design of this particular instrument, and some applications, are described by Robbins (1973). A prototype of another type of detector for mercury vapor in soil gas has been described by Bristow (1972).

Other, less expensive, portable mercury detectors are commercially available which operate on the principle of atomic absorption. With these instruments care must be taken to avoid, or compensate for, interferences caused by organic matter.

SUMMARY OF ANALYTICAL METHODS

From the descriptions and discussions of the various analytical methods one fact should have emerged, and that is that no one method can satisfy all exploration requirements. Also, it should be clear that, considering expense of equipment plus the necessity for specialists to operate the very sophisticated types, few laboratories are completely equipped to determine all elements of potential importance and at the necessary abundance levels.

Table 6-4.   Comparison of analytical methods commonly used in exploration geochemistry.

Name	Cost of Equipment	Detection Limits (see Table 6-3)	Advantages	Disadvantages	Cost of Analysis in Commercial Laboratories
Atomic Absorption	$4,000-20,000	Generally less than 10 ppm; some elements in ppb range	1. Rapid, sensitive, specific, accurate and inexpensive. 2. Several elements may be determined from same solution. 3. About 40 elements applicable to exploration geochemistry. 4. Partial or total analyses possible.	1. Accuracy suffers with high abundances. 2. Not satisfactory for some important elements such as Th, U, Nb, Ta and W. 3. Destructive.	1. Generally $1 per sample for first element, and 50¢ for each subsequent element determined on the same sample solution. 2. Au, Ag, Hg, Te, etc., from $1.50-$3.00. 3. Biogeochemical samples about twice regular rate. 4. Sample preparation (e.g. crushing, sieving) charges usually extra.
Colorimetry	Usually less than $1,000	Generally less than 10 ppm for elements commonly analyzed	1. Inexpensive, simple, sensitive, specific, accurate and portable. 2. Partial or total analyses possible.	1. Only one element (or a small group) determined at one time. 2. Not suitable for high abundances. 3. Some reagents unstable. 4. Tests not available for some important metals. 5. Destructive.	1. Generally 75¢ per element for cold-extractable. 2. Fusions or hot acid leaches generally $1.00 for first element, and 50¢ for each subsequent element determined on same solution. 3. Mo and W about $2.50 each.
Emission Spectrography	a. Visual detection $2,000-8,000 b. Photographic detection $15,000-30,000 c. Electronic (direct reader) $60,000-150,000	a. Usually only major and minor elements detected b. Generally from 1-100 ppm for most elements of interest. c. Same as (b) above.	1. Multi-element capabilities (for all instruments) 2. Only small sample required (for all instruments).	1. Complex spectra. 2. Requires highly trained personnel. 3. Generally slow (except for direct reader). 4. Sample preparation very critical and time consuming. 5. Destructive.	1. Qualitative or "semi-quantitative 30-element scans", $15-20. 2. Quantitative, about $5 for first element and $3 for each additional element on same film. 3. Direct-reader prices vary but probably about $30 for all elements recorded simultaneously.
X-ray Fluorescence	$30,000-40,000 (laboratory models)	50-200 ppm on routine basis; more sensitive with special procedures.	1. Simple spectra. 2. Good for high abundances of elements. 3. Uses relatively large sample. 4. All elements from fluorine to uranium are practical on modern equipment. 5. Certain liquids (e.g., brines) can be analyzed directly. 6. Excellent for rapid qualitative checks. 7. Non-destructive.	1. Sensitivities not as good as other methods for many elements. 2. Analyses slower than some other methods. 3. Analyses relatively expensive.	Generally $8-$12 per element per sample (quantitative); about half that price qualitative.

In order to assist those interested in the analytical aspects of geochemistry, Table 6-4 has been prepared which summarizes the main features of the four most important analytical methods used in exploration geochemistry: atomic absorption, colorimetry, emission spectrography and X-ray fluorescence. Included within Table 6-4 is a column which presents the approximate cost of an analysis by each method in North American commercial laboratories in 1973. Of course, it must be realized that costs will vary somewhat from one laboratory to another, with lower prices sometimes available in the winter season, but, in general, they are reasonably standard. The remarks made about the advantages and disadvantages of each method are, to some extent, based on individual preference.

Of great importance to those engaged in exploration geochemistry is the selection of the best method to use for the analysis of a specific element. Most persons concerned with the geological or collection aspects of exploration geochemistry, tend to avoid the analytical aspects by sending samples to their own or commercial laboratories, simply indicating the element(s) of interest, and leaving the decision-making to the analyst. If they do indicate a preference it will probably be atomic absorption, based on the current trend whether, in fact, it is the best method or not. Only occasionally will they discuss their problems with the analyst who might then be in a better position to obtain the information of most value to the geologist. He would, for instance, try to determine whether a quantitative, semi-quantitative or only a qualitative analysis is needed; what part of the sample (size fraction) should be analyzed; what interferences are likely to occur (if any); if a total or partial analysis is needed; the number of samples to be analyzed, as this will have a bearing on the method selected; how rapidly the results are really needed; the problems, if any, to be encountered in sample preparation and their effect on the results; and, finally, the accuracy and precision needed. Then, on the basis of all this information, he would be able to select the best analytical method. In practical situations, the lack of liaison between sample submitters and analysts has always been, and remains, a serious problem; it is found in all disciplines, not only exploration geochemistry, in which there is a submitter-analyst relationship. The lack of communication between the geologist and the analyst engaged in exploration geochemistry was deplored by several panelists engaged in a discussion entitled "What is a Geochemical Analysis" (Hansuld, 1969).

Notwithstanding the other factors which enter into the selection of the best analytical method, one which weighs very heavily is the lower limit of detection. Approximate limits of detection for atomic absorption, colorimetry and the emission spectrograph are presented in Table 6-3. Values are not included for X-ray fluorescence because, for all elements heavier than calcium, they are, on a routine basis, between 50 – 200 ppm. In fact they are predictable, in that the lower limit for calcium is about 200 ppm decreasing (i.e., becoming more sensitive) regularly to the transition metals where the limit of detection is about 50 ppm, and increasing again to about 200 ppm for barium, all based on the use of the K spectra. For the rare earths and all heavier elements, the L spectrum is used, with about 200 ppm the practical limit of detection for lanthanum on a routine basis, decreasing to about 50 ppm for uranium. The data in Table 6-3 comes from various sources, most of which are identified, but also include the author's experiences. It must be remembered that the limits of detection quoted are approximate, and will vary with changes in matrix and other factors mentioned above.

With several analytical methods available for a specific element, which one is preferable? The choice is especially difficult if two or more methods appear to have the same sensitivities. Another difficult choice occurs if one method has a lower limit of detection than another, but sample preparation is more difficult for the former method. The answer to the above question is very difficult indeed and because, for one reason or another, most analysts have their preferences, it is almost certain that uniformity of opinion will never be reached. Nevertheless, with full realization of the variables involved, graphical presentations are given in Figs. 6-15 and 6-16 which represent what appears to be a consensus of the first choice of many analysts. Because there is such a great difference in many cases between techniques for trace amounts less than 100 ppm, and trace and minor amounts above 100 ppm, two separate figures are presented using 100 ppm as the dividing point. The selection of techniques for the analysis of elements in Figs. 6-15 and 6-16 is based on the following assumptions and limitations.

1. The analyst is concerned with the analysis of one, or a few elements quantitatively, rather than a multi-element semi-quantitative orientation or reconnaissance survey, in which case emission spectrography might be recommended.

2. Where both X-ray fluorescence and emission spectrography are rated equal, preference is given to the former because many more North American and Australian commercial laboratories and universities have excellent XRF capabilities. However, for the rare earths and scandium, both of these methods are recommended.

3. Where two techniques appear to have equal capabilities, both are indicated.

4. Major quantities (e.g., over 5%) may, although not necessarily, require special assay methods (e.g., volumetric or gravimetric).

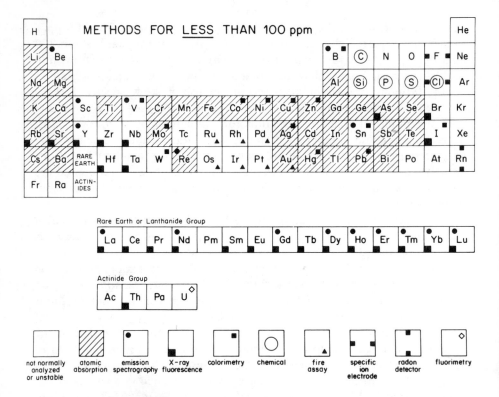

Fig. 6-15. Suggested analytical methods for the analysis of samples for purposes of exploration geochemistry, in which the elements to be determined occur in amounts of *less* than 100 ppm. If two methods are considered equally satisfactory, both are indicated. See text for details of the selection procedure.

5. Samples are typical of rocks, sediments, soils and vegetable matter submitted to commercial or company laboratories, or collected by university personnel for exploration geochemistry purposes (air and water are not included as they require special methods); no special chemical pre-concentration treatments, or unusual methods are considered.

6. The selection of the methods in Figs. 6-15 and 6-16 are first choices, but matrix effects, element interferences, or paucity of sample, may dictate the use of others. The selection assumes all analyses will be done in the laboratory, except for radon which should be measured at the sampling site.

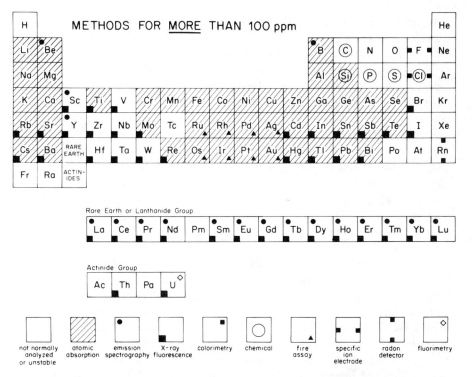

Fig. 6-16. Suggested analytical methods for the analysis of samples for purposes of exploration geochemistry, in which the elements to be determined occur in amounts of 100 ppm or *more*. If two methods are considered equally satisfactory, both are indicated. See text for details of the selection procedure.

COMMERCIAL LABORATORIES

With millions of samples being collected each year for exploration purposes, the need for good analytical facilities to analyze them is evident. In the last decade, the number of small mining and oil companies engaged in exploration work has increased, and the number of large companies concluding that it is cheaper to have some (or all) of their samples analyzed externally has also increased, particularly in those cases where specialized equipment is needed for one or a few projects (e.g., the analysis of niobium or thorium by X-ray fluorescence). The result has been the establishment of a significant number of commercial laboratories. This is now a big and very competitive business.

The main centres for commercial laboratories in Canada and the United States are in Toronto, Vancouver and Denver, with other laboratories of significance in many other places such as Ottawa, Calgary,

Tuscon and Salt Lake City. There are perhaps 10 major laboratories in Canada, and six each in the U.S. and Australia. The proliferation of commercial laboratories began about 1965, with the rise in the use of atomic absorption and the decline in the use of colorimetric field methods. Considering the relative youth of the industry, it is not surprising that the field has not yet stabilized, but still some larger, stronger companies are emerging, and will probably dominate the business in the future (but this must not be construed to necessarily equate size with quality). Whereas in 1965 a commercial laboratory could be operative with an atomic absorption unit, half a dozen lamps, and some inexpensive colorimetric capabilities, now most of the larger commercial laboratories have, or are considering, X-ray fluorescence, small direct reading emission spectrographs, and other types of sophisticated auxiliary equipment including computers, map plotters and other types of interpretive equipment, to extend their services.

The commercial laboratories, in several well-established cases, are supervised or run by personnel who received their training in government laboratories (e.g., the U.S. or Canadian Geological Surveys), or in mining company laboratories, and as such are eminently well qualified, assuming they have kept up with the latest developments. But under any circumstances, one particular aspect of the commercial laboratory worth noting is what may be described as an apparently higher-than-average turnover of key and intermediate level personnel. This, of course, can result in a change in the quality of analyses almost overnight. The problem of personnel changes and their effects is particularly noticeable during the summer months, when the commercial laboratories may be forced to hire temporary student help to cope with the great volume of samples. This is very true in Canada, where most samples are collected in the short summer season. During these peak seasons, commercial laboratories are prone to make mistakes, just like other humans working under strenuous conditions, and these errors may be traced to unexpected sources, for example, mislabelled samples, or typing errors in reports. It must be remembered that the commercial laboratories are geared to large volume projects which lend themselves to very routine methods and procedures, whether they be sample preparation or the actual analyses. The analysis during the summer season of a half dozen samples for one or two elements is actually a courtesy, except of course, to regular customers.

Being businessmen as well as analysts, the personnel of commercial laboratories are anxious to give the best service possible to clients, and are more than anxious to communicate with those collecting the samples and interpreting the data. Their experience, often based on information gained from the analysis of similar-type samples of other clients, is worth having. Almost all laboratories are willing to let clients observe the operation in order to better appreciate the nature of the work. Several of the larger commercial laboratories are associated with consultants experienced in geochemistry or economic geology and, should a problem be unusual, or there be a need for assistance in interpretation, they can normally give the necessary advice. Several laboratories also offer to do complete geochemical surveys (e.g., reconnaissance in remote areas) for clients.

In dealing with commercial laboratories it should be remembered that, although their literature may indicate the capability to analyze almost anything, by almost every method, the fact is that many of the smaller laboratories do not have all the necessary instrumentation, and samples may be sub-contracted to other laboratories with specialized equipment (e.g., X-ray fluorescence, fire assay), or with specially qualified personnel for difficult problems. This practice is to be commended as it ensures that the submitter obtains the best possible analytical data. Further, the commercial laboratories use two terms, "geochemical analysis" and "assay", with meanings somewhat different from what some geochemists might suspect. A geochemical analysis in a commercial laboratory, is one performed in the most routine, rapid and generally least expensive, manner. Duplicate samples are not normally analyzed, nor is any special effort made to account for matrix variations or other potential pitfalls. An assay, on the other hand, is a more accurate and precise analysis, possibly by a slower method, and in all likelihood duplicate samples would be run. Accordingly, it is perhaps twice (at the most five times) as expensive as the so-called geochemical analysis.

Finally, it is wise to spot-check any laboratory (commercial, company or university) by (a) submitting duplicate samples with different numbers to check the precision and (b) sending a certain percentage out to a second laboratory to check the accuracy, not only for samples with unusually high or low values, but on a routine basis. If this advice is taken, every effort must be made to ensure that the samples sent to the second laboratory are as similar as possible to the ones submitted to the first. The unused portions of identical samples, or a split of them, are preferred, and both laboratories, in the case of (b), must use the identical sample preparation and digestion procedures, analyzing the sample by the same method, whether it be instrumental or colorimetric. This should certainly be done before drilling or other expensive commitments are made. Usually the results on precision within the same laboratory will be satisfactory, and most often, it is hoped, the checks on accuracy between the two different laboratories will also be satisfactory. Agreement between two (or preferably three) laboratories can usually be considered as confirming both accuracy and precision, but on occasion the lack of agreement (accuracy) between several laboratories supposedly using identical methods of sample preparation, digestion and analysis can be shocking.

STANDARDS

The use of the word "shocking", which ended the preceding section, is purposeful. It represents the actual feeling which may develop if samples are sent for checks on accuracy or precision to two or more analysts. The situation is not, by any stretch of the imagination, limited to exploration geochemistry. Shocking results have originated from the elite of analytical society, but the magnitude of the problem only came to the forefront about 25 years ago.

In 1949 the U.S. Geological Survey sent two now-famous samples, G-1 and W-1, a granite and basalt, respectively, for analysis to many of the best laboratories in the world, for major, minor and trace element analyses, with the object of determining the accuracy and precision of analyses being obtained at that time. The results were, to say the least, disturbing. They showed that for major elements in a silicate analysis in which geologists had been inclined to accept the second decimal as significant, the first decimal place and in some cases the number to the left of the first decimal place, were to be treated with suspicion. The results indicated that, for the methods of the time, (many of which were gravimetric or volumetric and have since been replaced by instrumental techniques), the accuracy and precision decreased with decreasing content of the element within the rock, to the point where data on some trace elements were almost useless. The results have been discussed in many publications (e.g., Fairbairn, 1951). Many subsequent compilations of the data on G-1 and W-1 were made (both samples are now exhausted), the most recent, and probably the last of which, was by Fleischer (1969). Even to this day nobody knows the true value of the elements present in G-1 and W-1, and only "preferred" values, based on a combination of consensus and some subjective evaluations, are indicated by Fleischer (1969).

The results of the study of G-1 and W-1 emphasized the need for standards by which analysts may judge their own accuracy. Because there are many different types of natural and artificial materials with widely differing matrices and range of element contents, resulting in problems of inter-element effects, the need for many standards to attempt to cover or bracket the ranges of interest became evident. As a result, there are now well over 100 different standard samples of natural materials (including some ceramic and metallurgical materials of geochemical significance) which have been listed by Flanagan (1970). Some of the standards are free, and the charge for others is nominal.

As far as exploration geochemistry is concerned, the number of instrumental methods which can be used is increasing rapidly, and they are being continuously improved. Speed is a major factor, but this can well mitigate against high accuracy and precision. As the analytical laboratory increases its ability to turn out data faster and faster, the whole analytical process becomes more and more routine and impersonal, with the result that the recognition of normally obvious errors is becoming increasingly difficult. To be sure, most analytical laboratories will run one of their own standards with every tenth or twentieth unknown but, more than likely, they make no attempt to vary the matrix of their standards, or to analyze standards of varying concentrations. Yet the idea of running any standard as a check sample is extremely good, as among other things, it can detect such items as machine drift over a period of time, and the human errors which creep into an analysis. These types of errors, which tend to repeat themselves, are called bias, and are very difficult to detect. An alternative way to detect bias is to divide the sample being analyzed into two parts, each of which is analyzed by different methods.

The use of standards in the analysis of geochemical samples, and other natural materials, is essential. In the past few years several standards have been prepared primarily for exploration geochemistry in that they cover elements within the range and types of matrices (e.g., soils, sulfide ores) likely to be encountered by the analytical laboratory. These are especially welcome in view of the fact that instrumental methods are, by definition, relative methods — relative to a standard or group of standards. Table 6-5 presents a list of standards which have particular utility in the analysis of samples collected specifically for exploration. Several are quite recent, having become available between 1970 and 1972, and they are therefore not listed in Flanagan (1970). Nevertheless Flanagan's article should be consulted by the serious analyst interested in the broadest possible selection of standards. At this time the six standards (the last entry in Table 6-5) recently issued for geochemical purposes by the U.S. Geological Survey are being analyzed by many laboratories, and a compilation of the results is expected shortly. Probably the most ambitious standards preparation program of direct interest to exploration geochemists is being conducted by the Mines Branch, Department of Energy, Mines and Resources (Ottawa, Canada). The materials available are listed in Table 6-5, and are described by Faye (1972). Other standards, such as the glass standards containing 5, 50 and 500 ppm of many trace elements described by Myers et al. (1970), are not listed in Table 6-5, as they are not available for general distribution. Flanagan (1973) has compiled the average values for the international geochemical reference samples (mostly rocks) based on the most recent data available. Several additional standard samples are expected to be available from the U.S. Geological Survey in 1974. Abbey (1972, 1973) has reviewed the present state of the international reference samples of silicate rocks and minerals, and has assigned "usable" values, where possible.

Table 6-5.   Analytical standards of particular importance in exploration geochemistry.

Type	Designation	Source	Comments
Rocks (granite, andesite, granodiorite, basalt, peridotite, dunite)	G-2, AGV-1, GSP-1, BCR-1, PCC-1, DTS-1	1	U.S.G.S. rocks standards covering a wide range of matrices from granite to dunite. Compilation by Flanagan (1969, 1973). No charge to approved applicants.
Gold-quartz	GQS-1	1	2.61 ppm Au.
Molybdenite ore	PR-1	2	0.594% Mo, 0.793% S, 0.111% Bi, and 1.244% Fe (Faye), 1972). $50 for 200 grams.
Zinc-tin-copper-lead ore	MP-1	2	16.33% Zn, 2.50% Sn, 2.15% Cu, 1.93% Pb, 0.014% Mo, 0.071% In, 0.025% Bi, 0.791% As, and 59.5 ppm Ag (Faye, 1972). $50 for 200 grams.
Platiniferous black sand	PTA (formerly SSC-PT-1)	2	3.05 ppm Pt (0.089 oz/ton) (Faye, 1972). $50 for 400 grams.
Nickel-copper-matte	PTM-1	2	44.8% Ni, 30.2% Cu, 1.58% Fe, 21.6% S, 5.73 ppm Pt, 7.80 ppm Pd, 0.89 ppm Rh and 1.79 ppm Au (Faye, 172). $100 for 400 grams.
Sulfide (nickel-copper) ore	SU-1	2	Compilation of analyses by Sine et al. (1969) and Faye (1972). Provisional analyses: 1.3% Ni, 0.8% Cu, and 0.05% Co. $25 for 100 grams.
Syenite rocks	SY-2 and SY-3	2	Compilation of analyses by Sine et al. (1969) and Faye (1972); rocks are similar except SY-3 has higher U, Th and rare earth contents. Each $25 for 100 grams.
Ultramafic rocks	UM-1, UM-2, UM-4,	2	Sulfide-bearing rocks intended as standard for sulfur and ascorbic acid-hydrogen peroxide-soluble Cu, Ni and Co. (Cameron, 1972; Faye, 1972). 0.41-0.95% Cu, 0.19-0.83% Ni, and 0.007-0.029% Co. Each $25 for 100 grams.
Copper-molybdenum ore	HV-1	2	Nominal analyses: 0.5% Cu and 0.06% Mo (Faye, 1972). $50 for 200 grams.
Platinum metal-bearing flotation concentrate	PTC-1	2	5.0% Cu, 9.6% Ni, 0.3% Co, 24.2% Fe, and 24.1% S. Nominal values (in ppm): 3 Pt, 13 Pd, 1 Rh, 1 Ru, 1 Au and 6 Ag (Faye, 1972).
Radioactive ore standards	DHG-1, DLG-1, BL-1, BL-2, BL-3, BL-4	2	In preparation (Faye, 1972).
Uranium and thorium containing minerals, ores and counting standards; includes carnotite, monazite, uraninite, etc.	Numerous	3	About 2 dozen standards with the contents of most ranging from 1 ppm to 5%; several contain both U and Th. Prices generally from $2.50 - 6.00 each. Price list supplied by source.
Miscellaneous minerals (e.g. bauxite, lepidolite, spodumene, feldspars), ores (tin, zinc, manganese, iron), rocks (e.g., limestone), glasses, cements and refractories.	Numerous	4	Some are analyzed for major elements; other have data on rare elements (e.g., Cs, Rb, Li). Catalogue and price list supplied by source.
Miscellaneous ores (e.g. chrome, iron, manganese), alloys, slags, ceramic materials (e.g., fire brick), minerals (e.g., sillimanite, dolomite, magnesite), and spectrographic standards of steels for both emission spectroscopy and X-ray fluorescence.	Numerous	5	Wide variety with particular emphasis on materials and standards of value to the steel and ceramic industries. Catalogue and price list supplied by source.
Miscellaneous mixtures of synthetic standards; none of natural materials.	Numerous	6	Mixtures of a wide variety, with particular emphasis on standards applicable to emission spectroscopy (e.g. 49 elements in mixtures of varying concentration; noble metals). Catalogue and price list supplied by source.
1 jasperoid, 3 soils, 1 porphyry copper mill head, 1 Fe-Mn-W rich sample.	1, 2, 3, 4, 5 , 6	7	Specifically prepared as exploration geochemistry standards. These samples have been described in detail in J. Geochem. Explor. 2, 185-191 (1973).

Sources:
1.   Mr. F. J. Flanagan, U.S. Geological Survey, Washington, D.C. 20242.
2.   Standards Co-ordinator, Mineral Sciences Division, Mines Branch, 555 Booth Street, Ottawa, Ontario, Canada K1A 0G1. See Faye (1973) for the most recent prices, recommended values, and further details.
3.   Special Analysis Section, U.S. Atomic Energy Commission, Box 150, New Brunswick, New Jersey 08903.
4.   Office of Standard Reference Material, National Bureau of Standards, Washington, D.C. 20234.
5.   Bureau of Analyzed Samples, Ltd., Newham Hall, Newby, Middlesbrough, Teeside, TS8 9EA, England.
6.   Spex Industries Inc., Box 798, Metuchen, New Jersey 08840.
7.   Mr. H. W. Lakin, U.S. Geological Survey, Federal Center, Denver, Colorado 80225.

MOBILE LABORATORIES

Mobile laboratories are gaining in popularity and have replaced, to some extent, the base-camp type of laboratory. They vary in size, analytical instrumentation, capabilities and function. A very common type is housed in trailers and trucks of various sizes, whereas others which require the use of radioactive materials such as the portable neutron-activation equipment, are towed by a truck (e.g., Philbin and Senftle, 1971).

In general, mobile laboratories will contain either an emission spectrograph or an atomic absorption unit, and in addition, they have some colorimetric or wet chemical facilities because of the necessity for water, power and other services for the instruments. If the instrumental methods were not included, then the mobile laboratory would in essence, be a colorimetric laboratory or typical base-camp laboratory, for which no special housing (trailer) is necessary. As an indication of the remote areas in which they can operate with proper planning for long periods of time, mobile laboratories with atomic absorption have been flown to within a few hundred miles of the North Pole (in Canada) and have been shipped by sea to sites in the Solomon Islands. (Presumably, the decision was made that base-camp colorimetric methods would not suffice for the projects to be undertaken). The data obtained on the spot were necessary for the immediate direction of the projects being undertaken. The U.S. Geological Survey has a unique group of mobile laboratories, based upon the emission spectrograph, which are used as permanent laboratories during the winter months. They can be connected up to central sources of power and services when not in use in the field.

The use of mobile laboratories will probably continue to increase, and in fact, some commercial analytical laboratories now have them available for lease. In some cases, these mobile laboratories are used to service the needs of exploration groups in areas of high activity, so that immediate direction of a program is possible. Mobile laboratories, depending on the equipment installed and the type of trailer housing selected, can be outfitted for as little as $15,000 ranging up to as much as $40,000 or more. Airlifting of mobile laboratories to remote locations by helicopter appears to be increasing.

7/ *Primary Dispersion*
(Selected Topics)

INTRODUCTION

Various aspects of the primary environment, including primary dispersion and primary halos, were discussed in chapters 1 and 2. It was pointed out that primary halos may be classified on the basis of the time of their formation (syngenetic or epigenetic) or on the basis of their geometry (areal, leakage or wall rock patterns). In all cases, the primary halo must be produced at the same time, or nearly so, as the mineralization.

In the earlier discussions, following the precedent of Hawkes (1957) and Hawkes and Webb (1962), "primary" was defined as that part of the geochemical cycle embracing deep-seated processes of igneous differentiation and metamorphism, whereas "secondary" was related to the superficial processes of weathering, transportation and sedimentation at the Earth's surface. However, James (1967) observed that such a genetic classification, when extended to dispersion patterns, can result in difficulties and misunderstandings. For example, patterns existing in the walls of ore bodies formed by syngenetic sedimentary processes are not primary dispersion patterns in the strict sense but "fossil sedimentary patterns." He suggested that the term *primary dispersion pattern* should be used to describe the distribution of chemical elements in unweathered rock surrounding an ore body, no matter how or where the ore body itself was formed. A *secondary dispersion pattern* could then be defined as one brought about by the redistribution of primary patterns within the zone of oxidation and weathering. James' (1967) suggestions have been accepted by many exploration geochemists and, as a result, primary dispersion has gradually become essentially synonymous with the geochemistry of unweathered rocks and minerals.

In this chapter primary dispersion, in the James' (1967) sense, is discussed with the aim of illustrating how it may be used on a practical basis. Examples of the topics to be discussed include the use of rock geochemistry to delineate geochemical provinces, and how primary dispersion patterns may be able to assist in the discrimination between "barren" and "mineralized" environments. Additional comments are presented on the interpretation and use of primary dispersion halos of mercury and helium. Fluid inclusions and isotopic methods are discussed only briefly as their use in geochemical exploration is relatively limited at present. A voluminous literature now exists on many aspects of primary dispersion and it is possible only to refer to a few representative studies.

317

PRIMARY DISPERSION ON A REGIONAL SCALE: GEOCHEMICAL PROVINCES

Total Analysis Techniques

The use of rock geochemistry as an exploration tool has reached its most advanced state of refinement in the U.S.S.R., as the Soviets have conducted numerous such surveys for many years. Lithogeochemical surveys over extensive areas have been carried out to delineate geochemical (metallogenic) provinces, mineralized belts and other large targets of economic interest. Among the numerous examples which can be cited as typical of the approach used in the Soviet Union are those of Tauson and Kozlov (1973), and Beus and Sitnin (1972).

Tauson and Kozlov (1973) have been able to distinguish five main geochemical types of granites or granitoids (the following terms are rarely used in North America): (1) plagiogranites, which are terminal acid differentiates of gabbroic magma; (2) ultrametamorphic granites, which form by partial melting of highly metamorphosed rocks; (3) palingenic granites, which form as a result of complete remelting of different metamorphic rocks in the Earth's crust, and apparently can be subdivided into normal and subalkaline types; (4) plumasitic leucogranites, often termed "tin-bearing," which are genetically the latest acid differentiates of large chambers of normal palingenic granitic magma or of alkali basalts with a high content of potassium and volatiles; and (5) agpaitic leucogranites, often termed "rare-metal" or "columbite-bearing", which are considered to be differentiates of subalkaline palingenic magma or alkali basalt magma with a lower volatile content. As each of these granite types has different economic potential, it is desirable to be able to determine the economic potential of a granitic area on the basis of its geochemical characteristics. Each of the five types of granite has a distinctive assemblage of trace elements which assists in this task. Such assemblages are not only characteristic of the granites themselves, but also of the halos they form in surrounding rocks.

The identification of the geochemical types of granite and, in particular, of the potential ore-bearing varieties, is made possible by certain element concentration ratios. According to Tauson and Kozlov (1973), the most informative ratios are those given in Table 7-1. Especially noteworthy are the Ba/Rb ratio, and the (Li x 1000)/K ratio. The latter ratio, when high, identifies those granites high in volatiles (especially fluorine), which may possibly also contain tin, niobium and other trace elements concentrated as late differentiates. The Ba/Rb data shows that Rb has a strong affinity with fluorine and migrates with it. Thus, as a magma differentiates (roughly equivalent to the first five entries in Table 7-1), the barium content decreases while rubidium and fluorine generally increase. Very low Ba/Rb ratios, especially when confirmed by high (Li x 1000)/K ratios, are particularly favorable indications of potentially economic granitic geochemical provinces. Several other useful ratios, based on Soviet experiences, are given in Table 7-1.

Table 7-1. Concentration ratios in geochemical granitic types.
From Tauson and Kozlov (1973).

Concentration Ratios	1	2	3	4	5	6
K/Na	0.16	2.2	1.1	1.4	1.2	1.2
K/Rb	1250	385	240	100	290	165
Ba/Rb	45	11.5	5.3	0.5	0.34	4.1
(Li x 1000)/K	0.4	0.15	1.1	2.4	0.9	1.2
F/Li	75	16	16	31	29	20

1. plagiogranites
2. ultrametamorphic granites
3. palingenic granites

4. plusmasitic leucogranites
5. agpaitic leucogranites
6. "average" granite

Beus and Sitnin (1972) also applied this concept of indicator elements or ratios as criteria for identifying promising ore-bearing magmatic complexes by using different elements and ratios (Table 7-2). They established criteria for determining the economic potential of granites by comparing the distribution of the indicator elements in known ore-bearing complexes with those of barren rock complexes of the same type. Such studies should be the first stage of exploration for hidden deposits of rare metals (e.g., Li, Be, Sn) in granitic rocks, according to Beus and Sitnin (1972). At the same time, consideration of metasomatic alterations and other petrologic features of the rocks must be taken into account. Ratios of the indicator elements have been found to be more useful in many cases than the element abundances alone. From 40 to 75 well-selected samples appear sufficient to obtain meaningful relationships, based on the Soviet experiences.

The data in Table 7-2 are from rare metal deposits in various parts of the U.S.S.R. only, and are not universally applicable. In Australia, for example, several very detailed rock geochemistry studies, similar in many respects to those described above, have been made in the search for tin mineralization (e.g., Hesp, 1971; Flinter, 1971). These studies did not find any unambiguous correlation between the tin content of rocks and their economic potential. Thus, although a tin content of 20 ppm or more in a granitic rock is reasonably certain to indicate economic tin mineralization

in some parts of the Soviet Union, it does not necessarily follow that
economic deposits will be found elsewhere, even though such a figure is
unquestionably well above the 3 ppm average for such rocks (Table 2-1).
Examples are known of granites with less than 20 ppm Sn which do have
economic tin deposits, as well as those with more than 20 ppm Sn which do
not.

Table 7-2. Geochemical criteria for determining parent granites
of rare metal (Li, Be, Sn, W, Ta, Nb) deposits. Modified from
Beus and Sitnin (1972).

Indicator Element or Ratio	Average Abundance (ppm) or Ratio		Content (ppm) or ratio selected as criteria
	Ore-bearing	Barren	
Li	80 ± 20	37 ± 6	100 and more
Sn	15 ± 4	5 ± 1	20 and more
Mg/Li	75 ± 30	270 ± 80	30 and less
Zr/Sn	30 ± 10	76 ± 20	30 and less

In a more recent paper, Flinter et al. (1972) confirmed their earlier
conclusions that a purely geochemical approach is inadequate to establish a
relationship between the tin content and mineralization of the New
England (Australia) igneous complex. The same applies to tungsten, molybdenum and copper. They did find, however, that the ferromagnesian content (color index) of granitoids (a term used to denote the group of intermediate to acid igneous rocks of granitic composition and texture; Flinter, 1971, p. 323) is the most important feature to be considered and,
therefore, careful petrological and mineralogicial studies are possibly of
greater value than geochemical data in this particular area. Tin
mineralization in the New England complex is found in (1) lode and high
temperature veins in biotite-bearing leucogranites and (2) disseminated
within more mafic granitoids containing both hornblende and biotite.
Therefore, the relationship between rock type and mineralization is not a
simple one, which may account for some of the difficulties encountered in
using geochemical methods.

The reasons for the lack of correlation between the trace element content of rocks and their economic potential in some cases, and positive
correlation in others, are difficult to ascertain. Certainly, there are circumstances in which there is more than one controlling factor. In the case
of tin, the presence of biotite and other micas is probably important. Tin is
able to substitute in the structural positions normally occupied by ferric

iron (Fig. 2-6) and, should tin preferentially enter the mica structure, there may well be insufficient quantities remaining in the system to form economic deposits, even though the over-all abundance of tin in a particular granite may be high. In the case of Australian granites, the tin contents vary from 2.5 – 45 ppm in the bulk rocks and from 7 – 315 ppm in the biotites (Hesp, 1971). The biotites carry up to 45% of the total tin content of the rock and, therefore, play a major role in the mechanism of tin accumulation. On the other hand, Groves (1972) believes that the high tin content of biotites from some granites appears to be the result of a normal distribution of this element between liquid and crystals in a "tin-rich" liquid, and not to a selective substitution of tin for particular cations within the biotite structure. Discussion of these and other processes which may be operative in the formation of ore deposits, and which affect the relationship between trace element content of rocks and mineral deposits, are more properly in the area of economic geology, and the reader is referred to the vast literature in this field.

In the Canadian Shield, bedrock geochemistry has been used as a guide to areas of base-metal potential in the volcano-sedimentary belts (e.g., Sakrison, 1971; Descarreaux, 1973; Davenport and Nichol, 1973). These belts contain well-differentiated mafic and felsic extrusive pyroclastic rocks, but the massive zinc-copper sulfide deposits are found only within certain types of volcanic rocks. Descarreaux (1973) studied rocks extending from Matagami to a point near Rouyn (Ontario) and found that the ore bodies occur in calc-alkaline volcanic sequences but not within tholeiitic types (basalts poor in olivine and containing orthopyroxenes). Regionally, chemical analyses for the major components K_2O and TiO_2 permit a distinction to be made between these two types, and in a reconnaissance program this can mean considerable savings in costs and time. Locally, within the calc-alkaline sequences, certain types of anomalies are recognized only in association with the zinc-copper mineralization. These local anomalies are ascribed to hydrothermal alteration consisting of the introduction of MgO and removal of Na_2O, while enrichment of K_2O is also observed where zinc mineralization predominates. Enrichment of SiO_2, along with K_2O, may result in very extensive alteration halos (several thousand feet), but the dispersion patterns are significantly smaller than those of a geochemical province. Clearly, major element, as well as trace element, analyses provide important data for certain regional and local rock surveys. Similar, but not identical, results were reported by Davenport and Nichol (1973) from a different suite of rocks in a different part of the Canadian Shield (Birch-Uchi lakes, Ontario). They suggested that the high zinc content of felsic volcanic rocks containing economically significant mineralization may be a useful guide for distinguishing mineralized from unmineralized regions.

Many other primary dispersion studies involving numerous different rock types have been performed over large areas in attempts to use bedrock geochemistry as a tool for broad mineral reconnaissance and to establish correlations between geochemistry and mineral occurrences. These methods have a distinct advantage over secondary dispersion patterns in that the trace element patterns are not modified or complicated by the secondary environment. However, under certain conditions, the primary dispersion patterns may be reflected by soils in both regional and detailed studies, provided these soils are residual.

The modern approach to the study of primary dispersion on a regional basis involves the use of advanced statistical and computer techniques to aid in the elucidation of geochemical-geological relationships because these relationships are often difficult to decipher visually or by simple statistical methods of interpretation. For example, Hesp and Rigby (1973) used cluster analysis in their study of the New England (Australia) igneous complex. The results suggested that not only is this technique a useful tool for exploration geochemistry, but that it can also yield information which may be relevant to the process of ore formation. Garrett (1973) found that with the aid of advanced mathematical and statistical techniques (principal-component and multiple-regression analysis) on the data he obtained from a regional geochemical study of acid igneous rocks in east and central Yukon, it was possible to delineate those plutons known to be associated

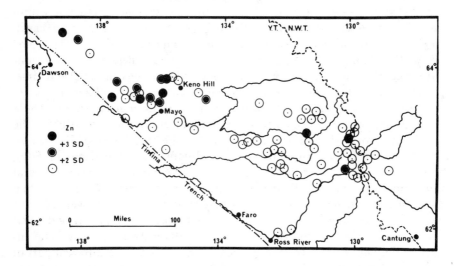

Fig. 7-1. Computer-generated results of a regional rock geochemistry study showing the maximum residual principal component scores for zinc in granitoids from the Yukon. From Garrett (1973).

with mineralization and, at the same time, eliminate others which appeared to be anomalous but which were not considered favorable on other grounds. The fact that these techniques seem to be capable of rejecting ground that does not warrant further expenditure is at least as important as the more positive aspects, since it is one objective of a regional survey. Some of the results of Garrett's (1973) study for zinc are shown in Fig. 7-1. Six plutons had maximum statistical (principal component) scores in excess of 3 standard deviation units (solid dots), and four of these are known to be associated with mineralization. Nine plutons exhibited maxima between 2 and 3 standard deviation units, and five of these are known to have mineralized showings in the vicinity.

Partial Extraction Techniques

The discussions to this point have been concerned with rock analyses in which the total element contents were determined. Partial extraction techniques have also been successfully applied to regional (and detailed) lithogeochemical studies. For example, when exploring for nickel and associated copper and cobalt in ultramafic rocks, it is important to know what proportion of the nickel is in sulfides and what is present in silicates and other generally uneconomic forms. Cameron et al. (1971) described a method (originally suggested in a Soviet paper) by which the sulfide nickel, copper and cobalt are preferentially dissolved from Canadian Shield ultramafic rocks, by an ascorbic acid-hydrogen peroxide mixture, whereas metals in non-sulfide forms are not attacked. (In certain situations, samples which are reacted with cold bromine and then shaken with 0.01M hydrochloric acid, give better partial extraction results, according to Davis, 1972). In sulfide ore-containing ultramafic rocks, leachable copper, nickel, cobalt and sulfur are considerably enriched compared with barren rocks of the same types. These relationships have been verified by studies of ultramafic rocks in Alaska (Dr. T. C. Mowatt, personal communication).

Hausen et al. (1973) pursued this approach further by studying mafic and ultramafic rocks from mineralized prospects in Canada and Australia. Using both partial extraction methods (ascorbic acid-hydrogen peroxide; water-bromine) and total analyses, they were able to show that log-log plots of nickel against sulfur enable evaluation of the mineralization potential of specific rock types. A comparison of serpentinites from mineralized and unmineralized localities suggests that concentrations and ratios of sulfur and nickel (S/Ni) near 1 to 1 are favorable indications of economic potential, with a broad range of 1:2 to 2:1 acceptable. If the ratios are significantly less than 1:2, the environment is sulfur-poor and a major part of the nickel will likely be in the serpentine (silicate) lattice. If the ratio is large, a sulfur-rich environment is indicated, and silicate nickel will be at a minimum.

Brabec and White (1971) determined aqua regia-extractable copper on 352 fresh rock samples from the Guichon Creek batholith (a porphyry copper province in British Columbia) and reported values ranging from 1 to 1600 ppm. Zinc extracted from 253 fresh rock samples by the same method showed a more limited range of from 5 to 152 ppm. The proportion of the total copper extracted by the acid attack was considerably higher than that of the zinc, indicating that the two metals may differ in their forms of occurrence in these rocks. Both copper and zinc show little or no correlation with the modes of the main rock-forming minerals, although the two metals correlated with each other. Of great economic significance is the fact that, on a regional scale, the central parts of this composite granitic intrusion, which contain the main porphyry copper deposits, are lowest in copper. From Fig. 7-2, it is clear that some of the major copper producers in this porphyry copper district (e.g., Lornex, Valley Copper) are located in areas where aqua regia-extractable copper values of rocks are low. Obviously, a relatively high aqua regia-extractable copper content of a granitic phase is not necessarily indicative of its superior ore potential.

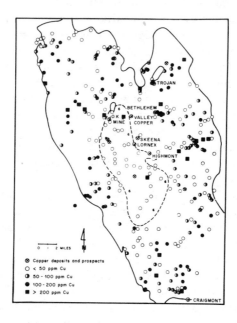

Fig. 7-2. Regional distribution of aqua regia-extractable copper in the Guichon Creek batholith, British Columbia. Note that the Highmont, Lornex, Valley Copper and other deposits are located where aqua regia-extractable copper values are low by comparison with the external zones. From Brabec and White (1971).

PRIMARY DISPERSION ON A LOCAL SCALE: DETAILED PATTERNS

Most primary halos actively investigated during geochemical surveys are smaller in size than those of a geochemical province (which can be from 10's to 100's of miles across), and they are the result of more localized geological processes. Although any distinction based on size is arbitrary, for the sake of discussion those dispersion patterns which can be detected up to a few miles, and occasionally a few tens of miles, from mineralization can be considered as useful for detailed studies. Bradshaw et al. (1972), based on Hawkes and Webb (1962), recognize several types of smaller dispersion patterns: (1) local syngenetic patterns, which are useful for detecting mineralized rock sequences (generally igneous), and for distinguishing between mineralized and barren sequences in the same rock type; (2) epigenetic wallrock anomalies, which are rarely more than a few hundred feet in width, and are useful in providing information on a detailed scale as to the localization of mineralization; and (3) leakage anomalies, which extend up to several hundred (or possibly a few thousand) feet above mineralization, and are of great potential value for locating hidden ore bodies. The difference between the first two types is based mainly on the time of emplacement and, since the trace element patterns of both are developed during endogenic processes and are often similar, the distinction is generally unnecessary for interpretation of the geochemical data. In addition, because of the lack of geochronological or detailed geological information, it is sometimes not clear whether the patterns are syngenetic, or were formed by the action of hydrothermal solutions introduced after the rock crystallized.

Patterns Due To Chemical Zoning

Most detailed studies of primary halos in the past have been directed towards the more limited patterns generated by mineralizing solutions or magmas. By way of example, Gott and Botbol (1973) analyzed 4000 rock and 8000 soil samples, and showed that zoning of major and minor metals in the Coeur d'Alene mining district of Idaho can be correlated with the location of mineralization. Lead-zinc-silver replacement veins in Precambrian rocks are the principal ore deposits in the district. The metal distribution shows the effect of different mobilities of metals during primary mineralization (Fig. 7-3) and, in addition, a temperature gradient is indicated. Many of the metals are zoned both vertically and laterally with respect to the center of mineralization, as well as to major geologic features of the entire district. Silver is found at shallow to moderate depths.

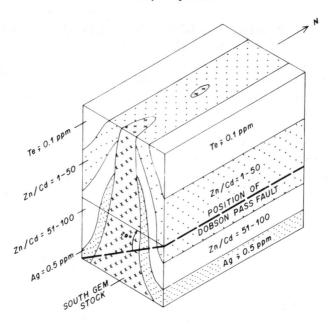

Fig. 7-3. Zoning of silver, cadmium relative to zinc (Zn/Cd), and tellurium around the South Gem stock, Coeur d'Alene district, Idaho. The east-west distance is about 3 miles; the vertical distance is about 7000 feet. The Te zone is 0.5 – 1 mile wide, the combined Zn/Cd zones are 0.3 – 1 mile wide, and the Ag zone is also generally 0.3 – 1 mile wide. (The zones are of variable width with some overlap of the adjacent zones). From Gott and Botbol (1973). Data on the scale of the figure and the width of the zones were supplied by G. B. Gott (personal communication).

Tellurium, on the other hand, appears to have been driven beyond the ore metals at the time the ore deposits were emplaced and, as it forms halos at considerable distances from the centers of mineralization, it is a useful pathfinder for buried deposits. Manganese (not shown on Fig. 7-3), like tellurium, forms extensive halos beyond the ore deposits.

The zoning is well defined by the dispersion of silver, tellurium, manganese and antimony (Mn and Sb are not shown on Fig. 7-3). However, zinc, cadmium and several other elements (e.g., copper, lead) by themselves do not define the ore deposits as well as ratios of these elements to each other (e.g., Zn/Cd). In Fig. 7-3 the ratio Zn/Cd is illustrated. Cadmium is more temperature-sensitive than zinc and the two metals tend to be fractionated from each other around the intrusive. Sphalerite in or near the intrusive is relatively depleted in cadmium, whereas it is relatively enriched in this element at greater distances. Thus the cadmium-rich halo around the stock can be used to indicate the relative position of mineralization. Cadmium is also depleted relative to zinc in the vein material, and enriched in the wallrock of the ore deposits. Geochemical data, in conjunction with geologic information, enabled the investigators to reconstruct the original position of the stock, as movement of about 16 miles had taken place subsequent to its emplacement along the Dobson Pass fault (Fig. 7-3).

Boyle and Smith (1968), Boyle et al. (1969), and Dass et al. (1973) have described the halos of trace elements formed by silver-bearing veins which cut several rock formations in the Cobalt area, Ontario. The dispersion of Sb, As, Hg, Co, Ni and Ag are illustrated in Fig. 7-4. Halos of these elements may be detected as far as 80-100 ft away from the veins in some formations. This is considered a broad halo for thin veins, but in other formations the halos are very narrow. The dispersion patterns of Cu, Pb and Zn (not shown) are also broad and coincident with the veins. The pronounced trace element dispersion patterns in the wallrocks can be used to locate the narrow silver veins in selected rock units. However, from Fig. 7-4 it is obvious that the dispersion pattern of each element is distinctive.

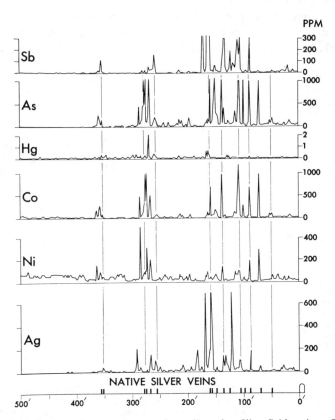

Fig. 7-4. Distribution of trace elements in wall rocks, Silverfields mine, Cobalt, Ontario. From Boyle and Smith (1968).

Numerous investigations have been made into the form and nature of primary dispersion patterns in wallrocks of various types. The results demonstrate widespread and irregular patterns in the vicinity of some ore deposits, and narrow regular patterns related to others. A very detailed study of the morphology and nature of primary halos around lead-zinc ore bodies in skarns has been described by Ovchinnikov and Grigoryan (1971). Their results, illustrated in Fig. 7-5, show the variations which typically are found in the lateral and vertical zonation of a number of pathfinder elements. Clearly, halos in wallrocks are very complicated, particularly in three dimensions. The non-symmetrical patterns are caused by any number of factors, such as variations in the mobility of elements as a function of changes in temperature, pressure and lithology. Thus, although primary halos for a number of elements may be valuable for prospecting purposes, their great variability in size, shape and relative position with respect to an ore body must be taken into account during the interpretation phase of a reconnaissance survey; while attempting to discover mineralization by drilling; and in any other situation in which data from primary dispersions are used. Perfectly symmetrical dispersion patterns, such as those illustrated in the idealized Fig. 7-3, have seldom been recognized.

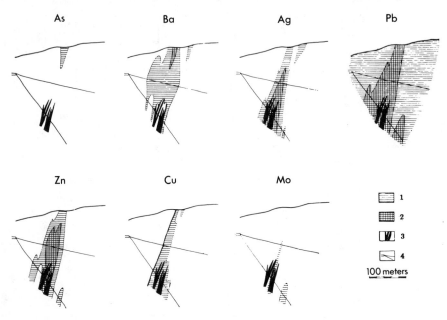

Fig. 7-5. Primary geochemical halos around lead-zinc ore bodies in skarns. The ore body is at a depth of 250 meters. Elemental contents (in ppm): 1. As, 15-30; Ba, 100-1000; Ag, 0.3-1; Pb, 30-100; Zn, 50-100; Cu, 30-500; Mo, 3-10. 2. Ba, 1000-2000; Ag, 1-8; Pb, 100-10,000; Zn, 100-10,000. 3. Ore body. 4. Position of drill holes. From Ovchinnikov and Grigoryan (1971).

Wallrock Alteration

Halos based on the trace or major element content of rocks are not the only patterns or products of primary dispersion processes. Hydrothermal solutions, in particular, affect rocks through which they pass by altering their mineralogical constitution. Mineralogical zoning patterns may result, which may or may not be accompanied by detectable trace or major element changes. These zones may be related to distance from the source of mineralization, or they may reflect temperature gradients, or both. For example, in the tin-bearing rhyolite region of central Mexico, halogen-containing mineralogical indicators are found consisting of apatite, secondary mica and fluorite. These secondary minerals represent a decreasing temperature gradient outward from the mineralized rhyolites. Traditionally, wallrock studies have been concerned with mineralogical and associated textural changes. As in the example just cited, they have attempted to identify the secondary minerals, such as clays (e.g., kaolinite, sericite, chlorite), other silicates (e.g., tourmaline, quartz), and carbonates which may be part of the gangue of the ore bodies, and to describe and map metasomatic zones. Wallrock alteration studies have been particularly useful in underground mine mapping and development, and in seeking extensions of known ore deposits through drilling. Hand specimen and microscopic studies have been, and continue to be, employed extensively in the mineralogical study of alteration zones although chemical methods are now widely used. Meyer and Hemley (1967) have published a very complete review on all aspects of wallrock alteration.

Lowell and Guilbert (1970) made a classic study of 27 porphyry copper-molybdenum deposits, most of which are located in the southwestern United States. They noted that these deposits exhibit many common characteristics, particularly the zonal pattern of lateral and vertical alteration and mineralization. Fig. 7-6, modified from Lowell and Guilbert (1970), is an idealized generalization of the distribution of the zones and it also emphasizes the influence of the level of erosion upon the zonal distributions observed. Although characteristic trace element variations are associated with these zones in many deposits, such geochemical distributions are not always easily identifiable. Further, it must be emphasized that these idealized concentric zones are not found in all porphyry copper-molybdenum deposits and in some deposits they are absent entirely.

Geochemical surveys for porphyry copper-molybdenum deposits conducted in the past few years have taken into account the probable presence of associated potassic and other alteration zones. Chemical analyses for potassium or trace elements (e.g., Cu, Mo, and possibly Rb) have not always unequivocally outlined the various alteration and ore-containing zones, nor have they always been successful in the recognition of such fine-

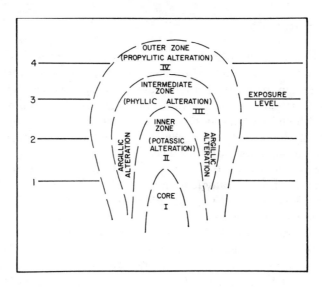

Fig. 7-6. Diagrammatic section through a porphyry copper-molybdenum deposit. At different levels of erosion (indicated by numbers 1–4), different alteration zones will be observed. The exposure levels are based on the following deposits: 1. Morenci, Butte; 2. Mineral Park, Silver Bell; 3. Bingham, Santa Rita; 4. Ajo. After Lowell and Guilbert (1970) by De Geoffroy and Wignall (1972). Reproduced from *Economic Geology*, 1972, v. 67, p. 657.

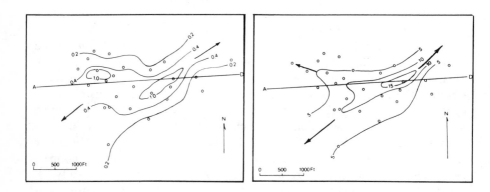

Fig. 7-7. *Left:* Percent copper in rock samples. *Right:* Percent sericite in rock samples as determined by X-ray diffraction analysis. All values are on drill hole composites from 1000-1900 ft elevation at Copper Creek, Arizona. Along the line A-D vertical profiles were taken (not shown here) which confirmed a similar close correlation between the sericite and copper contents. From Hausen and Kerr (1971).

grained minerals as sericite and secondary feldspar. Hausen and Kerr (1971) employed X-ray diffraction methods, which are ideally suited to the determination of mineralogical phases present in fine-grained altered rocks, and showed (Fig. 7-7) that it is possible to obtain a good correlation between trace elements (e.g., Cu and Mo) and the alteration products. In this case, high sericite values correspond with higher grades of copper. This technique, which is applicable to other types of alteration zones in addition to those associated with porphyry copper, permit the delineation of the limits of mineralization in a manner analogous to that of primary halos based on trace elements, and may permit the projection of inferred mineralization along trends that might otherwise go unrecognized.

The potassic alteration characteristic of many porphyry copper deposits should be readily detectable by airborne gamma-ray spectrometry techniques, as well as on the ground, as has been pointed out by Coope (1973), as well as by others previously. Similarly, Bennett (1971) described examples from Australia where anomalous potassium/uranium ratios, obtained by airborne gamma ray spectrometry, outlined areas where potassium was introduced into rocks during the formation of certain metasomatic copper deposits. Other airborne geophysical data discussed by Bennett (1971) indicate that various ratios of potassium, uranium and thorium may be used to deduce primary dispersion patterns.

Discussion

Geochemical surveys based on the analysis of rocks in the search for primary halos and other patterns indicative of primary dispersion may encounter difficulties during interpretation which are characteristic of these types of surveys. For example: (1) the anomalous values detected may have formed below the ore deposit which has since been eroded away; (2) metal-rich hydrothermal fluids which produced values suggestive of a primary or leakage halo may never have formed an ore deposit; (3) secondary dispersion resulting from groundwater movement or supergene enrichment, especially of mobile elements such as zinc and uranium, may form patterns which superficially appear as primary halos; (4) trace elements within certain minerals (e.g., tin or copper in biotite) may be the result of normal petrologic processes not necessarily associated with economic mineralization; (5) the halos may have formed either before or after mineralization, and thus be unrelated to it; this is especially likely to be encountered in the case of mercury, helium and other vapors which follow faults and similar structures; (6) vapors may produce non-significant leakage halos in unmineralized structures, or these halos may represent mineralization at greater depths. Finally, it is worth repeating again that there are several well-documented cases in which economic deposits are not reflected by elemental patterns in rocks.

At present, most geochemical methods used in the search for hidden ore deposits in igneous rocks are based on lithogeochemical methods, but such methods are only effective where primary halos are present in the rock and accessible for sampling. In certain stratiform deposits, such as some Pb-Zn-Ag ore bodies, primary dispersion patterns are absent and, in fact, this may be used as evidence that such deposits are syngenetic.

PRIMARY DISPERSION OF MERCURY, HELIUM AND RELATED ELEMENTS

Primary dispersion of mercury, helium and other elements which may result in gaseous leakage halos were discussed earlier (chapters 1 and 2), and it was stressed that these elements have great potential for the discovery of deeply buried deposits. Numerous studies of the dispersion patterns of mercury in rocks have been made in the past decade, and the results have often been interpreted on the assumption that the total amount of this element is present in the vapor form. Whereas this can be considered valid for mercury collected in atmospheric samples, it may not always be true for mercury in other media and, in fact, recent studies have shown that mercury in the vicinity of mineralization may be present in several different phases, none of which are gaseous. Native mercury is a primary constituent of some low-temperature hydrothermal ores, but Hg is more commonly found as cinnabar and within the structures of some sulfide minerals (particularly sphalerite; see Table 2-5), and in these forms it is most likely to be released only during weathering (or metamorphism).

Bradshaw and Koksoy (1968) pointed out that in rocks they studied from western Turkey, different proportions of mercury could be extracted when a sample was heated to different temperatures. They were able to relate the mercury released at various temperatures to the characteristic decomposition of HgS, $HgCl_2$, and Hg held in the lattice of rock-forming minerals. Significantly, they were unable to prove the presence of a mercury vapor phase. Thus, the mode of occurrence of mercury at this locality may be considered as an example of primary dispersion, but not of the gaseous variety (it is immaterial whether the mercury originally reached its present position in the vapor state; it is the present form of the mercury which is the determining factor).

Watling et al. (1973) published the thermal release curves of mercury vapor obtained by heating various mercury compounds to 550°C. These are illustrated in Fig. 7-8 and clearly show that each phase has its own characteristic thermal release pattern, and that if mercury exists in several phases

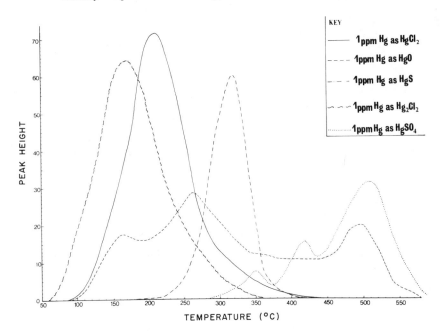

Fig. 7-8. Thermal release curves of mercury vapor from various mercury compounds. From Watling et al. (1973).

within a rock, identification of the individual compounds may be difficult. Elemental mercury (not shown in Fig. 7-8), gives a thermal release peak at about 80°C. Mercury within the structure of sulfide or silicate minerals will be released at temperatures generally above 450°C (pyrite at 450°C; sphalerite at 600-650°C), depending on such factors as the thermal stability of the lattice, and the bonding of the mercury. Applying this information to a zinc-cadmium deposit at Keel, Eire, Watling et al. (1973) found that the mercury thermal release pattern (Fig. 7-9) varies for different zones which are spatially related to mineralization, but elemental mercury was not detected (as there is no peak at 80°C). Mercuric sulfide (HgS; cinnabar) was found to be concentrated in the vicinity of the base-metal sulfides, whereas mercuric chloride is the predominant phase in rocks more remote from the sulfide ore. Therefore, the total mercury content of a rock (obtained by completely dissolving a sample, or heating it to a temperature sufficiently high to release all the mercury) (1) may originate from more than one source, none of which is a vapor phase, and (2) may not always be directly related to the proximity of ore deposits. Analysis of mercury compounds, by the continuous determination of mercury release (Figs. 7-8 and 7-9), may be of more value than that of total mercury. Bradshaw and Koksoy (1968) noted, and the studies at Keel tend to confirm, that primary

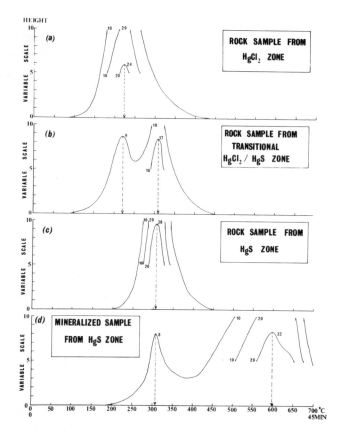

Fig. 7-9. Variations in the thermal release patterns in rock samples from different mercury compound zones, Keel, Eire. (Numbers such as 10, 20 and 24, which are indicated on the thermal release patterns, are related to changes in instrument settings, and are not pertinent to this discussion). From Watling et al. (1973).

mercury halos do not always form wider dispersion patterns than other elements, as would be expected on the basis of the gaseous nature of elemental mercury. They observed that in the western Turkey cinnabar and stibnite deposits they studied, antimony gave an anomaly of comparable distance (on the order of hundreds of meters) and strength to that of mercury. This, of course, is because mercury at this deposit does not exist in the elemental state with its high vapor pressure.

The importance of the physical state in which elements capable of occurring as gases are found is also emphasized in studies involving the use of helium for geochemical exploration. Eremeev et al. (1973) have discussed the importance attached to this element for mapping faults and for predicting the location of ore deposits (also earthquake forecasting) in the Soviet

Union. They recognize three forms in which helium may occur: (1) solid solution, with helium ions occupying structural sites within a crystal; (2) atomic inclusions, with neutral atoms in crystal lattices or in closed pore spaces; and (3) free atoms, which are exposed at the surface of minerals, and in pore spaces. The first two are "immobile", and would only be detected when a rock is completely decomposed, or heated to an extremely high temperature. The "mobile" helium, presumed to be mostly from abyssal sources, is capable of forming primary dispersion patterns, provided the rocks involved are reasonably permeable. Fault zones are particularly favorable sites for such dispersion patterns to develop and it is for this reason that helium is used for structural mapping (Fig. 7-10). Eremeev et al. (1973) suggested that groundwater sampling is the most simple and efficient method of helium surveying (at depths of 20-30 meters atmospheric factors are eliminated). Water from springs and wells is also suitable. Except in the search for uranium and a few other specialized types of minerals (Table 1-2), helium is not a pathfinder. Rather its use in the search for most ore deposits is for locating porous zones, and especially intersecting fracture patterns, which are favorable locations for mineralization. Should the movement of gases (such as mercury or helium) along fractures take place after the ore-forming events, dispersion patterns which may result are

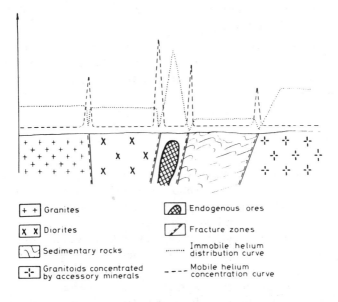

Fig. 7-10. Diagrammatic representation of the distribution of "mobile" and "immobile" helium in the outer part of the Earth's crust. Mobile helium (free atoms) occurs in anomalous concentrations above fracture zones which may contain mineralization. From Eremeev et al. (1973).

not true primary dispersion patterns, even though they may occur in the primary environment. In the Soviet Union, great emphasis is being placed on the value of vapors for structural (fault) mapping on the assumption that ore deposits are related to such features. The Soviets have established that, although in some cases the gases which are generated at depth and move up along faults may be predominantly of one component, most are mixtures of several gases in varying proportions.

TRACE ELEMENT CONTENT OF MINERALS

The trace element contents of various mineral phases have been used for geochemical exploration. These include mica, feldspar, magnetite, tourmaline, apatite, chalcopyrite, sphalerite, pyrite, and, less commonly barite, fluorite and several other minerals. One essential requirement is that the host mineral must have structural sites into which the trace and minor elements, which may act as pathfinders or indicators of mineralization, may substitute. A discussion of the reasons for trace element associations and diadochy was presented in Chapter 2, and several points are illustrated in Table 2-2 and Fig. 2-6. A second essential requirement is that the mineral being studied must be part of, or related to, the mineralization history (that is, the primary dispersion). In actual practice this relationship may not always be easy to establish.

The more important minerals whose trace element contents have been used in geochemical exploration will now be briefly discussed, and reference made to selected representative sources of additional information.

Mica

The muscovite formula, $KAl_2(Si,Al)_4O_{10}(OH)_2$, may be considered representative of the mica group. The muscovite structure contains positions for (a) large cations; these sites are normally occupied by K, but may contain Rb, Cs, Ba and Na; (b) large anions; these positions are normally occupied by hydroxyl, but may contain Cl or F; and (c) smaller sites normally occupied by Al, but which may contain Fe, Mg, Li, V, Zn, Cr, Sn, Mn, Ni, Cu, Sr, Ga and Nb. The elements listed above generally occur in muscovite in trace or minor quantities but, under certain conditions, these small quantities may be very important from the point of view of exploration. The use of the tin content of micas in exploration for that element has been mentioned above, and is typical of the approach that is used; high abundances of tin in mica are assumed, as a first approximation, to indicate a tin-bearing geochemical environment. Similarly, from studies of pegmatitic muscovites, it has long been known that rare-element pegmatites are generally reflected by high amounts of the rare-alkalis (Li, Rb, Cs). Heinrich (1962) observed that large amounts of niobium and beryllium in muscovite could be used as a prospecting guide for pegmatites containing economic amounts of minerals of these elements.

Most studies involving the micas in recent years have been concerned with exploration for porphyry copper mineralization. Parry and Nackowski (1963) found that the copper content of biotites from intrusions in porphyry copper areas tended to be relatively high. In the Sierrita and Santa Rita mountains of southern Arizona, Lovering et al. (1970) found that rocks from igneous intrusives genetically associated with copper deposits contain as much as 0.3% copper; however, biotite separated from these rocks contain as much as 1% copper. Rocks from igneous intrusives in the same area that are not associated with copper deposits contain very little copper, and the biotites separated from them contain at most 200 ppm copper. In one large composite stock there is a well-defined increase in the copper content of biotite from a few parts per million in the northern part to as much as 1% near copper deposits at its southern end. Lovering et al. (1970) concluded (1) that copper anomalies in biotite in the rocks in this area provide a more sensitive and extensive indication of associated copper mineralization than do those in the whole-rock samples, and (2) that the copper content of biotite may be useful in recognizing plutons which are genetically related to copper mineralization, as well as serving as a geochemical guide to copper ore deposits within such plutons. However, there have been other studies which appear to contradict the foregoing, or at least indicate that the relationships may be considerably more complicated in other situations.

Similar studies, utilizing the chlorine and fluorine content of biotites to distinguish geochemically between barren and productive intrusives, are also reported from porphyry copper and other deposits from the southwestern United States and British Columbia. By way of example, even though Parry (1972) found no abolutely clear distinction between the chlorine content of biotites and their occurrence in mineralized or barren plutons (Fig. 7-11), with one exception, those biotites with less than 0.2% chlorine came from plutons with little or no mineralization. Parry (1972) did find a close correlation between the chlorine and copper contents of biotite, and although insufficient information is presently available to reach any firm conclusions, the evidence seems to indicate that the chlorine and fluorine content of micas will become useful in exploration. Studies such as those of Stollery et al. (1971) mentioned previously (p. 48) have stressed the importance of chlorine and fluorine as complexing agents assisting in the transport of base metals and, on the basis of this, it is

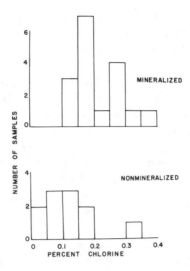

Fig. 7-11. Histograms of the chlorine content of biotites from various plutons in the
Basin and Range Province, Utah and Nevada. From Parry (1972). Reproduced from
Economic Geology, 1972, v. 67, p. 974.

logical to assume that these elements will be found in the vicinity of
mineralization and can be used as possible pathfinders. Darling's (1971)
preliminary study of the distribution of numerous trace and minor elements
(e.g., F, Zn, Cu, Li, Na, Y, Yb, Sc, B, Bi, Cr, V) in biotites from contact
metasomatic W-Mo-Cu ore deposits in California showed two independent
trends for fluorine. Fluorine (also Zr, Cu and Li) is highest in biotites from
plutons associated with certain ore deposits, and thus may be used to
distinguish between mineralized and barren environments. However, as
ore-bearing contacts are approached, the fluorine contents of the biotites
decrease, but the amount of other elements, including B, Bi, Cu, Sc, Ag, Sr,
Y, Cr and V may increase.

Feldspar

Slawson and Nackowski (1959) showed that potassium feldspars from
intrusives associated with lead mineralization have higher trace lead contents than feldspars from barren intrusives. Bradshaw and Stoyel (1968)
found that the mean contents of Sn, Pb and Zn in feldspars from
mineralized granites in Great Britain are significantly higher than those
from non-mineralized rocks. However, the copper contents of the feldspars
showed no essential differences.

Magnetite

Hamil and Nackowski (1971) studied magnetites from 14 monzonitic intrusives in ten districts of the Basin and Range Province, Utah and Nevada, and found that each stock or district has its own individual trace-element population. In general, accessory magnetites from major lead-zinc districts have high trace element abundances, whereas magnetites from major copper-producing districts have trace element abundances near the mean for all districts. High abundances of Ti and Zn in magnetite correlate with major lead-zinc mineralization, but low abundances of Ti and Zn correlate with major copper mineralization. High concentrations of Cu and Zn were found by de Grys (1970) in magnetite from intrusives associated with porphyry-copper mineralization in Ecuador. However, Huff (1971) found no significant copper anomalies in magnetites from drainages down-stream of mineralization in the Lone Star district, Arizona. Data on the major element composition of magnetites as related to occurrence have been compiled by Fleisher (1965). Theobald et al. (1967) have presented the results of an extensive study of detrital magnetite from the Inner Piedmont belt of North and South Carolina. They discussed the various ramifications of mineralogy, geochemistry, and geologic relationships in some detail, and furnish valuable background information relative to any attempt to utilize magnetite in geochemical exploration.

Tourmaline

Boyle (1971b) briefly mentioned that analyses of tourmaline obtained in heavy mineral surveys suggest that this mineral may be indicative of certain types of deposits. Thus, tourmaline from lithium pegmatites is enriched in Li, Cs and Rb, and tourmaline from tin deposits is enriched in Li and Sn. However, in areas where the country rock contains abundant tourmaline, difficulties in interpretation may be encountered. Fisher (1971) reported similar studies involving the use of trace elements in tourmaline as ore indicators in Australia. This mineral is regarded as particularly suitable for exploration purposes because its high chemical and mechanical stability permits it to be preserved in deeply leached soils, laterite profiles and stream sediments.

Apatite

This mineral, ideally $Ca_5F(PO_4)_3$, has great potential in geochemical exploration because a large number of cations and anions may substitute within its structure, it is common in many different types of rock, and it is relatively stable in the secondary environment. McConnell (1973) has compiled the most complete summary of the crystal chemistry, mineralogy and other aspects of this mineral including all major chemical sub-stitutions, although his treatise is not designed primarily for exploration purposes. Among the better-known substitutions are rare-earth elements, U, Mn and Sr for Ca; Cl and (OH) for F; and CO_3, SiO_4 and SO_4 for PO_4. Several studies relating the composition of apatite to geological occurrence and mineralization have been reported in the Soviet literature, but the publicized use of this mineral elsewhere has been limited. The relative proportions of (OH), F and Cl may prove to be particularly valuable parameters to ascertain.

Chalcopyrite, Sphalerite and Pyrite

Trace elements in these and other sulfide minerals have been used on a limited basis for exploration purposes. In a classic paper, Burnham (1959) studied the compositional zoning of ore minerals in the south-western United States and northern Mexico. He found that the Sn, Ag, and combined Co-In-Ni-Ag-Sn content of chalcopyrite and sphalerite defined three broad zones which corresponded with both regional tectonic features and previously established metallogenic provinces. He postulated that the relationship of zonal patterns to chemical variations originated with a deep-seated source.

Geochemical studies by Johnson (1972) using pyrite suggested that the Co, Ni, Ti and As contents of this mineral, together with Co/Ni ratios, are useful in the identification and genetic interpretation of different types of mineralization in Cyprus. In this locality, the pyrite is typically cobaltian, and the Co content increases at depth within the ore bodies. Arsenian pyrite is confined to secondary sulfide ores, whereas the Ni content is highest in disseminated pyrite in certain lavas most distant from the sulfide ore bodies.

Discussion

Johnson's (1972) study, as well as that of Gott and Botbol (1973) mentioned above (Fig. 7-3), illustrates the manner in which trace elements in minerals may possibly be used for interpretation of the temperature of formation, and additionally for the reconstruction of temperature gradients or zones, relative to mineralization (in most cases the effect of pressure is small compared to that of temperature). From a knowledge of the cobalt content of pyrite in Cyprus, for example, Johnson (1972) was able to infer relative temperatures and, when combined with Ni

and As data, it might be possible to determine the most probable localization of ore. By means of the temperature-sensitive Zn/Cd ratio in
sphalerite (Fig. 7-3), it is possible to infer the position of this mineral
relative to the stock and to mineralization, at least in one major mining
district in Idaho. At one time it was thought that the iron content of
sphalerite could be used for purposes of geothermometry, but it is now
known that variations in this parameter are controlled by other factors.
Therefore, not all elements may be used in all minerals, and caution is
necessary when attempting to determine temperature gradients, as well as
other geologic information, from trace element data alone.

Before concluding this section it is desirable to stress two important
points. First, for all geochemical studies involving the use of minerals it is
essential that the mineral concentrates be absolutely pure. Microscopic
confirmation of purity is the minimum requirement. Minerals which are
assumed to be pure may, in fact, have inclusions of other phases.
Muscovite, for example, upon microscopic study may be found to have
minute crystals of rutile, tourmaline, cassiterite and various other minerals
between the mica sheets. Sulfide ore minerals often contain minute phases
of other sulfide minerals, and magnetite commonly contains exsolved
ilmenite. In the case of the opaque minerals, X-ray diffraction, reflected
light microscopy or electron probe studies may be necessary to confirm
purity, but this is essential for confidence in some interpretations. Second,
as Bradshaw et al. (1972) pointed out, when undertaking whole rock
(lithogeochemical) analyses, caution is required because of the variations
which are possible in the mineral content of rocks (modal variations).
Using the example of Bradshaw et al. (1972) from granites in southwest
England, the average zinc concentration in feldspar for both mineralized
and non-mineralized granites is 6-14 ppm, and in biotite and muscovite
300-350 and 130-240 ppm, respectively. A 10% change in the modal content of the micas will, therefore, result in a significant shift in the zinc content of the whole rock. Such shifts could be interpreted as related to
mineralization if modal variations were not considered. Bradshaw et al.
(1972) also pointed out, based on data from the same locality, that it is
common for some metals in multi-metal mining districts to show a positive
correlation of trace metal concentrations in minerals (also igneous rocks)
with mineralization, while others show no association, either positive or
negative.

FLUID INCLUSIONS

When crystals grow or recrystallize in any kind of fluid medium,
small portions of the fluid, or gas, or both, are trapped in minute cavities

within the solid crystal. By careful examination, these fluids can be found in the minerals of most rocks, including ores and gangue, and can give evidence regarding the chemical nature of the fluid phase, the temperature at which the crystal grew, and many other facts of interest to economic geology and exploration geochemistry, only a few of which can be mentioned here. There are, however, several uncertainties connected with the interpretation of fluid inclusions, such as whether the trapped fluid is truly representative of the parent solution, the time of the trapping and, particularly in the case of gases, the amount of leakage from the cavity. Such problems notwithstanding, fluid inclusions remain our most direct means of attempting to sample hydrothermal fluids. This point was stressed by Roedder (1967) in his review of the voluminous literature, some of which dates back about 100 years. In a more recent review, Roedder (1972) showed that qualitative (mainly microscopic) methods may yield surprisingly large amounts of useful data.

Analyses of these fluids have shown them to be mainly dilute chloride-rich water containing Ca, Na, Mg, Fe, K, SO_4 and other common elements and radicals. The principal gas reported is CO_2, along with minor quantities of other gases such as H_2S, N_2 and H_2. The amounts of heavy metals (Cu, Mn, Zn) in solution may be high (100-10,000 ppm) in some cases. Fluid inclusions are examples of primary dispersion and can be expected to be detected in halos corresponding to the area impregnated by hydrothermal solutions.

Notwithstanding the value of knowing the composition of ore-forming fluids, and in using this information for exploration purposes, the chemical analysis of fluid inclusions has not been used to any great extent in geochemical exploration up to the present time. The reason, in part, has been the great difficulty of separation and analysis of the fluids, whose volume seldom comprises more than a few tenths of 1 percent of a sample, even though the inclusions may be exceedingly numerous. Modern methods of analysis (e.g., neutron activation) have overcome some of the problems inherent in working with minute quantities of fluid, but problems in interpretation are still encountered. It is important to realize that fluid inclusions are not restricted to the minerals formed during hydrothermal or mineralization processes. Any given sample will probably contain inclusions which are the result of different processes, and are of different ages. Nevertheless, certain studies (e.g., Roedder, 1971) have shown that some properties and measurements (e.g., temperature of formation, salinity) determined from fluid inclusions can be correlated with the mineralogical zonation. For example, the highest indicated temperatures and most saline solutions would be found in the cores of mineralized areas.

Kesler et al. (1973) and Van Loon et al. (1973) used selective ion electrodes to determine the water-soluble fluoride and chloride content of several granodioritic intrusives (which, after fine-grinding, can be considered a measure of the fluid inclusion content) from the Caribbean and Central America, only some of which are mineralized. Their results suggested that, within individual pluton complexes, high copper values are associated with high chloride values in the case of (presently) marginal or sub-ore grade mineralization. However, high copper values are associated with low chloride values in rocks that are host to ore grade mineralization. Fluoride values, on the other hand, tend to be high in mineralized intrusions.

If a mineral crystallizes near the boiling point of a fluid, the cavity will be full, or nearly full, of liquid, but at lower temperatures the fluid will occupy less space in the cavity. If the assumptions are made that the fluid contents of these cavities were deposited from the original ore-bearing fluids, and there has been no gain or loss of material, it is possible to estimate the minimum temperature of the original fluid. This is done by heating the sample until the liquid and gas expand to fill the cavity, the required temperature (a pressure correction should be made) representing an estimate of the temperature of formation of the mineral. These techniques are well established and are described by Roedder (1967), and others. Using this method, Bradshaw and Stoyel (1968) found cavity-filling temperature to be of value in the evaluation of some ore bodies in south-west England and, in fact, the approach was superior to soil and stream sediment surveys. Thus, in a hypothetical example, economic concentrations of a specific mineral, say sphalerite, may occur where the cavity-filling temperature of the fluid inclusions ranges from 200-220°C. A knowledge of the temperature gradient in an area obtained from fluid inclusion studies might possibly be used to determine the location of mineralization.

Whether fluid inclusions are used to determine the composition, temperature of formation, or any other physical (e.g., fluid density) or chemical (e.g., isotopic abundances) parameters of ore fluids, it must be remembered that these studies are often very complex and generally require the services of well-trained personnel. Sampling, analysis and interpretation are relatively difficult, and costs are high, by comparison with techniques generally employed in geochemical exploration. The greatest potential for fluid inclusion studies in geochemical exploration would appear to be in detailed surveys of very promising areas.

ISOTOPE METHODS

Isotopic methods are potentially of great importance in geochemical exploration but, at the present time, they are not widely used. There are a number of reasons for this, one being that these analyses require the use of a mass spectrometer and are generally expensive by comparison with other techniques. They also require personnel with special skills, and in many cases equivalent information may be obtained from alternative methods. Many studies of isotopic abundances of sulfur, oxygen, carbon and lead, as well as geochronological investigations, are described in the literature and, in fact, isotopic geochemistry is currently one of the most active and important areas of theoretical and applied (e.g., groundwater studies) geochemistry. The methods of analysis and the procedures for interpretation are well established and, therefore, it is probable that isotopic methods will be used with greater frequency in geochemical exploration in the future. The potential uses in geochemical exploration of stable isotopes (particularly oxygen and sulfur) have been discussed by Jensen (1971) and by Shima and Thode (1971). The suggested uses of lead isotopes in exploration have been described by Cannon et al. (1971). In the following paragraphs a brief summary is presented of the more important applications of isotope geochemistry to geochemical exploration, but, the reader is referred to the above references, or any of the numerous textbooks presently available, for the theoretical bases of these methods.

Geochronology

Age dating techniques, chiefly by the K-Ar and Rb-Sr methods, have potential for assisting in selecting economic plutons, areas of hydrothermal alteration, and so forth. For example, most porphyry copper deposits in the southwestern United States and northwestern Mexico are 58 – 72 m.y. (million year) in age, whereas those in British Columbia are about 50 m.y., and those in the southwest Pacific region are considerably younger (about 25 m.y.). Age dating of rock formations can be of great value; several studies have confirmed the similarity in age of porphyry and other types of mineralization in important mining districts.

Lead Isotopes

In several papers, Cannon and Pierce (1969) and Cannon et al. (1971) have suggested the manner in which lead isotopes may be used as guides for Mississippi Valley-type lead-zinc deposits. They found that in searching for "big" ore deposits of this type, a ratio of Pb^{206}/Pb^{207} in galena near 1.40 is extremely favorable. They refer to lead (galena) specimens with this ratio as "J-type lead", because this ratio was first detected in galena from Joplin, Missouri. In Fig. 7-12, the relationship of the Pine

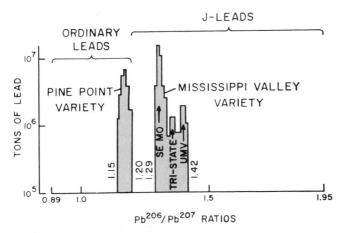

Fig. 7-12. Isotopic composition versus tons of lead in world-wide Phanerozoic stratiform lead-zinc deposits. SE Mo (southeast Missouri); Tri-State (the Tri-State mining region of Missouri-Kansas-Oklahoma); UMV (Upper Mississippi-Valley district). From Cannon and Pierce (1969).

Point and Mississippi Valley-type mineralization to the above-mentioned ratio is illustrated. Deposits containing more than 100,000 tons of lead are generally grouped about a ratio of $1.29 - 1.42$ (Pine Point is lower), as has been corroborated in studies by Angino et al. (1971), whereas low-tonnage deposits have a wide range of values for this ratio (but not $1.29 - 1.42$). The use of lead isotopes in this manner is encouraging and, if confirmed by studies in other parts of the world, may be extremely important in delineating target areas. Even a single lead isotope analysis may reveal information on the potential of a mineral occurrence. Further, Cannon et al. (1971) have observed lead isotope gradients in the Leadville, Colorado district that appear to point toward the focus of mineralization. Although thus far almost all isotopic information with respect to exploration has been obtained from galena, it is possible that geochemical lead anomalies in soil and water may also be analyzed isotopically for significant ratios.

Boyle and Garrett (1970) suggested that lead isotopic data may be useful in some areas in the search for lead-silver deposits. Silver isotopic data appears to be uniform in all types of deposits, and there does not appear to be any way to use such data in exploration at present. However, they noted that certain large silver-bearing deposits, such as the Sullivan and Broken Hill mines, are characterized by relatively uniform lead isotope ratios indicating Precambrian (or somewhat younger) ages, whereas satellitic deposits in the general area of major deposits have variable isotopic ratios that are usually anomalous. They suggested that if anomalous lead isotopic ratios are found in lead veins in an area, consideration should be given to the possibility of the occurrence of large massive lead deposits with uniform ratios in the same mineralized belt.

Sulfur Isotopes

Variations in the S^{34}/S^{32} ratio enable economic geologists and geochemists to determine whether sulfide minerals form at low or high temperatures. This ratio may also aid in the distinction between sulfides of hydrothermal and sedimentary origin. Boyle et al. (1970) mentioned that sulfur isotopes in the lead-zinc-silver deposits of the Keno Hill-Galena Hill area, Yukon, differ markedly from those of the adjacent country rock (the sulfur in the ore is enriched in S^{32}). Gold-bearing and other mineralized areas may be differentiated from barren areas by means of sulfur isotope studies. Boyle et al. (1970) also pointed out that the sulfur isotope ratio in the sulfate of waters leaching mineralized zones reflects the ratio in the ores, and this fact may be useful in hydrogeochemical prospecting.

Oxygen Isotopes

Variation in the O^{18}/O^{16} ratio has particular application in geothermometry. Although other methods for geothermometry have been mentioned in this chapter, specifically the use of trace element abundances and ratios in sulfides (Fig. 7-3), and fluid inclusion techniques, oxygen isotope methods permit fairly accurate determinations of the temperature that existed during the crystallization of specific silicate and oxide minerals. Oxygen isotopic distributions are temperature dependent, and are not affected by pressure. The temperature of formation of hydrothermal, metamorphic, and magmatic minerals may be determined. Oxygen isotopic methods have the potential for assessing the distance of migration of hydrothermal fluids, as well as the distance from mineralization, as indicated by a gradient of changing isotope ratios, possibly over distances of thousands of feet.

8/Secondary Dispersion
(Selected Topics)

INTRODUCTION

The fundamental principles of the secondary environment were discussed in Chapter 3. In Chapter 11 the principles of geochemical exploration in those secondary environments of particular interest in Canada and other areas in northern latitudes (e.g., permafrost, muskeg, glacial) are considered separately. This chapter is concerned with miscellaneous concepts and principles related to secondary dispersion which are not covered in either chapters 3 or 11. As in the case of the preceding chapter on primary dispersion, the emphasis here will be on practical applications, including how secondary dispersion patterns may be recognized and interpreted. The term secondary dispersion pattern is used in the sense of James (1967) discussed previously (p. 317), and is defined as the distribution (or redistribution) of chemical elements in the surface zone of oxidation and weathering. Even though secondary dispersion may take place on or near the Earth's surface where geological processes are well understood, it is still a very complex phenomenon which, at times, is difficult to interpret.

As most geochemical surveys are conducted in materials of the secondary environment (Table 1-1) a huge literature, comprising perhaps several thousand titles, has evolved concerning various aspects of secondary dispersion. Obviously, it is not feasible to attempt to refer to more than a few representative topics. Extremely thorough, well-illustrated discussions on various aspects of the secondary environment and secondary dispersion are found in Hawkes and Webb (1962) and Granier (1973). From these and other sources (e.g., Bradshaw et al., 1972) it can be concluded that the same fundamental principles of geochemical exploration are valid in widely separated secondary environments throughout the world, with due allowance for differing geologic conditions. For example, the techniques and methods of interpretation useful in soil surveys in tropical environments are also applicable to the more complex glaciated areas, provided the procedures are altered in detail as dictated by local environmental, geological and other conditions.

These alterations in detail, however, require an appreciation of all of the processes which may affect secondary dispersion and the resulting patterns. In some cases, just one process may control the dispersion, but very often several processes are influential simultaneously. From the discussions in Chapter 3 and elsewhere in this book, four major factors affecting secondary dispersion are recognized (chemical, biological, mechanical and environmental) and, within each category based on these factors, there are many processes operative. Some of the more important processes and points to be considered within each category are:

Chemical factors: (1) original bulk elemental composition and distribution in bedrock and in ores, and the mineralogical character of the material; (2) supergene processes responsible for the changes in the mineral composition of the rocks and ores (e.g., pH, extent of oxidation); (3) mobility characteristics of the element or its compounds in aqueous and other media in the presence or absence of organic substances (e.g., complex ion formation, precipitation of secondary minerals, adsorption, organometallic complexes); (4) presence of media capable of precipitating migrating metals and thus limiting the extent of dispersion patterns (e.g., limestone, certain bacteria).

Biological factors: (1) vegetation (amount, type, and depth of root penetration); (2) microorganisms (role in the production of humus, role in oxidation-reduction, assimilation and concentration of metals).

Mechanical factors: (1) gravity movement (slope angle, soil creep, solifluction, presence of vegetation to stabilize slopes); (2) dispersion in surface water (runoff, streams); (3) dispersion in groundwater (e.g., downward movement in rain water, upward movement by capillary forces); (4) glacial dispersion (see Chapter 11); (5) aeolian transfer of material (effect generally limited to the surface); (6) action of burrowing animals (generally restricted to the upper few feet of the surface).

Environmental factors: (1) climate (e.g., amount and distribution of rainfall, with its resultant effect on the availability of vegetation and on soil formation); (2) topography; (3) geological (e.g., thickness and origin of overburden, hydrologic environment, extent of denudation of ore-bearing bedrock, size of ore body); (4) time (in general, the longer the period of weathering and soil formation, the larger the secondary dispersion pattern, although under favorable circumstances, significant dispersion can occur in a few centuries).

In view of the multitude of variables just enumerated, it is not surprising that secondary dispersion patterns are complex, and that an extensive literature exists describing them.

DISPLACED ANOMALIES

Discussions of anomalies in the secondary environment beginning with Fig. 1-1 (left) have indicated that secondary dispersion patterns very commonly are displaced from the source of mineralization. Such displacements in stream sediments and water, for example, are extremely useful, as they are the basis for many regional reconnaissance surveys. However, these anomalies may assume many sizes and shapes, and the distances of displacement from mineralization may vary considerably. Secondary dispersion patterns developed in residual soils may be found directly over mineralization, but hydromorphic anomalies originating from the same mineralization may be found many miles away. Recognizing, interpreting and tracing secondary dispersion patterns to their source can be one of the more difficult aspects of geochemical exploration.

Fig. 8-1 illustrates seven of the many different displaced anomalies which may be encountered in geochemical surveys. Figs. 8-1A and B are

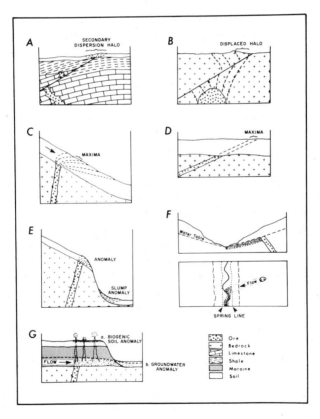

Fig. 8-1. Displaced anomalies (structural, hydromorphic, gravity and biogenic). A. Fault controlled secondary dispersion. B. Primary halo displaced by fault. C. Anomaly displaced by creep in residual soil. D. Vestigial ore band with offset anomaly at the surface. E. Anomaly displaced by slumping. F. Typical drainage anomaly produced by groundwater intersecting stream bed. G. Anomalies in transported overburden. From Andrews-Jones (1968).

examples of dispersion controlled by structure, specifically faulting. The secondary dispersion pattern in A formed after the faulting, possibly by means of rising fluids or vapors. The dispersion pattern on the surface in B, as shown, is an exposed primary dispersion pattern but within a short period of time soil formation would be expected to result in a superimposed secondary dispersion pattern. That such post-mineralization faulting can displace primary halos a considerable distance is illustrated by the Dobson Pass fault in the Coeur d'Alene district, Idaho (Fig. 7-3) along which movement was 16 miles. Clearly, very detailed studies are essential for the interpretation of anomalies such as those illustrated in Figs. 8-1A and B. On a practical basis, many anomalies of this type cannot be traced to their source and, as a result, they may be erroneously considered as false anomalies.

Figs. 8-1C and E illustrate anomalies displaced mechanically by gravity movement. Most gravity movement takes place near the surface in loose material, such as in the zone of soil and rock fragments (regolith) above bedrock. Slow displacement by soil creep may occur on almost any slope, and it is assisted by such factors as abundant moisture, alternate freezing and thawing, and permeable soil formations. With an increase in the angle of the slope, accelerated rates of movement are possible resulting in slumping, landslides, debris avalanche and other types of downslope movement. Vegetation tends to slow and stabilize gravity movements and, where it is thick and the root systems shallow, maximum movement of the soil and the dispersion pattern may take place below the root system. In general, however, the rate of gravity movement in soils on slopes decreases from the surface to the bedrock, and anomalies take forms such as those shown in Figs. 8-1C and 8-2. This results in mineralized fragments on the surface being transported farther from the source of mineralization than those in the lower soil horizons that are closer to bedrock. One characteristic feature of many soil anomalies which are displaced mechanically from mineralization is that soil profile samples taken along traverses at some distance from mineralization — perhaps starting at 600 feet away and working uphill toward the mineralization, as in Figs. 8-1C and 8-2 — will show a decrease in element concentration with depth. However, as the mineralization is approached the depth at which anomalous values are encountered will increase, while samples taken directly over the mineralization may show little change with depth. This characteristic is very useful in interpreting soil anomalies, and in locating the mineralization exactly, but such an approach assumes that the anomaly is continuous. Applying this concept to slump and other disrupted anomalies can result in misinterpretations.

When slumping (Fig. 8-1E) or landslides occur, the dispersion pattern is disrupted as well as displaced, a feature which must be considered in the

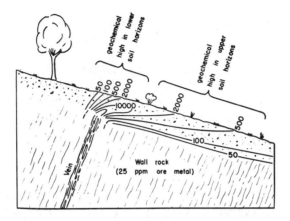

Fig. 8-2. Residual soil anomaly displaced mechanically by soil movement on a slope. Note the change of anomalous values with depth. From Huff (1952).

interpretation of anomalies in mountainous areas. It is possible that a residual soil anomaly will develop on top of a slump anomaly but it is more than likely that in mountainous areas the slump anomaly will be buried by barren colluvium (transported fragmental material built up at the base of slopes), or that it will be carried away in the drainage system. In addition to having a low cold extractable to total metal ratio, mechanically transported anomalies (e.g., slump anomalies) can usually be recognized from field studies of the topography, and also by the recognition of angular fragments containing sulfide minerals.

Displaced soil anomalies often have a "fan shaped" distribution pattern, especially if they occur on slopes (Figs. 1-3, 8-1C, 8-2), that is, they spread outward from a source. Fan shaped anomalies may also develop in groundwater (Fig. 3-23) and in glacial till. Displaced soil anomalies can also have the shape of a "train" if the soil movement takes place in a restricted channel, or if the soil is developed on anomalous material which had formed a linear dispersion pattern resulting from movement along well-defined drainage channels.

Except in certain circumstances, such as stagnant groundwater, most hydrogeochemical anomalies are displaced. This is illustrated in Figs. 8-1F and G where the displaced groundwater anomalies may not be recognized until they intersect the surface at the spring line (break-in-slope; see also Fig. 1-3). Depending upon the mobility of the elements involved, dispersion patterns of soluble elements may be detected either at considerable distances from the spring line, or they may form very small patterns restricted to the vicinity of the break-in-slope. Dispersion in surface water (Fig. 8-1F) may also take place in colloidal or adsorbed form, as discussed in Chapter 3, and the extent of the dispersion pattern will depend on such factors as the particle size of the colloid or adsorbing mineral phase and the velocity of the flowing water.

Displaced biogenic anomalies are often reflected in the soil (Fig. 8-1G), as well as in the vegetation itself. These may take a variety of forms, and may be considerably displaced from mineralization. Fig. 8-1D illustrates an offset anomaly in soil formed from a vestigial ore band which at one time cropped out on the surface. In the case illustrated, soil movement has been minimal.

PHYSICAL FORM AND CLASSIFICATION OF SECONDARY DISPERSION PATTERNS

Mention was made above of fans and trains, but these are just two of a great number of physical forms which secondary dispersion patterns may assume, many of which are difficult to describe in simple terms because of their complex origins and irregular dimensions. This is particularly true for displaced anomalies. Other terms which have been used to describe the form or position of secondary dispersion patterns include (from Hawkes and Webb, 1962): *superjacent,* a pattern developed more or less directly over the bedrock source; *lateral,* a pattern displaced to one side and entirely underlain by barren bedrock; and *halo,* a superjacent, nearly symmetrical pattern disposed about the source. To these the descriptive terms *intense* (if values rise sharply to well-defined peaks), or *diffuse* (if the pattern is subdued and lacks pronounced peaks), may be added to describe the distribution of anomalous metal values. Secondary dispersion patterns may also be classified as *clastic, hydromorphic,* or *biogenic patterns* depending upon the dispersion mechanism.

The terms syngenetic and epigenetic are also applied to secondary dispersion patterns although, on a practical basis, the distinction between these two types is not as significant as it is in the case of primary dispersion patterns. This is because dispersion patterns formed in the secondary environment are likely to result from several processes. For example, a soil or stream sediment anomaly may be partly mechanical (syngenetic) and partly hydromorphic (epigenetic). Nevertheless, it is important to attempt to determine the time of formation of secondary patterns (or of the individual components in the case of a multiple mechanical-hydromorphic anomaly) relative to the host geological materials, as such information may be of great value during the interpretive phase of a survey.

Syngenetic patterns in the secondary environment are generally formed mechanically, such as in the case of heavy minerals or sulfide grains in alluvium. Other examples include mechanically transported mineralized fragments in colluvium or glacial till. Dispersion patterns in residual soil are also generally classified as syngenetic. In the examples cited, the matrix and anomalous material were deposited at the same time and from the same source (or approximately so), which results in dispersion patterns which are relatively easy to interpret. Dispersion patterns of biogenic origin in soils, such as those developed in the humus layer, are considered syngenetic if they formed at the same time as the soil.

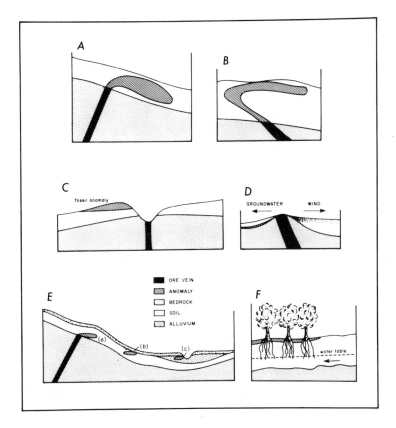

Fig. 8-3. Miscellaneous secondary dispersion patterns developed in soils. A. Anomaly in residual soil with no surface expression. B. Anomaly in residual soil with surface expression. C. Fossil soil anomaly developed before erosion formed the valley. D. Anomalies formed in opposite direction due to wind and groundwater (Note that the wind anomaly is limited to the surface). E. Multiple, disrupted anomalies in soil (a and b) and alluvium (c) originating from the same deposit. F. False anomaly resulting from the uptake by vegetation of trace metals from groundwater.

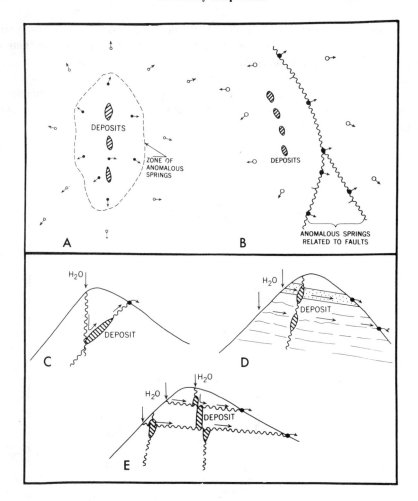

Fig. 8-4. *Top.* Some secondary dispersion patterns of springs. A. Idealized surface pattern of anomalous springs forming a halo around a mineralized zone. B. Idealized representation of anomalous springs forming a linear and sinuous pattern near a mineralized zone.
Bottom. Sketches illustrating some relationships of springs to mineralization and the difficulty in interpreting some dispersion patterns which may form. C. Spring at outcrop of mineralized fault. D. Springs along surface traces of porous beds and bedding planes. E. Springs along surface traces of cross faults. From Boyle et al. (1971).

Epigenetic patterns would include all patterns presently detectable in stream sediments, soils, glacial deposits, bogs and so forth, which have resulted from transport in solution. For the sake of uniformity and simplicity, dispersion patterns of elements currently in transport, such as those detected in spring water, groundwater, lakes and rivers, are here considered to be epigenetic, since they significantly post-date the host rock. (However, Hawkes and Webb, 1962, p. 146, consider the soluble load of groundwater and surface water as another kind of syngenetic dispersion pattern, in which water is the matrix rather than soil or some other solid material.) Elements forming epigenetic hydromorphic halos may be: loosely adsorbed onto organic matter, mineral or iron-manganese oxide surfaces; precipitated as insoluble compounds (e.g., $PbCO_3$); or co-precipitated within mineral phases (e.g., iron-manganese oxides). Such patterns are superimposed (1) upon the local background levels of element abundances, or (2) upon syngenetic anomalies, such as mechanically transported sulfide grains. In general, the most important epigenetic patterns are those formed by the mobile metals (e.g., Zn, Cu, Mo, U), as these elements are readily moved in solution, and subsequently adsorbed by fine-grained materials.

Figure 8-3 illustrates several additional examples of secondary dispersion patterns, most of which can be described and classified by a combination of terms. For example, in Fig. 8-3D, the wind (aeolian) pattern is displaced, clastic and lateral, as well as also possibly fan-shaped, intense and syngenetic. The other sketches in Fig. 8-3 are self-explanatory.

In Fig. 8-4 a selection of secondary dispersion patterns (A and B) emanating from springs is shown. According to Boyle et al. (1971), the use of springs is particularly effective in reconnaissance work in areas in which they are widespread, and where the surface pattern of anomalous springs assumes a halo about a mineralized zone, as shown in Fig. 8-3A. However, more generally, the patterns of anomalous springs are non-symmetrical with respect to mineralization, being linear and sinuous in plan (Fig. 8-3B) and related to the locations of mineralized faults and shear zones, bedding planes, strata of porous rocks, etc. Figs. 8-4C, D and E are cross-sectional sketches which illustrate some of the complex relationships which may exist between springs and mineral deposits, accounting for some of the complex forms hydromorphic dispersion patterns may take. Tolman (1937) has described many different types of springs, and from his discussions and illustrations, numerous additional types of dispersion patterns originating from springs can be imagined.

From the examples shown in Figs. 8-1 to 8-4, it is clear that in the secondary environment, even though it may represent only a few tens or hundreds of feet of material close to the surface, dispersion patterns may be displaced considerable distances, and they may appear in many different forms. The extent of the dispersion and the shape of the dispersion pattern formed are not only governed by factors of weathering and erosion, and by the inherent mobility in solution of the elements concerned, but also by geological, hydromorphic, and numerous other factors which must be considered for each particular area. It is always worthwhile to map and

study the form of secondary dispersion patterns as thoroughly as possible
with the objectives of (1) determining the mechanism of dispersion; (2)
selecting optimum sampling and analytical methods; and (3) ultimately
locating the source of anomalous values. The achievement of these objectives will be aided by a knowledge of the chemical, biological, mechanical
and environmental factors which are influential in each specific area.

MECHANICAL DISPERSION

Anomalous concentrations of metals in the secondary environment
may be derived from a mineralized source, or from non-mineralized
country rock, and moved either mechanically (in solid form) or hydromorphically (in solution). The anomalies formed by mechanical or hydromorphic processes differ markedly and each gives a different geochemical
response. Interpretation of geochemical data is often dependent upon the
recognition of these two types of dispersion (as well as biogenic) within an
area. Usually, the most effective way to distinguish between mechanically
and hydromorphically transported anomalies is by the use of different
chemical extractions. Metals moved in solution are generally loosely
bonded and have a high ratio of cold, or dilute acid, extractable to total
metal.

Metals moved mechanically typically occur as resistant primary
minerals, such as sulfides in some environments, in metal-containing
accessory silicate minerals, or as secondary minerals and precipitates (e.g.,
iron and manganese oxides precipitated from solution on the surface of
clastic fragments). Such minerals usually have a high specific gravity (beryl
is an exception) and they concentrate in the heavy mineral fraction of
stream sediments. In such cases, the reduction in the intensity of the
anomaly is largely through dilution with barren alluvium (detrital material,
often of distant origin, deposited by running water). Alternatively, dilution
may be with colluvium, collapsed bank material, sheet wash, and other
materials of local derivation. The dilution may be extremely rapid,
especially in the case of colluvium or collapsed bank material, and the dispersion pattern can be short. In order to accentuate mechanically dispersed
anomalies, the heavy minerals indicative of mineralization may be
separated from alluvium or colluvium prior to analysis. This technique has
been used successfully by Huff (1971) in alluvium in Arizona, by Garrett
(1971) in glacial tills in Quebec, and in numerous other situations. In all
cases, the aim is to improve or enhance the contrast between background
and anomalous samples.

In the simplest of examples, all metals of interest will occur in one
mineral phase in the heavy mineral fraction. However, several compre-

hensive studies on the mechanically transported metal content of stream sediments have shown that anomalous metal values may actually be distributed among several size and density fractions, each with distinctive geochemical and mineralogical characteristics. Thus, to obtain the most useful information and the most intense anomalies, the mechanically transported components must be separated into convenient, geologically significant fractions, each of which must be analyzed and interpreted separately.

By way of example, Huff (1971) studied the alluvium in several desert washes of the Lone Star porphyry copper district near Stafford, Arizona, and prepared seven different subsamples. Some of the washes drain areas of copper mineralization and several drain barren areas. The types of subsamples are indicated on Fig. 8-5. In view of the desert environment, all evidence of mineralization can be considered to be mechanically transported. The four size fractions on the left side of Fig. 8-5 were obtained by sieving, whereas those on the right side required special treatments, that is, the separation of the heavy fraction into a magnetic and a non-magnetic fraction, and obtaining a concentrate by ultrasonic treatment. The ultra-

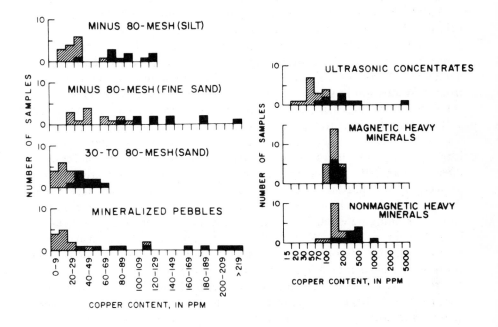

Fig. 8-5. *Left:* Samples of four alluvial sediment fractions classified according to their copper contents. *Right:* Heavy mineral and ultrasonic concentrates classified according to their copper contents. Samples from washes draining mineralization are indicated by the dark pattern. From Huff (1971).

sonic concentrates are assumed to be coatings on grains (iron-manganese oxides), as such coatings are generally released by this treatment.

It is clear from inspection of Fig. 8-5 that large differences are found among the seven types of alluvial sediments. Some, such as "mineralized pebbles" and "ultrasonic concentrates," show a substantial range of values, and samples from washes draining mineralization are clearly defined. Values above 200 ppm in the "nonmagnetic heavy minerals" are indicative of mineralization, whereas the magnetic heavy fraction cannot be used to discriminate between anomalous and background washes in this area. The main disadvantage of using the non-magnetic heavy fraction in this case would be the labor involved in preparing the samples when, in fact, "mineralized pebbles" or either of the minus 80 mesh fractions (Fig. 8-5, left) give distinct anomalies related to mineralization. This is an excellent example of a situation in which an orientation survey can enable the proper size and density fraction of material to be selected for analysis.

Although Fig. 8-5 is not a particularly good example, it does illustrate that size fractions coarser than 80 mesh (specifically, "mineralized pebbles") may contain anomalous values indicative of mineralization. The minus 80 mesh fraction has been found to be generally useful in those secondary environments where the dispersion patterns are hydromorphic in origin. However, where mechanical dispersion is dominant, and particularly where the rate of oxidation is significantly slower than the rate of erosion, analysis of fractions coarser than 80 mesh will often result in anomalies of greater size and contrast. One complicating factor results from the reduction in size of minerals with good cleavage (e.g., galena and sphalerite) with increasing distance from their source. Such minerals often will be found concentrated in the coarse size fractions in stream head-waters, but they are found in the finer size fractions at greater distances from mineralization because of mechanical disintegration. In the case of those mechanically transported ore and heavy minerals which are resistant to mechanical abrasion and reduction in size because they lack good cleavage (e.g., pyrochlore), sorting during alluvial transport can result in anomalies being detected in various size fractions depending on distance from the source. In fact, Watts et al. (1963) found that the size distribution of niobium from pyrochlore-bearing carbonatites in Northern Rhodesia could give a clear indication of proximity to the bedrock source. In those environments (e.g., deserts, glacial) where mechanical dispersion is pre-dominant, the dispersion patterns in alluvium, streams, till and other materials tend to be short by comparison with those environments in which oxidation and hydromorphic dispersion is predominant.

In a very thought-provoking paper, Brown (1970) has pointed out that there are cases in which lead mineralization in areas of mechanical weathering is not reflected in the stream sediments; presumably the same would apply to other metals as well. It is not practical to attempt to completely review his assumptions and calculations here, but suffice it to say that at Silver Creek (drainage area about one square mile) in the Slana-Ahtell district of Alaska, there is a quartz vein with 100 square meters of net exposure containing 4% galena (this galena has high silver values which justified the interest). Erosion of this deposit would yield only 37 grams of galena along with 200 million grams of waste annually. Taking into account the specific gravity and other characteristics of galena and the waste material, Brown (1970) calculated that the probability of getting one grain of galena in a one gram sample for analyses is less than one in 50 million (if the sample is collected at the mouth of the drainage basin). Even assuming that one grain of galena did get into the sample to be analyzed, it would only result in an additional 13 ppm lead. Clearly, it is quite possible to have a drainage area with a geologic anomaly in the form of mineralized veins without necessarily having an associated geochemical anomaly in the stream sediments. Ideally, mechanical dispersion will yield significant anomalies in those instances where the mineralization is large by comparison with the drainage basin, which was not the case in the example cited by Brown (1970). Although he made suggestions as to how this problem may possibly be minimized (e.g., replicate sampling; attempt to measure some mobile ion), consideration should also be given to analyzing the heavy mineral fraction. As these generally comprise only a few percent of most stream sediment samples, the resulting concentration factor (perhaps 20 to 50 times) will tend to make the detection of anomalous samples easier. Nevertheless, there is no doubt that Brown's (1970) suggestion that extreme caution be used in areas of high mechanical dispersion is justified. This would apply to glacial as well as to fluvial environments. Similar situations exist in areas of hydromorphic dispersion in which a high dilution factor makes the detection of mineralization virtually impossible using routine methods.

DISPERSION PATTERNS DETECTABLE BY PARTIAL EXTRACTION TECHNIQUES

Having considered mechanically-formed secondary dispersion patterns, it is now appropriate to discuss some aspects of those formed by hydromorphic processes. First, however, it is worthwhile to point out a few of the differences that exist between secondary dispersion patterns formed by the two processes. These differences include: (1) clastic patterns are generally composed of solid mineral phases and, although hydromorphic anomalies may consist of solids precipitated from solution, they are more than likely to be represented by metals loosely adsorbed on fine-grained mineral surfaces; (2) dispersion by mechanical methods generally takes place very near the surface, but hydromorphic patterns may be found at much greater depths in soils, glacial till, groundwater or lakes; (3) hydromorphic anomalies generally are more uniform than those produced from mechanically dispersed heavy minerals or other solid phases; (4) where both mechanical and hydromorphic anomalies are produced from weathering and erosion of the same deposit, the hydromorphic patterns will usually be found at greater distances from the mineralization. The hydromorphic patterns produced by the more mobile elements can be considered as being particularly useful in locating mineralized areas (regional survey), whereas those of the less mobile elements, or mechanically-formed dispersion patterns, are useful in locating the ore body (detailed survey).

Partial extraction techniques, as discussed in Chapter 6, can be of great value in the recognition and interpretation of hydromorphic anomalies, as well as in distinguishing them from mechanically-derived anomalies. The use of partial extraction techniques has been practiced for many years in geochemical exploration, and has met with particular success in tropical and temperate areas of the world. More recently these methods have been applied in colder environments such as the bog areas of Canada and in glaciated terrain (Fig. 11-22). However, it must always be kept in mind that hydromorphic anomalies can only occur in those environments in which there is oxidation and the release of elements into solution (gases from deep-seated sources dispersed in groundwater would be one of the usual exceptions). Therefore, partial (or cold) extraction methods are not as likely to be successful in deserts or other areas where mechanical erosion is predominant.

The objective of partial extraction techniques applied to secondary dispersion is to determine those metal values in a sample which are formed by epigenetic hydromorphic or biogenic processes. These readily extractable metals may be a better guide to mineralization than total metal values. In essence, the aim is to exclude background values which can often conceal metal patterns derived from mineralization. If this is achieved, the contrast between anomalous and background values may be sufficiently enhanced to find the source of mineralization. In general, the percentage of cold extractable, or easily extractable metal, to total metal in a sample is high for hydromorphic (and biogenic) anomalies, and low for mechanically transported anomalies.

Simple as the concept is, there are, occasionally, unusual aspects of partial extraction techniques. A case in point is the well-documented study by Coope and Webb (1963) of copper in stream sediments near mineralization in Cebu, Philippine Republic. In the upper left part of Fig. 8-6(a), the cxCu to total copper values (within circles) are more or less typical of what would be expected in stream sediments draining soils with a high copper content (Fig. 8-6b). The cxCu values account for about 5-20% of the total copper content. However, other tributaries (right side of Fig. 8-6a) draining similar copper-containing soils, have a very low percentage of cxCu (less than 5%). The explanation for this situation lies in the fact that very acid stream and groundwater conditions are generated by the oxidation of pyrite (Fig. 8-6b). Under acid conditions copper is mobile and not readily adsorbed onto mineral surfaces. The total copper values in the stream sediments, some of which are as high as 2500 ppm (Fig. 8-6a), still reflect anomalous conditions in the highly acid tributary. As cold-extractable values may not reveal the presence of mineralization in acid environments, the pH of water and soil should be determined before attempting to use cold extractable techniques.

Bradshaw et al. (1972, p. 45) pointed out, "A facet which can be of critical importance in exploration geochemistry, and one all too frequently ignored by geologists, is the choice of the correct metal extraction method during sample analysis." They described, in decreasing order of severity, three broad ranges of chemical extraction: (1) "total" perchloric acid extraction; (2) "weak acid" hydrochloric acid (0.5N HCl) extraction (hot); and (3) "cold" (EDTA) extraction. They, and others previously, have observed that the weak (2.5% or less), cold EDTA extraction is effective in

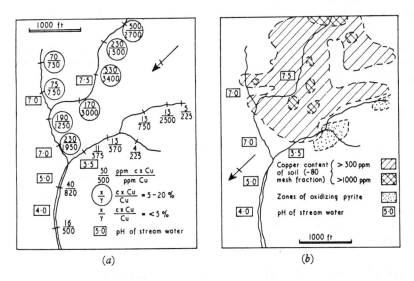

(a)   (b)

Fig. 8-6. Stream sediment and soil anomalies in the Luay drainage area, Cebu, Philippine Republic. (a) Relationship between pH of water and the ratio of cold-extractable copper (cxCu) to total copper (Cu) in the sediment. (b) Copper content of soil, and location of oxidizing pyrite responsible for acid pH of the stream water. From Coope and Webb (1963).

detecting hydromorphic anomalies, particularly in tropical environments, whereas the hot hydrochloric acid treatment will not only remove adsorbed elements, but also will dissolve some precipitated salts and possibly even attack the less resistant silicates, such as some clay minerals. (In certain cases, EDTA is not sufficiently strong to remove adsorbed metals, and this is attributable to the fact that some metal-mineral bonding is stronger than others.) Sodium or ammonium citrate solutions (buffers) are even weaker extractions than the EDTA treatment.

Bradshaw et al. (1972) have demonstrated that by means of the three classes of extractions described above, good success can be obtained in the interpretation of bog, break-in-slope, glacial, soil, stream sediment and other types of anomalies. This is illustrated in Fig. 8-7 which shows the results obtained by using the three extractions on stream sediments from the Northwest Territories. The "total" perchloric ($HClO_4$) extraction shows a distinct difference between creek D to the northwest and the remainder of the area, and this probably reflects bedrock geology. The same attack also identified an anomaly at the headwater of creek B, but only two anomalous samples from tributaries near the headwater of creek A. However, those tributaries closest to the known chalcopyrite mineralization do not show strong anomalies. The hydrochloric acid extractable copper, however, does show strong (first and second order) anomalies in five of the eight tributaries draining the mineralization. The EDTA extractable copper shows anomalous values in all the tributaries near the mineralization, as well as numerous others in creeks A and C. A further indication of fundamental differences between the various chemical digestions can be seen by comparing the histograms in Fig. 8-7. The EDTA and hydrochloric acid distributions appear to be normal, and in these two cases it is easy to select a threshold and to recognize anomalous samples. The distribution from the perchloric acid treatment appears to be the result of several parameters (polymodal).

There is no certain way of knowing which extraction method will produce the most desirable results. Only experimentation to determine optimum acid strength, digestion period and temperature (see Fig. 6-2), preferably as part of an orientation survey, will permit attainment of the ultimate objective of partial extraction methods, which is to obtain the maximum enhancement of contrast between anomalous and background samples. Based upon the experiences of Bradshaw et al. (1972), it is reasonable to suggest that the three classes of chemical extraction described above are useful starting points for the more common types of geochemical surveys. However, the suggested acid types and strengths will not necessarily yield the optimum results in every instance.

Although the discussion to this point has considered partial extraction techniques as they apply to hydromorphic patterns, by a judicious choice of chemical reagents, it is also possible to distinguish certain clastic patterns, particularly those formed by acid-soluble sulfide minerals, from background values. In actual practice, however, syngenetic clastic sulfide grains found in stream sediments or glacial till may be weathered, and their metals released only to be adsorbed by clays and other fine-grained materials, with the result that these elements are indistinguishable from those forming epigenetic hydromorphic anomalies.

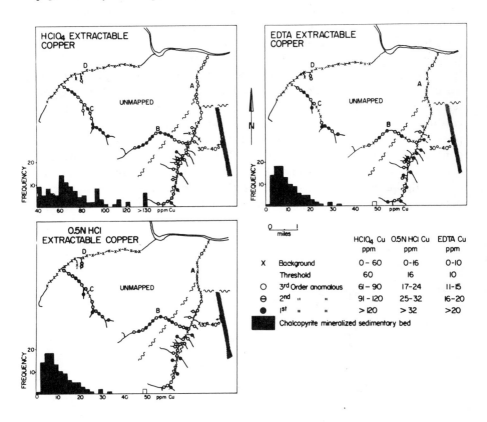

Fig. 8-7. Comparison of the results obtained from copper in stream sediments by three analytical extractions, Northwest Territories, Canada. From Bradshaw et al. (1972).

GEOTHERMAL PATTERNS

The secondary dispersion of elements has been used in recent years in several interesting ways which are unrelated to mineral exploration. Dispersion patterns in both the secondary and primary environments have been employed in environmental, agricultural and medical studies, in earthquake forecasting, and in the search for geothermal energy resources. The latter will be discussed briefly here, as it alone is directly concerned with appraisal of natural resources.

Of the eleven chemical indicators which White (1973) listed as having been used to determine temperatures in natural hot water (geothermal) systems, thus giving an indication of potential areas for geothermal energy resources, those obtained from hydrogeochemical samples have been especially useful. The dissolved SiO_2 content of thermal (e.g., hot spring) waters is presently regarded as the best indicator of subsurface temperatures. The silica geothermometry method is based upon the variation in the solubility of quartz as a function of temperature. Specifically, the silica content of thermal water increases with increasing temperature, assuming that equilibrium of quartz with water is effected at elevated temperatures and there has been no dilution or precipitation of silica upon cooling. Thus, if the SiO_2 content of a natural water is determined, it is possible to deduce the temperature of the rocks with which it has been in contact. Fournier and Truesdell (1970) presented the relationship in graphical form (Fig. 8-8). Curve A is probably the more reliable for springs of low discharge on the margins of certain active geothermal areas, whereas curve B has proved highly reliable for indicating temperatures of deep

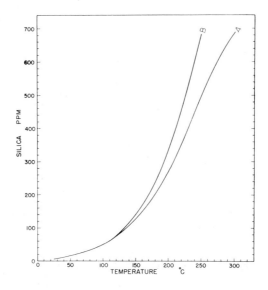

Fig. 8-8. Silica concentration in geothermal water versus estimated temperature of last equilibrium. Curve A applies to waters cooled entirely by heat conduction. Curve B applies to waters cooled entirely by adiabatic expansion at constant enthalpy. See text for further details. From Fournier and Truesdell (1970).

springs of high temperature and high discharge near the centers of geothermal activity (White, 1970).

Fournier and Truesdell (1973) recently described an empirical geothermometer for natural waters based on the (molar) concentrations of Na, K and Ca which is suited for the range of 4 to 340°C. There is also the possibility that mercury may be of value in exploring for geothermal resources as White et al. (1970), and others, have noted high contents of mercury in the water and gases of hot springs and associated sediments, as well as in volcanic fumaroles.

The use of geochemical methods in the search for geothermal energy is just one of many techniques that have been and are being employed (others include geophysical and remote sensing). Geochemical methods of sampling, analysis and interpretation of thermal waters differ significantly from those applied to cool, near-surface water used in the search for mineral deposits, but the specialized nature of the subject does not permit further discussion here. Suffice it to say that geochemical methods have been used throughout the western and northwestern parts of the United States, and elsewhere, where there is surface evidence of geothermal activity, such as New Zealand, Alaska and Yellowstone National Park. Hydrogeochemical methods are chiefly applicable in the exploration for the near-surface geothermal fields (to depths of about 4000 ft). The thermal waters are composed mostly, if not entirely, of meteoric water which has been in contact with subsurface heated rock, and is re-cycled to the surface. In some respects, thermal waters are analogous to oil seeps, as they are both indicative of buried energy. New geothermal fields are being developed at depths of 8000 ft and more, and at these depths, geophysical methods appear to be superior to those based on present geochemical techniques. However, a considerable amount of research is in progress relative to the application of geochemistry to the exploration for, and development of, geothermal resources.

9/ Regional and Detailed Surveys

INTRODUCTION

A general discussion of regional and detailed surveys was presented in Chapter 1 and, in subsequent chapters, continual reference has been made to these two types of surveys in various contexts. It was pointed out that regional and detailed surveys differ in objective, size of the area being surveyed, sampling density, type of material sampled, and so forth. It is unrealistic to attempt to discuss the many variables in this chapter, and some type of selection is therefore necessary.

The method adopted here consists of briefly describing how geochemical methods have been and are being used for regional reconnaissance surveys and then describing those surveys conducted on smaller scales until, finally, detailed surveys are discussed. This procedure was followed by Ginzburg (1960) who recognized three scales of geochemical mapping and surveys depending on purpose and objective: (1) reconnaissance; (2) prospecting; and (3) detailed. The sampling grid for each type of survey differs as befits the greatly differing objectives. Similarly, Bradshaw et al. (1972) found it convenient to discuss regional surveys on the basis of the number of stream sediment samples taken per square mile. In decreasing order, the sampling densities they considered were: (1) one sample per $40-80$ sq mi; (2) one sample per $5-20$ sq mi; and (3) one sample per $1-2$ sq mi. Each category is specific for a particular objective, and reflects different geochemical features.

The choice of sampling media in regional and detailed surveys is very important and, on a practical basis, is often dictated by environmental conditions. In arid or desert areas, soil and rock are used for surveys on all scales, and here the sampling stations would likely be much more closely spaced than in humid areas where stream sediments are available. Where reasonably well-defined surface drainage systems are developed, and conditions are favorable to the dispersion of metals derived from the weathering of mineralization, stream sediments are preferred, especially at the regional reconnaissance or "prospecting" level. For this reason, throughout this chapter examples based on the use of stream sediments have been selected where possible. It must be remembered, however, that rocks, soil, water, vegetation and vapors may also be used in all types of surveys, and on the detailed level, these are generally preferred to stream sediments even where the latter are available. Nevertheless, the principles involved in the use of stream sediments are equally applicable to the other media, with appropriate modifications.

For the purposes of discussion in this chapter it will be assumed that the minus 80 mesh fraction of all the sediment and soil examples discussed was analyzed unless otherwise stated. This assumption is appropriate because, for the sake of simplicity, preference has been given to examples from tropical or other areas in which chemical weathering is dominant. As indicated previously, in areas of chemical weathering the minus 80 mesh fraction is generally satisfactory for most surveys, especially those involving the use of active stream sediments. However, in certain areas stream sediments of any size fraction may be inferior to other sampling media. Brundin and Nairis (1972), for example, found that although inorganic stream sediments have been used extensively in the glacial till regions of Sweden for regional geochemical prospecting, such materials suffer from certain disadvantages. Their study of alternatives for regional geochemical prospecting in that terrain established that, for certain metals, organic mud and water appear to offer distinct advantages.

In a series of papers (Webb et al., 1968; Webb, 1970) it has been pointed out that regional geochemical reconnaissance surveys by means of stream sediment sampling can yield other types of valuable geological information in addition to evidence of ore deposits. The data can often be applied to geological mapping, stratigraphy, paleogeography and metamorphic geology. Results from stream sediment reconnaissance surveys, and to some extent more detailed surveys, are of particular value in areas with extensive glacial and alluvial cover, as they tend to reflect variations in the underlying bedrock, provided the surficial deposits are not exceptionally thick. In recent years stream sediment data have also been applied to agricultural and public health problems, such as the recognition of trace element deficiencies or excesses which may affect both crops and animals. Although these are certainly legitimate areas of applied geochemistry, they are only indirectly related to geochemical exploration and are not considered further here. Of course, should areas be observed where livestock, for example, have disorders attributable to the presence of excesses of trace elements obtained from water, vegetation or soil, such information is clearly suggestive of a geochemical province or rock formations which may have economic significance.

Before continuing with the main theme of this chapter it is important to consider the nature of the objective of the search in a regional geochemical survey. Erickson (1973) has presented the concept which is illustrated in Fig. 9-1. In this very simple diagram, A represents an exposed deposit, whereas the other three deposits are concealed either by a pre-mineral or by post-mineral cover such as volcanics or alluvium. The search for each type of deposit requires a different geochemical technique.

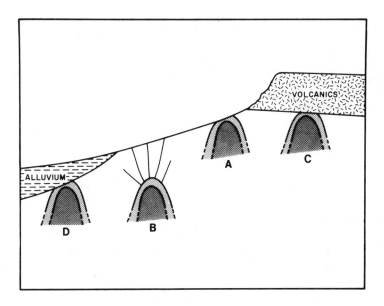

Fig. 9-1. Sketch of four categories of ore deposits only one of which (A) is exposed and directly amenable to stream sediment surveys. The detection of leakage halos may result in the location of B, but concealed deposits such as C and D are not likely to be detected by geochemical methods as now generally practiced. After Erickson (1973).

If the exposed deposit (A) is cut by a drainage system, there is an excellent chance of it being discovered during a regional reconnaissance geochemical survey by means of the collection and analysis of stream sediments for one or a few elements (usually Cu, Mo, Zn, Pb). This is a powerful technique in the circumstances indicated and has been successfully used in various parts of Africa, South America, North America, islands of the circum-Pacific region, and elsewhere. The main requirements here are that the target be exposed (or thinly veneered), and it must contribute mineralized material to the drainage basin in sufficient amounts for the elements of interest to produce a detectable anomaly. Areas of exposed porphyry copper mineralization in humid regions are often ideally suited to discovery by stream sediment surveys. This is because: (1) they are large targets; (2) they are commonly fractured and thus chemical weathering proceeds easily and releases the weathering products to the drainage basin; and (3) the abundance of pyrite gives rise to acid conditions which facilitate the mobilization of the disseminated copper and associated elements (e.g., molybdenum).

In the case of B (Fig. 9-1), which is concealed by barren rock or pre-mineral cover, the probability of finding this deposit is greatly reduced. Here one must look for leakage halos, and also be aware of the concepts of primary zones (dispersion) characteristic of different types of deposits (e.g., Figs. 7-3 to 7-6) which might be projected onto the present surface. Should a stream sediment reconnaissance survey be attempted in the search for this type of deposit, the minus 80 mesh fraction would likely yield negative results. The analysis of stained pebbles or crude pan concentrates would probably be of much greater value. Evidence for this approach was presented by Erickson et al. (1966) in their geochemical reconnaissance study of a 250 sq mi area of Elko County, Nevada where conventional stream sediment sampling and analysis failed to reveal any anomalous metal content (with one exception). Here the most successful method of geochemical reconnaissance was the analysis of iron-stained, discolored, or slightly altered float cobbles and pebbles, collected in the major drainages. These materials contained anomalous concentrations of several metals (chiefly Zn, Pb and Hg) which may have been a reflection of leakage halos.

Deposits C and D (Fig. 9-1) are concealed by post-mineral cover and geochemical exploration techniques on either a regional or detailed basis, as presently practiced, offer little promise of discovery. In the case of D, there has been some success with the sampling of groundwater, vegetation, basal caliche layers and vapors (e.g., Hg) but, in general, the vast majority of deposits of types C and D remain to be discovered. The resource potential of the concealed deposits (B, C and D) is undoubtedly greater than those of the exposed deposits (A), and development of methods for their discovery is the real challenge for the present and future.

LOCATION OF GEOCHEMICAL PROVINCES

Several examples of the use of low density stream sediment surveys (one sample per 20 – 100 sq mi) for outlining geochemical (metallogenic) provinces, mineralized belts and similar large scale geological features in both explored and unexplored terrain have been described, particularly from Africa.

Armour-Brown and Nichol (1970) carried out a regional reconnaissance survey in Zambia which covered over 80,000 sq mi at a sampling density of one sample per 75 sq mi. This study was specifically designed to investigate the usefulness of low density drainage sampling as a procedure for identifying regional background variations of minor element contents associated with different metallogenic zones. The samples were analyzed for 16 elements. By way of example, the results obtained for copper are presented in Fig. 9-2, where it can be clearly seen that the distribution of this element in stream sediments is consistent with the distribution of areas of known copper mineralization, although the correlation is by no means perfect. Many of the other minor elements revealed similar high back-

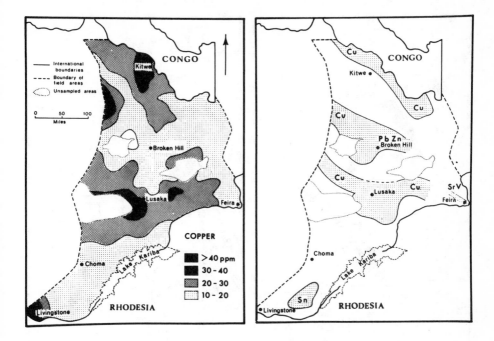

Fig. 9-2. Distribution of copper in stream sediments (left) in relation to known mineralized districts (right) in Zambia. Based on a sample density of 1 per 75 square miles. From Armour-Brown and Nichol (1970). Reproduced from *Economic Geology,* 1970, v. 65, p. 314 and p. 319.

ground concentrations within or surrounding metallogenic zones. Some elements, in particular Cu and Co, are enriched up to ten fold, and Sr and V two to four fold, in the mineralized bedrock formations by comparison with those in unmineralized areas, and this is reflected in the stream sediments. This study, therefore, also established the essential role of bedrock geochemistry in controlling the minor element distribution in stream sediments. This latter point is one which cannot be overemphasized, with regard to geochemical exploration.

A similar study of the 15,000 sq mi Basement Complex in Sierra Leone by Garrett and Nichol (1967) also employed widely spaced stream sediment samples (1 sample per 75 sq mi). Fig. 9-3 is typical of the results they obtained and it shows a pattern of chromium values which enclose all known chromite occurrences in the southeast corner of the country. In addition, the results of their work suggested some additional areas of possible chromium mineralization which may represent an extension of the known mineralized belt. Actually, from the stream sediment data alone it was not possible to discriminate unequivocally between chromium derived

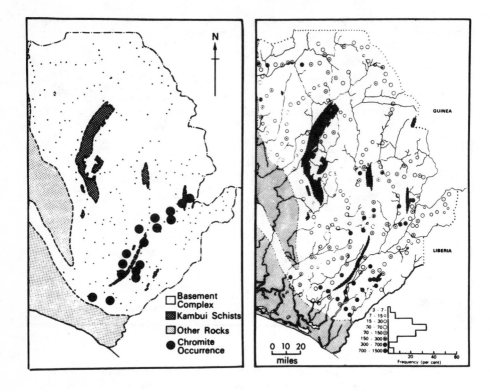

Fig. 9-3. Relation between chromite occurrences (left) and chromium in stream sediments (right) in the Basement Complex, Sierra Leone. From Garrett and Nichol (1967).

from chromite mineralization and that from ultramafic rock units within the Basement Complex. Nevertheless, the geochemical reconnaissance did reveal the existence of marked areal variations in the distribution of anomalous values which correspond with the chromium mineralization and a minor chromium province. The pattern is not evident in rock samples collected at a comparable density because of the heterogeneous distribution of chromite as discrete bands within the Basement Complex. Thus, these drainage basin geochemical methods, crude as they may appear to be, are nevertheless capable of producing results of great practical value.

It must be emphasized that at this sampling density there is no intention of identifying specific areas of mineralization, such as those of the order of several square miles. Rather the object is to identify belts, or minor provinces, of the order of perhaps many hundreds or a few thousands of square miles, worthy of more detailed examination. This type of survey is particularly useful in the preliminary exploration of large unknown areas where there is insufficient geological or other data to permit a reliable selection of zones of potential economic interest. In such situations the region is too large to be explored on a detailed level, at least initially, and some low cost method of exploration is necessary.

The sampling density in both examples cited above is nominally an average of one sample per 75 sq mi but, in actual practice because of logistics problems and the fact that stream drainages are not uniformly distributed, it may, of necessity, vary considerably. In those regions in which the rocks are essentially homogeneous over areas as large as 100 sq mi the problem is minor, as any one sample will be representative of a large area. However, in regions of complex and variable geology in which significant variations may be expected in areas as small as one to five sq mi, difficulties in interpretation are likely to be encountered. Bradshaw et al. (1972) have discussed this problem and have pointed out that in areas of complex geology, samples representative of larger catchment areas may have to be used but, in such cases, the sediments are likely to show little contrast. Consequently, they recommend that when the geology permits, it is better to sample smaller catchment areas (say 10 to 15 sq mi) and achieve only 15 – 30% coverage of the entire region, and then to interpolate over the gaps. This is preferable to using larger catchments (say 40 to 60 sq mi) and running the risk of all the sediments appearing very uniform. By reference again to Fig. 9-3, it is quite probable that the same regions of chromium enrichment would have been detected with a 50%, or possibly even a 20%, coverage.

The achievement of the objectives of a regional reconnaissance survey on this scale requires the extraction of a large amount of data from a relatively small number of samples. In this regard, it is generally advisable to perform multi-element analyses (16 elements were determined in the case of the Zambia survey and 15 for the Sierra Leone survey) and to use statistical methods to obtain the most significant interpretations. Armour-Brown and Nichol (1970) found that rolling mean (moving average) analysis of the Zambian stream sediment samples yielded improved definition of trends in the minor element distribution. Factor analysis served to establish the principal element associations related to the various bedrock types contributing to the metal content of the drainage samples, thus effectively permitting differentiation between anomalies related to mineralization and non-significant anomalies. Nichol et al. (1969) reached a similar conclusion based on the statistical study of the Sierra Leone data described above. Data from the Sierra Leone sediments are discussed further in connection with Fig. 12-9 to illustrate the use of the rolling mean (moving average) type of map. By use of statistical methods it is possible to recognize some of the more subtle, but nonetheless significant, information which may remain undetected when the data are interpreted by less rigorous means only. Such factors as the influence of weathering, erosion and secondary processes tend to modify stream sediment-bedrock-soil relationships in a subtle manner. The recognition of these factors, and the anomalies they produce, is particularly important in a low sample density survey.

It must not be assumed that only sophisticated statistical methods such as factor analysis and rolling mean or trend surface analysis are required for the interpretation of geochemical anomalies in a regional survey. For example, even such simple statistical parameters as Ni/Co and Ni/Cr ratios may enable the true source of anomalous nickel values to be determined. Enrichment of cobalt implies nickel sulfide mineralization whereas high chromium suggests that the nickel concentrations may be related to unmineralized ultramafic rocks. Although cobalt and nickel are elements which may themselves have economic potential, highly significant statistical information may also be obtained from elements with no direct economic significance (e.g., B, P, Ba, Na, Ce). Woodsworth (1971) stressed that in the geochemical drainage survey he conducted in the central Coast Mountains of British Columbia relatively simple statistical techniques, specifically cumulative frequency curves, permitted him to interpret the geochemical data, whereas more sophisticated methods, such as factor analysis and rolling mean analysis, could not be usefully applied.

SEDIMENT COLLECTION AT ONE SAMPLE PER 5 – 20 SQUARE MILES

Bradshaw et al. (1972) recognized stream sediment collection with an average density of approximately one sample per 5 – 20 sq mi as a method which may be used for: (1) the initial stage of an exploration program in regions suspected of containing mineralized belts or broad features similar to, but smaller than, those of a geochemical province, and (2) the follow-up to a low density survey designed to locate geochemical provinces (e.g., one sample per 75 sq mi).

In the first case, the intent and methods of execution of the survey would be essentially identical to those undertaken for the detection of a geochemical province, and the sampling density would probably be in the range of one sample per 10 – 20 sq mi. There are those who may argue that most, if not all, important geochemical provinces are known and, therefore, that low density sample surveys such as those carried out in Zambia and Sierra Leone, described above, are unnecessary. However, very few will disagree with the concept that many minor provinces and mineralized belts throughout the world have yet to be identified and, in such situations, samples collected at a density of one per 10 – 20 sq mi have great potential. In the second case, the sampling density would be close to one sample per 5 sq mi, and the work would assume many aspects of more detailed surveys (see below).

In any survey within the category being discussed here (one sample per 5 – 20 sq mi) the drainage sediments may, but more than likely will not, reflect individual mineral deposits. Only those deposits which are large or produce long dispersion trains with high contrast, may possibly be detected.

SEDIMENT COLLECTION AT ONE SAMPLE PER 1 – 5 SQUARE MILES

Regional geochemical reconnaissance and follow-up surveys using stream sediments collected at a density of one sample per 1 – 5 sq mi are now well established, and are possibly the most widely used geochemical reconnaissance technique where the objective is to recognize specific areas of mineral interest. They have the advantage of giving reasonably good coverage of an area and can frequently be sufficiently detailed to provide valuable geological information, such as the location of contacts and the nature of underlying rock types. Samples collected at a density of one per 1 – 2 sq mi can often indicate smaller features such as individual deposits and the location of alteration zones. As this category of survey merges into those conducted with a sample density of one per sq mi or less, the distinction between reconnaissance and detailed surveys may become obscure.

There are many examples which could be used to illustrate a survey conducted at a sampling density of one per 1 – 5 sq mi. The study by Govett and Hale (1967) in the Philippines is particularly apt for illustrating the effectiveness of stream sediment sampling in determining an anomalous dispersion train from a porphyry copper deposit for as much as 13 miles from the source (Fig. 9-4). The area is one of high relief with a humid climate and, because of these factors, both chemical and physical weathering are important processes. The geochemical data from which the map in Fig. 9-4 was prepared was obtained from unsieved stream sediment material analyzed at the sample site by a cold-extractable dithizone method. Hot extraction (hx) methods (not illustrated) based on the digestion of the samples with 1:1 HCl for 30 minutes also outlined the mineralization, although the contrast to threshold was higher for cxCu than hxCu (indicating a higher proportion of hydromorphic copper). The ratio of total to threshold values for streams K2 and K3 was in the range 16 – 32 for the cxCu method, and 4 – 8 for the hx method. Detailed follow-up work on stream K3 using sediments accurately outlined the mineralized areas, and

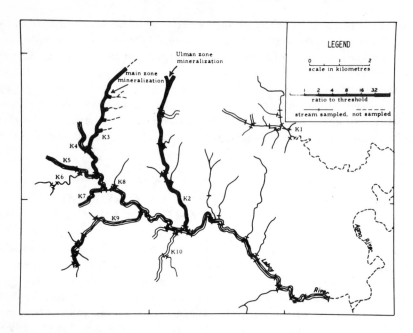

Fig. 9-4. Reconnaissance stream sediment survey in an area of porphyry copper mineralization, Luzon, Philippines. Results are based on cold-extractable copper analyses of unsieved samples. Background by this technique was 0.8 ppm Cu and threshold was 1.5 ppm Cu as determined in stream K1 (streams are designated K1 to K10). From Govett and Hale (1967).

the determination of cxCu in the minus 80 mesh residual soil fraction defined the mineralized source rock. This demonstrates that reconnaissance stream sediment geochemistry is capable of outlining areas with mineral potential which can later be located more exactly by detailed follow-up methods.

Some of the first geochemical reconnaissance surveys in North America were conducted at the density of one sample per 1 – 5 sq mi in eastern Canada, particularly in New Brunswick and the Gaspé Peninsula of Quebec. Hawkes et al. (1960) described the results of the collection and analysis of a total of about 15,000 stream sediment samples representing over 4,900 sample sites in a 27,000 sq mi area. In this part of Canada relief is flat to moderate, drainage is well developed, rainfall is sufficient for weathering to have formed dispersion trains, and the effects of glaciation have not seriously limited geochemical methods. The regional geochemical patterns they obtained clearly outlined terrains with geochemical characteristics favorable for the occurrence of base metals, and located specific drainage areas with outstanding promise. By reviewing the geologic setting of known deposits in the Bathurst (New Brunswick) mining district,

Hawkes et al. (1960) concluded that, with only minor exceptions, no important deposits occur (1) in areas underlain by predominantly sedimentary rocks, (2) within bodies of intrusive rocks, and (3) in the areas of low magnetic relief shown on the aeromagnetic maps of the Geological Survey of Canada. A composite map showing these unfavorable features was compiled (Fig. 9-5) and used as a guide in selecting areas for geochemical reconnaissance. Subsequent follow-up of favorable reconnaissance anomalies (dots in Fig. 9-5) led to the discovery of the Charlotte prospect (which carries over 12% combined lead and zinc, plus other values in gold, silver and copper), as well as other mineralized properties. This example also illustrates the advantage of combining geological and geophysical data with that obtained from geochemistry.

Numerous other geochemical studies have been conducted in this same area of New Brunswick and Quebec, and there are active geochemical programs in progress at this time. Not only stream sediments, but also stream and spring waters have been shown by Boyle et al. (1966) to give anomalous values which are related to mineralization. Stream sediments in this area are commonly enriched in manganese oxides and hydroxides, and all elements determined in their survey were found enriched in these minerals.

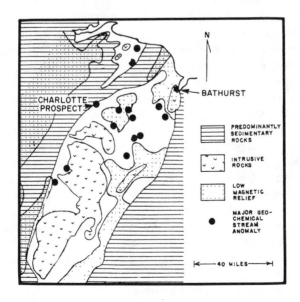

Fig. 9-5. Geologic favorability map of the Bathurst, New Brunswick, mining district. From Hawkes et al. (1960).

DETAILED SURVEYS

Detailed stream sediment surveys are generally carried out at a sampling density of greater than four or five per sq mi, and where soil, rock or vegetation are collected, the density may be as much as one sample per 5 sq ft. The areas in which detailed surveys are conducted may vary from a few square miles to some tens of square miles and are usually selected on the basis of results from regional geochemical surveys, favorable geology or geophysical evidence. Whereas the objective of a reconnaissance survey is to outline favorable areas, the detailed survey attempts to locate mineralization as exactly as possible preliminary to drilling and other more costly exploration methods. The results of detailed geochemical surveys are usually integrated with geophysical data during the interpretation phase and before drilling is started. It is at this stage that the geochemist must be aware of displaced anomalies (Figs. 8-1 to 8-3), as well as the many other factors which enter into interpretation. Should drilling reveal economic mineralization, the work of the geochemist is generally completed, although in some cases his opinion is required during drilling to interpret, for example, the significance of primary dispersion patterns and wall rock alteration. Should the drilling fail to outline a deposit, the geochemist may be called upon to re-evaluate the original geochemical data in the light of the drilling results and, because additional data would then be available in the third dimension, a completely different interpretation may result.

In favorable situations, stream sediments may be used in certain phases of a detailed survey, in addition to the regional reconnaissance surveys discussed above. This really entails increasing the sampling density, and is the best method to ensure the detection of an anomaly regardless of the material being sampled. Mention was made above of the work by Boyle et al. (1966) in the Bathurst district of New Brunswick which, in fact, was chiefly a detailed stream sediment survey. Stream sediments were collected along drainages at about 1500 ft intervals in an area of approximately 600 sq mi. A total of 3,550 such sediments were collected in addition to about an equal number of water samples and 170 heavy mineral concentrates. This study by Boyle et al. (1966) pointed out numerous anomalies which merited further investigation. Thus, in this particular region, detailed surveys (about 6 samples per sq mi) by means of stream sediments enabled relatively small areas to be isolated after which soil and other types of surveys (e.g., vegetation, spring water) could be used to pinpoint actual deposits.

Based on the reports by Hawkes et al. (1960), Boyle et al. (1966), and others, there is now a reasonably large and well-documented amount of information on the application of geochemical methods in the Bathurst district. Bradshaw (1973) has described some of the geochemical conditions encountered in this area and these provide an excellent opportunity to discuss examples of the practical aspects of detailed surveys. Much of what follows in the remainder of this chapter is based on Bradshaw's (1973) work.

On a regional scale, the Bathurst region can be divided into two principal zones: (1) areas with residual soil (some soils can be treated as residual, even though much of the area has been glaciated, because the amount of glacial movement was small); (2) areas covered with transported material such as alluvium. The length and intensity of anomalous dispersion patterns in the stream sediments and soils differ greatly in these two zones, and it is these variations which are explained by means of Figs. 9-6 to 9-9. Noteworthy is the fact that Presant (1967) found that in the Bathurst district, trace element contents are up to 100 times higher in soils above sulfide deposits than in soils above non-mineralized rocks. Zn, Cu and As are the elements most enriched above sulfides, whereas Mn and Fe show the least enrichment.

Fig. 9-6 illustrates the general patterns of stream sediment and soil anomalies in areas where the soil is of local derivation. From this figure it can be seen that: (1) there is a strong stream sediment anomaly which shows only a slow decrease in strength downstream in the stream draining the ore body, but once this joins the main stream, the dilution is relatively rapid; (2) a soil anomaly is developed directly over the mineralization and spreads out in a downslope direction similar to that shown in Fig. 8-2; and (3) there is a strong seepage anomaly developed at the break-in-slope which is locally higher in metal content than the residual soil anomaly (Fig. 1-3). One important exception (not illustrated) to this model would occur if there was a bog directly downslope from the mineralization and, in this case, a strong anomaly would be encountered in the bog material which, in fact, could be stronger than any other related anomaly from the same source (see Fig. 5-2). Fig. 9-7 is a three-dimensional view taken along A-B of Fig. 9-6, and it shows that the residual soil anomaly corresponds approximately to the dimensions of the ore body, as well as the size and position of the seepage (break-in-slope) soil anomaly, and the dilution of the stream sediment anomaly downstream.

Fig. 9-8 illustrates the general geochemical conditions found in those areas largely covered with transported material. In such cases: (1) the stream sediment anomaly is both shorter and weaker than the comparable anomaly in the residual soils, in part due to the influx of barren alluvium from the bank; (2) there is no surface soil anomaly directly over the mineralization, unless the overburden is thin and plant roots are able to extract metals from the mineralization; and (3) there is a seepage soil anomaly of only moderate strength built up at the break-in-slope. These features are illustrated in greater detail in Fig. 9-9. Similar conditions are

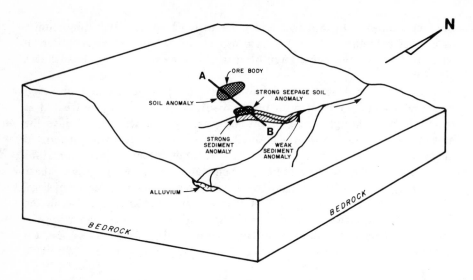

Fig. 9-6. Diagram showing the general patterns of stream sediment and soil anomalies in areas of essentially residual soil, Bathurst district, New Brunswick. Note that river flows north. Redrawn and modified from Bradshaw (1973).

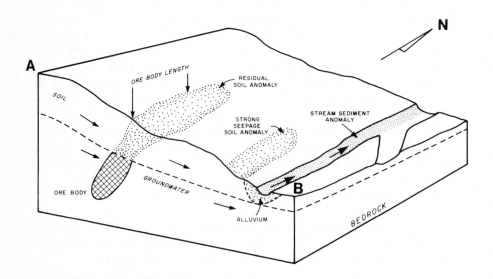

Fig. 9-7. Diagram based upon section along line A-B of Fig. 9-6. Redrawn and modified from Bradshaw (1973).

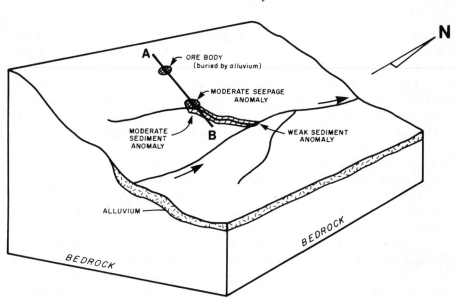

Fig. 9-8. Diagram showing the general patterns of stream sediment and soil anomalies in areas of transported overburden, Bathurst district, New Brunswick. Note that the river flows north. Redrawn and modified from Bradshaw (1973).

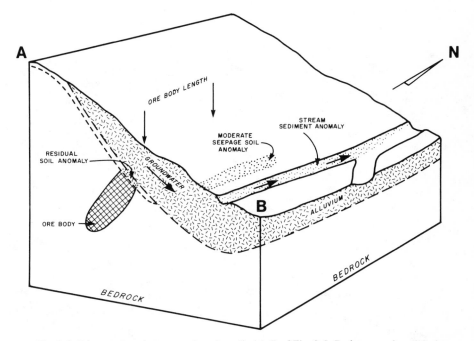

Fig. 9-9. Diagram based upon section along line A-B of Fig. 9-8. Redrawn and modified from Bradshaw (1973).

likely to be encountered in areas of glacial till, lacustrine clay, and other types of transported overburden. Strong bog anomalies also may be found in these situations.

Figs. 9-6 to 9-9 serve to illustrate the manner in which stream sediments may be used in the initial stages of a detailed survey, after which they are followed up by surveys employing soil, vegetation and any other materials which may assist in the exact location of mineralization. These figures also show the necessity for the careful determination of geological factors, local conditions and other variables (e.g., nature of overburden, topography, dilution mechanisms), particularly during the execution and interpretation of a detailed survey. In these diagrams the interrelationships of anomalies formed in or by stream sediments, soil and water within a small area are shown. From these figures it is also clear that the determination of the source of natural enrichments (anomalies) can be very complicated at times. Orientation surveys run before, or in conjunction with, detailed surveys are generally very rewarding. In summary it is apparent that, if properly undertaken and interpreted, detailed surveys can provide specific and valuable information leading to the location of mineralization.

10/*Vegetation Surveys*

INTRODUCTION

In Chapter 1 a brief introduction was given to the use of vegetation in geochemical exploration. In this chapter the subject is treated in further, but by no means exhaustive, detail. There is now a voluminous literature on this subject and the reader is referred to Brooks (1972), Malyuga (1964) and Chikishev (1965) for books directly concerned with the use of vegetation in exploration, and to Mortvedt et al. (1972), Sauchelli (1969) and Bowen (1966) for discussions on the roles of micronutrients and trace elements in relevant aspects of agriculture and biochemistry. Cannon (1971) has published an excellent review article on the use of plant indicators (geobotany) in groundwater surveys, geologic mapping and mineral prospecting.

As indicated previously, the detection of buried mineral deposits by the chemical analysis of vegetation (biogeochemistry) or the visual observation of plant cover (geobotany) is based on fundamentally simple principles. The root systems of vegetation obtain aqueous solutions from moist ground below the surface and act as an efficient sampling mechanism by providing a composite sample of a specific, often large volume of soil and, possibly, sub-soil. These aqueous solutions are a potential source of ore-associated metals that may ultimately either be concentrated in specific parts of plants, or result in a geobotanical anomaly. Certain plant species even have a limited ability to break down primary minerals and extract elements but normally vegetation surveys only detect those elements which move in solution. Elements of interest in exploration, therefore, must exist in the soil in a readily available form for uptake by the root system. Mechanical dispersion patterns, such as those in alluvium and glacial till in areas where mechanical weathering is predominant, are not as likely to be reflected in vegetation surveys because the metals in clastic grains are not available for uptake by the vegetation.

Mobilization Of Trace Elements

The total content of trace elements in a soil, especially a young soil, gives little information on the amount that is likely to be present in a form capable of being taken up by plants. Of most significance is the extent to which the elements in the primary minerals have been mobilized by chemical, and to a lesser extent biological, weathering and stored in the soil in an available form. Mitchell (1964, 1972) has summarized the forms in which mobilized (available) trace elements are held in soil for uptake by plants. These include: (1) in solution in ionic or combined form, in which case they can be removed from the soil by water extraction; (2) as readily exchangeable ions in inorganic or organic exchange-active complexes, extractable by neutral salts such as ammonium acetate; (3) as more firmly bound ions in the exchange complexes, extractable by dilute acetic acid or a chelating agent such as EDTA; (4) in insoluble organic or organo-mineral complexes, extractable by EDTA; (5) incorporated in precipitated oxides or other insoluble salts, extractable by more vigorous reagents such as acid ammonium oxalate; and (6) in secondary minerals in a fixed form. From the above descriptions, it can be seen that availability and mobility are measured by the amount of an element released during differing chemical treatments. In the laboratory, the effects can be determined by the amount of an element removed from soil under controlled conditions.

Mitchell (1972) pointed out that the proportion of a trace element present in any particular form depends on the nature and amount of the clay minerals and organic matter, and on the pH and Eh of the soil, as well as on the properties of the element in question. For instance, when copper is released during weathering, it is quickly chelated by organic complexes in the soil and, in acid soils, held in a form largely available to plants. In some soils over 90% of the soluble copper is present in organic combination, as compared with less than 75% of the zinc. Cobalt, on the other hand, appears to be present largely as firmly bound, but still exchangeable, ions in exchange complexes. In some soils, however, cobalt may be rapidly removed from the soil solutions by manganese oxides (by scavenging or coprecipitation) and will be fixed in a form unavailable to plants. Thus, in areas where the manganese content of soils is high, as in parts of Australia, the formation of cobalt anomalies in vegetation may be inhibited. In general, the lower horizons of freely drained soils show little mobilization of trace elements (because chemical and biological weathering is greatly reduced where water is freely drained away, and because weathering is reduced away from the daylight surface). However, nearer the surface there is often sufficient weathering to mobilize some elements. These features are illustrated in Fig. 10-1. The practical effect of this situation is that biogeochemical anomalies developed in a soil with impeded drainage, such as that shown in Fig. 10-1, might be ten times larger, or more, than ones developed over a freely drained soil that is similar in all other respects.

That it is possible to have trace elements in a soil without plants absorbing them in significant quantities is well known to agronomists and

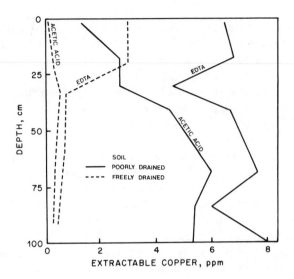

Fig. 10-1. Mobilization of copper in adjoining freely drained and poorly drained soil profiles developed on similar parent materials derived from olivine gabbro. This difference in mobilization is demonstrated by extraction with both 2.5% acetic acid and 0.05 molar EDTA. After Mitchell (1972).

those concerned with plant nutrition. Sauchelli (1969), for example, discussed the subject with particular emphasis on deficiencies of trace elements in soils and how such deficiencies may be corrected by means of the addition of fertilizers and specific chemical additives. The serious student interested in the use of vegetation for exploration purposes can gain much from careful study of the pertinent literature in the fields of agriculture and forestry. Characteristic and specific geobotanical anomalies related to deficiencies of specific elements (e.g., Fe, Mn, B, Zn, Mo, Co, Cl) are well established and these may, in some cases, be of as much value in exploration as those anomalies resulting from excesses of certain elements.

Soil-Plant Transfer of Elements; Plant Nutrition

The uptake of elements by plants, which is reflected in either geobotanical or biogeochemical anomalies, is a complex process. The plant root draws its supply of nutrients from the neighboring soil (Fig. 10-2) assuming, of course, that the elements are present in an available form. Once this zone is depleted in trace elements by plant uptake, several processes are involved in the repletion. According to Mitchell (1972) the two main processes appear to be: (1) mass flow of the soil solution to replace the liquid directly absorbed into the root, and (2) diffusion of the constituents of the soil solution toward the root zone down the concentration gradient created by the absorption of ions, salts or other constituents of the solution into the root. Opinions differ regarding the significance of the process known as root interception. Some workers postulate that the root develops and penetrates into untapped areas as a result of local nutrient depletion.

Although these processes account for the uptake of many major and trace elements, for example Mn and Mo, for others, the processes necessitate a degree of mobility in the soil which, in fact, may not exist. For those elements which are firmly held in adsorbing complexes or other forms, it would appear that increased concentrations of hydrogen ions (acidity) and organic chelating agents secreted by the roots (Fig. 10-2) or certain organisms, may be important factors in increasing the transport of some ions to the roots. Many workers postulate a highly acid, corrosive local environment at the root tips and in the adjacent soil, such that plants are

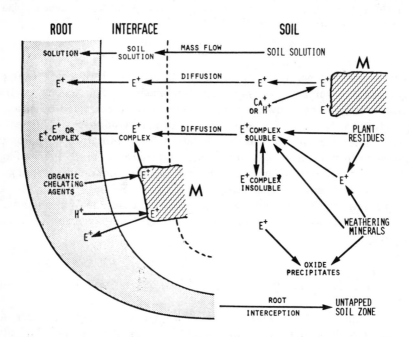

Fig. 10-2. The principal processes operative in soil during the transfer of trace elements between weathered minerals or decomposed organic residues and plant roots. The general symbol E^+ is used to represent any trace element cation that can participate in the particular process illustrated. M designates mineral fragments. From Mitchell (1972).

able to extract (dissolve) elements from minerals which are normally considered to be relatively stable in soils, possibly even some silicates and sulfides.

The discussion to this point has not considered the fact that uptake of trace elements can vary from species to species, with the stage of growth of a plant, and within different parts of the same plant. These are extremely important facts which cannot be ignored. The species effect is illustrated in the following example from Mitchell (1972) which shows the uptake of trace elements by three different constituent species of a mixed herbage (values in ppm in dry matter):

	Mo	*Mn*	*Mn/Mo*
clover	7.3	86	12
cocksfoot	1.3	142	109
crested dogstail	0.5	114	228

Such species relationships for the three constituents is not necessarily constant, but can vary with the nature of the soil and other factors. (The Mn/Mo ratio has no special significance other than to illustrate the behavior of the elements.) When different parts of a plant are compared, the situation becomes even more complicated because roots, stems, leaves and other plant parts will have differing trace element contents (see below). Further, effects of this type will vary from element to element and from species to species. In some cases one element will exhibit concentrations in a specific plant part whereas a geologically associated element will show no enrichment. In other words, plants may be selective in their relationship with specific elements. Elements such as uranium, lead and selenium are often preferentially excluded from some plants in which they are toxic. It follows that for exploration purposes not only must an element be in a form available for incorporation into a plant, but the physiology of the plant must be such that the element will be accepted for a positive vegetation anomaly to result. Clearly, the transfer of elements from the soil to a plant, the uptake of mineral matter by plants, and the movement of such elements from the root tips to the upper extremities, are extremely complex processes the discussion of which is beyond the scope of this book.

Biogeochemical Classification Of Elements

Chemical elements found in plants may be classified into three groups (Fortescue, 1971): macronutrients, micronutrients and non-essential elements. Sufficient amounts of macronutrients and micronutrients are needed by plants to complete their life cycle, but, the quantities must be within specific limits for each plant (Table 10-1). If one or more of these elements is available in too large or too small a quantity, growth rates will generally be reduced, or abnormalities of various types will take place in plant organs (e.g., leaves). Non-essential elements (e.g., silicon in grasses) are those taken up by the plant, either from the soil or atmosphere, which have no known role in the nutrition process. Some elements are not required by some plants, but are essential for others. Other elements are essential in certain quantities but are toxic at higher levels. Boron is an example of an element whose concentration range is very critical. Depending upon the element and the plant, critical concentration ranges may be either narrow or wide.

Table 10-1. Chemical elements found in plants. From Fortescue (1971), and based primarily upon data from Bowen (1966).

Elements essential for healthy plant growth		Elements which may, under certain conditions, reduce the growth rate of plants		
Macro-nutrients	Micro-nutrients	Very Toxic	Moderately Toxic	Scarcely Toxic
Hydrogen	Iron	Silver	Fluorine	Bromine
Oxygen	Copper	Beryllium	Sulfur	Chlorine
Nitrogen	Manganese	Copper	Arsenic	Iodine
Phosphorus	Zinc	Mercury	Boron	Germanium
Sulfur	Boron	Tin	Bromine	Nitrogen
Chlorine	Molybdenum	Cobalt	Chlorine	Phosphorus
Carbon	and possibly	Nickel	Manganese	Sulfur
Potassium	others	Lead	Molybdenum	Silicon
Calcium		(and	Antimony	Titanium
Magnesium		probably	Selenium	Calcium
		others)	Tellurium	Cesium
			Vanadium	Potassium
			Tungsten	Lithium
			(and others)	Magnesium
				Rubidium
				Strontium

Bowen (1966) divided elements into three classes based on their toxicities (Table 10-1). These are: (1) very toxic elements which are harmful to plants at concentrations in an available form of less than 1 ppm; (2) moderately toxic elements which produce toxicity symptoms at levels between 1 and 100 ppm; and (3) scarcely toxic elements which produce symptoms at higher concentrations that are not normally encountered in soils. Again it is important to recall that the chemical form, or availability, of the particular element is of primary importance in determining toxicity. Thus, under normal conditions, many of the elements listed as toxic in Table 10-1 are present in soils in amounts greater than the levels mentioned above (see Fig. 3-13), but they are generally tied up in silicates and other unavailable forms. In some cases, nutrients are accepted by plants only up to the toxic level after which they are rejected. In other words, the plant's nutritional requirements are the controlling factors in the movement of elements through the plant system.

Metal Uptake By Plants

The fundamental premise underlying biogeochemical surveys is that some chemical expression, however minor, will be reflected in vegetation growing over mineralization. This may be obtained from available elements in soil moisture, or from groundwater. As perhaps 95% or more of plant material is carbon, oxygen and hydrogen, trace element analyses are generally made on the dry residue (ash) from which these abundant and non-significant elements have been removed.

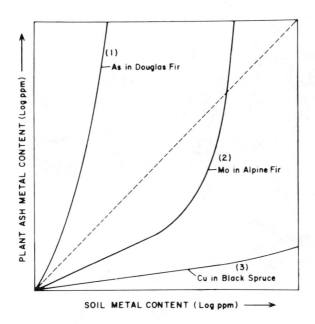

Fig. 10-3. Types of metal uptake behavior illustrated diagrammatically by relating metal concentrations in plant ash to metal concentrations in the underlying supporting soil. Dashed line represents a one-to-one relationship which is not commonly observed. Based upon Canadian examples. From Wolfe (1971).

As mentioned above, metal uptake by plants varies widely according to the particular species and element. Certain species have the ability to concentrate metals from soils selectively, whereas others appear capable of also excluding certain metals selectively. Three common types of metal uptake behavior are illustrated in Fig. 10-3. Curve (1) represents a case in which arsenic concentrations in the ash of first-year Douglas Fir twigs consistently exceeds the arsenic levels in related soils. In cases such as this, biogeochemical anomalies showing high contrast relative to soil anomalies would be expected. Curve (2) illustrates a case of partial exclusion of molybdenum by Alpine Fir at lower concentrations but at a certain threshold level the exclusion mechanism breaks down and the plant takes up large quantities of molybdenum. Curve (3) shows an extreme case of partial exclusion of copper by Black Spruce. Tenfold to 100-fold increases in the copper concentrations of soil are reflected by only 2- to 3-fold increases in the copper content of the ashed twigs from coniferous species growing on the soil. In this case, very poor biogeochemical anomalies would be anticipated.

Metal uptake is also strongly dependent upon the depth of root penetration and particularly the relationship of this to the water table. In

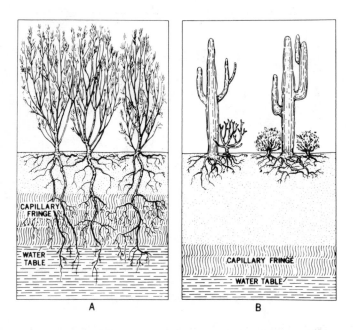

Fig. 10-4. Distinction between phreatophytes (A) and xerophytes (B) shown by their relationships to the water table. From Robinson (1958).

desert regions, in general, the flora is sharply divisible into two classes (Fig. 10-4). *Phreatophytes* are plants which obtain their water supply from the zone of saturation, either directly or through the capillary fringe. Phreatophytes (more than 70 species in different plant families are known) in desert areas are deep-rooted plants, some examples of which are capable of obtaining groundwater and yielding evidence of mineralization often at depths of 30 to 50 ft, and occasionally as much as 150 ft or more. *Xerophytes* are plants which occur where the water table is out of reach, and the vegetation is thus forced to depend upon the rains for a scant and extremely irregular water supply. *Mesophytes* are desert plants intermediate between the other two classes and are not particular in the source of their water. Clearly, the latter two, and especially the xerophytes, are of limited value in the search for buried mineralization. The distinction between phreatophytes and xerophytes is less noticeable in regions of high precipitation and it may be entirely absent in humid regions (Robinson, 1958). Deep-rooted phreatophytes (e.g., juniper, mesquite) have been successfully used for exploration in alluvium and other types of transported overburden, as well as in residual soil. Chaffee and Hessin (1971) detected anomalous metal patterns by applying biogeochemical techniques to desert plants whose roots penetrated as much as 100 ft of post-mineralization alluvium concealing a buried porphyry copper deposit in Arizona. Such biogeochemical techniques which utilize deep-rooted plants have great potential in many desert areas of the world. However, unusual examples have been described (e.g., Malyuga, 1964) where deep-rooted plants may not be used to detect an anomaly near the surface owing to the fact that metals are taken up, in some species at least, only at or near the end of the root system.

In the following pages discussions are presented on selected aspects of geobotany and biogeochemistry. These are followed by brief comments on the use of vegetation surveys, with particular reference to Canada. Other biogeochemical techniques, such as those based on the use of bacteria as indicators of the location of areas with high sulfide or sulfate contents, are not discussed because of the limited amount of pertinent information. Similarly, bryophytes (mosses and liverworts), even though they are capable of absorbing large amounts of trace elements from the rocks upon which they grow, are not discussed because they are rare and are extremely difficult to identify and classify except by a bryologist. Brooks (1972) has reviewed the state of knowledge concerning the use of bryophytes (including aquatic types) for exploration purposes. Details of the elemental composition of bryophytes have been published by Shacklette (1965). He (Shacklette, 1967) also reported on a study of copper mosses in Alaska that grow on substrates which contain large amounts of metals, and discussed the use of these plants in mineral exploration.

GEOBOTANY

Introduction

   As discussed previously, geobotany involves the visual identification
of vegetation. Stated in another way, geobotanical prospecting may be
considered as the visual investigation of plant cover types, or communities
which may indicate mineralization in the bedrock. In its simplest form,
geobotanical methods merely require the use of human eyes. Brooks (1972,
p. 11) observed, "Paradoxically, these methods are among the easiest to
execute and yet the most difficult to interpret of all the methods of exploration available at the present time." Interpretation requires some knowledge of such disciplines as botany, ecology, geology, biochemistry and
plant physiology and, for this reason, considerable expertise is generally
required to execute and interpret such surveys. It is true that some
botanical associations with mineralization have been known since at least
the 8th or 9th centuries, but these are examples of the very obvious associations which are the exceptions to the rule. Modern geobotanists must be
concerned about (Brooks, 1972): (1) the nature and distribution of plant
communities; (2) the nature and distribution of indicator plants; (3) the
recognition of morphological changes in vegetation; and (4) the application
of the above factors to botanical data obtained from aerial photography.
   Brooks (1972) credits the Russian geologist Karpinsky in 1841 with
being one of the first to recognize that different plants can be indicators of
the rocks and rock formations on which they grow. From this not only was
it natural to conclude that plants and plant associations could be used to
characterize the geology of an area, but also that plants are related to the
geologic environment. Although this is generally true, at least as a starting
point for discussion, it must be realized that there are many chemical,
environmental, and biological factors unrelated to geology which can
influence plant distribution, growth and health. These include the amount
of sunlight, precipitation, length of the growing season, altitude, latitude,
and the presence of insects. Notwithstanding the potential difficulties
involved, geologists have used the distribution of plant species in an
intuitive way as an aid in the mapping of geological formations, as well as
in locating fresh or saline aquifers and mineral deposits. The term *indicator
plant* (or geobotanical indicator) has been used previously (p. 24) for a
plant that shows, by its presence, the occurrence of an element in the soil
upon which it grows. However, in modern usage the term may also include
those plants which indicate rock types, groundwater and even hydrocarbons, as well as ores. Some authors also consider plants with characteristic variations (e.g., physiological changes) as indicator plants.

Indicator Communities

Brooks (1972) observed that indicator communities, or characteristic floras, will not necessarily indicate mineralization, but may permit the characterization of regions in which mineralization may occur. This is roughly analogous to the use of the low density drainage basin surveys described in Chapter 9 to outline favorable terrain. Examples of indicator communities include the study of serpentine flora to locate chromite deposits, and the use of selenium flora to locate uranium mineralization in such places (e.g., Colorado) where selenium is a pathfinder for uranium. The use of characteristic flora is particularly valuable where geologic maps are unavailable or of poor quality.

Geobotanical studies of serpentine furnish particularly good evidence that rock units of unusual chemistry support distinctive floras. Cannon (1971) pointed out that because serpentine (formed from the alteration of ultramafic rock) has high magnesium, nickel, chromium and iron contents and correspondingly low calcium, the growth of plants is affected. From an ecologic point of view serpentine areas are characterized as: (1) being sterile and unproductive; (2) having unusual floras of narrowly endemic species; and (3) producing strikingly physiognomic vegetation. Serpentine areas are commonly characterized by dwarf species of pine and shrubs, mosses, lichens, ferns, and certain genera of the laurel, chickweed and borage families. Plants living on serpentine-rich rocks must have the ability to extract sufficient calcium from these calcium-poor rocks. These plants must also be tolerant of high nickel and chromium. Ten percent NiO has been reported in the ash of leaves of endemic *Alyssum bertolonii* growing on serpentines in Italy.

In calcium-rich environments, such as soils developed on limestone and dolomite, other characteristic plant communities will flourish. The recognition of these flora often assist in geologic mapping. Limestone and dolomite, however, do not produce flora with stunted or peculiar characteristics. As the calcium content of soil is an important factor in controlling pH, plant communities which have different tolerances or requirements for calcium may also be used as indicators of strongly acid or alkaline soils. Some floras have been used as indicators of gypsum at White Sands, New Mexico (Cannon, 1971).

The selenium flora of the western United States furnishes a particularly good example of the use of floras in the search for uranium deposits (Cannon, 1971). Selenium commonly accompanies uranium in the sedimentary carnotite deposits of the Colorado Plateau (Table 2-3), and it is for this reason that selenium indicator plants are particularly useful, since there are no known species characteristic of uranium itself. The most widely known selenium indicators are certain species of *Astragalus* (locoweed). Although species of *Stanleya* and *Oonopsis,* and a species of wood aster (*Aster xylorrhiza*) require selenium, their requirements are less than many *Astragalus* species some of which contain as much as 4.6% selenium in the ash. The most useful species on the Colorado Plateau are *A. preussi* and *A. pattersoni.* In the Yellow Cat area of Utah, the occurrence of the latter corresponded with 81% of the ore that was within 32 ft of the surface. The presence of *Astragalus pattersoni* is also credited with the discovery of the Poison Canyon deposit near Grants, New Mexico. In addition, *Astragalus garbancillus* has been used to locate ore in Peru (Cannon, 1971).

Numerous other characteristic floras have been described in the literature which are indicative of elements such as zinc, while some are associated with saline soils containing sodium chloride, sodium carbonate, or sodium sulfate in Australia, the western United States, and the steppe region of the Soviet Union (Brooks, 1972).

Indicator Plants

In the preceding section, plant communities which indicate either rock type or mineralization were discussed although a few specific indicator plants representative of these communities, such as *Astragalus pattersoni,* were singled out for special mention. This section is concerned with such specific indicator plants. Indicator plants have the advantage over plant communities in that they are likely to enable the mineralization to be located more exactly.

Cannon (1971) observed that plant indicators of ore deposits may be classified into three groups: (1) those that have adapted to living exclusively on rocks or soils which supply unusual amounts of a particular element; (2) those that have acquired an immunity to large amounts of a particular metal by being able to reject the metal at the root site; and (3) those species of wide distribution that favor mineralized ground under certain local conditions because of acidity, water conditions, or availability of major plant nutrients. Thus there is the suggestion that indicator plants become adjusted to the environment by some type of natural selection process. Those plants which always indicate the presence of a definite element are called *universal indicators,* whereas those that can be used as indicators only within the limits of a certain district are called *local indicators* (Cannon, 1960).

Since indicator plants for mineralization have been known and studied for about a thousand years, it is not surprising that many such plants have been recognized. For example, a listing by Cannon (1971) contains the names of about 120 plants. Brooks (1972) listed 62 species from Cannon's

(1960) original compilation which he knew to be used for prospecting and also pointed out that the claims of usefulness for many indicator plants are unsubstantiated. In some cases these "indicators" have been found only in contaminated mine areas (soil and dumps), and are not from virgin areas. For the purposes of this book, Table 10-2 has been prepared which lists indicator plants known and used in the United States (no indicator plants are reported from either Canada or Mexico, although some of the copper indicator plants are likely to be found in northern Mexico). On a world-wide basis, of the 62 species considered by Brooks (1972) to be significant, 17 are indicators of copper. This includes the most successfully used of all copper plants, *Becium homblei,* a member of the mint family found in Zambia, which requires soil of more than 100 ppm copper to grow and can tolerate up to 5000 ppm. Discovery of several important anomalies have been attributed to geobotanical surveys based on this plant. Other elements not listed in Table 10-2 for which plant indicators are available in various parts of the world include (number of indicators in parentheses): boron (3, all in the U.S.S.R.); cobalt (2, in Katanga); iron (5, in New Caledonia, Venezuela and Germany); manganese (5, all in the U.S.S.R.); molybdenum (1, U.S.S.R.); nickel (3, in Italy, Norway and U.S.S.R.); and phosphorus (1, Spain). The large number of selenium indicators in Table 10-2 represents a

Table 10-2. Indicator plants used in the United States. Based on data in Cannon (1960, 1971) and Brooks (1972).

Element	Species*	Common Name	Location
Copper	*Eschscholtzia mexicana* (L)	California poppy	Arizona
	Mielichhoferia macrocarpa (U)	Copper moss	Alaska
	Mielichhoferia mielichhoferi (U)	Copper moss	North America
	Merceya ligulata (U)	Copper moss	North America
Lead	*Baptisia bracteata* (L)	Wild indigo	Wisconsin
	Erianthus giganteus (L)	Beardgrass	Tennessee
Selenium	*Aster venusta* (U)	Woody aster	Western U.S.A.
(and	*Astragalus* spp. (U)	Poison vetch	Western U.S.A.
uranium)	*Oonopsis* spp. (U)	Goldenweed	Western U.S.A.
	Stanleya spp. (U)	Princesplume	Western U.S.A.
Silver	*Eriogonum ovalifolium* (L)	Buckwheat	Montana
Vanadium	*Astragalus bisulcatus* (U)	Poison vetch	Western U.S.A.
Zinc	*Philadelphus* spp. (L)	Mock orange	U.S.A.

* U = Universal indicator;   L = Local indicator

special case, and only one other selenium indicator has been reported from elsewhere in the world (from Queensland). Most of the zinc indicators (six others in addition to the one listed in the table) are confined to Europe.

Brooks (1972) pointed out that indicator plants have certain characteristics in common. Most are herbaceous (no woody stem) rather than trees or shrubs. Cannon (1960) noted that true copper indicators ·belong mainly to three plant groups: the Caryophyllaceae (pink), or the Labiaceae (mint) families, and the mosses. Another characteristic of all indicator plants is that the elemental content of their ash is high. Mention was made above of the fact that *Astragalus pattersoni* has as much as 4.6% Se in its ash. Maximum concentrations of several other elements are presented in Table 10-3.

Table 10-3. Elemental contents in the ash of some indicator plants (ppm). From Cannon (1960) and Brooks (1972).

Element	Species	Normal Content	Maximum Content	Location
Cobalt	*Crotalaria cobalticola*	9	18,000	Katanga
Copper	*Becium homblei*	183	2,500	Zambia
Copper	*Gypsophila patrini*	183	500	U.S.S.R.
Manganese	*Fucus vesiculosus*	4,815	90,000	U.S.S.R.
Nickel	*Alyssum bertolonii*	65	100,000	Italy
Zinc	*Thlaspi calaminare*	1,400	10,000	Germany

Surveys using indicator plants give the best results in unglaciated areas; there do not appear to be any known indicator plants in Canada, and only three are listed from Scandinavia by Brooks (1972). Usik (1969a), after studying the distribution and morphological features of 165 plant species, could find no plants indicative of mineralization at two copper prospects and one molybdenum prospect in central British Columbia. The molybdenum prospect is the same one (Lucky Ship) which has been discussed in connection with Fig. 1-9, in which pronounced biogeochemical and soil anomalies have been detected.

Morphological, Physiological and Mutational Changes

Excesses (and deficiencies) of trace elements in soils in which plants grow can cause a wide variety of visible morphological and mutational changes. In addition, physiological effects, particularly diminished or inhibited growth, may result, exemplified by the dwarfed flora of serpentine discussed above. The subject has been considered in various degrees of detail by Cannon (1960, 1971) and Brooks (1972) who recognize such morphological changes related to mineralization as chlorosis of the leaves, dwarfism, gigantism, abnormally shaped fruits, changes in color in the flowers, and many more. Only a few will be discussed here.

Chlorosis results from an excess of many metals (e.g., Cu, Zn, Mn, Cr, Ni) which interfere with iron metabolism of the plant and produce a yellowing (chlorosis) of the leaves. Chlorosis is one of the most common types of geobotanical indicators of mineralization having been observed, for example, in vegetation growing on zinc-rich peat in New York, and on zinc-bearing dolomite in Wisconsin. Chlorosis has been used as an exploration technique for zinc in the Tri-State district of Missouri. Copper chlorosis is reportedly common in the Michigan copper belt (Cannon, 1971).

Changes in the color pattern of flowers are one of the more obvious indicators of mineralization. Geobotanical studies at an Armenian copper-molybdenum deposit showed changes in the petals of the poppy *Papaver commutatum* (Fig. 10-5). In the modified plants, the black spots are elongated, reaching the edge of the petal so as to form a cross. Numerous other color changes due to trace elements have been reported. In specific

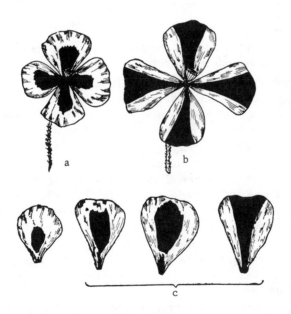

Fig. 10-5. Changes in the color of petals (as indicated by shading) of *Papaver commutatum* under the influence of copper-molybdenum mineralization. a. Normal flower. b. Modified flower. c. Degree of mutability of petals of the corolla. From Malyuga (1964).

instances manganese, nickel and cobalt cause pale foliage but boron, on the other hand, may darken foliage. Color changes due to radiation also have been reported. Both color and morphological changes have been described by Shacklette (1964) in *Epilobium angustifolium* affected by radiation emanating from the Port Radium, N.W.T. uranium deposits. However, similar color effects have been observed in *Epilobium angustifolium* growing on substrates derived from basaltic-gabbroic rocks in other regions of the Yukon Territory and Alaska which do not have uranium mineralization (Dr. T. C. Mowatt, personal communication).

Dwarfism (stunted growth) is a relatively common indicator of mineralization or unusual rock-types, such as serpentine mentioned above. The actual cause is often difficult to ascertain, but in some cases boron or radioactive minerals are present in large amounts. Gigantism, likewise, is usually associated with boron or radioactivity, but some examples are apparently related to the presence of bitumen. The best example of the relation of gigantism to bitumen (petroleum) has been reported in a work from the Soviet Union (summarized by Brooks, 1972) which showed gigantism in 29 species. Overall, however, gigantism is a rare phenomenon.

Discussion

From the above instances it should be clear that the use of geobotany in the exploration for minerals (including hydrocarbon) deposits has many limitations, the most obvious of which is the necessity for an expert in botany and/or ecology. Another limitation is that some surveys, such as those based on morphological, mutational or color changes in flowers, must be conducted during specific seasons of the year. In some cases, the indicator plant may be uncommon, or even rare, and this makes the geobotanical survey particularly difficult. Further, experience has shown that data obtained from geobotanical surveys are not necessarily universally applicable but may have only local significance. Where geobotany does work, the advantages are great. These include low cost, the fact that geologic formations, as well as mineralization can be mapped, and that sometimes geobotany may indicate buried mineralization where other methods fail. Although exact figures are not available, it is fair to state that geobotany comprises only a very minor proportion of the total effort included within the category of vegetation surveys.

Cole (1971) discussed several examples in which geobotanical techniques were successful and others in which they were not. She pointed out that an understanding of the environmental factors is essential. These include the interacting effects of geomorphology, climate, soil genesis, soil-water relationships, geology, and other variables. For example, the success of geobotany in the low tree and shrub savanna country of South West Africa is attributed to: (1) a semi-arid climate, with rainfall concentrated in the summer period of active growth when plants are particularly susceptible to toxic conditions in the soil, and (2) a direct relationship between vegetation and bedrock. On the other hand, where the presence of transported sand complicates the plant/bedrock relationships, soil sampling and biogeochemical sampling of deeply rooted plants offer more promise. In tropical rain forests, difficulties of access and of species recognition necessitate prior reconnaissance from low-flying aircraft. In mountainous areas, relief is responsible for altitudinal variations in the species composition of the forest but on Bougainville Island, Cole (1971) was able to find that, independent of height above sea level, a distinctive tree association developed in soils with the highest copper values over a porphyry copper deposit. Cole (1971) described numerous other examples to illustrate the effect of environmental factors in vegetation surveys, based particularly upon her experiences in Africa and Australia. These are of great value to those contemplating the use of such techniques in those or any other tropical or sub-tropical areas.

For the geologist or geochemist who has little training in botany, probably the most practical fact to keep in mind is that many ore deposits (e.g., the copper deposits in Rhodesia) occur in areas that are bare of vegetation in an otherwise forested region. Such areas, or terrain with a flora greatly reduced or distinctly different by comparison with the surrounding area, may result from poisoning of the vegetation by unusual concentrations of certain elements in solution with sulfuric acid generated by the oxidation of pyrite. Thus areas denuded of vegetation deserve special attention.

Finally, it is important to note that remote sensing techniques from aircraft and satellites are now being applied to the study of vegetation from agricultural, ecological and exploration points of view. Metal-rich soil and associated vegetation can, under favorable circumstances, be found by these methods, generally by multispectral techniques. For example, chlorosis in vegetation has been detected by the use of various color films in remote sensing experiments. At present it seems likely that these techniques may be quite useful in outlining geochemical provinces, and in the recognition of large, low-grade mineral deposits.

BIOGEOCHEMISTRY

Sample Collection

Biogeochemical surveys are based upon the analysis of plant material which has accumulated trace elements without producing any visible morphological or physiological effects. It is, therefore, necessary to collect and chemically analyze plant material in order to detect any evidence of mineralization. However, there is much more to biogeochemical techniques than the above generalizations would indicate. Brooks (1972, p. 99) succinctly stated, "It should be reemphasized that the main requirement of successful biogeochemical prospecting is that the plant species selected for the project should accumulate the element concerned in a *reproducible* manner and that the degree of accumulation should be in direct proportion to the concentration of the element in the soil." Unfortunately, there are probably about 20 different variables which can affect the accumulation of elements by vegetation (Brooks, 1972, p. 99), although some are relatively minor. Several of the more important variables which affect elemental accumulation by plants have been mentioned previously. They include: type of plant studied; plant organ sampled; age of plant or plant organ; pH of the soil; depth of the root system; and drainage conditions. These factors must always be considered during the collection of material for biogeochemical prospecting.

The method of collection, the particular plant, and the specific part of that plant which should be collected, will vary from area to area and with each survey. Therefore, it is essential to conduct an orientation survey before proceeding with any biogeochemical survey in a new area. The need for an orientation survey has been stressed by every expert in the field. Brooks (1972) has described in some detail the recommended procedures for carrying out such an orientation survey. Some of the main points he considered are: (1) selection of a site; (2) selection of plants for testing; (3) selection of a sampling pattern; (4) methods of sampling; (5) storage of plant and associated soil material; (6) preparation of material for analyses; and (7) evaluation of the survey. It is not possible to discuss all these points in detail in this chapter and, therefore, only representative topics will be considered. It must be remembered that all information on the geology of the area, as well as data on any mineralization, should be collected and integrated during the interpretation phase of the orientation survey.

Fortescue and Usik (1969), Fortescue and Hornbrook (1969), and Fortescue (1970) have described in detail the procedures they have found satisfactory in conducting biogeochemical surveys in Canada and, with appropriate modifications, the concepts are applicable elsewhere. Fortescue (1970) stressed that, for the most reliable results with biogeochemical surveys, it is necessary to obtain background information on bedrock geology, surficial soil cover, and plant cover types, these features being part of the overall geochemical landscape. An example of a plant cover map is illustrated in Fig. 10-6 from which it can be concluded that the type of vegetation likely to be found in any one area may vary greatly as

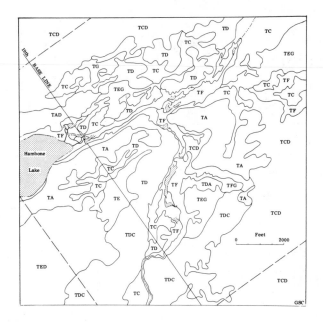

Fig. 10-6. General plant cover map at Hambone Lake, near Thompson, Manitoba. From Fortescue and Usik (1969).

TA   Deciduous forest
TC   Muskeg spruce-sphagnum forest
TD   Mineral soil spruce-hypnum forest

TE   Mixed muskeg spruce-shrub and spruce-sedge forest
TF   Shrub-sedge sites associated with water
TG   Muskeg shrub/sedge moss sites

a result of topography, drainage and other factors. Where the plant cover types are simple and uniform, the interpretation of biogeochemical data is also relatively simple, but where the plant cover types are complex, interpretation is difficult (Fortescue and Usik, 1969). As the ability of different plants to reflect mineralization varies greatly, the importance of obtaining as uniform a sampling medium as possible becomes apparent. Of course, it is not always possible to sample only one species and in such a situation it will be necessary to sample what is available and attempt to make appropriate compensation during the interpretation stage. One should no more sample vegetation indiscriminately any more than one should sample different soil horizons. As a rule, deep-rooted plants, if they are known, should be sampled preferentially in a new area.

Once the plant cover has been mapped, the actual sampling can begin, although in some cases both mapping and sampling may be achieved simultaneously. Inasmuch as one never knows beforehand what part of the vegetation will give the most significant geochemical response, several parts of

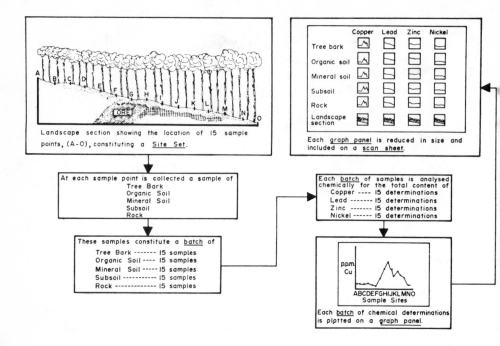

Fig. 10-7. Graphical display of the technique employed in obtaining biogeochemical data, taking into account many important aspects of the geochemical landscape. From Fortescue (1970).

the vegetation, the organic and mineral soils, and the bedrock should be sampled if an orientation survey was not run beforehand. Bedrock appraisal is essential so that the efficiency of the vegetation in extracting or accumulating metals can be judged. The technique employed by Fortescue (1970) is illustrated in graphic form in Fig. 10-7. The results of an actual survey conducted over a mineralized copper deposit in Quebec are shown in Fig. 10-8. In this study mineral and organic soils, as well as bark, twig, needle and current growth material from individuals of Balsam Fir *(Abies balsamea)* were sampled at 30 stations. Inspection of the "scan sheet" shows the great variations which were found for certain elements (e.g., lead) in various parts of the vegetation. This demonstrates the necessity for orientation surveys and consideration of all aspects of the geochemical landscape. It is because of variations such as these that most experts recommend that various parts of vegetation (e.g., twigs, leaves and stems) be collected and analyzed separately for several elements (e.g., Zn, Cu, Mo, Pb). In the example illustrated (Fig. 10-8), biogeochemical methods would have failed to locate the deposit (for unknown reasons), whereas B soil horizon sampling for copper would have been successful.

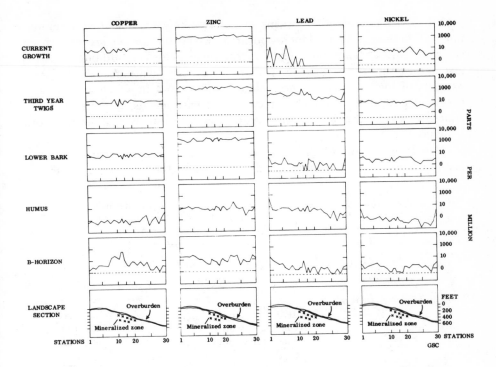

Fig. 10-8. "Scan sheet" showing minor element content of soil and vegetation and the landscape section at Pekan Creek, Gaspé Park, Quebec. Horizontal dashed lines represent analytical detection limits. The sampling interval at all stations is 50 feet except between stations 10 – 20 where it is 100 feet. From Fortescue and Hornbrook (1969).

Ideally, a common and uniformly occurring species is desirable in a biogeochemical survey. Further, as a rule, it is advisable to select a line (or grid) and to collect at a series of stations along it. This enables the results to be correlated with observable geological and mineralogical features, and the results for different species can be plotted to give a biogeochemical profile. On occasion it can be demonstrated that two or more species will give identical results in which case the problem of choice of sampling medium is greatly simplified.

Element Distribution In Vegetation

Although some plants can concentrate many elements to a striking degree, the accumulations are not uniformly distributed throughout the plant. Many studies have shown that, in general, first or second year growth of twigs and other organs are more likely to yield better (more intense) anomalies than older wood or bark and the former are generally easier to collect. Thus the metal content varies with age, although this is just one of the many variables (others have been mentioned above). An example of the age effect is presented in Table 10-4 in which arsenic, an excellent pathfinder for many precious and base metals, is preferentially concentrated in the ash of first-year growth of stems and needles of Douglas Fir.

Different organs of the same tree may vary in their ability to concentrate metals (also shown by the data in Table 10-4). Carlisle and Cleve-

Table 10-4. Variations in the arsenic content of the ash of various organs of different ages from Douglas Fir trees growing at two Canadian localities. From Warren et al. (1968).

Locality	Sample	Organ	Arsenic (ppm)
H. B. Mine	64 #3	first-year stems	510
		first-year needles	120
		second-year stems	70
		second-year needles	25
H. B. Mine	64 #5	first-year stems	780
		first-year needles	450
		second-year stems	280
		second-year needles	60
Bralorne Mine	64 #2	first-year stems	2110
		first-year needles	1060
		second-year stems	1390
		second-year needles	180

land (1958) found that for North American vegetation the elemental contents of tree organs decrease in the sequence: leaves, twigs, cones, woods, roots and bark. Among many examples in the literature which could be cited to illustrate this phenomenon, the data presented by Warren (1972) is typical. The cones of *Tsuga mertensiana* (Mountain hemlock) over the Britannia Mining and Smelting property in British Columbia contained 34,560 ppm copper in the ash, whereas the ash of the bark of the same tree contained only 2060 ppm. Both values are anomalous as background values are 1540 ppm and 210 ppm, respectively. Not to be overlooked, in addition, is the fact that there are seasonal variations in the metal content of the several organs of vegetation. Studies of this feature have established that the metal uptake by twigs is less likely to vary seasonally than that of leaves, fruit, needles and other plant parts. Guha (1961) produced the data illustrated in Fig. 10-9 which shows the seasonal variation of a number of elements in leaves of deciduous trees located away from mineral deposits in Scotland. As Fortescue (1967) pointed out, these data are of particular interest because they emphasize the fact that the amounts of some minor elements (e.g., Mn, B) tend to increase in concentration during the growing season whereas the amounts of others (e.g., Zn, Cu) decrease. In some cases the magnitude of the seasonal variation may be as much as sevenfold (e.g., boron in sycamore leaves). These variations also demonstrate the necessity for (1) collection of all samples in a biogeochemical survey within as small an interval of time as is possible, and (2) recognition of the fact that because of seasonal variations, anomalies may be missed by the collection of leaves (and other plant parts) at an inappropriate time of year. Brooks (1972) noted that some of Guha's findings are in contradiction with those

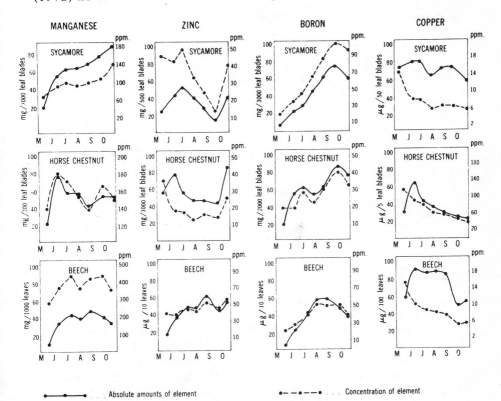

Fig. 10-9. Seasonal variations in the contents of some essential minor elements in the leaves of three species of deciduous trees. MJJASO refer to the months of the year in which samples were collected. ("Absolute amount" is the number of mg in a specific number of leaves; "concentration" is the ppm abundance in the leaves on an oven-dry basis.) From Fortescue (1967), and based on data from Guha (1961).

of other workers, but this may be caused by species differences. Also, the seasonal variations indicated in Fig. 10-9 appear to be characteristic of deciduous leaves but not necessarily of other types of vegetation.

Notwithstanding the inherent differences in the metal contents of vegetation as functions of plant species, plant age, and the many other variables, Table 10-5 has been prepared from data presented by Cannon (1960). This gives the reader some indication of the variations in the metal contents of the ash of selected types and parts of vegetation. The values presented in Table 10-5 were obtained from vegetation growing on unmineralized ground and are thus representative of background values.

Table 10-5. Average metal contents in the ashes of selected types of vegetation growing in unmineralized ground (in ppm). From Cannon (1960).

Element	Average	Grasses (above ground)	Other herbs (above ground)	Shrub (leaves)	Deciduous trees (leaves)	Conifers (needles)
Zn	1400	850	666	1585	2303	1127
Cu	183	119	118	223	249	133
Pb	70	33	44	85	54	75
Ni	65	54	33	91	87	57
V	22	25	23.5	25	16	21
Mo	13	34	19	15	7	5
Cr	9	19	10	14	5	8
Co	9	10	11	10	5	<7

Notes:
1. Above values are based on the analysis of from 4 to 2047 samples, mostly from the United States.
2. Other averages include (values in ppm): Al (8610); Fe (6740); Mn (4815); B (700); U (0.6); Sn (<5); Be (<2); Ag (<1); Au (<0.007).

Discussion

Biogeochemical techniques have been used successfully in many parts of the world. The history of biogeochemical prospecting, as reviewed and discussed by Brooks (1972), Malyuga (1964) and Warren (1972) vividly portrays viable, proven techniques. It is essential, however, that those contemplating the use of these techniques be aware of the fact that some reports may tend to be misleading in that they engender the impression that such techniques are generally relatively straightforward and easy. It cannot be stressed too strongly that the successful utilization of these methods involves a knowledge of many disciplines (e.g., botany, ecology, plant chemistry, plant physiology). Brooks (1972, p. 69) clearly stated, "It is not claimed that biogeochemical prospecting will be a universal panacea for the prospector in his search for minerals; under some conditions it is unwise to expect too much from the method." In the remainder of this chapter a few examples will be presented which point to some of the limitations which should be considered. By the same token, these comments must not be misconstrued as implying that biogeochemical prospecting is being down-graded; all techniques have characteristic limitations as well as advantages.

Not all elements are equally effective in biogeochemical surveys. Brooks (1972) mentioned that several elements have achieved the reputation of being "easy" (that is, they give good, reliable geochemical responses in most situations) whereas others are "difficult." Molybdenum and uranium are examples of the former whereas copper and zinc are examples of the latter. This is not to say that copper and zinc are always unreliable because there are numerous proven cases in which the contents of these elements in vegetation do, indeed, reflect mineralization. However, from Fig. 10-3 it can be seen that copper, at least in the example shown, tends to be excluded by the Black Spruce. Thus, the behavior of this particular plant species towards copper, is one reason why the success rate for biogeochemical surveys will vary with the element being determined. Another reason is related to the fact that copper and zinc are essential elements for the growth of plants (Table 10-1) and, as these elements are normally taken up by plants, it is necessary to find a suitable species which will reliably reflect anomalous contents of these elements. Brooks (1972) showed that certain statistical procedures are of potential value for assisting in the interpretation of "difficult" elements used in biogeochemical surveys.

There can be no doubt that anomalies and positive results described by Warren (1972), Hornbrook (1969), and others, for Cu, Zn, Mo, Ag, Au, Pb and Hg establish the value of biogeochemical methods in various parts of Canada. However, there are unquestionably examples where biogeochemistry does not work. For example, although Hornbrook (1969) described a classic example of a molybdenum anomaly in ash from needles of second year Alpine Fir (Fig. 1-9) which perfectly outlined the Lucky Ship deposit in west-central British Columbia, this might be classified as exceptional. In fact, some 20 deposits in British Columbia were considered, and seven "visits" completed, before the Lucky Ship was chosen to be the best one to illustrate this type of anomaly (Professor J. A. C. Fortescue, personal communication). That a good biogeochemical response was obtained by Hornbrook (1969) at this particular deposit is a reflection of almost ideal environmental (e.g., precipitation), geological and ecological conditions.

One of the main drawbacks to the successful use of biogeochemical methods in Canada is the presence of certain glacial deposits (e.g., tills and lacustrine deposits; see Chapter 11) which effectively mask or "blank out" geochemical responses. This problem, of course, is present in other parts of the world which have experienced glaciation in recent geological times (e.g., Scandinavia, Ireland, U.S.S.R.). By way of example, Fortescue and Hornbrook (1969) analyzed numerous samples of vegetation of various kinds at the Kidd Creek massive copper-zinc sulfide ore body near Timmins, Ontario, and found no significant variations in the concentration of any element that could be directly related to the mineralization. Similar disappointing results were obtained from soil samples, and are no doubt due to the fact that the area is covered by till, varved clay and other glacial deposits which range from about 20 – 60 ft in thickness. Another problem faced in some parts of Canada is based on the fact that although certain species of trees may reach heights of 60 ft or more, their root systems penetrate no further than two or three feet vertically into the glacial over-burden after which the roots grow laterally. This characteristic has been observed in trees in the vicinity of the porphyry copper deposits in the Babine Lake area, British Columbia, and elsewhere.

Wolfe (1971) concluded from a study of biogeochemical prospecting in glaciated terrain of the Canadian Precambrian Shield that (p. 79) "where plant root systems penetrate thick deposits of transported Pleistocene cover, biogeochemistry can provide an effective mechanism of sampling deep transported material and can be a useful exploration tool *if* the upper hori-zons of podzolic soils do not adequately reflect chemically anomalous conditions at depth. The biogeochemical method may be most useful in evaluating geophysical anomalies and/or deep geochemical anomalies in lower tills identified by overburden drilling." Wolfe (1971) also noted that apparently Mo, U, Pb, Co, and probably Fe and Mn, may show moderate to high anomaly contrast in the common deep-rooted tree species ("easy" elements), but Cu, Zn, Ni and Ag responses in vegetation of the Canadian Shield generally show low contrast ("difficult" elements).

The lack of biogeochemical responses in regions of transported over-burden is not limited to areas covered by glacial deposits. Similar problems are faced in humid tropical environments with transported soils. In addi-tion, masking of biogeochemical responses can result from the presence of caliche (calcrete) deposits in drier environments unless the vegetation receives drainage circulating from mineralization beneath the caliche (Cole, 1971). Cole (1971) also mentioned that in some cases, such as in the search for nickel in Australia, one major problem is distinguishing false anomalies from significant anomalies over ultramafic rocks (see Fig. 8-3F). Therefore, unless the above mentioned constraints are noted, and numerous others could be cited, the biogeochemical method of prospecting might appear too easy and, as a result, misleading to the reader.

$11/$ *Geochemical Exploration In Canada*

INTRODUCTION

This chapter is concerned with geochemical exploration in Canada, with particular emphasis placed on techniques to be employed in the various environments. Warren et al. (1967, p. 253) pointed out "that some of the best known workers in the field of applied geochemistry developed their techniques for use in the U.S.A., the U.S.S.R., Africa, and Australia. These techniques, although excellent for the purposes for which they were developed, are not always equally applicable elsewhere". By way of example, Warren et al. (1967) mentioned that many Canadian soils are transported and most are geologically young, often even where they are morphologically mature. Also, vegetation studies, which have proved very rewarding in some parts of the world, are not likely to prove equally rewarding in Canada because soils have not had sufficient time to develop with a distinctive plant ecology since the last glaciation.

In many respects attempting to discuss geochemical techniques which should be used in the world's second largest country in area is a challenge because there are several problems and features peculiar to this region for which very little of the necessary basic information is available. Hence, much of what is presented in this chapter is in a sense preliminary, even though it has been obtained from the best possible sources. The four topics which will be discussed concern geochemical exploration in:

1. *Permafrost:* This underlies about 50% of Canada and, until recently, it had been thought that rock materials in permafrost regions were essentially unaffected by such fundamental processes as weathering, leaching and secondary dispersion.

2. *Water:* About 20% of the world's fresh water is found in Canada and about the same percentage of area of the Northwest Territories, as well as the northern parts of Ontario, Manitoba and Saskatchewan are covered with lakes. This great abundance of surface water requires greater emphasis on hydrogeochemical methods here than anywhere else in the world. Lake sediments, as well as lake waters, will be discussed.

3. *Muskeg and Bogs:* Widespread organic materials called muskeg, bog, peat, muck, and various other names, cover much of the poorly drained, glaciated areas of Canada. Relatively little is known about the optimum geochemical techniques to be used in these materials.

4. *Glaciated Areas:* Most of Canada was covered by glaciers during the Pleistocene. Many glacial deposits are unrelated to their presently associated bedrock, and others (specifically glacial lake deposits) are impervious to element transport. However, with proper techniques, exploration geochemistry can be successfully used in glaciated areas.

In many parts of Canada all of these four features are encountered simultaneously, whereas in others only one or two are likely to be found. But everywhere an appreciation of at least one of the above four topics is essential to obtain the maximum value from a geochemical survey. The problems faced in Canada are similar in many respects to those faced in other localities in the northern latitudes, particularly in the Soviet Union, Scandinavia, Greenland and Alaska, where a large percentage of the bedrock is hidden beneath several kinds of younger deposits, such as glacial deposits and muskeg, and also lakes. Where possible, available information from these areas has been incorporated into the following discussions.

PERMAFROST

Introduction

Permafrost, or perennially frozen ground, underlies about 20% of the Earth's land surface and is widespread in North America, Eurasia and Antarctica. About 50% of the land mass of Canada (Fig. 11-1) and the Soviet Union is underlain by permafrost, as is most of Alaska and Greenland, and it occurs at high elevations in the mountainous regions of British Columbia and other parts of the world. At present, only a small percentage of the Earth's mineral resources are obtained from permafrost areas and, therefore, the near future will likely see much more exploration in these regions.

By comparison with geochemical techniques in the tropical and temperate environments, relatively little is known about the applicability, limitations and pitfalls of these methods in permafrost areas, particularly in North America, as only a few fundamental studies have been made. In the zone of continuous permafrost in Canada, for example, only studies in the Coppermine River area (Hornbrook and Allan, 1970; Allan and Hornbrook 1971; Allan et al., 1972) and in the Kaminak Lake area (Hornbrook and Jonasson, 1971; Shilts, 1973), both in the Northwest Territories, are sufficiently detailed in scope to be considered orientation studies, although numerous geochemical surveys, primarily by mining companies, are known to have been carried out in northern Canada, including the Arctic Archipelago. Lake sediment geochemistry east of Great Bear Lake (partly in the zone of continuous permafrost) has been studied by Allan et al. (1973). In

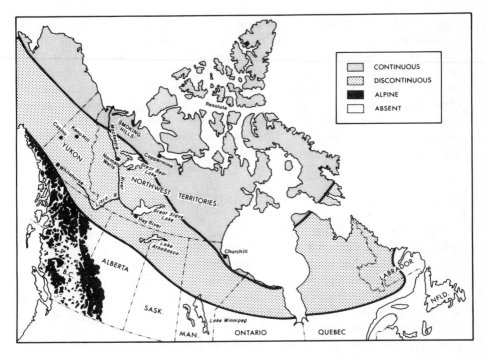

Fig. 11-1. Permafrost areas of Canada and Alaska. In British Columbia, the alpine permafrost areas indicated are at high altitudes (approximately 4800 ft at the Yukon border, and 7200 ft in the south). Not shown are patches of permafrost, for example in Saskatchewan, which are south of the permafrost limit, and mainly restricted to the drier portions of peatlands and peat bogs. Based mainly on R. J. E. Brown (1967).

the discontinuous permafrost zone, more studies have been completed, such as those in the Keno Hill (e.g., Boyle, 1958; 1965), Casino (Archer and Main, 1971) and Whitehorse (Smith, 1971) areas of the Yukon. A few studies pertinent to exploration geochemistry have been reported from Alaska (e.g., Sainsbury, 1957; Sainsbury et al., 1968; Burand, 1968; Brosge and Reiser, 1972), and from Antarctica (Claridge, 1965). Very fundamental and important work has been done in the Soviet Union, and those papers available in English which are of particular interest include the studies of Shvartsev (1965), Ivanov (1966) and Pitulko (1968). Shvartsev (1972) summarized the Soviet experiences and techniques in a most significant paper (also in English). Other studies from Canada and Scandinavia (e.g., Kvalheim, 1967) have been conducted in permafrost regions, but more attention has been paid to the problems resulting from glaciation than to those from permafrost.

Before proceeding further, it is necessary to describe a few of the more fundamental aspects of permafrost in order to appreciate its relationships to exploration geochemistry. The most modern and relevant discussion on permafrost, at least in Canada, is by R. J. E. Brown (1970), who also (1967) prepared a permafrost map of Canada. Most of what follows in the next section is based on these two works. Williams (1970) has described the permafrost regions of Alaska with particular emphasis on groundwater resources. Similarly, I. C. Brown (1967) has discussed groundwater occurrences, flow systems and chemistry in permafrost regions of Canada. Both studies have fundamental information applicable to exploration geochemistry.

Nature, Distribution And Occurrence Of Permafrost

Permafrost is defined exclusively on the basis of temperature. It refers to the condition of earth materials, such as soil, clay, silt, gravel, peat and rock, when their temperature remains continuously below 32°F. Permafrost includes ground which freezes in one winter, and remains frozen through the following summer and into the next winter. This one-year period is the minimum time limit for the duration of permafrost, and it may result in permafrost only a few inches thick. On the other hand, permafrost may be thousands of years old and hundreds of feet thick. It is important to realize that nowhere in the definition or in the descriptions of permafrost which follow, is there a requirement that it contain ice — the only factor of importance is the thermal condition of the earth materials. Ice is certainly an important component of permafrost in many areas, but it is an accessory, rather than a requisite component. Bedrock, for example, often contains no ice and may still, by definition, be permafrost. (This is sometimes referred to as "dry permafrost").

Even a small negative heat imbalance each year results in a thin layer being added annually to the permafrost. This process does not continue indefinitely, but a quasi-equilibrium is reached whereby the downward penetration of frozen ground is balanced by heat from the unfrozen ground below. Permafrost is not "permanently" frozen, as changes in climate and topography can cause it to thaw and disappear.

The permafrost region of the Earth is divided into two zones, *continuous* and *discontinuous*. These are illustrated in Fig. 11-2, in the form of profiles. The division between these zones was chosen arbitrarily by Soviet permafrost investigators as the − 5°C (23°F) isotherm of mean annual ground temperature, measured just below the zone of seasonal variation (depth of zero annual amplitude). This criterion has been adopted in North America. In the discontinuous zone there are areas ("islands") and layers of unfrozen ground (Figs. 11-2, 11-3) between the frozen layers. In the southern fringe of this zone, permafrost occurs in

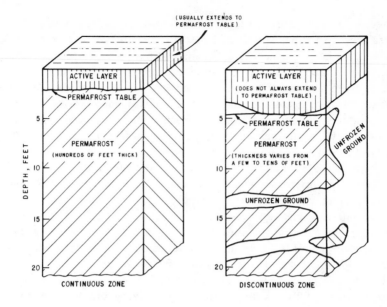

Fig. 11-2. Typical profiles in permafrost regions. From R. J. E. Brown (1970).

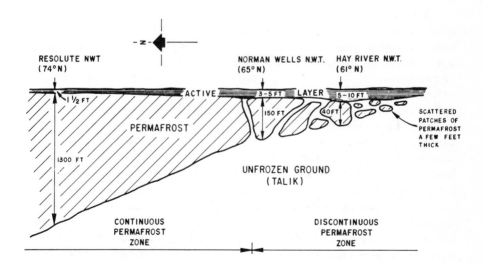

Fig. 11-3. Typical vertical distribution and thickness of permafrost in Canada. Locations at Resolute, Norman Wells and Hay River are shown on Fig. 11-1. From R. J. E. Brown (1970).

scattered patches, more or less like "Swiss cheese", a few square feet to several acres in area, and is confined to certain types of terrain, mainly peatlands. Other occurrences in the discontinuous zone are associated with either north-facing slopes, or forested stream banks where increased shading from summer thawing, and reduced snow cover, enhance permafrost development. On the other hand, a heavy snow in the autumn and early winter will inhibit winter frost penetration. In general, vegetation, because of its insulating properties, shields permafrost from the thawing effects of summer air temperatures. However, like snow, vegetation effects are quite variable, and the details are beyond the scope of this summary. Northward, permafrost becomes increasingly widespread and is associated with a greater variety of terrain types.

Permafrost varies in thickness from a few inches or feet at the southern limit, to about 200 ft at the boundary of the continuous zone (Fig. 11-3). In the northern part of the continuous zone, permafrost reaches a thickness of over 2000 ft. Unfrozen ground is found below all permafrost regardless of its thickness. The *active layer* is that area above the permafrost table (top of the permafrost), which is subject to annual freezing and thawing (Figs. 11-2 to 11-4). It generally varies in thickness from about $1^{1}/_{2}$ to 3 ft where it is insulated by mosses and vegetation. In exposed rock, the active layer will vary from about 15 ft, for example at Churchill, Manitoba, to about 7 ft on the islands of the Arctic Archipelago. The active zone is an important sampling medium in exploration geochemistry.

Variations in permafrost occurrences are governed predominantly by local variations in microclimate and such features of terrain as relief, vegetation, drainage, snow cover and soil type. However, the existence of large bodies of water also greatly influence the distribution and thermal regime of permafrost. Possibly of most significance in geochemical exploration in permafrost areas, at least in North America, is the abundance of large and medium-sized lakes (deeper than perhaps 8 ft). An unfrozen zone exists beneath bodies of water which do not freeze to the bottom (Fig. 11-5). The extent of this thawed zone (also called a talik) will vary with a number of factors which include the area and depth of water, water temperature, thickness of winter ice, snow cover and composition of the bottom sediments. The most important factor is the temperature of the water at the bottom of a lake. In the example shown (Fig. 11-5), even with a 4 feet-thick ice cover, about 6 ft of water in the 10 ft deep lake will remain unfrozen, and have a bottom temperature of 35°F. The area and depth of the thaw zone also depends on the geothermal gradient under a lake, which is quite variable. Nevertheless, as a generalization, 100 ft (vertical) of thaw will be found for each degree the bottom water temperature exceeds 32°F; in this case 300 feet of thawed permafrost, with the shape as illustrated in Fig. 11-5, could be expected. Not only does this thawed zone permit, by various mechanisms not completely understood, the mobility of metals in the direction of the lake water, but it is possible to have a thawed zone which extends through the entire permafrost layer, so that metal-containing fluids (and metal vapors) can migrate to the lake from deeply buried deposits. Similar thawed zones are found beneath rivers, and migrating fluids can discharge their metal contents into stream sediments, or to the surface as springs, in a manner analogous to groundwater discharge in more temperate climates. Larger taliks may be interconnected thus permitting even greater dispersion of elements in groundwaters. Obviously, permafrost is by no means "continuous", even in the continuous zone.

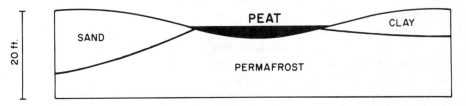

Fig. 11-4. Diagram illustrating differences in depth to permafrost in the continuous zone, and in those parts of the discontinuous zone underlain by permafrost, resulting from differences in types of surface materials. Because peat beds are good insulators, particularly when they are dry, permafrost is closest to the surface beneath them, and the active layer in them is usually thin. The active layer is thickest in well-drained sandy soil, where the ground freezes and thaws more deeply than in clay.

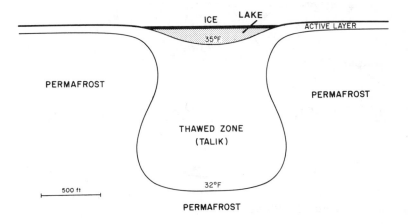

Fig. 11-5. Cross section of an ice-covered lake and thawed zone in the permafrost region in northern Canada. The ice is 4 ft thick; the active layer is 2 ft thick; and the lake is 10 ft deep (scale distorted in diagram). Bottom water temperature in the lake is 35°F.

Lakes on the Arctic Islands of Canada commonly range from 2 to 12 ft in depth, and have bottom water temperatures of from 32 to 35°F. Therefore, thawed zones as deep as several hundred feet are possible in the deeper lakes which do not freeze to the bottom. In more southerly parts of the permafrost region, bottom waters of 39°F have been reported, and in these cases, the thawed zone can extend through the permafrost. Clearly, lake waters, and bottom sediments in which adsorption is likely, are potentially good geochemical sampling media. In these southern areas, many connected areas of unfrozen ground (Figs. 11-2, 11-3) are found, and these enable migrating fluids to discharge at or near the surface in various forms, such as in springs, rivers or lakes, although it may be more difficult to determine the source of these waters than is the case with the more familiar hydrologic regimes of temperate climates.

Materials in the active layer, as mentioned above, are also favorable sampling media. This is a result of the fact that during the freeze-thaw cycle, material and moisture in the active zone will move, in a more or less specific manner, as illustrated in Fig. 11-6, and *frost boils* will result by a process called cryoturbation. These features are also known as mud boils (Shilts, 1973), nonsorted circles (Washburn, 1969), clay boils, tundra circles, and medallions or medallion patches in the Soviet literature. Frost boils, which are sorting phenomena restricted to the active layer, are probably formed because of the differences in thermal conductivity of

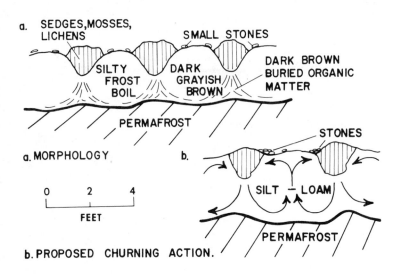

Fig. 11-6. Idealized cross section of silty frost boils developed on Arctic Brown soil (from Hornbrook and Allan, 1970). An equivalent diagram depicting the mechanism for the formation of mud boils on a till drumlin is illustrated by Shilts (1973, Fig. 3).

materials of various sizes. This suggestion is consistent with the observation of Shilts (1973) that frost (mud) boils develop on poorly-sorted, silt or clay-rich deposits such as till or fine-grained marine sediments, as opposed to well-sorted eskers. They are characterized by a segregation of clays, silts and other fine-grained materials in the center of the boil, whereas coarser material (stones) are pushed to the side and are found under the vegetation-covered rims, as is illustrated in Fig. 11-6. Thus sampling from the center of the boil will provide sufficient fine-grained (e.g., <80 mesh) material for analysis. Where frost boil formation is an active process, normal soil horizons are quickly disrupted and destroyed.

Hornbrook and Allan (1970) studied frost boils in detail in the Coppermine River region and found that the silt material in the center of the boil was a good sampling medium thus confirming the theory that the "churning" action should bring to the surface a homogenized sample of mineralization from within the active zone, or from the top of the permafrost, as well as any mineralized basal till float of glacial origin. Similar results were obtained by Shilts (1973) on tills, but not on eskers in the Kaminak Lake area. Others have found that cold extractable techniques (e.g., the Bloom Test (THM); Appendix B) will yield positive results on silty material from frost boils, indicating not only that frost boils are good sampling media, but that chemical weathering processes must be operative, at least to some degree, to produce loosely-held metal ions. The small amount of geochemical information available on frost boils appears to be limited to permafrost areas in till and soil, but there is no reason to suspect the process will not be operative, at least to a modest degree, in physically weathered, poorly sorted rock in non-glaciated areas as well. The factor which most limits the geochemical usefulness of frost boils is the thickness of the active layer (depth of summer thaw), which is rarely more than 3 ft in areas covered with glacial deposits or soil. Recent investigations by Pitulko (1968) in the Soviet Union have shown that the trace element (Cu, Zn, Ni, Pb, Nb and Y) content of frost boils is very close to that of the underlying permafrost. In essence, frost boils replace the classical "B" soil zone sampling methods of areas with residual podzolic soils. Frost boils and solifluction are the two characteristic examples of clastic dispersion in permafrost areas.

Weathering And Mobility Of Elements In Permafrost

The application of geochemical methods in permafrost terrain has, until recently, been largely neglected because of the assumptions that chemical weathering and leaching were greatly inhibited, and that permafrost would act as an impermeable barrier to water movement. However, physical and chemical dispersions in permafrost, particularly within the discontinuous zone and in the active layer, are now well documented. Mention was made above that cold-extractable methods are applicable to the fine fraction of frost boils and that this suggested that chemical processes were operative. Other evidence of chemical weathering in permafrost regions include:

1. Hydrogeochemical anomalies in areas of mineral deposits: Anomalies from metal deposits have been studied in stream sediments and waters from the Keno Hill area (Boyle 1958, 1965), and more recently, Boyle et al. (1971) have described the application of hydrogeochemical methods to the entire Canadian Shield, most of which is underlain by permafrost. The lake water study by Allan et al. (1973a), and the use of snow as a sampling medium (Jonasson and Allan, 1973), are also examples of the successful use of hydrogeochemical methods in permafrost areas. These studies showed that water from springs, rivers, and lakes, and groundwater and snow, are all potentially valuable media for exploration purposes. The elements present in solution can only have been released initially by chemical weathering.

   Allan and Hornbrook (1971) observed that only water from lakes at least 1000 ft in length should be sampled. This can probably be explained by the fact that these lakes are sufficiently deep (Fig. 11-5) for a large thawed zone to exist beneath them, and for copper-bearing groundwater or springs to discharge into them.

2. Iron and manganese oxide precipitates: These precipitates are known to occur in lakes, rivers and swamps from many localities in the northern latitudes such as Canada, Scandinavia, Alaska and the Soviet Union. The iron and manganese are leached from bedrock and subsequently are precipitated as a result of local environmental conditions (Eh-pH). Their presence indicates that chemical weathering does indeed take place.

3. Depletion of trace elements in near-surface soil layers: This condition has been reported by several writers, particularly when the soils are acid, and is interpreted as evidence of chemical weathering. When soil surveys are conducted in such areas, the possibility of metals being leached from clastic sulfides at the top of the soil should be considered, and the problem overcome by taking samples from as great a depth as possible. Leaching is likely to be strong in some cases, owing to the low pH of decaying organic materials. Alternatively, frost boils should be sampled as they do not suffer depletion to any appreciable extent because of the continual supply of new material from below.

4. Evidence of oxidation and element dispersion in frozen rock: Oxidation in perennially frozen rock may seem difficult to imagine, based

on our usual concept of chemical weathering (Chapter 3), yet there is very convincing evidence that just such a process does occur. Shvartsev (1972) suggested that the mechanism of freezing and thawing in the active layer promotes chemical weathering by causing a considerable amount of moisture to be drawn up towards the freezing front (called a "piston effect" or "freezing-out effect"). This solution movement is said to be the principal agent of weathering and removal of chemical elements in the top layers of the permafrost. The Soviet geochemists believe that a zone of oxidation is produced in the area of solution movement which can reach to depths of 150 – 200 meters.

Jonasson and Allan (1973) concur with Shvartsev (1972) and other Russian geochemists who believe that metal ions can be moved through permafrost to the surface. The mechanism they envisage is presented in Fig. 11-7. Once metals are at or near the surface, they may be dispersed by both groundwater and surface water. The elements may reach streams and subsequently lakes, where they may be adsorbed by sediments. This is one of the reasons why Allan et al. (1973a) were successful in using the fine inorganic sediments (less than 250 mesh fraction) of lakes in northern Canada. Of course, where sulfide deposits occur in the immediate vicinity of a lake, sulfides may move into the lake sediments in particulate form by physical dispersion. The model in Fig. 11-7 also explains the reason for the often-observed spring and fall hydrogeochemical anomalies. Such anomalies are related to a "flushing" of metals and certain gases, such as radon, after an accumulation of several months, in permafrost and snow covered areas. The use of snow as a sampling medium in permafrost areas is discussed later in more detail.

Ivanov (1966) reported his results of a study of copper-nickel (pyrrhotite-chalcopyrite) and cassiterite-sulfide deposits within the Arctic Circle in the U.S.S.R., and concluded that chemical weathering does take place in frozen rock. Other types of sulfide deposits are either not deeply oxidized, or show no evidence of oxidation at all. The oxidized sulfide-bearing rocks typically contain abundant readily soluble sulfates of iron (e.g., melanterite, $FeSO_4 \cdot 7H_2O$; fibroferrite, $FeSO_4(OH) \cdot 5H_2O$) which are, somewhat surprisingly, stable mineral phases in the permafrost environment. Ivanov (1966) reported that at the Ege Khaya deposit the oxidized zone is approximately 100 m deep, and that more than 50% of this zone is composed of the readily soluble iron sulfates. The mineral content of the oxidation zone varies with depth, and in descending order, the main sub-zones are: (1) limonite (to 5 m), (2) fibroferrite (to 30 m) and melanterite (to 100 m). However, galena, stibnite and other sulfides in close association with the nickel-copper and cassiterite-sulfide minerals are not oxidized and Ivanov (1966) explains this on the basis of the solubility of the various metal sulfates. The sulfates of zinc, copper and iron are highly soluble,

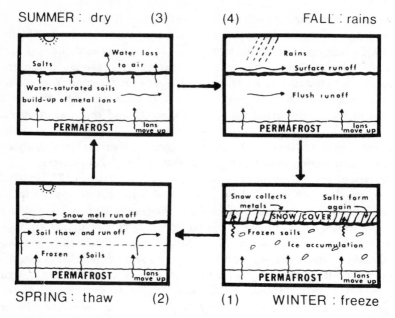

Fig. 11-7. Explanation of metal migration in permafrost. The basic process by which ions move upwards is probably ionic diffusion in solution. There is sufficient moisture in permafrost to allow capillary type migration to take place. (1) In winter the advance of the freeze front down from the surface and up from the permafrost causes metal cations (and anions) to concentrate in the frozen soil and subsequently to be moved to the surface where carbonate and sulfate crusts may form. The metals will also be moved into overlying snow. Thus there is a continuous accumulation over the winter months in the upper soil layers. (2) In spring the thaw flushes out the ions in the early runoff. As the thaw occurs only from the surface down, the process may take several weeks to leach the frozen soils. (3) In summer surface runoff has ceased, but subsurface migration may continue and carry dissolved material to the surface, where an ion build-up occurs again especially if the summer drought is long. (4) Fall rains complete the cycle, resulting in a flush runoff which may be as intense as the spring thaw under certain conditions. From Jonasson and Allan (1973).

whereas those of lead, antimony, arsenic and molybdenum are only slightly soluble. Support for this observation can be found in the report of B. H. Brown (1970) who noted that in the Alaska Range no gossans are found over galena and tetrahedrite veins, and concluded that in a sub-arctic environment, galena is relatively stable. (Others, however, claim to have seen galena oxidized to anglesite in sub-arctic regions in Alaska). The soluble sulfates which are produced during weathering in the permafrost zone can form hydromorphic halos which are detectable in waters of various types (springs, lakes, streams), as well as in the sediments associated with them, and in frost boils. Because of the definite tendency for the released elements to reach the active layer (also lakes, etc.) the possibility certainly exists of discovering deeply buried deposits in permafrost.

Both Soviet and Canadian geochemists have shown that the geochemical anomalies are greater, and the halos broader, in the first half of
the summer, because flushing and leaching reduces the contrast of the
anomalies by the later half of the summer (Fig. 11-7), particularly if the
active layer lies on a well-drained slope. However, there seems to be a lack
of agreement between Ivanov (1966) and Pitulko (1968) on the details of
relative mobility of the various ions in the permafrost regions. More work
is necessary to determine whether, in fact, a generalized order of mobility
can be obtained. In Fig. 11-8 trace element distribution curves illustrating
the results of natural leaching are presented in an idealized manner for Ni,
Pb, Cu, Zn, Nb and Y in the active zone overlying a mineral deposit in a
dolomite (modified from Pitulko, 1968). Three types of distributions have
been recognized: A. leached dispersion halo in an active layer undisturbed
by frost boil formation or other mechanisms; B. the leach pattern in frost
boils, and; C. the pattern obtained in active zone material moved by solifluction. Halos which have moved by solifluction characteristically have
lower amounts of the elements by comparison with the other two types, and
they are essentially thin, surface phenomena.

Oxidized sulfide deposits in permafrost have been reported from time
to time by various authors, but most have been attributed to an origin in
pre-glacial times. However, in addition to the Soviet interpretation of the
mode of formation of some of their gossans, and the evidence for secondary
halos given above, there is other confirming evidence that oxidation can

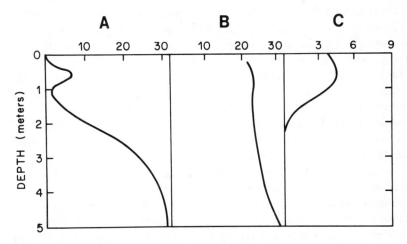

Fig. 11-8. Idealized trace element distribution illustrating three types of dispersion
halos related to mineralized sulfide deposits in permafrost areas. A. Undisturbed active
zone. B. Frost boil. C. Halo displaced by solifluction. Scale of concentration is
arbitrary. Modified from Pitulko (1968).

occur in permafrost regions. For example, in Antarctica Claridge (1965) noted crusts of iron stains and sulfides altered to various secondary minerals. Gossans of considerable size are reported from the northwest coast of Greenland, which has only been deglaciated in recent geological time, and where a pre-glacial goassan is not likely to have escaped erosion. Actually, this is not surprising in view of the studies by Washburn (1969) who reported that chemical weathering is locally important on the north-eastern coast of Greenland as evidenced by his observations of oxidation, solution and deposition of calcium carbonate, development of Arctic Brown soil, and other phenomena. Tedrow (1970) also described the formation of soils in northwest Greenland. He observed that secondary carbonates form in well-drained sites and, in general, he believed that a "desert-like" type of soil formation can occur in this dry, extremely cold environment. Salt efflorescences (mainly composed of Na, Ca, Mg and Cl) on certain soils were noted during the dry periods. Tedrow (1970) also presented an interesting short review of soils in the high Arctic, in which he summarized information not readily available in most libraries. On the northern coast of Canada, what is interpreted as oxidation is taking place at the Smoking Hills (Fig. 11-1). Rocks at this locality are black shales with bitumen, pyrite, and/or marcasite which intermittently "smoke" (oxidize), emitting heat, smoke and sulfurous fumes (Yorath et al., 1969). This example admittedly may be merely a surface phenomenon, but it does confirm that oxidation can take place under extreme climatic conditions.

Although more research is certainly needed on the problem of oxidation at sub-zero temperatures, it appears to be a workable assumption for geochemists interested in exploration in permafrost areas. The formation of oxidation zones at these temperatures is controlled by at least five factors (Ivanov, 1966): (1) depth of annual temperature fluctuations; (2) free energy and heats of reaction; (3) the absence or low content of free water as a liquid phase; (4) alteration of volume in the reactions; and (5) solubility of the reaction products in water. Oxidation of ore bodies in areas of discontinuous permafrost, for example at Keno Hill, Yukon (Boyle, 1958), is especially easy because there are sufficient "windows" in the permafrost to allow oxidizing waters to penetrate veins and to issue at lower elevations with their dissolved heavy metals.

It has been mentioned on numerous occasions that mercury can pene-trate permafrost (e.g., Jonasson, 1970; Hornbrook and Jonasson, 1971) to produce measurable dispersion halos which can be detected in water, till and soil at the surface in various parts of the Canadian Shield. Mercury halos on the surface near a tetrahedrite – sphalerite deposit in Greenland are also reported by Lehnert-Thiel and Vohryzka (1968). Vapor move-ment, including that of water vapor, through permafrost is a distinct possi-bility and possibly other types of metals, either as vapors or by ionic diffusion, may be transported but this has yet to be established. Shvartsev (1972) quotes two Soviet sources who suggest that chloride and sulfate anions, and calcium, lithium, sodium and potassium cations, can migrate through permafrost.

Geochemical Methods In Permafrost Terrain

When conducting geochemical surveys in permafrost areas it is necessary to take into account numerous special factors such as the fact that soils and soil profiles as known in the more temperate climates are either absent or poorly developed. In the northern parts of the Canadian Shield, permafrost inhibits free circulation of groundwater, especially when it reaches a thickness of 500 ft or more, except in thawed zones under lakes and rivers (see above). Near surface solution, leaching and transport of elements from metalliferous deposits may, however, occur in the active zone during the summer months. This process is likely to be more noticeable on the southern slopes of mountains. Archer and Main (1971) noted that at the Casino, Yukon porphyry copper deposit, permafrost did not appear to have an appreciable effect on the soil response for the metals they determined. However, the permafrost did have the effect of directing copper-rich spring waters to unfrozen "windows" on south-sloping drainage areas. Sediments associated with such springs have higher metal contents than sediments on the north-facing slopes.

Vegetation surveys have not been employed to any great extent in permafrost areas owing to numerous complications. Among these are the fact that plant roots are limited to the depth of the active layer and consequently do not penetrate very deep; the roots of many plants in northern Canada grow laterally. Further, the variety of vegetation, including mosses and lichens, is very limited. Also, the slow rate of decay in the cold climates creates a layer of humus which builds up to a considerable depth. Such organic material concentrates certain elements (e.g., Hg, Zn, Cu) during decay, causing erratic background values and non-significant anomalies in living vegetation. Boyle and Cragg (1957) mention such an effect at Keno Hill. Allan and Hornbrook (1971) analyzed the ash of lupines, heather and three types of lichens in the Coppermine region. Although the ash of these materials did show copper concentrations, which correlated with anomalous areas, the results were erratic, and not as reliable as those using lake water and stream sediment samples. However, the small amount of work so far reported on vegetation surveys in permafrost regions does not justify eliminating this technique from further consideration. It may well be very useful in some forested areas, such as in parts of the Yukon.

Clearly, permafrost regions are characterized by a number of special features which result in a diversity of geochemical halos. Based on experiences in the Soviet Union, Shvartsev (1972) has been able to distinguish five generalized situations:

1. Ore bodies with a thick oxidation zone outcropping on the surface: In this case, a thick active layer can be observed during the summer within the oxidation zone. The geochemical anomalies are "extremely sharp in character", cover a considerable area, and may be detected by analyzing water, soils, bottom sediments, vegetation, or overlying sediments.

2. Ore bodies completely concealed in permafrost, but at shallow depths: Such deposits may be delineated fairly distinctly by the use of data from springs in the active layer, particularly in the first half of the summer, or by analyzing soils or loose materials, taking into account the possibility of leaching of the active layer.

3. Ore bodies partially or completely below the permafrost zone: These deposits can be found by locating discharge points of the sub-permafrost waters into taliks. During the winter, "ice bodies accompanying these discharge points at the surface are good criteria in mapping the location of the discharge points" (Shvartsev, 1972, p. 383).

4. Ore bodies concealed under a peat layer in swampy areas: Halos are found by study of the water, peat and any associated materials.

5. Ore bodies overlain by unconsolidated sediments of appreciable thickness (up to 100 m): Such conditions are found in glacial materials, and in river valleys filled with alluvium. Thaw zones are commonly developed in these cases, and significant groundwater movement would be expected. Geochemical dispersion patterns will be revealed by the analysis of water and loose sediment (basal tills) at the boundary with the bedrock. Apparently the effects of glaciation do not add any complicating factors to geochemical exploration in permafrost areas of this type.

In conclusion it can be said that, in general, as was pointed out by Boyle in 1958, the occurrence of permafrost appears to present no insurmountable problems for exploration geochemistry except in obtaining soil samples and possibly in using vegetation. To overcome these difficulties, frost boils or hydrogeochemical methods (including lake sediments) should be emphasized, or it may be necessary to drill close to bedrock. The most effective use of the standard geochemical techniques requires an understanding of the origin and nature of permafrost, including the movement of water within and associated with it. In addition, it is necessary to recognize the fact that many of the waters in northern areas will have a higher organic matter content than those in other parts of the world, with a resultant interaction of the heavy metals and organic substances. Also, these waters commonly have relatively low values of pH and Eh, which produce a diversity of halos. As more case histories are published, understanding of the many complexities and ambiguous aspects should become clearer.

Contrary to popular belief, mining operations in the zone of continuous permafrost, especially if it is thick, have some distinct advantages which tend to offset the negative aspects of remoteness and extremely cold winter temperatures. Permafrost in these areas affords excellent ground support for construction, provided the accepted methods are followed. Also, mining operations which would normally be plagued by water flooding, as in the case of a vuggy, permeable host rock below the water table, may not be of concern in the continuous permafrost area because the ground, and its contained water, are frozen.

Snow

Jonasson and Allan (1973) have demonstrated that snow may be used successfully as a sampling medium in exploration geochemistry. Although this is technically a hydrogeochemical method, it is conveniently discussed with permafrost. The mechanism by which metals are moved through the permafrost and accumulated in snow is illustrated in Fig. 11-7. Such metals as Hg, Zn, Cu, Pb, Ni, Cd and Mn have been detected in snow samples over buried mineralization (atmospheric contamination was eliminated as a possible source). Jonasson and Allan (1973) concluded that the metals which constitute the bulk of the sulfide ore were the most useful indicators of the mineralization in their test sites. Although the dispersion halos were quite broad (several hundred feet) the technique appears to be most useful for detail geochemical surveys.

The best anomalies were obtained from samples of clean snow taken from 2 to 6 inches above the soil-ground ice layers. Studies of the metal content variations with depth of snow revealed a distinct increasing gradient of concentration from the snow surface downward, of up to 50-fold. Typical trace element contents found in snow, including snow over mineralization, are (all ppb): Hg, 0.01 – 1.69; Cu, 2.8 – 161; Zn, 1 – 53; Pb, 2 – 15; Ni, 3.6 – 40.8; and Cd, 0.4 – 7.2. In those areas which are not underlain by permafrost, snow sampling is also useful provided there is continuous snow cover for at least two to three months, without freeze-thaw cycles. This time interval is necessary to permit elements which move in vapor form, such as mercury or organo-metallic compounds generated in the soils, to infiltrate the snow and develop a measurable anomaly.

As a result of the spring melting, elements accumulated during the winter will be flushed out rapidly, and anomalies of great magnitude can result (Fig. 11-7). Jonasson and Allan (1973, p. 173) reported that the effects of precipitation and flush runoff can increase the concentration of mercury in the waters of lakes and streams by factors up to 500 times the concentrations measured during dry periods, that is, from about 0.01 ppb to about 5 ppb (this latter value is very high). Similar observations have been made by others with respect to radon anomalies. Obviously, caution is warranted in interpretation of hydrogeochemical anomalies in permafrost and snow covered areas during the period of the spring thaws, as false anomalies are very common.

WATER

Introduction

Hydrogeochemical methods are extremely important in Canada in view of the large areas of this country covered by water. That this fact has been realized is evidenced by the great interest in these methods since about 1969 by the Geological Survey of Canada and many exploration companies, although groundwater and spring water have been used in the Yukon and New Brunswick since the middle 1950's (e.g., Boyle et al., 1955). In this section the use of groundwater, river water, lake water, and lake sediments will be discussed from the point of view of what has been accomplished in the past, the value of these sampling media, and how they are best employed in the generally cold temperate to arctic Canadian climate. The general principles relating to the use of water sampling for exploration purposes have been discussed in chapters 1 and 3.

Groundwater

The most complete report on groundwater in Canada has been prepared by I. C. Brown (1967) primarily for hydrologic and water resources purposes. In this report he described Canada's six major hydrologic regions: (1) the Northern Region which includes, in general, all of Canada north of the southern limit of the discontinuous permafrost (Fig. 11-1); (2) the Cordilleran Region (most of British Columbia); (3) the Interior Plains Region (the southern parts of Alberta, Saskatchewan, and Manitoba); (4) the Canadian Shield Region (parts of Ontario and Quebec); (5) the St. Lawrence Lowland Region (the southernmost part of Ontario and along the St. Lawrence River); and (6) the Appalachian Region (the Maritime Provinces and Newfoundland). Each region tends to have distinctive rock types, climate, and hydrologic characteristics. More recent data of interest in exploration geochemistry may be found scattered in reports from various sources, such as the study by Rozkowski (1969) on the groundwater and surface water in one area of southern Saskatchewan, and another by Parsons (1970) on groundwater movement in a glacial complex (Clay Belt) in the Cochrane area, Ontario. Boyle et al. (1971) have discussed the use of groundwater obtained from drill holes and wells (Fig. 3-25) in some mineralized areas of the Canadian Shield.

In general, groundwater has not been used extensively for geochemical purposes in Canada. The reason for this is twofold. Firstly, groundwater data and wells are only abundant in the heavily populated St. Lawrence Lowlands Region of southern Ontario and Quebec, and the Interior Plains Region, where groundwater is used extensively for municipal and agricultural purposes. Although many oil wells have been drilled in the Interior Plains, they are usually deep, and no effort has been made in the past to collect waters suitable for base metal exploration (although tantalizing occurrences of sphalerite, galena and fluorite have been reported from time to time in well cuttings and cores). Saline waters associated with oil-producing formations have, however, been used for hydrocarbon exploration (Chapter 13). Secondly, very little information of any type on groundwater is available from the other regions, especially the Northern Region which comprises over one-half of Canada.

Groundwater is extremely important, nevertheless, in base metal (including uranium) exploration. The main functions of groundwater are: (1) it supplies streams and lakes with a large percentage of their water (e.g., Figs. 1-3, 3-22); (2) it appears as springs and spring precipitates at the surface (e.g., Fig. 1-4); and (3) it can be used to sample terrain covered by glacial deposits.

With regard to the Northern Region, permafrost is an important factor in the distribution and movement of groundwater, especially where the permafrost is thick. Hydrologists recognize three types of groundwater in permafrost areas (I. C. Brown, 1967): (1) *suprapermafrost water,* melted from the upper permafrost layers in the summer; (2) *intrapermafrost water,* within thawed zones (taliks) of permanently frozen ground, and commonly found associated with river channels, alluvium near rivers, and in glacio-fluvial material covering the floor of wide river valleys such as portions of the Pelly River in the Yukon; and (3) *subpermafrost water* found in the thawed zones beneath permanently frozen ground. In Alaska subpermafrost groundwater is used for municipal water supplies, such as in the Fairbanks area, where many wells encounter such water at depths of from 40 to 200 ft. There is little information on subpermafrost water in Canada, but it certainly exists, especially in the discontinuous permafrost areas (Figs. 11-1 to 11-3). This water, which may carry metals for considerable distances, probably supplies water to lakes through taliks and springs. I. C. Brown (1967, p. 182) observed that the mineral content of subpermafrost water is frequently high. The possibility of obtaining subpermafrost water in some permafrost areas is relatively poor because the permafrost frequently extends down to impervious bedrock.

Underground waters have been investigated thoroughly by Boyle et al. (1955, 1956), and Boyle (1965), in the Keno Hill-Galena Hill area of the Yukon. These studies have shown that groundwater can be used effectively in the search for metal deposits in discontinuous permafrost areas. At these particular localities, zinc values from 0.001 – 80 ppm and lead values from 0.002 – 1 ppm were recorded, the high values of both elements being anomalous. In the Cobalt area, Ontario, Boyle et al. (1969) found that groundwater and their precipitates (Table 3-2) in the vicinity of silver deposits carry higher than average amounts of the ore and gangue elements (e.g. Ag, Ni, Co, As, Sb, Cu, Pb). This information was a general guide for locating favorable areas for prospecting, but water analyses were not sufficiently specific to pinpoint individual deposits. In this part of the Canadian Shield region, the water is probably channeled by faults, fractures and joint systems because of the general impermeable nature of the rocks.

In New Brunswick manganese is moved by groundwater and may form bog manganese deposits upon reaching the surface, as described by Brown (1964) from a deposit near Moncton. High manganese, as well as many other trace element values, in the stream waters have been described from the Bathurst-Jacquet River district, New Brunswick, by Boyle et al. (1966), and these originate from groundwater seepages.

Other groundwater or related surveys have been reported from various parts of Canada, including hot springs in western Canada, but they are few in number and do not add materially to what has been said above.

River Water

River water has been used successfully for exploration purposes in New Brunswick and the Yukon at the same localities discussed above in connection with groundwater. The rivers in these locations obtain their trace metal contents from groundwater and springs (e.g., Figs. 1-3, 3-22). In British Columbia numerous hydrogeochemical surveys have been attempted, and although moderate success has been recorded in some cases, published reports are scarce. Mention was made earlier (Chapter 3) of the problems involved with the use of river water (e.g., fluctuations in the water flow, composition, and groundwater component), and these are certainly applicable to British Columbia. Further, in the limestone areas of the Rocky Mountains, the pH of river water will often be 7.5 or 8. Thus many elements of interest will not be soluble (Table 3-6). On the west coast, the high rainfall and steep topography make chemical weathering relatively less significant in comparison with physical weathering, thus transport of metals dissolved in river water may be negligible. Nevertheless, many river waters have been analyzed for molybdenum, mercury, copper and zinc, in the more favorable parts of the Province, particularly from smaller streams draining felsic rocks.

In the Canadian Shield only a limited amount of work has been done to assess the effectiveness of analyses of stream and river water in delineating mineralized zones (Boyle et al., 1971). One reason for this is that the streams are very poorly developed, often serving mainly as short interconnecting links between the innumerable small lakes. Further, they are often disrupted by glacial deposits, beaver dams and muskeg, which features make sampling and interpretation of the data difficult. As a result, stream and lake sediments generally give better, more easily interpreted indications of mineralization.

Boyle et al. (1971) mentioned that studies were in progress on the dispersion patterns of several elements (e.g., Cu, Pb, Zn, Fe, Ag, As, Sb, Co) in stream waters in the vicinity of gold, silver, nickel-cobalt and other deposits in the Canadian Shield (Yellowknife, Cobalt, northwestern Ontario). Their work showed that dispersion patterns in stream water do exist, but they are relatively restricted due to the strong adsorption by humic matter in the bottoms of nearly all streams, beaver ponds, and other organic-rich materials such as muskeg and organic silts. The trains seldom exceed 1000 ft in length and, in most cases, are measurable only for a few hundred feet. Such short dispersion trains are not favorable for reconnaissance purposes but they may be useful for detailed work, especially during the interpretation phase of a survey. Based on their work, which was not completed at the time, Boyle et al. (1971) suggested that sampling of water be done at or near the heads of main streams, and particularly at the heads of all tributary streams and creeks (which often start from groundwater seepages). All sites should be carefully scrutinized on aerial photographs, and all stream and spring waters in the vicinity, including the merest trickles, should be analyzed for the elements chosen as indicators of the deposits sought.

On the edge of the Canadian Shield, and particularly at the eastern edge of the Western Canada Sedimentary Basin between northeastern Alberta and Great Slave Lake, many rivers have been sampled for indications of lead-zinc mineralization similar to that at Pine Point. Reeder et al. (1972) have presented data for Zn, Pb, Cu and other metals from several of these rivers some of which have reasonably high values (e.g., the Firebag River had 10 ppb Zn at the time of sampling which is high for this region). This information, plus much more which is available from government agencies such as the Inland Waters Branch, certainly suggest that river water in the Canadian Shield and other parts of Canada have potential for exploration purposes which has yet to be effectively exploited. Except for the major rivers, very little sampling has been done in the winter through the ice, at which time the groundwater component would be most significant.

In the case of uranium, good results have been obtained with stream samples in several localities. As mentioned previously, uranium forms complexes with carbonate and sulfate ions, and the resulting complex is soluble under the alkaline pH conditions characteristic of many surface waters. Chamberlain (1964) was the first in Canada to report on the use of the uranium content of stream water as an indicator of mineralization. Based on work in the Bancroft area, he found that a regional plot of the hydrogeochemical data outlined the general configuration of the uranium mineralization, and that changes in background levels were spatially related to the character of the bedrock. Anomalously high uranium contents of up to 500 ppb in creek waters were associated with known ore bodies, but these values decreased to near-background levels within a few hundred feet because the uranium was extracted from solution due to the reducing action of decaying organic matter. Thus, the analysis of water samples offered only restricted guides to specific uranium targets in this area.

In a series of papers Dyck (1969, 1972) and Dyck et al. (1971) have expanded on the use of uranium and radon in surface waters (streams and lakes) from several areas in the Canadian Shield, particularly in the Beaverlodge area of northern Saskatchewan and the Bancroft area of Ontario (Fig. 11-9). Morse (1971) also described the results of his work in the Bancroft area, extending the well-known uranium and radon techniques to include the determination of radium. In most of the examples cited, soil samples and sediments were included with the stream water studies. The detailed results of these studies cannot feasibly be discussed here, but the indications are that surface water surveys for uranium and radon are quite helpful in uranium prospecting. In some situations the uranium content of the surface water proved better than the radon content for outlining favorable areas, and in others the reverse was true (and neither is infallible).

Dyck (1972) explained some of the reasons for difficulties and discrepancies in the use of both uranium and radon. For example, uranium in the Beaverlodge area is strongly bound by organic matter in the stream sediments, and the degree of mobility is dependent, at least in part, on the amount of organic matter in these sediments. The relative strength by which trace elements are bound to organic matter in the Beaverlodge area is Cu>Pb>U>Zn>Ni>Ra. The radon content of water is affected by another set of factors. These include the distance between the water sample

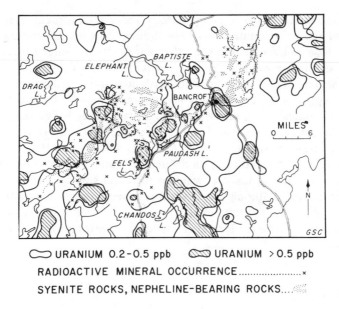

CD URANIUM 0.2-0.5 ppb   CD URANIUM > 0.5 ppb

RADIOACTIVE MINERAL OCCURRENCE ....×

SYENITE ROCKS, NEPHELINE-BEARING ROCKS....

Fig. 11-9. Geochemical map showing the uranium content of lake and stream waters, Bancroft area, Ontario. From Boyle et al. (1971).

which was collected and the sediment, temperature of the water, and to a lesser degree, flow, turbulence and other factors. Dyck (1972) recommends that when possible both uranium and radon should be determined (radon is determined on-site by means of a radon counter; Fig. 6-13). He concluded that uranium in lake water gives a better regional picture, whereas radon pinpoints the uranium sources more clearly. The selection of sample sites is particularly important in the case of radon. Sampling small lakes near inlets and outlets of streams entering and draining them gives best results (Boyle et al., 1971). Morse (1971) found that the radium content of stream and lake waters in the Bancroft area is too low to be an effective indicator in prospecting for uranium, and recommended radon or uranium. Radium is, however, effective in stream and lake sediments.

The results from radon surveys conducted in surface waters in Canada and elsewhere have been difficult to interpret on many occasions. One cause of false anomalies which occur seasonally during spring has been explained in Fig. 11-7. However, cases have been reported where radon has not been detected over known uranium mineralization. In a very carefully documented study Andrews and Wood (1972) explained that the release of radon is dependent on the extent to which rocks are fractured, and on the ability of groundwater to circulate through such rocks. Percolating groundwater, according to these authors, removes radon from fractured porous rocks by a transport, rather than a diffusion, mechanism.

Thus groundwater from various porous sedimentary rocks may acquire significant radon concentrations even though the uranium content of such rocks is close to average. On the other hand, stream water flowing over a low porosity or unfractured rock is unlikely to acquire significant radon concentrations, except when admixed with circulating groundwater. Therefore, for exploration purposes any use of radon must be made with caution because its content in groundwater and stream water is determined by the physical nature of the rock and aquifer to a much greater extent than by the uranium content of the formation. Andrews and Wood (1972) concluded that uranium in groundwater and surface water is the most promising technique for regional surveys, whereas for local surveys, radon in soil gas is most promising. Although their work is based on problems in the United Kingdom, it does suggest explanations for some of the difficulties experienced in Canada. However, specific confirmation remains to be presented.

Lake Water

The future emphasis on hydrogeochemical methods in Canada will be centered on lake water and sediments. Approximately 10% of Ontario, 15% of Manitoba and 2% of British Columbia are covered by lakes, as well as a significant percentage of the land surface of most of the other Provinces. The Northwest Territories and the Yukon contain an estimated 250 lakes exceeding 25 sq mi (65 sq km) in area and there are "hundreds of thousands of smaller ones within the tundra region of Canada alone" (Frey, 1963, p. 561). Black and Barksdale (1949) estimate that in some parts of the flat tundra of the Arctic Coastal Plain of northern Alaska 50 – 75% of the surface is covered with lakes, and it is reasonable to extrapolate this figure at least to the northern coast of Canada near the Alaskan border.

Because of the differences in climate, precipitation, topography, vegetation and biota, many different types of lakes are found in Canada and a complete discussion is not feasible here. In British Columbia alone for example, at least 12 limnological regions are recognized. Frey (1963) presented an excellent summary of the state of the science of limnology in North America as known at that time, much of which has application to exploration geochemistry. More recent information can be obtained from publications of the Canada Center for Inland Waters (Burlington, Ontario), journals connected with fisheries (e.g., *Journal of the Fisheries Research Board of Canada*), various publications in the fields of limnology, environmental science, water research and resources, as well as from studies by the Geological Survey of Canada.

The relationship between lakes and their environment, particularly temperature, precipitation, and the rock type on which they are found, is illustrated in Fig. 11-10 for the Provinces of Manitoba, Saskatchewan and Alberta. The limnological units in these Provinces can be conveniently discussed in terms of their total dissolved solids. The Northern Lakes are characterized by a low total dissolved solids content (less than 200 ppm and often less than 50 ppm), and are found in areas underlain by Precambrian rocks (mostly igneous and metamorphic). Many lakes in the Northwest Territories, including Great Bear Lake, and the eastern end of Great Slave Lake, are of this type also. The Northern Lakes are oligotrophic (see discussion of lakes in Chapter 3) even when small and shallow owing to the lack of nutrients and, as a result, they have a relatively small biota. A special type of oligotrophic lake, known as a *dystrophic* lake, is also common in this region. Dystrophic lakes are found in muskeg regions (see Fig. 11-14), and are characterized not only by a low total dissolved solids

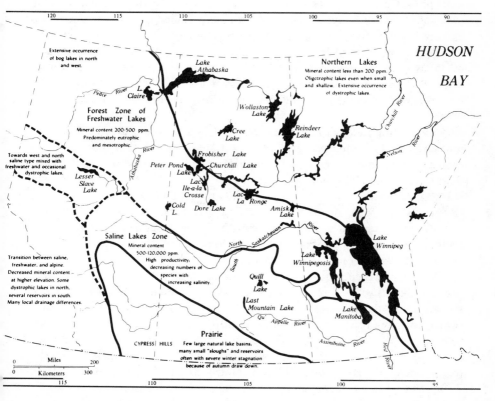

Fig. 11-10. Limnological regions of Manitoba, Saskatchewan and Alberta. From Frey (1963).

content, but also by the fact that they are acid and their water is brown (or yellow). Most oligotrophic lakes are alkaline and their water is clear. Waters in muskeg regions are discussed more fully below.

The lakes in the Forest Zone contain mineral contents of 200 – 500 ppm, or two to four times the average of the Northern Lakes on the Canadian Shield. Lakes in this zone overlie sedimentary deposits, and are in the parkland and the northern coniferous forest. Nutrients are abundant, and this results in a large and varied biota. In general, the characteristics of the larger lakes in this region are classed as eutrophic but there are exceptions, such as some deep ones which are oligotrophic. Mean depths of less than 18 m are typical, and thermal and oxygen stratification have been observed, although in the shallower examples wind may thoroughly mix the water and inhibit summer stratification. Several lakes in Saskatchewan and Manitoba straddle the boundary between the Northern and Forest zones. Lac la Ronge, for example, is oligotrophic at Hunter Bay which is deep, cold and on Precambrian rocks, whereas the main body of water is eutrophic. Similarly, Lake Winnipeg has oligotrophic characteristics on the east shore where water is received from the Canadian Shield, but it has eutrophic characteristics on the west shore where water comes from sedimentary rocks. One characteristic feature of all shallow (less than about 8 m) lakes in the Prairie Provinces is the phenomenon referred to as "winter kill". Because of the ice cover, all oxygen is used up, the lakes become reducing during the winter, and this causes fish and certain other organisms to die.

The Saline Lake Zone is characterized by a total dissolved solids content of 500 – 118,000 ppm, and a high sulfate content rather than the more typical bicarbonate or chloride. The high salinity of these waters is a product of the semi-arid climatological conditions and the chemistry (e.g., high sulfate), depends mainly on the lithology and groundwater flow conditions (Rozkowski, 1969). In the western part of Alberta, the lakes become less readily characterized as they merge with the more complex limnological regions of British Columbia. In the southern parts of Alberta and Saskatchewan (Prairie Zone), lakes are scarce.

With the exception of Great Slave Lake and a few of the other large northern lakes (e.g., Lake Athabasca and Great Bear Lake), very little is known about the lakes in the Northwest Territories and the Yukon. With regard to lakes in this region (and including Alaska and Greenland) Frey (1963, p. 559) states, "There can be few places where so little is known about so many lakes." However, studies which have been made in Alaska (e.g., Black and Barksdale, 1949; Livingstone et al., 1958; Boyd, 1959; Frey, 1963) and Greenland (e.g., Bocher, 1949; Hansen, 1967) can be extrapolated to northern Canada. These studies have established that the two primary factors influencing the chemical composition of river and lake water are (1) the amount of precipitation, and (2) the nature of the country rock. In the humid parts of southeastern Alaska, for example, the concentration of dissolved material in lake and river water is generally about 25 ppm, but in the Yukon basin where precipitation is much less, the concentration is higher, usually 100 – 200 ppm. These higher concentrations will be found in some waters of the Arctic drainage of Alaska and Canada. Superimposed upon the pattern of precipitation control, are the effects of the country rock. Thus in northern Alaska and northwestern Canada, where relatively soluble sedimentary rocks (e.g., limestone, gypsum) are exposed on the surface, the total dissolved solids content of surface water rises. However, in the eastern part of the Northwest Territories, where Precambrian igneous and metamorphic rocks are exposed, the effect of country rock on the composition of rivers and lakes is negligible. Clearly, the effects of climate, precipitation, type of country rock, topography, vegetation and biota cannot be overlooked in any discussion or explanation of the chemistry of lake and river water, whether it be for limnological or geochemical purposes. Because of variations in these parameters from one region to another, the basic information obtained for one area will not necessarily be applicable in another area.

The very cold arctic climate produces an ice cover about two meters thick in northern Alaska (Livingstone et al., 1958). This can greatly affect the composition of unfrozen water below the ice, if the lakes are shallow in relation to their ice cover. This is illustrated in Fig. 11-11 which shows, for example, how the chloride and magnesium contents of a shallow arctic lake will increase by the "freezing out" of salts into the water that remains under the ice. The increase in ice thickness is accompanied by an almost linear increase in the dissolved mineral content until a maximum is reached in May. Similar results have been obtained in Greenland (e.g., Hansen, 1967) and in arctic lakes in other parts of the world. By analogy, Canadian lakes will show similar variations and, therefore, the total dissolved solids and trace elements will vary with the time of sampling. In some cold lakes in arctic climates it is possible that precipitation of some salts will take place during the winter as the water temperature falls below the saturation temperature of some materials in solution. However, during the summer these materials will tend to re-dissolve.

In Chapter 3 it was pointed out that water stored in many lakes is poorly mixed, especially in lakes deeper than perhaps 30 ft. Therefore, single samples from stratified lakes can be assumed to represent only the location within the body of water from which they come unless information is available to indicate that the lakes are not stratified, that the water was collected during a period of overturn, or that the water collected (e.g., from the bottom) represents groundwater seepage which is the objective of the survey. For exploration in Canada, the additional factors of changing

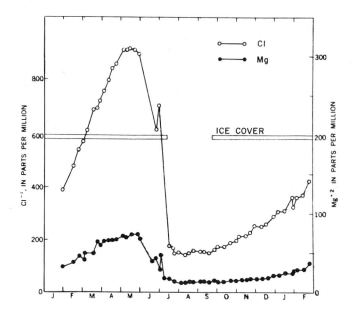

Fig. 11-11. Seasonal changes in the chemical composition of a small lake (over 8 feet deep) in Alaska, near the Arctic Ocean. Modified from Boyd (1959) by Livingstone (1963).

limnological regions, variations of lake characteristics (e.g., a change from oligotrophic to eutrophic) of individual lakes within one region, and the effects of ice cover are among the complexly interrelated parameters always to be considered. One very important simplifying assumption can be made with respect to the lakes in the northern part of Canada, i.e., that the biological influence is low or negligible. These lakes are oligotrophic, and they have a simple biota with a relatively short season of growth and, as a result, the bottom sediments are generally inorganic. In the more southern parts of Canada, especially in areas of abundant muskeg, and where the lakes are dystrophic (Fig. 11-14), influence of peat and mosses may become quite significant.

The orientation study by Arnold (1970) of lakes on the Precambrian Canadian Shield in the Flin Flon and La Ronge areas of Manitoba and Saskatchewan is one of the few (if not the only) published studies in which surface to bottom profiles of trace metals in lake waters were obtained specifically for exploration purposes. He determined the concentrations of Cu, Zn, Ni, Fe, Mn and Co, as well as pH, dissolved oxygen and temperature in several of the lakes which were generally stratified. Some lake sediments were also analyzed for these same metals. An example of the vertical water profiles obtained is shown in Fig. 11-12, and it illustrates

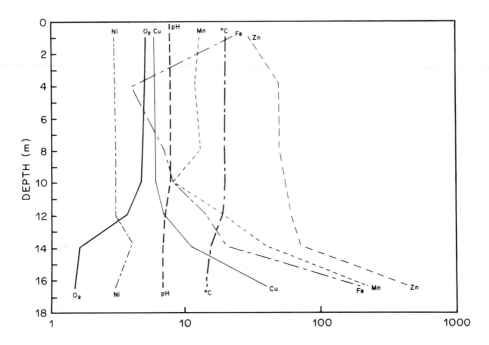

Fig. 11-12. Vertical profiles (obtained during the summer) of dissolved oxygen (cc/l), pH, temperature (°C), and metal contents (ppb) in water at Lake Athapapuskow, Saskatchewan. Stratification is recognized by means of the oxygen and temperature profiles (compare with Figs. 3-28 and 3-30). From Arnold (1970).

that changes in the metal concentration in water correlate approximately with changes in oxygen, pH and temperature. The hypolimnion (below 14 m) is reducing, and the concentration of all metals studied except nickel is significantly higher in the deep, cold and oxygen-poor water.

Although Arnold (1970) indicated that his results were too sparse to be generally applicable to all situations, he was able to suggest that, when mineralization occurs adjacent to a lake, the metals could be detected by a random sample in both the water and sediment. All metals studied were strongly partitioned into the sediments relative to water with partition ratios (concentration of metal in sediment to concentration in water) ranging from 1000 to 1,000,000, depending on the metal species and the environmental factors. As a result, Arnold (1970) concluded that lake sediments were particularly valuable for the detection of anomalous concentrations of metals. It appears, however, that in the examples studied, physical transport of sulfides into the sediments was predominant and that mineralization was adjacent to the lakes in all cases. Furthermore, only very small amounts (0.3 – 4.0%) of the total metal content of the samples were put into solution by cold extractable methods. Hence, Arnold's conclusions are certainly applicable to areas of physical weathering but they must be confirmed for areas in which chemical weathering occurs, in which case the metals would tend to be loosely bonded, before the conclusions can be generally applicable to large areas of the Canadian Shield.

Lake Sediments

Regional collection and analysis of lake sediments as a means of exploration is now receiving very serious attention in Canada, primarily as a substitute for sampling soils and stream sediments which is not practical in the Canadian Shield as both are poorly developed and meaningful samples are difficult to obtain. Many studies (e.g., Livingstone et al., 1958; Allan et al., 1972; Brunskill et al., 1972; Allan et al., 1973) have shown that lake sediments in the Canadian Shield and Alaska contain a clastic component composed of soils, silts, clays and possibly sulfide grains in some cases, which are derived from the watershed and physically transported to the lake by rivers and surface runoff (Fig. 3-22). In addition, groundwater may supply elements to the bottom sediments.

Lakes in muskeg areas (Fig. 11-14) are likely to have organic bottoms. Based on work in the southern part of the Canadian Shield, Timperly et al. (1973) recognize two types of organic materials in these lakes. One comprises "organic sediments" which represent varying mixtures of organic and inorganic materials. The second are "organic gels" which are found in organic-rich lakes in those parts of the Shield with a mixed deciduous-coniferous forest cover. These gels are almost completely organic in content, smell strongly of reducing conditions (H_2S), and are thixotrophic. They are thought to be mixtures of organic precipitates, residual organic matter, pollen, and some inorganic materials. The work of Timperly et al. (1973) is preliminary, therefore, further discussion of the methods of separation and analysis of these gels, and especially their potential advantages relative to other sampling media, must be deferred until additional results are published.

Lake sediments have been used by private companies for the initial reconnaissance stage of exploration programs since the early 1960's in Saskatchewan, Manitoba and other parts of Canada, both within and outside of the Canadian Shield, but very little of the results has been published. The papers from the Geological Survey of Canada (e.g., Allan, 1971; Allan et al., 1972, 1973) are the first published comprehensive studies on lake sediments. These studies have been concerned with two locations, the Coppermine area (Allan et al., 1972), and a 36,000 sq mi area of the northwestern Canadian Shield in the Bear and Slave geological provinces (Allan et al., 1973). In both cases the studies can be considered

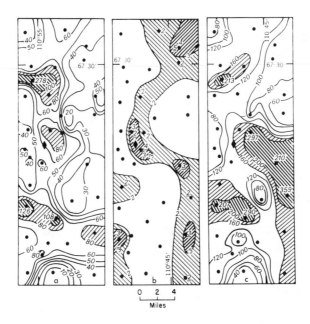

Fig. 11-13. Results of lake sediment and lake water surveys at High Lake, Northwest Territories. *Left:* copper in lake sediment. *Center:* copper plus zinc in lake water. *Right:* zinc in lake sediment. Sediment values in ppm; water values in ppb. From Allan et al. (1973).

as orientation and reconnaissance in nature. Fig. 11-13 illustrates some of the results obtained from lake sediments and, for comparison, lake water data are included. The maps clearly indicate areas which have higher concentrations of certain elements than others, although the results do not necessarily coincide in each case.

In both studies a sample density of one site per 10 sq mi was selected, and this is considered feasible for regional exploration of most of the Canadian Shield. Allan et al. (1973, p. 41) pointed out however, "that if anomalous amounts of ore elements within a given area were confined to one or a few ore deposits, it is improbable that the resulting geochemical signal could be detected at a sample density of one per 10 square miles." In another study in the Northwest Territories, Nickerson (1973) used a sampling density which was related to the rock type. For areas underlain by volcanic rocks he collected one sample per sq mi, but used a pattern density of one sample per two sq mi in sedimentary areas and one per ten sq mi in granitic areas. Nevertheless, all these studies suggest that known major deposits in the areas studied can be detected by low sample density lake sediment techniques.

Many facets of lake sediment geochemistry in the northern part of the Canadian Shield have yet to be elucidated. For example, the sediment types in the various lakes differ to some degree and research is needed to determine their ability to retain trace elements. In the Coppermine area Allan et al. (1972) used <80 mesh samples, whereas <250 mesh samples were used in the Bear-Slave area (Allan et al., 1973). Although these northern lakes are probably oligotrophic and the sediments probably have a low organic content, a knowledge of the trace element concentration in the organic phase, as well as in clays and any iron-manganese oxides, is necessary for a complete understanding of the geochemical processes in Canadian lake sediments. Variations in the proportions of these components are likely to cause changes in the overall trace element content of lake sediments and, consequently, in the contours illustrated in Fig. 11-13. Allan et al. (1972, 1973) discussed these points, as well as others such as sorting phenomena, the most desirable location(s) in a lake to collect sediments, oxidation-reduction processes and their effects, adsorption by iron-manganese oxides and organic matter, the various mechanisms of element transport, biological activity, and other subjects considered in this book. Considering the large number of lakes in northern Canada, and the variations which are likely to be found in them, it may well be many years before a reasonably complete understanding of all the geochemical variables is forthcoming. However, it will be worth the effort as the results ought to be applicable to large areas in Canada, Sweden, Norway, Finland and the Soviet Union.

MUSKEG AND BOGS

Definition And Occurrence

Following the usage of MacFarlane (1969), the word *muskeg* is defined as "organic terrain." Organic terrain, in turn, is a tract of country comprising a surficial layer of living vegetation and a sublayer of peat (fossilized or partially decomposed plant debris) existing in association with various hydrologic conditions and underlying rock formations. Included in the category of muskeg and organic terrain are such things as bog, organic swamp, muck, peat marsh and peatland (moors in European terminology). Muskeg forms a dense covering over bedrock which makes prospecting difficult.

In all localities where these materials are found there is an excess of water (water-saturated) and restricted drainage. As a result, decay of organic matter is inhibited and peat accumulates because, first, oxygen is depleted (anaerobic conditions), and second, decay-producing bacteria are repelled by the acid conditions.

Marshes (and swamps), on the other hand, are not considered muskeg, by definition, as they do not have substantial accumulations of peat. Although marshes occur in areas which are more or less permanently flooded, there is some water movement, and this results in a supply of oxygen sufficient to cause decay of organic matter. Marshes may have some organic matter in the process of decay (e.g., Fig. 3-18) and some may contain small amounts of peat which formed during earlier periods of restricted drainage. Emphasis in this chapter is on muskeg.

Fig. 11-14 shows location of muskeg in Canada. Canada has more muskeg, 500,000 sq mi (or about 12% of its total area), than any other country including the U.S.S.R. and, for reasons which have to do with both climate and terrain, most of it is on the Canadian Shield. Beyond the tree line in northern Canada muskeg becomes less abundant along with an increasing scarcity of vegetation. Muskeg covers extensive areas, such as those in northern Ontario and particularly in the Hudson Bay Lowland. Finland, however, has the greatest percentage of its land area covered by muskeg (peat). One-third of the country is mantled by this material which averages 2 meters in thickness.

The term *bog* is used for areas of confined muskeg or peat formation, usually in depressions of limited area where water can accumulate. A bog,

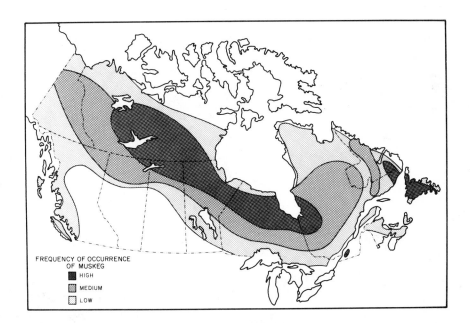

Fig. 11-14. Distribution of muskeg in Canada. The frequency of occurrence is indicated as "high", "medium" and "low", based upon engineering experiences. Modified from MacFarlane (1969).

therefore, is differentiated from muskeg mainly in terms of area. Thus, in those areas in Fig. 11-14 where muskeg abundance is indicated as "low" (e.g., British Columbia and the Maritime Provinces), the term bog would probably be appropriate as the organic terrain is generally not sufficiently extensive to warrant the name muskeg. In Fig. 11-15 two types of bogs are illustrated, a bog lake and a raised bog. Both have the required layer of accumulated peat formed in areas of restricted drainage and are deficient in oxygen. Much, but not all, of the dispersion in muskeg, swamps and bogs is chemical, as opposed to physical, owing to the sluggish water movements in these environments.

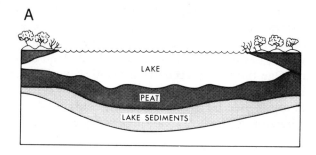

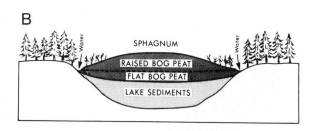

Fig. 11-15. Cross section of (A) bog lake, and (B) raised bog.

Bog Water Chemistry

One of the features of bogs is that their contained water is markedly different from any other type of natural water. Three significant characteristics of bog water are (Ruttner, 1963): (1) it has an unusually low total dissolved solids content, being especially deficient in lime (CaO), and its composition corresponds approximately to that of rain water; (2) it gives a strongly acid reaction; the pH in raised bogs usually lies between 3.5 and 4.5, and in flat bogs between 5 and 6; and (3) it has a high content of humic materials which impart a yellow to brownish color to the waters. In other geochemical environments (e.g., soils, groundwater) high acidity generally implies a large amount of dissolved solids, and possibly a sulfide-bearing source. Other features include a high carbon dioxide content which is released during decomposition proceeding in the peat, and a low oxygen content which results in a reducing environment. Gorham (1967) has reviewed various aspects of the chemistry of bogs with particular emphasis on the major elements. Clymo (1964) has discussed the origin of acidity in bogs.

Bog waters are classified as dystrophic (Gr. *dys* badly, *trophein* to nourish), a special type of oligotrophic water, and are very poor in fauna and flora primarily due to the low nutrient content. One reason for the low total dissolved solids content of bog water is that much of the supply is from relatively pure precipitation (Table 3-1). A second reason is that the organisms characteristic of bogs, particularly the *Sphagnum* moss, extract the elements (especially Ca) which seep in from groundwater or enter with the occasional surface water movement. In many parts of Canada up to 90% of the bog mass is sphagnum, or peat composed of sphagnum remains. Sphagnum establishes itself in a thick, floating mat of sod across the bog (Fig. 11-15A). The upper part grows from year to year but, as the lower parts die and disintegrate, the pieces sink through the water and eventually add to the growing peat layer on the bottom. For adsorbency, sphagnum is the next best thing to a sponge. It is logical to assume that at least some trace metals would be absorbed in the same way as calcium and other major elements, although there does not appear to be any study which has specifically investigated this possibility from the point of view of exploration. Peat adsorbs (also absorbs) elements about as well as sphagnum. (If tap water is filtered through a sufficiently thick layer of peat or sphagnum, it undergoes a reduction in lime content and becomes acid in reaction). Thus, a low copper content, for example, in bog water does not necessarily mean an absence of copper mineralization in the vicinity. An analysis of the peat material, at least, will be necessary to obtain an unambiguous answer. Brooks (1972) has discussed other aspects of the adsorption of elements by peat and stressed that the principal agents for element fixation are the humic acids in the peat.

Origin Of Bogs

Two conditions must be satisfied for the formation of bogs: (1) an abundant supply of precipitation (or groundwater seepage) — more than can be disposed of by drainage; and (2) a production of plant substance exceeding decomposition. These conditions apply on some parts of the

Canadian Shield (Fig. 11-14) where the precipitation is not readily drained because of the effects of glaciation, the cold climate inhibits the decay of organic matter, and in addition, sphagnum grows well. In these areas a lake may be converted to a bog with a small change in environmental conditions. Extensive bog development may be found in numerous other places in the world, such as in the tropics in Indonesia and Guyana where there are heavy rainfalls on low-lying, flat areas and extensive production of plant material. In British Columbia bogs may be found at higher elevations in areas where ponding (e.g., from glacial deposits) has restricted drainage, and in which the two conditions mentioned above are satisfied.

As undecomposed organic matter progressively fills a basin from the bottom upwards (Fig. 11-15A), there is usually at the same time a continual encroachment of shore flora towards the center of the basin. When the basin fills up, a *flat bog* (Fig. 11-15B) is formed, the water-saturated bottom of which has a pH of about 5 to 6. Owing to the continual adsorption of lime by sphagnum, the water of the flat bog becomes more and more acid. Specific types of sphagnum which continue to grow on top of dead sphagnum have the ability to draw groundwater and rainwater upward and, in fact, the water table rises. This results in a *raised bog* (Fig. 11-15B). Growth of the bog stops when evaporation and surface drainage balance rainfall and the upward flow of water.

Geochemical Studies

Usik (1969) has prepared the most complete review on geochemical and geobotanical prospecting in muskeg and bog areas (she prefers the term "peatland") and has concluded that only a few detailed studies on the metal distribution in these organic materials have been made in North America. Most of the studies on the subject, including the determination of the factors controlling the metal distribution, have been made in Scandinavia and the U.S.S.R. For several discussions on the use of peat and humus for exploration in the glaciated areas of Scandinavia, Kvalheim (1967) is highly recommended.

The occurrence of high concentrations of metals, particularly copper, zinc, uranium, vanadium, iron and manganese in organic materials has been known since 1824 when certain bogs in Ireland were reported to contain sufficient copper for extraction. At about the same time, peat bogs in Scotland occurring over lead-zinc veins were found to be enriched in these elements. In the United States, native copper has been reported in peat in Montana, copper-bearing peat bogs have been described in Montana and Colorado downstream from copper mineralization, and as much as 16% zinc is present in peats overlying the zinc-bearing Lockport dolomite in

New York (Cannon, 1955). Zinc carried into bog and muskeg by surface water is concentrated in abnormal amounts in the Keno Hill area, Yukon Territory, and at Chibougamau, Quebec (Boyle and Cragg, 1957). Horsnail and Elliott (1971) have reported several examples of similar accumulations in bogs (and swamps) in British Columbia. The occurrence of metal-containing bogs and swamps in New Brunswick has been mentioned previously. In all areas mentioned, there are numerous examples of bogs directly down-drainage from where even very weak mineralization results in an exceptionally strong anomaly being built up in the bog organic matter (Fig. 5-2). These anomalies (both significant and non-significant) are usually higher in metal content than any other anomaly (e.g., soil or stream sediment) related to the same source. Although some bogs have been mined for their metal content in the past, and some are presently being mined (e.g., in Finland), technological difficulties are great.

Gleeson and Coope (1967) made a detailed study of two bogs on the Canadian Shield of eastern Canada (Ontario and Quebec) and one in New-foundland, all of which are in glaciated areas. The Ontario and Quebec bogs are underlain by compact glacio-lacustrine clays. All three bogs are similar in that certain trace elements tend to become regularly distributed in the bog profiles, and the same relative (but not absolute) metal distribution is present in bogs from both background and mineralized areas. One such profile from a non-mineralized area is illustrated in Fig. 11-16. The important features of this figure are (1) the highest values for the copper (zinc and nickel are similar) are found in clay about 2 ft below the peat, and (2) the pH increases with depth in the peat section of the profile. Gleeson and Coope (1967) stated that it is very unlikely that metals present in the profile have been derived from upward moving groundwater (because when a lower basal till was intersected at depth, artesian flow was encountered). The increase in metal content with depth appears to be typical of bogs and parallels the increase in pH and humification of the peat. This has also been observed in other parts of the world. (e.g., Fin-land). They concluded that the metals in the peats have been derived from the lake and swamp water with which the peats were in contact at one time. The high metal content in the clay below the peat is also partially explained by adsorption of metal from swamp (bog) water when the clays were in contact with the water. In the Newfoundland bogs which are associated with mineralization, some lateral movement of metal-bearing groundwater supplied metals to the clay and peat, as the bog areas are interspersed with freely-drained overburden and bedrock and there are no impervious glacio-lacustrine deposits.

Salmi (1955, 1967) studied peat (bog) deposits in the vicinity of sul-fide ore bodies in Finland and found more copper, zinc and nickel in these by comparison with bogs remote from the mineralized zones. Maximum

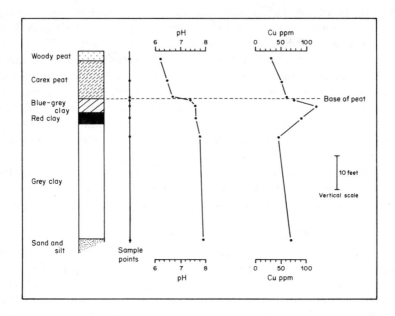

Fig. 11-16. Vertical variations in pH and copper content of a bog profile, Black Bay bog, near Kenora, Ontario. The highest concentrations of copper are immediately below the base of the peat. The pH in the peat layer of the bog is slightly acid (pH 5.5 – 6.8), whereas the underlying clays are slightly alkaline (pH 7.1 – 8.0). Redrawn from Gleeson and Coope (1967).

concentrations occur over or near the suboutcrop of the ore bodies which can be detected even where the peat is underlain by as much as 40 ft of sand and till. The mechanism by which the anomalous quantities of metals entered the peaty horizons was upward movement by capillary action, resulting from evaporation and transpiration of large quantities of moisture from the bog surface. Thus in Finland the highest metal values are found within the lower peat layers. Salmi (1967), therefore, recommends that the peat material be ashed and the metal values determined on a dry ash basis.

Although several other examples could be cited, the studies of Gleeson and Coope (1967) in Canada, and Salmi (1955, 1967) in Finland suffice to point out that sampling of profiles in bogs and other organic terrain in heavily glaciated areas can be used as a geochemical exploration technique, but that the specific horizons in which the trace elements are concentrated will vary. In some situations the highest concentrations may be within the peat, but in others they may be in the clay below it (Fig. 11-16). The exact mechanisms of metal transport will also differ; in some cases the metals may be transported by groundwater from below or from the surrounding higher ground, whereas in other situations they may represent metals adsorbed from lake water. Some anomalous values could be washed in directly from the surrounding highlands. During peat formation (humification) the metal values will increase along with pH (Fig. 11-16).

Partial extraction techniques have been applied to the study of a few bogs and related materials but the results are not conclusive. Hawkes and Webb (1962) reported that the best contrast for copper between anomalous and background dambos (seasonal swamps; not true bogs) in Rhodesia was obtained by using cold extraction rather than total methods for the determination of this element. On the other hand, Gleeson and Coope (1967) found that total rather than cold extractable zinc gave the best contrast from several bogs in eastern Canada. This they attributed to the formation of strong complexes between zinc and the organic matter. Maynard and Fletcher (1973) obtained similar results in bog samples containing copper from northwestern British Columbia. In the peat samples they studied the greatest contrast between anomalous and background samples was found with nitric-perchloric acid (total) digestions. Cold extraction with either dilute hydrochloric acid or dilute EDTA generally released 70 – 75% of the total copper, but the contrast between anomalous and background samples was not as good as that obtained using the total method. Cold extraction with ammonium citrate liberated a maximum of 8% of the copper.

The interpretation of geochemical data from bogs can be quite difficult. To outline base metal deposits by the analysis of peat or related materials (e.g., clays beneath them), it is necessary to obtain numerous profiles to find those bogs which contain higher abundances of the elements of interest. Thus, for example, in Newfoundland Gleeson and Coope (1967) found zinc values in bogs unrelated to mineralization (background bogs) to range from 20 – 110 ppm, whereas zinc values in bogs near mineralization range generally from 140 – 660 ppm (a few very low and high values are excluded). Several authors, including some in the Soviet Union, recommend the use of cold extractable/total metal ratios to assist in the interpretation. Similarly, others recommend the use of metal ratios and assemblages of elements in bog profiles (and water) in preference to individual determinations of any single specific element.

No discussion of muskeg, peat, bog or humus, would be complete without at least passing reference to the uranium concentrations commonly associated with these organic materials since their capacity to retain uranium is great. Uranium is transported in water (probably as the bicarbonate or sulfate complex), and is either adsorbed by the organic matter, or the uranium complexes can be reduced and the uranium precipitated as uraninite (pitchblende), UO_2. The uranium may originate from rocks with normal or somewhat higher than normal values (e.g., Fig. 3-24), or it may represent uranium from economic deposits. Uraniferous bogs have been reported from many places in the world (e.g., California, Scandinavia, U.S.S.R). In some localities they represent a local phenomenon and are related to an underlying bedrock source, but in others, the occurrences appear to be a regional phenomenon.

Among the numerous studies of uraniferous bog deposits, that by Armands (1967) in northern Sweden is particularly interesting. Values as high as 3.1% uranium were obtained from some peat bogs (on a dried sample basis), and the average for the organic component of the peat was 900 ppm. Springs averaging 100 ppb (0.1 ppm) uranium continually supply the bog. This indicates a concentration factor by the peat of 9000 times that of the associated spring water. Armands (1967) concluded that the supply of uranium (and radon which is also detected in the bog) emanates from below the covering of post-glacial sediments, and that the organic component (humus) of the peat serves as a collector for the uranium. The process probably began soon after the sediments were deposited, or about 5000 years ago. Calculations show that one spring can supply about 6 kg of uranium per year to the bog, and there are about 230 springs in the 35 sq mi area investigated. Biogeochemical prospecting was also successful, with twigs having much higher uranium contents than leaves. The highest uranium values detected in plant ash were in willow: twigs had 860 ppm uranium, whereas leaves had 450 ppm.

Usik (1969) has suggested that for a thorough detailed local investigation of bogs, peat, or muskeg, the following five steps should be involved:

1. Studies of physical conditions such as groundwater flow, upwelling of groundwater, and diffusion of elements.
2. Studies of element distribution in vertical profiles including the underlying clays and tills.
3. Studies of the vertical and areal distribution of pH, Eh, and elemental concentrations.
4. Studies of the nature of the binding of elements with organic matter (e.g., adsorption, chelation).
5. Botanical and ecological studies.

The first four factors have already been discussed, and next the botanical aspects will be considered briefly. Mention was made of the study by Armands (1967) of the uranium content of vegetation in the uraniferous bogs of northern Sweden. According to Usik (1969), in general, relatively few studies of this type have been carried out in muskeg (bog) regions. She mentioned that in Finland the twigs of Labrador tea contain more Zn, Cu and Pb than the peat in which they grow, and that this plant is thus more useful for prospecting than the peat itself. In addition, black spruce and tamarack, as well as Labrador tea, are considered useful in the muskeg regions of Canada.

Usik (1969) also pointed out that peat can be classified according to its vegetation cover (see also Radforth, 1961; MacFarlane, 1969). The vegetation, in turn, is an indication of the pH of the peat, and some types of muskeg vegetation can be used to indicate the character of the geologic substrata beneath the peat, provided the organic layer is not too thick. "Copper mosses" have been reported in peat bogs in New Brunswick and from peatlands in Alaska (Shacklette, 1967) and these have potential as geobotanical indicators. Vegetation may also be used to interpret hydrologic movement (including the occurrence of springs) in favorable circumstances. Airphoto interpretation has been used in muskeg and bog classification (e.g., MacFarlane, 1969) and in mineral reconnaissance, and Usik (1969) suggested that color and infrared photography offers great possibilities for future work in muskeg areas.

At present, regional studies in muskeg areas are hampered by many factors. The principal difficulty, according to Usik (1969), is to obtain an adequate sampling of these materials on a regional basis. Until this problem is resolved, possibly by using air photography to delineate favorable areas, most emphasis will continue to be placed on local exploration, primarily involving the use of bogs.

Bog Iron And Manganese

Reference has been made previously to "bog manganese" and "bog iron" ores. These occurrences are common in some areas with a cold humid climate, high water table, and poor drainage, similar to localities in which muskeg and bogs are found. Often they are not true bogs, but rather marshes (swamps), as peat is not a constituent.

In a marsh environment, reducing conditions are often present, as well as muck, an organic soil or marsh deposit which develops on periodically flooded or permanently wet ground, but not under stagnant water. This environment (decaying vegetation and reducing conditions) leads to the formation of abundant CO_2. Any iron (manganese will act similarly) in the surface or vadose (soil) water will be reduced to the ferrous state, and ferrous bicarbonate will form. If freely-moving oxygenated groundwater interacts with this descending water, the iron will be oxidized, and limonite will form in the vicinity of the water table. Such a deposit is sometimes called a marsh, or bog, iron ore (Fig. 11-17a). If there are lakes in the area, as is common in parts of Canada, the ferrous bicarbonate may move underground and interact with oxygenated lake water, and limonite will accumulate on the lake margin and bottom (Fig. 11-17b). In the example shown, the iron ore would be concentrated at the margin of the lake and would be absent in the center. In Fig. 11-17c, porous limestone acts as a conduit for

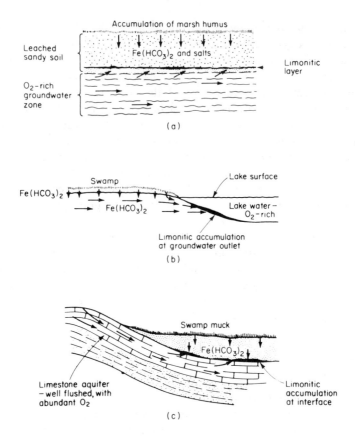

Fig. 11-17. Development of iron concentrations in marsh and bog environments. (a) limonite concentrations close to the water table. (b) deposition of iron oxides on a lake margin from groundwater interacting with oxygenated lake water. (c) deposition of iron oxides resulting from the interaction of oxygenated groundwater in a porous limestone aquifer with marsh water. From "Ore Petrology" by R. L. Stanton. Copyright, 1972. Used with permission of McGraw-Hill Book Company.

relatively fast-moving oxygenated water which interacts with descending ferrous bicarbonate at the bottom of the swamp, the ferrous iron is oxidized to ferric, and limonite precipitates. These reactions, or variations thereof, can also take place in bog environments (which are reducing, have decaying organic matter, and CO_2).

Base metals may be found within the iron or manganese deposits for any one of several reasons. They may have been mobilized by the same mechanism as the iron and manganese, and subsequently transported and precipitated with them. Alternatively, these metals may be extracted from the lake or marsh water during the formation of the iron or manganese deposits, or they may be adsorbed from circulating waters at some time after the formation of the marsh or bog deposits. Recognizing significant from false base metal anomalies in iron-manganese marsh or bog deposits usually is difficult.

GLACIATED AREAS

Introduction

During the Pleistocene, glaciers covered about 95% of Canada. Fig. 11-18 outlines the areas once beneath continental and valley (alpine) glaciers, as well as the unglaciated areas. Bradshaw et al. (1972, p. 34) stated that "The major drawback to use of conventional geochemical exploration techniques in Canada is the mechanical movement and removal, or mixing, of soil and rocks by glacial events." Nichol (1971, p. 37) concluded, "The problems facing geochemical exploration in Canada are largely those posed by the glacial history." Numerous studies in recent years have shown that

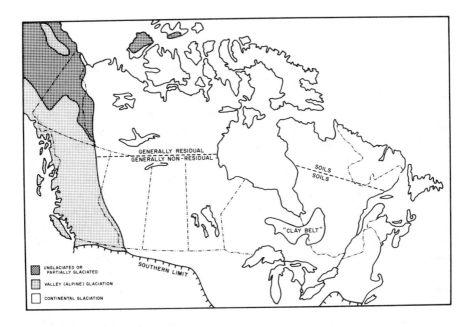

Fig. 11-18. Pleistocene glaciation in Canada. Generalized mainly from "Glacial Map of Canada" (Geological Survey of Canada Map 1253A) and from Flint (1971). The "Clay Belt" is just one of several important areas of glaciolacustrine deposits in Canada. The distribution of residual (including partially residual) and non-residual soils is taken from Bradshaw et al. (1972).

geochemical methods can be used successfully in many glaciated areas of Canada provided that there is a careful selection and adaptation of the well-established sampling and analytical procedures; above all, an understanding of the glacial history of the area is essential. This includes a knowledge of the distribution, fluctuations, erosional features and deposits of glaciers. At the present time great strides continue to be made in understanding the optimum procedures which must be employed, and in developing methods of interpretation of geochemical data from glaciated areas. Many of the advances which are being made in Canada are based upon the results of earlier experiences in Scandinavia, the U.S.S.R., Ireland and Great Britain. Among the recent pertinent discussions and summaries of geochemical exploration in glaciated areas are those of Kvalheim (1967), Bradshaw et al. (1972), Nichol and Bjorklund (1973), and Jones (1973).

Definitions And Basic Principles

Before proceeding further, it is necessary to define some terms commonly used in glacial geology and to discuss a few of the more fundamental aspects of this science which are of importance in exploration geochemistry. Terms are not used consistently in all parts of the world, or even within Canada, and in view of this it is not always possible to be certain as to what each author has in mind. For the purposes of this book, the terminology of Flint (1971) has been adopted. Easterbrook (1969) uses essentially identical terminology and is the source of a good general summary on glacial geology. Prest (1968), and Scott and St-Onge (1969) discuss aspects of glacial geology with particular reference to Canada.

The materials transported and deposited as a result of glaciation are collectively known as *glacial drift.* The term embraces all rock material in transport by glacier ice, all deposits made by glacier ice, and all deposits predominantly of glacial origin made in the sea or in bodies of glacial meltwater (including outwash), whether rafted in icebergs or transported in the water itself (Flint, 1971, p. 147). As defined, drift is an all-inclusive term. Glacial drift covers large parts of the glaciated regions of the world and even extends beyond them onto the floors of lakes and the sea, and into stream valleys, wherever water could carry material beyond the margins of the glaciers themselves.

Two general types of glacial drift are recognized, *till* (non-stratified) and *stratified drift,* but, no sharp dividing line separates one from the other as they tend to intergrade. Till is deposited directly from the ice without the direct aid of meltwater, and, as water plays a minimum role in the deposition, it is dominantly unsorted with respect to grain size. In former years this material was often referred to as "boulder clay," but this term is now seldom employed. Till is the most characteristic surficial deposit, and parent soil-type, in southern Canada. Its most outstanding characteristic is its lithologic and physical heterogeneity, and in fact, it is probably more variable than any other sediment. It may be cohesive and compact, or loose and friable, depending upon its texture, mineral composition and post-depositional history. The proportions of the various sizes in the mixture will vary greatly according to the character of the local bedrock and unconsolidated material over which the glacier has moved and the distance of transport. Till may consist principally of clay particles, or of large boulders, or any combination of these and intermediate sizes. Till sheets commonly are extensive and the thicker ones, 100 to 300 ft and in some places 1000 ft thick (usually over buried valleys), may extend for hundreds of miles while the thinner ones tend to be patchy. The average thickness of drift in central Quebec-Labrador and in Finland is less than 3 m (Flint, 1971). It is between 5 and 15 m thick in the Great Lakes region, south-western Alberta and Sweden. However in all glaciated localities till may be absent (e.g., in many parts of northern Canada in those areas where soils, if present, are generally residual; Fig. 11-18).

Two types of till are recognized: *lodgment till* and *ablation till* (Fig. 11-19). Lodgment till is deposited at the base of a glacier. The particles in this type of till are 'lodged' in the accumulating drift, under pressure, on the subglacial floor. Crushing and abrasion of the particles is intense, and the resulting till is compact. It is usual to think of lodgment till as being deposited during expansion or advance of a glacier, but there is no reason why it can not be deposited during shrinkage, if the terminal zone is actively flowing during shrinkage. Lee (1971, p. 32) stated, "The basal till of retreat ice is now recognized as a favourable sampling environment and as such has been used to successfully discover hidden orebodies."

Ablation (defined as the process by which substances, including ice, are lost from a glacier during evaporation and melting) till is deposited from drift in transport upon or within the terminal areas of a thinning and melting glacier (Fig. 11-19). Thus, as ice melts inward from any or all directions (top, terminus, base), the drift is moved towards the ground by any one of several processes (e.g., dumping, sliding, subsidence), and the resulting till is called ablation till. It is usually found on top of a lodgment till. This till is much looser, less compact, and less abraded than lodgment till which is deposited beneath the weight of an overriding glacier. During the process of settling, the fines are selectively washed away by glacial meltwater. Because the material available for the formation of ablation till is limited to the load within or upon the glacier at the time of ablation, it is generally thinner than lodgment till. However, ablation till may attain considerable thickness locally by means of sliding or flow movements of

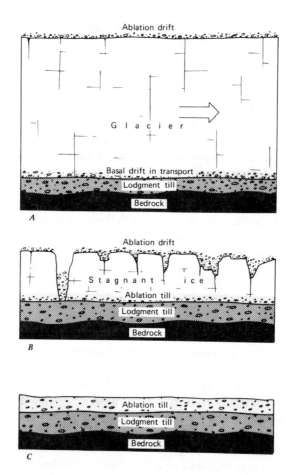

Fig. 11-19. Origin of lodgment and ablation till. A. Basal drift in transport over bedrock giving rise to a lodgment till. B. During period of ablation, ice near a glacial margin melts beneath a cover of glacial drift and an ablation till is deposited. C. Post-glacial condition. A layer of ablation till is found over a lodgment till. From Flint (1971).

various types. In theory it should be relatively easy to recognize an upper till as an ablation till but in practice difficulties are common. A so-called ablation till may actually be the surficial zone of a lodgment till reworked by frost action, or it may be a layer of a second lodgment till deposited during later glaciation. In many cases only one type of till is apparent at a particular locality and it is difficult to determine which type it is.

    Flint (1971, p. 183) recognized two classes of *stratified drift*. One, called ice-contact stratified drift, forms upon or immediately adjacent to glacier ice and, except for the stratified deposits of an esker, is of limited importance in exploration geochemistry. The second, called proglacial sediments, are stratified deposits built up beyond the glacier itself, such as in streams, lakes and in the sea, and is more likely to cover a large area. *Outwash sediments* (or simply *outwash*) is the term used for stratified proglacial sediments built by glacial streams much like streams of today, and they consist mainly of sand and gravel. The average grain size of outwash diminishes downstream, and eventually the finer fractions, which often travel a considerable distance from the glacier, are deposited. *Glaciolacustrine* (glacial lake) sediments are another type of stratified proglacial drift. These lake clays and silts are deposited in standing water dammed by ice and they may cover great areas. The "Clay Belt" of Ontario and Quebec represents lacustrine clay material deposited in such a glacial lake (Fig. 11-18). These deposits have very low permeability and consequently movements of metal-bearing fluids are limited (Parsons, 1970) which results in masking (p. 209). Clearly, glacial materials form a complete sequence from those deposited directly from ice to those deposited from running and standing water. Consequently, there is a complete gradation of sizes from rock flour and clay through sand and gravel to huge blocks, as well as a complete gradation of shapes from angular to perfectly round fragments.

The *deposits* discussed above, which are collectively known as glacial drift, accumulate within and adjacent to glaciers, and also at considerable distance from glaciers, to form various types of depositional *landforms*. Included in the landform category are such features as ground and end moraine, and various glaciofluvial landforms (e.g., valley train, esker). With regard to *moraine,* Flint (1971, p. 199) stated, "Accordingly we now think of moraine as an accumulation of drift deposited chiefly by direct glacial action, and possessing initial constructional form independent of the floor beneath it. This definition is preferable to that implied in the literature of Scandinavia and some other parts of Europe, in which *moraine* is a synonym of *till*." Thus ground moraine, which is defined as an accumulation of drift of low relief lacking transverse linear ridges, can contain deposits of lodgment and/or ablation till. Further detailed discussions on moraine may be found in Flint (1971) and Prest (1968). For a discussion of the methods to describe the essential features of till, the report of Scott and St-Onge (1969) is recommended.

Glacial Dispersion And Provenance

A rock fragment that has been glacially transported from its place of origin and deposited elsewhere, either on bedrock or till, is called an *erratic.* A *boulder train* consists of a series of erratics that have come from the same source, and they often have some characteristic which facilitate the recognition of their common origin. The trains may appear either as lines of erratics, such as are often found in the down-valley direction of a valley glacier, or in a fan-shaped pattern with the apex at the place of origin. By mapping erratics and boulder trains the source of the transported rock debris, or its *provenance,* can be determined in many cases, as well as giving an indication of the direction of ice flow. Grip (1953) stressed that the use of boulders provides a very sensitive and important exploration technique in Sweden, even when the sulfide metal content of the till is in the 50 ppm range. *"Float"* is a more generalized term signifying fragments transported by streams, frost action or gravity as well as by a glacier. Float has been used by prospectors for centuries and is a time-tested method of exploration. In Canada, float has been used in the discovery of at least 21 mines, including the Sullivan mine in British Columbia, the largest producer of lead and zinc in this country (Lee, 1971). One potential value of dogs in exploration is that they may be able to assist in tracing float (p.30). Determination of provenance is aided by use of various morphological features that indicate the orientation of ice flow. Such features include bedrock striations, large grooves and flutings, drumlins, and stoss-and-lee hillocks with smooth up-glacier faces and rough, steep lee faces. The latter features indicate the sense of movement as well as orientation of the flow. Till fabrics, especially orientations of large clasts, are also useful as directional indicators in some cases.

Interest has been shown recently in the use of *heavy clasts* for exploration purposes. These include rock fragments with pieces of attached ore, fragments of ore minerals, and certain pathfinder minerals (such as magnetite, zircon, garnet and rutile) diagnostic of favorable host rocks (Lee, 1971). A concentrate of heavy clasts requires a separation of the lighter minerals from the heavier ones, usually with a heavy liquid such as bromoform. Generally, the pieces considered as heavy clasts are from − 60 to + 120 mesh size. Other techniques (e.g., sluicing) may be used to separate the clasts in which case the size range will differ. In some respects the use of heavy clasts is similar to the use of "heavy minerals" except that clasts are coarse and contain a large proportion of light silicate minerals as a constituent of the fragments.

Clasts, as well as heavy minerals, are now frequently separated from basal till and other glacial deposits (e.g., the constituents of eskers) and analyzed chemically for base metals. The clasts may also be studied microscopically, primarily to provide information pertaining to geology, or to the nature of an EM anomaly (by identification of sulfides or graphite). However, visual microscopic identification of small amounts of the sulfides can often be very difficult, especially in those cases in which only traces of the elements of interest are present, the sulfide minerals have undergone partial leaching and alteration, or where hydromorphic dispersions are being sought. The use of chemical methods in tracing float, heavy clasts or heavy minerals in glacial deposits to the source of mineralization may be of value in either reconnaissance or detailed exploration programs.

Geochemists working with glacial drift, and particularly with till, must be concerned with the distances the constituent materials may have travelled. There is no doubt that some erratics have travelled 500 km or more but, in general, the distances are significantly less. Numerous studies, some dating back almost a century, have established that much of the coarser fraction of till is predominantly of local origin (for the sake of discussion, less than one mile). Presumably, these studies were made on lodgment till, as opposed to ablation till, because the latter, being formed of material within and on top of the glacier, would probably have travelled further. The fine fractions (e.g., clays) found in both types of till, on the average, travel the furthest. Pawluk and Bayrock (1967), and Bayrock and Pawluk (1969) have shown that the lithology of Alberta till closely reflects that of the underlying bedrock. Larsson and Nichol (1971), however, found that chemical analyses of till from Ireland did not permit them to identify the nature of the underlying bedrock. On the other hand, the mineralogical composition of the silt-sized fraction of the till, as determined by X-ray diffraction analysis, was found to be specific to the underlying bedrock.

Lee (1963, 1971), Garrett (1971), Gleeson and Cormier (1971), Bradshaw et al. (1972), and others, have produced evidence, such as that presented in Fig. 11-20, which shows that the distance travelled by sulfide grains mechanically moved by a glacier and deposited in the lodgment till

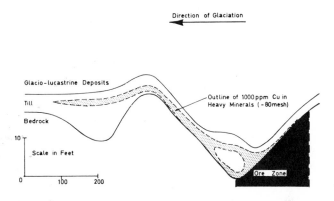

Fig. 11-20. Vertical profile of a metal (copper) dispersion train in till. The lateral dispersion of the sulfide grains from this deposit in the basal (lodgment) till is about 600 ft. Based on data from the Louvem deposit, Val d'Or, Quebec. From Garrett (1971).

is relatively limited. (In three dimensions such dispersions are often fan-shaped). Thus sampling of basal (lodgment) till is now an accepted exploration method in those parts of the Canadian Shield covered with glacial overburden. An analysis of the components of lodgment till formed from either an advancing or retreating glacier has resulted in good success (Lee, 1971). Similar favorable results have been reported from Scandinavia.

The constituents of eskers (long, narrow ice-contact ridges, commonly sinuous and composed chiefly of stratified drift; Flint, 1971) have also been used to trace the source of mineralization. Lee (1963, 1965, 1968, 1971) using the "glaciofocus method", was able to discover a kimberlite body in the Kirkland Lake area, Ontario, by analysis of pyrope garnet in the massive gravel unit of the Munro esker. The transport distance for the pyrope was about eight miles. A gold-bearing pyrite in the same massive gravel unit probably travelled only about one mile.

Shilts (1973) compared the effectiveness of sampling eskers and adjacent till in the continuous permafrost region of the Northwest Territories. He confirmed the general value of the esker method as a previously known area of copper-nickel mineralization was clearly outlined by means of anomalies in various fractions of both till and esker samples. However, he observed that some potentially interesting ore and associated minerals (e.g., sulfides and carbonates) are removed from the active zone of both the esker and till by chemical weathering. Because clays and secondary iron-manganese oxides can scavenge the released cations, the clay-containing, minus 250 mesh fraction from the thawed zone of eskers (at least in permafrost regions) is the best size fraction to analyze, but this fraction must be used with caution in the till. Heavy mineral and heavy clast separates from near-surface esker and till samples in regions of continuous permafrost are not recommended for routine analysis. Shilts (1973) also pointed out that offsetting some of the apparent advantages of sampling eskers is the fact that they generally only occur at intervals of several miles, and that the samples only represent relatively small areas. Further, the sampling and preparation techniques are time consuming and, therefore, relatively costly. Thus, till is a more useful prospecting medium.

Bradshaw et al. (1972) considered the various types of glacial sediments (e.g., till, glaciofluvial, glaciolacustrine) and the forms in which they are found (e.g., moraines, drumlins, outwash plains, kames, terraces, eskers) and concluded that till is preferred for exploration purposes. Some, such as drumlins, are too restricted in geographic extent to be generally useful, a conclusion similar to that reached by Shilts (1973) with respect to eskers. Outwash plains, kames, and glaciolacustrine deposits are of little use because of the distance of provenance, as well as the fact that they are sorted during transportation.

Exploration Using Till

The relationship between the composition of the coarse fragments (or heavy clasts) in drift and local bedrock, as illustrated in Fig. 11-20, has resulted in a great amount of effort being placed on research into the best method(s) of sampling and analyzing till samples, and in interpreting the data. Some of the most serious constraints on using till as a drift prospecting medium are: (1) the lack of understanding of the modes of till deposition in individual areas; (2) the analytical and interpretation problems created by the extreme variations from coarse to fine particle size; and (3) the presence in some cases of multiple till layers representing multiple ice advances, not all of which necessarily travelled in the same direction. The first usually manifests itself in the difficulty of distinguishing ablation facies from lodgment facies. The second is more serious, and appears during the sample preparation and interpretation phases. Problems with the third can be circumvented by distinguishing different till sheets on the basis of lithology and geochemistry, fabric, or mechanical distribution.

Till can be considered as being composed of four fractions (Shilts, 1971): coarse (> 60 mesh), fine sand (< 60, > 230 mesh), silt (< 230, > 2 micron), and clay (< 2 micron). The mineralogical and chemical composition of each size class is usually radically different from the other three. As with other types of geochemical samples, interpretation is greatly facilitated when samples are all of the same type. Therefore, it is imperative that the relative contributions of minerals and elements in each class be known when several classes are combined for analysis. Alternatively, a single class, or portion thereof (e.g., a magnetic fraction), should be analyzed separately. The optimum size fraction will almost certainly vary from area to area, depending on weathering and other environmental factors and, in addition it will depend on geological factors, such as the size and stability of the metal-containing mineral being sought. Thus, Shilts (1971) found that in the tills he studied, the clay fractions had to be avoided. They have a high adsorptive capacity for elements present in groundwater, as well as those made available by weathering and leaching during soil formation, and can either mask true anomalies or create false anomalies. On the other hand, Garrett (1971), and Gleeson and Cormier (1971), obtained good results with the total minus 80 mesh fraction, which would include the clay fraction. Even under the best of circumstances, significant anomalies in till have relatively low values in the immediate vicinity of ore bodies as compared to those developed by such conventional processes as residual soil formation. Orientation surveys are necessary to determine the proper combination of size or mineralogical fraction (e.g., the heavy clasts of the plus 80 mesh fraction), and the method of chemical treatment, for exploration in tills.

Methods of sampling till at depth are now well established. Various designs of drills are available, such as those illustrated in Fig. 5-3. Skinner (1972) compared the methods and costs of rotary and percussion methods in penetrating overburden in the Abitibi Clay Belt. Costs will depend on the depth of the hole (the drill should have the capability of drilling at least 100 ft), weather, type of equipment used, characteristics of the overburden, as well as many other factors, and they will vary widely. Gleeson and Cormier (1971) found their costs varied from 50 cents to two dollars per foot, whereas Bradshaw et al. (1972) estimated costs at 3 to 5 dollars per foot, and Skinner's (1973) more recent estimates varied from about 4 to 13 dollars per foot (average $7.30 for rotary drills).

In overburden surveys, samples are generally collected from the till-bedrock interface, as experience has shown that in most parts of the glaciated areas of Canada covered with till the best anomalies are obtained at this location. However, bedrock topography is a very important factor to be considered. Where the surface of the bedrock is flat, or nearly so, mineralized fragments will be found at the base of the till (Fig. 11-19). However, the effects of topography are illustrated in Fig. 11-20, where it can be seen that the base of the till is not always the location of the anomalous values. In some cases, owing to topographic and other effects, anomalous values in till which represent buried mineralization can reach the surface, and even be reflected in a soil anomaly. To overcome the possibility of missing an anomaly above the base of the till, continuous profile sampling and analysis, perhaps using composite 2 or 5 foot samples, is recommended. Profile sampling is not generally necessary for evaluating geophysical anomalies (such as gravity or EM conductors), as these anomalies presumably originate in the bedrock.

Two important conditions must be satisfied for successful overburden drilling. First, the hole must not be stopped because of the drill encountering an erratic, coarse gravel or permafrost. Ten percent of the holes drilled by Shilts (1973) were stopped by gravel. The possibility of being stopped by an erratic can be checked by drilling two or more holes within a few (or ten) feet of the initial hole, but the other cases are more difficult to interpret and counteract. Second, the samples must be taken in lodgment till. Samples taken in glaciolacustrine deposits or ablation till resting on bedrock can give misleading results as they will only rarely contain discernible evidence of nearby mineralization. Glaciolacustrine deposits are recognized by their high clay content. Ablation till can, however, be very difficult to recognize, especially when the identification must be made on a small amount of drilling sample (generally less than one-half pound is obtained) which may not be truly representative of this highly heterogenous deposit, and whose physical properties are difficult to determine on an unconsolidated sample.

Drift (till) prospecting has been carried out on two scales in Canada (Shilts, 1971). The first is a detailed scale where samples are collected from sites spaced from several tens to hundreds of feet apart (e.g., Fig. 11-20). This scale is often used where defined geophysical targets are to be further investigated, where there is other evidence of mineralization in an area, or where extensions of known deposits under overburden are sought. The second scale is regional, and sample sites are spaced thousands of feet to miles apart. At this scale, drift prospecting may be carried out over broad, unexplored or drift-covered regions, preferably where the drift is thin and drilling costs are modest. The disadvantages of regional sampling are that anomalies are generally weak (highly diluted) and difficult to interpret. At present, most drift prospecting has been concerned with using till on a detailed scale.

In addition to dispersion trains formed mechanically by glacial movement (syngenetic anomalies), secondary hydromorphic anomalies may also occur in glacial deposits. From the idealized profile shown in Fig. 11-21, it can be seen that solution and transport of metals in groundwater may result

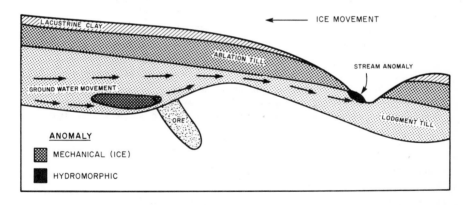

Fig. 11-21. Hydromorphic and mechanical (clastic) dispersions in till overlain by lacustrine clay. The hydromorphic dispersion results in an anomaly in stream sediments (or a bog) at a greater distance than the mechanical dispersion, and in this example, in an opposite direction.

in a laterally displaced hydromorphic anomaly at a considerable distance from buried mineralization, certainly at distances greater than those produced by mechanical dispersion. Further, the anomaly may be in a direction opposite to that of the mechanical (ice transported) anomaly. Many variations of this concept are possible, such as the hydromorphic anomaly being extended in the same general direction as the syngenetic mechanical anomaly (Fig. 11-22). Groundwater movements eventually may appear at the surface as springs or seepages, or within bogs, lakes and streams. These examples serve to show the importance of a knowledge of the glacial history, topography and groundwater movements in interpreting geochemical anomalies in a glaciated area.

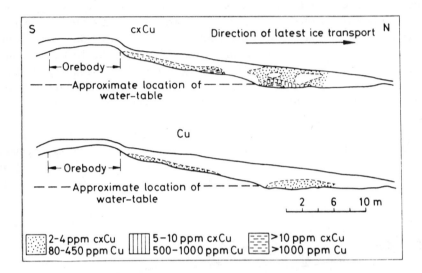

Fig. 11-22. Profile showing the distribution of anomalous cxCu and total Cu in over-burden (till) from Tverrefjellet, Norway. Note that the cold extractable anomalies form a more extensive halo than the total copper anomalies. From Mehrtens et al. (1973).

Evidence for hydromorphic dispersions in glacial areas has been accumulating for several years. Garrett (1971), for example, noted that at the Louvem deposit in Quebec (Fig. 11-20) there has been limited hydro-morphic ("saline") dispersion of zinc. At this locality the clastic dispersion train is controlled by the sum of the ice movements, but dominantly by the last readvance of the ice sheet. The small zinc hydromorphic dispersion pattern is controlled by the local hydrologic regime which, in turn, is related to bedrock topography. Garrett (1971) concluded that more often than not there is insufficient information on these two features to allow a thorough interpretation.

Mehrtens et al. (1973) obtained somewhat different results from Garrett (1971) in their study of the use of geochemical techniques in glaciated terrain in Norway, Wales and British Columbia (Fig. 11-22). They concluded that ore elements are dispersed from their bedrock source beneath glacial till dominantly in shallow groundwater and, to a lesser extent, by mechanical (ice) transport or biogeochemical processes. The mineralized groundwater is thought to be generated at or near the interface between till and the mineralized bedrock, and to travel at or just below this interface. Metal dispersion patterns related to sulfide mineralization in these and similar environments may be readily detected on a broad regional scale by sampling groundwater seepage sites, in agreement with several other earlier studies (e.g., Boyle et al., 1969, 1971). Where seepages occur in lakes as, for example, in parts of British Columbia, Mehrtens et al. (1973) recommend sampling of the organic-rich bottom sediments in the central deeper parts of these lakes for the detection of bedrock mineralization. Parslow (1972), however, found that in a mineralized area in northern Saskatchewan secondary dispersion in till was negligible. He concluded that in this locality the low temperature has inhibited chemical weathering processes and the oxidation of the sulfide minerals. On the basis of the differing results obtained by Garrett (1971), Mehrtens et al. (1973) and Parslow (1972), caution must be used in attempting to apply any one exploration philosophy to all situations.

As with secondary dispersion halos in residual soils and other types of transported overburden, the cold-extractable techniques can also assist in the recognition of hydromorphic dispersions in glacial deposits, and in distinguishing them from mechanical dispersions (Fig. 11-22). This is the procedure used by Parslow (1972), Mehrtens et al. (1973), and numerous others. Copper and zinc appear to form particularly well-defined secondary dispersion patterns. By taking profiles through till, and with the use of both total and partial (e.g., cold extractable) analytical methods, more information on the nature and significance of the dispersion patterns will be obtained than by taking samples only at one position within the till, or by using only one type of chemical treatment.

Comparison Of Continental And Valley Glaciers

In their discussion of geochemical exploration in Canada, Bradshaw et al. (1972) found it convenient and practical to consider areas of *continental* and *valley* (also called alpine or mountain) glaciation separately. The locations of the respective areas are shown on Fig. 11-18 (not shown at this scale are small areas of valley glaciation in the Maritime Provinces). Although the discussion to this point has mainly been concerned with exploration in areas of continental glaciation, the basic principles and terminology are equally applicable to valley glaciation. Such differences as do exist between the two types are mainly those related to location and size.

Valley glaciers are restricted to mountainous regions and flow in long sinuous arms from areas of high snow accumulation. They often have tributaries like rivers. Their widths are generally small in proportion to their lengths. Lengths can be measured from hundreds of meters to 112 km in the case of the Hubbard Glacier in the Yukon Territory and Alaska. The direction of movement of valley glaciers is easily determined and will be essentially the same for all advances. The direction of movement of continental glaciers, on the other hand, may fluctuate even during one period of glaciation, and this may complicate the determination of the source of the mechanically dispersed mineral grains. A *cirque* glacier is a small glacier that is wholly or largely restricted to a cirque, and it is normally found at the head of a glaciated valley. Debris deposited by such a glacier may be sampled at the cirque outlet, and this sampling location is generally representative of the entire cirque basin.

Three features of valley glaciers are of particular importance in exploration geochemistry. First, the principal sources of the load transported by valley glaciers are avalanches and rock falls onto the surface of the ice (Easterbrook, 1969). Material carried on the surface of the ice is called superglacial debris (that carried within the ice is called englacial), and such material is particularly abundant in valley glaciers in comparison with continental glaciers, because the valley sides rise above and immediately adjacent to the glacier. Second, ice from a tributary glacier may join the main glacier, and become a physical part of it, but the ice streams do not mix (Fig. 11-23). Thus, each tributary may maintain its own identity

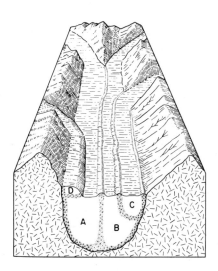

Fig. 11-23. Idealized diagram of a valley glacier complex showing that the two tributaries (C and D) do not mix with the two segments (A and B) of the main glacier. From Sharp (1960).

down-valley, much like some rivers mentioned previously (p. 164). The debris carried by each generally will not be mixed with that of the other tributaries until outwash, glaciolacustrine or certain other deposits are formed during melting or ablation. The lack of mixing creates severe sampling problems in tills deposited from valley glaciers, and it becomes particularly serious toward the source of the valley glaciers. A sample profile across a valley is often necessary. Third, valley glaciation can produce much alluvium, and many large "U" – shaped valleys in British Columbia and the Yukon are filled with 200 ft or more of this material. Soil produced on this alluvium is transported, and bears no genetic relationship to underlying bedrock, yet soil profiles developed on this material can easily be mistaken for those developed in residual soils. Further, if the alluvium is sufficiently thick, it may completely mask any geochemical response from bedrock, especially if the alluvium has a high clay content. Careful study of the topography is necessary to recognize such valley alluvium. Examination of profiles in available road cuts, pits, or stream beds may also be of assistance.

The three features just discussed clearly show that in some circumstances movement of materials for considerable distances is possible in areas of valley glaciation, and that local disturbances can be severe. However, at least in terms of reconnaissance surveys, and generally in terms of follow-up surveys in the valley glaciated areas of Canada, Bradshaw et al. (1972) state that samples may generally be treated as of local derivation.

Masking

Mention was made above of the masking effect of alluvium and now passing mention of other causes of masking are in order, although they are not necessarily related to glaciation. Glaciolacustrine deposits, such as those found in the Clay Belt of Ontario and Quebec (Fig. 11-18), are notorious for their impermeability to groundwater movement and other processes which would permit a surface geochemical anomaly to form (see Parsons, 1970). Thus, for example, the Kidd Creek deposit of the Texas Gulf Sulphur Company, near Timmins, Ontario, gave negative results for a geochemical soil survey because of the presence of overlying clay tills and varved clays. In contrast, strong geochemical anomalies were obtained from analyses of the coarse-textured lower till material found lying on the glacially paved bedrock surface below the clay (Fortescue and Hornbrook, 1969). Many other examples could be cited in which glaciolacustrine deposits, or till with a very high clay content, from both continental and valley glaciers, result in masking. Profile sampling to obtain samples below the impervious layers is necessary in many cases to overcome this problem. Alternatively, metal-containing groundwater may move laterally and form displaced anomalies at the break-in-slope, in stream sediments, lakes or bogs (Fig. 11-21). These anomalies are more likely to be found in areas of moderate to high relief, as in valley glaciated areas, than in flat areas.

Masking may also be caused by a covering of bentonite or volcanic ash which Bradshaw et al. (1972) report is present in northwestern British Columbia and in the southwestern Yukon. The position and thickness of the masking material is always a factor to be considered. Should the root systems of plants be able to penetrate the impervious materials, weak, but positive, anomalies may be detected in the soil. Lacustrine deposits unrelated to glaciation may also prevent geochemical anomalies from forming on the surface.

Soils In Glaciated Areas

Large parts of Canada which were covered by continental glaciers are presently devoid of any glacial drift. This is specifically true in the northern regions where glacial action has scoured the land surface to expose fresh bedrock and where soil has not had sufficient time to develop to any appreciable extent. It is in these areas that rock geochemistry, as well as lake and stream sediment geochemistry, is particularly applicable. Such soils as are present in the areas under discussion are residual or partly residual (Fig. 11-18). Conventional methods of soil sampling using the B horizon have been successful here, although the soils are generally thin, with poor profile development. Muskeg, lakes and permafrost are complicating factors as well. Organic material and A horizon samples are often collected, as in fact, these may be the only materials present in Arctic soils. In many respects, such sampling has much in common with biogeochemical prospecting. In the unglaciated parts of northern Canada, well-developed soil profiles may be found which were developed in pre-Pleistocene time and techniques applicable in the tropics (e.g., cold extraction methods) work well.

In the area of non-residual soils in southern Canada (Fig. 11-18), even where the A, B and C horizons are developed, soil sampling generally is not reliable because of the transported nature of the overburden. In addition, although till is the main parent material for most soils developed in the southern half of Canada, the great variations in temperature, precipitation and other factors discussed in Chapter 3 have resulted in the development of many different soil types (Fig. 3-5). Therefore, even where soil samples can be used in areas of previous continental glaciation (see below), careful attention must be paid to sampling and analytical methods. No one set of sampling and analytical procedures is generally applicable as it might be in large areas covered with residual soils.

Soil sampling can be used in some specific circumstances within areas covered by non-residual soil. Two such situations may be encountered occasionally. First, where the drift cover is thin, vegetation may be able to sample anomalies in lodgment till or other deposits. Such anomalies may be detected in the vegetation or in the A soil horizon. In addition, mercury in the A_3 horizon has been found to reveal the presence of underlying sulfides, whereas base metals did not, in some soils in Manitoba and elsewhere developed on transported overburden (Bradshaw et al., 1972). Second, inspection of Figs. 11-20 to 11-22 will show that it is possible for either mechanical or hydromorphic anomalies in till to reach the surface (or come sufficiently close) for soil profiles developed in such till to reflect mineralization. In these cases, the B or C horizon may be successfully sampled, but it is essential that the till be exposed (not covered by lacustrine deposits, for example). In either of the two situations mentioned, anomalies are likely to be displaced and difficult to interpret. Further, false anomalies, especially in the sampling of vegetation, are commonly encountered. Greater success has been obtained using soil sampling techniques in areas of valley glaciation by comparison with areas of continental glaciation because soils in the former are often partly residual.

12/ *The Statistical Treatment of Geochemical Data*

RICHARD B. MCCAMMON
Department of Geological Sciences
University of Illinois at Chicago Circle
Chicago, Illinois 60680

INTRODUCTION

The analysis of exploration geochemical data has for the most part been empirical. Because of its essentially numeric character coupled with the lack as yet of well defined mathematical models describing the source, migration and accumulation of minor and trace elements in rocks, it is not surprising that geochemical data have been subjected mainly to statistical analysis.

The use of statistics in exploration geochemistry, as in geology and geophysics, has increased significantly since the introduction of electronic computers in the early 1950's. Indeed, it is now sometimes difficult to separate statistical methodology from its computational aspects. There is a core of statistical knowledge of which the student or the professional engaged in the search for mineral deposits should be aware. It is therefore essential that basic statistical principles be firmly fixed in the mind of the explorationist. The present chapter should not be considered a substitute for a basic course in statistics. It should be accepted rather as a guide as to how current statistical methods can be used to enhance the interpretation of data gathered in geochemical surveys. Methods of statistical analysis of geologic data have received more detailed attention in texts by Miller and Kahn (1962), Krumbein and Graybill (1965), Griffiths (1967), Vistelius (1967), Koch and Link (1970, 1971) and Davis (1973).

Almost without exception the main objective in exploration geochemistry, to quote Hawkes and Webb (1962, p. 1),

" . . . is the discovery of abnormal chemical patterns, or *geochemical anomalies,* related to mineralization."

As we shall see, the definition of a "geochemical anomaly related to mineralization" implies more than just unusually high (or low) values of the content of an element or elements in the sampled media; it is related more specifically to the departure of a set of values from what is considered to be the normal background variation in a geochemical landscape. Many geochemical anomalies are unrelated to mineralization (see false anomalies, Chapter 4) and one of the main tasks of statistical analyses is to identify anomalies resulting from different causes. Success follows only if proper attention is paid to the nature of the sampling media, to the size, number and spacing of samples taken, and to the methods used to depict the spatial variation of the collected data. In addition this includes information on geologic as well as geochemical effects. Although many statistical problems remain, a number of techniques have been developed which have led to the more efficient collection, more detailed analysis and hopefully improved interpretation of the geochemical data concerned. In this chapter we discuss these techniques.

BACKGROUND VALUES

One of the major objectives of regional geochemical surveys is to establish the normal or background variation of an element considered important in the detection of economic mineral accumulations. This derives from the assumption that any such concentrations formed around an accumulation will be revealed by anomalous readings. Thus, if the distribution of a key element or a group of elements is defined, the presence of a hidden accumulation will be revealed by unusually high values of the elements in question. Consider, for example, Fig. 12-1 which is a histogram of the distribution of zinc in spring waters based on 3800 samples collected from an area of over 400 square miles in southwest

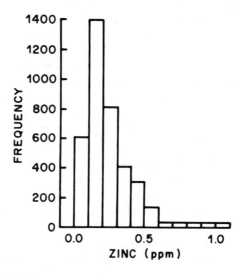

Fig. 12-1. Histogram of the zinc content in spring waters, southwest Wisconsin. Based on data from DeGeoffroy et al. (1968).

Wisconsin. The higher zinc contents are due presumably to the action of groundwater moving through halos associated with the zinc-lead ore bodies scattered throughout the area. The histogram exhibits a pronounced positive skewness towards these higher values which is a characteristic common to many trace element distributions. In the past, this tendency toward positive skewness has been attributed to the data following the lognormal distribution as shown in Fig. 12-2. This distribution has the property such that when the values are transformed by taking logarithms, the result is the normal distribution which is shown also in Fig. 12-2. On the assumption that the anomalous values for a given distribution of trace element contents are also the highest, the mode of the lognormal distribution can be taken as the background value. The mode rather than the mean is preferred since the mode is less sensitive to extreme values in a distribution. For grouped data, the mode is defined as the interval containing the greatest number of values. To avoid having to decide on the size interval for grouping a given set of data, however, an alternative is to take the median value of the distribution as the background value. The median is also less sensitive than the mean to extreme values. The median can be determined by plotting the data in the form of a cumulative frequency curve as shown for the example in Appendix C and taking as its value the 50th percentile. From the constructed cumulative curve, it is not difficult to select other percentile values (10th, 25th, 75th, etc.) and to use these to establish regional threshold values which were discussed in Chapter 4.

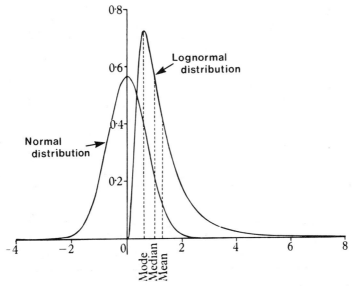

Fig. 12-2. Frequency curves of the normal and lognormal distributions. From Aitchison and Brown (1957).

The close similarity between the distribution of trace element contents and the lognormal distribution should not be regarded as proof that element concentration values follow any kind of lognormal law. Other distributions fit equally well; for example, Jeffreys (1970, p. 360) stated that the Pearson Type III distribution was appropriate for certain trace element distributions. For discrete bodies distributed randomly over an area, Griffiths and Ondrick (1970) have shown by sampling experiments that depending on the sampling strategy employed, a variety of observed frequency distributions can be obtained. Finally, Cameron and Baragar (1971) suggest that frequency distributions of elements reflect only whether or not the ore mineral is strongly or weakly segregated from the parent rock material. Thus, while it may not be possible to fit all empirical distributions to one type of probability distribution, the observed distribution does provide a method for establishing background variation.

The observed values for a given element are dependent upon both the sampled media and the parent material. Thus, systematic variation in background values is to be expected for different sampled material such as soils, stream sediments and glacial materials just as differences are to be expected for different rock types. Nichol (1971) has presented a chart shown in Fig. 12-3 in which the various mean minor-element contents of stream

	All data	Contrast	ORDOVICIAN						SILURIAN		DEVONIAN				CARB.	
			Acid volcanics	Augen schists	Basic volcanics	Sediments	Elmtree Grp.	Basic intrusives	Sediments	Volcanics	Sediments	Basic volcanics	Basic intrusives	Acid intrusive	Bathurst seds.	Bonaventure seds
Mn	1620	3	●	◉	◉									○	○	○
Ni	34	4			●	◉	◉					●	◉	○	○	◉
Cu	11	2	c	◉	◉	◉						●	●	○		
Zn	120	3					◉					●	●	○	○	
As	3	13	○				◉							●		
Mo	1·5	5			◉							●				◉
Ag	·3	2			◉							●	○	○		◉
Ba	400	2						○			◉	●	○	○		◉
Sb	3	2	◉			◉					○					●
W	2	1														
Pb	18	5	●				◉	○	○				●	○	○	
CxHM SS	3	2	●					○					○	○		○
HMW	<·001	3						●								
n	713		21	9	26	40	16	7	55	26	17	18	2	10	24	2

KEY
● Highmean content
◉
○
ᴑ Lowmean content

| ASSOCIATION | | | cxHM SS Mn Sb As | Ag Cu Mn | Ni Mn Cu Sb | Zn Cu Ni Pb As | HMW | Ba Pb cx HMSS | Pb | | Ni Ba | Zn Ag Ba Ni | Zn As Pb | Mo | Cu Ba cxHMSS | Mn Ni Cu Zn Ba Pb cxHMSS | Mn Zn Pb | Sb Ni Mn Ba | Mn cx HMSS |

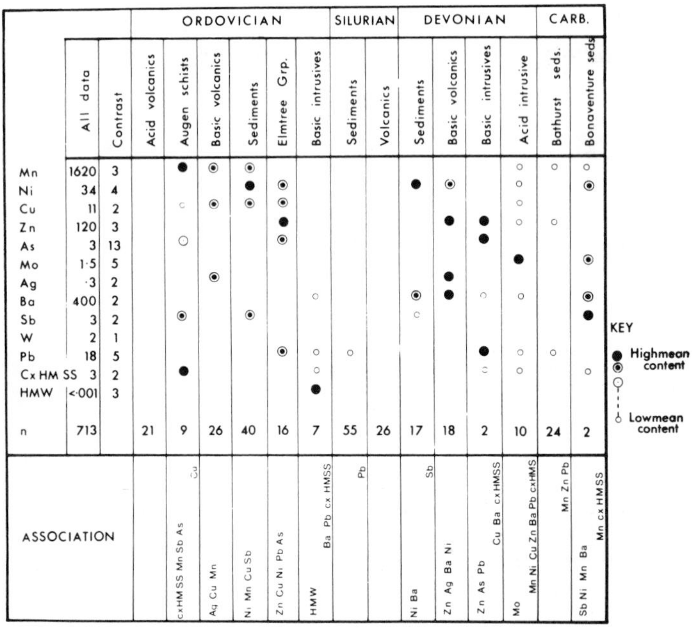

Fig. 12-3. Mean minor-element contents of stream sediments associated with different rock types. From Nichol (1971).

sediments associated with different rock types were compared. The chart was based on a random selection of 713 samples taken from multi-element data reported by Boyle et al. (1966) for a drainage reconnaissance survey in New Brunswick. The chart shows that regional background values vary with bedrock type. For background values established from observed frequency distributions of trace element contents therefore, care must be taken to insure that the samples involved are representative of the same or similar parent material.

The importance of establishing background values against which anomalous readings are contrasted is perhaps best exemplified by the recent discovery of a copper-molybdenum porphyry deposit in Yukon Territory, described by Archer and Main (1971). The location of the deposit was determined by stream-sediment sampling followed by grid sampling of soils and geological mapping. The map is given in Fig. 12-4. The map shows the deposit along with the surrounding anomalous samples assayed for copper for which the values were well above the regional threshold values established earlier during the reconnaissance investigations. We can summarize this discussion by emphasizing the importance of determining the background value of a given trace element and that when this is accompanied by geological mapping, the background content becomes a simple but powerful tool to aid in exploration.

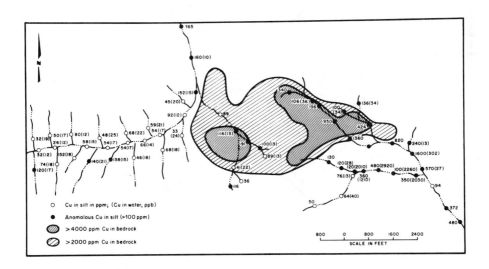

Fig. 12-4. Copper in silts and water, Yukon Territory. From Archer and Main (1971).

REGIONAL VARIATION

As important as the determination of background value is for the conduct of regional geochemical surveys, it is the prelude to the main task which is to ascertain the form and nature of the geochemical landscape. This task, however difficult, inevitably involves the geographic representation of the data in the form of maps. The use of maps is perhaps the most powerful single tool in the interpretation of spatially distributed data. Different maps show different things depending on the nature of the data, the amount of data, other information such as the geology, the cultural features present or whatever mathematical restraints exist for portraying data in two dimensions. Ideally, a map should be viewed as being either: (1) purely objective, or (2) largely interpretative. As an example of a purely objective map, Fig. 12-5 depicts the distribution of nickel in stream sediments in samples collected over an area covering 16,000 square miles in Sierra Leone. This map gives not only the location of each sampling point in the survey but also, based on the histogram in the lower right corner in the figure, the classification of each point according to the content of nickel observed in each sample. Furthermore, the regional geology has been ad-

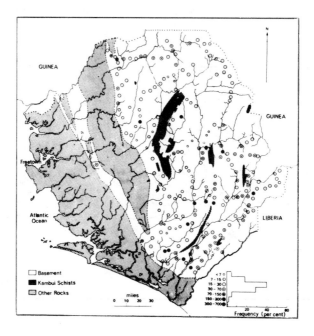

Fig. 12-5. Distribution of nickel in stream sediments over the Basement Complex, Sierra Leone. From Nichol et al. (1969). Reproduced from *Economic Geology,* v. 64, p. 205.

ded. Such a map reflects as nearly as possible the total information available for interpretation. Contrast this with Fig. 12-6 in which a set of statistical maps consisting of the linear, quadratic and cubic trend surfaces derived from the data in Fig. 12-5 are presented. The contours are expressed in ppm. These two examples allow a clear distinction to be made between the two types of map display. Maps of the type shown in Fig. 12-5 preserve the data at each sampling location but camouflage the general trend in the data whereas largely interpretative maps of the type shown in Fig. 12-6 reveal regional trends but fail to supply detailed information about a specific location. These two types of maps exemplify what is meant by local and regional variation. In practice, both types of variation are significant and considerable time and effort has been spent by geochemists and geologists in developing suitable quantitative methods of mapping. It is useful to consider general aspects of these methods.

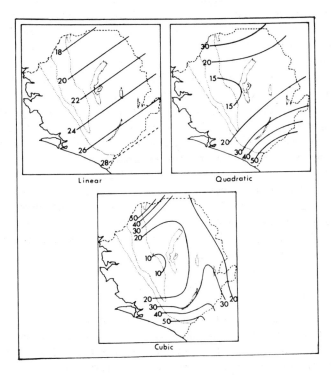

Fig. 12-6. Linear, quadratic and cubic trend surfaces of nickel distribution (ppm) in stream sediments over the Basement Complex, Sierra Leone. From Nichol et al. (1969). Reproduced from *Economic Geology,* v. 64, p. 206.

Trend Surface Method

The functional relationship between a value z_{xy}, say the content of some trace element designated z and the corresponding geographical coordinates (x,y), can be represented as

$$\bar{z}_{xy} = f(x,y)$$

where $\bar{z}_{xy}$ represents the average or expected value of z_{xy} and where $f(x,y)$ is given by some polynomial function or else by some Fourier series approximation. Most often, however it is the former. Based on least squares, the coefficients of these functions are calculated and trend surfaces are produced similar to those in the examples of Fig. 12-6. In general, the fitting of a trend surface does not depend upon the location of the data points and because of this, low-order polynomial surfaces have found wide application for establishing regional trends for geochemical data in which the sampling points are irregularly spaced. In deciding upon the significance of a given trend surface, however, it is important to consider the density and distribution of such points. A particularly good illustration of the effects of sample distribution and density in determining trend surfaces in geochemical prospecting is provided by Doveton and Parsley (1970). Fig. 12-7 depicts the progressive effect of clustered data on low-order polynomial trend surfaces. Clustering of data points tends to produce surfaces elongated parallel to the pattern of points. This effect can be minimized by insuring that the sample points in a survey are distributed as uniformly as possible within the survey area.

Another use of trend analysis is to select variables other than the geographical coordinates and to account for regional and local effects of an observed distribution of a given trace element or elements. Dahlberg (1968) has proposed for instance that the variation in trace metal content of stream sediments (T) can be characterized as a function of potential controlling factors using as a model

$$T = f(L,H,G,C,V,M,\epsilon)$$

in which L represents influences of lithologic units, H stands for hydrological effects, G for geologic features, C for cultural influences, V for type of vegetational cover, M for the effects of mineralized zones (an effect which may be present or absent, but which hopefully, is present) and ϵ for the error plus effects of factors not explicitly defined in the model. The basic idea is that by isolating and separating those factors unrelated to mineralization and by removing them from the regional trend, the possibility of recognizing anomalies due solely to mineralization should

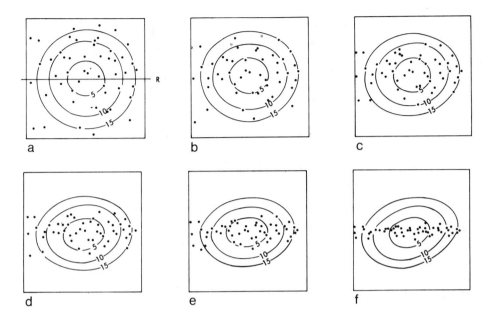

Fig. 12-7. Progressive effect of clustered data on quadratic trend surface. Illustration (a) shows the original surface and the reduced major axis (marked R) of the undeformed distribution of data points. Illustrations (b) to (f) show the progressive effect of narrowing the band width of data points about the reduced major axis. From Doveton and Parsley (1970).

become greater. In his paper, Dahlberg (1968) was able to demonstrate that the regional distribution of the copper content of stream sediments for an area in Pennsylvania was made up of local anomalies due to such effects as different lithology, known faults in the area, and cultural influences related to human activities. As an illustration, Fig. 12-8 shows the regional distribution of copper following the removal of simulated effects of cultural influences. The contours show the amount of the copper content which was removed. The results of Dahlberg's study are summarized in Table 12-1. Clearly, effects unrelated to mineralization in an area can and do affect the regional variation of trace element content but such effects if recognized can be removed. Similar studies of this kind are described by Rose et al. (1970), Rose and Suhr (1971) and Chatupa and Fletcher (1972).

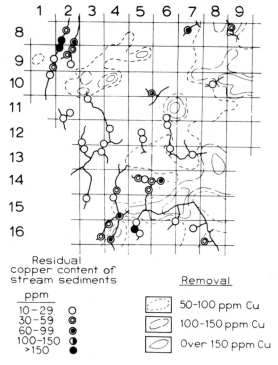

Fig. 12-8. Regional distribution of copper following removal of simulated effects of cultural influences. The contours represent the values removed. From Dahlberg (1968). Reproduced from *Economic Geology,* v. 63, p. 413.

Table 12-1. Percentage of samples reduced to less than 10 ppm by subtraction of simulated values for individual factors. From Dahlberg (1968).

Effect	Copper Content of Original Sample	
	30 to 50 ppm	*>50 ppm*
Cultural Features	57%	51%
Faults	40%	14%
Metabasalts	47%	34%
Prospects	42%	43%
	N=87	*N=41*

% = n/N where "N" = number of samples with greater than 30 or 50 ppm copper. "n" = number of samples reduced to less than 10 ppm copper by subtraction of simulated values.

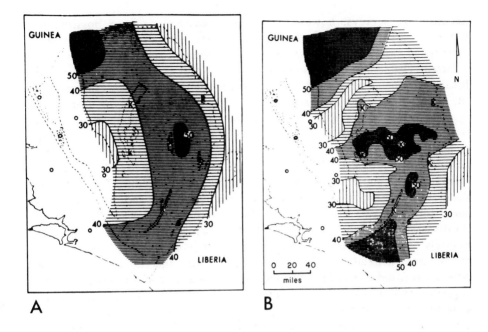

LEGEND

○ OTHER ROCKS
BASEMENT GRANITE
K KAMBUI SCHIST
〜 < 1·0 } CONTRAST
   < 1·0 }
● CONTRAST > 5·0
○ CONTRAST < 0·2

Fig. 12-9. Regional distribution of nickel content in soil samples over the Basement Complex, Sierra Leone. (A) cubic trend surface (B) moving average surface. From Nichol et al. (1969). Reproduced from *Economic Geology,* v. 64, pp. 207, 211.

Moving Average Method

A similar but slightly different approach to the problem of establishing regional trends which reside within geochemical data is to compute moving averages. In this method, the mean of the set of observations for samples contained within a constant-size window is computed as the window is moved progressively over the area covered by the data. For the rare cases where the data points are regularly spaced, relaxation methods described by Paul (1967) can be employed. In most geochemical surveys, however, it is not practical or even possible to sample on a regular grid so that the data points are more often irregularly distributed. The latter necessitates that a window of some arbitrary size be selected. The size of this window is chosen so that as nearly as possible the local variation is minimized relative to the regional variation. In general, the larger the window the greater is the smoothing effect on the data. This is analogous to choosing the degree of the polynomial function used to fit a trend surface to the data. As an indication of how the results obtained by trend surface methods are similar to those obtained by moving averages, consider the two maps in Fig. 12-9 for data obtained by Nichol et al. (1969) on the content of nickel in soils from Sierra Leone. The regional distribution of nickel is basically similar in both maps and in each map a small zone of high nickel content in the soils of the central zone is defined. As discussed by Nichol et al. (1969), the trends in the data were more accurately reflected by the moving average method which is not limited to a predetermined number of inflection points unlike the cubic trend surface which was fitted to the data. The moving average method, moreover, can be regarded as a weighted linear regression where the weights are applied to data points nearest the point of approximation to the regional trend. This approach has in fact been developed more formally within the mining industry by those whose interests lie mainly in the estimation of ore grade and production quality control. References to this literature may be found in Koch and Link (1970) and Davis (1973).

CORRELATION BETWEEN ELEMENTS

In a geochemical survey, it is unusual to analyze only a single element in a given set of samples. More likely, the contents will be determined for a number of elements. Since certain groups of elements respond more or less similarly to a given set of environmental conditions, it follows that mutual correlations between different elements will serve to identify more clearly the variations present in the geochemical landscape. Associations of some elements may lead directly to an interpretation of the type of deposit likely to be present in the area. On the contrary, certain associations of elements may indicate a false anomaly. Whatever their cause, correlations among elements provide additional information.

The use of simple correlation can be illustrated by an example in which it was desired to evaluate the use of different sampling media for a particular trace element. Timperley et al. (1970) reported a study in which they investigated the correlation between the nickel content in the soil and the nickel content in ashed leaves of *Nothofagus fusca,* the latter being much easier and speedier to sample. In their study, they observed a

statistically significant correlation in nickel content between the two different media sampled. Based on forty nine samples, the correlation coefficient was 0.82. The spatial association is shown in Fig. 12-10. They concluded from this study that sampling of ashed leaves of this particular plant
species was just as effective as taking soil samples. This type of information
is important in planning large scale orientation surveys.

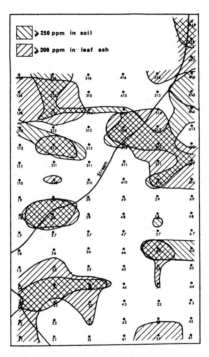

Fig. 12-10. Nickel anomalies in the Riwaka Basic Complex, South Island, New Zealand,
indicated by soil sampling and by sampling the ash of leaves from *Nothofagus fusca*.
From Timperley et al. (1970). Reproduced from *Economic Geology,* v. 65, p. 507.

Another example in which simple correlation proved effective was a
study of sedimentary lead-zinc deposits described by Michard and Treuil
(1969) who examined the correlations between selected trace elements of
mineralized and non-mineralized zones. Using the Spearman rank coefficient (for a discussion of rank correlation, the reader is referred to Kendall (1973, p. 494)), they observed the correlation between zinc and cadmium in the mineralized zone to be 0.672 based on nineteen samples,
whereas the correlation in the non-mineralized zone was found to be only
0.039 for twenty three samples. They attributed this marked difference in
correlation to the dissimilar physico-chemical environments between the
mineralized and non-mineralized zones.

It is not sufficient usually to consider a few correlations. The sources of variation of chemical composition in the sampled material are commonly diverse and in many cases have multiple origins. It is just as easy statistically to consider all the correlations between the element contents in the samples and to look for patterns among them.

The most simplistic procedure for correlating multi-element data is to generate the matrix which contains the correlations between all possible pairs of the elements being considered. In calculating these correlations, it is usual to transform the data first by taking the logarithms of the contents. This has the effect of removing much of the skewness of the original distribution of each element and greatly reduces the erratic fluctuations in the correlations calculated between different elements. Fig. 12-11 is a correlation matrix taken from a paper by Nichol (1971). The main advantage in presenting the correlations in a form of this type is that the patterns within the data are more readily apparent. Thus, for the 713 samples used in the study, the significant correlations of minor element contents of sediments and water can be summarized as follows: Zn is positively correlated with cx-HM, Pb, As, Cu and Ni; Pb with cx-HM, As and Zn; As with Pb and Zn. Unexpectedly, cx-HM is uncorrelated with W.

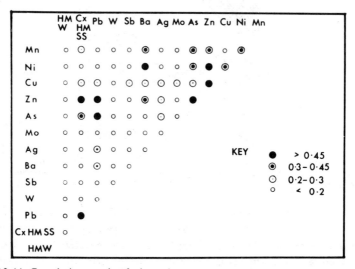

Fig. 12-11. Correlation matrix of minor-element contents of sediments and water based on 713 samples taken from drainage reconnaissance survey in the Bathurst-Jacquet area, New Brunswick. From Nichol (1971).

Rather than select mutually correlated groups of elements in this arbitrary manner, it is preferable more often to select them according to objective criteria. For instance, it may be more desirable to choose those mutually correlated elements which exhibit the greatest within-group correlation relative to the between-group correlation, taking into account all possible combinations of the given elements. This approach to correlation is recognized as cluster analysis. Cluster analysis is especially well adapted for situations in which the underlying causes of correlations among the variables are unknown or inadequately understood but where it is desirable to group the variables according to the mutual correlations. An extensive literature exists on cluster analysis and the reader is referred to Sokal and Sneath (1963), Parks (1966), and Gower (1967) for more details. Of particular interest is one special type of graphic output of cluster analysis. An example of this type is taken from a paper by Obial (1970) in which the results of clustering for the observed correlations of major, minor and trace element contents based on 170 stream sediment samples collected in a mineralized area underlain by Carboniferous limestone in Great Britain are represented. A dendrogram is shown in Fig. 12-12. It is a treelike diagram which depicts the mutual correlations among a

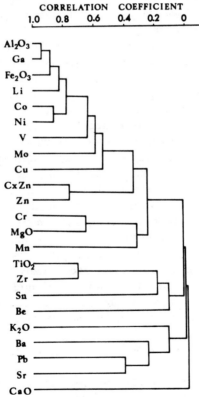

Fig. 12-12. Dendrogram showing result of R-mode cluster analysis based on the weighted pair-group average-linkage method of clustering for 170 samples. From Obial (1970).

given set of objects. The method of clustering may vary however and in Fig. 12-12 the results are based on the weighted-pair group average-linkage method described by Sokal and Sneath (1963). For this example, the following element groupings were noted. (1) Al_2O_3, Ga, Fe_2O_3, Li, Co, Ni, V, Mo and Cu; (2) CxZn and Zn; (3) Cr, MgO and Mn; (4) TiO_2, Zr, Sn and Be; (5) K_2O, Ba, Pb and Sr; and (6) CaO. When the elements with high correlations in each group were plotted as single element maps and compared, a closer similarity of their dispersion patterns was noted. Similarly, an increasing dissimilarity in the map patterns was observed for those elements in different groups or clusters. Thus the groupings established by correlation appear valid. A newer version of the dendrogram termed the dendrograph developed by McCammon (1968) promises even better visualization of the correlation structure of multi-element data.

Thus far, our concern has been with whole-correlations which exist among selected trace elements. We can ask, however, what the underlying factors are which give rise to an observed set of correlations. Furthermore, we can inquire as to how many of these factors are present and how their influence varies across any given area.

The subject of factor analysis is one which developed in the field of educational psychology and as a technique is one which has been applied mainly in fields of inquiry in which quantitative assessments of underlying causes are to be derived from a series of observations involving inter-correlated variables. Factor analysis is a method for reducing the complexity of a given set of intercorrelated data by accounting for the observed correlations among the variables in terms of the fewest possible number of underlying factors. Factor analysis is embraced by a very extensive literature and the reader should be forewarned before attempting to inquire after particular methods and procedures. The most complete modern treatment of factor analysis is a textbook on the subject by Harman (1967). A statistical treatment including the most up-to-date method for deriving factors can be found in a short monograph by Lawley (1971). As an introduction to factor analysis, the two-part article by Cattell (1965) is still the best starting point. In geology, factor analysis has been applied to a variety of problems and the interested reader should refer to articles by Imbrie (1963), Klovan (1966) and Miesch et al. (1966).

The basic factor analytic model may be expressed as

$$z_j = a_{j_1}F_1 + a_{j_2}F_2 + \cdots + a_{jm}F_m + d_jU_j \quad (j=1,2,\cdots,n)$$

where each of the n observed variables is described as a linear combination of m common factors and a unique factor. The reader should keep in mind that a factor is a mathematical concept and is not necessarily directly observable, although it could be, in certain situations. More often it is considered as some general characteristic of the samples being examined. It is generally assumed that the number of common factors, m, is much smaller than the number of observed variables, n. The coefficients (a_{jk}) of the factors are referred to as the loadings and are determined from the data. The coefficients (a_{jk}) for the common factors (F_k) are computed in a way which as nearly as possible will account for the observed correlations among the variables while the coefficients (d_j) of the unique factors (U_j) account for the remaining variance of each variable. The factors are considered initially at least as being independent variables with zero mean and unit variance. The problem in factor analysis is to determine the number of factors and the factor loadings which best reproduce the observed correlations. A number of computer programs are now available to carry out the necessary computations and the reader should refer to Cameron (1967), Garrett (1967), and Klovan and Imbrie (1971) for more details.

Factor	Factor Model				
	3	4	5	6	7
1	Ni Zn Ba Mn As Cu	Ni Ba Zn Mn As Cu Pb	HMW HMSS Pb Zn As	Ag Cu Pb	HMSS Zn Pb As Mn
2	Ag Pb HMW Sb Zn	HMW HMSS Pb Zn As	Ni Ba Zn Mn Cu As	Ni Ba Zn Mn Cu As Pb	Ni Ba Zn Mn Cu As
3	Mo W Mn	Mo W Mn	Mo Mn Sb As	Mo Mn As	Mo Mn As
4		Sb Ag Cu	Sb Ag Cu	Sb Cu	Sb
5			W	W	W
6				HMSS HMW Zn Pb	Ag Cu Pb Zn
7					HMW

Fig. 12-13. Metal associations of different factor models of the minor-element correlation matrix given in Fig. 12-11. From Nichol (1971).

Returning to the correlation matrix of Fig. 12-11, Nichol (1971) went ahead and performed a factor analysis and the results are shown in Fig. 12-13. In order to carry out the analysis on this particular data, Nichol performed the varimax rotation of Kaiser (1958) on the initial factor solution and followed this with the promax oblique rotation of Hurley and Cattell (1962). The details of these transformations are beyond the scope of this chapter; however, these operations have the effect of reducing the factor loadings to a simple structure, namely, placing as high loadings as possible on the fewest number of variables while preserving the independence of each factor and reducing the complexity further by allowing the factors to become slightly correlated. The metal associations given in Fig. 12-13 result from different factor models ranging from three to seven factors. On the basis of the geology, mineralization and surface environmental effects determined in the area, the seven factor model was judged the most significant. Taking into account all the information available, Nichol concluded that HMSS-Zn-Pb-As and Mn, the Ag-Cu-Pb and Zn, the Sb and the W factors were related to geology and mineralization and that the Ni-Ba-Zn-Mn, Cu and As, and the Mo-Mn-As associations were related to interactions of the geology, mineralization and dispersion processes in the surface environment. In this instance, a factor analytic approach allowed a more detailed interpretation of the correlations to be made.

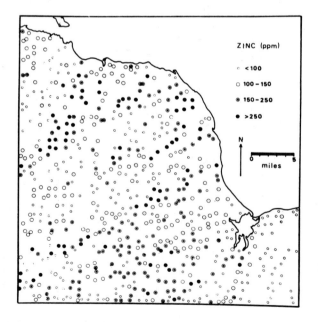

Fig. 12-14. Distribution of zinc (ppm) in stream sediments in the Bathurst-Jacquet River area, New Brunswick, after Boyle et al. (1966). From Nichol (1973).

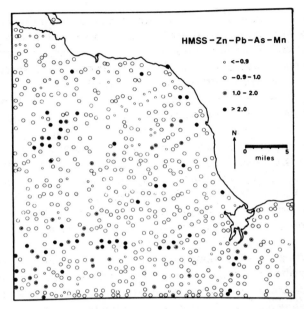

Fig. 12-15. Distribution of the HMSS-Zn-Pb-As-Mn association (factor scores). HMSS is heavy metals in stream sediments. From Nichol (1973).

In regard to mapping, a score or value for each factor can be calculated at each sampling point. Based on a method described by Kaiser (1962), a resulting set of factor scores can be used to map the spatial distribution of any given factor. Nichol (1973) provides an interesting example of this for geochemical data from New Brunswick. The areal distribution of zinc in the stream sediments collected in this study is shown in Fig. 12-14. The map indicates several areas where anomalous concentrations of zinc are located. In Fig. 12-15, the areal distribution of factor scores for a HMSS-Zn-Pb-As and Mn derived factor is given. It is seen that the areas of high factor scores of this factor match with sites of anomalous zinc content and furthermore that the pattern of zinc concentration is more apparent. Nichol went on to show that factor analysis in this instance served to identify the provenance of zinc content on the basis of element associations.

CLASSIFICATION OF SAMPLES

In a manner which is analogous to the search for groupings among trace elements, it is also possible to search for natural groupings among samples characterized by similar or dissimilar multi-element contents. In orientation geochemical surveys, it often happens that individual members of a suite of samples can be described as combinations of some number of end member components. In these instances, it is necessary first to identify the end member components and then to determine the proportion of each end member component present in each sample.

One approach adopted under these circumstances is Q-mode factor analysis. From a geometrical point of view, each sample is regarded as a p-dimensional vector where p is the number of elements. Samples collected in an area are represented as p-variate vectors distributed in a p-dimensional Euclidean space. For values of p greater than three, a representation is beyond ordinary graphical methods and the reader is referred to Van Andel (1964), Cattell (1965), Gower (1966) and Rayner (1966) for more complete details. An introduction to Q-mode analysis applied in geology can be found in Davis (1973), and a computer program has been provided by Klovan and Imbrie (1971).

An application of Q-mode factor analysis to exploratory geochemical data is described by Nichol et al. (1969) who studied the variability of trace element contents from Sierra Leone. A generalized geological map of the area is shown in Fig. 12-16A. For the thirteen elements analyzed (As, Co, Cr, Cu, Ga, Mn, Mo, Ni, Pb, Sn, Ti, V and Zn) using the Q-mode method, the first four vector factors extracted from the matrix of pairwise similarity coefficients were found to account for 91.9 percent of the total variation. Each extracted factor represented a particular end member composition. Scaled factor scores for each factor were calculated and these were used to prepare the maps shown in Fig. 12-16B-E. In their study, the authors associated vectors 1 and 3 with a granitic composition, vector 2 with an ultrabasic composition with somewhat lower values of Cu, Co, Ni, Cr, V, Mn and Zn, and vector 4 with ultramafic affinities but also with anomalous contents of As and Mo. In this way, the multi-element data groupings were explained in terms of variation of the bedrock. In regions where the bedrock composition is essentially unknown or the bedrock contacts poorly located, a Q-mode factor analytic approach is useful for separating the spatial effects of parent material composition on the observed distributions of trace element contents.

Another consideration in classification is the problem of assigning a sample to one of two or more known groups. This is the case for instance when the trace element contents of the bedrock are known or when the contents of selected trace elements in ore-bearing zones have been established relative to the contents in the country rock. In either case, what is desired is a separation between previously defined groups. This separation of groups can be effectively accomplished by discriminant analysis, a method first introduced by Fisher (1936) for the use of multiple measurements in taxonomic problems. Discriminant analysis has been used widely in geology and there have been a number of recent applications in geochemistry, in particular, those by Klovan and Billings (1967), Howarth (1971a, b), and Rose (1973). An introduction to the methods and techniques of discriminant analysis in geology can be found in McCammon (1969), Koch and Link (1972), and Davis (1973). The method most widely used is the linear discriminant function.

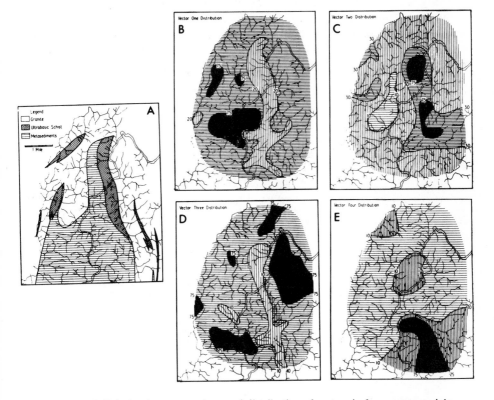

Fig. 12-16. Relation between geology and distribution of vectors in four vector models for trace element content data from Sierra Leone. (A) geology of the area. (B) to (E) distribution for each model vector. From Nichol et al. (1969). Reproduced from *Economic, Geology*, v. 64, p. 216.

Given a sample on which p measurements $(t_1, t_2, \cdots, t_p)$ have been taken, a single index value z is substitued according to

$$z = l_1 t_1 + l_2 t_2 + \cdots + l_p t_p)$$

The linear coefficients $(l_1, l_2, \cdots, l_p)$ are calculated from data which are available for the specified groups. The coefficients are chosen so that the differences in the values of z for the different groups are maximized relative to the differences in z within groups. Thus, for some unassigned sample having p measurements, the calculated z-value can be used to assign the sample to one particular group.

In a recent geochemical survey in Cyprus, reported by Govett (1972), it was observed that anomalous dispersion patterns for Cu, Zn, Ni and Co in rock samples could be detected in pillow lavas for a maximum distance of only 20 m from a sulfide deposit. Frequency distributions of the elements showed small systematic differences between background areas and groups of samples from traverses extending more than 2 km from mineralization. Govett (1972) demonstrated how discriminant analysis was successful in locating mineralized zones in this region of weak anomalies.

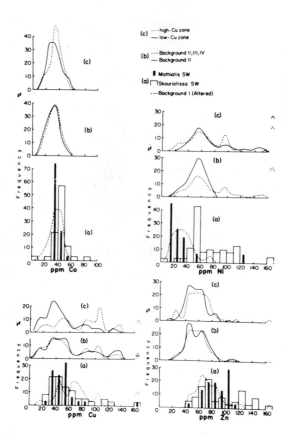

Fig. 12-17. Frequency distributions of Cu, Zn, Ni and Co in various groups of background, altered and mines samples collected in Cyprus. From Govett (1972).

The success of the method depended on the co-variation of all the elements involved. No single element as shown for the frequency distributions in Fig. 12-17 was found to discriminate uniquely between a rock 100 m away from mineralization and a rock 10 km from mineralization. It was considered likely however, that some linear combination of the four elements could be found which would vary as a function of proximity to mineralization and which, therefore, could be used to discriminate between anomalous and background conditions. The function derived was of the form

$$z = aCu + bZn + cNi + dCo + e$$

where a, b, c, d, and e were coefficients calculated from the data. Govett (1972) conducted a number of tests assuming different groups for different elements. Some of his results are shown in Table 12-2. Based on these and similar results, he was able to conclude that linear functions of the trace element concentrations enhanced the discriminatory power for detecting anomalous dispersion patterns in the rock sample data.

Table 12-2. Efficiency of discrimination between background and anomalous samples in standard type groups. From Govett (1972).

Test	Number of variables	Sample groups	Number of samples	% Correctly classified
1	4	1 - Background II	61	87
		2 - Skouriotissa SW	37	84
2	3	1 - Background II, III, IV	111	82
		2 - Skouriotissa SW and Mathiati SW	68	81
3	3	1 - High-copper zone	48	77
		2 - Skouriotissa SW and Mathiati SW	68	87
4	3	1 - Low-copper zone	63	84
		2 - Skouriotissa SW and Mathiati SW	68	81
5	4	1 - Altered Class	57	33
		2 - Background II	61	79
		3 - Skouriotissa SW	37	76
		4 - Mathiati SW	31	77
6	3	1 - Altered Class	57	23
		2 - Background II	61	64
		3 - Skouriotissa SW	37	73
		4 - Mathiati SW	31	71
7	4	1 - Altered Class	57	85
		2 - Background II	61	84

SURVEY SAMPLING

A major decision which has to be made at the outset of any large scale reconnaissance geochemical survey is the approximate density of sampling. The size of the sample, the method of sampling and the material to be sampled are all related to the analytical precision of the geochemical methods employed and can be treated separately. The greatest uncertainty lies in the unknown nature of the geochemical landscape. Invariably, prior knowledge of this surface is limited. The problem is to determine within reasonable limits how such a landscape can be exposed by means of point sampling at some sufficient number of locations. The scope of geochemical surveys varies so widely that sampling densities can be expected to range anywhere from one sample every few hundred feet, as along a stream bed for instance, to one or two samples per hundred square miles in areas where major changes in bedrock lithology are encountered only infrequently. Obviously, the density of sampling should be related to the inherent variability of the parent materials in question, which in turn is related to the size of the anomalies being sought. Such relations are never known with certainty in advance of a survey. At present, the guidelines used for determining the sampling density in a survey are based more on intuition than on sampling theory. There are some indications this viewpoint is changing.

In recent years, serious efforts have been made to identify the various sources of error in geochemical surveys and as a result computer simulation experiments have been conducted in order to understand better the effects of such errors for different sampling arrangements. There have been few instances where actual field studies have been carried out specifically to evaluate these effects. The notable exceptions are the studies by Baird et al. (1964) and Miesch (1964). The best account to date on the theory of error in geochemical data is given by Miesch (1967).

Sampling error in regional geochemical surveys is conveniently thought of in terms of a variance. This variance can be regarded as a measure of the uncertainty associated with a single measurement or group of measurements from a given locality. Over a given area, the sampling error will reflect the local relief in the geochemical landscape. Classic studies have shown, however, that if it is the mapping of an unknown surface which is desired, then the sampling strategy which is most efficient is the one which distributes the sampling points as uniformly as possible, that is, systematic sampling. The theoretical justification for this type of sampling has been given by Matern (1960) and Switzer (1968). To the extent that it is possible, orientation geochemical surveys should be conducted in this manner. Fig. 12-14, for instance, shows the distribution of sampling locations for zinc contents in New Brunswick as reported by Nichol

(1973). It can be seen that there is a fairly uniform coverage over the area of investigation.

In designing a survey sampling plan, the selection of the sampling interval is of primary concern. To check the effect of sampling density on the statistical significance of departures from a linear trend using artificially generated data, Miesch and Connor (1964) performed a series of experiments in which they divided an area into cells and selected at random a specified number of samples per cell. The sample observations were generated from a first-order trend surface defined for the entire area and to the trend value calculated at each location a random error component was added. From these data, a computed trend and a deviation about the trend was calculated and the results contoured as shown in Fig. 12-18. The contours, representing copper values, were based on cell means, each computed from a given number of samples. The number of samples per cell ranged from 2 in Fig. 12-18A to 16 in Fig. 12-18C. The sample size for each larger size sample was determined by first equating the mean within-cell variance to the between-cell variance and solving for n, the number of

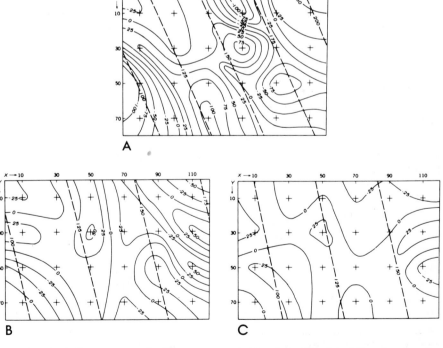

Fig. 12-18. Contour maps of cell means, computed trends and trend deviations (contour interval equivalent to 25 ppm). (A) 2 samples per cell. (B) 8 samples per cell. (C) 16 samples per cell. From Miesch and Connor (1964).

samples per cell required to balance the two variances. This method of comparing variances to determine sample size was based on work described earlier by Youden and Mehlich (1937). The sampling density was increased until the two variances were approximately equal. Clearly, the sampling error was reduced as the number of samples was increased. The comparison between the two variances provided a means of determining the sample size necessary to insure that trend deviations if they did occur could be considered to be significant. As Miesch and Connor (1967) point out however, in order to take a variance component approach, it is necessary beforehand to insure that stratified random samples (random samples drawn from homogeneous sub-populations) are taken and that proper attention is paid to the sample design so that the basic assumptions which underlie the analysis of variance as stated by Eisenhart (1937) are preserved.

In actual field studies which involve possibly the collection of thousands of samples, it may not always be practical to carry out a complete replicate sampling for the purpose of establishing the level of sample control needed to insure adequate coverage of an area. To this end, Garrett (1969, 1972) has proposed a compromise and has suggested that of the total sample sites visited, a subset of sites, preferably chosen randomly, be selected for more intensive subsampling for the purpose of determining local variability. (The reader should take special note to read both articles by Garrett as the first article contains an error which is subsequently corrected in the second article.) Ideally, the variance of the content of an element observed for all sites should not differ significantly from the variance calculated by taking a single value from each of the sites selected for more intensive subsampling. If this is the case, the mean within-site variance based on subsets of samples can be determined and the sample size adjusted accordingly. A comprehensive method for determining sample size has been proposed by Michie (1972) and applications of sample size determinations have been described by Howarth and Lowenstein (1972) and Plant (1971).

Apart from the formal approach for determining the sampling density, and therefore the sampling efficiency, it is possible also to view in retrospect a geochemical anomaly map constructed from a large number of samples and to consider how nearly that same map might have appeared had fewer samples been taken. In a study involving the distribution of Cu, Pb and Zn contents, Reedman and Gould (1970) have reported on some 10,000 stream sediment samples collected from over 7000 square miles of an area underlain by basement rocks in northeast Uganda. The geochemical results in this study were incorporated into single-element moving average maps and are shown in Fig. 12-19 (a-c). By dividing the map area into 75 cells in which each cell is one square mile in area and taking one sample

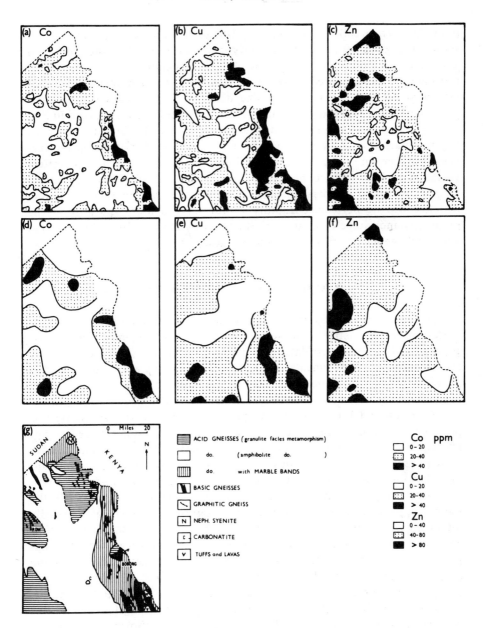

Fig. 12-19. Contoured moving average maps of cobalt, copper and zinc content of stream sediments from northern Karamoja, Uganda. (a), (b) and (c), maps based on 10,000 samples; (d), (e) and (f), maps based on 100 samples; (g) lithological map of northern Karamoja. From Reedman and Gould (1970).

point from each cell, maps based on 100 samples were prepared and are shown in Fig. 12-19 (d-f). In this study it was estimated that the collection and analysis of these 100 samples would have occupied approximately six man-months as compared to the approximately ten man-years which was estimated for the original study. The different sets of maps can be compared with each other and with the generalized geologic map shown in Fig. 12-19g. By visual inspection it can be seen that the broad pattern of regional variation is adequately reflected in the low-density sampling survey. Furthermore, the local variations in background levels reflect variation in bedrock geochemistry. Thus, where ore deposits are characterized by broad primary dispersion halos, low-density sample surveys which could be performed rapidly would serve a useful purpose at the reconnaissance stage of mineral exploration by drawing attention to areas with high geochemical background values.

ANOMALY DETECTION

As stated at the outset of this chapter, the main objective in exploration geochemistry is the detection of abnormal chemical patterns or geochemical anomalies, related to mineralization. As we have seen, the distribution of trace elements in an area often reveals a diverse and complex character of the geochemical landscape with the result that the search for anomalous zones is not always straightforward.

As Koch and Link (1971) have pointed out, there are essentially two possible conceptual approaches to geochemical exploration. In the first approach, trend surface methods or other regional smoothing techniques are used to seek a maximum or local maxima since it is likely that as one approaches the object being sought, the values of the indicator variables will increase. This represents a gradient or hill-climbing type of search. In the second approach, anomalies which are geographic clusters of unusually high (or rarely, low) values are defined as halos. This is the nugget effect and represents a hit-or-miss type of search. Mostly, however, these two approaches are combined. Orientation surveys are conducted mainly for the purpose of establishing the background variations and the regional trend pattern. Anomalous readings however can be identified by subtracting the regional values from the values observed at each sampling locality. In this way, residual maps can be produced. Such a map is shown in Fig. 12-20 which depicts anomalous copper content in soil samples taken from Bihar, India (Rao and Rao, 1970). The residual map in Fig. 12-20A was produced by subtracting a fitted trend surface value evaluated at each sampling point from the observed value at each sampling point. In their study, Rao and Rao (1970) noted the similarity between their residual map and

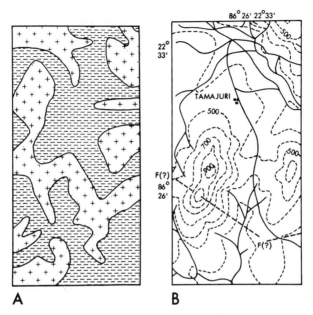

Fig. 12-20. Local variation in copper content in soil samples from Bihar, India. (A) residual map obtained by subtracting low-order trend surface from observed values. (B) topographic map of the area. From Rao and Rao (1970).

the topographic map for the area which is shown in Fig. 12-20B. It was observed that, in general, the hills were situated within the positive anomaly zones and that most of the tributaries of streams started within the negative anomaly zones, suggesting copper leaching. Positive anomalies not associated with favorable sites for leaching were considered to be indicative of the possible presence of an ore deposit. Such maps are useful in providing pictures of departures from regional trends. They do not indicate in general which anomalies are to be regarded as significant in terms of potential mineralization. The latter is best accomplished by follow-up surveys.

A similar but slightly different approach in recognizing anomalies is to regard values greater than some specified threshold value as anomalous. Hawkes and Webb (1962), for instance, suggested that values which are exceeded by not more than 2.5 percent of the total number of observations from an area could be considered anomalous. This is equivalent to saying that only 1 in 40 background samples is likely to exceed the threshold value. Since, in locating anomalies, the relative rather than the absolute values of the observations is important, such a rule seems valid. A difficulty arises, however, if the distribution of background values differ from the distribution of values associated with the anomalous zones. The extent of mineralization, furthermore, is expected to vary from one area to the next so that cutoff values determined in one region may not be appropriate in another. If more than a single group of values represents the difference between background and a given anomaly, the most appropriate threshold value should somehow divide the two groups of data. Bolviken (1971) has demonstrated how such a separation can be accomplished. He first constructed a hypothetical compound distribution composed of a background population and an anomaly population. As an example he compared this hypothetical distribution with a cumulative distribution for a set of copper values obtained from a stream sediment survey in Finnmark County, Norway. In this particular instance the situation was confounded by the fact that the distributions represented by the samples were derived from separate geological units and for his example it did not follow necessarily that the population with the highest mean content was associated with mineralization. To facilitate the dissection of empirical curves Bolviken (1971) prepared charts which permit the dissection of any curve into its component distributions. There are computer programs available however with the same purpose and two of these have been described by McCammon (1969) and Jones and James (1972).

Apart from establishing threshold values based on empirical distributions of selected trace element contents for the purpose of defining anomalies, it is possible also to consider the spatial geometry of anomalies along with their map expression. A new approach described by Puecker and Johnston (1972) identifies surface-specific points corresponding to given anomaly shapes. On the assumption that a sufficient number of samples have been collected during a survey so that the variation in the trace amount of an element can be represented as a continuous surface, then the neighboring values about any point on the surface can be used to classify the local surface shape. Puecker and Johnston (1972) have recognized shapes which correspond to a peak, pit, pass, ridge, ravine, slope, break or flat. These shapes are shown in Fig. 12-21. For each location on the surface, the value of the central point is subtracted from those of its neighbors in either clockwise or counterclockwise sequence. The result is an array of positive and negative differences from which each point can be classified as a surface-specific feature. The graphs on the right in Fig. 12-21 show the differences in the values for each type of feature. For this example, eight neighboring values are used. In practice, any number of values located at some specified distance away from a central value could be used. The method described by Puecker and Johnston has been used to catalogue the spatial characteristics of topographic maps but there is no reason why a similar approach could not be used to classify anomalies in geochemical maps. It could be a step closer toward automatic pattern classification. Clearly the method requires more investigation.

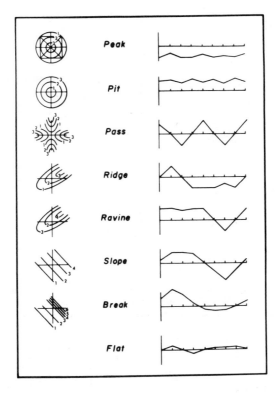

Fig. 12-21. Sequence of differences between central and neighboring elevations for various types of surface-specific points. From Puecker and Johnston (1972).

Having recognized a number of anomalous sample points on a geochemical map it is now necessary to determine if such groups of points are isolated or clustered. Obviously, a single anomalous point is suspect, whereas a cluster of anomalous points, reasonably close geographically, is regarded as strongly indicating an anomaly. Schuenemeyer et al. (1972) have developed a method for delineating clustered points in two dimensions by measuring perimeters of convex hulls. This is a fairly complex topic and the reader should refer directly to Schuenemeyer et al. (1972) for more complete details. Using their method, it is possible to determine which of the point distributions shown in Fig. 12-22 are clustered statistically and which are not. The method is based on the theory of points distributed randomly on a grid and involves the calculation of the probability that the measured perimeter of the convex hull defined for a given set of points is less than would be expected by chance. As an example in geochemical prospecting, Schuenemeyer et al. (1972) presented a map shown in Fig. 12-23 in which they plotted the locations of 21 out of 280

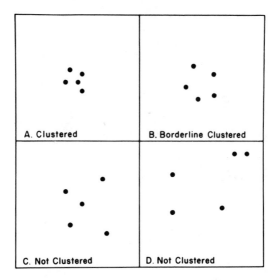

Fig. 12-22. Four sets of 5 points on a plane. Group A is clustered according to the method of convex hulls. Group B is on the borderline between the clustered and the not clustered. Groups C and D are not clustered. From Schuenemeyer et al. (1972).

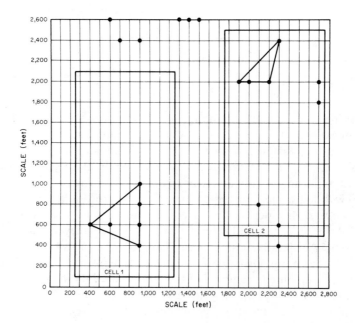

Fig. 12-23. Location of cells 1 and 2 used in analyzing geochemical data from Cape Rosier, Maine, for clustering and several selected convex hulls. From Schuenemeyer et al. (1972).

Table 12-3. Analysis of clustering in cell 1 of Fig. 12-23 using the method of convex hulls. For an explanation of a cutting perimeter for a specified risk level, see Schuenemeyer et al. (1972).

(1)	(2)	(3)
Risk level, percent	Observed number of subsets with perimeter less than the cutting perimeter at risk levels in column (1)	Probability that a cluster is nonrandom at risk level in column (1), percent
SUBSET SIZE = 3		
1	0	0.0
5	2	66.7
10	5	71.4
SUBSET SIZE = 4		
1	1	87.0
5	5	87.0
10	5	76.9
SUBSET SIZE = 5		
1	3	98.0
5	3	90.9
10	3	83.3
SUBSET SIZE = 6		
1	1	99.0
5	1	95.2
10	1	90.9

sample points for which the zinc contents were 133 ppm or higher based on a geochemical survey on Cape Rosier, Maine. Of the 364 intersections shown on the grid, zinc values were available for 280, with the others falling on roads, lakes and so forth. In their analysis, two subcells each containing 100 grid points were taken. The results of their analysis are summarized in Table 12-3 and Table 12-4. For cell 1, a convex hull enclosing 6 points had a 99 percent probability of being non-randomly clustered at a 1 percent risk level whereas for cell 2, a convex hull enclosing 4 points had only a 22 percent probability of being non-randomly clustered at the 5 percent risk level. Thus, if only one cluster or anomaly were to be selected for further sampling on the basis of zinc values, clearly the first location would have been chosen.

Essentially we have been discussing pattern recognition techniques for anomaly detection. The problem of detection begins with the choice of features of the spatial pattern which are to be measured. Associated with any anomalies related to mineralization, however, are other unsought patterns which may give rise to the same set of features. The basic problem is to devise a feature set which will differentiate the signal (anomaly) from the noise (background). Recently, Howarth (1972) described three-color map-

Table 12-4. Analysis of clustering in cell 2 of Fig. 12-23 using the method of convex hulls.

(1)	(2)	(3)
Risk level, percent	Observed number of subsets with perimeter less than the cutting perimeter at risk levels in column (1)	Probability that a cluster is nonrandom at risk level in column (1), percent
SUBSET SIZE = 3		
1	0	0.0
5	0	.0
10	2	26.3
SUBSET SIZE = 4		
1	0	0.0
5	1	22.2
10	4	36.4
SUBSET SIZE = 5		
1	0	0.0
5	1	26.3
10	4	41.7
SUBSET SIZE = 6		
1	0	0.0
5	1	41.7
10	1	26.3
SUBSET SIZE = 7		
1	0	0.0
5	0	.0
10	0	.0
SUBSET SIZE = 8		
1	0	0.0
5	0	.0
10	0	.0

ping as a display tool for visual pattern analysis of spatially distributed data. Clearly, these and other pattern recognition techniques will receive closer attention in the future.

COMPUTERIZED DATA SYSTEMS

In this chapter we have discussed various statistical methods and their role in the handling and analysis of geochemical data. We have not discussed in any detail the problems of data handling or the exact nature of the calculations involved in the use of these methods. This is because most of the processing and statistical treatment of geochemical data can now be performed by computer. To a large extent, the computer has rendered obsolete the monotony formerly associated with the calculations necessary for many of the statistical methods described earlier and has actually made possible the processing of large amounts of data on a fairly routine basis. It has freed the explorationist from the drudgery of having to perform endless calculations either manually or on a desk calculator and has expanded significantly the scope of data analysis. The numerical treatment of multi-element data before computers was barely possible. New problems have developed, however, and as Nichol (1973) has pointed out, the successful application of computer related procedures for data analysis in geochemical exploration requires close cooperation between geologists, geochemists, statisticians and programmers. The geologist and geochemist must be aware and have an understanding of the factors which have a bearing on minor element distribution in a search area and must be able to convey this information to the statistician who in turn can indicate the most powerful statistical tools available for planning a survey and in analyzing the data. Finally, to insure that the data are properly handled in the computer, a programmer is essential. The greatest potential in this teamwork of scientists lies in the ability to provide highly unique information which might otherwise be lost. Unquestionably, the use of computers has aided significantly in the development and implementation of more appropriate sampling, analytical and interpretative techniques in exploration geochemistry.

A consequence of the increased capability of information processing by computers has been the development of systematized field data recording systems. These systems ensure, firstly, the standardized recording of field data by a number of personnel, something which otherwise is difficult to obtain. Secondly, as we have seen, a number of factors, other than mineralization, may affect the element distribution in a sampled material and therefore it is necessary to take into account the local environment. As an example, Nichol (1973) has described an early system developed by the

Geological Survey of Canada designed for a geochemical drainage reconnaissance in New Brunswick which included the recording of information on bedrock type, character of the sample and the sample environment. In terms of the latter, this included such factors as flow rate of water, water level, turbidity and color of water, color of precipitate or stain, sample location in stream profile, color of sediment and sediment size analysis. To expedite the recording of this information the observations were assigned simple numeric values and were recorded on a field card. The information recorded in this manner was transferred later to computer cards for subsequent processing.

Many geochemical surveys are conducted for the sole purpose of locating a specific type of deposit whereas other surveys are conducted mainly to determine the regional variation of element contents for subsequent investigations. In the latter case particularly, it is important that information collected at an earlier date be subsequently retrievable at a later date and that the information obtained from a number of areas be directly comparable. With this objective in mind, the United States Geological Survey over the past several years, has developed an information system called the Rock Analysis Storage System (RASS). This system has been described by Botbol (1971). The system has provision for all analytical results and brief geological descriptions for every sample collected by survey personnel. Conceptually, RASS is a large rectangular array with a fixed number of columns and an infinite number of rows. The columns represent geological and chemical variables, and each sample occupies one row. Originally, the coded descriptive information on samples analyzed were recorded on standard 80-column punch cards and stored in a card file. Beginning in 1968, a computer-based file was initiated in which the data and descriptive information on samples were entered on magnetic tape and retrieved by means of computer processing techniques. This is the file system in use at the present time (instructions for submitting samples to the Branch of Analytical Laboratories and laboratories of the Field Services Section of the Branch of Exploration Research, and descriptions of procedures and computer systems for processing geochemical data are contained in a Sample Submittal Manual, Third Edition, 1969, prepared by the Data Processing Group, Branch of Geochemical Census, United States Geological Survey, Denver, Colorado). The collection of data, stored in such a way which allows for computer processing at a later date makes it more likely that variations in metal content due to factors other than mineralization will ultimately be recognized.

In handling the large volume of analytical data collected in geochemical surveys, and in retrospective studies involving the retrieval of large amounts of data from magnetic tape files, it is only natural that a system of computer programs designed specifically for the analysis of these data would be devised. The statistical methods and techniques described in this chapter form the basis of such a system.

Coupled with RASS, the United States Geological Survey has developed STATPAC, a collection of computer programs designed specifically for the reduction and statistical analysis of data which can be represented in the form of a two-dimensional matrix. For a more complete description, the reader should refer to Eicher and Selner (1971). The STATPAC system has been written to accommodate qualitative as well as quantitative data. STATPAC is general and can be useful in petrology, geophysics and hydrology in addition to geochemistry. A flow chart depicting the general scheme of STATPAC is given in Fig. 12-24. Although this is a fairly complex diagram, the main essentials consist of (1) a data source, (2) data preparation, (3) data reduction and analysis and (4) data display. A variety of options at each stage of processing allows a user maximum flexibility in the direction of data flow from beginning to end. The decisions which are required reflect the level of understanding on the part of the investigator as to the degree of complexity involved in the factors which interacted to produce the observed variation in content of the elements determined in a given geochemical survey.

This brings us to perhaps the most difficult question in the analysis of exploratory data and that is how to evaluate the results of a survey. As stated already, the purpose of a geochemical survey is to locate anomalies related to mineralization. The fact is indisputable. What is disputable however are the actions taken or decisions made with respect to the data collected during a survey.

There is, of course, a certain logic in geochemical exploration which for each survey gets translated into decisions made regarding the number and type of samples to be collected, the spacing between samples, the analytical techniques and the precision of measurements in the determination of elements in the samples, the statistical methods for analyzing the data and the criteria judged to be significant in evaluating the results of a survey.

To aid in the decision making process, Dahlberg (1971) has proposed an interesting algorithm. It is intended to assist the explorationist in deciding whether to continue, alter or discontinue the effort at each stage of exploration. The algorithm is shown in Fig. 12-25 in the form of a flow chart. The underlying principle is that of feedback. Exploration moves ahead when the relevant criteria specified at the preceding stage are satisfied. The four major aspects of the algorithm involve the precision, relevance, persistence and reliability of the observed data. These are seen as the decision steps represented by the diamond-shaped boxes in the diagram. Such an algorithm of course does not guarantee exploration success. It does insure, however, that proper attention will be given to factors

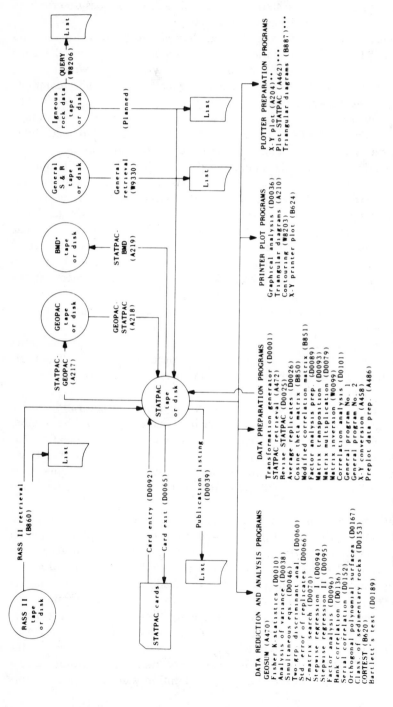

Fig. 12-24. System of computer programs for geochemistry and petrology RASS-STATPAC. Prepared by Branch of Regional Geochemistry, Geologic Division in cooperation with Denver Branch of Computations, Computer Center Division. U.S. Geological Survey (1970).

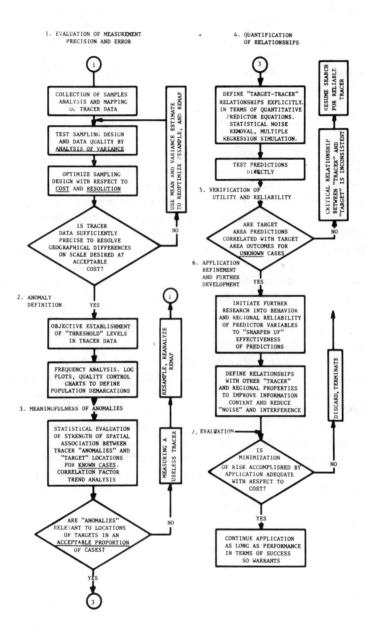

Fig. 12-25. Algorithmic development of a geochemical exploration program. From Dahlberg (1971).

which in the past have contributed to the discovery of deposits. Perhaps most important, as Dahlberg (1971) states, "it provides an integrated framework within which the entire exploration program can be mapped, prior to execution, from start to finish."

We have been concerned in this chapter with admittedly some of the more advanced statistical techniques for analyzing geochemical data. As pointed out elsewhere in this book, in the vast majority of cases, the straightforward approach of calculating the mean and standard deviation, preparing histograms and cumulative frequency curves and other simple graphical plots will provide the exploration geochemist with the basic information he needs in order to interpret his data, at least as a first approximation. The reason for presenting the more advanced statistical techniques is to make the reader aware of certain new developments in data analysis and to indicate something of its potential value. There is no doubt that as the geochemical methods of data acquisition and analysis become increasingly sophisticated, more, not less data, will be gathered. Moreover, as older areas are resurveyed (as for example, Great Britain), newly acquired data will have to be integrated with existing data. What this means is that the explorationist will have to contend with significantly larger volumes of data of a more diverse nature and will be expected to make correct interpretations within relatively shorter periods of time. Simultaneously, as the cost of exploration increases, the explorationist will need to extract greater meaning from the available data. It is likely that in future greater emphasis will be placed on the planning and design of orientation surveys, more careful sampling for follow-up surveys and more formal methods of assessing the errors of observation. It follows from this that the use of statistics and the role of statistical analysis in the proper treatment of exploration geochemical data will increase significantly in the coming years and that many of the techniques described in this chapter will be applied on a more routine basis. It is important to gain insight into the nature of these techniques now so that in future these techniques can be used without any delay caused by lack of understanding of the underlying principles.

13/ *Application Of Geochemistry To The Search For Crude Oil And Natural Gas*[1]

BRIAN HITCHON

Senior Research Officer

Alberta Research, Edmonton, Alberta, Canada

INTRODUCTION

The application of geochemistry to the search for crude oil and natural gas differs in some fundamental aspects from geochemical prospecting for mineral deposits. This difference exists despite similarities in origin between oil fields and some types of mineral deposits, more specifically those mineral deposits associated with unmetamorphosed sedimentary rocks. All mineral deposits and oil fields possess some common features. For example, all are aggregates of once widely dispersed matter subsequently accumulated at specific sites where physical and chemical changes in the aqueous carrier fluid caused deposition. These sites are often structurally controlled. As with oil fields, the fluid carrier for mineral deposits is governed by a well recognized system of energy potential fields (fluid, thermal, electroosmotic and chemicosmotic). Further, neither the components of the mineral deposit, especially the common base metals, nor crude oil are usually very soluble in the aqueous carrier. An important exception is natural gas. One major difference between mineral deposits and oil fields is the mobility of crude oil and natural gas in their own right. However, both the mineral deposits and oil fields are subject to dissipation. Another link between mineral deposits and oil fields is related to the phenomenon of metal-organic reactions (reviewed by Saxby, 1969) within the major geochemical cycles.

Comparisons and contrasts between geochemical exploration for oil fields and mineral deposits are summarized in Table 13-1. Succinctly, as

[1] Contribution No. 643 from Alberta Research.

509

Table 13-1. Comparisons and contrasts between geochemical exploration
for oil fields and mineral deposits.

Oil fields	Mineral deposits
Use both direct (organic) and indirect (inorganic) indicators	Essentially use only inorganic indicators
Predominantly deep exploration; shallow prospecting equivocal	Essentially only surface and shallow prospecting
Environment of exploration work in difficult conditions, e.g., high PT and H_2S	Surface conditions generally
Knowledge of hydrodynamics essential — but not applied sufficiently often	Knowledge of hydrodynamics important — but seldom applied
Use of microbiological indicators increasing, but still considered equivocal	Microbiological indicators almost never used
Biogeochemical and geobotanical techniques using living macro organisms never used	Minor use of biogeochemical and geobotanical (vegetation) techniques using living macro organisms
Accumulation sought is mobile, and may be seeping to the surface	Accumulation sought is static
Can examine kerogen (the source of the crude oil) and determine the type and amount of hydrocarbons evolved	Cannot examine the source of the ore — source often unknown, but fluid inclusions provide some information about carrier fluid

might be anticipated from knowledge of the fundamental differences between the deep organic deposits (oil fields) and the shallow inorganic deposits (mineral deposits), the dominant successful techniques used in the search for oil fields relate to the detection and study of organic matter encountered during drilling, and for mineral deposits to the detection of inorganic components at or close to the surface. This dichotomy of search techniques is not exclusive however. In the case of the search for oil fields much research has been carried out and there is increasing interest in the use of inorganic, as well as organic, indicators, especially those associated with formation waters encountered during drilling. Work on shallow prospecting techniques using both organic and inorganic indicators has generated considerable controversy in the petroleum industry regarding its efficacy and, at best, surface techniques of prospecting for oil fields are believed by many explorationists to be equivocal. Prospecting for mineral deposits is commonly carried out close to or at the surface, but the same techniques are now being developed to search for new buried deposits or extensions of existing deposits at depth. However, it should be stressed that the major value of petroleum geochemistry, in its widest sense, is not in direct prospecting but rather regional evaluation of the petroleum potential of sedimentary basins, including individual stratigraphic intervals.

One critical feature in common to the surface techniques applied both by the petroleum industry and by the mining industry is the general lack of knowledge of the near-surface formation water flow regime. This feature will be dealt with in more detail elsewhere in this chapter, but it is relevant to note here that no published study exists on surface prospecting, either for oil fields or mineral deposits, that is supported by a rigorous examination of the near-surface formation water flow regime.

Geochemistry may be applied at several stages during the exploration of sedimentary basins for crude oil and natural gas. During the initial stages, when few wells have been drilled and no commercial accumulations discovered, proper application of modern geochemical techniques provides reliable information on the petroleum generating potential of the basin and allows selection of the major source rock groups. Once hydrocarbons have been discovered it is important to classify the crude oils and natural gases into geochemical families having distinctive sources, hydrologic histories and timing of generation, migration and accumulation. In this manner geochemically related hydrocarbons may be sought in unexplored areas and different stratigraphic units. In the final stage of exploratory sophistication one may visualize computer modeling of all available geological, geophysical, geochemical and hydrologic data to seek out the last commercial accumulations.

In this chapter we will first describe the successful techniques used to determine the petroleum generating potential of sedimentary basins. This is followed by a review of processes which may alter crude oil once it has accumulated within the reservoir and how knowledge of geochemistry assists the search for high quality crude oils. The classification of crude oils and natural gases into geochemical families is difficult and the analytical techniques with which the individual organic marker compounds are determined are complex and sophisticated. Guidelines by which the problem of classification may be approached will be outlined. As the most common fluid recovered during drilling is formation water it is appropriate, next, to consider the present and potential value of knowledge of the geochemistry of formation water as applied to the search for crude oil and natural gas. The reasons why surface prospecting is still considered equivocal will be noted, and although these reasons do not apply to offshore prospecting, there are other phenomena to take into consideration in these regions. Finally, a personal view will be offered regarding the ideal application of geochemistry to the search for crude oil and natural gas.

REGIONAL EVALUATION OF PETROLEUM POTENTIAL

Crude oil and natural gas originate via a kerogen pathway from organic matter deposited with the sediments. This is illustrated simplistically in Fig. 13-1. The original organic matter comprises a small amount of allochthonous material inherited from pre-existing sedimentary rocks (mainly more stable hydrocarbons and NSO compounds), with the major portion being autochthonous material. This includes a dominant amount of high molecular weight biopolymers, defined on the basis of their solubility in acids and bases (Fig. 13-2), and a smaller amount of lipids, carbohydrates and amino acids. Within the first few tens of meters of burial the organic matter is subjected to intensive biochemical degradation within an increasingly anaerobic environment. Deamination, decarboxylation, saponification of esters, and denitrification are some of the processes

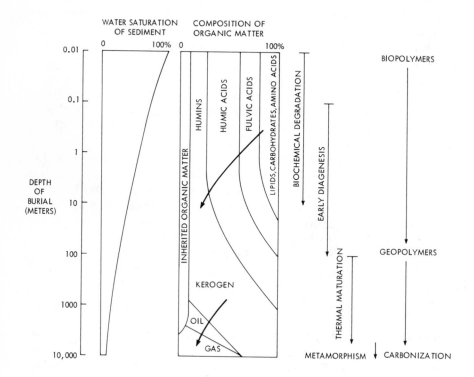

Fig. 13-1. Model for generation of crude oil and natural gas via a kerogen pathway from organic matter deposited with the sediments (after Tissot, unpublished lectures to Ecole Nationale Supérieure du Pétrole, Paris).

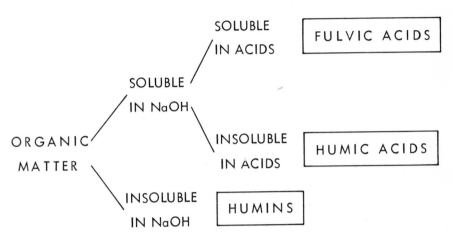

Fig. 13-2. Classification of biopolymers based on solubility in acids and bases (after Stevenson and Butler, 1969).

operating in this regime and they produce increasing amounts of complex, polycondensed, high molecular weight kerogen. The production of kerogen is represented by the upper arrow in Fig. 13-1.

Kerogen may be thought of as a geopolymer created by biochemical degradation and early diagenesis of original biopolymers. Its molecular architecture is incompletely deciphered at this time and future work will undoubtedly show a range of structures, the variations of which may well play an important role in some of the variations in crude oil composition. Basically, it consists of nuclei of condensed polycyclic structures (mainly aromatic, but with some naphthenic and heterocyclic molecules), crosslinked by alkyl chains, functional groups and heteroatoms, the whole forming a 3-dimensional network (Fig. 13-3). The lower arrow in Fig. 13-1 represents the pathway by which kerogen is converted to crude oil and natural gas during thermal maturation.

A considerable amount of work has been done on the geochemistry of kerogen but it is not necessary to determine its detailed molecular architecture in order to evaluate its petroleum generating potential. A number of procedures have been developed for the isolation of kerogen from sedimentary rocks (Forsman and Hunt, 1958; Forsman, 1963; Robinson, 1969; Saxby, 1970; Durand et al., 1972), and Fig. 13-4 shows a simple flow sheet. Private and published information (McIver, 1967; Long et al., 1968; Durand et al., 1972; Espitalié et al., 1973) on several sedimentary basins demonstrates a relation between the elemental composition of kerogen and the products of its maturation (crude oil and natural gas). The chemistry of kerogen pathways during petroleum generation parallel those of the maturation of coal macerals (Fig. 13-5) and basically amount to a loss of

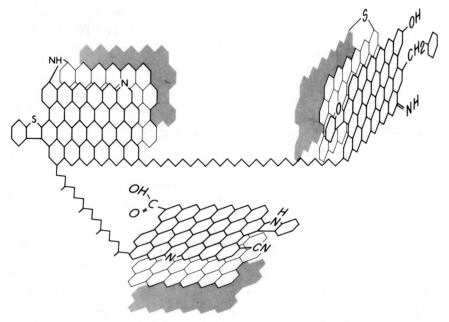

Fig. 13-3. Schematic molecular architecture of kerogen showing three dominantly aromatic nuclei, cross-linked by alkyl chains, and a variety of functional groups and heteroatoms.

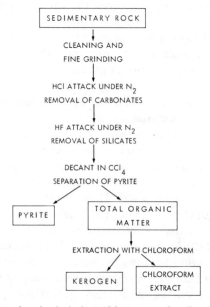

Fig. 13-4. Flow sheet for the isolation of kerogen (after Durand et al., 1972).

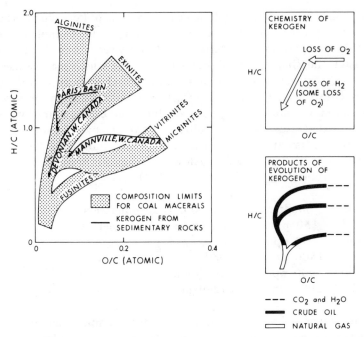

Fig. 13-5. Kerogen pathways during maturation (after Durand et al., 1972).

oxygen (as carbon dioxide and water) followed by a loss of hydrogen (mainly as hydrocarbons, with some additional loss of oxygen) as maturation proceeds. The products of evolution of kerogen are first immature natural gas, then crude oil with some natural gas, then natural gas alone. Changes in composition of kerogen with maturation also include a reduction in the amount of sulfur (Table 13-2), which may occur as H_2S in the more mature natural gas (Le Tran, 1971; Connan et al., 1972). Thus, a relatively simple elemental analysis can indicate the degree of maturation of kerogen and the favorability for crude oil or natural gas.

Table 13-2. Change in composition of kerogen with maturation (Toarcian shales, Paris basin, France). After Durand et al. (1972).

Depth (m.)	Elemental composition (wt %, ash-free basis)					Atomic ratios			
	C	H	O	N	S	H/C	O/C	N/C	S/C
Outcrop	74.9	7.9	10.4	2.0	4.8	1.26	0.104	0.023	0.024
1065	77.8	8.3	8.9	1.9	3.1	1.28	0.086	0.021	0.015
2450	85.4	7.1	5.0	2.3	0.2	1.00	0.044	0.023	0.001

Although the isolation and elemental analysis of kerogen are relatively simple procedures, they are time consuming and cannot be applied profitably to all samples of cuttings and cores recovered during drilling. Two faster and simpler techniques are available which may be used more widely. The extraction of soluble organic matter with an organic solvent such as chloroform provides information on the quantity of potentially mobile organic matter, though not on the degree of maturation. McIver (1962) has suggested the use of ultrasonics to remove the soluble organic matter from sediments. Source-rock quality scales have been developed and that of Philippi (1956) is given below, although Philippi used diisopropyl (a hexane isomer) as a solvent:

0 - 50 ppm	Very poor (noncommercial)
50 - 150 ppm	Poor (marginal commercial)
150 - 500 ppm	Fair
500 - 1500 ppm	Good
1500 - 5000 ppm	Very good
> 5000 ppm	Excellent

The second technique concerns the microscopic examination of the kerogen. Coal petrographers have developed the principles used for the identification and classification of kerogen although the comminuted and disseminated nature of kerogen, compared to coal, preclude use of the more refined techniques. However, reflectance of some types of vitrinitic organic matter is rather widely used as a technique to characterize the maturation of kerogen, and may be of value in determining the relation of coalification and deformation, as explained briefly by Hacquebard (1972). The relation between organic matter and petroleum has been recognized for many years but it is only recently (Correia, 1969; Staplin, 1969; Burgess, 1974) that scales have been developed to relate the degree of alteration of kerogen to the associated hydrocarbons (Table 13-3). Thermal maturation modifies the kerogen locked in the sedimentary matrix and thus the kerogen reflects the maximum paleo-temperature. For mapping, a thermal-alteration index of five progressive degrees of alteration is recognized. Where the two highest degrees prevail, prospects for wet gas and crude oil are negligible unless there has been post-thermal migration from outside the thermally affected area. In terms of practical exploration this has been evaluated for the western Canada sedimentary basin by Evans and Staplin (1971).

Two interesting techniques which are still in their development stages and which show promise of eventual practical use differ in major respects from the conventional isolation and analysis of kerogen. In the first, the soluble organic matter is extracted with chloroform, the rock sample is

Table 13-3. Kerogen thermal-alteration scale. After Staplin (1969).

Thermal-alteration index	Degree of alteration	Colour of organic matter	Associated hydrocarbons
1	None	Fresh, yellow	Crude oil, wet or dry gas
2	Slight	Brownish yellow	Crude oil, wet or dry gas
3	Moderate	Brown	Crude oil, wet or dry gas
4	Strong	Black	Dry gas
5	Severe	Black with additional evidence of rock metamorphism	Dry gas to barren

then pyrolysed and the degradation products of the kerogen identified by gas chromatography (Giraud, 1970). It is hoped that the relative contents of the light hydrocarbons and the nature of the degradation products, especially those from high temperature pyrolysis, will permit the characterization of the source rock and the determination of its degree of evolution. In the second technique the rock sample is examined by electron spin resonance spectroscopy (see Table 13-5 for explanation of this technique) and the paramagnetic characteristics of the kerogen determined (Pusey, 1973). The technique can be used on small samples (cuttings), is rapid, inexpensive and nondestructive, can be used with a wide variety of rock matrices, and is effective throughout the temperature range expected. It is claimed that, like the kerogen thermal-alteration scale (Table 13-3) it reflects the maximum paleotemperature but it should be possible with the ESR method to determine the maximum paleotemperature fairly accurately. Once the relation between the ESR and petroleum-generating characteristics of kerogen are known, this technique will be a useful adjunct in the arsenal of techniques used to determine petroleum generating potential.

We have already discussed the various methods by which the potential of kerogen for generating petroleum can be evaluated, and have observed that thermal degradation is the controlling mechanism. The generation of immature natural gas is followed by that of crude oil and wet gas — or as expressed by Vassoyevich et al. (1970) and Kartsev et al. (1971), the principal phase of oil formation — and finally by the generation of dry gas as the temperature increases. If the time available for oil generation is held constant, then the depth at which we may expect the maximum of the principal phase of oil generation will depend on the geothermal gradient — more strictly on the maximum geothermal gradient. With a high

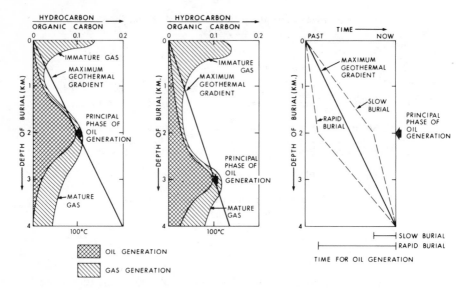

Fig. 13-6. Relation of principal phase of oil generation to geothermal gradient and time.

maximum geothermal gradient the maximum of the principal phase of oil generation will be at a shallower depth (Fig. 13-6, left hand diagram) than with a low maximum geothermal gradient (Fig. 13-6, centre diagram). For example, in the Los Angeles and Ventura sedimentary basins of California described by Philippi (1965), the geothermal gradients are 39.1°C. per kilometer and 26.6°C. per kilometer, respectively. Both basins comprise a thick sequence of Middle and Upper Tertiary sedimentary rocks, with production from Upper Miocene/Lower Pliocene strata. We may normalize the productive capacity of the shales of a basin by expressing the content of their extractable hydrocarbons in terms of unit quantity of their total organic carbon. Both the hydrocarbon content and the hydrocarbon/organic carbon ratio increase strongly with depth of burial and age of the rocks. However, the increase is quite slow. It is not until well into the Upper Miocene shales, which are about 15 million years old, that enough hydrocarbons have been generated to raise the hydrocarbon/organic carbon ratio to the range of 0.03-0.12 which is often observed in source sediments. The maximum hydrocarbon/organic carbon ratio reported in the Los Angeles basin is 0.113 at nearly 3.5 km. A threshold ratio of about 0.015 is reached at a depth of 2.4 km in the Los Angeles basin and 3.6 km in the Ventura basin. The temperatures at these depths are 94°C. and 96°C., respectively. It is thus temperature, and not depth, which controls the generation of hydrocarbons.

If we now hold the maximum geothermal gradient constant (Fig. 13-6, right hand diagram) it is clear that rapid burial to the temperature of the maximum of the principal phase of oil generation will allow more time for oil generation than slow burial. Thus, a long time of burial can compensate for lower amounts of organic matter provided the geothermal gradient is high enough to allow subsurface temperatures to reach about 75-100°C., the level at which significant hydrocarbon generation takes place. By means of time-depth charts Deroo et al. (1973) have shown that the Upper Devonian shales of the western Canada sedimentary basin have been buried to a depth corresponding to present temperatures of more than about 75°C. only since early Cretaceous time, some 120 million years ago, despite an age for the shales of more than 350 million years. The same authors have reported organic contents in the range 0.1 - 0.5%, or about one tenth those of the shales of the Los Angeles basin. Further, the geothermal gradient of the western Canada sedimentary basin (36.5°C. per kilometer) is slightly lower than that of the Los Angeles basin. Yet, despite these disadvantages, the shale productivities are sufficiently similar in terms of the ratio of hydrocarbon/organic carbon (about 0.8 for the Upper Devonian shales) that much of the high shale productivity of the Upper Devonian of the western Canada sedimentary basin, when compared to the Los Angeles basin, must be attributed to the greater length of time (about ten times) that the Devonian shales have been at hydrocarbon generating temperatures.

Thus variations in amount of organic matter, geothermal gradient and time available for oil generation explain most of the differences in petroleum potential between sedimentary basins (Philippi, 1965; Tissot, 1966; Albrecht and Ourisson, 1969). The importance of these observations means that it is only by using good basic geological information on the history of a sedimentary basin, together with data on the geochemistry of the organic matter, specifically kerogen, that the true petroleum potential can be determined.

SEARCH FOR HIGH QUALITY CRUDE OIL

The quality of crude oil accumulating during the maximum of the principal phase of oil generation is probably in the range 20-30° API. Contrast between iron-rich clastic environments and iron-poor carbonate environments, the presence of polar clay surfaces, and the common association of carbonates and anhydrite all tend to result in initial crude oils towards the lower end of the gravity range in carbonate rocks and towards the higher end of the gravity range in clastic rocks. Once crude oil has accumulated in the reservoir its quality may be drastically changed by several processes (Fig. 13-7). Evans et al. (1971) have offered the first well-documented examples of these processes, which may be so severe as to relegate the differences in the initial quality of the crude oils from carbonate and clastic environments to secondary economic importance.

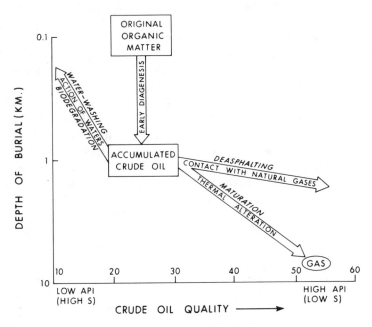

Fig. 13-7. Model for processes affecting crude oil quality (after Evans et al., 1971).

Two processes enhance the quality of crude oil. They are thermal maturation and gas deasphalting. We have already noted that the effect of thermal alteration on kerogen is first to produce immature natural gas, then crude oil with some natural gas, then natural gas alone (Table 13-3). A more detailed examination shows that within the principal phase of oil generation the effect of increasing temperature is to increase gradually the content of compounds containing less than 15 carbon atoms, at the expense of the C_{15+} heavy liquids, which themselves become increasingly paraffinic. This effect is illustrated diagrammatically in Fig. 13-8, and the detailed variations in hydrocarbon type between a thermally immature and a thermally mature crude oil can be seen in Table 13-4. The result of thermal maturation within the principal phase of oil generation is a lighter crude oil, with lower content of sulfur, and enhanced content of paraffins, and therefore a more valuable product.

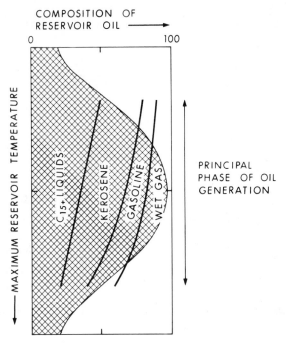

Fig. 13-8. Relation of crude oil composition to increasing maturation during principal phase of oil generation (based on Evans et al., 1971).

In the subsurface, there is an increase in pressure as well as in temperature. The increase in pressure has the effect of forcing the gases generated during maturation into solution in the crude oil. The result is a change in the phase equilibria and consequently the precipitation of asphaltic material. The process is analogous to propane deasphalting in refineries and to the routine laboratory procedure for separating asphaltenes from crude oils by the addition of pentane. The results are difficult to distinguish from the more mature thermally altered crude oils, except for the presence of bitumen plugging within the reservoir. In addition to a study of the solution-gas/crude oil ratio (GOR), Evans et al. (1971) postulate that the ratio of methane to nitrogen in the natural gas provides a clue as to whether natural gas had migrated into a previously under-saturated oil pool because they believe that nitrogen is for the most part an indigenous gas. Hence a change in the methane/nitrogen ratio indicates the addition of methane to the system. The effect of gas deasphalting is illustrated schematically in Fig. 13-9. Both thermal maturation and gas deasphalting are economically beneficial, and knowledge of the geochemical processes involved should assist the practical petroleum explorationist.

Table 13-4. Variation of crude oil properties with history of evolution and alteration. After Evans et al. (1971), Bailey et al. (1973b) and Speight (personal communication).

Property	Evolution and alteration			
	Biodegraded and water-washed	Thermally immature	Thermally mature	Deasphalted
Depth (m.)	1185	1300	2640	2590
Present reservoir temperature (°C.)	45	48	92	104
GOR (scf/bbl)	50	200	1780	500
°API	15.2	25.8	44.0	40.6
Sulfur (wt %)	2.99	2.77	1.0	0.16
C_{15+} oil fraction (BP>270°C.)				
Saturates (%)	19.1	33.4	66.0	88.8
Aromatics (%)	43.3	37.9	24.8	10.8
Eluted NSO's (%)	21.3	18.2	7.6	—
Asphaltenes (%)	16.2	10.4	1.7	0.3
Normal paraffins as % of saturate fraction	6.9	28.7	36.7	33.5
Asphaltenes (elemental composition on ash-free basis)				
C (%)	81.9	80.4	85.5	—
H (%)	8.0	7.8	9.4	—
O (direct) (%)	1.1	2.0	2.0	—
N (%)	1.0	2.6	1.2	—
S (%)	8.0	7.2	1.9	—

Two processes decrease the quality of crude oil: water-washing and biodegradation. During water-washing fresh water recharging from the surface washes away the light hydrocarbons in amounts approximately proportional to their solubility in the undersaturated fresh water moving past the reservoir. Although water-washing affects essentially all hydrocarbons, the results are most noticeable in the gasoline-range (C_4-C_7) fraction of crude oil, particularly with the one-ring aromatics (benzene, toluene, xylenes), which are the most soluble in water, but which are also the most resistant to biodegradation. Knowing the content of individual gasoline-range hydrocarbons in the original and the water-washed crude oils in the Carboniferous, Mission Canyon Formation of southeastern Saskatchewan, Bailey et al. (1973b) have calculated the relative volumes of formation water needed to effect the alteration, using solubility data of McAuliffe (1967). In terms of (units of formation water/unit of altered reservoir crude

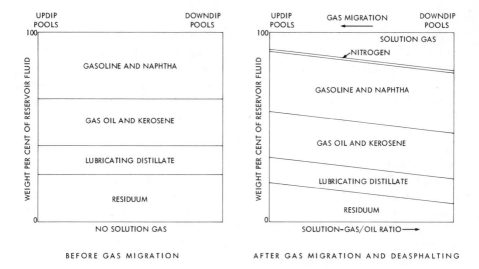

Fig. 13-9. Schematic model for gas deasphalting (based on Evans et al., 1971).

oil) x 10^3, the smallest volumes are needed for the most soluble compounds, benzene (0.75) and toluene (2.3), and the requirements increase
through cyclopentane (4.1) and cyclohexane (9.4) to n-pentane (19.8), n-
hexane (120.4) and n-heptane (492.2). Considering only the C_4-C_7
hydrocarbons and not those both lighter and heavier which also are affected by water-washing, this process has reduced the gravity by about 5°
API and increased the sulfur content about 0.3 per cent. The same authors
also point out that the efficient removal of n-hexane and n-heptane cannot
be explained in terms of water-washing, as these compounds are several orders of magnitude less soluble than benzene and toluene. Therefore,
another process in addition to water-washing must be operative, and it is
one which becomes increasingly efficient from nC_5 up to nC_7 and probably
also affects the branched paraffins, but to a lesser extent. They suggest
biodegradation.

The relations between microbes (principally bacteria, actinomycetes,
filamentous fungi and yeasts) and petroleum have been studied for over 50
years. Attention has been directed principally at the role microbes play in
the generation of petroleum, at their use in prospecting, and at the
detrimental effects they create, for example by corrosion of metals, reservoir plugging during water flooding and by generally decomposing organic
drilling fluid additives and petroleum products (Davis, 1967). However,
well-documented evidence became available only relatively recently concerning their importance in altering crude oil composition in the reservoir

(Winters and Williams, 1969). This process is termed biodegradation, and results from either the introduction of bacteria into the reservoir, presumably primarily with fresh water recharging from the surface, or from reactivation of dormant pre-existing bacteria, speculatively through reduction of salinity and addition of nutrients. Bailey et al. (1973a) have provided field and laboratory data which demonstrates that normal paraffins through to at least nC_{34} are severely depleted by biodegradation, with lower-ring naphthenes and aromatics being attacked at the same time as the lighter normal paraffins but before the heavier ones. The isoprenoids, pristane and phytane, were metabolized after the disappearance of the n-paraffins. The more condensed cyclic hydrocarbons were apparently unaffected. Additional NSO compounds and asphaltenes were formed by the metabolism of the hydrocarbons. In the Carboniferous of Saskatchewan, cited previously (Bailey et al., 1973b), the combined effects of water-washing and biodegradation have been to reduce crude oil gravities by more than 20° API and increase the sulfur content from less than 1 per cent to more than 3 per cent. In fact, the combined effects have resulted in a conventionally unproducible "tar".

The two processes producing crude oil with decreased quality have been well documented in the Powder River basin (Winters and Williams, 1969), southeastern Saskatchewan (Bailey et al., 1973b), and the Lower Cretaceous, Mannville Group of Alberta (Deroo et al., 1974). As the last-named report is concerned with the origin of crude oil in the Athabasca oil sand deposit, which is now economically recoverable by standard mining methods, it is pertinent to cite this region as an example of the effect of extensive destruction of reservoired crude oil, which, because of the size and situation of the deposit, is nevertheless a valuable asset.

Deroo et al. (1974) have demonstrated that the change in crude oil quality in going from the unaltered to the biodegraded crude oil is the result essentially of a decrease in n-alkanes, and concomitant relative increase in phytane and pristane, until only the latter isoprenoids are left — and finally even they disappear in the fully biodegraded "tar". These changes can be seen in the GC results presented in Fig. 13-10. Their study of the aromatic and sulfur compounds confirmed this observation. The change in n-alkane content coincides with a decrease in salinity of the associated formation water (Fig. 13-11), which suggests that water-washing accompanies biodegradation. Hitchon et al. (1969) have provided a hydraulic head cross-section through the Athabasca oil sand deposit showing the position of the freshwater recharge from the surface, and so this aspect of water-washing will be illustrated here (Fig. 13-12) by a hydraulic head and salinity cross-section through the Peace River oil sand deposit, also in northern Alberta. The close relation between hydraulic head, salinity and the oil sand deposit is very clear and further substantiates the genetic relation of hydraulic flow, water-washing, biodegradation and heavy oil deposits.

Connan (1972) carried out an interesting experiment with asphalts from the Aquitaine basin, France, which were similar in appearance and

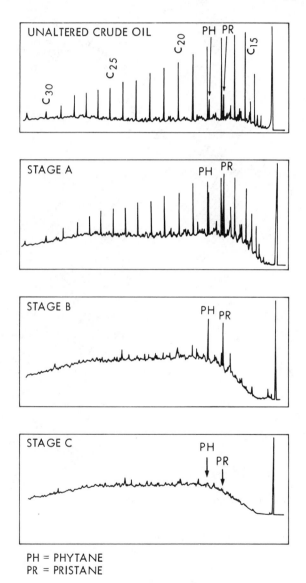

PH = PHYTANE
PR = PRISTANE

Fig. 13-10. Gas chromatograms showing stages in biodegradation of crude oil, Lower Cretaceous, Mannville Group, Alberta, Canada (after Deroo et al., 1974). For location of samples chromatographed see Fig. 13-11.

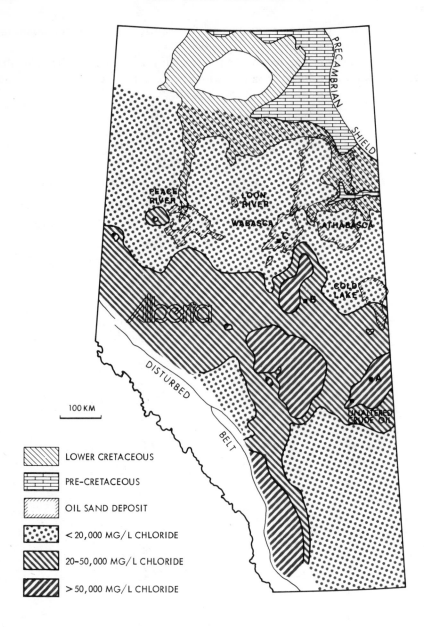

Fig. 13-11. Regional variations of chloride content of formation waters from Lower Cretaceous, Mannville Group, Alberta, Canada, in relation to oil sand deposits (based on unpublished data of Hitchon and Horn).

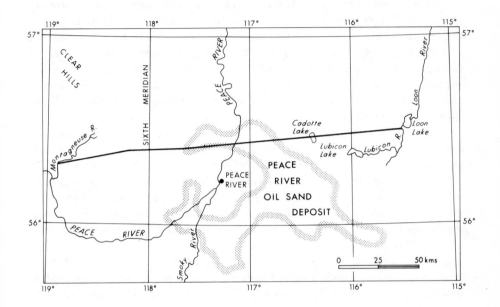

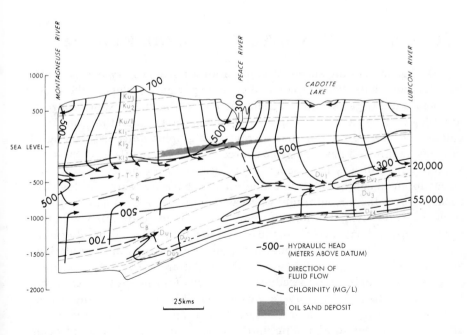

Fig. 13-12. Hydraulic head and salinity cross-section, Peace River oil sand deposit.

consistency but genetically different. Thermal alteration of the thermally immature, bacterially unaltered crude oil produced a thermally mature crude oil, whereas identical thermal alteration of biodegraded crude oil produced an oil having a gross composition comparable to that of unaltered immature, i.e., unproducible, oil. Thus simulated evolution is a useful technique for distinguishing between unaltered, immature crude oils and biodegraded crude oils. The importance of this observation for oil exploration is obvious.

Summarizing guidelines for the search for high quality crude oil, we can observe that geochemistry has a very important part to play in determination of the degree of thermal alteration and hence the quality of crude oil to be found, and certainly a knowledge of geochemistry is valuable in deciphering the effects of gas deasphalting. Hydrodynamics, in conjunction with geochemistry, and even simply mapping salinity variations, will assist the explorationist in avoiding regions in which water-washing and biodegradation have operated, but a caveat must be issued indicating that avoidance of these regions carries with it the penalty of overlooking a heavy oil deposit, which though not economically producible at the time of discovery, may become economically viable as mining and in-situ recovery techniques improve.

CLASSIFICATION INTO CRUDE OIL FAMILIES

During the middle stages of exploration of sedimentary basins, when hydrocarbons have been discovered, it is important to classify the crude oils and natural gases into geochemical families having distinct sources, hydrologic histories, and timing of generation, migration and accumulation. Two crude oils in different stratigraphic units may belong to the same family and have the same source, or two crude oils from the same stratigraphic unit may have separate sources — for example, in strata adjacent to an unconformity one crude oil may have its source in downdip rocks beneath the unconformity whereas the other crude oil may originate from shales overlying the unconformity. Analysis of crude oils yields two general types of parameters — those that are little affected by alteration (thermal maturation, gas deasphalting, water-washing and biodegradation) and can be used for correlation, and those that are affected by alteration processes and can be used to deduce the history of the deposit. No easy way is known of distinguishing these various possibilities without the use of hydrologic and geochemical techniques, and certainly exploration into new areas and different stratigraphic units will benefit from this type of prior knowledge.

The classification of crude oils into geochemical families is difficult.

The U.S. Bureau of Mines has pioneered methods of crude oil characterization based on USBM routine distillations, and their methods are capable of fairly detailed application, for example, the work of Jones and Smith (1965) on crude oils in the Permian basin of west Texas and New Mexico. In many areas where abundant USBM crude oil distillations are available, or when the wildcat well or pool can no longer be resampled, considerable useful information can be obtained from this type of data. However, the development of modern sophisticated analytical techniques, with which individual hydrocarbon compounds can be identified and their content determined, means that more reliable correlations can be made using specific marker compounds. It is beyond the scope of this chapter to describe in detail the various analytical techniques which have been applied to crude oils. Some of the more important are summarized in Table 13-5, and the interested reader is referred to Burlingame and Schnoes (1969), Douglas (1969), Murphy (1969), and Speight (1971), as well as to standard textbooks on analytical chemistry, for more details.

The vicissitudes of thermal maturation, gas deasphalting, water-washing, and biodegradation mean that minimal reliance can be placed on variations and concentrations of n-alkanes and most branched alkanes when attempting to group crude oils into genetically related families; nor are any of the light hydrocarbons likely to be useful for obvious reasons. There are no published reports on the value of the components of the asphaltene fraction in correlation, and in part this is probably due to difficulties in analyzing such high molecular weight material. Probably the most satisfactory compounds for tracing the genesis of crude oils are the so-called biological markers, of which the most important are the acyclic isoprenoid alkanes, steranes, triterpanes and porphyrins (reviewed by Hills et al., 1970; Hitchon, 1971). The basic carbon skeleton of representative biological markers is shown in Fig. 13-13. The acyclic isoprenoid alkanes were first reported from crude oil by Bendoraitis et al. (1963) and have now been identified in many crude oils, and in one crude oil all 12 isoprenoid hydrocarbons from C_{14} to C_{25} as well as head-to-tail conformations have been found (Han and Calvin, 1969). They are identified mainly by GC-MS methods and Welte (1970) has used them to solve correlation problems among crude oils in the Gifhorn trough of northwestern Germany, where their relative distributions differ with their source. A combination of capillary GLC and GC-MS was used by Henderson et al. (1969) to identify steranes and triterpanes but because of the complex procedures required to identify individual compounds there are no published reports of crude oil correlations which make use of the amount and proportions of the individual steranes and triterpanes. However, there are a few papers in which the steranes and triterpanes as a group have been used to characterize crude oils or to relate specific crude oils to their source rock (Tissot et al., 1971; Welte, 1972). Other studies of crude oil correlations with either hydrocarbons or stable isotopes as the markers include Silverman and Epstein (1958), Bray and Evans (1961), Kvenvolden and Squires (1967), Poulet and Roucache (1969), Thode and Monster (1970), Vredenburgh and Cheney (1971), and Monster (1972).

The classification of crude oils into families by means of biological markers is limited at present only by the complexity of the analytical procedures. It is very likely that in the near future we will have the

Table 13-5. Summary of analytical techniques for study and identification of kerogen, crude oil and natural gas.

Technique	Measurement	Molecular parameter determined	Identification	State during examination	Minimum quantities required (gm)
Chromatographic					
Gas-liquid chromatography (GLC or GC)	Time taken to emerge (RT)	Molecular size, solubility, and adsorption effects	Individual organic compounds	Gas phase/ solution in liquid phase	10^{-9}
Thin layer chromatography (TLC)	Distance travelled on plate (R_f value)	Solubility and adsorption behavior, molecular size	Individual organic compounds	In solution	10^{-9}
Linear elution adsorption chromatography (LEAC)	Retention volume	Adsorptive characteristics	Individual organic compounds and narrow groups of compounds	In solution	10^{-4}
Gel permeation chromatography (GPC)	Retention volume	Molecular size	Gross separation and molecular size identification	In solution	10^{-4}
Spectrometric					
Visible and Ultraviolet (UV)	Adsorption maxima (λ max) and intensity (ϵ)	Excitation of non-bonded electrons and conjugated systems	Individual organic compounds and ligands	In solution	10^{-3}
Infrared (IR)	Complex spectrum	Adsorption of energy of vibration of inter-atomic bonds	Ligands and radicals	In solution liquid state, or solid	10^{-3}
Mass spectrometry (MS)	Complex spectrum	Ease of fragmentation of molecule and stability of fragments	Individual organic compounds	Gas phase	10^{-4}
Nuclear magnetic resonance (NMR)	Complex spectrum	Interaction of nuclear spins of protons	Basic carbon skelton and radicals	In solution or liquid state	10^{-2}
Electron paramagnetic resonance (EPR)	Complex spectrum	Magnetic spin properties of free radicals and metal ligands	Free radicals	In liquid or solid state	10^{-3}
Optical rotary dispersion (ORD) and circular dichroism (CD)	Single value or curve	Degree of asymmetry of molecule and consequent effect on polarized light	Specific biogenic compounds	In solution or liquid state	10^{-2}
Other					
X-ray diffraction (X-ray)	Complex intensity pattern	Electron density in unit cell	Molecular size and spacing for platy molecules	Solid	10^{-2}
Stable isotope	Single value	Proportions of isotopic species	Assists search for source of elements	Gas phase	10^{-3}
Neutron activation	Complex spectrum	γ-Ray decay spectrum from neutron irradiation	Wide range of individual elements	Liquid or solid phase	10^{-2}

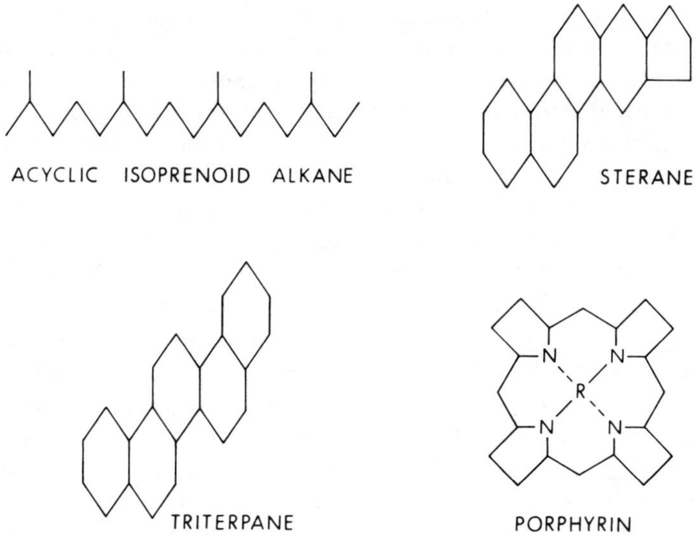

ACYCLIC ISOPRENOID ALKANE

STERANE

TRITERPANE

PORPHYRIN

Fig. 13-13. Carbon skeletons of representative biological markers.

capability, not only of grouping the crude oils by common sources, but also of indicating the type of organisms from which the crude oils originated because of the chemotaxonomic significance of biological markers (Hitchon, 1971). However, at present, stable carbon isotope ratios on various crude oil fractions (saturates, aromatics, eluted NSO compounds, asphaltenes) are probably the best tool for classification into crude oil families (Rogers et al., 1971). The application of modern statistical techniques such as stepwise multiple discriminant function analysis and factor analysis, which have been applied to crude oils and condensates by Connor and Gerrild (1971), and Hitchon and Gawlak (1972), respectively, will then be of even greater importance in interrelating the many individual biological-marker compounds and the stable isotope data.

PETROLEUM OCCURRENCE INDICATORS IN FORMATION WATERS

The commonest fluid recovered during drilling is formation water. It is most often found alone, but even when hydrocarbons are recovered they are frequently accompanied by formation water. This is to be expected because the pore space of sedimentary rocks is usually a water-wet system and formation water is the carrier fluid for crude oil and natural gas.

During the process of migration from the source rock the hydrocarbons and formation water are intimately linked in a physical-chemical system as yet incompletely understood, and we might therefore anticipate that even after the hydrocarbons have accumulated, the former intimate association would still be reflected in the formation water in some manner. What is surprising is that this line of reasoning has not been pursued more diligently in view of the frequent recovery of formation water. We will first briefly outline the origin and geochemistry of formation waters, then note some modern applications of geochemistry to the study of petroleum occurrence indicators, before finally suggesting possible future applications.

All formation waters comprise two entities, water (H_2O) and dissolved components, which may be genetically different. The term "dissolved" is used in its widest sense to include the micelles of Baker (1967) and the "accommodated" material of Hodgson and Hitchon (1965). In order to determine the origin of formation waters it is necessary to understand the geochemistry of both the water and dissolved components, and since the concentration of the latter is very variable (fresh water to brines) it is important to know the pore volumes and hydrodynamics for use in mass balance calculations. The western Canada sedimentary basin will be used as an example because it is the only basin for which there is published data on stable isotopes in the water (Hitchon and Friedman, 1969), the geochemistry of the dissolved inorganic components (Billings et al., 1969; Hitchon et al., 1971), pore volumes of the sedimentary rocks (Hitchon, 1968) and hydrodynamics (Hitchon, 1969a, b).

About 85 per cent of the strata in the western Canada sedimentary basin were deposited under marine conditions, with the remaining 15 per cent under brackish water, or possibly fresh water, conditions. The ancient sea water is the source of the majority of deuterium, oxygen-18 and dissolved salts. The deuterium of the original sea water is gradually being mixed with deuterium from surface water under the present hydraulic system. The distribution of deuterium has been derived through mixing of the diagenetically modified sea water with not more than 2.9 times, and possibly as little as 1.13 times, as much fresh water at the same latitude. Movement of this fresh water through the basin has redistributed the dissolved salts into the observed salinity variations. There has been fractionation of deuterium on passage of water through micropores in shales (confirmed recently in laboratory experiments by Coplen and Hanshaw, 1973), and extensive enrichment of oxygen-18 through exchange of oxygen isotopes between water and carbonate minerals.

The volume-weighted mean formation water has been calculated and shown to have a composition similar to sea water. The salinity is 1.3 times that of sea water with the major part of the gain resulting from solution of Middle Devonian bedded halite. With the exception of Mg and SO_4 all components show a net gain, compared to sea water. Relative to the present pore volume there has thus been negligible removal of dissolved salts, although there has been a major redistribution. Excepted is the large loss of water during initial compaction, but that water is believed to be essentially unaltered sea water. Saline springs at the regional discharge of the basin result from solution of halite and gypsum by meteoric water, and not from loss of formation water (Hitchon et al., 1969). The major factor causing concentration of the dissolved salts is membrane filtration (confirmed recently in laboratory experiments by Hanshaw and Coplen, 1973). Together with dilution by fresh water recharge this results in a chemical population ranging from fresh water to membrane concentrated brines. Dolomitization caused by formation waters has not been an important factor in controlling the relative proportions of Ca and Mg in the formation waters, and neither has bacterial reduction of SO_4 played a major role in the loss of SO_4 from the formation waters. Rather, Mg may have been used in the generation of chlorite and the amount of SO_4 controlled by the solubility of $SrSO_4$ and $BaSO_4$. Further, the gain in Sr in the volume-weighed mean formation water is possibly from the aragonite to calcite diagenetic recrystallization. Redistribution of some alkali metals and alkaline earth metals between rock matrix and formation water has taken place through exchange on clays. Desorption of Br and I from clays occurred, but there has also been a contribution of these halogens, principally I, from organic matter. $CaCO_3$ solubility equilibria have been demonstrated. The trace metals, Cu, Fe, Mn and Zn have all been concentrated in formation waters but their mode of concentration remains equivocal.

This summary of the various processes presently known to be operating in the western Canada sedimentary basin will provide some idea of the extreme complexity of water-rock interaction. With obvious modifications, similar interrelated processes are operating in all sedimentary basins — and it should not be forgotten that there is an organic phase to consider, which was not taken into account in the studies cited. The perceptive reader can now appreciate that when dealing with multivariate natural systems there is decreased possibility that determination of single components or study of simple ratios will lead to definitive petroleum occurrence indicators.

There seems to be only one published report that considers the chemical components in formation waters in a multivariate sense and which is directed at the determination of petroleum occurrence indicators, and that is the recent paper by Hitchon and Horn (1974), which, again, refers to a study of the western Canada sedimentary basin. In this study the analyses were divided into two populations depending on whether the formation water sample was from a stratigraphic interval initially producing oil and/or gas, or was from a non-producing interval. Succinctly, the two populations reflect the economic success of the exploration rather than the technical success. Statistical considerations limited further subdivisions of the two populations into two groups, Paleozoic (dominantly carbonate reservoirs) and Mesozoic (dominantly sandstone reservoirs). The entire data set was then subjected to discriminant function analysis. With 95 per cent assurance, the Alberta formation waters associated with economically producible hydrocarbon occurrences are chemically different, in a multivariate sense, from the formation waters from non-producing stratigraphic intervals. Iodide and magnesium are the most important discriminators in the Paleozoic group, accounting for 60 per cent of the relative contribution of the variables to the discriminants in the 8-component system analyzed. Formation waters from Paleozoic strata have a general background of iodide (and bromide) from argillaceous rocks (including associated organic matter). The occurrence of source sediments, and the resultant hydrocarbon accumulations, are accompanied by enhanced contents of iodide in the associated formation waters (Fig. 13-14),

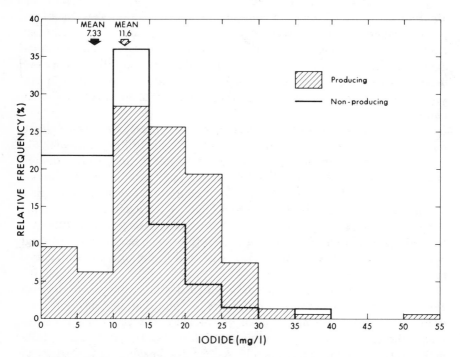

Fig. 13-14. Comparison of the relative frequency of iodide in formation waters from producing and non-producing stratigraphic intervals in the Paleozoic rocks of Alberta, Canada (Hitchon and Horn, 1974).

which originates from the enhanced contents of organic matter in the source sediments. The following indirect but inferred relationship may explain the importance of magnesium to the discrimination: Mg enrichment → increased dolomitization → increased effective porosity → increased probability of hydrocarbon occurrence. In the Mesozoic group the most important discriminators are chloride and sodium (49 per cent relative contribution to the discriminants). This reflects the fact that in the Mesozoic sequence in Alberta, which contains only 60 per cent marine strata and lies mostly within the regime of local and intermediate flow systems, the most probabilistic situation for both hydrocarbon occurrence and high salinity formation water is within the marine clastic section in regions subject to the minimum hydrodynamic flushing, which are also those situations in which the reservoired crude oils have been preserved from biodegradation and water-washing, and hence are economically producible by conventional techniques.

The discriminant function is a direct measure of the relative importance of the individual components of the multivariate system in distinguishing formation waters from producing and non-producing stratigraphic intervals. The statistical nature of the discriminant function calculation results in the discriminant index not being sharply focused due to the nature of the data, and consequently the chance of error increases as the difference between the computed discriminant score of the unknown sample and the discriminant index decreases. This is shown diagrammatically in Fig. 13-15 in terms of probability of misclassification. For example, if some future Paleozoic formation water sample from Alberta is analyzed for the 8 components considered in this study, and if the computed discriminant score is -14.5, we can say the chances are 9 out of 10 that the formation water was associated with hydrocarbons in the subsurface. This is a potentially powerful method which is capable of taking into account a great variety of components in the formation waters, which may be associated with the occurrence of hydrocarbons through diverse processes, and it can also provide an indication of the probability of economic occurrence.

At first appraisal, the most obvious petroleum occurrence indicators in formation waters would seem to be those organic compounds which are also found in crude oil and natural gas. For most of the low molecular weight material, which is normally gaseous at the surface, this involves collection of the formation waters in high-pressure bottom-hole samplers in order to prevent loss of the dissolved gases on reaching the surface. Studies of this nature have been carried out (Buckley et al., 1958), which suggest that local enrichment of the formation waters in dissolved gaseous hydrocarbons may occur in the vicinity of some oil or gas fields. However, the difficulties in sampling have limited the widespread use of this method.

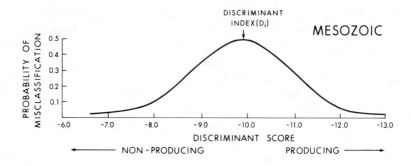

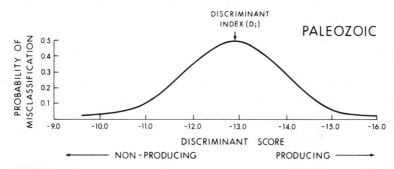

Fig. 13-15. Probabilities of misclassification of formation waters as a function of discriminant score for the Mesozoic and Paleozoic of Alberta, Canada (Hitchon and Horn, 1974).

Another technique which has found considerable favour concerns the analysis of benzene in formation waters (Zarrella et al., 1967; Blazejczak and van der Weide, 1967). Benzene is the most soluble of the aromatic hydrocarbons in formation water and with only elementary precautions may be completely retained in samples recovered to the surface. Zarrella et al. (1967) present several instances in which the concentration of benzene dissolved in formation waters decreases with increasing horizontal distance from the nearby oil field. There seem to be no published applications of the use of the other one-ring aromatics in exploration. The demonstration by Bailey et al. (1973b) of the process of water-washing and its dramatic effect on the benzene content of crude oil, and consequently of the formation water, suggests that extreme caution should be exercised in interpreting results of analysis of any low molecular weight material in formation waters, especially those which may be subject to water-washing.

In summarizing this section on petroleum occurrence indicators in formation waters it should be noted that many components present in formation waters have been suggested as potential indicators. Some, like sulfate, simply do not contribute to discriminating producing from non-producing stratigraphic intervals — at least for Alberta (Hitchon and Horn, 1974). For most, there has been insufficient rigorous statistical testing to be absolutely sure that what is being distinguished is in fact the presence or absence of a commercial hydrocarbon occurrence, and it is primarily for this reason that more papers are not cited from the extensive literature on hydrogeochemistry. The geochemical reasons for the choice of many of the suggested indicators are probably sound, and some of the more likely ones are naphthenic acids (Davis, 1969), fatty acids (Cooper and Kvenvolden, 1967), phenols and the ammonium ion. In the case of organic components caution must be exercised regarding the part these same components play in the various processes which may alter the crude oil in the reservoir. In view of our patently inadequate knowledge of petroleum occurrence indicators in formation waters, of the frequent recovery of formation water during drilling, and the relative ease with which most of the likely organic and inorganic indicator components can be determined by modern analytical methods, it is clear that the most effective way to utilize formation waters is multicomponent analysis and subsequent rigorous statistical analysis by a method such as discriminant function analysis. The maximum information available from formation waters is not being utilized, and this is particularly true for water-rock interaction phenomena as they relate to inorganic, organic and stable isotope reactions, and such important parameters as the prediction of porosity. Only by such an approach can we hope to determine the relative importance of a variety of indicators which are related to the presence of economic hydrocarbon occurrences by several probably unrelated processes, and at the same time obtain some indication of the probability of hydrocarbon occurrences.

SURFACE PROSPECTING

The tremendous expense of geophysical surveys and exploratory drilling has encouraged consideration of surface prospecting techniques. Success in surface prospecting relies on the principle that hydrocarbons or inorganic components associated with the oil field are continuously being lost to the surface by diffusion. It has not been satisfactorily proved that vertical diffusion of components is taking place. In addition, authors who have studied vertical diffusion of hydrocarbons either with laboratory models or by mathematical analysis have apparently failed to take into account fluid flow within the near surface region, which may range up to several meters per year.

In order to understand the significance of this argument, it is necessary to digress into an explanation of the phenomena that control fluid flow in the subsurface. As noted in the Introduction, fluids are governed by a well recognized system of energy potential fields (fluid, thermal, electroosmotic

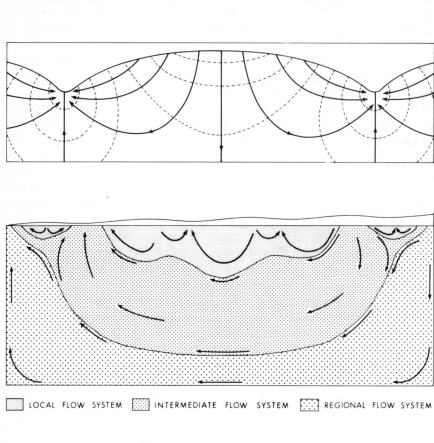

LOCAL FLOW SYSTEM   INTERMEDIATE FLOW SYSTEM   REGIONAL FLOW SYSTEM

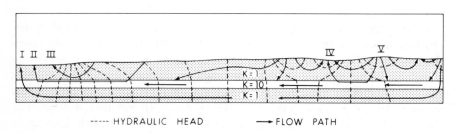

----- HYDRAULIC HEAD   ⟶ FLOW PATH

Fig. 13-16. Development of fluid flow models. Top: Hubbert (1940); Centre: Tóth (1962, 1963); Bottom: Freeze and Witherspoon (1966, 1967, 1968).

and chemicoosmotic). Of these, the fluid potential is the most important and the best known. The fluid potential is defined as the amount of work required to transport a unit mass of fluid from an arbitrary chosen datum (usually sea level) and state to the position and state of the point considered. The classic work of Hubbert (1940) on the theory of groundwater motion was the first published account of the basinwide flow of fluids that considered the problem in exact mathematical terms as a steady-state phenomenon. His concept of formation fluid flow is shown in the top model in Fig. 13-16. Expanding on the work of Hubbert, Tóth (1962, 1963) introduced a mathematical model in which exact flow patterns are obtained as solutions to formal boundary value problems. Specifically, his model (Fig. 13-16, centre diagram) was isotropic and homogeneous and, like Hubbert's, was based on the Laplace equation, but by superposition of a sinusoidal surface on the regional slope of the basin, Tóth accounted for local topographic relief. Local changes in topography control an upper zone of a local fluid flow system, and this system is separated by an intermediate flow system from the main regional fluid flow regime which extends around the extremities of the model and is controlled by the regional topographic trend. Freeze and Witherspoon (1966, 1967, 1968) pointed out a number of restrictions to the Tóth model and developed both analytical and numerical methods of analysis (Freeze, 1966, 1969) for multilayer (nonhomogeneous), nonisotropic successions with a general configuration for the topographic surface and a sloping basement (Fig. 13-16, bottom illustration).

The complex models generated by Freeze and Witherspoon are dimensionless, and thus the fluid flow nets apply to sedimentary basins of all sizes. The Tóth-Freeze-Witherspoon model is meaningful when applied to an actual sedimentary basin, as exemplified by recent studies of the western Canada sedimentary basin (Hitchon, 1969a, b). These studies show that the main variables affecting the fluid potential distribution are topography and geology (lithology and permeability). The dominant fluid potential in any part of the basin corresponds closely to the fluid potential at the topographic surface in that part of the basin. Major recharge areas correspond to major upland areas, and major lowland regions are major discharge regions. Large rivers often exert significant drawdown effects (Hitchon, 1971, Fig. 4). Relatively highly permeable beds, if sufficiently thick, can significantly affect the regional fluid potential distribution by drawing down the fluid potentials, although the dominant features are topographically controlled. These observations apply to all sedimentary basins because of the dimensionless nature of the model.

In itself, the model may be used as an exploration tool in regions with a minimum of fluid pressure data (Hitchon and Hays, 1971). However, it is the manner in which the fluid flow regime affects the application of geochemistry to the search for hydrocarbons that is most relevant here. Examination of the lower model in Fig. 13-16 shows that at location I, any hydrocarbons detected at the surface would be representative of hydrocarbons present in the lower layer only, whereas at location II they would be representative of hydrocarbons in most of the upper two layers. By contrast, at location III, only a relatively short distance from the other two locations, hydrocarbons detected at the surface would be representative of hydrocarbons present in only a small volume of the upper layer. These observations are made while cognizant of the fact that the basis of the argument for surface geochemical prospecting for hydrocarbons lies in the belief of the vertical diffusion of hydrocarbons from the oil field to the surface. As this same model shows, vertical diffusion sometimes has a vector in the same direction as the flow of the fluids (location IV), and sometimes in the opposite direction to the fluid flow (location V), all in a very short horizontal distance.

No proponent of surface prospecting for petroleum has published a study or case history taking into account these observations. In the opinion of the author, it is this lack of such a rigorous examination of the possible paths by which hydrocarbons may be carried to the surface that renders surface prospecting for oil fields equivocal and not any fundamental failure in the logical sequence of events between the occurrence of the indicator in the oil field and its appearance at the surface. Williams (1970) has pointed out the applicability of fluid flow studies to prospecting for mineral deposits and a number of the caveats considered above are pertinent.

The electroosmotic and chemicoosmotic potentials have been evaluated as mechanisms causing fluid movement (van Everdingen, 1968). The electroosmotic fluid movement is generated by electric potentials resulting from the existence of telluric currents and is assumed to have a negligible effect on fluid movement. The chemicoosmotic potential is more important, however, in those sedimentary basins with both a wide range in concentration of dissolved material in formation waters and shales which act as semipermeable membranes. A large literature has built up on the subject of the semipermeable membrane capacity of shales and it is now well documented that many fluid pressure anomalies in sedimentary basins, including closed fluid potential lows, owe their origin to chemicoosmotic forces (Berry, 1958; Hill et al., 1961; Coustau and Sourisse, 1967; Hanshaw and Hill, 1969; Hitchon, 1969b). The thermal potential is probably negligible as a force to move fluids in most sedimentary basins but it may be of significance in geothermal regions. However, the movement of formation waters may have an appreciable effect on the temperature distribution in the subsurface, for example, the effects from cool recharge water, which can affect the solubility of components dissolved in the water and the regime for biodegradation.

One aspect of fluid flow that is of direct importance in the application of geochemistry to the search for crude oil is flow resulting from compaction of sediments. Under normal conditions most compaction which takes place soon after deposition results in the movement of fluids upwards out of the sediments towards the sea bottom because porosity and permeability tend to decrease downwards. However, in thick sequences of argillaceous sediments fluid expulsion may be inhibited because of low permeability, thus compaction will not be great, and a high porosity — high fluid pressure situation will result. This is termed under-compaction and is a temporary state. The higher-than-normal fluid pressures cause limited downward fluid flow in sections where thick permeable sandstones underlie the zone of under-compacted shale, with the boundary surface between the zones of upward and downward fluid flow existing in the lower middle part of the shales (Magara, 1969). Examples of the relation between the zone of downward fluid flow and hydrocarbon occurrences have been described from Japan (Magara, 1968) and the western Canada sedimentary basin (Magara, 1973). In addition to the effect on fluid flow, abnormally high fluid pressures have been suggested as important in diagenesis and in the movement of some components dissolved in the formation waters resulting in deposition of cements (Rumeau and Sourisse, 1972). Certainly this aspect of fluid flow should be taken into account during subsurface exploration when using petroleum indicators but its most pertinent application is in offshore exploration.

As on land, hydrocarbons seeping from the sea floor may provide information about the location of deep oil fields — but unlike exploration on land there is no near-surface zone of local flow. The directions of fluid flow in a thick offshore wedge of sediments is shown schematically in Fig. 13-17, including effects due to an undercompacted zone. The presence of submarine fresh water springs in some nearshore regions suggests that fluid flow from the land may affect compaction flow in some nearshore areas, but the extent of this influence is not known. Direct detection of the shallower oil field in Fig. 13-17 is possible at the sea bottom, but due to the phenomenon of downward flow associated with the undercompacted zone such direct detection might not be possible with respect to the deeper oil field unless any movement due to vertical diffusion can overcome the downward flow due to undercompaction. Complicating factors include dispersal of the petroleum indicators by ocean currents, contamination of the indicator hydrocarbons by recent biologically produced hydrocarbons, and pollution. The work of Frank et al. (1970) and Brooks and Sackett (1973) on low molecular weight hydrocarbons in sea water in the Gulf of Mexico supports the feasibility of the method, although Gérard and Feugère (1969) suggested that examination of shallow cores for low molecular weight hydrocarbons might be more successful. Sea water and

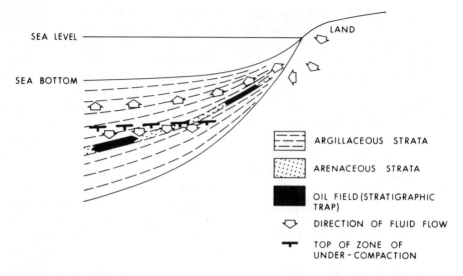

Fig. 13-17. Fluid flow directions in a thick offshore wedge of sediments.

bottom sediment sampling is a valuable reconnaissance technique, especially if proper statistical data-reduction, such as trend surface and residual mapping methods, are applied. The technique of sea-bottom water sampling is in current use by exploration companies and although there are no detailed published reports of the viability of this method it is known to have been used successfully.

We may summarize the flow of fluids in sedimentary basins by noting that fluids in porous media move from regions of high energy to regions of low energy. The energy gradient to which the fluids react comprises an aggregate of potentials resulting from elevation and pressure, thermal, electroosmotic and chemicoosmotic forces coupled in a system as yet incompletely described mathematically (van Everdingen, 1968). The elevation and pressure potential can be completely described by the Tóth-Freeze-Witherspoon model. It is only when the thermal, electroosmotic and chemicoosmotic potentials influence the kinetic state of the fluid that their effect can be seen in terms of hydraulic head in the model, and then probably only a portion of that potential is observed. In sedimentary basins with thick sequences of argillaceous sediments in which compaction is incomplete, the fluid flow system may be modified by high fluid pressure zones. Without adequate knowledge of the vicissitudes of fluid flow it is not possible to obtain the maximum understanding from geochemical exploration for crude oil and natural gas. Indeed, in the case of surface prospecting on land, the technique may be invalid.

Of the more than two hundred papers which have been published on surface geochemical prospecting for petroleum, a few are notable for the comprehensive manner in which they have summarized available knowledge (Kartsev et al., 1959; Sokolov et al., 1959; Horvitz, 1959, 1968; Chilingar and Karim, 1962; Davidson, 1963; Kroepelin, 1967), have outlined new or modified procedures (Smith and Ellis, 1963; Neglia and Favretto, 1964; Debnam, 1965; Horvitz, 1972) or presented detailed reports of project areas (Housse, 1961; Pomeyrol et al., 1961; Zak, 1961; Karim, 1964; Karim and Chilingar, 1965; Kvet and Michalicek, 1965; Nyssen et al., 1966, Debnam, 1969; Housse, 1971; McCrossan et al., 1972). A few have used modern statistical techniques to validate their conclusions and an outstanding example of a comprehensive study which considered most of the relevant factors which might affect the anomalies found is that of McCrossan et al. (1972) on the western Canada sedimentary basin. These authors concluded that syngenetic hydrocarbons from material in the glacial drift interfered so extensively that the method was of questionable value in such glaciated areas. Their conclusion is probably pertinent to most glaciated terrains. Although they considered the deep hydrodynamic situation, they did not attempt to relate the soil gas anomalies they observed to the nearsurface flow systems, possibly because of the extensive interference they found. Housse (1971) studied gases sorbed on sedimentary rocks recovered from shallow seismic shot holes, and has reported that in the Paris basin the gas anomalies are displaced a few kilometers to the north because of the flow of water towards the centre of the basin in shallow sandstones. Recent work by Leythaeuser (1973) on the effects of weathering of organic matter in shales indicates the care that must be taken when reporting data on organic matter in sedimentary rocks from shallow depth — therefore extreme caution must be exercised with sorbed gases.

From a careful evaluation of the literature on surface prospecting for petroleum, including the cited papers, it is the author's opinion that in offshore areas there is considerable merit to the technique if care is exercised for the various factors which may modify anomalies. On land, the case for the validity of surface prospecting for petroleum is not proven. It seems highly likely that in terrains with transported soils it will be impossible to make use of the method. In regions with residual soils the technique may be feasible, provided the near-surface flow regime is adequately known. In view of the fluid nature of crude oil and natural gas it is difficult to understand why geochemists have not been more concerned with the composition and flow of the near-surface and surface fluids and less with soil and weathered bedrock. It seems obvious that this should be the correct approach and until such a comprehensive study is undertaken, with rigorous statistical control, surface prospecting for petroleum on land will remain equivocal.

IDEAL APPLICATION — A PERSONAL VIEW

The prodigious problems of preparing a review paper carry with them the privilege of presenting a personal view of the subject being discussed. It is difficult not to be categorical when covering such a comprehensive and complex topic as the application of geochemistry to the search for crude oil and natural gas, and the readers of the book of which this chapter is a part must be cognizant of the fact that a book of similar size could have been written on this topic alone. The author has selected what he considers are the most important publications appearing in the 1960's and early 1970's (as late as October, 1973), as well as a few earlier ones, primarily in the English language, in easily accessible journals and publications, and which appear to him to be most directly applicable to *successful* geochemical exploration. Many hundreds of references could have been cited but their inclusion would not clarify this complex topic; indeed, they might only serve to confuse. Many suggestions have been made for the application of specific geochemical techniques to the search for petroleum, a considerable number of which are presented without adequate foundation in fact. Some may have merit as exploration methods but the characteristic univariate approach must surely fail in a multivariate system.

Crude oil and natural gas, together with formation water, are complex multicomponent fluids flowing in a multivariate sedimentary environment, with time an important factor. Only by using rigorous statistical techniques and large modern computers can the exploration geochemist hope to evaluate such a multivariate time-dependent system. It is hopeless to expect simple, "black-box" solutions to the application of geochemistry to the search for hydrocarbons. A tentative, though successful, beginning has been made in simulating geological development of sedimentary basins and processes (Merriam and Cocke, 1968; Ojakangas, 1970). Our data-reduction techniques such as factor analysis, cluster analysis and discriminant function analysis are fairly well advanced but insufficiently used. Capabilities exist for complex trend surface analysis and this method has already been applied to crude oil gravities and hydrodynamics (Harbaugh, 1964a, b). We have passed from the stage of being able to plot simple formation water quality diagrams to the use of computer programs for resolving complex multicomponent solution-mineral equilibrium computations (Kharaka and Barnes, 1973) which are relevant to water-rock interactions, porosity plugging and mineral solution in the underground environment. Three-dimensional modeling of crude oil quality is already being carried out. Schwartz and Domenico (1973) have developed as yet simple techniques of relating water flow and water chemistry, although extensions of the type of programs developed by Freeze and Witherspoon (1966, 1967, 1968) to models of actual sedimentary basins seem to be limited only by computer capacity. Kvenvolden and Weiser (1967) and Tissot (1969) have expressed organic geochemical reactions under earth conditions in terms of mathematical models. Multiple well log analysis using digital techniques has already been applied successfully (Rudman and Lankston, 1973) and if the correct parameters are logged can be a valuable geochemical exploration tool in conjunction with study of cuttings, cores and fluids.

All these advances in computer technology, together with the vast amount of geochemical data now available to industry and government will form the input to the computer programs of the future. These programs will be capable of delineating hydrocarbon trends by modeling the geological, geophysical, geochemical and hydrologic history of the basin using today's parameters, and will be capable of seeking the last commercial accumulations for an energy-short world. However, we are not obtaining all the analytical data necessary so that we can take maximum advantage of future computer programs, which will be compiled by governments, industry and universities, whether the basic data is available or not.

It is my personal view that maximum practical application of geochemistry to the search for crude oil and natural gas can only be accomplished by extensive multicomponent analysis of fluids and rocks, together with a knowledge of hydrodynamics and a thorough understanding of the basic geology of the sedimentary basin, coupled with computer modeling of the geological, geophysical, geochemical and hydrologic history of the basin. Geochemistry can never replace good geology, but it is axiomatic, in my view, that if the integrity of the skills of the geochemist are applied properly they will enhance economic success in the search for crude oil and natural gas.

A *CONVERSIONS*

Conversion of Elemental Content Units

Definitions

1 milligram	=	10^{-3} gram	=	1 mg
1 microgram	=	10^{-6} gram	=	$1\,\mu g$
1 nanogram	=	10^{-9} gram	=	1 ng
1 picogram	=	10^{-12} gram	=	1 pg
1 part per million	=	$1\,\mu g$ per gram	=	1 ppm
1 part per billion	=	1 ng per gram	=	1 ppb
1 part per trillion	=	1 pg per gram	=	1 ppt

Equivalents

weight %		*ppm*			*ppb*		
1.0%	=	10,000 ppm					
0.1%	=	1,000 ppm					
0.01%	=	100 ppm					
0.001%	=	10 ppm					
0.0001%	=	1 ppm	=	1,000 ppb	=	$10^{-4}\%$*
0.00001%	=	.1 ppm	=	100 ppb	=	$10^{-5}\%$
0.000001%	=	.01 ppm	=	10 ppb	=	$10^{-6}\%$*
0.0000001%	=	.001 ppm	=	1 ppb	=	$10^{-7}\%$
0.00000001%	=	.0001 ppm	=	.1 ppb	=	$10^{-8}\%$

For waters (dilute solutions)

1 mg/l	=	10^{-3} g/l	=	1000 ppb or 1 ppm
		10^{-4} g/l	=	100 ppb
		10^{-5} g/l	=	10 ppb
$1\,\mu g/l$	=	10^{-6} g/l	=	1 ppb
		*10^{-7} g/l	=	0.1 ppb

For gases (atmosphere, soils, exhalations)

1 ng/m^3	=	10^{-3} ng/l
''	=	$10^{-6}\,\mu g/l$*
''	=	10^{-9} mg/l
''	=	10^{-12} g/l
100 ng/m^3	=	10^{-7} mg/l*

*These values so designated are in common use in literature from the USSR.

Modified from a tabulation obtained from Dr. I. R. Jonasson,
Geological Survey of Canada.

Appendix A

CONVERSIONS
TROY, AVOIRDUPOIS AND METRIC WEIGHTS

Unit	Troy Grains	Troy Dwts.	Troy Ounces	Troy Pounds	Avoir. Ounces	Avoir. Pounds	Milli-grams	Grams	Kilo-grams
Troy Grain.........	1.	.0417	.0021	.00017	.00229	.00014	64.8	.0648	.000065
Troy Dwt...........	24.	1.	.05	.00417	.05486	.00343	1555.2	1.5552	.001555
Troy Ounce........	480.	20.	1.	.08333	1.0971	.0686	31103.5	31.103	.031103
Troy Pound........	5760.	240.	12.	1.	13.1657	.8229	373241.9	373.242	.37324
Avoir. Ounce.......	437.5	18.23	.9115	.076	1.	.0625	28349.5	28.350	.02835
Avoir. Pound.......	7000.	291.67	14.5833	1.215	16.	1.	453592.6	453.59	.45359
Milligram..........	.0154	.00064	.000032	.0000027	.000035	.0000022	1.	.001	.000001
Gram.............	15.432	.643	.03215	.00268	.03527	.0022	1000.	1.	.001
Kilogram..........	15432.3	643.01	32.1507	2.6792	35.2739	2.205	1000000.	1000.	1.

LENGTHS

Unit	Inches	Feet	Millimeters	Centimeters	Meters
Linear inch..........	1.	0.0833	25.399	2.5399	0.025399
Linear foot..........	12.	1.	304.788	30.4788	0.304788
Linear millimeter......	0.03937	0.00328	1.	0.1	0.001
Linear centimeter.....	0.3937	0.0328	10.	1.	0.01
Linear meter.........	39.37	3.2809	1000.	100.	1.

AREAS

Unit	Inches	Feet	Millimeters	Centimeters	Meters
Square inch.........	1.	0.00694	645.11	6.4511	0.000645
Square foot.........	144.	1.	92895.	928.95	0.0929
Square millimeter.....	0.00155	0.00001075	1.	0.01	0.000001
Square centimeter....	0.155	0.001075	100.	1.	0.0001
Square meter........	1550.	10.75	1000000.	10000.	1.

VOLUMES

Unit	Inches	Feet	Millimeters	Centimeters	Meters
Cubic inch..........	1.	0.000578	16385	16.385	0.00001638
Cubic foot..........	1728.	1.	28311000	28311.	0.028311
Cubic millimeter......	0.000061	0.000000035	1	0.001	0.000000001
Cubic centimeter......	0.061	0.000035	1000	1.	0.000001
Cubic meter.........	61023	35.316	1000000000	1000000.	1.

Courtesy of Engelhard Industries of Canada Ltd.

Appendix

B

SELECTED COLORIMETRIC METHODS

ZINC IN SOILS, SEDIMENTS AND ROCKS

A. Hot Acid Extractable Zinc

Procedure
1. Weigh or scoop 0.1 gm of a fine-grained sample (sieved to desired mesh size) into a test tube (16 x 150 mm).
2. Add 3 ml of 1:3 nitric acid (1 acid to 3 water). Disperse from a buret or an automatic pipettor. (When the organic matter content of the sample is high, the sample should be ignited at $450° - 550°C$ for 10 minutes before digestion to destroy the organic matter which tends to affect the colored compounds formed in this test.)
3. Simmer in a hot water bath for $^1/_2$ hour.
4. Allow to cool.
5. Adjust to 10 ml with metal-free water. (If the test tube is not calibrated to 10 ml, add 7 ml of water since loss on digestion is minor.) Mix well and allow the residue to settle.
6. Pipet an aliquot of 2 ml into another test tube (16 x 150 mm).
7. Add 5 ml of zinc buffer to (6) above.
8. Add 5 ml of 0.001% dithizone solution to (6) above.
9. Cap the tube and shake it vigorously for 30 seconds.
10. Compare the color of the upper colored layer with a standard zinc series.
11. If the color of the solvent phase in the test solution exceeds that of the highest standard, repeat from Step 6 using a smaller (perhaps 0.5 ml) aliquot.

12. Zinc (ppm) =

$$\frac{\text{micrograms } (\mu g) \text{ of matching standard}}{\text{weight of sample (gm)}} \times \frac{\text{volume of sample solution (ml)}}{\text{aliquot (ml)}}$$

Example: If 3 micrograms found: $\text{Zn (ppm)} = \frac{3}{0.1} \times \frac{10}{2} = 150$

Standards
1. To each of five test tubes (16 x 150 mm), add 5 ml of buffer solution.
2. Add respectively 0, 0.1, 0.2, 0.3 and 0.4 ml of a 10 microgram per ml standard zinc solution.

3. Add 5 ml of 0.001% dithizone solution to each.
4. Cap the tubes and shake vigorously for 30 seconds.
5. Color series should be:
   0 microgram = green
   1 microgram = blue-green
   2 micrograms = blue
   3 micrograms = purple
   4 micrograms = red
6. Store in a cool, dark place when not in use. Under ideal conditions these standards may last a few days, but they should be prepared daily.

Reagents

Nitric acid solution. 1 part of concentrated acid to 3 parts metal-free water.
Dithizone stock solution (0.1%).
    Dissolve 0.1 gm of dithizone in 100 ml of chloroform and shake for 5 minutes. Store in a thermos bottle away from sunlight in a cool place (refrigerator). This can be kept for a month provided no discoloration appears.
Dithizone test solution (0.001%).
    Dilute 1 ml of stock solution to 100 ml with toluene. Prepare fresh daily. NOTES: (1) Yellow discoloration of dithizone solution indicates oxidizing conditions and the dithizone solution should not be used. (2) Toluene is used here in preference to chloroform or carbon tetrachloride because it is lighter than the water-acid-buffer mixture, hence it will float. Unreacted soil or sediment, particularly in the cold extractable procedure, will be on the bottom of the test tube and will not, therefore, interfere with color interpretation.
Zinc buffer solution.
1. Weigh 125 gms of sodium thiosulfate into a beaker.
2. Dissolve by adding 400 ml of water.
3. Transfer the solution to a large separatory funnel (1 liter).
4. Add 300 grams of sodium acetate to a second beaker and dissolve by adding 400 ml of warm metal-free water.
5. Add 60 ml of glacial acetic acid to (4) above.
6. Add this solution to the separatory funnel.
7. Extract any heavy metals from (6) above with dithizone solution as follows:
   (a) Add about 50 ml of 0.01% dithizone-carbon tetrachloride solution (0.01 gm dithizone in 100 ml of carbon tetrachloride) to the separatory funnel and shake vigorously for three minutes. Allow the dithizone to settle to the bottom. If this layer shows any color except characteristic green, then heavy metals are present and forming colored dithizonates.
   (b) Drain off the lower layer of colored dithizone and discard.
   (c) Repeat using other 50 ml portions of dithizone stock solution until lower layer remains green.
   (d) For a very sensitive test for heavy metal contamination, repeat as in (a) and (b) using a diluted dithizone solution (0.001%) and note any color change. Repeat, if necessary, until no color change occurs. NOTE: A dithizone-chloroform or dithizone-carbon tetrachloride solution is preferred in the purification procedure, as they are heavier than the buffer solution; they will settle to the bottom of the separatory funnel and will be easy to discard.
   (e) When all heavy metals have been extracted, the excess dithizone solution dissolved in the buffer must be removed with 50 ml portions of carbon tetrachloride, discarding the lower layers. Extract with successive 50 ml portions of carbon tetrachloride until the bottom layer is colorless. NOTE: Be sure to release pressure on separatory funnel correctly — invert and slowly open stopcock.
8. Dilute to 2 liters and store (keeps indefinitely).
Standard zinc solutions.
a. 100 micrograms of Zn per ml (μg/ml): Dissolve 0.100 gm of reagent grade (30 mesh) zinc metal in 10 ml of conc. HCl acid and dilute to 1 liter with metal-free water.

SELECTED STANDARDS TO ILLUSTRATE COLORIMETRIC TESTS

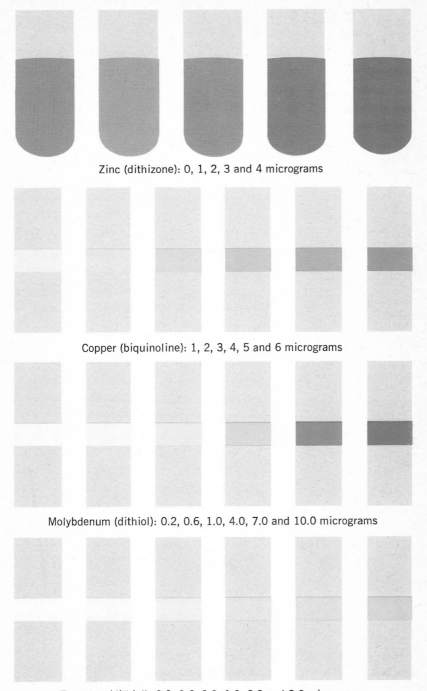

Zinc (dithizone): 0, 1, 2, 3 and 4 micrograms

Copper (biquinoline): 1, 2, 3, 4, 5 and 6 micrograms

Molybdenum (dithiol): 0.2, 0.6, 1.0, 4.0, 7.0 and 10.0 micrograms

Tungsten (dithiol): 0.2, 0.6, 0.8, 1.0, 2.0 and 3.0 micrograms

b. 10 micrograms of Zn per ml (μg/ml): Pipet 10 ml of the 100μg/ml solution into a 100 ml volumetric flask and dilute to volume with metal-free water.

B. Cold Extractable Zinc (Buffer Only)

Procedure

1. Scoop or weigh 0.2 gm of sample (fine fraction only) into a test tube (16 x 150 mm).
2. Add 5 ml of zinc buffer solution (same buffer as used in the hot extractable method).
3. Add 2 ml of dithizone test solution (0.001%).
4. Cap the tube and shake vigorously for 1 minute. Allow the dithizone layer to separate.
5. A color in the upper layer will be developed between green and red. If the color in the dithizone layer is green, record as 0μg; if it is blue-green, record as 1μg, and if it is blue, record as 2μg. In these cases the test is over. If the color developed is purple, red (or pink) add more dithizone solution. Do *NOT* shake. Mix gently by tipping the test tube back and forth, and rotating. Add dithizone in measured increments until blue (2μg) is obtained. Record the total volume of dithizone solution required. If the sample requires more than 20 ml of dithizone solution to achieve a blue color, record as "20+".

6. Zinc (ppm) =

$$\frac{\text{micrograms Zn (2}\mu\text{g for blue)}}{\text{weight of sample (0.2 gm)}} \times \frac{\text{no. ml dithizone added}}{\text{no. ml of dithizone used in standards}}$$

$$\text{Zn (ppm)} = 10 \times \frac{\text{no. ml dithizone added}}{5}$$

NOTE: In the field it will be sufficient to record and map the volume of dithizone solution added above 2 ml as an index of the amount of Zn reacting.

Standards; Reagents

Prepared as outlined in the hot extractable method for zinc, including preparation of standard zinc solutions.

* * * * *

COPPER IN SOILS, SEDIMENTS AND ROCKS

A. Hot Acid Extractable (Biquinoline)

Procedure

1. Scoop or weigh a 0.1 gm sample of soil, sediment, or finely ground rock (sieved to desired mesh size) into a test tube (16 x 150 mm).
2. Dissolve in hot HNO_3 as in steps (2) and (3) of the procedure for zinc (hot acid extractable method).

OR: Add 0.5 gm of potassium pyrosulfate, mix, and fuse for two minutes after the flux melts. Allow to cool and add 3 ml HCl (1:1); wait until "cake" is dissolved. (It may be necessary to place the test tube in a hot water bath to hasten the dissolution).

3. Dilute to 10 ml with metal-free water. Mix well and allow the residue to settle.
4. Pipet a 2 ml aliquot into another test tube (16 x 150 mm).
5. Add 10 ml of copper buffer to (4) above.
6. Add 1 ml of biquinoline solution to (4) above.
7. Cap the tube and shake it vigorously for 30 seconds.
8. Allow the isoamyl alcohol layer to separate and compare the pink color with a standard copper series.
9. If the intensity of the color in the solvent phase exceeds that of the highest standard, repeat from Step 4 using a smaller aliquot.

10. Cu (ppm) =

$$\frac{\text{micrograms } (\mu g) \text{ of matching standard}}{\text{weight of sample (gm)}} \times \frac{\text{volume of sample solution (ml)}}{\text{aliquot (ml)}}$$

$$\text{Cu (ppm)} = \frac{\text{micrograms } (\mu g) \text{ of matching standard}}{0.1} \times \frac{10}{2}$$

$$\text{Cu (ppm)} = \text{micrograms } (\mu g) \text{ of matching standard} \times 50$$

Standards

1. To each of seven test tubes (16 x 150 mm), add 10 ml of buffer solution.
2. Add respectively 0, 0.1, 0.2, 0.3, 0.4, 0.5 and 0.6 ml of a 10 microgram per ml standard copper solution.
3. Add 1 ml of biquinoline solution to each.
4. Cap the tubes and shake them vigorously for 60 seconds.
5. Allow the isoamyl alcohol layer to separate. A series of pink colors (a monocolor) of increasing intensity will be obtained.
6. Tightly cap or seal the test tubes (to prevent evaporation of isoamyl alcohol) and the color should remain stable for months.

Reagents

Nitric acid solution. 1 part of concentrated acid to 3 parts metal-free water.
Hydrochloric acid solution. 1 part of concentrated acid to 1 part metal-free water.
2,2′ – biquinoline solution (0.02%).
1. Dissolve 0.2 gm of 2,2′ – biquinoline in 900 ml of isoamyl alcohol.
2. Warm gently until biquinoline dissolves. CAUTION: requires a well ventilated room; keep away from flames.
3. Allow to cool and dilute to 1 liter with isoamyl alcohol. The solution must be clear; if it is yellow the reagent is impure and should not be used.
Copper buffer solution.
1. Dissolve 200 gm of sodium acetate, 100 gm of sodium tartrate and 20 gm of hydroxylamine hydrochloride and dilute to 1 liter with metal-free water.
2. Check pH of buffer solution with pH indicator paper. If pH is 6 ± 0.2 proceed to next step. If not, adjust the pH with HCl or NaOH.
3. Check for Cu contamination by placing 10 ml of buffer solution into a test tube; add 1 ml of biquinoline solution; shake for 2 minutes. Development of a pink color indicates contamination. If copper contamination is detected, purify the copper buffer with dithizone solution, until it is free of copper, as outlined in the zinc method.
Standard copper solutions.
a. 100 micrograms of Cu per ml (μg/ml): Dissolve 0.200 gm copper sulfate ($CuSO_4 \cdot 5H_2O$) in 500 ml of dilute HCl (8 ml conc. HCl in 1 liter of metal-free water).
b. 10 micrograms of Cu per ml (μg/ml): Pipette 10 ml of the 100 μg/ml solution into a 100 ml volumetric flask and dilute to volume with the above dilute HCl solution.

B. Cold Acid Extractable (Biquinoline)

Procedure

1. Scoop 0.1 gm of sample into a test tube (16 x 150 mm).
2. Add 1 ml of 1:1 HCl.
3. Add 10 ml of copper buffer solution.
4. Add 1 ml of biquinoline solution.
5. Cap tube and shake it vigorously for 30 seconds.
6. Allow the isoamyl alcohol layer to separate and compare the pink color with a standard copper series.

7. Cu (ppm) =

$$Cu\ (ppm) = \frac{\text{micrograms } (\mu g) \text{ of matching standard}}{\text{weight of sample (gm)}} \times \frac{\text{volume of sample solution (ml)}}{\text{aliquot (ml)}}$$

$$Cu\ (ppm) = \frac{\text{micrograms } (\mu g) \text{ of matching standard}}{0.1} \times \frac{1}{1}$$

$$Cu\ (ppm) = \text{micrograms } (\mu g) \text{ of matching standard} \times 10$$

Standards: Reagents

Prepared as outlined in the hot extractable method for copper, including preparation of standard copper solutions.

* * * * *

COLD EXTRACTABLE "TOTAL HEAVY METALS" IN SOILS AND SEDIMENTS (BLOOM TEST)

Procedure

1. Scoop 0.1 gm of the finest fraction of a soil or stream sediment into a test tube calibrated at 5 ml, and then at 2 ml intervals (or use a 25 ml glass stoppered, graduated cylinder).
2. Add 5 ml of buffer solution.
3. Add 1 ml of dithizone test solution (0.001%).
4. Shake vigorously for 5 seconds.
5. If the top layer is green, blue-green or blue, then the analysis is complete, and record as 0, $^1/_2$ and 1 ml respectively.
6. If the layer is purple or red, add 1 ml of dithizone test solution and gently mix. (Do not shake as this will remove more loosely-held metal from the sample). Repeat until a blue color is obtained. If the sample requires more than 20 ml to achieve a blue color, record as "20 +".
7. Record the volume of dithizone solution used to reach the end point (a blue color). These dithizone volumes are an index of the heavy metal content and can be "mapped" as such. The dithizone volume obtained is often equated with the "total heavy metal" content, but usually it is chiefly zinc.

Reagents

Dithizone stock solution (0.01%).
Dissolve 0.01 gm of dithizone in 100 ml toluene and mix for 10 minutes. Store in a thermos bottle.

Dithizone test solution (0.001%).
Dilute 10 ml stock solution to 100 ml with toulene.

Buffer solution.

1. Dissolve 50 gm of ammonium citrate and 8 gm of hydroxylamine hydrochloride in a beaker containing 600 ml of metal-free water.
2. Add a small amount of thymol blue to the solution. The solution will be pink if the pH is lower than 2.8, yellow if the pH is 2.8 – 8.5 and blue if the pH > 8.5. Add NH_4OH until the solution turns blue indicating pH = 8.5.
   OR: Add NH_4OH and test with pH paper until pH = 8.5.
3. Transfer the solution to a large separatory funnel and remove the heavy metals by extraction with 20 ml portions of a 0.01% dithizone-carbon tetrachloride solution (0.01 gm dithizone in 100 ml carbon tetrachloride) until the organic layer remains green. Extract the dissolved dithizone with 25 ml portions of chloroform until the chloroform is colorless; any yellow color in the buffer should disappear at this stage. Wash 2 or 3 times with carbon tetrachloride to remove dissolved chloroform.
4. Dilute to 1 liter with metal-free water.

*   *   *   *   *

HEAVY METALS IN WATER

Procedure

1. Place 50 ml of water in a 100 ml glass stoppered, graduated cylinder.
2. Add 5 ml of buffer solution.
3. Add 5 ml of dithizone test solution.
4. Stopper the cylinder and shake vigorously for 1 minute.
5. Allow the layers to separate and observe the color of the solvent layer.
   (a) If the color is a mixed color between green and red, estimate the amount of metal present by reference to the zinc content in the following table (modified from Ward et al., 1963, Table 4):

Micrograms (μg) of metal in 50 ml of water	Dithizone color		
	Zinc	Copper	Lead
4	red	blue	blue-green
3	purple	blue-green	bluish green
2	blue	bluish green	bluish green
1	blue-green	green	green
0	green	green	green

   Although copper and lead are capable of reacting with dithizone under the conditions of this test, from the above tabulation it can be seen that they are not as effective as zinc in forming colored dithizonates.
   (b) If the color is red, repeat the test using a smaller volume of sample (e.g., 25 ml); in this case the number of micrograms found by comparison with the above tabulation would be multiplied by 2. Alternatively, it may be necessary to increase the sensitivity of the test; in this event use a smaller volume of dithizone (e.g., 2.5 ml), and make an appropriate adjustment for the micrograms found (in this example, divide by 2).
6. Total heavy metals (as zinc equivalent) in water (ppb) = micrograms found (μg) x 20.

Reagents

Acetic acid solution.
   Dilute 114 ml of glacial acetic acid to 1 liter with metal-free water.
Dithizone stock solution (0.01%).
   Dissolve 0.01 gm dithizone in 100 ml of toluene and mix for 10 minutes. Store in a thermos bottle.
Dithizone test solution (0.0015%).
   Dilute 15 ml of stock solution to 100 ml with toluene.
Sodium acetate solution.
   Dissolve 164 gm of sodium acetate in metal-free water and dilute to 1 liter.

Buffer solution.
1. Mix 900 ml of the above sodium acetate solution with 100 ml of the above acetic acid solution.
2. Purify with dithizone-carbon tetrachloride, or dithizone-chloroform as outlined in the zinc procedure.

* * * * *

MOLYBDENUM AND TUNGSTEN IN SOILS, SEDIMENTS AND ROCKS

A. Molybdenum

Procedure

1. Weigh 0.25 gm of sample into a nickel crucible (15 ml).
2. Add 1.5 gm of flux and mix thoroughly.
3. Heat in a muffle furnace at 800°C. for 8 minutes (or in a gas flame about 10 minutes) until frothing has stopped.
4. Cool and add about 3 ml of water; cover the crucible and let it stand for several hours (preferably overnight).
5. Loosen the melt on the bottom of the crucible with a pyrex rod; decant into a test tube (16 x 150 mm) calibrated at 5 ml, and adjust to volume (exactly 5 ml) with water washings from the crucible.
6. Heat the test tube in a boiling water bath for 10 minutes shaking the tube occasionally; make sure that the melt has completely broken up.
7. Mix thoroughly; allow to cool and settle (or centrifuge the tube).
8. Pipet a 2 ml aliquot of the clear, supernatant solution into a second test tube. (If tungsten is to be determined on the same sample, pipet 0.5 ml of this clear solution into a third test tube and proceed with the tungsten determination described below).
9. Add 2 ml of hydroxylamine hydrochloride solution cautiously. Shake the test tube to liberate carbon dioxide. Then cool to below 25°C.
10. Add 0.5 ml of dithiol solution and occasionally shake gently over a period of 20 minutes.
11. Compare the color of the upper layer with a standard series. (It may be necessary to centrifuge some samples in order to separate the organic phase.)
12. If the color of the upper layer exceeds that of the highest standard, repeat from step 8 using a smaller aliquot of solution.

13. Mo (ppm) =

$$\frac{\text{micrograms } (\mu g) \text{ of matching standard}}{\text{weight of sample (gm)}} \times \frac{\text{volume of sample solution (ml)}}{\text{aliquot (ml)}}$$

Example: If a 2 ml aliquot was used (as suggested in step 8):

$$\text{Mo (ppm)} = \frac{\text{micrograms } (\mu g) \text{ of Mo}}{.25} \times \frac{5}{2}$$

Mo (ppm) = micrograms (μg) Mo x 10

Standards

1. To each of 12 test tubes (16 x 150 mm), add 2 ml of hydroxylamine hydrochloride.
2. Pipet 0, 0.2, 0.4, 0.6, 0.8 and 1.0 ml of a standard solution containing 1 microgram of molybdenum per ml, and 0.2, 0.3, 0.4, 0.5, 0.7 and 1.0 ml of a standard solution containing 10 micrograms of molybdenum per ml, into the test tubes.
3. Dilute each to 4 ml with water.
4. Add 0.5 ml of dithiol solution to each. Shake gently at frequent intervals over a period of 20 minutes. The colors of the standard series range from colorless through varying intensities of yellow-green, corresponding to a molybdenum content of 0 to 10 micrograms.

Reagents

Flux mixture.
Mix 500 grams of sodium carbonate, 400 grams of sodium chloride and 100 grams of potassium nitrate until homogeneous.

Hydroxylamine hydrochloride (2.5%).
Dissolve 5 gm of hydroxylamine hydrochloride in 20 ml of water; add 180 ml of concentrated hydrochloric acid. Store in a glass stoppered bottle.

Dithiol solution (1%).
Weigh 0.5 gm of zinc dithiol into a 50 ml glass stoppered cylinder; add 1 ml of hydrochloric acid (concentrated) and shake for 10 seconds. Dilute to 50 ml with amyl acetate and mix thoroughly.

Standard molybdenum solutions.
a. 100 micrograms of Mo per ml ($\mu g/ml$): Dissolve 125 mg of sodium molybdate (dihydrate) in water and dilute to 500 ml.
b. 10 micrograms of Mo per ml ($\mu g/ml$): Pipet ml of the 10 100 $\mu g/ml$ solution into a 100 ml volumetric flask and dilute to volume with water. Prepare fresh weekly.
c. 1 microgram of Mo per ml ($\mu g/ml$): Pipet 10 ml of the 10 $\mu g/ml$ solution into a 100 ml volumetric flask and dilute to volume with water. Prepare fresh daily.

B. Tungsten

Procedure

1. Decompose the sample as in Steps 1 – 7 of the above procedure for molybdenum.
2. Pipet a 0.5 ml aliquot of the clear, supernatant solution into a test tube (16 x 150 mm).

3. Add 5 ml of stannous chloride solution.
4. Heat in a boiling water bath for 10 minutes.
5. Add 1 ml of 0.5% dithiol solution and continue heating for 10 minutes, shaking occasionally.
6. Cool, and add 1 ml of the Stoddard-ethanol solution; stopper the tube and shake vigorously for 30 seconds.
7. Compare visually with a standard series. If the color obtained exceeds that of the highest standard, add a 1 ml aliquot of the Stoddard-ethanol solution and shake. Continue this procedure until the color intensity lies within the range of the standards. Alternatively, repeat from Step 2 using a smaller aliquot.
8. Record the volume of test solution (aliquot) used, the volume of solvent (Stoddard-ethanol solution) used, and the number of micrograms of tungsten from the matching standard series.

9. $$W \text{ (ppm)} = \frac{\text{micrograms } (\mu g) \text{ of matching standard}}{\text{weight of sample (gm)}} \times \frac{\text{solvent volume (ml)} \times 5}{\text{aliquot (ml)}}$$

Example: If a 0.25 gm of sample was fused, and only 1 ml of Stoddard-ethanol mixture was required (solvent volume) with a 0.5 ml aliquot, then the equation would reduce to:

$$W \text{ (ppm)} = \frac{\text{micrograms } (\mu g) \text{ of } W}{0.25} \times \frac{1 \times 5}{0.5}$$

Note: The factor of 5 appears in the equation by virtue of the fact that after the original fusion, the fused product was adjusted to a 5 ml volume with water (see Step 5 in the molybdenum procedure).

Standards

Pipet 0, 0.1, 0.2, 0.3, 0.4, 0.6 and 0.8 ml of a 1 microgram of tungsten per ml solution, and 0.1, 0.2 and 0.3 ml of a 10 microgram of tungsten per ml solution into test tubes (16 x 150 mm) and proceed as in Steps 3 to 6 above. The colors of the standard series range from colorless through varying shades of blue (often referred to as blue-green), corresponding to a tungsten content of 0 to 3 micrograms.

Reagents

Stannous chloride solution (10%).
Dissolve 50 gm of stannous chloride (dihydrate) in 500 ml of concentrated hydrochloric acid. Store in a glass stoppered bottle.

Stoddard-ethanol solution.
Mix 50 ml of Stoddard solvent and 50 ml of absolute ethanol. Store in a plastic bottle. (Shellsol, kerosene, petroleum spirit or white spirit may be substituted for Stoddard solution).

Dithiol solution (0.5%).
Dissolve 0.5 gm of zinc dithiol in 25 ml of sodium hydroxide solution; warming may be necessary. Dilute to 100 ml with water. Store in a glass stoppered bottle in a refrigerator.
Sodium hydroxide solution.
Dissolve 8 gm of sodium hydroxide pellets in 100 ml of metal-free water. Store in a plastic bottle.
Standard tungsten solutions.
a. 100 micrograms of W per ml (μg/ml): Dissolve 0.090 gm of sodium tungstate in 50 ml of water and then dilute to 500 ml with water.
b. 10 micrograms of W per ml (μg/ml): Pipet 10 ml of the 100 μg/ml solution into a 100 ml volumetric flask and dilute to volume with water. Prepare fresh weekly.
c. 1 microgram of W per ml (μg/ml): Pipet 10 ml of the 10 μg/ml solution into a 100 ml volumetric flask and dilute to volume with water. Prepare fresh daily.

* * * * *

Appendix

C

WORKED STATISTICAL EXAMPLES

In this section we consider some examples which illustrate the simplest but yet the most important types of statistical calculations performed on exploration geochemical data. We will not consider the more advanced statistical techniques described in Chapter 12 as these require a more detailed explanation than the present space allows. For the interested reader examples in geochemistry utilizing these techniques can be found in texts by Koch and Link (1970, 1971) and Davis (1973). We restrict our attention in this section to those types of calculations carried out on nearly every set of exploration geochemical data as the first step towards an interpretation. Furthermore, these calculations can be performed without aid of a computer.

Consider the group of values given in Table C-1. These represent the content of copper expressed in ppm for 59 stream sediment samples collected from the Casino prospect area in the Yukon Territory, Canada as reported by Archer and Main (1971). A map with the location of the samples was presented as Fig. 12-4 in Chapter 12. This particular set of data serves as a useful illustrative example because the samples are from an area in which an ore deposit has been found. Both the background values and the anomalous values reside within the data.

From an examination of the values in Table C-1, it is not readily obvious what proportion of these values would be greater than some specified

Table C-1. Copper content of silts (ppm). From Archer and Main (1971)

32	138	160	950	50
32	68	165	1360	64
20	54	89	100	76
74	46	116	136	120
26	59	41	424	360
50	54	116	1020	480
152	66	91	820	100
80	62	100	240	350
58	33	89	1600	94
140	92	340	570	372
48	45	106	130	480
54	152	196	120	

value, say for instance, a threshold value of 50 ppm. It would be difficult also to estimate by inspection the local background value for this set of data. The first step is to rank the values in order of increasing magnitude. The rank-ordered values of Table C-1 are given in Table C-2. Answers to the two preceding questions are now readily obtained. By counting the number of values starting from the lowest value of 20, it is seen there are 11 values less than or equal to 50 which means 48 out of the 59 values are greater than 50. If the regional threshold value was 50 ppm, and assuming that the collected samples were representative of the area as a whole, it follows that the level of localized concentration represented by these data would be well above the regional level.

To establish the local background value for a given set of data, it is usual to take as an estimate the *median* value of the data distribution. The median is defined as the value for which one half the values in the distribution are less and one half are greater. For the set of 59 values, this would correspond to the 29th rank-ordered value. Referring to Table C-2, the median value is 100. Had the number of values been an even number, the median would have been defined as the midpoint of the two values which divided the distribution in half.

It is important also to determine which of the values or range of values occur most frequently in these data. To accomplish this, the data can be subdivided into specific intervals and thus, cast in the form of a frequency table. To do this, we should note that the size of interval along with the number of intervals required to represent the data adequately is not always an easy choice. Too coarse an interval will unduly smooth the data distribution whereas too fine an interval will result in little improvement over the original data.

Table C-2. Copper content from Table C-1 arranged in ascending rank order

20	54	89	130	360
26	54	91	136	372
32	58	92	138	424
32	59	94	140	480
33	62	100	152	480
41	64	100	152	570
45	66	100	160	820
46	68	106	165	950
48	74	116	196	1020
50	76	116	240	1360
50	80	120	340	1600
54	89	120	350	

A common formula for choosing the number of intervals into which data are to be subdivided is given by

$$[k] = 10 \cdot \log_{10} N$$

where $[k]$ is the largest interval contained in the right hand expression and N is the number of observations. This formula is valid for N greater than 30. For the data in Table C-2 in which there are 59 values, the number of intervals is calculated as

$$[k] = 10 \cdot \log_{10} 59 = 17$$

To obtain the interval width, we can divide the largest value in Table C-2 by 17 and upon rounding to the nearest multiple of 10 obtain a value of 90. With this accomplished, we proceed to tabulate the data in the form of a frequency table shown in Table C-3.

It is possible now to gain an insight of the nature of the underlying distribution. The cumulative as well as the class frequencies are listed in

Table C-3. Distribution of grouped data for copper content values
in Table C-2

Class Interval	Midpoint of Interval	Class Frequency	Relative Frequency	Cumulative Frequency	Cumulative Frequency (pct)
0-90	45	25	.42	25	42
91-180	135	19	.32	44	74
181-270	225	2	.03	46	77
271-360	315	3	.05	49	82
361-450	405	2	.03	51	85
451-540	495	2	.03	53	88
541-630	585	1	.02	54	90
631-720	675	0	.00	54	90
721-810	765	0	.00	54	90
811-900	855	1	.02	55	92
901-990	945	1	.02	56	94
991-1080	1035	1	.02	57	96
1081-1170	1125	0	.00	57	96
1171-1260	1215	0	.00	57	96
1261-1350	1305	0	.00	57	96
1351-1440	1395	1	.02	58	98
1441-1530	1485	0	.00	58	98
1531-1620	1575	1	.02	59	100
		59	1.00		

Table C-3 and later we will see how the cumulative frequencies can be used in the interpretation.

What is clear by inspection of the class frequencies in Table C-3 is that the distribution of copper values is skewed towards the larger values. In other words, there are many more low copper values than there are high values. Actually this is typical of trace element distributions in general, since in most cases, the portion of a total area characterized by an anomalous content of a given element is usually very small and it is not surprising that this type of distribution is encountered. We can prepare a histogram for the data shown in Fig. C-1. A histogram is constructed by plotting class frequency against the midpoint of each class interval.

We can now answer an earlier question, that is, which interval contains the most number of values? From the histogram, we see it is the interval from 0 to 90. The midpoint of this interval is the *mode* of the distribution, in this case, 45. We now compare this with the median, remembering it was found to have a value of 100, and because of this difference, we can conclude that the distribution of copper contents is indeed skewed. As a general rule, the median rather than the mode is a more reliable estimate of the background value of an element because the mode depends on the interval chosen for subdividing the data into frequency classes. For any estimate of a background value, however, care must be taken to insure that the sampled data are representative of the same or similar parent populations.

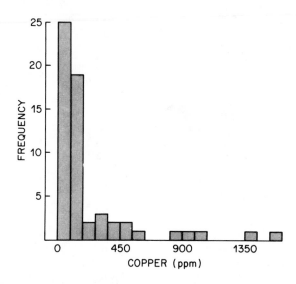

Fig. C-1. Histogram of copper contents based on fifty-nine samples.

For data with a large number of observations, the median can also be estimated from the cumulative frequency plot. If we take the cumulative frequency values in Table C-3 and prepare a plot of cumulative frequency against the midpoints of the class intervals, we obtain a median value equal to 112. Note that this value differs slightly from the median value obtained for the ungrouped data. In most cases, as in this case, the difference between the median value estimated from ungrouped data and the median value estimated from grouped data can be ignored.

Having constructed the cumulative frequency curve, it is not difficult to read off other percentile values which may be of interest. For instance, it may be desired to determine the value for which 95 percent of the data are less than or equal, so that any value above this value can be considered as anomalous. From Fig. C-2 it is seen, for instance, that the value for the 95th percentile is 1035. Referring to Table C-2, we see that only two of the original values are greater than 1035 which suggests that these two values alone are indicative of anomalous contents with a high degree of probability. Again, caution must be exercised in forming such conclusions in light of what has been emphasized in this book regarding the interpretation of anomalies.

The disadvantage of the median, or for that matter the mode, in describing a distribution is that neither of these two estimators takes into account very large (or very small) values and therefore, both are relatively

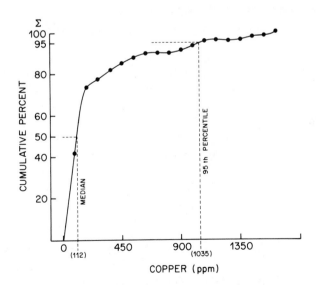

Fig. C-2. Cumulative frequency plot of copper data. The values for the median and the 95th percentile are given in parentheses.

insensitive to large fluctuations in the data. Consequently, neither the median or the mode provides a very efficient measure of the central tendency of a given distribution. If we are interested in the mean content of a sampled material, we are much better off to calculate the sample *mean*.

The sample mean, $\bar{x}$, is defined as

$$\bar{x} = \frac{\sum x_i}{n}$$

where $\sum x_i$ refers to the sum of a given set of values and n is the number of values. For the data in Table C-2, the sample mean is given by (20 + 26 + $\cdots$ +1600)/59 which is equal to 13260/59 or 224.8. We note that the value for the sample mean is considerably greater than either the median value or the mode derived from the same data. This difference is due to the presence of a few very high values.

Where a large number of observations are involved, it is often more convenient to calculate the sample mean from grouped data as shown in Table C-3. For the case of grouped data, the sample mean is calculated as

$$\bar{x} = \frac{\sum f_i x_i}{n}$$

where f_i is the ith class frequency, x_i is the midpoint of the ith class interval and n is the total number of values. From the frequency data given in Table C-3, we calculate the same means as

$$[(25 \times 45) + (19 \times 135) + \cdots + (1 \times 1575)]/59$$

which is equal to 13275/59 or 225. This value is very close to the value obtained for the sample mean calculated from the original data. In general, we can ignore such a small difference.

Another useful measure of a distribution is the sample *variance*, or taking its positive square root, the sample *standard deviation*. The sample variance is a measure of dispersion and has applications in checking on the homogeneity of values in any particular set of data. The sample variance is calculated as

$$s^2 = \frac{\sum (x_i - \bar{x})^2}{(n-1)}$$

where $\bar{x}$ is the sample mean, x_i is the ith value and n is the total number of values. The denominator for the sample variance is (n-1) rather than n however. The sample standard deviation is calculated as

$$s = +\sqrt{s^2}$$

The standard deviation is preferred more often to the sample variance for characterizing the dispersion about the mean because it is expressed in the same units as the mean, unlike the sample variance which is expressed in terms of squared units of the mean. The sample variance is calculated first, however.

In calculating the sample variance, particularly when using a desk calculator where there is little risk of truncation or roundoff error, it is convenient to use the computing formula given by

$$s^2 = \frac{n(\sum x_i^2) - (\sum x_i)^2}{n(n-1)}$$

Thus, one needs only to calculate $(\sum x_i)$ and $(\sum x_i^2)$ from the data in order to compute the sample variance. For the data in Table C-2, the sample variance is calculated as

$$\frac{59(8971424) - (13260)^2}{59(58)} = 103298.2$$

and by taking the positive square root, we obtain the sample standard deviation equal to

$$s = 321.4$$

If the calculations are to be performed by computer however, because of truncation and roundoff error, it is advisable to calculate the sample variance using the expression given first.

For grouped data, the sample variance is calculated as

$$s^2 = \frac{\sum f_i (x_i - \bar{x})^2}{(n-1)}$$

where f_i is the frequency of the ith class interval, x_i is the midpoint of the ith class interval, $\bar{x}$ is the sample mean calculated for grouped data and n is the total number of values. Unlike the sample mean, however, the sample variance calculated for grouped data is invariably greater than is the sample variance calculated from ungrouped data. In general, this difference can be approximated by $h^2/12$ where h is the interval width. In our example, h is equal to 90 so that the sample variance calculated from the grouped data in Table C-3 ought to be greater than the sample variance calculated for the ungrouped data in Table C-2 by an amount of $90^2/12$ or approximately 675. Actually, the calculated sample variance for the grouped data in Table C-3 is equal to 105020.7. This is 1722.5 greater than the sample variance calculated for ungrouped data. The predicted correction accounts for roughly one third of this difference. Comparing the two sample standard deviations, however, the first value is equal to 321.4 while the second value is equal to 324.1. After making the correction to the second value, the latter becomes 323.0. Despite the difference in the two calculated variances, the effect of this difference on the two standard deviations is negligible as shown by this example and in general, the correction applied to the sample standard deviation for grouped data can be ignored.

The sample mean and variance (or standard deviation) describe jointly the two main characteristics of frequency distributions, namely, the central tendency and the dispersion about the mean. As we have seen, however, the distributions of trace element contents indicate that such distributions are generally skewed toward the higher values. As a measure of this skewness, particularly for positively-valued variates, as in the case for element contents, we can calculate the *coefficient of variation* defined as

$$c = s/\bar{x}$$

where s is the sample standard deviation and $\bar{x}$ is the sample mean. The coefficient of variation is considered a measure of relative variability which takes into account both the mean and the standard deviation. For instance, as the mean is increased, the coefficient of variation for most of the observed geochemical data distributions tends toward zero. Conversely, for quantities present in very small amounts, the coefficient of variation for the observed data distributions tends toward infinity. In particular, values of the coefficient of variation above 2.0 or 2.5 are usual only for substances in trace amounts. For the copper data, dividing the sample standard deviation in which the value was 321.4 by the sample mean in which the value was 224.7, we obtain a coefficient of variation equal to 1.4 This suggests that copper is concentrated in amounts greater than expected for trace elements in general and therefore we might anticipate that some anomalous copper content values are included in the data. In addition, based on the comments of Koch and Link (1972), any value of the coefficient of variation which exceeds 0.5 for positively-valued variates makes it unlikely that such data are derived from a normal distribution but that instead, the data are drawn from some different distribution as for example, the lognormal distribution. We turn finally to consider this problem.

The most direct way to determine if a set of data values are approximately lognormally distributed is to plot the cumulative frequency distribution expressed in logarithms on arithmetic probability paper. If the cumulative frequency distribution is approximated by a straight line, it is safe usually to assume that the data follow a lognormal distribution. From the data plotted as a cumulative frequency curve in Fig. C-2, we obtain the result shown in Fig. C-3. The cumulative frequency distribution expressed in logarithmic units has been plotted on arithmetic probability paper. To a first approximation, the data fall along a straight line. The tendency toward lognormality is largely a function of mean content and the spatial variation in content for a given trace element in relation to the surrounding bedrock geochemistry. It remains to be seen that such variation is indicative solely of anomalies associated with ore bodies.

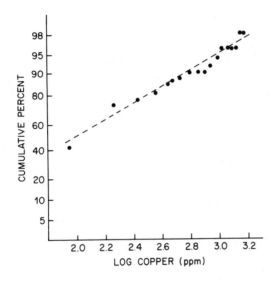

Fig. C-3. Cumulative frequency plot of log copper values on arithmetic probability paper. The dashed line represents the approximate best-fit straight line through the data.

D

CHEMICAL SYMBOLS
AND ELEMENTS

Ac . . . . . Actinium	He . . . . . . Helium	Pt . . . . . . Platinum
Ag . . . . Silver	Hf . . . . . Hafnium	Ra . . . . . . Radium
Al . . . . . Aluminum	Hg . . . . . Mercury	Rb . . . . . Rubidium
Ar . . . . Argon	Ho . . . . . Holmium	Re . . . . . Rhenium
As . . . . . . Arsenic	I . . . . Iodine	Rh . . . . . Rhodium
At . . . . . . Astatine	In . . . . Indium	Rn . . . . . . Radon
Au . . . . Gold	Ir . . . . Iridium	Ru . . . . Ruthenium
B . . . . Boron	K . . . . . Potassium	S . . . . Sulfur
Ba . . . . . . Barium	Kr . . . . . . Krypton	Sb . . . . . Antimony
Be . . . . . Beryllium	La . . . . Lanthanum	Sc . . . . . Scandium
Bi . . . . . . Bismuth	Li . . . . . . Lithium	Se . . . . . Selenium
Br . . . . . . Bromine	Lu . . . . . Lutetium	Si . . . . Silicon
C . . . . Carbon	Mg . . . . Magnesium	Sm . . . . . Samarium
Ca . . . . . Calcium	Mn . . . . Manganese	Sn . . . . Tin
Cd . . . . . Cadmium	Mo . . . Molybdenum	Sr . . . . . Strontium
Ce . . . . . . Cerium	N . . . . . . Nitrogen	Ta . . . . . Tantalum
Cl . . . . . . Chlorine	Na . . . . . . Sodium	Tb . . . . . Terbium
Co . . . . Cobalt	Nb . . . . Niobium	Tc . . . . Technetium
Cr . . . . Chromium	Nd . . . Neodymium	Te . . . . . Tellurium
Cs . . . . . . Cesium	Ne . . . . Neon	Th . . . . . Thorium
Cu . . . . . . Copper	Ni . . . . Nickel	Ti . . . . . Titanium
Dy . . . . Dysprosium	O . . . . . . Oxygen	Tl . . . . . Thallium
Er . . . . . . Erbium	Os . . . . . . Osmium	Tm . . . . . Thulium
Eu . . . . . Europium	P . . . . Phosphorus	U . . . . . Uranium
F . . . . . . Fluorine	Pa . . . Protactinium	V . . . . . Vanadium
Fe . . . . Iron	Pb . . . . Lead	W . . . . . Tungsten
Fr . . . . . Francium	Pd . . . . . Palladium	Xe . . . . . . Xenon
Ga . . . . . Gallium	Pm . . . Promethium	Y . . . . . . Yttrium
Gd . . . . Gadolinium	Po . . . . Polonium	Yb . . . . Ytterbium
Ge . . . . Germanium	Pr . . Praseodymium	Zn . . . . Zinc
H . . . . . .Hydrogen		Zr . . . . . Zirconium

THM or HM	total heavy metals (principally Zn, Cu, Pb)
Me or M	generalized symbol for any metal
cx	cold extractable (e.g., cxCu is cold extractable copper)
hx	hot extractable
RE or REE	a collective term for the rare earth elements with atomic number 57—71, which includes La, Ce, etc.

Actinium	Ac	Neodymium	Nd
Aluminum	Al	Neon	Ne
Antimony	Sb	Nickel	Ni
Argon	Ar	Niobium	Nb
Arsenic	As	Nitrogen	N
Astatine	At	Osmium	Os
Barium	Ba	Oxygen	O
Beryllium	Be	Palladium	Pd
Bismuth	Bi	Phosphorus	P
Boron	B	Platinum	Pt
Bromine	Br	Polonium	Po
Cadmium	Cd	Potassium	K
Calcium	Ca	Praseodymium	Pr
Carbon	C	Promethium	Pm
Cerium	Ce	Protactinium	Pa
Cesium	Cs	Radium	Ra
Chlorine	Cl	Radon	Rn
Chromium	Cr	Rhenium	Re
Cobalt	Co	Rhodium	Rh
Copper	Cu	Rubidium	Rb
Dysprosium	Dy	Ruthenium	Ru
Erbium	Er	Samarium	Sm
Europium	Eu	Scandium	Sc
Fluorine	F	Selenium	Se
Francium	Fr	Silicon	Si
Gadolinium	Gd	Silver	Ag
Gallium	Ga	Sodium	Na
Germanium	Ge	Strontium	Sr
Gold	Au	Sulfur	S
Hafnium	Hf	Tantalum	Ta
Helium	He	Technetium	Tc
Holmium	Ho	Tellurium	Te
Hydrogen	H	Terbium	Tb
Indium	In	Thallium	Tl
Iodine	I	Thorium	Th
Iridium	Ir	Thulium	Tm
Iron	Fe	Tin	Sn
Krypton	Kr	Titanium	Ti
Lanthanum	La	Tungsten	W
Lead	Pb	Uranium	U
Lithium	Li	Vanadium	V
Lutetium	Lu	Xenon	Xe
Magnesium	Mg	Ytterbium	Yb
Manganese	Mn	Yttrium	Y
Mercury	Hg	Zinc	Zn
Molybdenum	Mo	Zirconium	Zr

References

(Note: CIM = Canadian Institute of Mining and Metallurgy.
IMM = Institution of Mining and Metallurgy.)

ABBEY S. (1968) Analysis of rocks and minerals by atomic absorption spectroscopy. *Geol. Surv. Canada Paper* 68-20.

ABBEY S. (1972) "Standard samples" of silicate rocks and minerals — a review and compilation. *Geol. Surv. Canada Paper* 72-30.

ABBEY S. (1973) Studies in "standard samples" of silicate rocks and minerals – Part 3: 1973 extension and revision of "usuable" values. *Geol. Surv. Canada Paper* 73-36.

ADLER I. (1966) *X-ray Emission Spectrography.* Elsevier.

AGRICOLA G. (1546) *De Re Metallica.* (translation by H. C. and L. H. Hoover, 1912). Dover.

AHO A. A. and BROCK J. S. (1972) Handicaps to mining development in Yukon and the Far North. *Western Miner* 45 (No. 11), 14-31.

AHRENS L. H. (1954) The lognormal distribution of the elements. *Geochim. Cosmochim. Acta* 5, 49-73; 6, 121-131.

AHRENS L. H. and TAYLOR S. R. (1960) *Spectrochemical Analysis* (2nd edition). Addison-Wesley.

AITCHISON J. and BROWN J. A. C. (1957) *The Lognormal Distribution.* Cambridge University Press.

ALBRECHT P. and OURISSON G. (1969) Diagénèse des hydrocarbures saturés dans une série sédimentaire épaisse (Douala, Cameroan). *Geochim. Cosmochim. Acta* 33, 138-142.

ALEXANDER M. (1961) *Introduction To Soil Microbiology.* Wiley.

ALEXANDER M. (1971) *Microbial Ecology.* Wiley.

ALLAN R. J. (1971) Lake sediment: a medium for regional geochemical exploration of the Canadian Shield. *CIM Bull.* 64 (Nov.), 43-59.

ALLAN R. J., CAMERON E. M. and DURHAM C. C. (1973) Reconnaissance geochemistry using lake sediments of a 36,000-square mile area of the northwestern Canadian Shield. *Geol. Surv. Canada Paper* 72-50.

ALLAN R. J., CAMERON E. M. and DURHAM C. C. (1973a) Lake geochemistry — a low sample density technique for reconnaissance geochemical exploration and mapping of the Canadian Shield. *Geochemical Exploration 1972,* pp. 131-160. IMM. London.

ALLAN R. J. and HORNBROOK E. H. W. (1971) Exploration geochemistry evaluation study in a region of continuous permafrost, Northwest Territories, Canada. *Geochemical Exploration.* CIM Spec. Vol. 11, 53-66.

ALLAN R. J., LYNCH J. J. and LUND N. G. (1972) Regional geochemical exploration in the Coppermine River area, District of Mackenzie; a feasibility study in permafrost terrain. *Geol. Surv. Canada Paper* 71-33.

ANDREWS J. N. and WOOD D. F. (1972) Mechanism of radon release in rock matrices and entry into groundwaters. *Trans. IMM* (Sect. B, Appl. Earth Sci.) 81, B198-B209.

ANDREWS-JONES D. A. (1968) The application of geochemical techniques to mineral exploration. *Colo. School Mines, Mineral Indust. Bull.* 11, No. 6.

ANGINO E. E. and BILLINGS G. K. (1972) *Atomic Absorption Spectroscopy in Geology* (2nd edition). Elsevier.

ANGINO E. E., GOEBEL E. D. and WAUGH T. C. (1971) Lead isotopes and metallic sulphides as exploration guides in Mid-Continent Paleozoic rocks. *Geochemical Exploration.* CIM Spec. Vol. 11, 453-456.

ANTHONY L. M. (1967) The discovery of the Keystone gold mine, Cleary Hill area, Fairbanks district, Alaska – a geochemical prospecting case history. *Geol. Surv. Canada Paper* 66-54, 3-12.

ARCHER A. R. and MAIN C. A. (1971) Casino, Yukon – a geochemical discovery of an unglaciated Arizona-type porphyry. *Geochemical Exploration.* CIM Spec. Vol. 11, 67-77.

ARMANDS G. (1967) Geochemical prospecting of a uraniferous bog deposit at Masugnsbyn, northern Sweden. In, *Geochemical Prospecting In Fennoscandia* (editor A. Kvalheim), 127-154. Interscience.

ARMOUR-BROWN A. and NICHOL I. (1970) Regional geochemical reconnaissance and the location of metallogenic provinces. *Econ. Geol.* 65, 312-330.

ARMSTRONG F. A. J. and SCHINDLER D. W. (1971) Preliminary chemical characteristics of waters in the Experimental Lakes Area, northwestern Ontario. *J. Fish. Res. Bd. Canada* 28, 171-187.

ARNOLD R. G. (1970) The concentrations of metals in lake waters and sediments of some Precambrian lakes in Flin Flon and La Ronge areas. *Saskatchewan Res. Council Geol. Div. Circ.* 4.

BAILEY N. J. L., JOBSON A. M. and ROGERS M. A. (1973a) Bacterial degradation of crude oil: comparison of field and experimental data. *Chem. Geol.* 11, 203-221.

BAILEY N. J. L., KROUSE H. R., EVANS C. R. and ROGERS M. A. (1973b) Alteration of crude oil by waters and bacteria — evidence from geochemical and isotope studies. *Amer. Assoc. Petroleum Geologists Bull.* 57, 1276-1290.

BAIRD A. K., MCINTYRE D. B. and WELDAY E. E. (1967) A test of chemical variability and field sampling methods, Lakeview Mountain tonalite, Lakeview Mountains, Southern California batholith. *Calif. Div. Mines Geol., Spec. Rept.* 92.

BAKER E. G. (1967) A geochemical evaluation of petroleum migration and accumulation. *Fundamental Aspects of Petroleum Geochemistry* (editors B. Nagy and U. Colombo), 299-329. Elsevier.

BAKER W. E. (1973) The role of humic acids from Tasmanian podzolic soils in mineral degradation and metal mobilization. *Geochim. Cosmochim. Acta* 37, 269-281.

BARAKSO J. J. and BRADSHAW B. A. (1971) Molybdenum surface depletion and leaching. *Geochemical Exploration.* CIM Spec. Vol. 11, 78-84.

BARNES I. and HEM J. D. (1973) Chemistry of subsurface waters. *Ann. Rev. Earth Planet. Sci.* 1, 157-181.

BAYLISS P. (1972) An interpretation of the deeply weathered profile. *Clay Minerals* 9, 438-440.

BAYROCK L. A. and PAWLUK S. (1967) Trace elements in Alberta tills, *Canad. J. Earth Sci.* 4, 597-607.

BEAR F. E. (1964) *Chemistry Of The Soil.* Amer. Chem. Soc., Mono. Ser. 160. Reinhold.

BENDORAITIS J. G., BROWN B. L. and HEPNER L. S. (1963) Isoprenoid hydrocarbons in petroleum. *Anal. Chem.* 34, 49-53.

BENNETT R. (1971) Exploration for hydrothermal mineralization with airborne gamma-ray spectrometry. *Geochemical Exploration.* CIM Spec. Vol. 11, 475-478.

BERRY F. A. F. (1958) Hydrodynamics and geochemistry of the Jurassic and Cretaceous systems in the San Juan basin, northwestern New Mexico and southwestern Colorado. Ph.D. Thesis, Stanford University, California.

BEUS A. A. and SITNIN A. A. (1972) Geochemical specialization of magmatic complexes as criteria for the exploration of hidden deposits. *Intern. Geol. Congr. 24th,* Montreal, Vol. 6, 101-105.

BILLINGS G. K., HITCHON, B. and SHAW D. R. (1969) Geochemistry and origin of formation waters in the western Canada sedimentary basin. 2. Alkali metals. *Chem. Geol.* 4, 211-223.

BLACK R. F. and BARKSDALE W. L. (1949) Oriented lakes in northern Alaska. *J. Geol.* 57, 105-118.

BLANCHARD R. (1968) *Interpretation Of Leached Outcrops.* Nevada Bur. Mines Bull. 66.

BLAZEJCZAK J. and VAN DER WEIDE B. M. (1967) Dosage de faibles quantités d'hydrocarbures aromatiques dans les eau de formation. *Bull. Centre Rech. Pau-SNPA* 1, 199-202.

BLEACKLEY D. (1964) Bauxites and laterites of British Guiana. *Geol. Surv. Brit. Guiana Bull.* 34.

BLOOM H. (1955) A field method for the determination of ammonium citrate-soluble heavy metals in soils and alluvium. *Econ. Geol.* 50, 533-541.

BLOOM H. (1963) Toxic properties of several organic solvents used in exploration geochemistry. *Econ. Geol.* 58, 1000-1002.

BLOOMFIELD K., REEDMAN J. H., and TETHER J. G. G. (1971) Geochemical exploration of carbonatite complexes in eastern Uganda. *Geochemical Exploration.* CIM Spec. Vol. 11, 85-102.

BOCHER T. W. (1949) Climate, soil, and lakes in continental west Greenland in relation to plant life. *Medd. om Gronland* 147, No. 2, 1-63.

BOLVIKEN B. (1971) A statistical approach to the problem of interpretation in geochemical prospecting. *Geochemical Exploration.* CIM Spec. Vol. 11, 564-567.

BOLVIKEN B., LOGN O., BREEN A. and UDDU O. (1973) Instrument for *in situ* measurements of pH, Eh and self-potential in diamond drill holes. *Geochemical Exploration 1972,* pp. 415-420. IMM. London.

BOLVIKEN B. and SINDING LARSEN R. (1973) Total error and other criteria in the interpretation of stream sediment data. *Geochemical Exploration 1972,* pp. 285-295. IMM. London.

BORCHERT H. (1960) Genesis of marine sedimentary iron ores. *Trans. IMM* 69, 261-279.

BOTBOL J. M. (1971) Geochemical exploration data processing techniques utilized by the U.S. Geological Survey (abs.). *Geochemical Exploration.* CIM Spec. Vol. 11, 569.

BOWEN H. J. M. (1966) *Trace Elements In Biochemistry.* Academic Press.

BOWIE S. H. U., DARNLEY A. G. and RHODES J. R. (1965) Portable radio-isotope X-ray fluorescence analyzer. *Trans. IMM* 74, 361-379.

BOWIE S. H. U., SIMPSON P. R. and RICE C. M. (1973) Application of fission-track and neutron activation methods to geochemical exploration. *Geochemical Exploration 1972,* pp. 359-372. IMM. London.

BOYD W. L. (1959) Limnology of selected arctic lakes in relation to water supply problems. *Ecology* 40, 49-54.

BOYLE R. W. (1958) Geochemical prospecting in permafrost regions of Yukon, Canada. Symposium de Exploration Geoquimica (Primer Tomo), 175-188. *Intern. Geol. Congr. 20th,* Mexico (T. S. Lovering, coordinator).

BOYLE R. W. (1965) Geology, geochemistry and origin of the lead-zinc-silver deposits of Keno Hill-Galena Hill area, Yukon Territory. *Geol. Surv. Canada Bull. 111.*

BOYLE R. W. (1967) Geochemical prospecting — retrospect and prospect. *Geol. Surv. Canada Paper* 66-54, 30-43.

BOYLE R. W. (1968) A source for metals and gangue elements in epigenetic deposits. *Mineral. Deposita* 3, 174-177.

BOYLE R. W. (1969) Elemental associations in mineral deposits and indicator elements of interest in geochemical prospecting. *Geol. Surv. Canada Paper* 68-58.

BOYLE R. W. (editor) (1971) *Geochemical Exploration.* Canadian Institute of Mining and Metallurgy (CIM) Special Vol. 11, 1-594.

BOYLE R. W. (1971a) Geochemical Prospecting. In, *Encyclopedia of Science and Technology* (3rd edition). McGraw-Hill.

BOYLE R. W. (1971b) Boron and the boron minerals as indicators of mineral deposits (abstract). *Geochemical Exploration.* CIM Spec. Vol. 11, 12.

BOYLE R. W. and CRAGG C. B. (1957) Soil analyses as a method of geochemical prospecting in Keno Hill-Galena Hill area, Yukon Territory. *Geol. Surv. Canada Bull.* 39.

BOYLE R. W. and DASS A. S. (1967) Geochemical prospecting – use of the A horizon in soil surveys. *Econ. Geol.* 62, 274-276.

BOYLE R. W. and DASS A. S. (1971) The geochemistry of the supergene processes in the native silver veins of the Cobalt – South Lorraine area, Ontario. *Canad. Mineral.* 11, 358-390.

BOYLE R. W., DASS A. D., CHURCH D., MIHAILOV G., DURHAM C. C., LYNCH J. J. and DYCK W. (1969) Research in geochemical prospecting methods for native silver deposits Cobalt area, Ontario, 1966. *Geol. Survey Canada Paper* 67-35.

BOYLE R. W. and GARRETT R. G. (1970) Geochemical prospecting — A review of its status and future. *Earth – Sci. Rev.* 6, 51-75.

BOYLE R. W., HORNBROOK E. H. W., ALLAN R. J., DYCK, W. and SMITH A. Y. (1971) Hydro-geochemical methods — application in the Canadian Shield. *CIM Bull.* 64 (Nov.), 60-71.

BOYLE R. W., ILLSLEY C. T. and GREEN R. N. (1955) Geochemical investigation of the heavy metal content of stream and spring waters in Keno Hill-Galena Hill area, Yukon Territory. *Geol. Surv. Canada Bull.* 32.

BOYLE R. W. and LYNCH J. J. (1968) Speculations on the source of zinc, cadmium, lead, copper and sulfur in Mississippi Valley and similar types of lead-zinc deposits. *Econ. Geol.* 63, 421-422.

BOYLE R. W., PEKAR E. L. and PATTERSON P. R. (1956) Geochemical investigation of the heavy metal content of streams and springs in the Galena Hill-Mount Haldane area, Yukon Territory. *Geol. Surv. Canada Bull.* 36.

BOYLE R. W. and SMITH A. Y. (1968) The evolution of techniques and concepts in geo-chemical prospecting. In, *The Earth Sciences in Canada* (editor E. R. W. Neale), Royal Soc. Canada Spec. Public. No. 11, 117-128. University of Toronto Press.

BOYLE R. W., TUPPER W. M., LYNCH J., FRIEDRICH G., ZIAUDDIN M., SHAFIQULLAH M., CARTER M. and BYGRAVE K. (1966) Geochemistry of Pb, Zn, Cu, As, Sb, Mo, Sn, W, Ag, Ni, Co, Cr, Ba, and Mn in the waters and stream sediments of the Bathurst-Jacquet River District, New Brunswick. *Geol. Surv. Canada Paper* 65-42.

BOYLE R. W., WANLESS R. K. and STEVENS R. D. (1970) Sulfur isotope investigation of lead-zinc-silver-cadmium deposits of the Keno Hill-Galena Hill area, Yukon, Canada. *Econ. Geol.* 65, 1-10.

BRABEC D. and WHITE W. H. (1971) Distribution of copper and zinc in rocks of the Guichon Creek Batholith, British Columbia. *Geochemical Exploration*. CIM Spec. Vol. 11, 291-297.

BRADSHAW P. M. D. (1973) The use of strategic models in exploration geochemistry: data for mobile elements from Bathurst area, New Brunswick. *Res. Rept. Ser.* No. 17, *(Studies in Landscape Geochem.* No. 8), pp. 36-56. Brock Univ., Dept. Geol. Sci., (St. Catharines, Ontario).

BRADSHAW P. M. D., CLEWS D. R. and WALKER J. L. (1972) *Exploration Geochemistry*. A series of seven articles reprinted from *Mining In Canada* and *Canadian Mining Journal*. Barringer Research Ltd., 304 Carlingview Dr., Rexdale, Ontario, M9W 5G2.

BRADSHAW P. M. D. and KOKSOY E. M. (1968) Primary dispersion of mercury from cinnabar and stibnite deposits, W. Turkey. *Intern. Geol. Congr. 23rd*, Prague, Vol. 7, 341-355.

BRADSHAW P. M. D. and STOYEL A. J. (1968) Exploration for blind ore bodies in southwest England by the use of geochemistry and fluid inclusions. *Trans. IMM* (Sect. B, Appl. Earth Sci.) 77, B144-B152.

BRAY E. E. and EVANS E. D. (1961) Distribution of n-paraffins as a clue to recognition of source beds. *Geochim. Cosmochim. Acta* 22, 2-15.

BRISTOW Q. (1972) An evaluation of the quartz crystal microbalance as a mercury vapour sensor for soil gas. *J. Geochem. Explor.* 1, 55-75.

BRISTOW Q. and JONASSON I. R. (1972) Vapour sensing for mineral exploration. *Canad. Mining J.* 93 (No. 5), 39-44, 47, 85.

BRITTON H. T. S. (1955) *Hydrogen Ions* (4th edition). Chapman and Hall. London.

BROCK J. S. (1972) The use of dogs as an aid to exploration for sulphides. *Western Miner* 45 (No. 12), 28-32.

BROOKS J. M. and SACKETT W. M. (1973) Sources, sinks, and concentrations of light hydro-carbons in the Gulf of Mexico. *J. Geophys. Res.* 78, 5248-5258.

BROOKS R. R. (1972) *Geobotany and Biogeochemistry In Mineral Exploration*. Harper and Row.

BROSGE W. P. and REISER H. N. (1972) Geochemical reconnaissance in the Wiseman and Chandalar districts and adjacent region, southern Brooks Range, Alaska. *U.S. Geol. Surv. Prof. Paper* 709.

BROWN A. C. (1964) Geochemistry of the Dawson Settlement bog manganese deposits, New Brunswick. *Geol. Surv. Canada Paper* 63-42.

BROWN B. W. (1964) A statistical case study in geochemical prospecting for copper. *Econ. Geol.* 59, 492-500.

BROWN B. W. (1970) Error in lead anomalous stream sediments. *Econ. Geol.* 65, 514-515.

BROWN G. (1961) *The x-ray identification and crystal structures of clay minerals.* Mineralogical Society (London).

BROWN I. C. (1967) Groundwater in Canada. *Geol. Surv. Canada Econ. Geol. Report* No. 24.

BROWN R. J. E. (1967) *Geol. Surv. Canada Map* 1246A.

BROWN R. J. E. (1970) *Permafrost in Canada.* University of Toronto Press.

BRUNDIN N. H. and NAIRIS B. (1972) Alternative sample types in regional geochemical prospecting. *J. Geochem. Explor.* 1, 7-46.

BRUNSKILL G. J., POVOLEDO D, GRAHAM B. W. and STAINTON M. P. (1972) Chemistry of surface sediments from sixteen lakes in the Experimental Lakes Area, northwestern Ontario. *J. Fish. Res. Bd. Canada* 28, 277-294.

BUCKLEY S. E., HOCOTT C. R. and TAGGART M. S. (1958) Distribution of dissolved hydrocarbons in subsurface waters. *Habitat of Oil* (editor L. G. Weeks), 850-882. American Association of Petroleum Geologists, Tulsa, Oklahoma.

BUCKMAN H. O. and BRADY N. C. (1969) *The Nature And Properties Of Soils* (7th edition). Macmillan.

BURAND W. M. (1968) Geochemical investigations of selected areas in the Yukon-Tanana region of Alaska 1965 and 1966. *Alaska Dept. Nat. Res., Div. Mines Minerals, Geochemical Rept.* No. 13.

BURGESS J. D. (1974) Microscopic examination of kerogen (dispersed organic matter) in petroleum exploration. *Geol. Soc. Amer. Symposium, Carbonaceous Materials as Indicators of Metamorphism* (in press).

BURLINGAME A. L. and SCHNOES H. K. (1969) Mass spectrometry in organic geochemistry. *Organic Geochemistry* (editors G. Eglinton and M. T. J. Murphy), 89-160. Springer-Verlag.

BURNHAM C. W. (1959) Metallogenic provinces of the southwestern United States and northern Mexico. *New Mexico Bur. Mines Mineral Res. Bull.* 65, 1-76.

BURNS R. G. (1970) *Mineralogical Applications Of Crystal Field Theory.* Cambridge University Press.

CADEK J., MALKOVSKY M. and SULCEK Z. (1968) Geochemical significance of subsurface waters for the accumulation of ore components. *Intern. Geol. Congr. 23rd.,* Prague, Vol. 6, 161-168.

CAMERON E. M. (editor) (1967) Proceedings, Symposium On Geochemical Prospecting. Ottawa, April, 1966. *Geol. Surv. Canada Paper* 66-54, 1-282.

CAMERON E. M. (1967) A computer program for factor analysis of geochemical and other data. *Geol. Surv. Canada Paper* 67-34.

CAMERON E. M. (1972) Three geochemical standards of sulphide-bearing ultramafic rock: UM-1, UM-2, UM-4. *Geol. Surv. Canada Paper* 71-35.

CAMERON E. M. and BARAGAR W. R. (1971) Distribution of ore elements in rocks for evaluating ore potential: frequency distribution of copper in the Coppermine River Group and Yellowknife Group volcanic rocks, N.W.T., Canada. *Geochemical Exploration.* CIM Spec. Vol. 11, 570-576.

CAMERON E. M., SIDDELEY G., DURHAM C. C. (1971) Distribution of ore elements in rocks for evaluating ore potential: nickel, copper, cobalt and sulphur in ultramafic rocks of the Canadian Shield. *Geochemical Exploration.* CIM Spec. Vol. 11, 298-313.

CAMPBELL A. S., ADAMS J. A. and HOWARTH D. T. (1972) Some problems encountered in the identification of plumbogummite minerals in soils. *Clay Minerals* 9, 415-423.

CANADA DEPARTMENT OF AGRICULTURE (1970) The system of soil classification for Canada. Queen's Printer for Canada Catalogue No. A42-4069.

CANNEY F. C. (editor) (1969) *International Geochemical Exploration Symposium.* Colo. School Mines Quart. 64, 1-520.

CANNEY F. C. and NOWLAND G. A. (1964) Determination of ammonium-citrate-soluble cobalt in soils and sediments. *Econ. Geol.* 59, 1361-1367.

CANNEY F. C. and WING L. A. (1966) Cobalt: useful but neglected in geochemical prospecting. *Econ. Geol.* 61, 198-203.

CANNON H. L. (1955) Geochemical relations of zinc-bearing peat to the Lockport Dolomite, Orleans County, New York. *U.S. Geol. Surv. Bull.* 1000-D, 119-185.

CANNON H. L. (1960) Botanical prospecting for ore deposits. *Science* 132, 591-598.

CANNON H. L. (1971) Use of plant indicators in ground water surveys, geologic mapping, and mineral prospecting. *Taxon* 20, 227-256.

CANNON R. S. and PIERCE A. P. (1969) Lead isotope guides for Mississippi Valley lead-zinc exploration. *U.S. Geol. Surv. Bull.* 1312-G.

CANNON R. S., PIERCE A. P. and ANTWEILER J. C. (1971) Suggested uses of lead isotopes in exploration. *Geochemical Exploration.* CIM Spec. Vol. 11, 457-463.

CARLISLE D. and CLEVELAND G. B. (1958) Plants as a guide to mineralization. *Calif. Div. Mines Geol. Spec. Rept.* 50.

CATTELL R. B. (1965) Factor analysis: an introduction to essentials. Pt. 1 and 2. *Biometrics* 21, 190-215; 405-435.

CHAFFEE M. A. and HESSIN T. D. (1971) An evaluation of geochemical sampling in the search for concealed 'porphyry' copper-molybdenum deposits on pediments in southern Arizona. *Geochemical Exploration.* CIM Spec. Vol. 11, 401-409.

CHAMBERLAIN J. A. (1964) Hydrogeochemistry of uranium in the Bancroft-Haliburton region, Ontario. *Geol. Surv. Canada Bull.* 118.

CHATUPA J. and FLETCHER K. (1972) Application of regression analysis to the study of background variations in trace metal content of stream sediments. *Econ. Geol.* 76, 978-980.

CHAWLA V. K. (1971) Changes in the water chemistry of Lakes Erie and Ontario. *Bull. Buffalo Soc. Nat. Sci.* 25 (No. 2), 31-66.

CHÉTELAT E. DE (1947) La genèse et l'évolution des gisements de nickel de la Nouvelle-Calédonie. *Soc. Geol. France Bull.,* Series 5, 17, 105-160.

CHIKISHEV A. G. (1965) *Plant Indicators of Soils, Rocks, and Subsurface Waters.* Consultants Bureau.

CHILINGAR G. V. and KARIM M. (1962) Gaseous survey methods in exploration and prospecting for oil and gas: a review. *Alberta Soc. Petroleum Geologists J.* 10, 610-617.

CLARIDGE G. G. C. (1965) The clay mineralogy and chemistry of some soils from the Ross Dependency, Antarctica. *N.Z. Jour. Geol. Geophys.* 8, 186-220.

CLARKE W. B. and KUGLER G. (1973) Dissolved helium in groundwater: a possible method for uranium and thorium prospecting. *Econ. Geol.* 68, 243-251.

CLEMA J. M. and STEVENS-HOARE N. P. (1973) A method of distinguishing nickel gossans from other ironstones on the Yilgarn Shield, Western Australia. *J. Geochem. Explor.* 2, (in press).

CLOKE P. L. (1966) The geochemical application of Eh-pH diagrams. *J. Geol. Education.* 14, 140-148.

CLYMO R. S. (1964) The origin of acidity in *Sphagnum* bogs. *The Bryologist* 67, 427-431.

COLE M. M. (1971) The importance of environment in biogeographical/geobotanical and biogeochemical investigations. *Geochemical Exploration.* CIM Spec. Vol. 11, 414-425.

CONNAN J. (1972) Laboratory simulation and natural diagenesis. 1. Thermal evolution of asphalts from the Aquitaine basin (SW France). *Bull. Centre Rech. Pau-SNPA* 6, 195-214.

CONNAN J., LE TRAN K., RUMEAU J. -L., VAN DER WEIDE B. and COUSTAU H. (1972) Problèmes posés par les gisements de gaz à faible profondeur et par l'hydrogène sulfuré des gisements profonds. *Compt. Rend. du 89ᵉ Congrès de l'Indust. du Gaz, Paris*, 114-127.

CONNOR J. J. and GERRILD P. M. (1971) Geochemical differentiation of crude oils from six Pliocene sandstone units, Elk Hills U.S. Naval Petroleum Reserve No. 1, California. *Amer. Assoc. Petroleum Geologists Bull.* 55, 1802-1813.

COOPE J. A. (1971) The Association of Exploration Geochemists. *Geochemical Exploration.* CIM Spec. Vol. 11, 5-6.

COOPE J. A. (1973) Geochemical prospecting for porphyry copper-type mineralization — a review. *J. Geochem. Explor.* 2, 81-102.

COOPE J. A. and WEBB J. S. (1963) Copper in stream sediments near disseminated copper mineralization, Cebu, Philippine Republic. *Trans. IMM* 72, 397-406.

COOPER J. E. and KVENVOLDEN K. A. (1967) Method for prospecting for petroleum. U.S. Patent 3,305,317.

COOPER J. R. and HUFF L. C. (1951) Geological investigations and geochemical prospecting experiment at Johnson, Arizona. *Econ. Geol.* 46, 731-756.

COPLEN T. B. and HANSHAW B. B. (1973) Ultrafiltration by a compacted clay membrane — 1. Oxygen and hydrogen isotopic fractionation *Geochim. Cosmochim. Acta* 37, 2295-2310.

CORREIA M. (1969) Contribution a la recherche de zones favorables a la genèse du pétrole par l'observation microscopique de la matière organique figurée. *Rev. l'Inst. Francais du Pétrole* 24, 1417-1454.

COUSTAU H. and SOURISSE C. (1967) Pressions anormales dans les réservoirs Infralias a Paléozoique du sud Aquitan (sud-ouest). *Bull. Centre Rech. Pau-SNPA* 1, 143-152.

CRONAN D. J. and THOMAS R. L. (1970) Ferromanganese concretions in Lake Ontario. *Canad. J. Earth Sci.* 7, 1346-1349.

CURTIN G. C. and KING H. D. (1972) An auger-sleeve sampler for stony soils. *J. Geochem. Explor.* 1, 203-206.

CURTIN G. C., LAKIN H. W., NEUERBERG G. J., and HUBERT A. E. (1968) Utilization of humus-rich forest soil (mull) in geochemical exploration. *U.S. Geol. Surv. Circ.* 562.

DAHLBERG E. C. (1968) Application of a selective simulation and sampling technique to the interpretation of stream sediment copper anomalies near South Mountain, Pennsylvania. *Econ. Geol.* 63, 409-417.

DAHLBERG E. C. (1971) Algorithmic development of a geochemical exploration program. *Geochemical Exploration.* CIM Spec. Vol. 11, 577-580.

DALL'AGLIO M. and TONANI F. (1973) Hydrogeochemical exploration for sulphide deposits: correlation between sulphate and other constituents. *Geochemical Exploration 1972*, pp. 305-314. IMM. London.

DANSEREAU P. (1957) *Biogeography.* The Ronald Press.

DARLING R. (1971) Preliminary study of the distribution of minor and trace elements in biotite from quartz monzonite associated with contact-metasomatic tungsten-molybdenum-copper ore, California, U.S.A. *Geochemical Exploration.* CIM Spec. Vol. 11, 315-322.

DASS A. S., BOYLE R. W. and TUPPER W. M. (1973) Endogenic halos of the native silver deposits, Cobalt, Ontario Canada. *Geochemical Exploration 1972*, pp. 25-35. IMM. London.

DAVENPORT P. H. and NICHOL I. (1973) Bedrock geochemistry as a guide to areas of base-metal potential in volcano-sedimentary belts of the Canadian Shield. *Geochemical Exploration 1972*, pp. 45-57. IMM. London.

DAVIDSON M. J. (1963) Geochemistry can help find oil if properly used. *World Oil* 157 (No. 1), 94, 96, 100, 104-106.

DAVIS C. E. S. (1972) Analytical methods used in the study of an ore intersection from Lunnon Shoot, Kambalda. *Econ. Geol.* 67, 1091-1092.

DAVIS J. B. (1967) *Petroleum Microbiology.* Elsevier.

DAVIS J. B. (1969) Distribution of naphthenic acids in an oil-bearing aquifer. *Chem. Geol.* 5, 89-97.

DAVIS J. C. (1973) *Statistics And Data Analysis In Geology.* Wiley.

DEBNAM A. H. (1965) Field and laboratory methods used by the Geological Survey of Canada in geochemical surveys. No. 6. Determination of hydrocarbons in soils by gas chromatography. *Geol. Survey Canada Paper* 64-15.

DEBNAM A. H. (1969) Geochemical prospecting for petroleum and natural gas in Canada. *Geol. Survey Canada Bull.* 177.

DE GEOFFROY J. and WIGNALL T. K. (1972) A statistical study of geological characteristics of porphyry-copper-molybdenum deposits in the Cordilleran Belt — application to the rating of porphyry copper. *Econ. Geol.* 67, 656-668.

DE GEOFFROY J. G., WU S. M. and HEINS R. W. (1968) Selection of drilling targets from data in the southwest Wisconsin zinc area. *Econ. Geol.* 63, 787-795.

DEROO G., ROUCACHÉ J. and TISSOT B. (1973) Etude geochimique du Canada occidental, Alberta. *Geol. Survey Canada Open-file Report.*

DEROO G., TISSOT B., MCCROSSAN R. G. and DER F. (1974) Geochemistry of the heavy oils of Alberta. *Bull. Canadian Petroleum Geol.* (in press).

DERRY D. R. (1971) Geochemistry — the link between ore genesis and exploration. *Geochemical Exploration.* CIM Spec. Vol. 11, 1-4.

DESCARREAUX J. (1973) A petrochemical study of the Abitibi volcanic belt and its bearing on the occurrences of massive sulphide ores. *CIM Bull.* 66 (Feb.), 61-69.

DOUGLAS A. G. (1969) Gas chromatography. *Organic Geochemistry* (editors G. Eglinton and M. T. J. Murphy), 161-180. Springer-Verlag.

DOVETON J. H. and PARSLEY A. J. (1970) Experimental evaluation of trend surface distortions induced by inadequate data-point distributions. *Trans. IMM* (Sect. B, Appl. Earth Sci.) 79, 197-207.

DUNHAM K. C. (1972) Basic and applied geochemists in search of ore. *Trans. IMM* (Sect. B, Appl. Earth Sci.) 81, 1-5.

DURAND B., ESPITALIÉ J., NICAISE G. and COMBAZ A. (1972) Étude de la matière organique insoluble (kérogène) des argiles du Toarcien du bassin de Paris. Premiere partie: Étude par les procédés optiques, analyse élémentaire, étude en microscopie et diffraction électroniques. *Rev. l'Inst. Francais du Pétrole* 27, 865-884.

DYCK W. (1969) Development of uranium methods using radon. *Geol. Surv. Canada Paper* 69-46.

DYCK W. (1971) The adsorption and coprecipitation of silver on hydrous oxides of iron and manganese. *Geol. Surv. Canada Paper* 70-64.

DYCK W. (1972) Radon methods of prospecting. In, *Uranium Prospecting Handbook* (editors: S. H. U. Bowie, M. Davis and D. Ostle), pp. 212-243. The Institution of Mining and Metallurgy. London.

DYCK W. (1973) Feasibility study of geochemical sampling of Arctic coastal streams by helicopter based on a Department of Transport icebreaker. *Geol. Surv. Canada Paper* 72-42.

DYCK W., DASS A. S., DURHAM C. C., HOBBS J. D., PELCHAT J. C. and GALBRAITH J. H. (1971) Comparison of regional geochemical uranium exploration methods in the Beaverlodge area, Saskatchewan. *Geochemical Exploration.* CIM Spec. Vol. 11, 132-150.

DYCK W. and MEILLEUR G. A. (1972) A soil gas sampler for difficult overburden. *J. Geochem. Explor.* 1, 199-202.

EASTERBROOK D. J. (1969) *Principles of Geomorphology.* McGraw-Hill.

ECKSTRAND O. R. (1971) The nickel potential of serpentinized ultramafic rocks. *Canad. Mining J.* 1971, 40-45.

EISENHART C. (1947) The assumptions underlying the analysis of variance. *Biometrics,* 3, 1-21.

EREMEEV A. N., SOKOLOV V. A., SOKOLOV A. P. and YANITSKII I. N. (1973) Application of helium surveying to structural mapping and ore deposit forecasting. *Geochemical Exploration 1972*, pp. 183-192. IMM. London.

ERICKSON R. L. (1973) Presidential address given at the annual general meeting of The Association of Exploration Geochemists, Vancouver, April 13, 1973. *Newsletter* No. 9, The Association of Exploration Geochemists.

ERICKSON R. L., MARRANZINO A. P., UTEANA O. and JANES W. W. (1966) Geochemical reconnaissance in the Pequop Mountains and Wood Hills, Elko County, Nevada. *U.S. Geol. Surv. Bull.* 1198-E.

ESPITALIÉ J., DURAND B., ROUSSEL J., -C. and SOURON C. (1973) Étude de la matière organique insoluble (kérogène) des argiles du Toarcien du bassin de Paris. Deuxième partie: Études in spectroscopic infrarouge, en analyse thermique différentielle et en analyse thermogravimétrique. *Rev. l'Inst. Francais du Petrole* 28, 37-66.

EVANS C. R., ROGERS M. A. and BAILEY N. J. R. (1971) Evolution and alteration of petroleum in western Canada. *Chem. Geol.* 8, 147-170.

EVANS C. R. and STAPLIN F. L. (1971) Regional facies of organic metamorphism. *Geochemical Exploration*. CIM. Spec. Vol. 11, 517-520.

VAN EVERDINGEN R. O. (1968) Studies of formation waters in western Canada: geochemistry and hydrodynamics. *Canad. J. Earth Sci.* 5, 523-543.

VAN EVERDINGEN R. O. (1970) The Paint Pots, Kootenay National Park, British Columbia — acid spring water with extreme heavy-metal content *Canad. J. Earth Sci.* 7, 831-852.

VAN EVERDINGEN R. O. and BANNER J. A. (1971) Precipitation of heavy metals from natural and synthetic acidic aqueous solutions during neutralization with limestone. *Canada Dept. Energy, Mines and Resources, Inland Waters Branch, Tech. Bull.* No. 35.

FAIRBAIRN H. W. AND OTHERS (1951) A comparative investigation on precision and accuracy in chemical, spectrochemical, and modal analysis of silicate rocks. *U.S. Geol. Surv. Bull.* 980.

FAIRBRIDGE R. W. (1972) *The Encyclopedia of Geochemistry and Environmental Sciences*. Van Nostrand Reinhold.

FAYE G. H. (1972) Standard reference ores and rocks available from the Mines Branch. *Canada Dept. Energy, Mines and Resources, Mines Branch, Info. Circ.* IC294.

FAYE G. H. (1973) Standard reference ores and rocks available from the Mines Branch as of October 1973. *Canada Dept. Energy, Mines and Resources, Mines Branch, Info. Circ.* 309.

FEIGL F. (1958) *Spot Tests In Inorganic Applications* (4th edition). Translated by R. E. Oesper. Elsevier.

FETH J. H., ROBERTSTON C. E. and POLZER W. L. (1964) Sources of mineral constituents in water from granitic rocks, Sierra Nevada, California and Nevada. *U.S. Geol. Surv. Water-Supply Paper* 1535-I.

FICKLIN W. H. (1970) A rapid method for the determination of fluoride in rocks using an ion-selective electrode *U.S. Geol. Surv. Prof. Paper* 700-C, C186-C188.

FINLAYSON A. M. (1910) Problems of ore deposition in the lead and zinc veins of Great Britain. *Geol. Soc. London. J.* 66, 299-338.

FISHER N. H. (1971) Recent research in geochemical prospecting in Australia. *Geochemical Exploration*. CIM Spec. Vol. 11, 16-20.

FISHER R. A. (1936) The use of multiple measurements in taxonomic problems. *Annals Eugenics* 7, 179-188.

FLANAGAN F. J. (1969) U.S. Geological Survey Standards. II. First compilation of data for the new USGS rocks. *Geochim. Cosmochim. Acta* 33, 81-119.

FLANAGAN F. J. (1970) Sources of geochemical standards. II. *Geochim. Cosmochim. Acta* 34, 121-125.

FLANAGAN F. J. (1973) 1972 values for international geochemical reference samples. *Geochim. Cosmochim. Acta* 37, 1189-1200.

FLEISCHER M. (1955) Minor elements in some sulfide minerals. *Econ. Geol., Fiftieth Anniversary Volume,* 970-1024.

FLEISCHER M. (1965) Composition of magnetites as related to type of occurrence. *U.S. Geol. Surv. Prof. Paper* 525D, D-82-D84.

FLEISCHER M. (1969) U.S. Geological Survey Standards. I Additional data on rocks G-1 and W-1, 1965-1967. *Geochim. Cosmochim. Acta* 33, 65-79.

FLETCHER K. (1970) Some applications of background correction to trace metal analysis of geochemical samples by atomic-absorption spectrophotometry. *Econ. Geol.* 65, 588-589.

FLINT R. F. (1971) *Glacial and Quarternary Geology.* Wiley.

FLINTER B. H. (1971) Tin in acid granitoids: the search for a geochemical scheme of mineral exploration. *Geochemical Exploration.* CIM Spec. Vol. 11, 323-330.

FLINTER B. H., HESP W. R. and RIGBY D. (1972) Selected geochemical, mineralogical and petrological features of granitoids of the New England complex, Australia, and their relation to Sn, W, Mo and Cu mineralization. *Econ. Geol.* 67, 1241-1262.

FORSMAN J. P. (1963) Geochemistry of kerogen. In, *Organic Geochemistry* (editor I. A. Breger), 148-182, Pergamon Press.

FORSMAN J. P. and HUNT J. M. (1958) Insoluble organic matter (kerogen) in sedimentary rocks of marine origin. In, *Habitat of Oil* (editor L. G. Weeks), 747-778. American Association of Petroleum Geologists, Tulsa, Oklahoma.

FORTESCUE J. A. C. (1967) Progress report on biogeochemical research at the Geological Survey of Canada 1963-1966. Section A. Background, scope and objectives. *Geol. Surv. Canada Paper* 67-23, Part 1, 1-29.

FORTESCUE J. A. C. (1970) A research approach to the use of vegetation for the location of mineral deposits in Canada. *Taxon* 19, 695-704.

FORTESCUE J. A. C. (1971) Biogeochemistry, plant growth and the environment. *CIM Bull.* 64 (August), 77-82.

FORTESCUE J. A. C. (1973) Relationship between landscape geochemistry and exploration geochemistry. *Res. Rept. Ser.* No. 17, *(Studies in Landscape Geochem.* No. 8) pp. 1-21. Brock Univ., Dept. Geol. Sci., (St. Catharines, Ont.).

FORTESCUE J. A. C. and HORNBROOK E. H. W. (1967) A brief survey of progress made in biogeochemical prospecting at the Geological Survey of Canada 1962-65. *Geol. Surv. Canada Paper* 66-54, 111-113.

FORTESCUE J. A. C. and HORNBROOK E. H. W. (1969) Two quick projects, one at a massive sulfide ore body near Timmins, Ontario and the other at a copper deposit in Gaspé Park, Quebec. *Geol. Surv. Canada Paper* 67-23, Part II, 39-63.

FORTESCUE J. A. C. and USIK L. (1969) Geobotanical and soil geochemical investigations during 'visits' to eight landscapes with undisturbed mineral deposits. *Geol. Surv. Canada Paper* 67-23, Part II, 10-38.

FOSTER J. R. (1971) The reduction of matrix effects in atomic absorption analysis and the efficiency of selected extractions on rock-forming minerals. *Geochemical Exploration.* CIM Spec. Vol. 11, 554-560.

FOSTER J. R. (1973) The efficiency of various digestion procedures in the extraction of metals from rocks and rock-forming mineral. *CIM Bull.* 66 (August), 85-92.

FOURNIER R. O. and TRUESDELL A. H. (1970) Chemical indicators of sub-surface temperature applied to hot spring waters of Yellowstone National Park, Wyoming, U.S.A. In, *Proceedings U.N. Symposium on the Development and Utilization of Geothermal Resources. Geothermics* Spec. Issue 2, Vol. 2, Part 1, 529-535. Pisa, 1970.

FOURNIER R. O. and TRUESDELL A. H. (1973) An emperical Na-K-Ca geothermometer for natural waters. *Geochim. Cosmochim. Acta* 37, 1255-1275.

FRANK D. J., SACKETT W., HALL R. and FREDERICKS A. (1970) Methane, ethane, and propane concentrations in Gulf of Mexico *Amer. Assoc. Petrol. Geol. Bull.* 54, 1933-1938.

FREEZE R. A. (1966) Theoretical analysis of regional groundwater flow. Ph.D. Thesis, Univ. of California, Berkeley, California.

FREEZE R. A. (1969) Theoretical analysis of regional groundwater flow. *Canada Inland Waters Branch, Sci. Ser. 3.*

FREEZE R. A. and WITHERSPOON P. A. (1966) Theoretical analysis of regional groundwater flow. 1. Analytical and numerical solutions to the mathematical model. *Water Resour. Res.* 2, 641-656.

FREEZE R. A. and WITHERSPOON P. A. (1967) Theoretical analysis of regional groundwater flow. 2. Effect of water-table configuration and subsurface permeability variation. *Water Resour. Res.* 3, 623-634.

FREEZE R. A. and WITHERSPOON P. A. (1968) Theoretical analysis of regional groundwater flow. 3. Quantitative interpretations. *Water Resour. Res.* 4, 581-590.

FREY D. G. (editor) (1963) *Limnology In North America.* University of Wisconsin Press.

GALLAGHER M. J. (1967) Determination of molybdenum, iron, and titanium in ores and rocks by portable radioisotope X-ray fluorescence analyzer. *Trans. IMM* (Sect. B, Appl. Earth Sci.) 76, B155-B164.

GARRELS R. M. and CHRIST C. L. (1965) *Minerals, Solutions and Equilibria.* Harper and Row.

GARRETT R. G. (1967) Two programs for the factor analysis of geologic and remote sensing data. Northwestern Univ. Dept. of Geology Rept. 12, NASA.

GARRETT R. G. (1969) The determination of sampling and analytical errors in exploration geochemistry *Econ. Geol.* 64, 568-569; discussion 68, 281-283 (1973).

GARRETT R. G. (1971) The dispersion of copper and zinc in glacial overburden at the Louvem Deposit, Val d'Or, Quebec. *Geochemical Exploration,* CIM Spec. Vol. 11, 157-158.

GARRETT R. G. (1973) Regional geochemical study of Cretaceous acidic rocks in the northern Canadian Cordillera as a tool for broad mineral exploration. *Geochemical Exploration 1972,* pp. 203-219. IMM. London.

GARRETT R. G. and NICHOL I. (1967) Regional geochemical reconnaissance in eastern Sierra Leone. *Trans. IMM* (Sect. B. Appl. Earth Sci.) 76, B97-B112.

GÉRARD R. E. and FEUGÈRE G. (1969) Results of an experimental offshore geochemical prospection study. In, *Advances in Organic Geochemistry, 1968* (editors P. A. Schenck and I. Havenaar), 355-372. Pergamon Press.

GIRAUD A. (1970) Application of pyrolysis and gas chromatography to geochemical characterization of kerogen in sedimentary rock. *Amer. Assoc. Petrol. Geol. Bull.* 54, 439-455.

GIBBS R. J. (1972) Water chemistry of the Amazon River. *Geochim. Cosmochim. Acta* 36, 1061-1066.

GIBBS R. J. (1973) Mechanisms of trace metal transport in rivers. *Science* 180, 71-72.

GILBERT M. A. (1959) Laboratory methods for determining copper, zinc and lead. *Geol. Surv. Canada Paper* 59-3.

GINZBURG I. I. (1960) *Principles of Geochemical Prospecting.* Pergamon. (Translation from the Russian).

GLEESON C. F. and COOPE J. A. (1967) Some observations on the distribution of metals in swamps in eastern Canada. *Geol. Surv. Canada Paper* 66-54, 145-166.

GLEESON C. F. and CORMIER R. (1971) Evaluation by geochemistry of geophysical anomalies and geological targets using overburden sampling at depth. *Geochemical Exploration.* CIM Spec. Vol. 11, 159-165.

GOLDSCHMIDT V. M. (1937) The principles of distribution of chemical elements in minerals and rocks. *J. Chem. Soc.* 1937, 655-673.

GOLDSCHMIDT V. M. (1954) *Geochemistry.* Oxford University Press.

GORDON M., TRACEY J. I. and ELLIS M. W. (1958) Geology of the Arkansas bauxite region. *U.S. Geol. Surv. Prof. Paper* 299.

584   References

GORHAM E. (1967) Some chemical aspects of wetland ecology. *Proc. Twelfth Annual Muskeg Conf. 1966, Nat. Res. Council Canada Tech. Memorandum* No. 90, 20-38.

GORHAM E. and SWAINE D. J. (1965) The influence of oxidizing and reducing conditions upon the distribution of some elements in lake sediments. *Limnol. Oceanog.* 10, 268-279.

GOTT G. B. and BOTBOL J. M. (1973) Zoning of major and minor metals in the Coeur d'Alene mining district, Idaho, U.S.A. *Geochemical Exploration 1972*, pp. 1-12. IMM. London.

GOVETT G. J. S. (1972) Interpretation of a rock geochemical exploration survey in Cyprus — statistical and graphical techniques. *J. Geochem. Explor.* 1, 77-102.

GOVETT G. J. S. and HALE W. E. (1967) Geochemical orientation and exploration near a disseminated copper deposit, Luzon, Philippines. *Trans. IMM* (Sect. B, Appl. Earth Sci.) 76, B190-B201.

GOVETT G. J. S. and WHITEHEAD R. E. (1973) Errors in atomic absorption spectrophotometric determination of Pb, Zn, Ni and Co in geologic materials. *J. Geochem. Explor.* 2, 121-131.

GOWER J. C. (1966) A Q-technique for the calculation of canonical variates. *Biometrika* 55, 588-589.

GOWER J. C. (1967) A comparison of some methods of cluster analysis. *Biometrics* 23, 623-627.

GRANIER C. L. (1973) *Introduction a la Prospection Géochimique Des Gîtes Métallifères* (in French). Masson et Cie. Paris.

GRIFFITHS J. C. (1967) *Scientific Method in Analysis of Sediments*. McGraw-Hill.

GRIFFITHS J. C. (1971) Problems of sampling in geoscience. *Trans. IMM* (Sect. B, Appl. Earth Sci.) 80, B346-356.

GRIFFITHS J. C. and ONDRICK C. W. (1970) Structure by sampling in the geosciences. *Random Counts in Scientific Work*. Vol. 3. Pennsylvania State University Press.

GRIMALDI F. S. and SCHNEPFE M. M. (1968) Determination of palladium and platinum in rocks. *U.S. Geol. Surv. Prof. Paper* 600-B, B99-B102.

GRIMES D. J. and MARRANZINO A. P. (1968) Direct-current arc and alternating-current spark emission spectrographic field methods for the semi-quantitative analysis of geologic materials. *U.S. Geol. Surv. Circ.* 591.

GRIP E. (1953) Tracing of glacial boulders as an aid to ore prospecting in Sweden. *Econ. Geol.* 48, 715-725.

GROVES D. I. (1972) The geochemical evolution of tin-bearing granites in the Blue Tier Batholith, Tasmania. *Econ. Geol.* 67, 445-457.

DE GRYS A. (1970) Copper and zinc in alluvial magnetites from central Ecuador. *Econ. Geol.* 65, 714-717.

GUHA M. (1961) A study of the trace-element uptake by deciduous trees. Unpublished Ph.D. thesis, University of Aberdeen, Scotland. (Cited in Fortescue, 1967).

HACQUEBARD P. A. (1972) Pre- and post-deformational coalification and its significance for oil and gas exploration (Abst.). *Twenty-fourth Internat. Geol. Congress,* Sect. 5, 18.

HAMIL B. M. and NACKOWSKI M. P. (1971) Trace-element distribution in accessory magnetite from quartz monzonite intrusives and its relation to sulfide mineralization in the Basin and Range Province of Utah and Nevada — a preliminary report. *Geochemical Exploration*. CIM Spec. Vol. 11, 331-333.

HAN J. and CALVIN M. (1969) Occurrence of C_{22}-C_{25} isoprenoids in Bell Creek crude oil. *Geochim. Cosmochim. Acta* 33, 733-742.

HANSEN K. (1967) The general limnology of arctic lakes as illustrated by examples from Greenland. *Medd. Om Gronland* 178, No. 3, 1-77.

HANSHAW B. B. and COPLEN T. B. (1973) Ultrafiltration by a compacted clay membrane — II. Sodium ion exclusion at various ionic strengths. *Geochim. Cosmochim. Acta* 37, 2311-2327.

HANSHAW B. B. and HILL G. A. (1969) Geochemistry and hydrodynamics of the Paradox basin region, Utah, Colorado and New Mexico. *Chem. Geol.* 4, 263-294.

HANSULD J. A. (1967) Eh and pH in geochemical prospecting. *Geol. Surv. Canada Paper* 66-54, 172-187.

HANSULD J. A. (moderator) (1969) What is a geochemical analysis? A panel discussion. *Colo. School Mines Quart.* 64, 5-26.

HARBAUGH J. W. (1964a) A computer method for four-variable trend analysis illustrated by a study of oil-gravity variations in southeastern Kansas. *Kansas State Geol. Surv. Bull. 171.*

HARBAUGH J. W. (1964b) Trend-surface mapping of hydrodynamic oil traps with the IBM 7090/94 computer. *Colo. School Mines Quart.* 59, 557-578.

HARDEN G. and TOOMS J. S. (1964) Efficiency of the potassium bisulphate fusion in geochemical analysis. *Trans. IMM.* 74, 129-141.

HARMAN H. H. (1967) *Modern Factor Analysis.* University of Chicago Press.

HARSHMAN E. N. (1972) Geology and uranium deposits, Shirley Basin area, Wyoming. *U.S. Geol. Surv. Prof. Paper* 745.

HAUSEN D. M., AHLRICHS J. W. and ODEKIRK J. R. (1973) Application of sulphur and nickel analyses to geochemical prospecting. *Geochemical Exploration 1972*, pp. 13-24. IMM. London.

HAUSEN D. M. and KERR P. F. (1971) X-ray diffraction methods of evaluating potassium silicate alteration in porphyry mineralization. *Geochemical Exploration.* CIM Spec. Vol. 11, 334-340.

HAWKES H. E. (1957) Principles of geochemical prospecting. *U.S. Geol. Surv. Bull.* 1000-F, 225-355.

HAWKES H. E. (1963) Dithizone field tests. *Econ. Geol.* 58, 579-586.

HAWKES H. E., BLOOM H., RIDDELL J. E. and WEBB J. S. (1960) Geochemical reconnaissance in eastern Canada. *Intern. Geol. Congr. 20th*, Mexico, Vol. 3, 607-621.

HAWKES H. E. and WEBB J. S. (1962) *Geochemistry In Mineral Exploration.* Harper and Row.

HEINRICH E. W. (1958) *Mineralogy and Geology of Radioactive Raw Materials.* McGraw-Hill.

HEINRICH E. W. (1962) Geochemical prospecting for beryl and columbite. *Econ. Geol.* 57, 616-619. (Discussion pp. 1127-1130).

HELGESON H. C. (1964) *Complexing and Hydrothermal Ore Deposition.* Pergamon Press.

HEM J. D. (1970) Study and interpretation of the chemical characteristics of natural water (2nd edition). *U.S. Geol. Surv. Water-Supply Paper* 1473.

HEM J. D. (1972) Chemical factors that influence the availability of iron and manganese in aqueous systems. *Bull. Geol. Soc. Amer.* 83, 443-450.

HENDERSON W., WOLLRAB V. and EGLINTON G. (1969) Identification of steranes and triterpanes from a geological source by capillary gas liquid chromatography and mass spectrometry. In, *Advances in Organic Geochemistry, 1968* (editors P. A. Schenck and I. Havenaar), 181-207. Pergamon Press.

HESP W. R. (1971) Correlations between the tin content of granitic rocks and their chemical and mineralogical composition. *Geochemical Exploration.* CIM Spec. Vol. 11, 341-353.

HESP W. R. and RIGBY D. (1973) Cluster analysis of rocks in the New England igneous complex, New South Wales, Australia. *Geochemical Exploration 1972*, pp. 221-235. IMM. London.

HILL G. A., COLBURN W. A. and KNIGHT J. W. (1961) Reducing oil-finding costs by use of hydrodynamic evaluations. In, *Economics of Petroleum Exploration, Development, and Property Evaluation. Proc. 1961 Inst. Internat. Oil Gas Educ. Center*, 38-69. Prentice-Hall.

HILLS I. R., SMITH G. W. and WHITEHEAD E. V. (1970) Hydrocarbons from fossil fuels and their relationship with living organisms. *J. Inst. Petroleum* 56, 127-137.

HITCHON, B. (1968) Rock volume and pore volume data for plains region of western Canada sedimentary basin between latitudes 49° and 60° N. *Amer. Assoc. Petrol. Geol. Bull.* 52, 2318-2323.

HITCHON B. (1969a) Fluid flow in the western Canada sedimentary basin. 1. Effect of topography. *Water Resour. Res.* 5, 186-195.

HITCHON B. (1969b) Fluid flow in the western Canada sedimentary basin. 2. Effect of geology. *Water Resour. Res.* 5, 460-469.

HITCHON B. (1971) Origin of oil: geological and geochemical constraints. In, *Origin and Refining of Petroleum* (editors H. G. McGrath and M. E. Charles), 30-66, *Advances in Chemistry Series* 103. American Chemical Society, Washington, D.C.

HITCHON B., BILLINGS G. K. and KLOVAN J. E. (1971) Geochemistry and origin of formation waters in the western Canada sedimentary basin. III. Factors controlling chemical composition. *Geochim. Cosmochim. Acta* 35, 567-598.

HITCHON B. and FRIEDMAN I. (1969) Geochemistry and origin of formation waters in the western Canada sedimentary basin. I. Stable isotopes of hydrogen and oxygen. *Geochim. Cosmochim Acta* 33, 1321-1349.

HITCHON B. and GAWLAK M. (1972) Low molecular weight aromatic hydrocarbons in gas condensates from Alberta, Canada. *Geochim. Cosmochim. Acta* 36, 1043-1059.

HITCHON B. and HAYS J. (1971) Hydrodynamics and hydrocarbon occurrences, Surat basin, Queensland, Australia. *Water Resour. Res.* 7, 658-676.

HITCHON B. and HORN M. K. (1974) Petroleum occurrence indicators in formation waters from Alberta, Canada. *Amer. Assoc. Petrol. Geol. Bull.* 58, (in press).

HITCHON B., LEVINSON A. A. and REEDER S. W. (1969) Regional variations of river water composition resulting from halite solution, Mackenzie River drainage basin, Canada. *Water Resour. Res.* 5, 1395-1403.

HODGSON G. W. and HITCHON B. (1965) Research trends in petroleum genesis. *Eighth Commonwealth Mining Metall. Congress Proc.* 5, 9-19.

HOLLAND H. D. (1972) Granites, solutions and base metal deposits. *Econ. Geol.* 67, 281-301.

HOLMAN R. H. C. (1963) A method for determining readily-soluble copper in soil and alluvium. *Geol. Surv. Canada Paper* 63-7.

HOLMES R. and TOOMS J. S. (1973) Dispersion from a submarine exhalative ore body. *Geochemical Exploration 1972*. pp. 193-202. IMM. London.

HORNBROOK E. H. W. (1969) Biogeochemical prospecting for molybdenum in west-central British Columbia. *Geol. Surv. Canada Paper* 68-56.

HORNBROOK E. H. W. and ALLAN R. J. (1970) Geochemical exploration feasibility study within the zone of continuous permafrost; Coppermine River region, Northwest Territories. *Geol. Surv. Canada Paper* 70-36.

HORNBROOK E. H. W. and JONASSON I. R. (1971) Mercury in permafrost regions: occurrence and distribution in the Kaminak Lake area, Northwest Territories. *Geol. Surv. Canada Paper* 71-43.

HORSNAIL R. F. and ELLIOTT I. L. (1971) Some environmental influences on the secondary dispersion of molybdenum and copper in western Canada. *Geochemical Exploration*. CIM Spec. Vol. 11, 166-175.

HORSNAIL R. F., NICHOL I. and WEBB J. S. (1969) Influence of variations in secondary environment on the metal content of drainage sediments. *Colo. School Mines Quart.* 64, 307-322.

HORVITZ L. (1959) Geochemical prospecting for petroleum. *Twentieth Internat. Geol. Congress*, Symposium de Exploración Geoquímica 2, 303-319.

HORVITZ L. (1968) Hydrocarbon geochemical prospecting after thirty years. *Unconventional Methods in Exploration for Petroleum and Natural Gas* (editor W. B. Heroy), 205-218. Southern Methodist University Press, Dallas, Texas.

HORVITZ L. (1972) Vegetation and geochemical prospecting for petroleum. *Amer. Assoc. Petrol. Geol. Bull.* 56, 925-940.

HOSKING K. F. G. (1971) Problems associated with the application of geochemical methods of exploration in Cornwall, England. *Geochemical Exploration*. CIM Spec. Vol. 11, 176-189.

HOUGH J. L. (1958) *Geology of the Great Lakes*. University of Illinois Press.

HOUSSE B. (1961) Résultats d'une étude géochimique en surface dans la zone du gisement de pétrole du Djebel Haricha (Maroc). *Rev. l'Inst. Francais du Pétrole* 16, 140-149.

HOUSSE B. (1971) Enseignements tirés de l'expérience d'un programme de prospection géochimique systématique des hydrocarbures. *Geochemical Exploration.* CIM Spec. Vol. 11, 523-528.

HOWARTH R. J. (1971) Empirical discriminant classification of regional stream sediment geochemistry in Devon and east Cornwall. *Trans. IMM* (Sec. B, Appl. Earth Sci.) 80, 142-149.

HOWARTH R. J. (1971) An empirical discriminant method applied to sedimentary rock classification from major element geochemistry. *J. Inter. Assn. Math. Geol.* 3, 51-60.

HOWARTH R. J. (1973) The pattern recognition problem in applied geochemistry. *Geochemical Exploration 1972,* pp. 259-273. IMM. London.

HOWARTH R. J. and LOWENSTEIN P. L. (1971) Sampling variability of stream sediments in broad-scale regional geochemical reconnaissance. *Trans. IMM* (Sec. B, Appl. Earth Sci.) 80, 363-372.

HUBBERT M. K. (1940) The theory of groundwater motion. *J. Geol.* 48, 785-944.

HUFF L. C. (1948) A sensitive field test for heavy metals in water. *Econ. Geol.* 43, 675-684.

HUFF L. C. (1952) Abnormal copper, lead, and zinc content of soil near metalliferous veins. *Econ. Geol.* 47, 517-542.

HUFF L. C. (1970) A geochemical study of alluvium-covered copper deposits in Pima County, Arizona. *U.S. Geol. Surv. Bull.* 1312-C.

HUFF L. C. (1971) A comparison of alluvial exploration techniques for porphyry copper deposits. *Geochemical Exploration.* CIM Spec. Vol. 11, 190-194.

HUFF L. C. and MARRANZINO A. P. (1961) Geochemical prospecting for copper deposits hidden beneath alluvium in the Pima district, Arizona. *U.S. Geol. Surv. Prof. Paper* 424-B, B308-310.

HUNT C. B. (1972) *Geology of Soils.* W. H. Freeman.

HURLEY J. L. and CATTELL R. C. (1962) The Procrustes program, producing direct rotation to test an hypothesized factor structure. *Behavioral Sci.* 7, 258-262.

HUTCHINSON G. E. (1957) *A Treatise on Limnology.* Vol. 1. Geography, Physics, and Chemistry. Wiley.

HUTCHINSON G. E. (1967) *A Treatise on Limnology.* Vol. 2. Introduction to Lake Biology and the Limnoplankton. Wiley.

HUTCHINSON G. E. (1973) Eutrophication. *Amer. Scientist* 61 (No. 3), 269-279.

IMBRIE J. (1963) Factor and vector analysis programs for analyzing geological data. *ONR Tech Rept.* 6.

IMBRIE J. and VAN ANDEL T. H. (1964) Vector analysis of heavy mineral data. *Bull. Geol. Soc. Amer.* 75, 1131-1155.

INGAMMELLS C. O., ENGELS J. C. and SWITZER P. (1972) Effect of laboratory sampling error in geochemistry and geochronology. *Intern. Geol. Congr. 24th,* Montreal, Sect. 10, 405-415.

IVANOV O. P. (1966) Basic factors in the development of sulfide deposit oxidation zones under permafrost conditions. *Geokhimiya,* 1095-1105, (In Russian). Translation: *Geochem. Intern.* (1966) 3, 875-884.

IUPAC (International Union of Pure and Applied Chemistry) (1964) *Reagents and Reactions For Qualitative Inorganic Analysis* (Fifth Report). Butterworths, London. (The contents of this book also appear in *Pure and Applied Chemistry,* 8, No. 1, 1964).

JAMES C. H. (1964) The application of white spirit as a solvent for dithizone in geochemical analysis for copper in tropical climates. *Econ. Geol.* 59, 1596-1599.

JAMES C. H. (1967) The use of the terms "primary" and "secondary" dispersion in geochemical prospecting. *Econ. Geol.* 62, 997-999.

JEFFREYS H. (1970) *The Earth.* Cambridge University Press.

JENKINS R. and DEVRIES J. L. (1967) *Practical X-ray Spectrometry.* Springer-Verlag.

JENNY E. A. (1968) Controls on Mn, Fe, Co, Ni, Cu, and Zn concentrations in soils and water: The significant role of hydrous Mn and Fe oxides. *Amer. Chem. Soc., Advances In Chemistry Series,* No. 73, 337-387.

JENNY H. (1941) *Factors of Soil Formation.* McGraw-Hill.

JENSEN M. L. (1971) Stable isotopes in geochemical prospecting. *Geochemical Exploration.* CIM Spec. Vol. 11, 464-468.

JOHNSON A. E. (1972) Origin of Cyprus pyrite deposits. *Intern. Geol. Congr. 24th,* Montreal, Sect. 4, 291-298.

JONASSON I. R. (1970) Mercury in the natural environment: a review of recent work. *Geol. Surv. Canada Paper* 70-57.

JONASSON I. R. and ALLAN R. J. (1973) Snow: a sampling medium in hydrogeochemical prospecting in temperate and permafrost regions. *Geochemical Exploration 1972,* pp. 161-176. IMM. London.

JONES M. J. (1973) *Prospecting in Areas of Glacial Terrain.* Institution of Mining and Metallurgy. London.

JONES M. J. (1973a) *Geochemical Exploration 1972.* Institution of Mining and Metallurgy. London.

JONES T. A. and JAMES W. R. (1972) MAXLIKE: FORTRAN IV program for maximum likelihood estimation. *Geocom. Bull.* 5, 187-202.

JONES T. S. and SMITH H. M. (1965) Relationships of oil composition and stratigraphy in the Permian basin of west Texas and New Mexico. *Fluids in Subsurface Environments* (editors A. Young and J. E. Galley), 101-224. American Association of Petroleum Geologists, Tulsa, Oklahoma.

KAISER H. F. (1958) The varimax criterion for analytic rotation in factor analysis. *Psych.* 23, 187-200.

KARIM M. F. (1964) Some geochemical methods of prospecting and exploration for oil and gas. Ph.D. Thesis, Univ. of Southern California. University Microfilms Inc. No. 65-1291.

KARIM M. F. and CHILINGAR G. V. (1965) Exploracion quimica para petroleo en la costa del Pacifico de Nicaragua. *Bol. Serv. Geol. Nacl. Nicaragua* 7, 97-136.

KARTSEV A. A., TABASARANSKII Z. A., SUBBOTA M. I. and MOGILVESKII G. A. (1959) *Geochemical Methods of Prospecting and Exploration for Petroleum and Natural Gas* (English translation Ed. by P. A. Witherspoon and W. D. Romey), University of California Press, Berkeley, California.

KARTSEV A. A., VASSOEVICH N. B., GEODEKIAN A. A., NERUCHEV S. G. and SOKOLOV V. A. (1971) The principal stage in the formation of petroleum. *Eighth World Petroleum Congress Proc.* 2, 3-11.

KAURANNE L. K. and NURMI A. (1967) The analysis of trace amounts of copper in soil with neocuproine. In, *Geochemical Prospecting In Fennoscandia* (editor A. Kvalheim), pp. 331-333. Interscience.

KENDALL M. G. and STUART A. (1973) *The Advanced Theory of Statistics.* Vol. 2. Charles Griffin, London.

KESLER S. E., VAN LOON J. C. and BATESON J. H. (1973) Analysis of fluoride in rocks and an application to exploration. *J. Geochem. Explor.* 2, 11-17.

KESLER S. E., VAN LOON J. C. and MOORE C. M. (1973) Evaluation of ore potential and granodioritic rocks using water extractable chloride and fluoride. *CIM Bull.* 66 (Feb.), 56-60.

KHARAKA Y. K. and BARNES I. (1973) SOLMNEQ: Solution-mineral equilibrium computations. *NTIS,* PB-215 899.

KLOVAN J. E. (1966) The use of factor analysis in determining depositional environments from grain-size distributions. *J. Sed. Petrol.* 36, 115-125.

KLOVAN J. E. and BILLINGS G. K. (1967) Classification of geological samples by discriminant function analysis. *Bull. Can. Petrol. Geol.* 15, 313-330.

KLOVAN J. E. and IMBRIE J. (1971) An algorithm and FORTRAN IV program for large-scale Q mode factor analysis and calculation of factor scores. *J. Inter. Assn. Math. Geol.* 3, 61-78.

KOCH G. S. JR. and LINK R. F. (1970, 1971) *Statistical Analysis of Geological Data.* Vols. 1 and 2. John Wiley.

KOCH G. S. JR. and LINK R. F. (1971) The coefficient of variation – a guide to the sampling of ore deposits. *Econ. Geol.* 66, 293-301.

KRAUSKOPF K. B. (1967) *Introduction to Geochemistry.* McGraw-Hill.

KROEPELIN H. (1967) Geochemical prospecting. *Seventh World Petroleum Congress Proc.* 1B, 37-57.

KRUMBEIN W. C. and GRAYBILL F. A. (1965) *An Introduction to Statistical Models in Geology.* McGraw-Hill.

KUNZENDORF H. (1973) Non-destructive determination of metals by radioisotope X-ray fluorescence instrumentation. *Geochemical Exploration 1972.* pp. 401-414. IMM. London.

KVALHEIM A. (editor) (1967) *Geochemical Prospecting in Fennoscandia.* Interscience.

KVENVOLDEN K. A. and SQUIRES R. M. (1967) Carbon isotopic composition of crude oils from Ellenburger Group (Lower Ordovician), Permian basin, west Texas and eastern New Mexico. *Amer. Assoc. Petroleum Geologists Bull.* 51, 1293-1303.

KVENVOLDEN K. A. and WEISER D. (1967) A mathematical model of a geochemical process: Normal paraffin formation from normal fatty acids. *Geochim. Cosmochim. Acta* 31, 1281-1309.

KVET R. and MICHALICEK M. (1965) Results and perspectives of the application of geochemical surface prospection for bitumin in the CSSR. *Geochem. v Ceskoslovensku* 1, 287-294.

LANG A. H. (1970) Prospecting in Canada. *Geol. Surv. Canada, Economic Geology Report* No. 7.

LARSSON J. O. and NICHOL I. (1971) Analysis of glacial material as an aid in geological mapping. *Geochemical Exploration.* CIM Spec. Vol. 11, 197-203.

LEARNED R. E. and BOISSEN R. (1973) Gold — a useful pathfinder element in the search for porphyry copper deposits in Puerto Rico. *Geochemical Exploration 1972.* pp. 93-103. IMM. London.

LEE H. A. (1963) Glacial fans in till from the Kirkland Lake fault: a method of gold exploration. *Geol. Surv. Canada Paper* 63-45.

LEE H. A. (1965) Investigation of eskers for mineral exploration. *Geol. Surv. Canada Paper* 65-14.

LEE H. A. (1968) An Ontario kimberlite occurrence discovered by application of the glaciofocus method to a study of the Munro esker. *Geol. Surv. Canada Paper* 68-7.

LEE H. A. (1971) Mineral discovery in the Canadian Shield using the physical aspect of overburden. *CIM Bull.* 64 (Nov.), 32-36.

LEEPER G. W. (1964) *Introduction to Soil Science* (2nd edition). Melbourne University Press.

LEGGET R. F. (1967) Soil – its geology and use. *Geol. Soc. Amer. Bull.* 78, 1433-1460.

LEHNERT-THIEL K. and VOHRYZKA K. (1968) Studies of mercury halos around polymetallic deposits in permafrost ground of East Greenland at latitude 72° north. *Montan-Rundschau* 16 (No. 5), 104-108 (Vienna). (Cited in Hornbrook and Jonasson, 1971).

LEPELTIER C. (1971) Geochemical exploration in the United Nations Development Programme. *Geochemical Exploration.* CIM Spec. Vol. 11, 24-27.

LE TRAN K. (1971) Etude geochimique de l'hydrogene sulfure adsorbe dans les sediments. *Bull. Centre Rech. Pau-SNPA* 5, 321-332.

LEVINSON A. A. and TAYLOR S. R. (1971) *Moon Rocks And Minerals.* Pergamon.

LEYTHAEUSER D. (1973) Effects of weathering on organic matter in shales. *Geochim. Cosmochim. Acta* 37, 113,120.

LIVINGSTONE D. A. (1963) Chemical composition of rivers and lakes. *U.S. Geol. Surv. Prof. Paper* 440-G.

LIVINGSTONE D. A., BRYAN K. and LEAHY R. G. (1958) Effects of an Arctic environment on the origin and development of freshwater lakes. *Limnol. Oceanogr.* 3, 192-214.

LOGANATHAN P. and BURAU R. G. (1973) Sorption of heavy metal ions by a hydrous manganese oxide. *Geochim. Cosmochim. Acta* 37, 1277-1293.

LONG G., NEGLIA S. and FAVRETTO L. (1968) The metamorphism of the kerogen from Triassic black shales, southeast Sicily. *Geochim. Cosmochim. Acta* 32, 647-656.

LOUGHNAN F. C. (1969) *Chemical Weathering of the Silicate Minerals.* Elsevier.

LOVERING T. S. (coordinator) (1958) Symposium de exploracion geoquimica. *Intern. Geol. Congr. 20th.,* Mexico, 1956 (in 3 volumes).

LOVERING T. G., COOPER J. R., DREWES H. and CONE G. C. (1970) Copper in biotite from igneous rocks in southern Arizona as an ore indicator. *U.S. Geol. Surv. Prof. Paper* 700-B, B1-B8.

LOWELL J. D. and GUILBERT J. M. (1970) Lateral and vertical alteration-mineralization zoning in porphyry ore deposits. *Econ. Geol.* 64, 373-408.

LOWENSTEIN P. L. and HOWARTH R. J. (1973) Automated colour mapping of three component systems and its application to regional geochemical reconnaissance. *Geochemical Exploration 1972,* pp. 297-304. IMM. London.

LYNCH J. J., GARRETT R. G. and JONASSON I. R. (1973) A rapid estimation of organic carbon in silty lake sediments. *J. Geochem. Explor.* 2, 171-174.

MACDONALD J. A. (1969) An orientation study of the uranium distribution in lake waters, Beaverlodge district, Saskatchewan. *Colo. School Mines Quart.* 64, 357-376.

MACFARLANE I. C. (1969) *Muskeg Engineering Handbook.* University of Toronto Press.

MACKAY J. R. (1970) Lateral mixing of the Liard and Mackenzie rivers downstream from their confluence. *Canad. J. Earth Sci.* 7, 111-124.

MACKAY J. R. (1972) Application of water temperatures to the problem of lateral mixing .n the Great Bear – Mackenzie River System. *Canad. J. Earth Sci.* 9, 913-917.

MACLIVER C. N., MCDONALD D., SAMPEY D. and SULLIVAN J. V. (1969) Multi-element analysis in geochemistry: a six-channel atomic absorption spectrophotometer (abstract). *Colo. School Mines Quart.* 64, 511.

MAGARA K. (1968) Compaction and migration of fluids in Miocene mudstone, Nagaoka plain, Japan. *Amer. Assoc. Petrol. Geol. Bull.* 52, 2466-2501.

MAGARA K. (1969) Upward and downward migrations of fluids in the subsurface. *Bull. Canad. Petrol. Geol.* 17, 20-46.

MAGARA K. (1973) Compaction and fluid migration in Cretaceous shales of western Canada. *Geol. Serv. Canada Paper* 72-18.

MAIGNIEN R. (1966) Review of research on laterites. *UNESCO Natural Resources Research IV,* 1-148.

MALYUGA D. P. (1964) *Biogeochemical Methods of Prospecting.* Consultants Bureau. New York. (Translation from the Russian).

MANSKAYA S. M. and DROZDOVA T. V. (1968) *Geochemistry of Organic Substances.* (Translated and edited by L. Shapiro and I. A. Breger). Pergamon.

MARMO V. and PURANEN M. (editors) (1960) Geological results of applied geochemistry and geophysics. *Intern. Geol. Congr. 21st.,* Copenhagen, Rept. Session Norden 2(2).

MASON B. (1966) *Principles of Geochemistry.* (3rd edition). Wiley.

MATERN B. (1960) Spatial variation. *Medd. Statens Skogsforskningsinstit* 49.

MAXWELL A. A. and LAWLEY D. N. (1971) *Factor Analysis as a Statistical Method.* Butterworth, London.

MAYNARD D. E. and FLETCHER W. K. (1973) Comparison of total and partial extractable copper in anomalous and background peat samples. *J. Geochem. Explor.* 2, 19-24.

MAZZUCCHELLI R. H. (1972) Secondary geochemical dispersion patterns associated with the nickel sulphide deposits at Kambalda, Western Australia. *J. Geochem. Explor.* 1, 103-116.

MAZZUCCHELLI R. H. and JAMES C. H. (1966) Arsenic as a guide to gold mineralization in laterite-covered areas of Western Australia. *Trans. IMM* (Sect. B, Appl. Earth Sci.) 75, 286-294; discussion: 76, 127-129.

McAuliffe C. (1967) Solubility in water of paraffin, cycloparaffin, olefin, acetylene, cyclo-olefin and aromatic hydrocarbons. *J. Phys. Chem.* 70, 1267-1275.

McCammon R. B. (1968) The dendrograph: a new tool for correlation. *Bull. Geol. Soc. Amer.* 79, 1663-1670.

McCammon R. B. (1969) FORTRAN IV program for nonlinear estimation. *Kansas Geol. Surv. Computer Contrib.* 34.

McCammon R. B. (1969) Multivariate methods in geology. In, *Models of Geologic Processes* (editor, P. Fenner). American Geological Institute (pages not numbered; McCammon's article is about 141 p. long).

McCarthy J. H. (1972) Mercury vapor and other volatile components in the air as guides to ore deposits. *J. Geochem. Explor.* 1, 143-162.

McConnell D. (1973) *Apatite.* Springer-Verlag.

McCrossan R. G., Ball N. L. and Snowdon L. R. (1972) An evaluation of surface geochemical prospecting for petroleum, Olds – Caroline area, Alberta. *Geol. Surv. Canada Paper* 71-31.

McIver R. D. (1962) Ultrasonics — a rapid method for removing soluble organic matter from sediments. *Geochim. Cosmochim. Acta* 26, 343-345.

McIver R. D. (1967) Composition of kerogen — clue to its role in the origin of petroleum. *Proc. Seventh World Petroleum Congress* 2, 25-36.

McKenzie R. M. (1967) The sorption of cobalt by manganese minerals in soils. *Austral. J. Soil Res.* 5, 235-246.

McKenzie R. M. and Taylor R. M. (1968) The association of cobalt with manganese oxide minerals in soils. *Trans. Ninth Intern. Congr. Soil Sci.*, Vol. 2, 577-584.

Mehrtens M. B., Tooms J. S. and Troup A. G. (1973) Some aspects of geochemical dispersion from base-metal mineralization within glaciated terrain in Norway, North Wales and British Columbia, Canada. *Geochemical Exploration 1972*, pp. 105-115. IMM. London.

Merriam D. F. and Cocke N. C. (editors) (1968) *Computer Applications in the Earth Sciences: Colloquium on Simulation.* Kansas Geol. Surv. Computer Contrib. 22.

Meyer C. and Hemley J. J. (1967) Wall rock alteration. In, *Geochemistry of Hydrothermal Ore Deposits* (editor H. L. Barnes), pp. 166-236. Holt, Rinehart and Winston.

Michard G. and Treuil M. (1969) Réflexions sur l'utilisation des corrélations entre éléments-traces en géochimie: exemples. *Bull. Soc. Geol. France* 11, 595-598.

Michie U. M. (1973) The determination of sampling and analytical errors in exploration geochemistry. *Econ. Geol.* 68, 281-284.

Miesch A. T. (1964) Effects of sampling and analytical error in geochemical prospecting. In, *Computers in the Mineral Industry* (editor G. A. Parks) Pt. 1, *Stanford Univ. Publ. Geol. Sci.* 9, 156-170.

Miesch A. T. (1967) Theory of error in geochemical data. *U.S. Geol. Surv. Prof. Paper* 574-A.

Miesch A. T., Chao E. C. T. and Cuttitta F. (1966) Multivariate analysis of geochemical data on tektites. *J. Geol.* 74, 673-691.

Miesch A. T. and Connor J. J. (1964) Investigation of sampling error effects in geochemical prospecting. *U.S. Geol. Surv. Prof. Paper* 475-D.

Miller R. L. and Kahn J. S. (1962) *Statistical Analysis in the Geological Sciences.* Wiley.

Mitchell R. L. (1964) Trace elements in soils. In, *Chemistry of the Soil* (2nd edition) (editor F. E. Bear). *Amer. Chem. Soc. Mono. Ser.* 160, 320-368. Reinhold.

Mitchell R. L. (1972) Trace elements in soils and factors that affect their availability. *Geol. Soc. Amer. Bull.* 83, 1069-1076.

Monster J. (1972) Homogeneity of sulfur and carbon isotope ratios S^{34}/S^{32} and C^{13}/C^{12} in petroleum. *Amer. Assoc. Petrol. Geol. Bull.* 56, 941-949.

Morse R. H. (1971) Comparison of geochemical prospecting methods using radium with those using radon and uranium. *Geochemical Exploration.* CIM Spec. Vol. 11, 215-230.

MORTIMER C. H. (1941-42) The exchange of dissolved substances between mud and water in lakes. *J. Ecol.* 29, 280-329; and 30, 147-201.

MORTVEDT J. J., GIORDANO P. M. and LINDSAY W. L. (1972) *Micronutrients in Agriculture.* Soil Science Society of America. Madison, Wisconsin.

MURPHY M. T. J. (1969) Analytical methods. *Organic Geochemistry* (editors G. Eglinton and M. T. J. Murphy), 74-88. Springer-Verlag.

MYERS A. T., HAVENS R. G. and DUNTON P. J. (1961) A spectrochemical method for the semi-quantitative analysis of rocks, minerals, and ores. *U.S. Geol. Surv. Bull.* 1084-I.

MYERS A. T., HAVENS R. G. and NILES W. W. (1970) Glass reference standards for trace element analysis of geologic materials. *Develop. Appl. Spect.* 8, 132-137.

NAKAGAWA H. M. and LAKIN H. W. (1965) A field method for the determination of silver in soils and rocks. *U.S. Geol. Surv. Prof. Paper* 525-C. C172-C175.

NEGLIA S. and FAVRETTO L. (1964) Study of an analytical method for the execution of surface geochemical prospecting for petroleum and natural gas. *Advances in Organic Geochemistry, 1962* (editors U. Colombo and G. D. Hobson), 285-295. The Macmillan Co.

NICHOL I. (1971) Future trends of exploration geochemistry in Canada. *Geochemical Exploration.* CIM Spec. Vol. 11, 32-38.

NICHOL I. (1973) The role of computerized data systems in geochemical exploration. *Bull. CIM* 66 (January) 59-68.

NICHOL I. and BJORKLUND A. (1973) Glacial geology as a key to geochemical exploration in areas of glacial overburden with particular reference to Canada. *J. Geochem. Explor.* 2, 133-170.

NICHOL I., GARRETT R. G. and WEBB J. S. (1969) The role of some statistical and mathematical methods in the interpretation of regional geochemical data. *Econ. Geol.* 64, 204-220.

NICHOL I. and HENDERSON-HAMILTON J. C. (1965) A rapid quantitative spectrographic method for the analysis of rocks, soils and stream sediments. *Trans. IMM* 74, 955-961.

NICHOL I., HORSNAIL R. F. and WEBB J. S. (1967) Geochemical patterns in stream sediment related to precipitation of manganese oxides. *Trans. IMM* (Sect. B, Appl. Earth Sci.) 76, B113-B115.

NICHOLLS G. D. (1971) Geochemical sampling problems in the analytical laboratory. *Trans. IMM* (Sect. B, Appl. Earth Sci.), 80, B299-B304.

NICKERSON D. (1973) An account of a lake sediment geochemical survey conducted over certain volcanic belts within the Slave structural province of the Northwest Territories during 1972. *Geol. Surv. Canada Open File* 129.

NORTON S. A. (1973) Laterite and bauxite formation. *Econ. Geol.* 68, 353-361.

NOWLAN G. A. (1971) A field method for the determination of cold-extractable nickel in stream sediments in soils. *U.S. Geol. Surv. Prof. Paper* 700-B, B177-B180.

NYSSEN R., VAN EGGELPOEL A. and GODARD J. M. (1966) Contribution a la geochimic de surface sur le permis de Colomb-Bechar (Sahara). *Advances in Organic Geochemistry, 1964* (editors G. D. Hobson and M. C. Louis), 303-315. Pergamon.

OBIAL R. C. (1970) Cluster analysis as an aid in the interpretation of multi-element geochemical data. *Trans. IMM* (Sect. B, Appl. Earth Sci.) 79, 175-180.

OJAKANGAS D. R. (1970) FORTRAN IV program for simulating geologic development of sedimentary basins. *Kansas Geol. Surv. Computer Contrib.* 49.

OVCHINNIKOV L. N. and GRIGORYAN S. V. (1971) Primary halos in prospecting for sulphide deposits. *Geochemical Exploration.* CIM Spec. Vol. 11, 375-380.

OVCHINNIKOV L. N., SOKOLOV V. A., FRIDMAN A. I. and YANITSKII I. N. (1973) Gaseous geochemical methods in structural mapping and prospecting for ore deposits. *Geochemical Exploration 1972,* pp. 177-182. IMM. London.

PARK C. F. and MACDIARMID R. A. (1964) *Ore Deposits.* W. H. Freeman.

PARKER R. L. (1967) Composition of the Earth's crust. *U.S. Geol. Surv. Prof. Paper* 440-D.

PARKS J. M. (1966) Cluster analysis applied to multivariate geologic problems. *J. Geol.* 74, 703-715.

PARRY W. T. (1972) Chlorine in biotite from Basin and Range plutons. *Econ. Geol.* 67, 972-975.

PARRY W. T. and NACKOWSKI M. P. (1963) Copper, lead and zinc in biotites from Basin and Range quartz monzonites. *Econ. Geol.* 58, 1126-1144.

PARSLOW G. R. (1972) Geochemical exploration in northern Saskatchewan. *Intern. Geol. Congr.* 24th, Montreal, Sect. 10, 394-401.

PARSONS M. L. (1970) Groundwater movement in a glacial complex, Cochrane district, Ontario. *Canad. J. Earth Sci.* 7, 869-883.

PAULING L. (1960) *The Nature of the Chemical Bond* (3rd edition). Cornell University Press.

PAWLUK S. and BAYROCK L. A. (1969) Some characteristics and physical properties of Alberta tills. *Res. Council Alberta Bull.* 26.

PEACHEY D., ROBERTS J. L. and SCOT-BAKER J. (1973) Rapid colorimetric determination of phosphorus in geochemical survey samples. *J. Geochem. Explor.* 2, 115-120.

PERCHAC R. M. and WHELAN C. J. (1972) A comparison of water, suspended solid and bottom sediment analyses for geochemical prospecting in a northeast Tennessee zinc district. *J. Geochem. Explor.* 1, 47-53.

PEUCKER T. K. and JOHNSTON E. G. (1972) Detection of surface specific points by local parallel processing of discrete terrain elevation data. *Univ. Maryland Compt. Contrib. Tech. Rept.* TR-206.

PHILBIN P. and SENFTLE F. E. (1971) Field activation analysis of uranium ore using ^{252}Cf neutron source. *Trans. Soc. Min. Eng.* 250, 102-106.

PHILIPPI G. T. (1956) Identification of oil source beds by chemical means. *Twentieth Intern. Geol. Congr.* Sect. III Geologia del petroleo, 25-38.

PHILIPPI G. T. (1965) On the depth, time and mechanism of petroleum generation. *Geochim. Cosmochim. Acta* 29, 1021-1049.

PITULKO V. M. (1968) Features of geochemical searches for rare-metal deposits in permafrost areas. *Geologiya i razvedka* 11, 43-52 (In Russian). Translation: *Intern. Geol. Rev.* (1969) 11, 1239-1246.

PLANT J. (1971) Orientation studies on stream-sediment sampling for a regional chemical survey in northern Scotland. *Trans. IMM* (Sect. B, Appl. Earth Sci.) 80, B324-B345.

PLANT J. and COLEMAN R. F. (1973) Application of neutron activation analysis to the evaluation of placer gold concentrations. *Geochemical Exploration 1972*, pp. 373-381. IMM. London.

POMEYROL R., BIENNER F. and LOUIS M. (1961) Exemple de prospection géochimique par l'analyse des gaz adsorbés en surface dans le bassin de Fort Polignac. *Rev. l'Inst. Francais du Pétrole* 16, 868-874.

POULET M. and ROUCACHE J. (1969) Etude géochimique des gisements du Nord-Sahara (Algérie). *Rev. l'Inst. Francais du Pétrole* 24, 615-644.

PRESANT E. W. (1967) A trace element study of podzol soils, Bathurst district, New Brunswick. *Geol. Surv. Canada Paper* 66-54, 222-232.

PREST V. K. (1968) Nomenclature of moraines and ice-flow features as applied to the glacial map of Canada. *Geol. Surv. Canada Paper* 67-57.

PUSEY W. C. (1973) The ESR-kerogen method . . .How to evaluate potential gas and oil source rocks. *World Oil* 176 (No. 5), 71-75.

PYRIH R. Z. and BISQUE R. E. (1969) Determination of trace mercury in soil and rock media. *Econ. Geol.* 64, 825-828; discussion, 65, 357-359.

RADFORTH N. (1961) Organic terrain. In, *Soils In Canada* (editor R. F. Legget). Royal Soc. Canada Spec. Public. No. 3. University of Toronto Press.

RANKAMA K. and SAHAMA T. G. (1950) *Geochemistry.* University of Chicago Press.

RAO S. V. L. N. and RAO M. S. (1970) A study of residual maps in the interpretation of geochemical anomalies. *J. Inter. Assn. Math. Geol.* 2, 15-24.

RAYNER J. H. (1966) Classification of soils by numerical methods. *J. Soil Sci.* 17, 79-92.

REED B. L. and MILLER R. L. (1971) Orientation geochemical survey at the Nixon Fork mines, Medra quadrangle, Alaska. *U.S. Geol. Survey Bull.* 1312-K.

REEDER S. W. (1971) Cross-sectional study of the effects of smelter wastewater disposal on water quality of the Columbia River downstream from Trail, British Columbia. *Canada Dept. Energy, Mines and Resources, Inland Waters Branch, Tech. Bull.* No. 39.

REEDER S. W., HITCHON B. and LEVINSON A. A. (1972) Hydrogeochemistry of the Mackenzie River drainage basin. I. Factors affecting inorganic composition. *Geochim. Cosmochim. Acta* 36, 825-865.

REEDMAN A. J. and GOULD D. (1970) Low sample density stream sediment surveys in geochemical prospecting: an example from northeast Uganda. *TRANS. IMM* (Sect. B, Appl. Earth Sci.) 79, 246-248.

RITCHIE A. A. (1969) Recent advances in the chromatographic analysis of geologic materials. *Colo. School Mines Quart.* 64, 427-435.

ROBBINS J. C. (1973) Zeeman spectrometer for measurement of atmospheric mercury vapour. *Geochemical Exploration 1972*, pp. 315-323. IMM. London.

ROBINSON T. W. (1958) Phreatophytes. *U.S. Geol. Surv. Water-Supply Paper* 1423.

ROBINSON W. E. (1969) Isolation procedures for kerogens and associated soluble organic materials. *Organic Geochemistry* (editors G. Eglinton and M. T. J. Murphy), 181-195. Springer-Verlag.

ROEDDER E. (1967) Fluid inclusions as samples of ore fluids. In, *Geochemistry of Hydrothermal Ore Deposits* (editor H. L. Barnes), pp. 515-574. Holt, Rinehart and Winston.

ROEDDER E. (1971) Fluid inclusion studies on the porphyry-type ore deposits at Bingham, Utah, Butte, Montana, and Climax, Colorado. *Econ. Geol.* 66, 98-120.

ROEDDER E. (1972) Composition of fluid inclusions. *U.S. Geol. Surv. Prof. Paper* 440-JJ.

ROGERS M. A., BAILEY N. J. L. and EVANS C. R. (1971) A plea for inclusion of basic sample information when reporting geochemical analyses of crude oils. *Geochim. Cosmochim. Acta* 35, 632-636.

ROMAN R. J. and BENNER B. R. (1973) The dissolution of copper concentrates. *Minerals Sci. Engng.* 5 (No. 1), 3-24.

ROSE A. W. (1972) Favorability for Cornwall-type magnetite deposits in Pennsylvania using geological, geochemical and geophysical data in a discriminant function. *J. Geochem. Explor.* 1, 181-194.

ROSE A. W., DAHLBERG E. C. and KEITH M. L. (1970) A multiple regression technique for adjusting background values in stream sediment geochemistry. *Econ. Geol.* 65, 156-165.

ROSE A. W. and SUHR N. H. (1971) Major element content as a means of allowing for background variation in stream-sediment geochemical exploration. *Geochemical Exploration.* CIM Spec. Vol. 11, 587-593.

ROSTAD O. H. (1967) Geochemical case history at the Little Falls molybdenite prospect, Boise County, Idaho. *Geol. Surv. Canada Paper* 66-54, 249-252.

ROZKOWSKI A. (1969) Chemistry of groundwater and surface waters in the Moose Mountain area, southern Saskatchewan. *Geol. Surv. Canada Paper* 67-9.

RUDMAN A. J. and LANKSTON R. W. (1973) Stratigraphic correlation of well logs by computer techniques. *Amer. Assoc. Petrol. Geol. Bull.* 57, 577-588.

RUMEAU J. L. and SOURISSE C. (1972) Compaction, diagenese et migration dans les sediments argileux. *Bull. Centre Rech. Pau-SNPA* 6, 313-345.

RUTTNER F. (1963) *Fundamentals of Limnology.* (3rd edition). University of Toronto Press.

SAINSBURY C. L. (1957) A geochemical exploration for antimony in southeastern Alaska. *U.S. Geol. Surv. Bull.* 1024-H.

SAINSBURY C. L, CURRY K. J. and HAMILTON J. C. (1973) An integrated system of geologic mapping and geochemical sampling by light aircraft. *U.S. Geol. Surv. Bull.* 1361.

SAINSBURY C. L., HAMILTON J. C. and HUFFMAN C. (1968) Geochemical cycle of selected trace elements in the tin-tungsten-beryllium district western Seward Peninsula, Alaska – a reconnaissance study. *U.S. Geol. Surv. Bull.* 1242-F.

SAKRISON H. C. (1971) Rock geochemistry — its current usefulness on the Canadian Shield. *CIM Bull.* 64 (Nov.), 28-31.

SALMI M. (1955) Prospecting for bog-covered ore by means of peat investigations. *Bull. Comm. Geol. Finlande* 169, 5-34.

SALMI M. (1967) Peat in prospecting: applications in Finland. In, *Geochemical Prospecting In Fennoscandia* (editor A. Kvalheim), pp. 113-126. Interscience.

SANDELL E. B. (1959) *Colorimetric Determination of Traces of Metals* (3rd edition). Interscience.

SAUCHELLI V. (1969) *Trace Elements in Agriculture.* Van Nostrand Reinhold Co.

SAXBY J. D. (1969) Metal-organic chemistry of the geochemical cycle. *Rev. Pure Appl. Chem.* 19, 131-150.

SAXBY J. D. (1970) Isolation of kerogen in sediments by chemical methods. *Chem. Geol.* 6, 173-184.

SCHNITZER M. and KHAN S. U. (1972) *Humic Substances In The Environment.* Marcel Dekker. New York.

SCHUENEMEYER J. H., LIENERT C. E. and KOCH G. S. Jr. (1972) Delineation of clustered points in two dimensions by measuring perimeters of convex hulls. *U.S. Bur. Mines Rept. Invest.* 7565.

SCHWARTZ F. W. and DOMENICO P. A. (1973) Simulation of hydrochemical patterns in regional groundwater flow. *Water Resour. Res.* 9, 707-720.

SCHWARTZ M. O. and FRIEDRICH G. H. (1973) Secondary dispersion patterns of fluoride in the Osor area, Province of Gerona, Spain. *J. Geochem. Explor.* 2, 103-114.

SCOTT J. S. and ST-ONGE D. A. (1969) Guide to the description of till. *Geol. Surv. Canada Paper* 68-6.

SENFTLE F. E., DUFFEY D. and WIGGINS P. F. (1969) Mineral exploration of the ocean floor by *in situ* neutron absorption using a californium-252 (^{252}Cf) source. *J. Marine Techn. Soc.* 3, 9-16.

SENFTLE F. E. and HOYTE A. F. (1966) Mineral exploration and soil analysis using *in situ* neutron activation. *Nucl. Instr. Meth.* 42, 93-103.

SHACKLETTE H. T. (1964) Flower variation of *Epilobium angustifolium* L. growing over uranium deposits. *Canad. Field Natural.* 78, 32-42.

SHACKLETTE H. T. (1965) Element content of bryophytes. *U.S. Geol. Surv. Bull.* 1198-D.

SHACKLETTE H. T. (1967) Copper mosses as indicators of metal concentrations. *U.S. Geol. Surv. Bull.* 1198-G.

SHARP R. P. (1960) *Glaciers.* University of Oregon Press.

SHAW D. M. (1961) Element distribution laws in geochemistry. *Geochim. Cosmochim. Acta* 23, 116-134.

SHERMAN G. D., MCHARGUE J. S. and HODGKISS W. S. (1942) Determination of active manganese in soil. *Soil Sci.* 54, 253-257.

SHILTS W. (1971) Till studies and their application to regional drift prospecting. *Canad. Mining J.* (April), 45-50.

SHILTS W. (1973) Drift prospecting; geochemistry of eskers and till in permanently frozen terrain: District of Keewatin; Northwest Territories. *Geol. Surv. Canada Paper* 72-45.

SHIMA M. and THODE H. G. (1971) A geochemical prospecting method using stable isotopes. *Geochemical Exploration.* CIM Spec. Vol. 11, 469-472.

SHVARTSEV S. L. (1965) Hydrochemical method of prospecting in northern swamp areas. *Geologiya i Geofisika* (No. 7), 3-10 (In Russian). Translation: *Intern. Geol. Rev.* (1966) 8, 1151-1156.

SHVARTSEV S. L. (1972) Geochemical prospecting methods in regions of permanently frozen ground. *Intern. Geol. Congr. 24th.,* Montreal, Section 10, 380-383.

SILVERMAN S. R. and EPSTEIN S. (1958) Carbon isotopic compositions of petroleums and other sedimentary organic materials. *Amer. Assoc. Petrol. Geol. Bull.* 42, 998-1012.

SINE N. M. and TAYLOR W. O., WEBBER G. R. and LEWIS C. R. (1969) Third report of analytical data for CAAS sulfide ore and syenite rock standards. *Geochim. Cosmochim. Acta* 33, 121-131.

SIVARAJASINGHAM S., ALEXANDER L. T., CADY J. G. and CLINE M. G. (1962) Laterite. *Advances in Agronomy* 14, 1-60.

SKINNER R. G. (1972) Drift prospecting in the Abitibi Clay Belt; overburden drilling program methods and costs. *Geol. Surv. Canada Open File* 116.

SLAVIN W. (1968) *Atomic Absorption Spectroscopy.* Interscience.

SLAWSON W. F. and NACKOWSKI M. P. (1959) Trace lead in potash feldspars associated with ore deposits. *Econ. Geol.* 54, 1543-1555.

SMALES A. A. and WAGER L. R. (editors) (1960) *Methods in Geochemistry.* Interscience.

SMITH A. Y. (1964) Cold extractable "heavy metal" in soil and alluvium. *Geol. Surv. Canada Paper* 63-49.

SMITH F. M. (1971) Geochemical exploration over complex mountain glacial terrain in the Whitehorse Copperbelt, Yukon Territory. *Geochemical Exploration.* CIM Spec. Vol. 11, 265-275.

SMITH G. H. and ELLIS M. M. (1963) Chromatographic analysis of gases from soils and vegetation, related to geochemical prospecting for petroleum. *Amer. Assoc. Petrol. Geol. Bull.* 47, 1897-1903.

SOKAL R. R. and SNEATH P. H. A. (1963) *Principles of Numerical Taxonomy.* W. H. Freeman.

SOKOLOV V. A., ALEXEYEV F. A., BARS E. A., GEODEKYAN A. A., MOGILEVSKY G. A., YUROVSKY Y. M. and YASENEV B. P. (1959) Investigations into direct oil detection methods. *Fifth World Petroleum Congress Proc.* 1, 667-687.

SOWER F. B., EICHER R. N. and SELNER G. I. (1971) The STATPAC system. *U.S. Geol. Surv. Comput. Contrib.* 11.

SPEIGHT J. G. (1971) The application of spectroscopic techniques to the structural analysis of coal and petroleum. *Applied Spectroscopy Reviews* 5, 211-264.

STANTON R. E. (1966) *Rapid Methods of Trace Analysis For Geochemical Application.* Edward Arnold. London.

STANTON R. E. (1970a) The colorimetric determination of molybdenum in soils, sediments and rock by zinc dithiol. *Proc. Austral. Inst. Min. Metall.* 235, 101-102.

STANTON R. E. (1970b) The colorimetric determination of tungsten in soils, sediments and rocks by zinc dithiol. *Proc. Austral. Inst. Min. Metall.* 236, 59-60.

STANTON R. E. (1971a) The colorimetric determination of thorium in soils, sediments, and rocks. *Proc. Austral. Inst. Min. Metall.* 239, 101-103.

STANTON R. E. (1971b) The determination of bismuth in soils, sediments and rocks. *Proc. Austral. Inst. Min. Metall.* 240, 113-114.

STANTON R. E. and HARDWICK A. J. (1971) The colorimetric determination of vanadium in soils, sediments, and rocks for use in geochemical exploration. *Proc. Austral. Inst. Min. Metall.* 240, 115-117.

STANTON R. E. and McDONALD A. J. (1966) The colorimetric determination of boron in soils, sediments, and rocks with methylene blue. *Analyst* 91, 775-778.

STANTON R. E., MOCKLER M. and NEWTON S. (1973) The colorimetric determination of molybdenum in organic-rich soil. *J. Geochem. Explor.* 2, 37-40.

STANTON R. L. (1972) *Ore Petrology.* McGraw-Hill.

STAPLIN F. L. (1969) Sedimentary organic matter, organic metamorphism, and oil and gas occurrence. *Bull. Canadian Petrol. Geol.* 17, 47-66.

STEVENSON F. J. and BUTLER J. H. A. (1969) Chemistry of humic acids and related pigments. *Organic Geochemistry* (editors G. Eglinton and M. T. J. Murphy), 534-557. Springer-Verlag.

STOLLERY G., BORCSIK M. and HOLLAND H. D. (1971) Chlorine in intrusives: A possible prospecting tool. *Econ. Geol.* 66, 361-367.

SWITZER P. (1967) Reconstructing patterns from sample data. *Ann. Math. Stat.* 38, 138-154.

TAUSON L. V. and KOZLOV V. D. (1973) Distribution functions and ratios of trace-element concentrations as estimators of the ore-bearing potential of granites. *Geochemical Exploration 1972*, pp. 37-44. IMM. London.

TAUSON L. V., OVCHINNIKOV L. N., POLIKARPOCHKIN V. V. and GRIGORYAN S. V. (1971). The use and development of geochemical prospecting methods for ore deposits in the U.S.S.R. *Geochemical Exploration.* CIM Spec. Vol. 11, 42-43.

TAYLOR R. M. (1968) The association of manganese and cobalt in soils – further observations. *J. Soil Sci.* 19, 77-80.

TAYLOR R. M. and MCKENZIE R. M. (1966) The association of trace elements with manganese minerals in Australian soils. *Austral. J. Soil Res.* 4, 29-39.

TAYLOR S. R. (1964) Abundance of chemical elements in the continental crust: a new table. *Geochim. Cosmochim. Acta* 28, 1273-1284.

TAYLOR S. R. (1966) The application of trace element data to problems in petrology. *Physics and Chem. of the Earth* 6, 135-213.

TAYLOR S. R. (1969) Trace element chemistry of andesites and associated calc-alkaline rocks. Proc. Andesite Conf., *Oregon Dept. Geol. Min. Resources Bull.* 65, 43-63.

TEDROW J. C. F. (1970) Soil investigations in Inglefield Land, Greenland. *Medd. om Gronland* 188 No. 3.

TENNANT C. B. and WHITE M. L. (1959) Study of the distribution of some geochemical data. *Econ. Geol.* 54, 1281-1290.

THEOBALD P. K., LAKIN H. W. and HAWKINS D. B. (1963) The precipitation of aluminum, iron and manganese at the junction of Deer Creek with the Snake River in Summit County, Colorado. *Geochim. Cosmochim. Acta* 27, 121-132.

THEOBALD P. K., OVERSTREET W. C. and THOMPSON C. E. (1967) Minor elements in alluvial magnetite from the Inner Piedmont Belt, North and South Carolina. *U.S. Geol. Surv. Prof. Paper* 554-A.

THODE H. G. and MONSTER J. (1970) Sulfur isotope abundances and genetic relations of oil accumulations in Middle East basin. *Amer. Assoc. Petrol. Geol. Bull.* 54, 627-637.

TIMPERLY M. H., BROOKS R. R. and PETERSON P. J. (1970) Prospecting for copper and nickel in New Zealand by statistical analysis of biogeochemical data. *Econ. Geol.* 65, 505-510.

TIMPERLY M. H., JONASSON I. R. and ALLAN R. J. (1973) Sub-aquatic organic gels: a medium for geochemical prospecting in the southern Canadian Shield. *Geol. Surv. Canada Paper* 73-1, Part A, 58-62.

TISSOT B. (1966) Problèmes géochimiques de la genèse et de la migration du pétrole. *Rev. l'Inst. Francais du Pétrole* 21, 1621-1671.

TISSOT B. (1969) Premières données sur les méchanismes et la cinétique de la formation du pétrole dans les sédiments. Simulation d'un schéma réactionnel sur ordinateur. *Rev. l'Inst. Francais du Pétrole* 24, 470-501.

TISSOT B., CALIFET-DEBYSER Y., DEROO G. and OUDIN J. L. (1971) Origin and evolution of hydrocarbons in early Toarcian shales, Paris basin, France. *Amer. Assoc. Petrol. Geol. Bull.* 55, 2177-2193.

TOLMAN C. F. (1937) *Ground Water.* McGraw-Hill.

TOOMS J. S. and WEBB J. S. (1961) Geochemical prospecting investigations in the Northern Rhodesian Copperbelt. *Econ. Geol.* 56, 815-846.

TÓTH J. (1962) A theory of groundwater motion in small drainage basins in central Alberta, Canada. *J. Geophys. Res.* 67, 4375-4387.

Tóth J. (1963) A theoretical analysis of groundwater flow in small drainage basins. *J. Geophys. Res.* 68, 4795-4812.

Trudinger P. A. (1971) Microbes, metals and minerals. *Minerals Sci. Engng.* 3 (No. 4), 13-25.

Trudinger P. A., Lambert I. B. and Skyring G. W. (1972) Biogenic sulfide ores: a feasibility study. *Econ. Geol.* 67, 1114-1127.

Tugarinov A. I. and Grigorian S. V. (editors) (1968) Geochemical and geophysical prospecting for deep-seated deposits (abstracts). *Intern. Geol. Congr. 23rd.*, Prague.

Turekian K. K. (1969) The oceans, streams, and atmosphere. In, *Handbook of Geochemistry* (editor K. H. Wedepohl), Chap. 10, pp. 297-323. Springer-Verlag.

Turekian K. K. and Wedepohl K. H. (1961) Distribution of the elements in some major units of the Earth's crust. *Geol. Soc. Amer. Bull.* 72, 641-664.

Turton A. G., Marsh N. L. McKenzie R. M. and Mulcahy M. J. (1962) The chemistry and mineralogy of lateritic soils in the south-west of Western Australia. *Commonwealth Sci. Ind. Research Organ. (CSIRO) Soil Public.* No. 20.

United Nations (1963) Proceedings of the seminar on geochemical prospecting methods and techniques. *Mineral Resources Develop. Series* No. 21.

United Nations (1970) Proceedings of the second seminar on geochemical prospecting methods and techniques. *Mineral Resources Develop. Series* No. 38.

United States Department of Agriculture (1957). *Soil.* The Yearbook of Agriculture.

United States Department of Agriculture (1960) *Soil Classification - a Comprehensive System - 7th Approximation.* (see also Supplement, March, 1967).

Usik L. (1969) Review of geochemical and geobotanical prospecting methods in peatland. *Geol. Surv. Canada Paper* 68-66.

Usik L. (1969a) Botanical investigations at three known mineral deposits. *Geol. Surv. Canada Paper* 68-71.

Valeton I. (1972) *Bauxites.* Elsevier.

Van Loon J. C., Kesler S. E. and Moore C. M. (1973) Analysis of water-extractable chloride in rocks by use of a selective ion electrode. *Geochemical Exploration 1972*, pp. 429-434. IMM. London.

Vassoyevich N. B., Korchagina Yu. I., Lopatin N. V. and Chernyshev V. V. (1970) Principal phase of oil formation. *Internat. Geol. Rev.* 12, 1276-1296.

Vinogradov A. P. (1959) *The Geochemistry of Rare and Dispersed Chemical Elements in Soils.* Translated by Consultants Bureau. New York.

Vinogradov A. P. (1963) Biogeochemical provinces and their role in organic evolution. *Geochemistry* (translation of *Geokhimiya*), 1963 (No. 3), 214-228.

Vistelius A. B. (1967) *Studies in Mathematical Geology.* Consultants Bureau. New York.

Vredenburgh L. D. and Cheney E. S. (1971) Sulfur and carbon isotopic investigation of petroleum, Wind River basin, Wyoming. *Amer. Assoc. Petrol. Geol. Bull.* 55, 1954-1975.

Wager L. R. and Brown G. M. (1960) Collection and preparation of material for analysis. In, *Methods In Geochemistry* (editors A. A. Smales and L. R. Wager), pp. 4-32. Interscience.

Wainerdi R. E. and Uken E. A. (editors) (1971) *Modern Methods of Geochemical Analysis.* Plenum.

Ward F. N., Lakin H. W., Canney F. C. and others (1963) Analytical methods used in geochemical exploration by the U.S. Geological Survey. *U.S. Geol. Surv. Bull.* 1152.

Ward F. N., Nakagawa H. M., Haims T. F. and VanSickle G. H. (1969) Atomic-absorption methods of analysis useful in geochemical exploration. *U.S. Geol. Surv. Bull.* 1289.

Warren H. V. (1972) Biogeochemistry in Canada. *Endeavor* 31, 46-49.

Warren H. V. and Delavault R. E. (1953) Geochemical prospecting finds widespread application in British Columbia. *Min. Eng.* 5, 980-981.

Warren H. V. and Delavault R. E. (1956) Pathfinding elements in geochemical prospecting (abstract). *Intern. Geol. Congr. 20th.*, Mexico City, p. 359.

WARREN H. V., DELAVAULT R. E. and BARAKSO J. (1968) The arsenic content of Douglas Fir as a guide to some gold, silver and base metal deposits. *CIM Bull.* 61, 860-868.

WARREN H. V., DELAVAULT R. E. and CROSS C. H. (1967) Some problems in applied geochemistry. *Geol. Surv. Canada Paper* 66-54, 253-264.

WARREN H. V., DELAVAULT R. E., FLETCHER K. and PETERSON G. R. (1971) The copper, zinc and lead content of trout livers as an aid in the search for favourable areas to prospect. *Geochemical Exploration.* CIM Spec. Vol. 11, 444-450.

WASHBURN A. L. (1969) Weathering, frost action, and patterned ground in the Mesters Vig district, northeast Greenland. *Medd. om Gronland* 176, No. 4.

WATLING R. J., DAVIS G. R. and MEYER W. T. (1973) Trace identification of mercury compounds as a guide to sulphide mineralization at Keel, Eire. *Geochemical Exploration 1972,* pp. 59-69. IMM. London.

WATSON J. P. (1972) Distribution of gold in termite mounds and soils at a gold anomaly in Kalahari sand. *Soil Sci.* 113, 317-321.

WATTS J. T., TOOMS J. S. and WEBB J. S. (1963) Geochemical dispersion of niobium from pyrochlore-bearing carbonatites in Northern Rhodesia. *Trans. IMM* 72, 729-747.

WEBB J. S. (1970) Some geological applications of regional geochemical reconnaissance. *Proc. Geol. Assoc.* 81 (Part 3), 585-594.

WEBB J. S. (1971) Research in applied geochemistry at Imperial College, London. *Geochemical Exploration.* CIM Spec. Vol. 11, 45.

WEBB J. S., NICHOL I. and THORNTON I. (1968) The broadening scope of regional geochemical reconnaissance. *Intern. Geol. Congr. 23rd.,* Prague. Vol. 6, 131-147.

WEDEPOHL K. H. (1969) *Handbook of Geochemistry.* Springer-Verlag. (Supplements have been issued in 1970, 1971 and 1972 in the form of loose-leaf volumes.)

WELTE D. H. (1970) Correlation problems among crude oils. *Advances in Organic Geochemistry, 1966* (editors G. D. Hobson and G. C. Speers), 111-127. Pergamon.

WELTE D. H. (1972) Petroleum exploration and organic geochemistry. *J. Geochem. Explor.* 1, 117-136.

WENNERVIRTA H., BOLVIKEN B. and NILSSON C. A. (1971) Summary of research and development in geochemical exploration in Scandinavian countries. *Geochemical Exploration.* CIM Spec. Vol. 11, 11-14.

WHITE D. E. (1973) Geochemistry applied to the discovery, evaluation and exploitation of geothermal energy resources. In, *Proceedings UN Symposium on the Development and Utilization of Geothermal Resources. Geothermics Spec. Issue 2.* Vol. 1, Part 2, (in press). Pisa, 1970.

WHITE D. E., HEM J. D. and WARING G. A. (1963) Chemical composition of subsurface waters. *U.S. Geol. Surv. Prof. Paper* 440-F

WHITE D. E., HINKLE M. E. and BARNES I. (1970) Mercury contents of natural thermal and mineral fluids. *U.S. Geol. Surv. Prof. Paper* 713, 25-28.

WHITTAKER E. J. W. and MUNTUS R. (1970) Ionic radii for use in geochemistry. *Geochim. Cosmochim. Acta* 34, 945-956.

WILLIAMS J. R. (1970) Groundwater in the permafrost regions of Alaska. *U.S. Geol. Surv. Prof. Paper* 696.

WILLIAMS R. E. (1970) Applicability of mathematical models of groundwater flow systems to hydrogeochemical exploration. *Idaho Bur. Mines and Geol., Pamphlet No. 144.*

WILSON H. N. (1966) *An Approach to Chemical Analysis.* Pergamon.

WINTERS J. C. and WILLIAMS J. A. (1969) Microbiological alteration of crude oil in the reservoir. *Symposium on Petroleum Transformations in Geologic Environments,* Div. Petrol. Chem., Amer. Chem. Soc., New York, Sept. 7-12, 1969, E22-E31.

WOLFE W. J. (1971) Biogeochemical prospecting in glaciated terrain of the Canadian Precambrian Shield. *CIM Bull.* 64 (Nov.), 72-80.

WOLLENBERG H. A. (1973) Fission-track radiography of uranium and thorium in radioactive minerals. *Geochemical Exploration 1972,* pp. 347-358. IMM. London.

WOLLENBERG H., KUNZENDORF H. and ROSE-HANSEN J. (1971) Isotope-excited X-ray fluorescence analyses for Nb, Zr and La + Ce on outcrops in the Ilímaussaq intrusion, South Greenland. *Econ. Geol.* 66, 1048-1060.

WOODSWORTH G. J. (1971) A geochemical drainage survey and its implications for metallogenesis, central Coast Mountains, British Columbia. *Econ. Geol.* 66, 1104-1120.

YORATH C. J., BALKWILL H. R. and KLASSEN R. W. (1969) Geology of the eastern part of the northern Interior and Arctic Coastal Plains, Northwest Territories. *Geol. Surv. Canada Paper* 68-27.

YOUDEN W. S. and MEHLICH (1937) Selection of efficient methods for soil sampling. *Boyce Thompson Inst. Contrib.* 9, 59-70.

ZAK I. (1961) Geochemical prospecting for oil in the Heletz-Brur area (First progress report). *Geol. Survey Israel, Rept. No. Chem.* /4/1961.

ZARRELLA W. M., MOUSSEAU R. J., COGGESHALL N. D., NORRIS M. S. and SCHRAYER G. J. (1967) Analysis and significance of hydrocarbons in subsurface brines. *Geochim. Cosmochim Acta* 31, 1155-1166.

ZEISSINK H. E. (1969) The mineralogy and geochemistry of a nickeliferous laterite profile (Greenvale, Queensland, Australia). *Mineral. Deposita* 4, 132-152.

ZEISSINK H. E. (1971) Trace element behavior in two nickeliferous laterite profiles. *Chem. Geol.* 7, 25-36.

Index

Benzene, 522, 523, 536
Beryllium, 14, 60
Beryllometer, 295
Biodegradation, 520, 522, 524, 535, 540
Biogenic dispersion patterns, 352
Biogeochemical classification, 387
    provinces, 58
Biogeochemistry, 5, 6, 23, 35, 383-408
    false anomaly, 194
Biological activity; effect on weathering and
    soils, 88, 109
Biological markers, 529-531
Biophile, 63
Biotite; decomposition, 76
    Sn content, 320
Biquinoline, 270
Biringuccio, 4
Birnessite, 135
Black shales, 60
    false anomalies, 207
Black Spruce; metal uptake, 389, 407, 448
Blank out; see masking
Blind River; leached uranium, 79
Bloom H., 7
Bloom test, 280
Bog; also see muskeg
Bog; definition, 441
    false anomalies, 205
    Fe-Mn deposits, 428, 449
    general, 31, 409
    lake, 442
    origin, 443
    profile, 446
    sampling, 229
    water chemistry, 442
Bolivia, 56
Boron; pathfinder, 54
    plant toxicity, 388
Bougainville, 34
    vegetation surveys, 399
Boulder-tracing, 30, 455
Bowen's reaction series, 41, 64
Boyle R. W., 7
Bradshaw P. M. D., 29
Braunite, 77
Break-in-slope anomaly, 13
British Columbia, 3, 6, 14, 30, 34, 210, 229
    bogs, 447
    false anomalies, 204
    Fe-Mn oxides, 137-138
    glacial, 462
    lake seepages, 463
    native Cu in lake sediments, 190
    organic matter adsorption, 139
    porphyry copper ages, 344
    regional survey, 374
    sediments in mountainous areas, 225
    vegetation surveys, 407

Broken Hill; Pb isotopes, 345
Brown forest soil, 108
Brown soils, 106
Bryophytes, 391
Buffer; extractions with, 248
Buried deposits, 34

Cadmium, 66, 69, 199
Calciomorphic soils, 108
Calculation of element content, 273
Caliche, 94, 103, 104, 106
Canada; also see individual provinces and
    territories
    Canada Dept. of Agriculture Soil
        classification, 92, 100
    drilling overburden, 230
    geochemical exploration, 409-467
    lakes, 181, 432-438
    soil map, 101
    ultramafics, 323
    vegetation surveys, 400-408
Canadian Shield, 14, 22, 32, 33, 57
    bedrock geochemistry, 321
    biogeochemistry surveys, 408
    hydrogeochemistry, 429
    lake sediments, 438
    permafrost, 411, 423
Cannon H., 7
Carbon; determined in lake sediments, 190
Carbon dioxide vapor surveys, 28
    rod furnace, 263
Carbonatite, 60, 202, 233, 358
Caribbean; porphyry copper, 343
Carlin, Nevada, 3
Cation exchange capacity, 96
Cebu, Philippines, 360
Central America; porphyry copper, 343
Cerargyrite, 82, 144
Cerussite, 78
Cesium, 59, 66
Chalcophile, 63
Chalcopyrite; trace elements in, 68, 340
Chattanooga shale, 57
Chemical weathering, 72-86
    mineral sequence, 72
    products, 81
Chemical zoning patterns, 325
Chemicals, safety, 249, 270
Chernozem soils, 106, 123
Chestnut soils, 106, 123
Chile, 55
Chinese, 5
Chloride in granites, 343
Chlorine, 48
    in micas, 338
Chlorosis, 397
Chromate fusibility, 252

ERRATA

p. 29 line 14. *for* over the Mt. Isa *read* in West Pilbara (but not over Mt. Isa)

p. 60 Table 2-4. Under bauxites, note that Nb is only concentrated in those bauxites derived from alkalic rocks.

p. 61 line 5. *for* never *read* rarely

p. 69 lines 25-26. In some places (e.g., Cornwall, Pennsylvania), Co and other elements in the pyrite structure have been recovered commercially.

p. 77 line 7. *for* birnessite *read* ramsdellite

p. 83 line 7 from bottom. *read* oxidation, leaching takes place near the surface, and the gossan is formed.

p. 84 to the caption of Fig. 3-1 *add* "The water table fluctuates in a gossan, as it does elsewhere, and is not meant to be permanently fixed as might be construed from the figure".

p. 92 last line in Table 3-4. The Canadian designation for bedrock is R (not D).

p. 118 line 8 from bottom. *for* Garnierite . . . in part) *read* Garnierite, a mixture of hydrous nickel silicates,

p. 128 first line of table heading. *After* dilute *add* (0.025 - 0.0025M)

p. 131 line 8 from bottom. *for* $FeO(OH)_3$ *read* $Fe(OH)_3$

p. 132 line 8 from bottom. *delete* molybdenite

p. 142 first line (heading of section). *Add* Chemical (so the title reads "Chemical Mobility In The Secondary Environment").

p. 146 line 12 from bottom. *for* ppb *read* ppm

p. 154 line 4. *after* district *add* where the soils are transported and the country is arid.

p. 183 to the caption of Fig. 3-31 *add* All values in ppb.

p. 193 line 16. *For* in the following chapter *read* in Chapter 6

p. 197 line 2 from bottom. *for* CaCl *read* $CaCl_2$

p. 204 line 14. *for* in the next chapter *read* on page 274

p. 213 line 18. *After* range *add* (not a value). The sentence now reads, "Background is defined as the normal range (not a value) of . . ."

p. 218 line 7. *After* is *add* sometimes

p. 223 lines 4-6 from bottom. Note that when wet sediments are sieved in the field, as they are collected, caution must be exercised to prevent loss of the metal-containing fine fraction.

p. 224 lines 7-9. Note that envelopes should contain water-proof glue, otherwise they fall apart.

p. 224 lines 12-13 from bottom. *delete* from as near the middle of the stream as possible. (This is because this is impractical in many cases, and in others the swift currents make it unlikely fine sediments will be found here).

p. 255 lines 5-9. In actual practice, crushing is rarely done on stream sediments or soils. For these sampling media, *for* crushing *read* disaggregate (when referring to lumps).

p. 256 lines 18-19. Similar to above, stream sediment and soil samples are not generally ground, especially if they have been sieved or screened.

p. 267 to caption for Fig. 6-8 *add* Based on design by Prof. H. Bloom.

p. 270 line 21. *After* gloves *add* (Some chemists feel that plastic gloves are likely to be dangerous because manipulative ability is reduced).

p. 281 line 20. *After* encountered. *add* Bloom (1955) and

p. 283 line 7 from bottom. *After* Ward et al. (1963). *add* The most important of these is the widely used thiocyanate method for molybdenum which is equally as effective as the dithiol procedure. It can also be adapted for the determination of Mo in water.

p. 420 line 2 of text. *for* B.H. *read* B.W.

p. 438 line 8. *delete* soils and

p. 460 line 10. *for* Skinner's (1973) *read* Skinner's (1972)

p. 466 line 6 from bottom. *after* overburden *add* (however, recent examples have established that soils developed on lodgment till do reflect mineralization.)

p. 481 lines 6-7 from bottom. *for* Kendall *read* Kendall and Stuart

p. 484 line 11 from bottom. *for* Lawley *read* Lawley and Maxwell

p. 488 line 12. *for* Van Andel *read* Imbrie and Van Andel

p. 489 line 3. *for* Rose (1973) *read* Rose (1972)

p. 489 line 4 from bottom. *after* z = *add* (

p. 492 lines 15-16 from bottom. *for* Baird et al. (1964) *read* Baird et al. (1967).

p. 492 line 4 from bottom. *for* Switzer (1968) *read* Switzer (1967)

p. 494 line 8. *for* Miesch and Conner (1967) *read* Miesch and Conner (1964).

p. 494 line 13. *for* Eisenhart (1937) *read* Eisenhart (1947)

p. 494 lines 18-19. *for* Garrett (1969, 1972) *read* (Garrett 1969, 1973)

p. 494 line 30. *for* Michie (1972) *read* Michie (1973)

p. 494 line 31. *for* Howarth and Lowenstein (1972) *read* Howarth and Lowenstein (1971)

p. 502 line 5. *for* Howarth (1972) *read* Lowenstein and Howarth (1973)

p. 519 line 18. *for* 0.8 *read* 0.08

p. 531 Fig. 13-13. The sterane molecule has three benzene rings instead of four as shown.

p. 549 Step 3. *for* "Simmer . . . hour" *read* Place in boiling water for ½ hour.

p. 553 bottom half of page between steps 2 and 3. *add* 2A. Slosh for ¼ minute.

p. 554 a. The cold extractable test as described on this page is a *modified* Bloom (1955) test.

　　b. Although this modified test works satisfactorily as described, if one wishes to make it identical to that described by Bloom (1955), in step 6 *delete* the sentence in parentheses, "(Do not shake . . . from the sample)" and *replace* it with, "Shake another 3 seconds".

p. 555 line 4 (Step 2). *for* ml *read* drops (so it now reads, "Add 5 drops of buffer solution").

p. 563 line 4. *for* interval *read* integer

p. 577 Burgess (1974) reference. *for* in press *read* Geol. Soc. Amer. Spec. Paper 153, 19-30.

p. 580 Deroo et al. (1974) reference. *for* Bull. Canadian Petroleum Geol. (in press) *read* In, *Oil Sands Fuel of the Future* (Ed. L. V. Hills), Canad. Soc. Petrol. Geol. Mem. 3, 148-167.

p. 586 Hitchon and Horn (1974) rerference. *for* (in press) *read* 464-473.

p. 598 Turekian and Wedepohl (1961) reference. *for* 641-664 *read* 175-192

p. 600 Add the following references:

KAISER H. F. (1962) Formulas for component scores. *Psychometrika* 27, 83-87.

LAWLEY D. N. and Maxwell A. E. (1971) *Factor Analysis As A Statistical Method*. Elsevier.

PAUL M. K. (1967) A method of computing residual anomalies from Bouguer gravity map by applying relaxation technique. *Geophysics* 32, 708-719.

Introduction to
Exploration Geochemistry
Second Edition

Introduction to
Exploration Geochemistry
Second Edition

The 1980 Supplement

A. A. Levinson
DEPARTMENT OF GEOLOGY AND GEOPHYSICS
UNIVERSITY OF CALGARY
CALGARY, ALBERTA, CANADA

Applied Publishing Ltd.
Wilmette

Photocomposed in Canada by Alcraft Printing Co. Limited,
Calgary, Alberta

Printed in the United States of America by
The University of Chicago Printing Department,
Chicago, Illinois.

Library of Congress Catalog Card Number: 79-54677

ISBN 0-915834-04-9

Preface
Second Edition

Six years have passed since the first edition was written and, in the interim, sufficient advances in exploration geochemistry have been published to warrant a second edition. In addition, numerous reviews of the first edition have appeared with suggestions for improvement and expansion.

When writing a second edition of any book, an author is always faced with the problems of how much and what material should be re-written, deleted and added. Most second editions of geological texts, in my experience, are only slightly modified versions of the first edition, although third and subsequent editions are often entirely new. Because it is only six years since this book first appeared, an unusual approach has been followed; it is to retain the first edition in its entirety without change and to add an additional several hundred pages as an expanded addendum entitled *"The 1980 Supplement"*. The first edition is used as the base upon which to add new advances in the field, and to discuss some topics which were not previously included. Even though the method will not appeal to all, there are considerable advantages, not the least of which will be the savings to purchasers (and particularly students). I am convinced, therefore, that the approach is best in this particular case.

A serious attempt has been made to comply with the numerous excellent suggestions contained in the reviews of the first edition. However, in some cases, this is impossible for the reviewers disagree among themselves on specific points. Whereas in the first edition North American and Australian examples were used wherever possible, here no such limitation is imposed. Further, every effort has been made to use examples from arid environments, particularly those in tropical latitudes, and uranium has been stressed in view of the great interest in exploration for this element at present. Also, when possible, reference has been made to the advances in the Soviet Union, particularly where their concepts and methods have potential application to problems in the West.

The format and topics discussed are all based upon those in the first edition, a consequence of the decision to use the addendum method for a second edition. Additional chapters are numbered 6A, for example ("A" signifies Addendum), and figures and tables are numbered as 6A - 1. All topics covered by second-order headings (such as "Types Of Surveys", p. 9), and by third-order headings (such as Soil Surveys, p. 11) in the first edition are considered. If such topics do not warrant additional discussion, such will be noted (for example, "Rock Surveys. No additional comments"; see p. 620). On the other hand, topics new to this second edition are so indicated by the statement, "A new topic", for example, "Lithogeochemistry In The Soviet Union (A new topic)", on p. 761.

Calgary, Alberta, Canada   *A. A. Levinson*
August, 1979

v

Acknowledgements

I would like to express my sincere thanks to a small group of friends and colleagues whose encouragement, assistance and advice have made this second edition possible. Most visible among them are Dr. R. B. McCammon and Dr. Brian Hitchon who wrote Chapter 12A and Chapter 13A, respectively. As in the first edition, both Dr. McCammon and Dr. Hitchon enjoyed complete freedom in the selection of their material and method of presentation.

Behind the scenes, so to speak, are Dr. K. Bloomfield and Dr. J. D. Appleton of the Institute of Geological Sciences (Overseas Division), and Dr. P. M. D. Bradshaw, of Placer Development Ltd. These gentlemen critcally reviewed the first eleven chapters in their entirety. Their recommendations on matters of content, augmented by their years of practical experience in the field of exploration geochemistry, as well as their suggestions on the manner of presentation and exposition, have been invaluable to me. Obviously, I am deeply indebted to all of them.

Others who have critically read various chapters, or sections thereof, include: Professor P. Bayliss (Chapter 3A), Professor C. J. Bland (Chapter 3A), Dr. R. R. Brooks (Chapter 10A), Dr. J. Plant (Chapter 6A), Professor S. B. Romberger (Chapter 3A), and Dr. W. W. Shilts (Chapter 11A). Their valuable contributions to this work is greatly appreciated.

Dr. McCammon wishes to acknowledge the assistance of Dr. R. G. Garrett for critically reading Chapter 12A. Dr. Hitchon's chapter was reviewed by Dr. J. A. Boon, Mr. C. W. D. Milner and Dr. E. I. Wallick to whom he expresses his thanks.

Contents

4A / Some Basic Principles   699

5A / Field Methods   711

6A / Analytical Methods   721

7A / Primary Distribution (Selected Topics)   747

8A / Secondary Dispersion (Selected Topics)   779

9A / Regional and Detailed Surveys   787

10A / Vegetation Surveys   801

11A / Geochemical Exploration In Canada   815

$1A\Big/$ *Introduction*

INTRODUCTION   (pp. 1-4)

The past several years have seen exploration geochemistry reach its highest degree of acceptance and respectability. Exploration geochemists, *sensu stricto,* sit in the highest councils of geological societies and on policy making boards, and at the 25th International Geological Congress held in Sydney, Australia in 1976, the "Geochemistry" section was divided into two parts, one theoretical and the other entitled "Exploration Geochemistry." Achievement of this status has not been easy. The battle has been an up-hill one, and still continues, for as Webb (1973) recently pointed out with respect to the acceptance of exploration geochemistry in the academic world (p. 24), "In the main, there has been a marked (though fortunately waning) reluctance in university circles to recognize the legitimate scientific basis of applied geochemistry." The reluctance to which Professor Webb refers continues to wane, albeit slowly, because academics finally are awakening to the scientific status of applied geochemistry which, for the most part at present, is concerned with exploration. The matter is succinctly phrased by Webb (1973, p. 24) who said, "In this connexion I cannot overstress that exploration geochemistry is not merely a technique; it is a demanding, highly challenging, branch of applied science in the strictest sense; it is a practical extension of fundamental geochemistry disciplined by the need to develop cost-effective methods of accumulating useful data followed by the application of interpretational criteria based on sound fundamental concepts; it is, in short, a body of knowledge in its own right."

Since exploration geochemistry is "a body of knowledge in its own right", it is not surprising that the majority of geochemical surveys are conducted today under expert guidance and most of the larger companies, governmental agencies, and the United Nations employ geochemical specialists in their exploration programs. A review of the literature originating from these specialists reveals that all their contributions make a major point of the vital necessity of the thorough understanding of the recent geological history (e.g., glacial history, weathering processes) of an area being prospected *prior* to surveys being undertaken, and certainly before reliable interpretation of the data can be attempted. The constant repetition of this apparently self-evident prerequisite over the years has now finally convinced everyone that geochemical techniques must be adapted to suit the particular environment in which exploration is being conducted.

As any body of knowledge, such as exploration geochemistry, matures, there is a tendency for the scientist concerned with this knowledge to see applications of his expertise to other disciplines. Some exploration geochemists are now applying their expertise to a wide range of environmental and related problems. By virtue of the job he is trained to do, the exploration geochemist has a unique breadth of experience in developing and using cost-effective methods for mapping the distribution of elements in many types of sampling media. The techniques are essentially identical to regional and detailed geochemical exploration methods. Webb (1975) has noted that this expertise can be used to advantage in the environmental field including such particular specialties as agriculture, fisheries, human and animal health and nutrition, pollution, and so forth. These subjects are rightly in the field of applied geochemistry but as this book is concerned only with the application of geochemistry to exploration for mineral deposits and hydrocarbon accumulations, they are not considered here.

Remote sensing received only scant mention in the first edition (p. 4 and p. 399) owing to the fact that the relationships between the complex techniques used and exploration geochemistry were only beginning to merge at that time. The interim has seen an explosion of remote sensing studies applied to many fields of endeavor including mineral exploration. This is particularly the result of data obtained from the Landsat-1 (previously called ERTS-1) satellite launched in July, 1972; data from this satellite now cover almost the entire land area of the Earth. Landsat-2, launched in January, 1975, and others to follow, promise further aid in mineral exploration although it should be noted that at the present stage of development, remote sensing is strongly dependent upon supportive field studies, the so-called "ground truth" verifications.

Numerous reviews of remote sensing methods and their results are now available in the recent literature of which that by Viljoen et al. (1975) is particularly informative for the exploration geochemist. This review is extremely well-documented, beautifully illustrated, and applicable to the subject matter of this book. Further, the authors take what might be called a conservative position with respect to the present ability of remote sensing techniques to locate mineral deposits (in conformity with the opinions of this writer). Viljoen et al. (1975, p. 150) state, "Exaggerated claims have been made about the ability of ERTS to locate new mineral deposits but to date there have been no verified reports of ore bodies discovered exclusively from the use of ERTS imagery."

Based on the conclusions of Viljoen et al. (1975) and other conservative sources, the practical uses of remote sensing (particularly Landsat) in mineral exploration can now be summarized. Thus, remote sensing:
1. Suggests areas of regional interest and sometimes focuses on specific structures and surface anomalies as possible exploration targets. These would include: (a) faults and fracture intersections, and other structures, which might provide likely sites for mineral localization; (b) gossans, zones of oxidation, and zones of hydrothermal alteration if sufficiently large and if they extend to the surface; and (c) identification of lithologies favorable as host or source rocks.
2. Provides important information necessary for planning practical aspects of an

exploration program, including the best exploration techniques to be used, thus reducing the time necessary for decision making.

3. Provides better geologic maps (occasionally with unique information) than those available in many parts of the world.

4. Assists geologists (including geochemists) in gaining a better understanding of regional geological factors, and a broader geological knowledge of the area, thus possibly inspiring more imaginative overviews.

5. Saves time and money.

As experience with the interpretation of remote sensing data grows, it may be possible to extend the above list of uses of these techniques, but for the present the time-tested sequence used in modern mineral exploration remains: (1) data compilation for the particular project or objective; (2) selection of target area; (3) basic mapping; (4) geochemical prospecting followed by selective geophysical work; and (5) drilling to determine the amount and extent of a deposit. Remote sensing contributes primarily toward phase (2) above, and to a lesser extent, to phase (3). It clearly has much to offer as a reconnaissance technique to be used as a first step in virgin territory enabling very large areas to be scanned at low cost in order to identify promising potential. A case in point would be the use of the Landsat imagery of eastern Bolivia where circular structures and other features were identified and these turned out to be a new alkaline intrusive province. The province has Nb-La rich residual soils developed over possible carbonatites, and also potential tin mineralization similar to the Rondonia area of western Brazil (Dr. J. D. Appleton, personal communication).

Specific examples of how remote sensing and geochemical and other types of surveys are integrated in the whole exploration program are now available from extremely different environments. For example, in northern Finland, Talvitie and Paarma (1973) showed that eskers tend to be concentrated along ancient fractures and that these fractures may be mineralized. Once these features are identified, in part by remote sensing methods, the geochemical effort can be properly directed. In the remote, heavily forested areas of the Amazon Basin in Brazil, radar imagery was used to record details of geology, soil, hydrology and vegetation conditions. Following this, technical personnel were moved into the area to collect geochemical samples (rock, stream sediments, soil, water and vegetation) at carefully selected sites in the project area (Anonymous, 1975). In both examples above, remote sensing was used for site selection during the early reconnaissance stage.

So far, the term remote sensing has been used for all types of data (usually photographs and various types of imagery) obtained from satellites and high flying aircraft; it has not included vapor surveys or the measurement of natural gamma radiation obtained near the surface or in the atmosphere by conventional aircraft. It is necessary, however, to be aware of the importance of the latter types of surveys, not the least of which are those which complement or integrate vegetation surveys. A few examples have now been published (properly classified as orientation surveys) which attempt to recognize and interpret the complex spectral signatures of vegetation which are obtained from aircraft and/or satellite imagery with the ultimate objective of using the results for exploration purposes. A case in point is the very detailed study by Cole et al. (1974). They described their work in western Queensland, Australia, in which spectral signatures of vegetation communities recognized on multi-spectral photography obtained from aircraft, were evaluated with reference to bedrock geology and mineralization, soil geochemistry, and geobotany, all determined on the ground. These, in turn, were also compared to Landsat imagery. As this was an orientation survey in an area with known mineralization, no discoveries were attempted or claimed, but their results did show that vegetation-oriented remote sensing techniques certainly assist in mineral exploration within the framework outlined above for satellites. It is almost inevitable that, sooner or later, remote sensing methods based primarily on vegetation will be given major credit for a discovery. However, it should be borne in mind that tonal vegetational changes are the basis of most orthodox photogeological interpretations. Also, the *absence* of vegetation caused naturally by heavy metal poisoning of soil and vegetation can sometimes be detected by remote sensing methods (see Figs. 10A-2 and 10A-3). Bolviken et al. (1977) showed that an occurrence of natural copper poisoning in northern Norway could be detected by Landsat imagery (resolution 0.45 hectare).

Noteworthy is the publication by Williams and Carter (1976) which, perhaps more than any other to date, points out the considerable advantage of simple pictorial presentation of Landsdat imagery under a large variety of conditions and for a large variety of uses, including mineral exploration. It is one of the most comprehensive, and certainly the best illustrated, publication to have appeared so far on the practical application of ERTS (Landsat) imagery. Of the numerous other publications concerned with remote sensing technology, those of Woll and Fischer (1977) and Smith (1977) deserve mention. The former is also concerned with the use of Landsat satellite data for mineral and petroleum exploration, whereas the latter stresses the cost-benefit aspect of remote sensing in mineral exploraton.

Finally, it is worth stressing that remote sensing techniques should be used in combination with aeromagnetics, gravity measurements, geochemistry, and other accepted geological and geophysical techniques appropriate to the area of interest, for maximum benefit.

HISTORY   (pp. 4-9)

Hawkes (1976a) has documented some historical aspects of exploration geochemistry, with particular emphasis on its development in North America, in a very enjoyable narrative-style presentation. It may be surprising to the present generation of mineral explorationists that most of the apparently very obvious techniques of exploration geochemistry are very new and, in fact, they were developed and proven in the lifetime of many still active explorers. Hawkes (1976a) noted that the very earliest publication on exploration geochemistry, as we know it today, was in 1935. The first important discoveries based directly on these methods, at least in the

Western World, resulted from surveys conducted in 1950. However, Bostrom (1978) stated that geochemical methods had previously been used "... at the turn of the century in Northern Sweden and possibly even earlier elsewhere, and incidentally, led to the discovery of Sandlidberget in 1906".

TYPES OF SURVEYS   (pp. 9-31)

The discussions presented under this heading in the first edition are of a very general nature and need no additional comment except for "Vapor Surveys" and "Other Types of Surveys". A new heading "Airborne Particulate Surveys", has also been added.

It is, perhaps, advisable to point out that in Continental Europe (e.g., France), the terms *strategic* and *tactical* are used as approximate synonyms for the terms *reconnaissance* and *detailed* (or *follow-up*) commonly used in English-speaking countries. The terms strategic and tactical have also been used on occasion by English-speaking scientists concerned with landscape geochemistry (pp. 202-203).

No suitable additional figures are available on the types of geochemical samples analyzed for exploration purposes and, therefore, the data in Table 1-1 (p. 9) cannot be revised. However, certain material, particularly water, is receiving very greatly increased attention primarily because of the interest in uranium exploration. Uranium, its daughter product radon, and certain pathfinders (particularly fluorine) are readily soluble and mobile in water and form extensive hydrogeochemical dispersion halos (see pp. 22, 156, 215, 299, 430). The use of water as a sampling medium will certainly increase as the Energy Research and Development Agency (ERDA) in the United States has initiated a nationwide hydrogeochemical reconnaissance program, to be completed in 1982, the objective of which is to produce "uranium favorability maps" incorporating geochemical, geophysical, remote sensing and geological data. A similar program was initiated in Canada (the Federal-Provincial Uranium Reconnaissance Program).

Although hydrogeochemical methods are being used extensively at this time for uranium exploration, it is interesting to note that historically these techniques have not been particularly successful in locating mineral deposits. Boyle (1979), quoting Soviet and other sources, noted that nearly all geochemical successes are attributable to surface lithogeochemical (soils, stream sediments, heavy minerals, and rocks) surveys and very few to hydrogeochemical, biogeochemical and atmogeochemical surveys.

Soil Surveys   (pp. 11-14)

No additional comments.

Rock Surveys   (pp. 14-16)

No additional comments.

Stream Sediment Surveys   (pp. 16-19)

No additional comments.

Water Surveys   (pp. 20-23)

No additional comments.

Vegetation Surveys   (pp. 23-27)

No additional comments.

Vapor Surveys   (pp. 27-31)

As indicated earlier, vapor surveys are among the most recent developments in exploration geochemistry. In view of this, it is not surprising that the exact mechanisms by which elements are transported to, and dispersed in, the atmosphere are not yet thoroughly known. Curtin et al. (1974) reviewed what is known about the migration of elements from vegetation and noted that the transport of elements in plant exudates had been demonstrated as early as 1944; in other words, elements can migrate into the atmosphere during transpiration. Apparently the first attempt to verify or use this fact for exploration purposes was made by Kolotov et al. (1965) in the Maritime Provinces of the Soviet Union. Specifically, they collected oak leaves along traverses over several base metal deposits, after which the leaves were rinsed with demineralized water and the water analyzed. The rinse water from the above leaves contained from 2 to 100 times the metal content of water used to rinse similar leaves collected several hundred meters away from the deposits. They attributed the metal content of the rinse water to airborne metals washed from the surface of the leaves; this was corroborated by the analysis of rainwater and snow which also showed higher metal contents over the ore deposits by comparison with that falling beyond the boundaries of the deposits. From this study, Kolotov et al. (1965) postulated that metals may migrate into the atmosphere and form "atmospheric dispersion aureoles" (although the exact form of the metals in the aureoles was undefined.)

In what was essentially an extension of the above study, Curtin et al. (1974) collected exudates from various types of trees in Colorado and Idaho, the purpose of which was to determine whether metals and other elements are actually contained in the volatile exudates of forest vegetation. The exudates were collected in plastic bags

(Fig. 1A-1) and after appropriate preparation (i.e., various filtration processes), their ashes were analyzed. At least 27 elements were detected in the ashed residues, many

Fig. 1A-1. Plastic bags attached to conifer trees to collect exudates for geochemical analysis. Alpine environment west of Denver, Colorado. From Chaffee (1975) but based on work described by Curtin et al. (1974).

of which are of interest to exploration geochemists (e.g., Li, V, Cr, Co, Ni, Cu, Zn, As, Mo, Ag, Pb, Bi, Cd, Sn, Sb, and Ba). They concluded that the presence of these elements suggests that true volatile exudates (as terpene-derived, organometallic aerosols) are a medium for the transport of elements in the biogeochemical cycle and that their collection and analysis may be a useful geochemical tool, for example, in areas with transported overburden.

Airborne Particulate Surveys
(A new topic)

The collection and analysis of airborne particulates for mineral exploration was pioneered by Weiss and first reported in the literature in 1971 (Weiss, 1971). His collection method depended on towing nylon thread, held around a frame, below an aircraft. The particles adhered to the thread by electrostatic charge. The thread and frame were recovered, and the particles physically removed and analyzed by conventional means. The exposure time of each frame was a minimum of one minute and so anomalies were normally in the order of several miles long, their width depending on the flight spacing.

More recently, Barringer Research has developed the AIRTRACE[R] system (Barringer, 1977) for more rapid collection and analysis of airborne particulates. This system employs large aerodyne collectors on either side of a helicopter for sample collection (Fig. 1A-2). These devices also have the advantage that they reject the fine

Fig. 1A-2. AIRTRACE — a patented airborne exploration system. Aircraft with collection system on its wings flying in the U.S.A. Courtesy of Barringer Research Ltd., Rexdale, Ontario, Canada.

particles and, as presently employed, collect only particles greater than 20 microns whose dwell time in the atmosphere is confined to a few minutes or less. This reduces the effect of wind smearing. The collected particles are impinged on a special adhesive tape where they are stored for later analysis. Analysis, for 20 elements, is achieved by the LASERTRACE[R] system. This system vaporizes the particles off the tape and the vapor is passed into a direct reading plasma emission spectrometer system (see p. 740). Trace elements detected down to background concentrations include Cu, Pb, Zn, Ni, Co, Cr, V, Cd, Mn, Sr, Be and P and the full range of major elements are also determined (i.e., Si, Al, K, Ca, Na, Fe, Mg and Ti). Uranium can only be analyzed using a special modification to the system. To date, this system has been employed for reconnaissance and semi-reconnaissance coverage of areas which are several tens of square miles or greater. A summary of recent use of this system is given by Barringer (1979). One distinct advantage of this system is that it is flown together with magnetics and gamma-ray spectrometry allowing a more integrated approach to airborne exploration.

A new system called SURTRACE[R] has been adapted for the collection of surface microlayer material, directly from the ground or from vegetation. The surface microlayer is the top monolayer of particulate matter on the Earth's surface, or particulate matter on vegetation, and is the material which is available for movement into the atmosphere. However, the particle size of this material is such that its movement is typically very limited, making the SURTRACE system better suited for detailed surveys than AIRTRACE. The SURTRACE system can be operated from a helicopter together with magnetics and gamma-ray, or on the ground using a backpack system. Analysis is for the same suite of elements as AIRTRACE. To date, this system has only been used commercially in arid and semi-arid areas. Some uses of this system and the surface microlayer are given by Barringer and Bradshaw (1979), and Bradshaw and Thomson (1979).

From the above discussion it would appear that for an ore deposit to be detected by means of a particulate survey it would have to outcrop on the surface, or it would have to have a soil anomaly with surface expression. (An exception might be volatile exudates originating with vegetation, and representing buried mineralization, which is adsorbed on to particulate matter and carried into the atmosphere.)

Finally, from the discussion of vapor surveys and airborne particulate surveys presented above, it should be obvious that surveys involving the collection, analysis and interpretation of gases or particles in the atmosphere ("atmogeochemical surveys") require special expertise and equipment. Thus, on a practical basis, these types of surveys are unavailable to most geochemists.

Other Types of Surveys (p. 31)

Among the several types of surveys mentioned on p. 31 at least one, the collection and analysis of heavy minerals (including "pan concentrates"), has received increased attention in recent years and these are often considered as useful sampling media. The remainder, for the most part, continue to be used for special purposes on a limited basis. Some, such as termite mounds are, however, believed to have significant potential particularly in tropical areas with overburden. In southern Africa (e.g., Botswana) termite mounds have been used successfully in the search for diamond pipes by means of identifying the heavy minerals, such a pyrope and chrome diopside, brought to the surface from beneath as much as 40 m of Kalahari sand. This technique may also have potential in the search for Au, W, Sn and other elements which form resistate minerals, as well as Cu in favorable situations (d'Orey, 1975).

Other materials which have been used, or whose use has been suggested, are listed below:
1. Human hair; based on the observation that certain elements, which may reflect the chemistry of soils and rocks in an agrarian society, may be concentrated in hair (Venkatavaradan, 1973).

2. Seaweed, shellfish and other marine organisms; based on the study of Bollingberg (1975) who found that the lead and zinc concentrations within seaweed and mussels of a West Greenland fjord increased toward an adjacent sphalerite-galena-pyrite deposit.

3. Ice in the form of naleds; these are ice sheets which form during the cold season in areas of discontinuous permafrost (Shvartsev, 1972).

4. Chaffee (1975) has suggested that the following may also be useful sampling media in appropriate situations: desert varnish, geodes, submarine springs, guano, microorganisms, animal tissue and waste products, and caliche (the latter is presently used routinely in certain parts of southern Africa and Australia in connection with uranium exploration). Hoffman (1977) has shown the value of sampling the fine fraction of talus in mountainous regions under certain conditions; talus had previously been used in an arid area in Chile (Maranzana, 1972).

5. Other biological materials which have been known to have been sampled for exploration purposes include fossil bone, the feathers of certain birds, trout livers, and wool. Wogman et al. (1976) have suggested that radium accumulations in animal thyroid glands may be a possible method for uranium and thorium prospecting. Czehura (1977) reported that color differences in a green-rimmed lichen (*Lecanora cascadensis* Magn.) can be used to discriminate between copper-bearing and barren rocks in a California locality.

6. Studies have been made to locate areas where trace element-associated diseases of animals and man exist as these diseases may aid in identifying regions that are enriched or depleted in various elements. (Strictly speaking, surveys in this category, analogous to geobotanical surveys, are not geochemical because samples are not chemically analyzed). This concept has been employed during the successful base metal exploration in Ireland; however, that success is certainly not universal as the work of Andrianova (1971) showed. She attempted to characterize a biogeochemical province (see p. 58) on the basis of "white muscle disease" which occurs in farm animals mainly living on floodplains in certain marshy valleys in the permafrost region of central Yakutia, USSR. Analysis of 11 elements (B, Cu, Mo, Sr, Ba, Mn, Fe, Co, Ni, Cr and V) failed to find any significant relationship, either excess or deficiency, between any of the elements in the soils, rocks or plants, and the disease (even though the B and Sr contents were elevated, and Mn, Cu, Mo, Fe and Co contents of plants were low). And, finally, it has been reported that certain areas of the Karoo District, South Africa, with a higher than normal incidence of respiratory cancer in humans, have been selected by at least one company as the starting point for uranium exploration; presumably the respiratory cancer would be caused by effects from the decay of radon and its daughter products.

Most of the above types of surveys are untried, and all must be considered unproved, even though a few have been used successfully in very specific and limited circumstances.

PRESENT STATUS   (pp. 31-33)

An up-date of the statistics on the use of geochemical methods, and the number of samples analyzed, is possible based on figures published by Webb (1973) and Beus and Grigorian (1977).

Webb (1973) stated that a survey (about 1973) showed that 90% of 150 companies surveyed in 11 countries made frequent use of geochemical methods, and that they spent about 12% of their exploration budgets for this purpose. He estimated that about 8,000,000 samples were collected annually in the Western World and 10 million per annum in the U.S.S.R. The figure for the Western World may be low because Coetzee (1974) estimated the sample volume in South Africa alone at 3 to 5 million (from which 12 to 15 million determinations are obtained). Based on information in Beus and Grigorian (1977, p. 169) it also appears that all previous estimates of the number of samples collected in the Soviet Union are low because in Kazakhstan alone more than 60 million geochemical samples were collected between 1947 and 1969 (and 20 new deposits were found).

GEOCHEMICAL SUCCESSES   (pp. 33-34)

It is becoming increasingly difficult to identify mineral deposits in which geochemistry alone is credited with a discovery although there are many published reports in which geochemistry is credited with being "part of the team". This is, indeed, a healthy situation and it tends to emphasize what has been stated earlier (p. 3); "... it must be stressed that geochemistry is just one method of exploration which should be used in conjunction with as many other techniques as possible." And (p. 3), "It is unusual for geochemistry to be the only technique used in an exploration project, and at some stage geophysics and geology are generally combined with geochemistry."

In a recent review, Boyle (1979) listed numerous examples in which geochemical methods have resulted in the discovery of a mineral deposit. These include (and many others could be cited): Carlin-type gold deposits, Nevada; the auriferous Muruntau deposit in Uzbek, U.S.S.R.; the Beltana and Aroona willemite deposits, South Australia; the McArthur River and Lady Loretta lead-zinc deposits, Australia; the Woodlawn copper-lead-zinc deposit, Australia; the Husky lead-zinc-silver deposit, Keno Hill, Yukon; the Island Copper porphyry deposit, British Columbia; and the Sam Goosely copper-silver-molybdenum deposit, British Columbia.

Boyle (1979) has also reviewed estimates of the number of anomalies found in the U.S.S.R. and in other countries, and the number of mineable deposits which have resulted from them. In the U.S.S.R., according to a source used by Boyle (1979), in the 20 year period before 1974 about 80,000 surface (metallic) halos and anomalies were identified by geochemical methods which resulted in approximately 220 deposits of numerous elements, and some 900 prospects. The percentage ratio of deposits to anomalies is, therefore, 0.3. This is not a very impressive figure. He estimated that about 100,000 geochemical anomalies have been located in Africa, Europe, North America, South America and Australia in the last 10 years by geochemical methods, and about 150 orebodies have been found associated with these anomalies, giving about the same order of magnitude for the ratio. However, the ratio may become larger as more of the prospects and anomalies are tested.

RESEARCH IN THE IMMEDIATE FUTURE   (pp. 34-35)

The six current trends in exploration geochemistry listed on pp. 34-35 are still pertinent for research in industry, government and universities. To those the following may be added:

1. Research into the use of different partial and total acid digestions, thermal release curves for vapors (see Fig. 7-8), and specific digestions for heavy minerals, in order to determine the source of elements in geochemical samples. Such studies will enable geochemists to distinguish between Ni or Ag, for example, in silicates and sulfides. They will also enable mechanical, hydromorphic, and vapor anomalies to be characterized. In effect, this area of research is an extension of the partial *vs* total extraction (analysis) techniques discussed on pp. 245-252 and 723-725.

2. Analytical advances are needed for certain elements in certain situations. These include: (a) lower detection limits for Au, Sn, W and Pt; (b) less expensive isotopic analyses for Pb, S and O (pp. 345-346), as well as for isotopes in the uranium decay series (e.g., $^{234}U/^{238}U$ activity ratios, ^{226}Ra); (c) less expensive methods for Se, Te, Cl_2, I_2 and certain other elements and vapors; (d) reliable and inexpensive field methods for certain gases, such as He, I_2, SO_2 and H_2S.

3. Research on primary geochemical halos in order to be able to recognize and locate geochemical zoning, channel ways, and leakage halos associated with buried deposits. Beus and Grigorian (1977) have summarized the Soviet concepts, techniques, and experiences in this area but they have yet to be applied in a similar manner in the Western World.

4. Continued research into methods of discovering ore bodies hidden by rock, as opposed to those concealed by overburden. An example would be an ore body covered by a volcanic flow such as that illustrated in Fig. 9-1 (C), p. 369. This is particularly a problem in well-known ore districts where the potential is high.

5. The delineation of geochemical provinces and their relationship to metallogenic provinces. Boyle (1979) considers this to be the most important problem for exploration geochemists.

6. Study of exploration techniques to be used in arid environments, particularly deserts, has been sorely neglected. As more interest is being shown in such areas

throughout the world, geochemical methods should be perfected for these environments.

7. Methods for the interpretation of geochemical data must be improved as analytical productivity and precision continues to outpace the ability to evaluate the data. Anomalies are often extremely subtle and, therefore, difficult to recognize. It appears that the relative costs of exploration, including the interpretation phases, should be increased before costly drilling is initiated.

Boyle (1979) has discussed numerous other desirable research projects which should receive immediate attention.

LITERATURE (pp. 35-37)

The literature of exploration geochemistry continues to expand. Additions to the literature presented on pp. 35-37 may be summarized as follows (numbers correspond with categories on pp. 36-37):

1. The book by Beus and Grigorian (1977) is the first in-depth discussion of Soviet methods and concepts to appear in English in about 20 years. It stresses rock geochemistry although many other topics are discussed. Rose et al. (1979) have prepared a second edition of the well-known, and highly respected, book on exploration geochemistry by Hawkes and Webb (1962).

4. Hood (1979) edited "Geophysics and Geochemistry in the Search for Metallic Ores" which is the Proceedings of an international symposium held in Ottawa, Canada in October, 1977. The symposium contains nine "State of the Art" reviews on topics of direct interest to exploration geochemists. It also has about 15 case histories, many of which integrate geochemical and geophysical results.

5. Jones (1975) and IMM (1977) have edited extremely important volumes concerned with exploration methods in areas of glaciated terrain. These volumes are essential for those working, or contemplating work, in that environment.

7. e. Fifth Symposium, Vancouver; Elliott and Fletcher (1975).
   f. Sixth Symposium, Sydney; Butt and Wilding (1977).
   g. Seventh Symposium, Denver; Watterson and Theobald (1979).

8. Several annual supplements to the *Bibliography of Exploration Geochemistry, 1965-1971* have been published as follows:
   a. 1972 Supplement: *J. Geochem. Explor.* 2, pp. 41-75.
   b. 1973 Supplement: *J. Geochem. Explor.* 3, pp. 89-128.
   c. 1974 Supplement: *J. Geochem. Explor.* 4, pp. 271-314.
   d. 1975 Supplement: Hawkes (1976 b).
   e. 1977 Supplement: Hawkes (1977).

$2A$ / The Primary Environment

INTRODUCTION (pp. 39-41)

There can be no doubt that, over the years, geochemists in the Soviet Union have made major contributions to the understanding of geochemical principles as applied to the primary environment. We are now fortunate to have available the translated text of Beus and Grigorian (1977) in which Soviet concepts and experiences in this particular environment are emphasized. This chapter, therefore, will concentrate on those concepts discussed by Beus and Grigorian (1977) which are new to the English literature (e.g., "geochemical specialization") or which justify further amplification (e.g., geochemical provinces). In Chapter 7, the application of some of the Soviet methods are illustrated with practical examples.

At the outset it is important to point out that in the Soviet literature the word "dispersion" is not used in connection with the primary environment as it is in Chapter 2 of this book and throughout the literature in English. Rather the word "distribution", as in primary distribution, is preferred and will be used herein. *Dispersion,* following the original definition by Fersman, refers to deviations toward *lower* values, as could be the case in the secondary environment where a deposit is being chemically or mechanically destroyed by weathering. *Concentration* (or accumulation) is the deviation of the average (background or clarke) value toward *higher* values which, in the primary environment, ideally results in a deposit. Therefore, in recognition of the fact that there is no dispersion in the sense of destruction in the primary environment, the term primary *distribution* is preferred. Both dispersion and concentration are due to movement of the elements, such movement being called *migration,* again adopting Fersman's original definition.

The extent of element migration is affected by many factors which Beus and Grigorian (1977) classify as either internal or external. Internal factors influencing the migration of elements include thermal, radioactive, and certain chemical properties (e.g., the tendency to form complexes) of the elements. External factors are characterized by the thermodynamic condition (environment) of the medium in which an element migrates and in which certain inherent properties of an element are expressed. Thus, external factors affecting migration include temperature, pressure, and the chemical environment of the migration. The last factor, which is of particular importance in the secondary environment, was dealt with at length in Chapters 2 and 3 of the first edition where such phenomena as pH-Eh (pp. 124-134) and adsorption (pp. 134-141) were discussed.

Of particular interest to exploration geochemistry is the concept of a *geochemical barrier*. This term, originally introduced by Perel'man (see Perel'man 1967, 1977), implies an abrupt change in the physicochemical environment in the paths of migration of elements causing the precipitation of certain elements from solution. Usually the term is applied to phenomena occurring in the surficial zone, such as changes in pH (pH of hydrolysis) or the presence of iron-manganese oxides which adsorb elements from solution. It may, however, also be used for endogenic processes, such as when an abrupt decrease in temperature and pressure cause the precipitation of certain minerals; accordingly, geochemical barriers may properly be introduced in this chapter. The important geochemical barriers of interest to exploration geochemists are listed in Table 2A-1. Numerous other types of geochemical barriers, and a discussion of their significance, may be found in Perel'man (1967, 1977).

Table 2A-1.   Selected geochemical barriers. From Beus and Grigorian (1977).

Barrier type	Characteristics
Temperature	This is very important for migration in endogenic processes. The role of this barrier in exogenic processes is insignificant.
Decompression	In endogenic processes an abrupt decrease in the pressure within the system plays a major role in the processes of mineral formation. It is much less significant in exogenic processes.
Acid-alkaline	The effect of changes in the acidity-alkalinity regime of a solution during endogenic processes is sometimes a decisive factor in the separation of many components in the solid phase, and in the concentration of ore substances. It is of less significance in exogenic processes; however, the alkaline barrier is responsible for the precipitation of iron, nickel, and other metals from solution when the solution comes into contact with limestones at the boundary of weakly acidic soil horizons and deeper levels rich in carbonate materials, etc.
Oxidizing-reducing	In endogenic, as well as in exogenic processes, a sudden change in the oxidizing-reducing environments in the paths of migration has a decisive effect on the precipitation of some metals:
a) Oxidizing	takes place as juvenile or ground waters, low in oxygen, come into contact with surface waters rich in oxygen. It is very important in the precipitation of the oxides of iron and manganese in surface waters;
b) reducing hydrogen sulfide	causes precipitation of the majority of metals in the form of sulfides;
c) reducing gley	causes precipitation of some anion-producing metals, such as uranium, vanadium, and molybdenum.
Sulfate and carbonate	Occurs at the initial interaction of sulfate and carbonate waters with other types of waters rich in calcium, strontium, and barium. Gypsum and celestine are formed.
Adsorption	Typically an exogenic geochemical barrier. It is of great importance in the precipitation of trace elements from surface and ground waters.
Evaporation	Occurs in regions of rapid evaporation of ground waters. It is accompanied by salinization, the formation of gypsum, etc.
Mechanical	Results from changes in the velocity of water flow (or air movement) and is responsible for the precipitation of heavy minerals. It plays a major role in the formation of placer deposits.

DISTRIBUTION OF ELEMENTS IN
IGNEOUS ROCKS AND MINERALS   (pp. 41-49)

Bowen's Reaction Series   (pp. 41-42)

No additional comments.

Distribution of Trace Elements   (pp. 43-47)

In Table 2-1 (p. 43) the average abundances of 65 elements are tabulated for the Earth's crust, various rock types, soil, and river water. Many compilations of this type can be found in the literature (e.g., Beus and Grigorian, 1977, Table 3) all of which will vary slightly one from another. Vernadskii took the average crustal abundances of the elements (Earth's crust values) and divided them into "decades", that is, into categories based on a 10-fold difference in their crustal abundances (Table 2A-2). In this manner, it is possible to readily observe useful aspects of element distribution, such as the fact that 25 elements average from 1 - 10 ppm, and that such well-known elements as Ag, Hg, and B, occur in much smaller amounts than relatively unknown elements such as Nd and Sc.

Table 2A-2.   Distribution of the average contents of 82 elements in the continental lithosphere based on Vernadskii's "decades". After Beus and Grigorian (1977).

Decade	Average content in decade (in weight %)*	Number of elements	Element
I	> 10	2	O, Si
II	$10^0 - 10^1$	6	Al, Fe, Ca, Mg, Na, K
III	$10^{-1} - 10^0$	4	Ti, P, H, C
IV	$10^{-2} - 10^{-1}$	9	Mn, S, F, Ba, Sr, V, Cr, Zr, Cl
V	$10^{-3} - 10^{-2}$	14	Ni, Rb, Zn, Cu, Co, Ce, Y, La, Nd, Sc, N, Li, Ga, Nb
VI	$10^{-4} - 10^{-3}$	25	Pb, B, Th, Sm, Gd, Pr, Dy, Er, Yb, Hf, Br, Cs, Sn, As, Be, Ar, U, Ge, Mo, Ho, He, Eu, Tb, W, Ta
VII	$10^{-5} - 10^{-4}$	8	Lu, Tl, I, In, Sb, Tm, Gd, Se
VIII	$10^{-6} - 10^{-5}$	5	Ag, Hg, Bi, Ne, Pt
IX	$10^{-7} - 10^{-6}$	4	Pd, Te, Au, Os
X	$10^{-8} - 10^{-7}$	3	Re, Ir, Kr
XI	$10^{-9} - 10^{-8}$	1	Xe
XII	$10^{-10} - 10^{-9}$	1	Ra

*See tabulation on p. 547 for conversion of exponential units (e.g., 10^{-4}%) to ppm and ppb units.

Earlier, the statement was made (p. 45), " . . . it is important to determine what proportion of the total amount of an element of interest is in a form amenable to mining, milling and metallurgical extraction and what quantity is in an economically

unrecoverable form, primarily in silicates or other minerals such as magnetite."
Unfortunately, no details were discussed nor were practical examples given of this
important concept. Beus and Grigorian (1977), however, have treated this subject in
detail, and they have recognized and discussed three principal modes of occurrence
of elements in rocks:

1. **Elements present in igneous rocks as independent structural components of
minerals.** This type includes the main rock-forming elements (O, Si, Al, Fe, Ca, Mg,
Na, K) of Vernadskii's first two decades, and they occur mainly as silicates. Several
of the minor elements in the third decade (Ti and P), and some trace elements under
certain conditions, may also form their own minerals which, in some varieties of
igneous rocks (especially those which have undergone metasomatic alteration), may
concentrate the bulk of the elements which are present in the rock (Table 2A-3).

2. **Isomorphic (ous) form of element occurrence.** This form of occurrence was discussed in detail earlier under such topics as dispersed elements (p. 61) and diadochy
(p. 65). For this form of occurrence to be studied properly, monomineralic fractions
must usually be isolated from a rock, their abundances determined by quantitative
mineralogic and petrographic methods, and the element(s) of interest must be determined quantitatively by chemical analysis. (In certain cases it may be possible to
determine the form of occurrence simply by using different chemical digestions).
With this information, it is then possible to make a *distribution balance,* which can be
defined as the determination of the amount of an element in the various mineralogical components of a rock. An example of such a distribution balance for lead in a granite is presented in Table 2A-4. A similar example for uranium is presented in Table
2A-5.

From Tables 2A-4 and 2A-5 it is possible to identify *mineral concentrators* and
mineral accumulators (the latter sometimes called "carriers"). The former contains
maximum amounts of the particular element (zircon in the case of both Pb and U),
whereas the latter contains the bulk of the element in the rock (feldspars in the case
of Pb; quartz in the case of U). Additional examples of mineral concentrators and
accumulators, with particular reference to the occurrence of uranium in rocks, may
be found in Larsen and Gottfried (1961).

Table 2A-6 lists important mineral concentrators and accumulators, based on
Soviet experiences. This table complements Table 2-2 (p. 46). Based on the data in
Table 2A-6, Beus and Grigorian (1977) concluded that biotite is the best indicator
mineral of the geochemical characteristics of siliceous and intermediate igneous
rocks because it isomorphously incorporates Li, Cs, Cu, Zn, Nb, Ta, Sn and W (thus
permitting its use in exploration). Of lesser importance are muscovite (Be, F, Ta,
Sn); plagioclase (Be, Mo); potassium feldspar (Pb, Rb, Mo); sphene, ilmenite and
zircon (Nb, Ta, Mo, Sn); magnetite (Mo, Pb, Cr); and pyroxenes and amphiboles
(Cu, Zn). In basic and ultrabasic rocks, the most interesting indicators of Cr, Ni, Co
and Cu, in addition to sulfides, are pyroxenes, amphiboles and, to a lesser extent,
olivine.

Table 2A-3.   Principal minerals formed by minor and trace elements in rocks of the lithosphere.
From Beus and Grigorian (1977).

Element	Granites and granodiorites	Intermediate rocks	Basic rocks	Alkaline rocks	Ultrabasic rocks
Titanium	ilmenite, sphene	ilmenite sphene	ilmenite, titanomagne-tite	ilmenorutile, sphene, complex titanium silicates	—
Phosphorus	apatite, monazite	apatite	apatite	apatite	—
Sulfur		sulfides of iron, copper, lead, zinc, etc.			
Boron	tourmaline	—	—	—	—
Zirconium	zircon	zircon	—	zircon, complex zirconium silicates	—
Lithium	lithium micas, amblygonite, spodumene (in metasomatically altered granites)	—	—	—	—
Beryllium	beryl, bertrandite, phenacite, chrysoberyl (in metasomatically altered granites)	—	—	—	—
Fluorine	fluorite, topaz	—	—	villiaumite, fluorite	—
Chromium	—	—	chromite	—	chromite
Manganese	secondary oxides	secondary oxides	secondary oxides	secondary oxides	secondary oxides
Copper, zinc, lead, nickel	sulfides	sulfides	sulfides	sulfides	sulfides
Arsenic	arsenopyrite	arsenopyrite	arsenopyrite	—	—
Molybdenum	molybdenite	—	—	molybdenite	—
Tin	cassiterite	—	—	—	—
Rare-earth elements	monazite, allanite, xenotime, rare-earth niobates	—	—	loparite, complex silicates	—
Niobium and tantalum	Columbite-tantalite, pyrochlore-microlite, rare-earth tantalo-niobates (in metasomatically altered granites)	—	—	loparite	
Thorium	monazite, thorite	—	—	thorite	—
Uranium	uranium oxides, phosphates, etc.	—	—	—	—

3. **Elements present in a soluble (mobile) form.** Included in this category are elements which occur (a) within gas-liquid inclusions, and (b) within capillary and pore solutions. After fine grinding, such elements can be removed and determined when rocks are treated with water or other solvents which do not destroy silicates and other insoluble minerals. Solvents may include, for example, weak solutions of hydrochloric acid, sodium carbonate or ammonium carbonate, but these may also

dissolve extremely small inclusions of sulfides, native elements and phosphates. Beus and Grigorian (1977) reported that if granitoid rocks are treated with a weak (1:50) solution of HCl containing small amounts of NaCl (1 gram per liter), 30 - 50% of the lead, and 70 - 90% of the Zn, may be expected to go into solution. (If Beus and Grigorian are correct about the amount of readily soluble lead, it follows that perhaps 50% of the lead indicated in Table 2A-4 as being within feldspars may actually be loosely held rather than in silicate structures.)

Table 2A-4.   Distribution balance of lead in a granite from the U.S.S.R.
From Beus and Grigorian (1977).

Mineral	Content of mineral in rock, %	Content of lead in mineral, (ppm)	Quantity of lead within rock due to mineral, %
Quartz	28.0	Not determined	—
Plagioclase	29.0	11	18.8
Potassium feldspar*	37.9	34	75.8
Biotite	3.7	40	8.7
Magnetite	0.2	11	0.1
Zircon**	0.004	110	0.02
Sphene	0.003	Not detected	—

Abundance of lead in rock, ppm:
1) calculated value: 17.6
2) determined by analysis: 17

 * Potassium feldspar — carrier of the majority of the lead in the rock.
** Zircon — concentrator of lead in the rock.

Table 2A-5.   Distribution balance of uranium in a granite from the Southern California batholith. Based on data from Larsen et al. (1956, Table 5).

Mineral	Content of mineral in rock, %	Content of uranium in mineral, (ppm)	Quantity of U within rock due to mineral, %
Quartz*	40	2.4	37
Orthoclase	36	1.3	18
Plagioclase	16	2.3	14
Muscovite	7	8.0	22
Biotite	0.1	5.2	trace
Garnet	0.2	5.8	trace
Apatite	.001	47.	trace
Zircon**	.002	4600.	4
Monazite	.004	2500.	4
Xenotime	.003	360.	trace
Magnetite	trace	3.0	trace

Abundance of uranium in rock, ppm:
1) calculated value: 2.59
2) determined by analysis: 2.7

 * Quartz — carrier of the majority of the uranium in the rock
** Zircon — concentrator of uranium in the rock

Table 2A-6.   Mineral concentrators and mineral accumulators (carriers) of trace elements in rocks.
From Beus and Grigorian (1977).

Element	Mineral
Lithium	Micas (in acid rocks), especially biotite; hornblende (in intermediate and basic rocks)
Rubidium	Biotite (concentrator of rubidium in intermediate and acid rocks); potassium feldspar (carrier in granitoid rocks); micas (in intermediate rocks)
Beryllium	Plagioclases (carrier of beryllium in intrusive rocks); muscovite (concentrator of this element)
Fluorine	Micas and hornblendes (in the absence of its own minerals)
Chromium	Pyroxenes and magnetite (in ultrabasic and basic rocks in the absence of chromite)
Nickel and cobalt*	Magnesium pyroxenes and olivine (in ultrabasic and basic rocks); biotite (in intermediate and acid rocks)
Copper*	Pyroxenes and amphiboles, as well as magnetite and biotite (in intrusive rocks)
Zinc*	Biotite and amphibole (in intermediate and acid rocks); magnetite (in basic rocks). The role of tourmaline is being studied.
Niobium and tantalum	Ilmenite, zircon, cassiterite, sphene (in granitoid rocks). Biotite is an interesting indicator. Other micas are of minor significance.
Molybdenum	Potassium feldspar and plagioclase (carriers in igneous rocks). The highest concentrations are found in magnetite, ilmenite, and sphene
Tin	Biotite (in granitoid rocks), muscovite (in muscovite granites), tourmaline
Cesium	Biotite and, to a lesser extent, muscovite (in granites)
Tungsten	Biotite (in granitoid rocks)
Lead	Orthoclases (in acid and intermediate rocks). Maximum concentrations are found in zircon (and in some other accessory minerals)

*Biotite and other minerals often contain extremely small inclusions of sulfides of these elements whose quantitative role is difficult to assess.

Post-magmatic (Hydrothermal) Ore Deposits   (pp. 47-48)

No additional comments.

Complex Formation   (pp. 48-49)

The literature, both experimental and descriptive, on the significance of complexes in the transport of metals in all types of fluids (e.g., magmatic, hydrothermal, groundwater) continues to grow at a rapid rate, underscoring the importance of this subject to economic geology and exploration geochemistry. Beus and Grigorian (1977) have discussed the subject as it is viewed by Soviet geochemists. However, most of these studies and discussions are beyond the scope of this book.

PRIMARY HALOS AND PRIMARY DISTRIBUTION
(pp. 49-53)

The techniques used in the Soviet Union for recognizing and interpreting primary halos and the primary distribution of elements, in large measure based on the concepts of hydrothermal zoning, constitute a large part of the book by Beus and Grigorian (1977). Some of the more important basic concepts are presented in this chapter, whereas the practical application of these concepts to mineral exploration will be found in Chapter 7A.

PATHFINDERS   (pp. 53-55)

In the Soviet literature the word pathfinder is not used. Rather, they use the word indicator, of which there are two types: *direct indicators* and *indirect indicators* (Beus and Grigorian, 1977). The former refers to those elements which concentrate in ores (e.g., the use of Cu in the search for copper deposits), whereas the latter refers to associated elements (e.g., the use of Li in the search for tantalum deposits because there are no good analytical techniques for the determination of tantalum).

GEOCHEMICAL PROVINCES   (pp. 55-58)

Geochemical provinces were defined (p. 55) as "relatively large, well-defined areas of the Earth's crust that have a distinctive chemical composition." Although correct as far as it goes, based on the discussion and level of understanding of geochemical provinces in Beus and Grigorian (1977), that definition is naive at best, and, at worst, misleading. For example, nowhere in the definition or the ensuing discussion in the first edition, was it stressed that within a geochemical province there may be considerable differences in the distribution of individual elements, or element associations, for rocks of the same type but of different ages. Nor was it emphasized that geochemical provinces may have formed over a considerable period of time requiring a combination of several complex geological processes. Accordingly, it is appropriate to quote the all-inclusive definition of geochemical provinces used by Beus and Grigorian (1977, p. 15): "large-scale crustal units characterized by common features of geological and geochemical evolution expressed in the chemical composition of the constituent geological complexes [formations], as well as in the endogenic and exogenic metalliferous and nonmetalliferous concentrations of the chemical elements."

Beus and Grigorian (1977) also pointed out, as would be expected upon reflection, that the content of trace elements in similar rock types differ from province to province. By means of the *concentration ratio,* CR, which is the ratio of the average content of an element in the material studied (rock type, pluton, igneous complex, etc.) to the regional (i.e., geochemical province) or global average content, these differences and their significance can be readily seen. For example, they observed that the lithium contents of certain granite plutons within two different geochemical provinces, the Ukrainian Crystalline Massif and the Eastern Transbaikalian Region, are about 100 ppm; this is much higher than the average content of this element in granites of the lithosphere (CR = 2.6). However, in the latter province the lithium contents of all the Mesozoic granites is 90 ppm and, therefore, CR for the specific pluton sampled is only slightly higher than unity: 100/90 = 1.1. On the other hand, the average lithium content in the former province is only 40 ppm, with the resultant CR = 100/40, or 2.5. Thus, the content of lithium in the specific granite sampled in the Ukrainian Crystalline Massif should be regarded as a strong local geochemical anomaly which merits further consideration.

Beus and Grigorian (1977) have discussed other more complicated methods of studying the rocks of geochemical provinces and, among other things, determining their ore potential. For example, instead of using individual elements such as lithium in the above example, data on two or more elements may be combined yielding what are termed multiplicative or additive (see p. 765) concentration ratios. Also, empirical procedures have been developed for comparing geological complexes on the basis of their trace elements being placed in a series of decreasing abundances. In such cases these techniques require a thorough understanding of the petrography and history of the rock, as well as the trace element geochemistry, and are beyond the scope of this book.

GEOCHEMICAL ASSOCIATIONS   (pp. 58-70)

General   (pp. 58-61)

No additional comments.

Goldschmidt's Geochemical Classification   (pp. 61-66)

Fig. 2-5 in the first edition (p. 62) illustrated four major categories in Goldschmidt's geochemical classification: siderophile, chalcophile, lithophile, and atmophile. It was noted in the accompanying text that there was a fifth type of chemical affinity, i.e., biophile, and in all cases the groupings were only generalized because some elements have characteristics common to two groups.

Other geochemists have compiled different geochemical classifications, although they are usually also based on Goldschmidt's original work. One such classification, that of Beus and Grigorian (1977), is presented in Table 2A-7. It should be noted that whereas the classification in Fig. 2-5 is based on elemental characteristics within the entire Earth (including the core and the mantle), that of Beus and Grigorian (1977) is based on elemental affinities only in the lithosphere — the part of the Earth of most interest to exploration geochemists. Thus, whereas iron and its associated elements form a specific group (the siderophile elements) based on their high abundances in the Earth's core, iron and certain associated elements (V, Cr, Co, Ni; but not Au and the Pt group elements) can also be classified as a sub-group of the oxyphile elements when only the lithosphere is considered (Table 2A-7). Beus and Grigorian (1977) also use two additional categories in their classification, i.e., noble and hydrophile elements, whose significance is self-explanatory.

Table 2A-7. Geochemical classification of elements in the Earth's crust. After Beus and Grigorian (1977).

Type	Major (>1%)	Minor (0.1-1.0%)	Trace elements (<0.1%) Mineral-forming	Dispersed
I. Oxyphile				
a. Lithophile	O, Si, Al, Fe, Mg, Ca, Na, K	Mn, Ti, P, (C)	Li, Be, B, F, Sr, Ba, Y and rare earth elements, Zr, Nb, Ta, Sn, Cs, W, Th, U	Ga, Ge, Rb, Hf, Sc, Ra, (Tl)
b. Siderophile	Fe	—	V, Cr, Co, Ni	—
II. Chalcophile	—	—	S, Cu, Zn, As, Se, Mo, Ag, Sb, Te, Hg, Pb, Bi	Cd, In, Re, Tl, (Ga), (Ge)
III. Noble	—	—	Pd, Os, Ir, Pt, Au	Rh, Ru
IV. Hydrophile	(O)	H	Cl, Br, I, (S)	
V. Atmophile	(O)	C	He, N, Ne, Ar, Xe, Rn	

Applications   (pp. 66-70)

Boyle (1974a) has revised and up-dated his earlier compilation on elemental associations in all types of mineral deposits, and discussed the use of various indicator (pathfinder) elements of interest in geochemical prospecting.

GEOCHEMICAL SPECIALIZATION (CHARACTERIZATION)
(A new topic)
Introduction

In the preceding sections of this chapter various aspects of trace element distribution, abundance, association, migration, and concentration were discussed. With these largely theoretical aspects now sufficiently developed, it is appropriate to ask, "What elemental characteristics of rocks are likely to make them favorable for the occurrence of undiscovered ore?" In other words, it is necessary to convert the theoretical concepts into practical criteria for the recognition of rocks likely to produce mineralization, and to be able to distinguish these from rocks that are barren. This is the subject referred to as *geochemical specialization* in the Russian literature, and which is very well explained by Beus and Grigorian (1977). The word "specialization" is a direct translation from the Russian but English-speaking geochemists will probably prefer the word *characterization*. Whereas the major element components of similar rock types (e.g., granites) are very similar in most parts of the lithosphere, this is not so in the case of the trace elements which can vary greatly in either direction (higher or lower) of an average value and, therefore, geochemical specialization is concerned primarily with trace elements. Thus, the term geochemical specialization can be defined as the "specific features in the distribution of one or several trace elements in rocks, expressed in appreciably higher or lower contents of these elements, or in the anomalous values of variance unusual for the given rock type" (Beus and Grigorian, 1977, p. 53). In addition to assisting in the determination of the ore-producing potential of rocks, these geochemical features may be used as indications of the similarity or dissimilarity of rocks classified petrographically as the same type, and in the determination of the genetic affiliation of rocks.

The distinctive features of geochemical specialization of igneous rocks are determined by a variety of factors, such as the chemical nature of the magma, the physical characteristics of its crystallization (including the effect of the enclosing rocks), as well as by post-magmatic alteration of the rocks. In sedimentary rocks, geochemical specialization is the result of the geochemical differentiation of material during the processes of migration and sedimentation, followed by the diagenetic transformation of the sediments. The term geochemical specialization can be used in connection with geochemical provinces (e.g., the porphyry copper province of the circum-Pacific region), as well as with a small intrusive body or a sedimentary suite. However, it should be stressed that geochemical factors alone, even though they may favor the concentration of ore, are not sufficient for the formation of economic mineralization; a large number of inter-related geochemical and geological factors must be considered. For example, in the case of the formation of primary ore concentrations associated with the development of a particular igneous complex, the geochemical factors are responsible for: the behavior of the ore elements in the primary magmatic melt; the features controlling the separation of the elements from the magma chamber during the magmatic or post-magmatic stage of its development; the transport mechanism of the ore elements; and the chemical reactions leading to the concentration of these elements in ores. The geological factors, in turn, control the separation and movement of the ore-forming melts and solutions by creating conditions, either favorable or unfavorable, for the formation of economic concentrations of ore elements. Thus, both geochemical and geologic factors must be favorable for the formation of deposits of economic significance.

Criteria for the Assessment of Geochemical Specialization

The main criteria for the assessment of the geochemical specialization of igneous, metamorphic and sedimentary rocks are based on the: (1) trace element distribution patterns in rocks; (2) trace element distribution patterns in rock-forming, minor and accessory minerals; (3) ratios of the contents of geochemically related elements in rocks and minerals. Examples of some of these criteria were discussed in Chapter 7 of the first edition.

1. **Trace element distribution patterns in rocks.** The trace element distribution patterns of most interest to exploration geochemists are those which show an appreciably higher or lower content of a particular element, or group of elements, as compared with similar rocks in the same region or elsewhere in the world. This is illustrated in Table 2A-8 and in Table 2A-9 for tin and lithium, respectively, where the differences in the average content of these elements in granitic rocks permits ore-bearing igneous complexes to be differentiated from those which are barren (see also p. 320). However, as shown in the case of lithium (Table 2A-9), the method is not unequivocal because there is an overlap in the range of variations between ore-bearing and barren complexes. Nevertheless, by statistical methods (not discussed here) Beus and Grigorian (1977) have shown that only 2.6% of the barren granites will have over 100 ppm of lithium, whereas this value will be equalled or exceeded in 23.6%, 40.5% and 63.7%, respectively, of the parent (ore-bearing) rocks containing various types of deposits.

Table 2A-8.   Average contents of tin (in ppm) in granites from tin-bearing and barren igneous complexes. After Beus and Grigorian (1977).

Region	Barren granites	Tin-bearing granites
Kalba, Gorny and Rudny Altai, U.S.S.R.	5	16 - 30
Chukchi Peninsula, U.S.S.R.	7 - 8	10 - 16
Transbaikalia, Far East, Kazakhstan, Urals, Middle East	5	15
Caucasus, Ukraine	2 ± 0.7	—
U.S.A., Canada, Japan	3	—
Great Britain	—	27
Yugoslavia, France	—	37

Table 2A-9.   Lithium in parent (ore-bearing) and barren granitic rocks in selected localities of the U.S.S.R. After Beus and Grigorian (1977).

Regions	Arithmetic mean of contents (ppm)	Range of contents (ppm)	Probability of Li content equal to or higher than 100 ppm (in %)
Granites of the Earth	38	8 - 130	2.4
Barren for Sn, W, Be and Ta	37.5	8 - 130	2.6
Parent rocks for pegmatite deposits of Li, Be and Ta	100	25 - 300	40.5
Parent rocks for greisen and quartz-vein type deposits of Sn, W and Be	80	4 - 540	23.6
Parent rocks for Ta-containing apogranites	130	23 - 440	63.7

Beus and Grigorian (1977) mention two cases, one involving chalcophile elements in igneous rocks and the other specifically involving beryllium in granites, in which the presence of ore is not represented by high contents of specific elements in rocks. Therefore, caution must be exercised in applying this method.

2. **Trace element distribution patterns in rock-forming, minor and accessory minerals.** The possibility of using the trace element composition of separated mineral fractions as guides to ores was first suggested by Fersman in the 1930's in the U.S.S.R. Micas have received particular attention (pp. 336-338). The first relationship to be noticed was between the tin content of pegmatitic muscovites and the presence of cassiterite in the pegmatites. Subsequently, increased contents of cesium in lepidolites were used as a positive indication of pollucite, followed by the discovery that high contents of Be, Nb and Ta in muscovites from pegmatites may indicate the presence of beryl and columbite-tantalite in these pegmatites. For example, Beus and Grigorian (1977, Table 19) reported that the tantalum content of muscovites from non-metalliferous pegmatites of the Soviet Union averages 4 ppm, whereas muscovites from rare-metal pegmatites containing columbite-tantalite average 35 ppm tantalum.

The use of the tin and chlorine contents of biotites, as well as the trace elements in several other minerals, has been discussed earlier (pp. 336-341; see also p. 774).

3. **Ratios of the contents of geochemically related elements in rocks and minerals.** An example of the use of ratios of geochemically related elements in the context of geochemical specialization, although this term was not specifically used, may be found in Table 7-1 and the ensuing discussion (p. 319). Clearly, these criteria are based on differences, rather than similarities, in the chemical properties of diadochic (isomorphous) elements. The reason for the variations in ratios of isomorphous pairs, such as K-Rb and Al-Ga, is that there are slight differences in the chemical properties of each member of the pair which results in slightly different migration characteristics in, for example, magmatic and post-magmatic hydrothermal or metasomatic environments. Thus, whereas Rb is typically dispersed in K minerals maintaining a K/Rb ratio of approximately 170 during the crystallization of most igneous rocks, this ratio falls to about 100 during the metasomatism of granites. Further, when individual minerals are considered, the quantitative relationships between trace and major elements can be different in the same mineral of different origin.

Table 7-2 (p. 320) illustrates the use of some indicator ratios based on various pairs of elements, not only isomorphous (diadochic) elements. Beus and Grigorian (1977) point out that the most informative ratios are those which are based on elements whose migration tendencies, such as Mg/Li, have opposing directions during the crystallization of a magma, or in other processes such as metasomatism.

An important aspect to consider in assessing the value of geochemical specialization is that the ore-bearing potential of a pluton can usually be determined from the analysis of only 10 to 30 properly selected samples. The procedures for sampling are discussed on p. 761.

3A / *The Secondary Environment*

INTRODUCTION   (pp. 71-72)

Notwithstanding the increased emphasis being placed on the primary environment in this edition, as evidenced by the greatly expanded discussions in chapters 2A and 7A, the secondary environment remains the one of most importance to exploration geochemists.

The organization of material, and the concepts, presented in Chapter 3 of the first edition, will be used as the basis of discussion in this edition. Because the concepts in the first edition are so fundamental in many instances, little or no new material need be added. The major additions will be found in those subjects in which significant new data of particular interest to exploration geochemists have become available within the past few years, such as: gossans; hydrous iron and manganese oxides; and water (hydrogeochemistry). This chapter ends with two new topics: "The Arid Environments"; and "Mobility of Uranium and Its Daughter Nuclides in the Secondary Environment".

CHEMICAL WEATHERING   (pp. 72-86)

Introduction   (pp. 72-73)

Krauskopf (1979) has written the second edition of his book *Introduction To Geochemistry* which discusses chemical weathering, soil formation, Eh-pH, and numerous other topics covered in this chapter. Brownlow (1979) also has written a book, entitled *Geochemistry,* which covers many of the same topics. These excellent textbooks are highly recommended to those interested in the more advanced aspects of this subject. On an even more advanced level, the book by Yariv and Cross (1979) can be recommended.

Some Chemical Weathering Reactions   (pp. 73-80)

The importance of ferrous sulfate and sulfuric acid generated by the oxidation of iron sulfides as powerful agents in the chemical decomposition of sulfide minerals was discussed in connection with equations 7 to 11 (p. 77). Dutrizac and MacDonald (1974) reviewed the importance of the ferric ion in acidic media as a leaching agent

643

for base metal sulfides and uranium oxides. Although their interest is mainly in connection with hydrometallurgical extraction processes, their discussions and review do have fundamental information applicable to the weathering of sulfide deposits.

Bacteria   (pp. 80-81)

The importance of bacteria in the geochemical cycle of iron and sulfur was discussed earlier (pp. 80-81). However, more emphasis should have been placed on the influence of bacteria in the oxidation of other elements, particularly manganese. For example, the largest manganese deposit in Russia is believed to have formed with the aid of manganese-oxidizing bacteria (Zajic, 1969). Bacteria are also believed to be essential to the formation of the manganese oxide precipitates at the thermal springs in the Badgastein area, Germany. These manganese oxides have subsequently adsorbed large quantities of radium which is the main source of the high radon content of the thermal waters.

In the field of microbiology it is well known that fermentative anaerobic hydrogen-producing bacteria are very abundant and that they can produce hydrogen by means of the decomposition of organic compounds. Such bacteria have been found in groundwaters, soils, muds and brines. Levinson (1977) suggested that hydrogen produced in this manner may be the agent for the reduction of U^{+6} to U^{+4}, which results in the precipitation of uraninite or other uraniferous minerals, when no other reducing agent can be postulated. Methane may also be produced by bacteria and it would act in a similar manner to hydrogen.

Products Of Chemical Weathering   (pp. 81-86)

In mineral exploration for base metal sulfides it is frequently necessary to evaluate the economic significance of gossans (pp. 85-86). This requires a distinction to be made between true gossans and false gossans; as defined here, the former represents economic (or potentially economic) mineralization whereas the latter does not. Blain and Andrew (1977) have thoroughly reviewed this subject, including detailed considerations of the mineralogical, textural and geochemical features of both types of gossans (see pp. 84-86 for a brief introduction to these topics). Moeskops (1977) reviewed the geochemical aspects of gossans with specific application to the nickel mineralization in the Yilgarn Shield, Western Australia.

The basic concept behind the geochemical methods of gossan interpretation assumes that as sulfide minerals are weathered, a component of the ore minerals, usually quite variable, is retained within the gossan-forming system. Of course, false gossans and other types of ironstones (e.g., laterites) may also contain significant concentrations of these same elements, but not usually in the same relative proportions.

The geochemistry of the true nickel gossans of Western Australia is now sufficiently well studied and documented to enable their chemical characteristics to be used to differentiate them from false gossans, as inspection of Table 3A-1 will reveal. Specifically, Moeskops (1977) found that true gossans compared with false gossans: (1) are generally enriched in Ni and Cu; (2) generally possess lower Ni/Cu ratios; (3) generally yield low Zn, Mo and Mn values; (4) are generally enriched in Bi; and (5) are invariably enriched in Pt, Pd and Se. To positively distinguish true gossans, Pt, Pd and Se must be determined, the reason being that these elements occur in sulfides, and not in silicates, and they have limited mobility under gossan-forming conditions.

Table 3A-1. Summary of selected geochemical data on nickel sulfide ores, true gossans, and false gossans, Yilgarn Shield, Western Australia. All values in ppm. After Moeskops (1977).

Element	Fresh Sulfide Ore		True Gossan		False Gossan	
(ppm)	"Range"	Median	"Range"	Median	"Range"	Median
Cu	2,000-5,000	2,100	800-3000	1000	80-150	90
Ni	25,000-40,000	31,000	1500-5000	3000	800-2500	1500
Zn	60-120	80	50-200	140	200-1500	800
Pb	30-60	40	30-40	35	10-30	20
Co	400-1800	1,200	200-400	280	200-800	480
Cr(s)	300-700	600	200-2000	750	100-1800	600
Mn	250-600	450	200-600	400	800-2000	1400
Mo	<5	<5	<5	—	~10	10
Pt	.08-.80	0.45	0.1-0.2	0.15	—	—
Pd	0.5-1.2	0.70	0.1-0.3	0.15	—	—
Au	~0.02	0.02	≤0.15	—	—	—
Ag	~3.0	3.0	≤3	2	—	—
As	<5-10	8	20-30	25	20-30	25
Bi	30-50	40	20-60	45	~15	15
Se	10-40	25	20-40	35	≤2	—
Sb	<4	<4	20-40	30	—	—
Ni/Cu	8-20	14.8	2-6	3.7	5-20	14.9

Notes:

1. "Range" is actually a "normal range" (i.e., 15-85 percentiles).

2. Cr(s) is acid soluble (HClO$_4$-HNO$_3$) chromium.

3. Blanks (—) indicate not detected (below detection limit which is 0.02 ppm for Pt, Pd and Au; 2 ppm for Ag; 5 ppm for Mo).

4. Based on analysis of 40 nickel ores, 103 true gossans and 69 false gossans from 32 localities, including most producing nickel mines in Western Australia.

Other geochemists, for example Wilmshurst (1975), reached conclusions similar to those of Moeskops (1977). Several earlier studies found that acid soluble chromium Cr(s) values have particular significance in gossan interpretation; however, Moeskops (1977, p. 257) believes that the nature and origin of the Cr(s) in true and false gossans needs further research.

Relatively little geochemical data have been published on other types of gossans, although such information undoubtedly exists in the files of some mining companies. Table 3A-2 summarizes data presented by Blain and Andrew (1977) on copper-zinc and lead-zinc gossans and associated ironstones in southern Africa and, in it, certain relationships can be seen, such as high Ba and Ag in Cu-Zn gossans by comparison with false gossans (i.e., pyrite gossans and ferricretes). However, much more work is needed on gossans from base metal deposits before broad conclusions, similar to those reached for the Western Australia nickel area, can be made with any degree of confidence. Further, it is important to recognize that schemes for discriminating between true gossans and false gossans are generally applicable only within a restricted area. Blain and Andrew (1977) pointed out that the classification methods used for nickel gossans in Western Australia misclassify many of the nickel gossans in Botswana.

Table 3A-2. Geochemical data on copper-zinc and lead-zinc gossans and associated ironstones (false gossans) from southern Africa. Numbers in parenthesis indicate the number of samples analyzed. All values are averages in ppm. After Blain and Andrew (1977).

Type	Ni	Co	Cu	Zn	Pb	Mn	Cr	Ba	Ag	Cd
Cu-Zn gossans (50)	27	70	1800	90	250	110	—	1800	8	2
Cu-Zn gossans (19) (siliceous)	32	28	370	580	105	125	75	5900	2	7
Pb-Cu gossans (5)	30	40	330	800	3800	3300	50	720	9	—
Zn-Cu gossans (19)	24	25	115	600	6200	360	65	260	7	7
Stratiform Fe-S gossans (Minor Cu-Zn) (22)	10	100	3900	920	390	1000	—	140	8	10
Copper gossans (11) (vein type)	270	240	15200	50	220	330	45	50	3	6
Pyrite gossans (12) (false gossans)	25	20	45	45	30	70	25	30	—	—
Ferricretes (laterites) (32)	75	—	15	105	65	1250	—	—	—	4

A few additional points, based mainly on experiences in Western Australia, merit comment:

1. The coarse fractions of both true and false gossans often yield higher trace element contents than do the fine fractions (see pp. 117, and 255-256, for examples from Western Australia).
2. Single element interpretation of gossans is not feasible (possible exceptions: Pt, Pd, Se, and probably Ag and Au).
3. Gossan interpretations based on mobile, scavengable elements like Zn, Cu and Ni, should also consider the amounts of Fe (and Mn) oxides present. A high ratio of, for example, Cu/Fe, or Cu/limonite, is more likely to indicate a true gossan than a high absolute value of Cu.
4. Gossan prospecting need not be confined to outcrops. Limonitic (gossan) pebbles can be collected from stream sediments and analyzed in the hope of recognizing mechanical dispersion patterns from source gossans.

Gossan formation and the resultant distribution of elements have recently been studied in the laboratory. Thornber and Wildman (1979), for example, weathered supergene Australian nickel ores in a circulating flow tank devised to simulate natural processes. Their initial results showed that the tendency of the base metals to coprecipitate with the Fe (in goethite) is: $Cu > Pb > CO \sim Ni > Zn$. Studies of this type are of great potential value for exploration geochemists. And, finally, it is important to recognize that most gossan and related laterite studies have been performed on materials from Western Australia which are old (mid-Tertiary) and, at present at least, in an arid environment. What is needed are more data on, and interpretations of, younger gossans in more humid and temperature environments.

PHYSICAL WEATHERING (pp. 86-87)

With respect to physical weathering, the following statements were made in the first edition (p. 86): "Overall, it plays a minor role in weathering processes on the Earth"; and, "Where chemical weathering is predominant, physical weathering plays a supplementary role".

Although these statements are true, one might be led to erroneously conclude that clastic (mechanical) fragments are not likely to be found in significant amounts in residual soil anomalies over mineralization in temperate, humid climates. However, such is not the case as Barbier and Wilhelm (1978) proved in their studies of surficial (secondary) dispersion around sulfide deposits in France. They showed that dispersion of anomalies in residual soils over mineralization is more predominantly mechanical than chemical (hydromorphic). Their work is discussed in more detail in Chapter 7A under "Mechanical Dispersion". On the other hand, in Antarctica, which might be considered a bastion of physical weathering, Boyer (1975) found evidence for chemical weathering in immature soils developed on plutonic rocks. The evidence comes from the discovery, by X-ray diffraction methods, of illite, montmorillonite, and certain mixed-layer clays characteristic of this type of weathering. He concluded that present meteorological data are compatible with an interpretation for slow, present-day chemical weathering. From the above two examples one can only conclude that caution must be exercised in predicting the type of weathering which may be predominant, or found, in any particular environment.

ENVIRONMENTAL FACTORS AFFECTING WEATHERING
(pp. 87-89)

Climate (pp. 87-88)

No additional comments.

Biological Activity   (p. 88)

No additional comments.

Parent Material   (p. 88)

No additional comments.

Topography   (pp. 88-89)

No additional comments.

SOIL   (pp. 89-124)

Introduction   (pp. 89-90)

The basic principles and discussions on soils in the first edition require no elaboration. Suffice it to mention that in the past six years many fine text and reference books have appeared on this subject. These include Buol et al. (1973), Birkeland (1974), Brady (1974), Bolt and Bruggenwert (1976), Marshall (1977), Dixon and Weed (1977), and Foth (1978). To those with a special interest in the soils of Canada, the two-volume compilation by Clayton et al. (1977) can be recommended as the first comprehensive attempt to describe them in some detail. UNESCO (1971) compiled an excellent review on soils and weathering in the tropics. Dregne (1976) has written a very comprehensive book on soils of arid regions which contains chapters on all the continents except Antarctica.

Soil Profiles   (pp. 90-96)

No additional comments.

Clay Minerals   (pp. 96-99)

No additional comments.

Classification Of Soils   (pp. 99-102)

Birkeland (1974) reviewed soil classification problems and presented a good exposition of modern classification schemes. Modifications to the Seventh Approximation (p. 100) system of soil classification, which is now officially called the "Soil Taxonomy", have been published by the U.S. Department of Agriculture (1975).

The most current state of soil taxonomy in Canada has been described and explained by Canada Department of Agriculture (1978).

These newer classifications are still not widely used by exploration geochemists primarily because they: (1) are apparently more difficult to understand; and (2) require laboratory tests in order for the soil to be classified unequivocally. Nevertheless, some of the terms and concepts of the Soil Taxonomy are being used with increasing frequency in numerous disciplines as they do emphasize certain chemical and physical features which are of value. Accordingly, it is appropriate to at least list the roots (usually Latin or Greek) for the names of the ten "orders" (the broadest classification level) of the Soil Taxonomy presented in Table 3-5 (p. 100) so that the reader will have a better understanding for the basis of this classification.

Soil Order	Root	Soil Order	Root
Entisol	Rec*ent*	Spodosol	Gr. *spodos* (wood ash)
Vertisol	L. *verto* (turn; inverted)	Alfisol	Al-Fe
Inceptisol	L. *inceptum* (beginning)	Ultisol	L. *ultimus* (last)
Aridisol	L. *aridus* (dry)	Oxisol	F. *Ox*ide
Mollisol	L. *mollis* (soft)	Histosol	Gr. *histos* (tissue)

Zonal Classification (pp. 102-107)

No additional comments.

Intrazonal Soils (pp. 107-108)

No additional comments.

Azonal Soils (p. 108)

No additional comments.

Factors Affecting Soil Formation (pp. 108-114)

No additional comments.

Laterites (pp. 114-120)

Many very important studies have been reported on the weathering, mineralogy, and trace element geochemistry of laterites in Western Australia. Among the more recent and important are those of Wilmshurst (1975), Nickel and Thornber (1977), Nickel et al. (1977), Smith (1977), Gedeon et al. (1977), and Thornber and Wildman (1979). From the point of view of exploration geochemistry, these studies form a useable data base by which non-economic ironstones (i.e., false gossans, laterites, ferricrete) can be distinguished from true gossans (Table 3A-1) although the objectives of the various authors were different in most cases. The subject of distinguishing between true and false gossans was discussed above and needs no further comment here other than to stress that, since laterites represent residual iron (and aluminum) oxide accumulations from barren rock, and because iron oxides (i.e., limonite) are excellent scavengers of mobile elements such as Cu, Zn and Ni, a similarity can result between the trace element abundances of laterites and true gossans. Also, it is worth mentioning again that the Western Australian laterites which are found on the Archean Yilgarn Shield are the result of deep weathering of ultramafic rocks which occurred between the Eocene and middle Miocene epochs (Nickel and Thornber, 1977) when more humid conditions prevailed. Clearly, conclusions based on these studies cannot have universal application. Many more geochemical studies on laterites from other parts of the world are needed, such as that of Kronberg et al. (1979) on Brazilian laterites, in order to assess the degree of variability in lateritic weathering processes and trace element contents under different environmental and geological conditions. At this time there is no successful routine method of prospecting for bedrock mineralization using laterite samples in a manner analogous to conventional soil sampling.

Trace Elements In Soil Profiles   (pp. 120-124)

The model for soil profile formation (pp. 90-96) used by most exploration geochemists is one in which the processes of removal, addition and transfer of elements, as well as mineral transformations, are active. According to this model, the relative intensities of certain of these processes, operating in different environments on rocks which have different element concentrations, are responsible for the variability of soils and their trace element contents from place to place, and within a soil profile (Fig. 3-14, p. 122). As a result of these variabilities, geochemists must always be aware of potential difficulties when sampling soils in general, and in studying soil profiles in particular. The most important factor is to be able to distinguish between soils developed on residual materials from those developed on transported materials. Prior to conducting a soil survey, this is the most important observation which can be made. In the latter case, the procedure of sampling the B soil horizon will be ineffective, and it may even by misleading. Among other more common pitfalls are: (1) failure to take into account changes in soil type within the survey area (Fig. 3-9, p. 113); (2) failure to sample consistently from one soil horizon; (3) incorrect interpretation of the nature of the layering in soil profiles; and (4) contamination and soil disturbance.

Recognizing the possible great variability is one thing, but to be able to properly use the trace element content of soils and of soil profiles in exploration, is another. The latter requires much basic knowledge of trace element concentrations, including ranges and means, for various soil types throughout the world (Fig. 3-13, p. 121). Although more data continue to become available, as for example, the recent data on the Australian laterites mentioned above, still more is needed.

One major step in this direction is the book by Aubert and Pinta (1977) which is an extensive compilation of selected trace element concentrations for various soil types and their parent rocks (when available), in regions throughout the world. (Unfortunately, most of the data are pre-1969.) The following elements are reviewed and discussed in some detail in separate sections: B, Cr, Co, Cu, I, Pb, Mn, Mo, Ni, Se, Ti, V and Zn. Other elements discussed in lesser detail include: Li, Rb, Cs, Ba, Sr, Bi, Ga, Ge, Ag and Sn. Of special value to geochemists are the separate sections for certain of the above elements which are subdivided to include concentrations for soils in: (1) temperature and boreal regions; (2) arid and semi-arid regions; and (3) tropical humid regions. Concentrations of trace elements in chemical extractants indicative of plant toxicity and/or plant availability are also discussed. Clearly, information of this type is extremely valuable but cannot possibly be summarized here (but see Appendix E). Other sources of valuable information include papers concerned with environmental aspects of soils. For example, Connor and Shacklette (1975) reported the geochemical summary statistics for 48 elements in samples of natural materials (rocks, soils, plants and vegetables) from 147 landscape units in the conterminous United States. There summaries demonstrated the wide diversity to be expected in elemental properties of landscape units, and suggested that published element abundances for broad categories, like "soil", may be misleading.

Profile sampling involves the analyses of samples from several horizons in a soil profile; ideally it is used on residual soils but the principle may also be applied to displaced soils in humid-temperature areas (Fig. 8-2, p. 351), soils in permafrost regions (Fig. 11-8, p. 421), lake sediments, and in other special situations. The objective is to assist in the interpretation of anomalies including determining: (1) the type of geochemical migration (i.e., hydromorphic or mechanical); (2) whether an anomaly is directly over mineralization or displaced from it; and (3) whether an anomaly may be unrelated to mineralization (as in the case of surface contamination). Profile sampling may involve one or several size fractions. Various partial extractions, as well as a total extraction procedure, usually can be advantageously employed. The application of soil profile sampling in exploration geochemistry has been lucidly discussed by Bradshaw et al. (1974).

APPLICATION OF pH AND Eh   (pp. 124-134)

Introduction   (pp. 124-125)

No additional comments.

pH   (pp. 125-128)

No additional comments.

Eh (pp. 129-131)

Seasonal variations in the Eh of soils can be extreme in some cases. Fig. 3A-1 shows such variations over a two-year period. Clearly, such changes will have a marked effect on the mobility of certain elements in soils over a period of time.

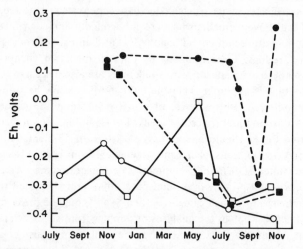

Fig. 3A-1. Seasonal variations in the Eh of five waterlogged soils (location not known) over a two year period. After a source cited by Carroll (1970, p. 115).

Eh—pH Diagrams (pp. 131-134)

No additional comments.

ADSORPTION (pp. 134-141)

Hydrous Iron And Manganese Oxides (pp. 134-138)

With respect to studies on adsorption of metals by hydrous oxides of iron and manganese, the statement was made (p. 135), "Literally hundreds of papers can be found, many of which report on the pH, Eh, and other conditions under which these minerals form, and also on their trace element contents." And the trend continues.

A selection of recent papers on various topics related to adsorption by hydrous iron and manganese oxides which are applicable to soils and sediments (both fresh water and marine) include: Murray (1975 a, b), Hem (1977, 1978), Means et al. (1978), and Eaton (1979). Among the many papers concerned with various aspects of exploration, the following may be cited as representative examples: Chao and Anderson (1974), Carpenter et al. (1975), Nowlan (1976), Bampton et al. (1977),

Wagner et al. (1978), Carpenter and Hayes (1978), and Butt and Nichol (1979). The entire subject of the significance of secondary iron and manganese oxides in geochemical exploration was reviewed by Chao and Theobald (1976).

Now that the importance of the ability of iron and manganese oxides to selectively scavenge more elements (primarily by adsorption or coprecipitation) is firmly established, it remains only to point out the important new findings of direct application to exploration. These include:

1. The findings of Whitney (1975) who showed that the prevalence and thickness of oxide coatings increases with particle size, and that the contents of Zn and Cd increase in parallel with the Mn content, but that the distribution of Pb is less regular, being related to Fe in some samples and Mn in others.

2. The documentation by Nolan (1976) of the selective scavenging of certain elements by Fe-Mn oxides, based on a multi-element correlation study of stream sediments and oxide coatings from Maine. He separated the elements into the following categories:

a.   elements not scavenged by Fe-Mn oxides: B, Cr, K, Mg, Rb, Sc, Ti, V and Zr;
b.   elements probably not scavenged by Mn-Fe oxides: Ag, Be, Ca, Ga, La, Sb and Y;
c.   elements scavenged weakly by Mn-Fe oxides: Cu, Mo, Pb and Sr;
d.   elements strongly scavenged by Mn oxides: Ba, Cd, Co, Ni, Tl and Zn;
e.   elements scavenged strongly by Fe oxides: As and In.

The above adsorption sequences appear to contain discrepancies by comparison with those reported previously, particularly with respect to Ag, Cu, Mo, and Pb (see discussions on pp. 135-137). Still other studies (e.g., Murray, 1975a) report the adsorption affinities of the selected metals for manganese oxides as: $Co > Zn > Ni > Ba > Sr > Ca > Mg$. For amorphous iron oxides, sequences such as $Pb > Cu > Zn > Ni > Cd > Co > Sr > Mg$, have been reported in the literature. Such variations in the affinities of metals for manganese and iron oxides are possibly the result of local environmental (or experimental) conditions, as well as the fact that different manganese and iron minerals may have different adsorption characteristics.

Of special importance to exploration geochemists is the necessity to determine which elements, and what proportion of these elements in a sample, are adsorbed (coprecipitated) with the Fe and/or Mn oxides, as this knowledge is extremely valuable during the interpretative phases of a program (see Fig. 3A-8). Such information can be gained from selective chemical fractionation techniques, a subject thoroughly reviewed by Chao and Theobald (1976), and also by the use of ratios, such as Zn/Mn (Nichol et al., 1975; Butt and Nichol, 1979). Over the years, many different extraction procedures have been used, for example, various concentrations of hydrochloric, nitric, acetic or oxalic acids. Chao and Theobald (1976) proposed the following extraction sequence in order to partition the Fe and Mn into specific fractions (ignoring the small percentage of exchangeable metals), and to sequentially release their adsorbed or coprecipitated metals:

a. Mn oxides: 0.1 M hydroxylamine hydrochloride in 0.01 M HNO_3 at room temperature, 30 minutes;

b. Fe oxides (amorphous): 0.25 M hydroxylamine hydrochloride in 0.25 M HCl at 70°C, 30 minutes;

c. Fe oxides (crystalline): sodium dithionite extraction at pH=4.75 at 50°C, 30 minutes;

d. Sulfide minerals:$KClO_3$ + (HCl − 4N HNO_3), 20 minute boil;

e. Silicate matrix: HF - HNO_3 digestion.

Needless to say, many geochemists will use different extraction procedures which will also lead to valuable information. For example, subsequent to the above recommendation by Chao and Theobald (1976), Peachey and Allen (1977) reported on the use of an ascorbic acid - ammonium acetate - hydrogen peroxide leaching system. They found that it dissolves significant amounts of S^{--}, SO_4^{--}, CO_3^{--}, PO_4^{--}, and MoO_4^{--} with little effect on secondary iron and clay minerals. Labile Mn oxides and organic matter are also dissolved but the relative importance of these phases can be quantified. In regions with Pb-Zn mineralization in northern England, this selective dissolution procedure helped to distinguish between anomalous samples related to mineralization and those related to adsorption onto secondary manganese oxides (Dr. J. D. Appleton, personal communication).

Organic Matter   (pp. 139-140)

The effect of organic matter on the adsorption of metals was documented earlier (pp. 139-140) and additional examples are presented below (for example, in connection with bogs; see Figs. 11A-1 and 11A-3). As in the case of the Fe-Mn oxides discussed above, there is no unanimity of opinion as to the relative affinity of metals for organic matter. Typical adsorption sequences for divalent elements might be: Ni > Co > Pb > Cu > Zn > Mn > Ca > Mg, or, Cu > Ni > Co > Pb > Ca > Zn > Mn > Mg. In addition to the diverse nature of materials falling within the category of organic matter, other factors, such as pH, Eh, inorganic mineralogy and humus content, will affect the retention mechanism. The entire subject of metal-sorption interactions with organic matter has been reviewed by Jackson et al. (1978).

Two additional points merit repetition. The first point is that, analogous to the Fe-Mn oxides, the organic content (or loss on ignition, LOI, which is a reflection of it) is frequently determined on soils and sediments which contain significant amounts of such material, prior to the interpretation of the trace element contents. This is the common procedure in the case of: (1) center-lake bottom sediments in Canada which have a high organic content (e.g., Garrett and Hornbrook, 1976); and (2) many uranium surveys employing soils or sediments. The second point is that organic matter can have the opposite effect to adsorption, that is, it may increase the mobility of elements primarily by forming metal-organic complexes. By way of example, Baker (1978) showed that humic acid can dissolve, complex and transport gold (as much as 330 ppb was taken up by solutions containing 550 ppm humic acid during a 50 day period). Jackson and Skippen (1978) found that organic acids are capable of increasing the solubility of Cu, Zn and Ni in the presence of clays under certain experimental conditions. The entire subject of increased mobility resulting from the formation of metal-organic complexes, is extremely complicated and beyond the scope of this book; it has been reviewed by Jackson et al. (1978).

Other Examples Of Adsorption (p. 141)

No additional comments.

MOBILITY IN THE SECONDARY ENVIRONMENT
(pp. 142-150)

From the earlier discussion of chemical mobility in the secondary environment, and from Fig. 3-19 (p. 143), it is clear that the mobility of any element in an aqueous medium varies considerably from one location to another. The factors affecting the aqueous migration of elements in the secondary environment which have received the most attention in this book, as well as in other studies, are pH (acidity-alkalinity) and Eh (oxidizing or reducing environment). These are certainly the most important of the numerous geochemical controls (see Table 2A-1). However, experience in the Soviet Union has shown that it is desirable and feasible to consider two main types of reducing environments (Table 2A-1): (1) reducing gley (i.e., without hydrogen sulfide); and (2) reducing with hydrogen sulfide.

In a *reducing gley environment* not only are free oxygen and strong oxidizing agents absent from the waters, but so also is hydrogen sulfide. Sometimes, the waters will contain abundant carbon dioxide or methane often, in part, at least, the result of bacterial action. Under gley conditions, such as in bogs, certain deep lakes, water-logged soils, and some aquifers, iron and manganese, in the form of Fe^{2+} and Mn^{2+} respectively, migrate easily. The waters may be either acid (environment 1 in Fig. 3-15, p. 130) as is usually the case, or alkaline (generally between environments 2 and 6, at an Eh of 0.15 volts or less, in Fig. 3-15). The rocks and soils are generally grey or green in color. If the pH decreases considerably, copper, vanadium and uranium, among other elements, can be reduced (Cu^{2+}-Cu^{+}-Cu^{0}; V^{5+}-V^{3+}; U^{6+}-U^{4+}) and insoluble minerals of these metals will form (see p. 190 for a discussion of the formation of native copper in British Columbia lakes).

In a *reducing hydrogen sulfide environment* the waters, like those in the gley environment, do not contain free oxygen or strong oxidizing agents, but they do sometimes contain methane and other hydrocarbons, in addition to hydrogen sulfide. Bacterial activity is often a very important factor in the formation of the hydrogen sulfide (pp. 80-81). Here, iron and many other metals do not migrate in solution since they form very insoluble sulfides (e.g., pyrite, sphalerite, galena). The waters in this environment are generally alkaline, and the Eh is frequently strongly negative (as in environment 2 in Fig. 3-15).

Soviet geochemists have quantified the relative *rate of migration* (mobility) of an element in water by means of the following equation, which relates the content of an element in the dissolved solids of a water and its content in associated rocks (Perel'man, 1977):

$$K = \frac{m \cdot 100}{a \cdot n}$$

where, K is the coefficient of aqueous (water) migration;

m is the content of an element in the surface or ground water which drains an area (g/l);

n is the content of an element in the rock over, or through, which the water flows (%);

a is the total mineral residue (dissolved solids) in the water (g/l).

The greater the coefficient K, the faster the migration (i.e., the greater the mobility). For example, K for Ca and Mg in the majority of landscapes ranges from 2 to 20, whereas for potassium it is one-tenth of that range. Fe^{3+}, whose K is generally less than 0.1, is a poor migrant (immobile) in the weathering zone. U and Mo, whose K values are generally greater than 1 in oxidizing environments, are considered good migrants (mobile).

Table 3A-3, based on coefficients of aqueous migration presented by Perel'man (1977), is an attempt to expand Fig. 3-19 (p. 143) by subdividing the reducing environment into reducing gley (without hydrogen sulfide) and reducing hydrogen sulfide categories. Only elements of particular value, or potential value, in hydromorphic (chemical) dispersion have been considered. Very acid conditions (pH < 4) such as exist at oxidizing sulfide deposits or their springs (see Table 3-7, p. 161) are not considered in Table 3A-3 because they are rare. The fact that many metals, such as Mo and As, may be scavenged by Fe-Mn oxides in the oxidizing environment, and by organic matter in the reducing environment, is also not considered.

As with any other classification, Table 3A-3 does not take into account all behavioral characteristics of the elements in the secondary (surficial) environment, only some of the major ones. Clearly, the classification in Table 3A-3, is another example of the use of the concept of geochemical barriers, that is, areas in which the rates of migration of the elements decrease sharply, and the elements consequently become concentrated. As the concept of geochemical barrier implies a change in the geochemical environment, rather than geochemical stability, it is entirely possible, for example, to have a reducing environment change to an oxidizing environment. Accordingly, an immobile element may become mobile, and *vice versa,* as is so common in lakes (see Fig. 3-29, p. 180). This phenomenon is referred to as re-solution at a geochemical precipitation barrier.

Table 3A-3. Relative aqueous mobilities of selected elements in waters of the secondary environment. Based mainly on data in Perel'man (1977, Tables 4 and 10).

RELATIVE MOBILITIES	ENVIRONMENTAL CONDITIONS		
	Oxidizing (pH > 4)	Reducing Gley (without hydrogen sulfide)	Reducing Hydrogen Sulfide
Very Mobile $K = 10 - 100$	S, Cl, Br, I, B, He, Rn	Cl, Br, I, B, He, Rn	Cl, Br, I, B, He, Rn
Mobile $K = 1 - 10$	Ca, Na, Mg, F, Sr, Zn, U, Mo, V, Se, Te, Re	Ca, Na, Mg, F, Sr, Mn^{2+}, Zn, Cu, Ni, Pb, Cd	Ca, Na, Mg, F, Sr
Slightly Mobile $K = 0.1 - 1.0$	Si, K, Mn, P, Ba, Li, Rb, Cs, Pb, Ni, Cu, Co, As, Cd, Tl, Ra, Hg, Ag	Fe^{2+}, Co, Hg, Ag, Si, K, P, Ba, Li, Rb, Cs, As, Tl, Ra, Hg, Ag	Si, K, P, Ba, Li, Rb, Cs, Tl, Ra
Immobile $K = <0.1$	Fe; see footnote 3	U, Mo, V, Se, Te, Re	S, V, Mo, Se, Te, U, Re, Mn, Zn, Cu, Ni, Pb, Cd, Fe, Co, As, Hg, Ag

Notes:
1. In the oxidizing and reducing gley environments:

    (a) Zn, Cu, Ni, Pb and Cd are mobile or slightly mobile as long as the pH is less than 7; these elements are precipitated in an alkaline environment (pH of hydrolysis, Table 3-6, p. 128).

    (b) Hg and (Ag?) are slightly mobile in both acid and alkaline environments.
2. Although most elements listed above are assumed to travel as ions, some (e.g., Mo, U, V, Se, Re) travel as complexes, such as, MoO_4^{2-} and $UO_2(CO_3)_2^{2-}$.
3. The following elements are considered chemically immobile in all common aqueous environments (pH > 4, oxidizing and reducing): Al, Ga, Cr, Ti, Zr, Hf, Y, rare-earths, Nb, Ta, Be, Th, Sn, Pt-group, Au.

Although chemical mobility has been primarily discussed in this section, it is important to emphasize again (also see Chapters 8 and 8A) that several kinds of mobility and resultant anomalies, i.e., both chemical and mechanical, can be represented in the same sampling medium of the secondary environment. A case in point can be illustrated by reference to the study by Butt and Nichol (1979) in which they found three different types of heavy metal anomalies in the stream sediments of Northern Ireland. Two were significant anomalies: (1) a hydromorphic anomaly identified on the basis of abnormal Zn/Mn ratios; and (2) a clastic Pb anomaly. One was a non-significant (false) hydromorphic anomaly derived by leaching of Zn and other metals from acidic and/or gley soils which were subsequently concentrated by coprecipitation with manganese oxides. Distinguishing between the various dispersion mechanisms and anomalies was achieved by different partial chemical extractions and the analysis of several size and gravity fractions.

Electrochemical Dispersion. It is, perhaps, appropriate at this point to mention that the exact mechanism by which elements migrate in the secondary environment is not always clear. Although most geochemists simply assume that the metals in solution have a minimum of difficulty in migration, others have been concerned with developing a mechanism to explain the transport of metals, i.e., dispersion of elements, which results in geochemical anomalies in overburden (e.g., till, colluvium) or rock overlying and post-dating mineralization. Toward this end, several respected researchers (e.g. Logn and Bolviken, 1974; Bolviken and Logn, 1975; Govett, 1975, 1976; Govett et al., 1976; Bolviken and Gleeson, 1979) have invoked the self-potential theory of Sato and Mooney (1960) in order to explain the migration of ions in an electrochemical field; migration of this type is called *electrochemical dispersion*.

In essence, the Sato and Mooney (1960) theory postulates that sulfide bodies can act as conductors connecting zones of differing oxidation potential. One of these zones would be above the water table (oxidizing) and the other would be below it (reducing), and a vertical gradient in oxidation potential (Eh) would result. The essential principles of electrochemical dispersion applied to this problem are illustrated in Fig. 3A-2. In this figure, electrons can be seen to flow from depth toward the surface within the ore body, a natural conductor. As a result, the top of the ore body

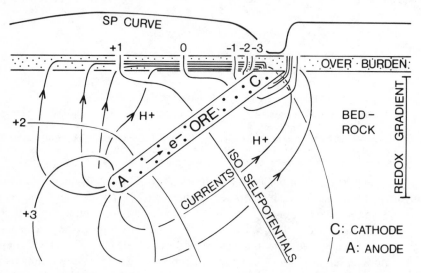

Fig. 3A-2. Simplified model illustrating principles of electrochemical dispersion. Courtesy of B. Bolviken.

is negatively charged with respect to its surroundings, whereas the bottom is positively charged. In the groundwater outside the conductor, cations, such as H^+ and the base metals (e.g., Cu^{2+}, Pb^{2+}, Zn^{2+},) move toward the surface as anions move to the deeper environments. Geochemical anomalies resulting from movements of metals upward into the overburden (note that arrows designated H^+ in Fig.

3A-2 also represent base metal cations) are likely to be weak and would probably be more easily detected by partial extraction procedures. However, in some cases there may be no detectable conventional geochemical response. In this situation, Govett (1975) has suggested that soil conductivity measurements may provide useful data as an adjunct to traditional soil geochemistry. (The determination of soil conductivity is simple and rapid. One gram of -80 mesh soil is weighed into 100 ml of deionized water and stirred for one minute after which the conductivity is measured with a conductivity meter.)

Webber (1975), however, suggests that ionic migration in an electric field is extremely slow and that the above mechanism is not nearly as powerful as either transport by flow in solution over long distances and times, or by diffusion over short distances and times. Additional research on this subject is justified in order to explain the observed weak geochemical anomalies which have formed through transported overburden and barren cap rock above mineralization.

WATER — GENERAL COMMENTS (pp.150-151)

From the discussions on element mobility in aqueous media in the previous section, it is clear that only those elements which have some degree of chemical (hydromorphic) mobility in natural waters, or are a component of a stable suspension (e.g., colloid), can be of value in hydrogeochemical surveys. When such elements are released from mineralization and are sufficiently mobile in solution for anomalous dispersion patterns to form in waters, *hydrogeochemical anomalies,* which are also called *aqueous dispersion patterns,* are produced. Aqueous dispersion patterns, like those in other sampling media, produce targets which are larger than the ore deposit and, as a result, the mineralization may be detected some distance away in both surface waters (Figs. 1-4 and 1-6) and groundwater (Fig. 3-23, p. 154). The halos always extend in the direction of water flow. The above statements apply to both surface and ground waters, as do all of the comments which follow in this section.

From the data in Table 3A-3 and in Fig. 3-19 (p. 143), it can be seen that there are at least 30 trace and minor elements which are sufficiently mobile under specified conditions in either surface or ground water, to be of possible use as pathfinders or indicator elements in hydrogeochemical surveys. Table 3A-4, based on work in the Soviet Union, lists some important element associations in aqueous dispersion halos originating from selected types of deposits. Beus and Grigorian (1977) note that such halos have a zonal structure, i.e., as the distance from the ore body increases, the concentration of elements varies, and the zones have different concentrations. Practically all the elements in the associations listed in Table 3A-4 are found immediately surrounding the specific type of deposit, in large part because of the more acidic (strongly oxidizing) nature of the waters. The intermediate zones are characterized by increased concentrations of Cu, Pb and Se. The peripheral zones are characterized by anomalous concentrations of a limited number of the more mobile elements, such as Zn, As, Mo and, sometimes, Hg. Numerous processes, such as dilution, adsorption, and oxidation-reduction barriers, lead to the "removal" of some of the elements listed in the associations in Table 3A-4 in different parts of the dispersion halo. (In connection with the discussion of Fig. 3-23, p. 154, hydrogeochemical zoning was implied inasmuch as the Mo halo was detectable 8 miles from the mine, but copper yielded a more restricted halo).

Beus and Grigorian (1977, p. 208) note, "A characteristic feature of hydrogeochemical anomalies (as compared to lithogeochemical anomalies) is the rather uniform distribution of the indicator elements within anomalous fields." It is

Table 3A-4. Hydrogeochemical indicator elements for different types of deposits. From Beus and Grigorian (1977).

Type of Deposit	Strongly oxidizing environment	Weakly oxidizing environment
Copper-pyrite	Cu, Zn, Pb, As, Ni, Co, F, Cd, Se, Ge, Ag, (Au), (Sb)	Zn, Pb, Mo, As, F, Se, (Ge)
Polymetallic	Pb, Zn, Cu, As, Ni, Ag, Cd, Se, (Mo), (Sb), (Ge)	Pb, Zn, As, Mo, Ni
Molybdenum	Pb, Cu, Zn, Be, F, Co, Ni, Mn, (Mo), (W)	Mo, Pb, Zn, F, As, Li
Tungsten-beryllium	Zn, Cu, As, F, Li, Rb, (W), (Mo), (Be)	Mo, F, Li, (W)
Mercury-antimony	Hg, As, Zn, F, B, Se, Cu, (Sb)	Zn, B, F, Ag, As
Gold ore	Ag, As, Se, Pb, Cu, Zn, Ni, Co, (Au), (Mo), (Sb)	As, Mo, Zn, Ag, (Sb)
Tin ore	Pb, Cu, Zn, Li, F, (Sn), (Nb), (Be)	Li, F, Zn, (Sn), (Be)
Titaniferous magnetite	Fe, Ni, Co, Cr, (Ti)	Ni, Fe
Spodumene	Li, Rb, Cs, Mn, Pb, Sr, F, (Nb), (Ga)	Li, Rb, Cs, F
Copper-nickel	Ni, Cu, Zn, Co, Ag, Ba, Sn, Pb, U	Ni, Zn, Ag, Ba, (Sn)
Beryllium-fluorite	F, Li, Rb, W, (Be)	F, Li, (Be)
Baritic-polymetallic	Ba, Sr, Cu, Zn, Pb, As, (Mo)	Sr, As, Mo, (Be)

Note:

The elements in brackets are either immobile under the specified conditions (e.g., Mo is only soluble under weakly oxidizing or alkaline conditions; see Fig. 3-20, p. 147), or there are no suitable analytical techniques for the determination of these elements (e.g., Sb, Ge, Sn) to the low detection limits required in waters. However, there may be conditions where it is desirable to analyze for these elements.

for this reason that one hydrogeochemical sample, particularly of groundwater, may be representative of the water chemistry over a large area. From this it follows that hydrogeochemistry can be profitably used in regional exploration under favorable environmental and geological conditions. However, in many cases stream sediments are equally, or more, effective because they are considerably less variable in trace element composition by comparison with stream waters. In the surface environment there are very few elements that are more stable in the water than in the inorganic sediments (uranium and molybdenum, which travel as anionic complexes in alkaline waters, as well as fluorine, are examples of elements which are more stable in specific waters). Therefore, hydrogeochemical sampling of streams is usually recommended as just one possible technique to be used during regional exploration (see Fig. 9A-5), particularly for the detection of subsurface discharges. Except in Canada and parts of Scandinavia where lake waters are important, groundwater surveys are, by far, the most valuable type of hydrogeochemical surveys. Such surveys can be used as an aid in finding concealed deposits when no surface methods, such as soil or stream sediment sampling, are applicable. The groundwater may be sampled directly by means of springs, seeps or wells, or indirectly by evaluating the base-flow component of stream water, particularly at periods of low flow.

In view of the increasing emphasis being placed on hydrogeochemical methods at the present time, it is necessary to expand upon the inter-relationships between trace element content and three fundamental parameters: total dissolved solids, conductivity, and discharge.

At the outset, it is important to emphasize that the absolute content of an element in an aqueous medium can be misleading for any number of reasons. One is that the element may be affected by variations in the *total dissolved solids* (TDS) content. This is illustrated in Fig. 3A-3 where a linear relationship can be seen between variations in the uranium content and the total dissolved solids of streams in the Caucasus which drain granodiorites containing 2 ppm U (Lopatkina, 1964). A similar linear relationship between uranium and total dissolved solids was demonstrated for streams draining granites containing 6 ppm U in other parts of the Soviet Union, and a plot of these values on Fig. 3A-3 is shifted "upwards" (e.g., waters with 100 ppm TDS would have about 1.5 ppb U compared to 0.5 ppb U for waters draining the granodiorites). Hence, this example illustrates the point that the trace element content of the rock over, or through, which a stream or groundwater flows must be taken into account (or at least that portion of the element content which is available for dispersion in solution) when interpreting the trace element content of water. Another example of the influence of rock type on the uranium content of surface waters may be seen in Fig. 4-8 (p. 216). Obviously, there is likely to be a poor correlation between the trace element content of a stream and its total dissolved solids content if there is a considerable variation in the trace element content of the rocks traversed by the stream. Accordingly, geochemists ideally should consider the following factors, among others, in their evaluation of hydrogeochemical data: (a) variations in the types of rocks within a drainage basin (groundwater flowing through aquifers will tend to be relatively uniform in composition); (b) the maximum amount of the element within the respective rock types which is soluble in the surface or ground water; and (c) the total dissolved solids content.

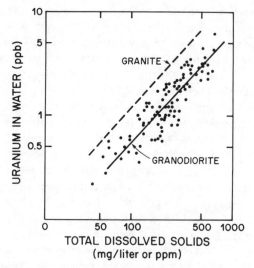

Fig. 3A-3. Relationship between uranium content and total dissolved solids of some surface waters (streams) in the Soviet Union. The solid line is the approximate "best fit" for data obtained from streams in the Caucasus draining granodiorites containing 2 ppm U. The dashed line represents the approximate "best fit" for similar data from streams draining granites containing 6 ppm U elsewhere in the Soviet Union. After Lopatkina (1964).

The total dissolved solids of a water sample is *approximately* directly proportional to its *conductivity* (specific conductance). Conductivity (an electrical property easily measured in the field or laboratory with a small portable meter) for a specified amount of total dissolved solids will vary with the chemical nature of the water, more specifically, with the relative percentages of the major components of waters, listed in Table 3-1 (p. 74). There are three major types of river water (Table 3-8, p. 171): Ca-HCO_3; Ca-SO_4; and Na-Cl. There are even more types of groundwater, some of which are unusual such as the $CaCl_2$ or $MgCl_2$ brines associated with oil fields. As a result, there is no unequivocal way in which conductivity readings (in micromohs/cm) can be accurately converted to total dissolved solids (and especially the content of a trace element) unless a significant amount of data on the chemistry of specific waters are available beforehand. For example, conductivity values (in micromohs/cm) multiplied by the empirically determined factor 0.55 yield approximate values for the total dissolved solids content of rivers (mostly the Ca-HCO_3 type) in western Canada south of the Saskatchewan River. The range of this factor for individual rivers in this region is 0.51 - 0.62 which, alone, can introduce a significant error. Nevertheless, north of the Saskatchewan River, such as in northern Saskatchewan and northern Alberta, the conversion factor increases to 0.68, and it approaches 0.80 in the northern part of the Northwest Territories. One reason advanced for the increased value of the conversion factor is that humic acids, which are abundant in the rivers of northern Canada (as in the tropics), add to the total dissolved solids content but do not increase the conductivity. Clearly, the conversion of conductivity values to total dissolved solids content for any type of water which may be used for exploration purposes is subject to error.

Usable conversion factors for specified localized areas, at least for most purposes, can be determined following detailed chemical analysis of samples collected for hydrogeochemical orientation surveys. They also may be calculated from published water analyses for specific areas or, alternatively, they may be obtained from appropriate governmental agencies. Because accurate conversion factors are difficult to obtain, most knowledgeable exploration geochemists simply use conductivity values *per se*, and normalize trace element data against these measurements (e.g., U/conductivity). This is based on the assumption that the only parameter which will affect such measurements in an exploration area will be the total dissolved solids content, specifically the major components listed in Table 3-1 (p. 74), and these are assumed to be essentially constant for a specific area. In most cases this empirical approach is satisfactory as long as the possible pitfalls are recognized, and the area to be explored is small.

Now that the approximate *directly* proportional relationship between total dissolved solids and conductivity has been established, it is appropriate to discuss a third variable — *discharge* (or flow rate). This parameter is *approximately inversely* proportional to the other two (Fig. 3-26, p. 162). Fig. 3A-4 illustrates the usual relationship

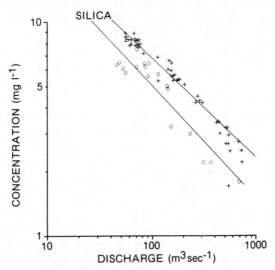

Fig. 3A-4. Relationship between discharge (cubic meters per second) and dissolved silica content (mg/1) in waters of the Squamish River, British Columbia. Regression lines are included for two statistically significant seasonal periods (represented by circles for December through June and by crosses for July through November). From Kleiber and Erlebach (1977).

between an element, dissolved silicon (H_4SiO_4, reported as silica) in this case, and discharge for the Squamish River, an unpolluted river in southwestern British Columbia. The decrease in dissolved silica is simply related to the dilution effect of increased discharge from precipitation or melting snow. However, statistical studies by Kleiber and Erlebach (1977) show that two seasonal periods exist with respect to the concentration-discharge data, and these are represented by two regression lines in Fig. 3A-4. Clearly, only an approximate inverse relationship between element content and discharge exists on the basis of the spread in the data, and the fact that there are two regression lines.

Although an inverse relationship of the type illustrated in Fig. 3A-4 can be expected for most minor and trace elements, as well as major elements, exceptions do exist. A case in point is illustrated in Fig. 3A-5, again from the Squamish River. Here, the phosphorus content *increases* with discharge. Kleiber and Erlebach (1977) suggest that the increase in phosphorus can be correlated with an increase in the concentration of phosphorus-containing fine-grained solids re-suspended from the river bed by increased turbulence. Similar relationships might also be expected for other

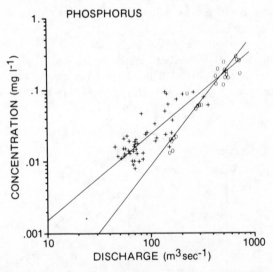

Fig. 3A-5. Relationship between discharge and phosphorus content, Squamish River, British Columbia. Explanation of symbols the same as in Fig. 3A-4. From Kleiber and Erlebach (1977).

trace elements adsorbed on the fine particulate matter. Although the analytical technique used for the Squamish River samples (Fig. 3A-5) detected "total" phosphorus (i.e., phosphorus in very fine particles as well as adsorbed phosphorus), which may not be applicable to all exploration situations, Fig. 3A-5 does show that caution is required before generalizations can be made with respect to element abundances, particularly of the trace elements, and physical parameters in natural waters. Further, it is appropriate to emphasize that *almost every surface and groundwater system has its own peculiarities,* a fact well known to chemical hydrologists. Accordingly, the direct relationship between phosphorus and discharge illustrated in Fig. 3A-5 may only be applicable to the Squamish River.

A question commonly asked when hydrogeochemical surveys are being contemplated is, "What field and laboratory measurements should be made, and what elements determined, in the waters collected?" Of course, there is no set answer to this question because of the diverse nature of: (a) the objectives of such a survey (e.g., type of mineralization sought; see Table 3A-4); and (b) the environmental and geological parameters in the exploration area (e.g., see Table 3A-3).

For the sake of illustration, let us assume that a reconnaissance hydrogeochemical survey is to be conducted for uranium in one of the semi-arid to humid regions of the western United States or British Columbia, and that both river and spring waters are available for sampling (but not groundwaters from wells). In this set of circumstances, the following parameters would be determined, ideally in the field, but if this is not possible, as soon as feasible in the laboratory to which the samples will be expeditiously sent:

a. *conductivity:* determined by means of a conductivity meter. Although conductivity is preferably determined in the field, it is a reasonably stable parameter and can be determined later in the laboratory.

b. *pH:* determined both in the field and later in the laboratory. pH values in the laboratory are likely to be higher (increased alkalinity) by comparison with those obtained in the field, the increase being a measure of CO_2-loss. Measurements should be made with a pH meter, as none of the commercially available pH indicator papers give accurate and reproducible results for all types of surface waters with low dissolved solids. Some of the color range pH indicator kits (based on the sensitivity of color indicators, such as phenol red) apparently give reasonable results.

c. *alkalinity:* this is effectively the HCO_3^- and CO_3 content. The necessary titrations should preferably be performed in a field laboratory.

d. *Eh:* determinations of the oxidation-reduction potential with meters, whether in the field or laboratory, yield results which are notoriously difficult to reproduce and interpret. Accordingly, Eh is determined indirectly, if at all, by observing any smell of hydrogen sulfide in spring waters, or by the determination of dissolved ferrous iron in these waters.

The following elements should be determined in a laboratory:

e. *Uranium:* may be determined in a field laboratory equipped with a fluorometer (Fig. 6-12) or a laser-induced fluorescence analyzer (Fig. 6A-2).

f. *Phosphate:* should be considered in view of the recently recognized importance of the uranyl-phosphate complexes (see Fig. 3A-10).

g. *Pathfinders* for uranium: pathfinders for the roll-front (sandstone) type of deposit include Cu, Mo, V, and Se. For certain other types of uranium deposits, perhaps F, Cu, Co and Ni may be the desired pathfinders (see Appendix E).

h. *others:* if pH is low (acid), SO_4 might be advantageously determined to establish the source of the acidity, such as oxidizing pyrite or sedimentary rocks (e.g., gypsum). Use of ratios, such as SO_4/conductivity, or SO_4/Cl, may be useful.

The above tabulation and very brief discussion on field and laboratory measurements is just the "tip of the iceberg". Factors which must also be considered before the survey is begun include: whether stream sediments should be collected along with waters; whether the sample should be acidified or not (some determinations may require acidified samples and others unacidified samples, so two bottles of water may have to be collected at each station); and, which analytical methods should be employed. All these topics are within the realm of orientation surveys, field methods, and analytical techniques, discussed elsewhere in this book.

GROUNDWATER (pp. 152-161)

Occurrence, Movement And Fluctuation (pp. 152-155)

No additional comments.

Groundwater Aquifers And Element Transport (pp. 155-157)

No additional comments.

Chemistry Of Groundwater (pp. 158-161)

Most exploration geochemists recognize the fact that surface waters may change in composition for any one of many reasons. The concentration of a trace element in surface waters may be decreased by dilution, or it may be increased by evaporation particularly as the water passes from a humid to an arid region. These changes can be monitored by determining the conductivity or the total dissolved solids content of water samples, as discussed above.

However, many geochemists fail to recognize that the chemistry of groundwater can also change significantly for a number of different reasons. Fig. 3-24 (p. 156) illustrates the changes in the chemistry of groundwater when the environment changes from oxidizing to reducing (an oxidation-reduction geochemical barrier) in a confined aquifer. Here, accompanying the precipitation of uranium, in some cases, selenium, vanadium and other elements also may precipitate Oxidation-reduction sequences in such groundwater flow systems have been discussed by Champ et al. (1979) who recognize three redox zones: (1) oxygen-nitrate; (2) iron-manganese; and (3) sulfide.

Other causes of change in the chemistry of groundwater are explained in terms of the sulfate-carbonate and the adsorption geochemical barriers (Table 2A-1), which have been discussed earlier (pp. 158-159). Well-defined wet and dry seasons may result in significant variations in the chemistry of groundwater particularly near the water table (p. 160). Seasonal springs may contain very high contents of metals when groundwater "flushes" oxidized mineralization at the onset of the wet season; these high values may also be reflected in the base flow of streams long after the wet season depending on the rate of groundwater movement, on the distance of the mineralization relative to discharge site, and on other factors mentioned on p. 163. In some cases, evaporation of groundwater may occur, particularly when the water table is close to the surface and the zone of aeration is very porous, and this process may result in higher trace element and dissolved solids contents. (This is the evaporation geochemical barrier listed in Table 2A-1). For evaporation to be a significant process in increasing the content of elements in groundwater, the following additional conditions also must be satisfied: (1) slow movement of the groundwater from the recharge area to discharge area; and (2) the "moisture gradient" must be sufficient to encourage the diffusion of water from the saturated to the unsaturated zone by capillary forces. Both Hambleton-Jones (1978, p. 184) and Thomson and Reed (1979, p. 315) mention that "evapotranspiration" of groundwater is a viable process in arid regions.

Finally, by way of example, solution of rocks (the aquifer) by slowly moving groundwater, as well as solution of their cements and secondary minerals and salts (e.g., those in halomorphic soils), particularly in arid and desert regions, can result in an increased content of individual elements and the total dissolved solids content (salinity). An example of solution of an aquifer (including secondary minerals) causing a spectacular change in the uranium content (from 4 to 400 ppb) is shown in Fig. 3A-6 which is based on an example from an arid region in southern Africa. As the high values in this figure do not represent mineralization, they are false anomalies. A general increase in the conductivity and total dissolved solids content of the ground-

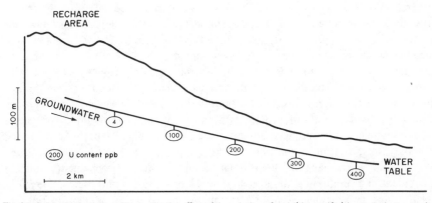

Fig. 3A-6. Schematic diagram illustrating the effect of the solution of rock (the aquifer) by groundwater on the uranium content (there is no known uranium deposit in the area). Based on an example from an arid region in southern Africa.

water accompanies the increase in uranium, and these suggest that the increase in uranium is related to rock solution. It is also possible that evaporation (discussed above) is the mechanism responsible for the increase in uranium; however, in the case shown in Fig. 3A-6, the groundwater is fairly deep which would make the evaporation process of minor importance. This is another example in which the abundance of an element is best expressed as the ratio between that element and the total dissolved solids content, or the conductivity. It also illustrates that the element contents of groundwaters, as well as surface waters, are influenced, sometimes strongly, by climatic and regional factors. In general, however, such factors are more important in explaining the chemistry of surface waters, e.g., the low pH of bog waters in the colder regions of Canada, as well as waters in high rainfall tropical regions, because both have high organic acid contents.

In cases where sequential values such as those shown in Fig. 3A-6 do, in fact, represent mineralization, following the collection and analysis of water samples from several wells, it is possible to calculate the maximum distance to the source of an anomaly. Hoag and Webber (1976a) showed, that in the specific region of eastern Quebec they investigated, travel distances of Cu and Zn in waters from mineralization (also from contamination) could be approximated as a function of hydraulic head, sodium concentration in the water, and a constant which had to be derived for the specific area. In another study on groundwaters from this same area, Hoag and Webber (1976b) have been able to evaluate the significance of the total sulfate content from the point of view of recognizing the proportion which originates from the oxidation of sulfide minerals, as well as from other sources. Specifically, in their study area, which contains no sulfate of sedimentary origin such as anhydrite or gypsum beds, sulfate concentrations of 28 ppm or less indicate no, or very little, surface oxidation, and no sulfides exposed to free oxygen. In this case, conditions may not be favorable for geochemical soil surveys, as these depend upon metal dispersion, either chemical or mechanical, at the surface from the weathering of sulfides. However, mineralization may exist below the zone of oxidation and be detectable by means other than sulfate, for example, by metals in the groundwater. Sulfate concentrations between 25 and 160 ppm indicate some surface oxidation in which both inorganic and bacterial processes are operative. Soil geochemistry should be useful in this case although the mineralization is probably disseminated or partially protected from oxygenated waters. Sulfate concentrations greater than 160 ppm are certain indicators of extensive bacterial surface oxidation; there should be a well developed gossan zone at the surface indicative of mineralization. Whether the sulfate content of groundwaters in other areas (containing no sedimentary sulfate deposits) can be used as successfully, is an open question.

For the serious student of the relationships between the chemistry and the movement of groundwater, the excellent books by Custodio and Llamas (1976) in Spanish, and by Schoeller (1962) in French, are highly recommended. In English, the books by Domenico (1972), and by Freeze and Cherry (1979), can also be recommended.

RIVER WATER   (pp. 161-173)

Sources And Fluctuations   (pp. 161-166)

No additional comments.

Mechanism Of Trace Element Transport In Rivers   (pp. 166-170)

No additional comments.

Chemistry Of River Water   (pp. 170-171)

For the student interested in a quantitative treatment of the variables which determine the composition of natural waters of all kinds (river, lake, groundwater, ocean water and estuarine water), the book "Aquatic Chemistry" by Stumm and Morgan (1970) is highly recommended. Their presentation draws upon the basic principles of physical chemistry in order to explain many of the topics treated qualitatively in this book.

Extent Of Geochemical Anomalies in River Water   (pp. 171-173)

No additional comments.

LAKE WATER AND SEDIMENTS   (pp. 173-191)

Introduction   (pp. 173-175)

Lerman (1978) has edited an advanced textbook on lakes which includes discussions on their chemistry, geology and physics. This book is particularly suited to the advanced student.

Stratification, Overturn And Classification   (pp. 175-179)

On p. 178, the following statement was made: "The wind is able to work effectively on a lake several miles long and mix it to a depth of some tens of feet . . .". The first published proof of this statement in the exploration literature can be found in a study by Cameron (1978). His report, based on hydrogeochemical exploration on small lakes (2 sq. km. or less) north of the tree line in the Northwest Territories, Canada, showed that the lake waters in this tree-less windy area are mixed when the lakes are ice-free. Therefore, surface lake water samples, such as those which can be rapidly obtained during a helicopter-supported survey, are satisfactory for exploration purposes. However, lakes further south in the Yellowknife area, N.W.T., and in northwestern Ontario, are stratified (Nichol et al., 1975).

Influence Of Oxidizing And Reducing Conditions   (pp. 179-182)

No additional comments.

Water Movement In Lakes   (pp. 182-184)

No additional comments.

Chemistry Of Lake Water   (pp. 184-185)

Not only are pH and Eh the main factors in controlling ability of a trace element to remain in solution in river water and groundwater, but they are also extremely important in lake water. Although this topic was discussed earlier (e.g., in connection with Fig. 3-29, p. 180), it deserves additional emphasis with an example from a different environment. Fig. 3A-7 shows the effect of increasing alkalinity (pH) on

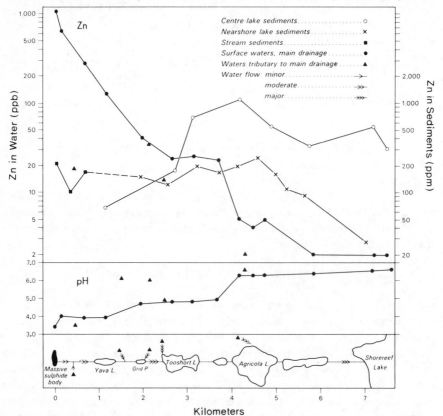

Fig. 3A-7. The distribution of Zn in lake water and stream sediments, Agricola Lake system, N.W.T., Canada. Note the coincident increase in pH in the waters and Zn in the lake sediments (and decrease of Zn in the waters). After Cameron (1977a).

restricting the dispersion of Zn (pH of hydrolysis = 7.0, Table 3-6, p. 128) in waters in the Agricola Lake area of the barren Canadian Shield (Cameron, 1977a).

Under the acid conditions generated by oxidation of the massive sulfide deposit whose location is shown in Fig. 3A-7, zinc is mobile in the surface waters and remains so until the pH of these waters approaches 7; the change in pH is caused mainly by the influx of higher pH waters from tributary streams. When the trace metals reach Agricola Lake (pH = 6.5) zinc becomes markedly concentrated in the lake sediments. Similar decay patterns were found for Cu and Pb in these waters but hydrogeochemical anomalies from these elements were shorter than the 4 km for zinc. The reduction (decay) in the metal contents of waters from mine drainages will appear similar to that shown in Fig. 3A-7 and will result in false anomalies.

Biological Aspects (pp. 185-187)

No additional comments.

Lake Sediments (pp. 187-190)

The observation was previously made that (p. 190), "Certain heavy metals, particularly those that are transported in solution such as zinc, copper and uranium, may be enriched in sediment samples containing organic matter, in a manner similar to the process described previously for soils. For this reason the organic matter content, as well as that of the iron-manganese oxides, is often determined along with the heavy metals." This is such an important concept, particularly with respect to the Fe-Mn oxides, that it bears re-emphasis.

The concept can be illustrated by reference to Fig. 3A-8 in which the highest values for arsenic are found in sediments at the north end of the lake where the manganese content also is highest. By plotting the arsenic to manganese ratio (multiplied by 100 to obtain integers as opposed to fractions), the proximity of the anomalous arsenic values to the mineralization can be immediately recognized. This is yet another example of: (1) the ability of Fe or Mn oxides to scavenge; and (2) the use of properly selected ratios of elements, as opposed to the content of an individual element, for data interpretation.

Summary (pp. 191)

The use of the sediments and waters of lakes is now an accepted geochemical exploration technique particularly in lakes of the Canadian Shield. The techniques are being refined and they are being applied in other areas (e.g., the lakes of British Columbia). The entire subject of lake sediments and lake waters applied to mineral exploration has recently been reviewed and discussed by Coker et al. (1979).

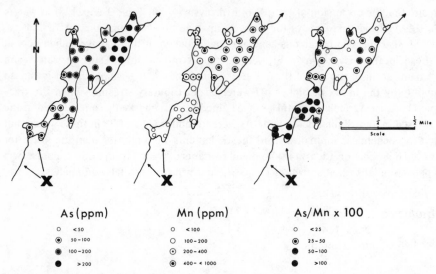

As (ppm)

○	< 50
◉	50 - 100
◉	100 - 200
●	> 200

Mn (ppm)

○	< 100
○	100 - 200
◉	200 - 400
◉	400 - < 1000

As/Mn x 100

○	< 25
◉	25 - 50
●	50 - 100
●	> 100

Fig. 3A-8. The distribution of As and Mn in lake sediments in relation to gold mineralization (indicated by X), Yellowknife, N.W.T., Canada. A plot of ratio As/Mn·100 shows the proximity of the As anomaly to the mineralization. From Nichol et al. (1975).

THE ARID ENVIRONMENTS
(A new topic)

Introduction

About one fifth of the Earth's land surface is desert (Holmes, 1978, p. 481). However, contrary to popular opinion, only about 20 percent of the desert areas are covered with sand. Deserts are usually considered to comprise the extremely arid and arid areas of the Earth which, according to Cooke and Warren (1973, p. 7), contain 4 percent and 15 percent of the land surface, respectively. If the semi-arid regions are also considered in this discussion, we are then concerned with another 15 percent, or a total of about one third of the Earth's land surface (Fig. 3A-9). (Extremely arid, arid, semi-arid, desert and other terms, are defined below).

Aridisols (from the Latin, *aridus*, see Table 3-5, p. 100) are the most abundant soil type (order) in the world covering 19 percent of the land surface (Foth, 1978, p. 269). In comparison, podzols (spodosols), which form the basis of many of the concepts used in connection with soil profiles and surveys in geochemical exploration (pp. 90-100), rank *eighth* among the soil orders covering only 5.4 percent of the Earth's land surface. Clearly, a study of the characteristics of desert and arid landscapes, and the geochemical techniques which are likely to be most successful in these areas, is justified. Regrettably, for one reason or another, serious attention to desert and arid environments in exploration geochemistry has been sorely neglected, at least in the published literature (including textbooks).

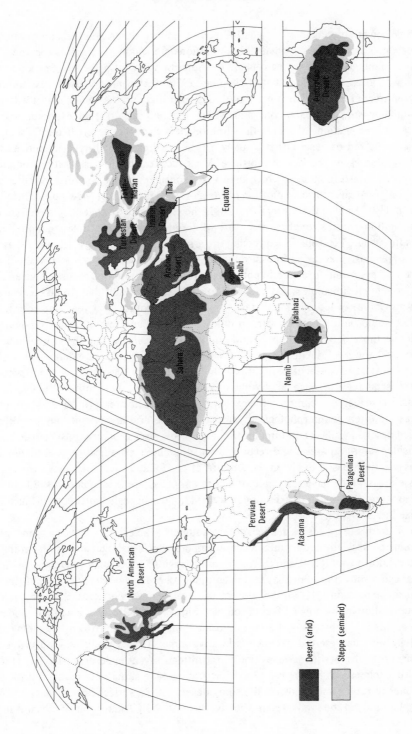

Fig. 3A-9. Map of the world showing the distribution of desert and semi-arid regions. From Zumberge and Nelson (1976).

When attempting to select a definition of the terms desert and arid environment, it becomes immediately evident that definitions are legion. A desert is regarded by some simply as a barren area capable of supporting few forms of life. For a definition with greater precision, that of Glennie (1970, p. 195) may be used: "An almost barren tract of land in which precipitation is so scanty and spasmodic that it will not adequately support vegetation, and where the potential rate of evaporation far exceeds precipitation". From both definitions it is clear that desert (from the Latin, *desertis,* barren or deserted) is actually a type of climate which is characterized by aridity, that is, by the lack of moisture. Most recent climatic classifications employ aridity or moisture indices, which involves a consideration of potential evaporation and water balance; the classification of Meigs (1953) is the most widely used. Briefly, Meigs (1953) noted that as temperature is the climatic factor that affects the rate of evaporation most strongly, some sort of ratio involving precipitation and temperature would be useful in expressing the aridity of an area. Meigs' scheme is based on a previously developed moisture index (by Thornthwaite), which relates the adequacy of precipitation to the hypothetical evapotranspiration of natural vegetation. Where the precipitation is just adequate to supply all the water that would be needed for maximum evaporation and transpiration in the course of a year, the moisture index would be 0. Climates with an index between 0 and minus 20 are *sub-humid,* whereas those between minus 20 and minus 40 are *semi-arid,* and those areas below minus 40 are *arid* (McGinnies et al., 1968). *Extremely arid* areas are defined as those where at least 12 consecutive months without rainfall have been recorded, and where there is no regular seasonal rhythm of rainfall. *Deserts,* as mentioned above, are usually considered to comprise the extremely arid and arid areas. However, semi-arid areas are often geomorphologically similar to more arid areas, and they are often called deserts locally. Therefore, for the sake of this discussion, we must occasionally consider semi-arid areas, and recognize that the above classification is gradational. Classifications based solely on the amount of precipitation (e.g., defining deserts as those areas which have less than 10 inches annually) are inadequate in the light of present knowledge. (In this discussion we will not consider the polar deserts, such as those in Antarctica.

The soils (e.g., desert, reddish desert, solonetz; Table 3-5) of the desert and arid regions possess numerous unique characteristics which distinguish them from their better-known counterparts in the more humid regions. According to Dregne (1976) arid region soils commonly have: (1) a low level of organic matter; (2) a slightly acid to alkaline reaction (pH) on the surface; (3) a calcium carbonate (caliche or calcrete) accumulation in the upper 5 feet of soil (see Figs. 3-7 and 3-8); (4) a weak to moderate profile development; (5) a coarse to medium texture; and (6) a low biological activity. In true deserts (extremely arid environment), as a result of wind erosion, transport, and subsequent deposition, three different types of desert surfaces are produced (Holmes, 1978): (1) the rocky desert, a desolate surface of bedrock with local patches of rubble and sand; (2) the stony desert, with a surface of rubble, gravel or pebbles; and (3) the sandy desert. Clearly, the extremely arid regions of the land surface have no soil cover at all, but are covered with exposed rock outcrops, gravel, rubble, and lastly sand. Complementary to the three types of desert surfaces is loess (eolian silt) deposited on the bordering steepes from dust-laden winds that blow outward from the desert.

Understanding the general concepts of landforming processes operative in extremely arid, arid, and semi-arid regions, as well as the soil (if present)-landscape relations, is essential for those attempting to use geochemical methods in these regions. Among the books which can be recommended for this purpose are those by McGinnies et al. (1968), Cooke and Warren (1973), Dregne (1976) and Mabbutt (1977) (most have been mentioned earlier in this discussion). Glennie (1970) has stressed ancient (from the Cambrian to the Recent) desert sedimentary environments. Numerous introductory geology and geography textbooks contain discussions on many of the essential geomorphological features of importance to exploration geochemists.

Arid environments are discussed in this chapter (The Secondary Environment) primarily because they represent the opposite extreme in precipitation from the water-related topics in the immediately preceding sections, even though fresh and altered rocks, which are usually associated with the primary environment, are frequently sampled for exploration purposes in regions with these climates.

Sampling Media In Arid Environments

Lithogeochemistry, biogeochemistry (phreatophytes, p. 390), groundwater, and soil gas become more important as sampling media with increasing aridity. At one extreme, in the case of a rock desert with thick transported soils, rock materials and deep groundwater may be the only sampling media possible, whereas in semi-arid areas with thin transported or residual soils the sampling of soils may be successful, as well as vegetation, and occasionally even surface water may be feasible. In an arid environment, as in all other environments, selection of the sampling media must be made so that the maximum information, which may lead to the discovery of mineralization, is obtained in the most cost-effective manner, i.e., the sampling media must offer the maximum chance of defining the exploration target at minimum cost.

Lovering and McCarthy (1978) have discussed the advantages and disadvantages of various sampling media used in the Basin and Range Province of the western United States. This region has climates ranging from desert, where outcrop (lithogeochemical) sampling is employed either alone or in combination with other media, to semi-arid where some of the more usual sampling media can be used. The remainder of this section will be concerned with a summary of the sampling media discussed by Lovering and McCarthy (1978) for the Basin and Range Province which, by analogy, can be used in other similar environments. Note also that many of the sampling media discussed below are *not* unique to the arid environments, but may be modified for these climates.

1. Fresh rock samples.

Application. recognition of: geochemical provinces; potentially mineralized plutons; primary halos; and ore bodies.

Advantages. generally only a small number of samples are needed.

Disadvantages. generally only applicable to intrusive rocks; anomaly contrast often low; statistical treatment of multi-element data may be needed.

Comments. see discussion on lithogeochemistry, primary halos and other related topics in Chapters 2, 7, 2A and 7A.

2. Altered rock samples.

Application. analysis of hydrothermally or pyrometasomatically altered rocks may indicate the presence of blind ore deposits related to these types of alteration.

Advantages. altered rocks: may define smaller targets than unaltered rocks; their location may indicate favorable zones for exploration (e.g., faults); trace element contents may be higher than in unaltered rocks.

Disadvantages. rock alteration generally limited to small areas in the vicinity of ore deposits, and it may result in fewer outcrops in areas with low to moderate relief because altered rocks chemically weather more easily; ore elements may be leached away during chemical weathering.

Comments: silicic (jasperoid), pyritic, and contact-metasomatic alterations, have been successfully used. See Fig. 9A-2.

3. Mineralized rock samples (includes vein fillings and fracture coatings).

Application. sampling primary metal concentrations (halos) of hydrothermal origin (vein fillings), or secondary dispersion of these elements (fracture coatings).

Advantages. provides information on channels followed by mineralizing solutions, and information on the more mobile elements which are found in the secondary fracture coatings (usually Fe-Mn oxides).

Disadvantages. anomalies largely confined to pre-ore conduits which were open at the time of mineralization.

Comments. because fracture coatings of Fe-Mn oxides (in deserts sometimes called "desert varnish") are so abundant in rock outcrops in some localities, such coatings can only be effectively used in selected circumstances, such as in the search for blind ore bodies under a relatively shallow cover of bedrock near other known deposits (see Figs. 9A-2 and 9A-3).

4. Mineral separates from rocks.

Application. trace elements in mineral separates (e.g., micas, magnetite) may be used to: discriminate between barren and mineralized plutons; define regional geochemical provinces; identify specific exploration targets (see discussions in Chapters 7, 2A and 7A).

Advantages. in some types of reconnaissance surveys certain mineral separates may yield more useful information than rock samples; primary halos based on some minerals (e.g., biotite) may be larger and more easily detected than other types of primary halos.

Disadvantages. requires extra sample preparation to separate the mineral phase; method still not widely tested and evaluated.

5. Stream sediment samples.

Application. when available, the same as stream sediments in more humid regions, i.e., detection of mineralization within a catchment area.

Advantages. the same as in more humid regions.

Disadvantages. not always available; the fine fractions (e.g., minus 80 mesh), so useful in more humid environments, are often diluted with fine windblown dust.

Comments. As in the more humid areas, water transported sediments in desert-arid regions may be composed of materials of numerous particle sizes (e.g., clay-size, sand-size) and mineral compositions (e.g., heavy minerals, resistates, light fractions), and without an orientation survey, the most effective type of stream sediment sample is often not known. One approach is to use two size fractions, one coarse (e.g., −4 to +16 mesh) to concentrate any gossan fragments and multiple grains cemented by metal-rich Fe-Mn oxides, and one fine (e.g., −80 mesh) to concentrate any hydromorphically transported metals (possibly moved upward by capillary action from groundwater), again with due caution to windblown dust. In some arid regions, coarse fractions (−10 to +30) have been found to be related primarily to the composition of the major rock units, whereas a heavy mineral (panned concentrate) fraction is related to mineralization (see Fig. 9A-3). Clearly, deposits can be missed if sediment sampling is conducted without due regard to the advantages and disadvantages of the various sizes and types of sediment fractions for the particular area.

Several studies in deserts have concluded that when two size fractions are selected for analysis, the coarser fraction should be subjected to a total extraction technique, whereas the finer fraction should be analysed following a weak partial (cold) extraction. In places like Saudi Arabia where perhaps 95 percent of the transport is by wind, but where it does rain occasionally at which time detritus is moved: (a) the minus 80 mesh size is of little value for exploration purposes; (b) the 30-80 mesh size fraction can be used for regional evaluation (e.g., determination of the rock type and background geochemistry); and (c) the 10-30 mesh size fraction can be used for detailed work to determine the local expression.

In some cases, the selective sampling of very coarse materials, such as stream cobbles and pebbles of hydrothermally altered or mineralized rock is recommended (Figs. 9A-1 and 9A-2); they are obtained where streams leave the mountains and there is a marked decrease in gradient. Such samples can provide important information on the presence of mineralization in the drainage basin, particularly in those regions where nothing else is available for sampling. If available, Fe-Mn oxides may be scraped off the rocks and these coatings analyzed separately for evidence of metals carried hydromorphically and subsequently absorbed or coprecipitated.

Lovering and McCarthy (1977, p. 261-262) note that in the Basin and Range Province, the silt-sized fraction of stream sediments is the most commonly used sampling medium for reconnaissance purposes, just as it is in the more humid regions. However, the geochemical anomalies result from finely ground particles of ore minerals and gossans transported mechanically, sometimes for many kilometers, as opposed to principally hydromorphic transport in the same size fraction in more humid regions. (See discussion of "Stream Sediments" as applied to arid regions in Chapter 5A).

6. Heavy mineral concentrates.

Application. primarily to recover ore minerals (e.g., cassiterite, gold), mineralized fragments, indicator minerals (e.g., magnetite), and oxidation products (e.g., gossan) which will lead to the primary mineralization.

Advantages. trace element anomalies are enhanced because the light fraction is removed; anomaly trains can be traced much further; mineralogical source of geochemical anomalies can often be determined by microscopic examination.

Disadvantages. requires extra effort either by panning in the field or by gravity separations in the laboratory. Panning requires water but a battery operated air separator or shaking table may be substituted. (See discussion of "Stream Sediments" as applied to arid regions in Chapter 5A).

Comments. Panning should only be carried out to the point where the light fraction (chiefly quartz, feldspar and mica) are removed; excessive panning may result in the loss of iron oxides, some sulfides, fluorite, barite, and other important minerals. Heavy minerals may also be advantageously divided into a magnetic fraction (chiefly magnetite which may, on occasion, concentrate Zn, Ti, Sn, W, Cr, Co and Ni) and a non-magnetic fraction. Further separation may be obtained, if warranted, by means of various settings on a magnetic separator.

7. Soil samples.

Application. generally for exploration on a detailed scale.

Advantages. anomalies are generally close to mineralization; fewer false anomalies because of the absence of organic matter to adsorb metals.

Disadvantages. soils in arid regions: generally are poorly developed, in part, because the decay products of vegetation are blown away and do not assist in the formation of soils; have about the same amount of dispersion as rock samples because they are formed by mechanical weathering; generally are not suited for reconnaissance programs; often have impermeable caliche (calcrete) layers which may inhibit the movement of water and thus mask underlying mineralization (however, Cox, 1975, and others as well, have found that the existence of a calcrete horizon overlying a Ni-Cu gossan in Western Australia did not impede soil geochemical surveys and that the calcrete itself could even be used as the equivalent of a soil sampling horizon).

Comments. In upland areas of predominantly desert regions, such as in the Basin and Range Province, precipitation and vegetation may be more abundant, and well developed soils and soil profiles may be found.

8. Groundwater samples.

Application. water from wells, springs and seeps represent a sample which may have leached large volumes of rock and may present clues to hidden mineral deposits; penetrate the base of alluvial fill on sediments.

Advantages. anomalous metal contents may be derived from concealed mineralization; even where specific metals are not soluble in groundwater, pathfinder elements, or a change in pH of the groundwater, may indicate proximity to ore.

Disadvantages. sources for the collection of groundwater are often very widely spaced; elements of interest may not be soluble; knowledge of hydrologic regime in an area may be essential for interpretation; changes in the chemistry of groundwater may be caused by evapotranspiration, or by solution of minerals in the aquifer, and this does not represent mineralization (see Fig. 3A-6).

Comments. The absence of a groundwater anomaly in a geologically favorable area should not be interpreted to indicate the absence of mineralization.

9. Biogeochemical samples.

Application. the detection of mineralization covered by soil, alluvium, or other transported materials by means of the analysis of deep-rooting vegetation (phreatophytes).

Advantages. detection of mineralization buried as much as 50 metres provided deep-rooted species are sampled.

Disadvantages. the extent to which elements are concentrated, and the variations in the concentration of the elements in different organs, make interpretation difficult (see Chapters 10 and 10A); only those elements which are mobile in groundwater can be detected by biogeochemical methods; chemical analyses are more expensive.

Comments. geobotanical anomalies can only occur over ore or pathfinder elements which are soluble in groundwater (and then under favorable conditions).

10. Others.

In favorable circumstances, vapor surveys (Hg, He, Rn, SO_2 and other volatile gases, etc.) may be used in arid environments and, in some cases, they work better than in humid regions because of the absence of soil or vegetation cover. The limited use of termite mounds and talus samples have been discussed elsewhere in this book.

Discussion

From the preceding section which considered the sampling media used in the various types of arid environments, it is clear that exploration geochemists face special difficulties in regions where this type of climate prevails. Nevertheless, in spite of the problems, when properly executed, geochemical surveys have been shown to be effective in these environments in southern Africa, Australia, southwestern United States, Saudia Arabia, Iran and Egypt.

Coetzee (1974) summarized the major difficulties in the arid regions of southern Africa as:

(1) narrow, low contrast anomalies over mineralized bodies;

(2) spurious anomalies produced by lithological differences; and

(3) problems in selecting the optimum size fraction for analysis.

Similar problems have been faced elsewhere, for example in Western Australia where Friedrich and Christensen (1977, p. 219) reported, ". . . the secondary dispersion halos of Scotia and Ringlock are narrow, restricted to a few tens of metres, and sampling at closer intervals is needed to locate the ore-bodies." Similarly, on the arid Stuart Shelf, South Australia, Rattigan et al. (1977) also found secondary dispersion halos very narrow and best detected by analysis of rock material obtained by deep (about 25 meters) drilling where permeable sequences allow the upward movement of mineralized waters from ore-deposits. Such problems are particularly serious in those arid areas covered with sands, whether they be stable (stationary) as in the Kalahari (Fig. 3A-9), or shifting as in parts of the Sahara.

Considering the sampling media which are available in arid environments in a positive manner we may conclude:

1. Soils, when residual or near-residual, give good geochemical responses, but may suffer from problems associated with dilution where there is a strong windblown component. Calcrete may reduce the intensity of a soil anomaly but it usually does not destroy it completely unless the calcrete is very thick.

2. Stream sediment (or wadi) sampling can be used with a considerable chance for success, however, the grain size which is selected is likely to be crucial.

3. Biogeochemical sampling may be successful provided deep-rooting plants (e.g., mesquite) are used and soluble elements are determined.

4. Lithogeochemical methods may be used in arid environments in essentially the same way they are used in more humid environments and, because of the absence of leaching (at least in recent times), superior results may even be obtained.

5. Hydrogeochemical methods, particularly groundwater, may be used but great caution must be exercised as the composition of the waters may be a reflection of various processes which are unrelated to mineralization, such as evapotranspiration or dissolution of minerals in the aquifer (Fig. 3A-6).

As is the case in the more humid environments, no substitute exists for a thorough, comprehensive, multi-media orientation survey over an undisturbed ore deposit of the same type, and in the same environment, as that being sought. As such an ideal circumstance is not usually encountered, the choice of sampling media, types of analyses to be done, and other parameters normally determined in an orientation survey, often are obtained by analogy to other known deposits in similar environments elsewhere in the world.

MOBILITY OF URANIUM AND ITS DAUGHTER NUCLIDES IN THE SECONDARY ENVIRONMENT
(A new topic)

Introduction

More has been written on the scientific aspects (e.g., physics, chemistry, geology) of uranium than on any other element. Therefore, any discussion of uranium within the confines of a textbook must, of necessity, be selective. This section will be primarily concerned with the mobility of uranium and its daughter nuclides in the secondary environment, because this topic is of utmost usefulness to exploration geochemists. The discussion will also attempt to integrate various fundamental geochemical principles found in other parts of this book, such as complexing and geochemical barriers. The specific aims of this section will be to emphasize that:

(1) exploration geochemists must recognize that in the search for uranium they are dealing with a multi-element (e.g., U, Ra, Rn) system;

(2) each element has its own chemical and physical properties which results in different mobilities so that different sampling and detection methods may be required; and

(3) in the chemical weathering environment, disequilibrium may occur (the definition of disequilibrium and its implications are discussed below).

There are many reviews and summaries of various aspects of uranium geochemistry and exploration, and only a few such as Rogers and Adams (1969), Rich et al. (1977), Gabelman (1977), and Dyck (1978) need to be mentioned at this point. In other sections of this book, the reader will find various special topics related to uranium exploration (for example, radon counters, high radium springs, and uranium false anomalies).

Before proceeding further it is advisable to digress into a short discussion of the geochemical behavior of uranium during igneous processes (the primary environment). One practical reason for this is that it permits the reader to understand the nature of the element associations which result in pathfinders in both the primary and secondary environments.

Geochemical Behavior Of Uranium During Igneous Processes

In the igneous environment, uranium is a mobile element and tends to accumulate in the late differentiates of igneous melts, primarily because the uranous (U^{4+}) ion has a high charge, which prevents it from substituting for any element its own size (e.g., Ca, Na) likely to be found in the crystal structures of early crystallizing major rock-forming minerals. Thus, the content of uranium in igneous rocks increases from an average of 0.001 ppm in ultramafic rocks to 4.8 ppm in granites. For the same reason, the thorium content of igneous rocks increases similarly from an average of 0.003 ppm in ultramafic rocks to 17 ppm in granites. From the following tabulation (using data in Table 2-1, p. 44):

	U(ppm)	Th(ppm)	Th/U
ultramafic	.001	.003	3.0
mafic (basalt)	.6	2.2	3.6
intermediate (granodiorite)	3	10	3.3
felsic (granite)	4.8	17	3.5

it can be seen that the Th/U ratio remains essentially constant (about 3.5) throughout the differentiation process, and that both elements accumulate in the residual magmas. Similar to U and Th, the potassium content of rocks increases with differentiation, and the K/U and K/Th ratios are also relatively constant at about 13,000 and 4,000, respectively. Therefore, there is, a geochemical association among U, Th and K in the igneous environment. However, in many alkalic rocks, the U and Th increase relative to K, with U contents of more than 15 ppm (up to 50 ppm) being common. Thus, many syenitic stocks and alkalic volcanic provinces are relatively high in U and Th, and as a result, they are good source rocks for subsequent uranium deposits in which formation by water transport following chemical weathering is required (for example, the roll front type; see Fig. 3-24, p. 156). Thorium, on the other hand, is not mobile in solution in the secondary environment and only thorium placer deposits can be expected.

In the late stages of differentiation, uranium may crystallize (substitute) in such primary minerals as zircon, allanite, sphene, pyrochlore, monazite, xenotime and apatite (see Tables 2-2, and 2A-5) or, where sufficiently concentrated, as a uranium mineral such as uraninite. (In this book, pitchblende is considered to be a fine-grained, poorly-crystallized, thorium-free variety of uraninite with a variable composition from $UO_{2.0}$ to $UO_{2.6}$; Rich et al., 1977, p. 35). Nevertheless, within late crystallizing rocks, the major rock-forming minerals (quartz, feldspar, mafic minerals) may carry the bulk of the uranium, even though these minerals contain very low amounts of the element. This is illustrated by the data in Table 2A-5 where it is shown that 91% of the uranium is found in the rock-forming minerals (normally much less, perhaps 50% of the uranium in granites, is found in rock-forming minerals). However, in some granites, the uranium may be dispersed as ionic or molecular disseminations. These disseminations, which generally are easily leached by dilute acid (or possibly even water when the rock is finely ground), may be found adsorbed on crystal surfaces, in crystal structure defects, along grain boundaries, in fluid inclusions, or in deuterically altered areas. This type of uranium, which exists outside of the crystal structure, is not firmly bonded and is called *labile* uranium. Labile uranium is relatively mobile, particularly if channels (e.g., microfractures) are available for the penetration of water, as is that uranium which is released by chemical weathering from the rock-forming minerals; both provide a source of uranium for subsequent secondary mineralization. Labile uranium in granite has been proposed as a source of many uranium deposits, such as the sandstone type in the United States and the vein type in France.

Thorium may also concentrate in some of the above-named minerals, particularly zircon and monazite, but it may also be found in other late stage minerals, such as thorianite, davidite and brannerite, with or without uranium. Both U and Th are often associated with Y and the other rare earth elements, Rb, Cs, Nb, Ta, Be and other dispersed or "incompatable" elements characteristic of pegmatites and other late stage differentiates.

Instead of crystallizing in the igneous rocks itself, the uranium may be separated from the magma chamber and precipitate as some of the above minerals (e.g., uraninite, brannerite) or other minerals (e.g., uranyl silicates) in late stage hydrothermal veins. In these late-stage fluid phases, the uranium commonly becomes separated from the thorium, in part, because of oxidation of the uranium from the uranous (U^{4+}) ion to the uranic (U^{6+}, or uranyl, UO_2^{2+}) ion. Because thorium is less efficiently separated into the late stage fluids (it does not go through a change in oxidation state), pegmatites and hydrothermal veins tend to have relatively higher uranium (lower Th/U ratios) than primary igneous rocks.

The mode of transportation of uranium in hydrothermal solutions has been the subject of several investigations. Because of the correlation between the uranium and carbon dioxide contents of fluid inclusions, high temperature transport as the uranyl carbonate complex has been proposed, and also questioned (e.g., Rich et al., 1977, p. 59) because of the decreasing stability of these complexes above 25°C. The association of uranium with the halogens, particularly fluorine (e.g., fluorite associated with uraninite in certain types of deposits), has suggested the formation of uranohalogen complexes. Complexes such as uranium hexafluoride, UF_6, are volatile and this may provide a mechanism for the escape of uranium from a magma chamber. Likewise, explanations for the precipitation of uranium from hydrothermal solutions are not entirely satisfactory. Based on the study of specific uranium deposits, some researchers have suggested that precipitation is promoted by the instability or break-up of the various complexes. Others, working on different types of deposits, which contain hematite, have suggested that the oxidation of Fe^{2+} to Fe^{3+} is accompanied by reduction of U^{6+} to U^{4+} and the subsequent precipitation of uraninite. These, and numerous other hypotheses, suggest that several mechanisms are possible to explain the formation of uranium deposits in late-stage igneous environments, and each situation must be considered individually.

Geochemistry Of Uranium In The Secondary Environment

The geochemistry of uranium in the secondary environment is governed primarily by three factors: (1) its variable oxidation state; (2) its tendency to form complexes with a large number of naturally occurring anions; and (3) its tendency to be adsorbed on organic matter and mineral surfaces. All of these points, which are essential to understanding and interpreting data from hydrogeochemical uranium surveys, involve a fluid medium (hence the commonly used term "solution chemistry"). In the secondary environment the fluid is either surface water or groundwater. (Mechanical transport and concentration of uranium minerals, such as uraninite, into placers is not considered in this discussion).

1. Variable oxidation state. Uranium, unlike thorium, can exist in a hexavalent (uranic; U^{6+}) state which, when it forms complexes (see below), is much more soluble than tetravelent uranium (uranous; U^{4+}) or thorium (Th^{4+}). In fact, the high solubility of uranium in the surficial environment is due to the oxidation of uranium to U^{6+} as UO_2^{2+} in aerated or near-surface oxidizing environments. The reaction for oxidation of tetravalent uranium in water can be written as:

$$U^{4+} + H_2O + \tfrac{1}{2}O_2 = UO_2^{2+} + 2H^+$$

or using uraninite:

$$UO_{2(solid)} + 2H^+ + \tfrac{1}{2}O_{2(aqueous)} = UO_2^{2+} + H_2O$$

2. Tendency to form complexes. Uranium forms complexes with many components in natural aqueous systems. Langmuir (1978b) and Romberger (1979) have reported a total of 43: fifteen with water; ten with fluoride; nine with phosphate; four with sulfate; three with carbonate; and two with chloride. The importance of each will depend on the pH and on the activities of the various anions. The most important complexes are those of the carbonates and several phosphates, particularly the uranyl acid biphosphate, $UO_2(HPO_4)_2^{2-}$. The importance of phosphate complexes has only recently been recognized and, in fact, they are probably more stable than the carbonate complexes under certain conditions (Langmuir, 1978 a, b; Romberger, 1979). The common occurrence of uranium in phosphate rock has long been known, and gives added weight to the importance of the strong bonding between uranium and phosphate.

The distributions of the several important aqueous phosphate species and carbonate complexes with uranium as a function of pH are illustrated in Fig. 3A-10. From this diagram it can be seen, for example, that if stream water had a pH of 7.5, about 55% of the phosphate would be present as acid biphosphate (forming a uranyl acid biphosphate complex with available uranium), and 45% would be present as $H_2PO_4^-$ which forms a weaker uranyl acid phosphate complex. Similarly, at a pH of 7.5, about equal amounts of the uranium complexed with carbonate would be as the uranyl dicarbonate (UDC) complex, $UO_2(CO_3)_2^{2-}$, and as the undissociated uranyl carbonate, $UO_2CO_3^0$. The uranyl tricarbonate (UTC) complex is only stable in more alkaline solutions, and above pH = 8.5 it is the predominant uranyl carbonate complex. UDC may be converted to UTC, and vice versa, the stabilities of each being markedly effected by the partial pressure of carbon dioxide. Because the complexes UDC, UTC and uranyl acid biphosphate are all very stable at pH values above 7.5 (Fig. 3A-10), uranium is most mobile in alkaline aqueous environments. In fact, uranium is more mobile than elements such as Zn, and much more mobile than Cu, because these elements have relatively low pH of hydrolysis values (Table 3-6, p. 128) and they do not form stable complexes soluble in alkaline waters.

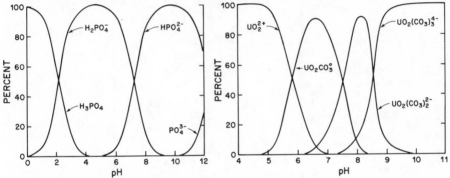

Fig. 3A-10. Distribution of aqueous phosphate species (left), and carbonate species (right), as a function of pH at 25°C. $PO_4 = 0.1$ ppm; $P_{CO_2} = 10^{-2}$ atm.; $U = 10^{-8}$M (2.4 ppb). After Langmuir (1978 a, b) and Romberger (1979).

Fig. 3A-11 illustrates the stability fields for the soluble uranyl carbonates, uranyl hydroxide, the uranyl ion, and the uranous (U^{4+}) ion, as well as the fields for insoluble uranium oxides (for the purposes of this book, UO_2, U_3O_8 and U_4O_9 are all considered to be uraninite, as their formulas are within the range $UO_{2.0}$ to $UO_{2.6}$). From this figure, as well as Fig. 3A-10 (right), it can be seen that in this system it is not possible to precipitate uranium from natural waters of the chemical weathering environment (see Fig. 3-15, p. 130) solely by a change of pH (because uranyl complexes are stable over the entire range of pH). Only by means of an environmental change in which the Eh becomes more reducing, will uranium precipitate. This figure also shows that the uranous ion is only soluble under very acid conditions in a very limited field.

With respect to uranyl complexes, the following generalizations can be made (Romberger, 1979):

1.   Carbonate complexes are very important in slightly acid to alkaline pH in the absence of phosphate, but their stability gradually decreases with increasing temperature because the solubility of CO_2 (and other volatiles) decreases with increasing temperature resulting in less CO_2 to complex with U. Hence, boiling of thermal waters is a good mechanism for the deposition of U, as the volatiles are partitioned into the vapor phase.

2. In the presence of phosphate, even as little as 0.1 ppm PO_4 at low termperatures, phosphate complexes dominate the slightly acid to slightly alkaline pH range.

Therefore, uranium is generally transported as some uranyl complex, and deposition takes place on reduction to the U^{4+} state.

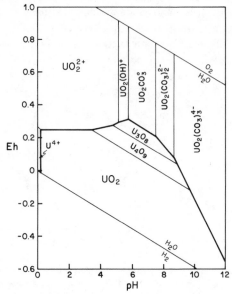

Fig. 3A-11. Eh-pH diagram for the system U-O-H-CO_2 showing the mobility of the uranyl complexes, and the stability fields of three solid uranium oxides (uraninite or varieties), at 25°C. U = 1 ppm; $P_{CO_2} = 10^{-3.5}$. After Langmuir (1978 a, b) and Romberger (1979).

Natural waters carrying uranium, carbonate and phosphate will also carry many other anions, such as the sulfate, fluoride and chloride which, as indicated above, may also form complexes with uranium. Uranium complexes of some of these anions are important under specific environmental conditions (Langmuir, 1978b): (a) uranous (U^{4+}) fluoride complexes in reducing water (e.g., bogs, muskeg) below pH 3-4; (b) uranyl fluoride, and the uranyl ion (UO_2^{2+}), in oxidized waters at Eh values of +0.2 to -0.1 volt, and pH values 1-7; and (c) hydroxide complexes in neutral to alkaline environments. (These conclusions all depend on the activity of the various anions, and must not be taken too rigidly at face value. For example, if the fluoride ion is low in a specific acid reducing water, the fluoride complex will not be important.)

It is now clear that uranium in (near-surface) natural waters is normally complexed. The nature of the complexing inorganic anions, such as carbonate, phosphate, hydroxide and fluoride, will depend on the relative activities of these anions and the specific conditions of Eh-pH. They are also affected by other variables such as the total dissolved solids content and the presence of other ions or anions (e.g., vanadate). Therefore, there is a trend toward analyzing for many of the variables found in water samples collected for uranium surveys in an attempt to assist in the interpretation of the results. These include: U, PO_4, HCO_3, SO_4, Cl, F, Ca, Mg, K, Fe, total dissolved solids, conductivity, pH, Eh, and pathfinder elements (e.g., Mo, Se).

The relationship between uranium and total dissolved solids (or conductivity) was discussed earlier (Fig. 3A-3). However, this simplified relationship does not always give satisfactory results. For example, in those cases where phosphate is available, the phosphate is perhaps ten times more significant to the mobility of uranium than is the dissolved solids content. In unpolluted northern Canadian lakes, phosphate is generally so low that it can be excluded from consideration, and only the bicarbonate content (effectively equated with total dissolved solids) need be considered. On the other hand, uranium complexing with phosphate (which begins to occur with as little as 0.007 ppm PO_4; Romberger, 1979) can result in false anomalies, if a specific water has significant phosphate in relation to other components, as the phosphate will tend to extract uranium from rocks, stream sediments, and other media. Sometimes simply studying the uranium to phosphate ratio may permit such false anomalies to be recognized. Along these lines, Culbert and Leighton (1978) were able to show that highly anomalous (up to several thousand ppb) uranium contents of alkaline lake waters in the Okanagan region, British Columbia could be correlated with the dissolved bicarbonate content.

Some exploration geochemists have attempted to take all possible variables into account by calculating a "solubility index" (also called a "saturation index"), which is capable of evaluating the true nature of the data obtained from a hydrogeochemical uranium survey. Such efforts, which are to be commended, are expensive, because they require analysis or determination of all the possible chemical and physical parameters (14 were mentioned above) likely to effect the solubility of uranium in water. On the basis of the analytical and physical parameters determined, a computer program determines the influence of perhaps a dozen aqueous species (e.g., UDC, UTC) in the particular system.

3. Tendency for uranium to be adsorbed. The importance of uranium adsorption by organic matter has been discussed earlier. For example, peat bogs in Sweden containing as much as 3.1% uranium were mentioned on p. 448. The adsorption of uranium on to the organic matter in lake sediments was mentioned on p. 190. It is for these reasons that loss on ignition (LOI), a reflection of the organic content of a sample, is often determined to aid in the interpretation of certain surveys (however, the correlation is not always good). The relationship between the amount of uranium adsorbed in the organic matter in streams and bogs is generally given in the Soviet literature as:

$$U(\text{in organic matter}) = 10,000\ U\ (\text{in water})$$

Following adsorption, the uranyl ion may be retained in an adsorbed form, or it may be reduced to U^{4+} (uraninite, or some other reduced phase).

Various other materials, such as clays, zeolites and iron oxides and hydroxides, which may be found in stream sediments or soils, may also be adsorbents of uranium. Some, such as freshly precipitated iron hydroxides, may remove uranium from solution almost quantitatively, however, they tend to release the uranium as they become better crystallized. Langmuir (1978 a, b) concluded that adsorption of the uranyl ion is maximal in pH range of 5-8.5. He further noted (p. 29), "Unfortunately, not enough is known of the detailed mechanisms of uranyl sorption from natural waters onto specific sorbents to allow accurate prediction of the role of sorption in low-temperature sedimentary environments." In theory, uranium is soluble in water over the entire range of pH (Figs. 3A-10, 3A-11). Yet, some Canadian geochemists have observed that when the pH is alkaline, waters should be sampled, but when the pH is acid, sediments should be sampled. Perhaps, in certain cases during the "competition" between adsorption by clay materials and solution as complex ions, adsorption is favored. To be absolutely certain, both waters and sediments should be sampled (at least in Canada) until the matter is resolved.

Precipitation Of Uranium In The Secondary Environment

To this point the discussion has been concerned with the mobility of uranium. Now, it is appropriate to briefly discuss the mechanisms by which mobility is reduced, including how uranium is precipitated.

There are three main mechanisms:

1. *Reduction.* The reduction of U^{6+} to U^{4+} will result in the formation of uraninite or other uranous phases. This mechanism explains the occurrence of uranium in roll front deposits (Fig. 3-24, p. 156), bogs, uraniferous logs, and coal beds. Methane has been postulated as the reductant in the formation of the Rabbit Lake and other large deposits in Saskatchewan (Hoeve and Sibbald, 1978). Other mobile reductants of importance include hydrogen sulfide, other hydrocarbons, and hydrogen. The oxidation of ferrous to ferric iron may result in the reduction of uranium in many deposits.

2. *Adsorption.* Adsorption of uranyl complexes by organic matter and other materials, e. g., clays and iron oxides, has been discussed above. These materials certainly reduce the mobility of uranium. Evidence appears to be emerging that following the initial adsorption of the uranyl complexes, there may be subsequent reduction and precipitation by mobile reductants, such as methane and hydrogen sulfide.

3. *Formation of insoluble compounds.* Uranyl ion (in an oxidizing environment) can be precipitated from solution when it combines with vanadate, phosphate, arsenate, silicate, carbonate or sulfate anions in association with Ca, Mg, K, Ba, Pb, Cu and several other elements. Minerals such as carnotite, $K_2(UO_2)_2(VO_4)_2 \cdot 3H_2O$, and soddyite, $(UO_2)_5Si_2O_9 \cdot 6H_2O$, are the result. Uranyl ions may also be precipitated from solution after the dissociation of uranyl carbonate complexes at elevated temperatures.

Decay Of Uranium

To this point, only selected geochemical characteristics of uranium itself have been considered. However, in the search for uranium ore deposits, explorationists may attempt to detect: (1) ^{214}Bi by means of a scintillometer or other types of gamma ray detectors; (2) ^{222}Rn in soils and waters by means of a radon counter (Fig. 6-13), Track Etch (Fig. 6A-4), or by some other method; (3) ^{226}Ra, particularly in waters, by the "sealed can" or some other method; and (4) ^{4}He, usually in soil gas or waters, by means of mass spectrometry. Where do these topics fit into our discussion of geochemical exploration for uranium? For the answer to this question we must consider, first, the decay of the most abundant isotope of uranium, ^{238}U, and second, the relative mobilities of the daughter nuclides* formed by the decay of ^{238}U.

There are three main radioactive elements in nature, i.e., uranium, thorium, and potassium (only the ^{40}K isotope, constituting 0.0119% of K, is radioactive). These elements are continually disintegrating to stable end products (e.g., lead, helium and argon) by means of nuclear decay, and their relative abundances are becoming less, but because of their very long half-lives, they have not yet totally disappeared.

Uranium has two abundant isotopes, ^{238}U and ^{235}U, constituting approximately 99.3% and 0.7%, respectively, of the total uranium in rocks, ores, and other natural materials. These abundances are remarkably constant although an exception is found in the uranium ores from Oklo, Gabon, West Africa, where a natural fission reactor about 1.8 billion years ago consumed some of the ^{235}U so that this isotope now constitutes as little as 0.3% of the ore from the locality; however for our purposes, the ^{238}U/^{235}U ratio (137.5 ± 0.5) in nature can be considered constant. (Cowan and Adler, 1976, have reported other minor variations in the ^{235}U content of uranium ores.) A third isotope, ^{234}U, results from the decay of ^{238}U, and in equilibrium its abundance is only 0.0054% (this isotope is discussed further below).

Fig. 3A-12 shows the principal decay products of the ^{238} decay series. The nuclides decay mainly be the emission of alpha (α) and beta (β) particles. In some cases, such as with ^{214}Bi, the product of the β particle emission is left in one of several excited states and it immediately releases gamma (γ) radiation of a particular energy (1.76 MeV being one allowed value); it is this radiation which is detected by scintillation counters. Gamma rays are electromagnetic radiation and can penetrate through matter relatively well (several hundred meters in air), but a uranium deposit covered with more than about 0.5 m of rock would not be detected at the surface. Beta particles have a range of about 3 m in air, and about 2 mm in rock. Alpha particles, which are helium nuclei (^{4}He), travel no more than about 50 mm in air, and

---

*The term "nuclide" rather than "atom" is favoured when speaking of nuclear reactions (e.g., radioactive decay). A nuclide is simply a species of nucleus characterized by a certain atomic number Z, and atomic weight A, such as $^{238}_{92}$U, $^{222}_{86}$Rn or $^{23}_{11}$Na. When one nuclide is changed to another, adjustments to the number and orbital distributions of the atomic electrons surrounding the nucleus are required. An isotope is a nuclide of a given Z (i.e., an element) characterized by a certain value of A, e.g. ^{238}U, ^{234}U and ^{235}U.

only a few microns in rocks. The adsorption of α, β and γ rays by air and other materials is a fundamental matter which must be understood by those prospecting for uranium. It explains, for example, the fact that in order for an alpha particle to be detected by any of the radon techniques, the radon atom must be within about 50 mm of the detector (because that is how far the alpha particle emitted by the decay of [222]Rn can travel in air).

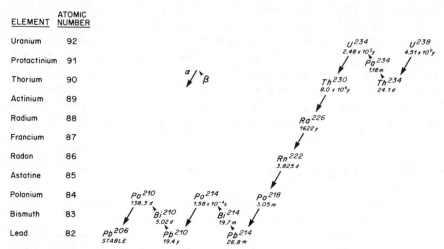

Fig. 3A-12. The principal members of the [238]U series decay chain. Note the position of [222]Rn and [214]Bi far down the decay chain.

It is not necessary for our discussion to consider the decay of [235]U. Suffice it to say that it decays in a manner similar to that of [238]U (α and β rays are emitted), and its stable end product is [207]Pb. Further, as we have assumed it is always found in a fixed ratio with [238]U (see above), the search for [238]U is effectively a search for [235]U. Of course, [235]U, which is fissionable, is the only uranium isotope of economic interest.

Concept Of Equilibrium

One may logically expect to detect the presence of [238]U by the analysis of any one of the daughter products in the decay series. For example, the detection of [214]Bi or [222]Rn will indicate the presence of [238]U. However, this is only true if the system (ore, sample) is in *equilibrium,* also called "secular equilibrium". Equilibrium can be defined as a state in which all nuclides of a decay series have fixed abundances to each other. Alternatively, it may be defined as a balanced, natural proportion of parent and daughter nuclides in which the daughter products(s) is (are) decaying as fast as the parent(s). The opposite condition is called *disequilibrium.* Inherent in the concept of equilibrium is a closed system — a condition in which a daughter nuclide is not fractionated from the parent. If all the nuclides shown in Fig. 3A-12 are in equilibrium with 1 gm of ^{238}U, there will be 3.42 x 10^{-7} g of ^{226}Ra and 1.1 x 10^{-21}g of ^{214}Po, i.e., about 3 atoms of polonium (Dyck, 1978, p. 58).

The time interval for equilibrium to be achieved in a closed system is 10 half-lives of the longest lived intermediate nuclide in the decay series. In the ^{238}U series, ^{234}U is the longest lived intermediate nuclide whose half life is 248,000 years. Thus, if fresh, pure ^{238}U could be obtained, it would require 10 times 248,000, or 2.48 my to achieve equilibrium. If, however, one were to start with ^{238}U in equilibrium with its daughter ^{234}U, which is approximately the case in most natural situations, only 10 times 80,000 years (the half-life of ^{230}Th), or 800,000 years, would be required for equilibrium to be attained. In actual practice, effective equilibrium for exploration purposes can be achieved in 4 or 5 half-lives as the following tabulation will show (based on the assumption that the half-life of the daughter is significantly shorter than that of the precursor; for an exact treatment see the Bateman equation which can be found in any standard textbook on nuclear physics):

Half-lives	Percent equilibrium
9.97	99.9
6.6	99.
5.0	97.
4.0	94.
2.0	75.
1.0	50.

If one peruses Fig. 3A-12, one will notice that the half-lives of the nuclides progressively become shorter (with exceptions) down the decay chain. Further, it should be apparent that the nuclides with very short half-lives are dependent for their existence on either (1) their immediate parent, or (2) their grandparent, and not to ^{238}U. For example, ^{222}Rn (half-life 3.825 days) is dependent for its existence upon its parent ^{226}Ra (half-life 1622 years) so that physicists will say that the ^{222}Rn is "supported" by the ^{226}Ra. Because the two nuclides (^{218}Po and ^{214}Pb) between ^{214}Bi and ^{222}Rn have short half-lives, it is proper to consider that ^{214}Bi is dependent upon ^{222}Rn; in other words, there must be ^{222}Rn in the vicinity (purposely vaguely stated) if the 1.76 MeV gamma radiation from ^{214}Bi is detected.

If the decay chain is broken, equilibrium no longer exists and a state of disequilibrium prevails. Disequilibrium in the secondary environment results from at least two causes: (1) the chemical properties of the various nuclides in the decay chain are different, e. g., uranium is mobile in oxidizing fluids (discussed above), whereas thorium is immobile, or in other words, because of differential leaching; (2) radon is a gas and, theoretically at least, it can escape from a crystal containing its parent ^{226}Ra and its precursors (e.g., ^{238}U).

A clear understanding and appreciation of disequilibrium is of extreme importance primarily because exploration for uranium is often based on detecting elements far down the decay chain, specifically ^{222}Rn and ^{214}Bi (the latter being the primary source of gamma radiation detected by scintillometers). In some cases these nuclides may become separated by great distances from a uranium ore body.

There are at least four positions in the decay series, where parent ^{238}U can be fractionated from its daughters resulting in disequilibrium: (1) ^{234}U; (2) ^{230}Th; (3) ^{226}Ra; and (4) ^{222}Rn. The first three are relatively long-lived nuclides allowing a considerable period of time for the chemical processes responsible for disequilibrium to operate. Cumulative fractionation over two or three of the above positions can make exploration for uranium a difficult and frustrating experience.

Disequilibrium In The ^{238}U Series

Disequilibrium occurs when one or more of the nuclides are chemically or physically (the gases) transported away from their immediate parent or from ^{238}U. Those nuclides with the longest half-lives are likely to be moved the farthest.

At this point, the mechanisms by which the four most important fractionations take place will be discussed, as well as certain other practical aspects of disequilibrium. The fractionation and use of helium in exploration was discussed earlier (p. 335 and p. 000) and will not be considered further. Various aspects of disequilibrium from the point of view of exploration geochemistry have been reviewed by Dyck (1978), Hambleton-Jones (1978), and Levinson and Coetzee (1978), where the reader is referred for more detailed treatments and additional references.

Fractionation of ^{234}U. Fractionation of ^{234}U from ^{238}U was discovered in 1953 in the Soviet Union by Cherdyntsev and associates (see Cherdyntsev, 1971; a translation in English). The fractionation is controlled by several factors (e.g., "alpha recoil"; also called the Szilard-Chalmers or "hot atom" effect), and the net result is that ^{234}U is more mobile in water than ^{238}U. The increased mobility is explained by the fact that ^{234}U is likely to be in the higher oxidation state (U^{6+}) by comparison with ^{238}U (U^{4+}) and, therefore, ^{234}U is more likely to form soluble uranyl complexes, such as uranyl dicarbonate.

Ratios of $^{234}U/^{238}U$ (actually "activity ratio") in waters range from less than the ideal of 1.0 to as much as 9, and in most natural waters it is 1.15. These ratios may be used for exploration in soils and waters in several ways. For example, Cowart and Osmund (1977) showed that groundwaters down dip from an actively forming uranium accumulation are enriched in ^{234}U. One disadvantage of using the $^{234}U/^{238}U$ ratio for exploration is the difficulty in analysis (extensive chemical separations followed by alpha spectrometry are necessary). However, the method has not yet been thoroughly evaluated and in the future it could be very important in exploration.

Fractionation of ^{230}Th. Fractionation of ^{230}Th from the two parent uranium isotopes, ^{234}U and ^{238}U, takes places because thorium is extremely insoluble in aqueous media by comparison with uranium (averages for river water: Th = 0.1 ppb and U = 0.4 ppb; Table 2-1, p. 44). Therefore, uranium moves away from thorium resulting in disequilibrium between ^{230}Th and ^{234}U.

Fractionation of ^{226}Ra. Radium is an alkaline-earth element and chemically it is similar to barium. Some compounds of radium, such as the sulfates, have very low solubilities (i.e., the solubility of $RaSO_4$ is about 20 ppb at 25 C). Nevertheless, the extremely low abundance of radium (each ton of uranium will contain only 0.334 g of ^{226}Ra) permits all that is produced by decay of ^{230}Th to be mobile. Radium also forms soluble complexes with the chloride ion as evidenced by the high Ra content of deep, chloride-type brines. However, even though radium can be leached from its parent ^{230}Th, there are several natural mechanisms by which its mobility can be reduced or stopped, at least temporarily, in the waters of the surficial environment: cocrystallization, coprecipitation, inorganic adsorption, and biological absorption.

Cocrystallization occurs when radium is able to substitute for barium in barite, resulting in the so-called "radiobarite". As neither uranium nor thorium can occupy such structural positions, the radium is "unsupported", and all evidence of ^{226}Ra will disappear within 10 half-lives, or about 16,000 years. The presence of radiobarite must indicate recent geological processes, such as active leaching of pyrite-containing uranium ores. In such cases, sulfuric acid is formed, uranium is leached away as a soluble uranyl sulfate complex, and radium will coprecipitate with any available barium as radiobarite (see discussion of the Blind River uraninite deposit, p. 79).

Coprecipitation of radium may occur where Fe-Mn oxides and hydroxides are being precipitated, such as at the orifices of springs. Radium, and U if present in spring water, may also be coprecipitated with barite and Fe-As precipitates (Cadigan and Felmlee, 1977). In actual practice, it is difficult to determine whether radium is coprecipitated or absorbed by Fe-Mn oxides.

Adsorption of Ra may be effected by Fe-Mn oxides, clay minerals or organic matter. Clay minerals preferentially adsorb divalent, alkaline-earth cations (p. 97), including radium. Differential adsorption of Ra and U is illustrated in Fig. 3A-13 based on an example from the Soviet Union. This figure shows the paths taken by U and Ra as they are leached from three uranium-bearing veins that have been weathered to depths of several dozens of metres below the surface in a humid environment; presumably ^{230}Th would remain immobile in the weathered zone of the ore veins. Radium is adsorbed by clays in the glacial moraine, whereas uranium is mobile until it is adsorbed and/or reduced by the organic matter in the peat bog.

Many springs and wells (groundwater) have high contents of radium. Some, such as certain wells in the Helsinki area, Finland, which contain as much as 256 pCi/l* of ^{226}Ra, have been discussed in connection with false anomalies (p. 000). In the Badgastein area of Austria, thermal spring waters containing about 10-20 pCi/l ^{226}Ra

---

*By original definition 1 curie (Ci) is the number of disintegrations per second occurring in 1 gram of radium. However, because of its low abundance, radium is always reported as picocuries (pCi), where 1 pico is defined as 10^{-12}. Radium abundances in water are reported as picocuries per litre, written as pCi/l (10^{-12} curies per litre). Subsequently, the definition of curie has been modified to be a measure of radioactivity in general. Today, one curie of a radioactive substance is that amount of the substance which gives 3.700×10^{10} dps (disintegrations per second).

have been reported. Manganese oxides (locally called "reissacherite", a radioactive wad; Palache et al., 1944, p. 569) may contain over 400,000 pCi/g of ^{226}Ra. This radium has been adsorbed from the spring waters by the Mn oxides over an extended period of time, and is the main source of the radon and other radioactivity in the area. Other less spectacular concentrations of radium-containing Mn (and other types) precipitates from springs are known from many localities, such as the western United States and more recently Saskatchewan (Dr. C. J. Bland, personal communication).

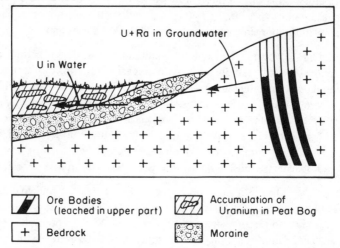

Fig. 3A-13. Differential dispersion of uranium and radium from a chemically weathered uranium deposit in a taiga landscape, USSR. After Perel'man (1977).

In connection with radioactive spring and well waters (usually containing ^{226}Ra and/or ^{222}Rn) it is important to note that they frequently contain little or no uranium and, therefore, the Ra and Rn are unsupported. In most cases, no uranium deposits have so far been found in association with these nuclides. Apparently, warmed meteoric waters have leached large volumes of granitic rock, which may or may not contain high contents of uranium, and these waters are then "channelled" to specific orifices. It has been suggested that at the elevated temperatures characteristic of thermal springs, radium carbonate is soluble whereas the uranyl carbonate complex is unstable and dissociates, following which the uranium is removed from solution (Hambleton-Jones, 1978).

Biological absorption of radium by vegetation is discussed in connection with Fig. 10A-1 and Fig. 10A-5.

Fractionation of ^{222}Rn. Various aspects of radon geochemistry and mobility have been discussed previously (e.g., pp. 300, 430, 431), and are also discussed below in connection with false anomalies associated with springs (p. 703) and with new methods of radon detection (e.g., Figs. 6A-3, 6A-4). In way of review: (1) ^{222}Rn is an inert gas with a half-life of 3.8 days whose detection may indicate the presence of uranium mineralization; and (2) the immediate parent of ^{222}Rn is ^{226}Ra, and the detection of ^{222}Rn only guarantees the presence of ^{226}Ra.

The second point can be illustrated with Fig. 3A-14, which is based on an example from an arid region in southern Africa, where the position of the water table has fluctuated in recent geological times. This has resulted in disequilibrium, as ^{226}Ra has been moved to point "A". A Rn anomaly detected at the surface at the present time represents only Ra, and not uranium. As this Ra is unsupported, it will have completely disappeared in about 16,000 years from the time of its original precipitation at "A". Rn generated at "B" is too deeply buried to be detected at the surface.

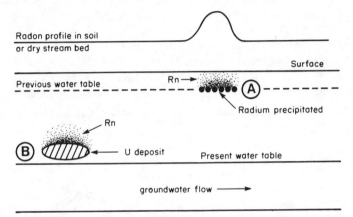

Fig. 3A-14. Idealized diagram showing a radon (^{222}Rn) anomaly over radium (^{226}Ra) at A, which was precipitated when the water table was higher. The uranium deposit is located at B where the radon anomaly presently being generated does not reach the surface. Such situations, or variations thereof, are believed to be common in certain desert environments, such as in southern Africa. After Levinson and Coetzee (1978).

Another point which merits at least passing comment is the *emanation coefficient,* defined as the ratio of liberated radon to the radon formed in a sample. Clearly, Rn cannot migrate toward the surface if it cannot escape (emanate) from a sample. The emanation coefficient for most uranium ore samples only averages 21%, but for rocks it is usually only a few percent. The factors controlling the emanation of ^{222}Rn are complex with the most important variables being the mineralogy, the particle size of the uranium minerals, and the porosity and permeability of the ore. Poorly crystallized, very fine-grained samples will likely have a higher emanation coefficient compared to samples with well-crystallized, coarse grained uranium minerals. The variability of the emanation coefficient in uraniferous materials has been discussed by Dyck (1978), and Levinson and Bland (1978). From their reports it is obvious that because of the variable and unpredictable amounts of Rn emanation, radon surveys can only be considered qualitative at best. Further, what Rn does emanate and is detected at the surface may only be a reflection of Ra, and not necessarily U (Fig. 3A-14).

Disequilibrium and gamma ray spectrometry. As indicated earlier, *gamma ray spectrometers (or scintillometers) detect ^{214}Bi and not uranium* (at least those field and laboratory models used for uranium exploration). Also, it was noted that the presence of ^{214}Bi indicates proximity to ^{222} Rn because, on a practical basis, ^{222}Rn is the parent of ^{214}Bi due to the short half-lives of the intermediate nuclides ^{218}Po and ^{214}Pb (Fig. 3A-12). Thus with reference to Fig. 3A-14, a gamma ray spectrometer survey should record anomalous values in those same areas where the high Rn anomalies were detected. However, there is no uranium occurrence corresponding to anomaly "A" as there is at "B" (Fig. 3A-14) as a result of disequilibrium between Ra and U at location A. The apparent amount of uranium calculated from the gamma ray emission of ^{214}Bi is called *equivalent uranium* (often abbreviated eU, or in the oxide form as eU$_3$O$_8$), and is based on the assumption that equilibtrium conditions prevail, which is not always the case.

The relation between actual uranium and equivalent (apparent) uranium shows marked deviations, similar to that in Fig. 3A-14, in many instances as a result of disequilibrium. Fig. 3A-15 illustrate the point further with plots of the apparent (equivalent) uranium from ^{214}Bi and actual uranium (which can be determined by

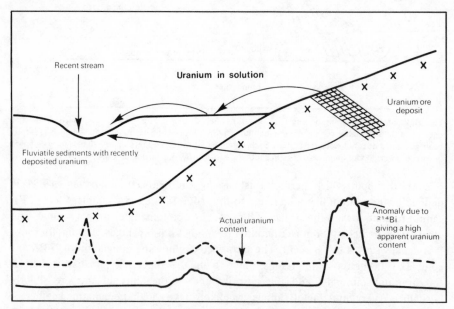

Fig. 3A-15. Diagrammatic representation of the relationship between actual uranium and apparent (equivalent) uranium (eU; based on gamma radiation from ^{214}Bi) in an environment with chemical weathering. After Hambleton-Jones (1978).

fluorometry, delayed neutron activation or X-ray fluorescence). In Fig. 3A-15, soluble uranium is leached from a deposit, transported as a complex such as uranyl dicarbonate, and deposited in stream sediments. Residual ^{226}Ra, in part generated from insoluble ^{230}Th in the ore deposit, will give an apparent uranium anomaly determined by gamma ray methods significantly above the actual uranium content (the emanation coefficient of ^{222}Ra is generally less than 20% and, therefore, sufficient ^{214}Bi will be available in the deposit). The actual uranium content of the stream sediments in Fig. 3A-15 may be high, but since there has been insufficient time for ^{214}Bi to have been produced, the gamma ray activity is negligible. In an analogous situation, Levinson et al. (1978) have shown that uranium hydromorphically transported into organic lakes sediments in the Seahorse Lake area, near the large Key Lake uranium deposit, Saskatchewan, yielded apparent uranium values of only 3-8% of the actual uranium present in most samples owing to the young age of the uranium (the area was glaciated during the Pleistocene, and the lakes are believed to be no older than 10,000 years). Any clastic uranium minerals in the stream sediments (Fig. 3A-15) or lake sediments will be equilibrium and equivalent uranium will correspond with actual uranium.

Concluding remarks. We may summarize the preceding discussions by recognizing that there are two basic types of disequilibrium, and both involve transport in solution:

(1) residual, i.e., where uranium is leached and the daughter nuclides remain, and

(2) new ("young"), hydromorphically transported uranium, which has not reached equilibrium with its daughters.

A point well worth emphasizing is that disequilibrium must not be viewed only in a negative light, because the dispersion of radioactive nuclides, which is the result of disequilibrium, can often be used to advantage. In Canada, for example, some exploration geochemists recommend that the uranium content of lake waters be used for regional surveys, whereas the radium and radon contents of these waters be used on a semi-detailed and detailed scale, respectively, in a manner analogous to the concept of pathfinder elements. Success in uranium exploration based on the use of disequilibrium will depend, in large part, on the degree to which the exploration geochemist understands the fundamental principles.

In actual practice, the determination and explanation of the degree and exact nature of disequilibrium at any one locality is likely to be difficult. It may require the measurement of several of the following nuclides: ^{238}U, ^{234}U, ^{230}Th, ^{226}Ra, ^{222}Rn, and ^{214}Bi. However, it may well be worth the effort. This may be illustrated with the case where a company is reported to have located its mine in the Colorado Plateau area over the zone which gave the highest radiometric (gamma ray) responses, and its processing plant over the zone with the lowest radioactivity. Unfortunately, the ore was out of equilibrium which, in practical terms, meant that the mine area was high in ^{226}Ra and ^{214}Bi, and correspondingly low in uranium (although some uranium was definitely present), whereas the processing plant was located above the highest uranium values.

4A / *Some Basic Principles*

CONTAMINATION (pp. 193-199)

New examples of contamination in materials which are often collected for geochemical purposes continue to be reported, not only in the geochemical literature, but also in other fields, such as limnology and environmental control. A selection of these is discussed below.

Mining Contamination (pp. 193-195)

Mining activity traditionally produces contamination which is detected in water, lake and river sediments, soil and air. However, this is not universally so. A case in point is the study by Allan (1974a) which reported on the As, Ni and Cu contents of lake sediment cores from three established mining areas in Canada. At each there are numerous large deposits but the absolute concentrations of the contamination had not reached levels that would reflect the ore occurrences.

Industrial Contamination (pp. 195-196)

Shacklette and Conner (1973) found that Spanish moss, an epiphyte which obtains all its element load from the atmosphere, can be used in an economical and rapid manner for determining the degree of local atmospheric metal pollution. A particular advantage of such plants is that their metal content is an integration of periodic fluctuations in the amount of metal pollution over relatively long periods of time. From their studies Shacklette and Conner (1973) found that tin could be detected in Spanish moss up to 50 miles from the only tin smelter in the United States, and that samples from industrial and highway locations contained high amounts of As, Cd, Cr, Co, Cu, Pb, Ni and V.

The effect of fallout on soil by windblown Zn, Pb, Cu and Cd from a fuming kiln in South West Africa that had been in operation for slightly over one year, has been reported by Marchant (1974). The area in which the study was conducted is semi-arid and underlain almost exclusively by calcareous rocks — both factors that would tend to limit chemical weathering. They found that the fallout was detectable for several kilometers, and that it was wind controlled. Metal contamination, however, even when present in thousands of ppm on the surface, had not yet pene-trated to a depth of 10 cm. These findings are in agreement with others from smelters in Rhodesia, Arizona, and Idaho (see pp. 194 and 196) which have recorded exten-sive lateral but little vertical contamination, and geometric patterns unrelated to geology. In a marginally related, but interesting study, Roberts et al. (1974) have documented the effects of lead contamination on humans (mostly children) living close to two smelters in different parts of Toronto. Accumulations of lead were found in their blood and hair. Contamination decreased exponentially with distance from the smelters and generally reached background levels within 300 meters of the discharge stacks.

Agricultural Contamination   (p. 196)

Agriculture requires water and, to induce precipitation in certain areas, seeding of clouds with silver iodide is sometimes employed. This has resulted in silver anomalies in some very unexpected places. A case in point is the high silver values (as high as a few ppm) found in the recent sediments of some lakes.

Constructional Contamination   (p. 196)

No additional comments.

Domestic, Municipal, Human and Animal Contamination   (pp. 197-198)

Contamination of lake sediments caused by local industry and municipal waste is well known, and a case where sediments have been contaminated by agricultural activity has been documented above. Allan and Timperly (1975), working with lake sediments, primarily in the glaciated terrain of Canada, have presented some fascinating data on industrial and urban contamination which they trace to local, regional and even global sources. As a consequence, they conclude that dredged lake sediment samples should be avoided, and that even with core samples, the upper 5-10 cm should be discarded. In practice, it will be necessary for the geochemist to determine if contamination of the type found by Allan and Timperly (1975) is likely to adversely affect his particular project. Factors to be considered include what ele-ment is being sought (uranium, for example, is not likely to be introduced by urban contamination), the type of element transport (clastic sulfide grains are not likely to be contaminants in most situations), etc.

That contamination of lakes can originate from distant sources has been con-firmed by numerous chemists and limnologists concerned with the acidity of lake water (e.g., Beamish and Van Loon, 1977). Acidification of lakes has become a serious problem leading to severe damage to fish populations, in places such as Sweden and Canada. Acid lake waters result from acid rain, the result of atmospheric pollution by sulfur dioxide (chiefly from the burning of fossil fuels) which oxidizes readily and dissolves in water to form sulfuric acid. The acid rainwater, whose pH is even lower than the theoretical and normal 5.7 (see p. 73), may sometimes be neutralized by reaction with limestone, or by other mechanisms (e.g., ion exchange with clays) but, if not, it may enter lakes where it tends to increase solubility and availability of the heavy metals normally found in sediments (Beamish and Van Loon, 1977).

Sewage from many cities, including both domestic and industrial waste, is subject to varying degrees of treatment before the sludge is dumped into near-shore disposal areas. Enhanced anomalies of several metals can be found in the sediments and water (and fish) of the immediate area. In addition to the elements which might normally be expected (Cu, Pb, Zn, Ni, Cr, Fe, Mn) some unexpectedly high abundances of rare elements have been reported. Silver, for example, is considerably enhanced (but still generally less than 1 ppm) in sludge from such areas as New York harbor and the Firth of Clyde (into which sewage from Glasgow is discharged). The studies of Halcrow et al. (1973) show that sediment contamination is restricted to a limited depth (40 - 50 cm in the Firth of Clyde), analogous to wind-transported contamination in soils.

Discussion (pp. 198-199)

Dijkstra et al. (1979) reported on the use of modern methods of exploration geochemistry in the Moresnet region of northeastern Belgium. In this old lead-zinc district there are several old mine and smelter dumps which have contaminated the stream sediments. The effect of the trace element dispersal patterns from these dumps can be considerably reduced by means of certain partial extraction techniques. Specifically, zinc contamination was determined by the difference between hydroxlamine hydrochloride extractable and sodium dithionate extractable metal. Similarly, lead contamination was determined by the difference between sodium citrate extractable and magnesium chloride extractable metal. Such successful studies give encouragement for geochemical exploration methods in other old mining districts characterized by contamination.

ORIENTATION SURVEYS (pp. 199-204)

Orientation surveys are now conducted as part of almost all carefully thought out, well-organized exploration programs in which geochemical methods are employed. It should be emphasized that orientation surveys are applicable anywhere in the entire range of geochemical activity, from the broad-scale reconnaissance level to the most detailed soil survey preliminary to drilling.

In some places, however, orientation surveys cannot be carried out at the beginning of a project because there are no appropriate deposits known in the exploration area. It is then necessary to use the exploration parameters obtained elsewhere. By way of example, if exploration were planned in the desert or semi-arid regions of Botswana or South West Africa, procedures found successful in similar environments in Egypt, southwestern United States, or Australia would be employed, at least initially, until an appropriate deposit or occurrence in the study area is found. It is important, also, to recognize that orientation surveys conducted over known mineralization found by other methods, such as geophysics, may show that geochemical methods do not work. In such cases it is necessary to determine the reason for the negative results; the explanation may be that the overburden contains a recent ash bed, that the wrong extraction or the wrong size fraction was used, or any one of a dozen other possibilities. However, if after a very detailed orientation survey all proposed geochemical techniques do not yield the desired results, i.e., similar mineralization would not be found within the exploration area, it is recommended that the use of geochemical techniques be deferred.

As pointed out by Buhlmann et al. (1975), in general, five sequential stages in an exploration program may be distinguished: (1) regional selection; (2) area selection; (3) area exploration; (4) follow-up exploration; and (5) detailed exploration. Development of interest in an area (regional or area selection above) is normally preceded by geologic studies in the field and office. In these initial stages, as well as others that follow, a combination of several methods will likely lead to better results than the use of one technique alone. Geological, as well as remote sensing (e.g. Landsat) and geophysical methods, should be considered at an early stage in regional and area selection. During or after the selection stage, an orientation survey should be run to determine the optimum field, analytical and interpretive parameters which can distinguish a regional or area anomaly from background in, for example, stream sediments. In actual practice, several orientation surveys, with different objectives in mind, are likely to be conducted sequentially during various stages of a continuing exploration program, as need and encouraging results dictate.

FALSE (NON-SIGNIFICANT) ANOMALIES   (pp. 204-210)

Awareness of the fact that anomalies unrelated to mineralization (false anomalies) can, and do, occur frequently, is essential for those engaged in geochemical exploration. In fact, publicizing false anomalies and their explanations can be of great assistance to the explorationist by saving time, effort and expense. Many examples of false anomalies (e.g., unexplained anomalies in soils, anomalies caused by adsorption by Fe-Mn oxides and organic matter) were discussed previously pp. 204-210). The recent literature contains numerous additional examples of false anomalies which can usually, be explained by one or more of the mechanisms discussed earlier. However, almost all the recent descriptions of false anomalies are from the northern latitudes, particularly areas which have Pleistocene glacial deposits, and those areas in which lake sediments are used for exploration purposes (e.g., Canada and Scandinavia). The many case histories concerned with geochemical exploration in the Canadian Cordillera and the Canadian Shield compiled by Bradshaw (1975), and those in glacial terrain edited by Jones (1973, 1975), include examples of these types of false anomalies. The best and most abundant false anomalies in these areas appear to be related to seepage anomalies into bogs, and to changes in lithology. With respect to the latter, Dr. P. M. D. Bradshaw (personal communication) states that the dominant pattern seen on all Geological Survey of Canada geochemical maps is lithology (not mineralization); thus the anomalies are examples of false anomalies (or perhaps "non-significant" anomalies in this case because they do relate to something).

Recent interest in uranium exploration has resulted in ever more examples of radon false anomalies (p. 431) being recognized. By way of example, radon false anomalies are now known to be caused by variations in atmospheric pressure, as illustrated in Fig. 6A-3. Actually, all gases tend to give erratic results (some of which result in false anomalies) because "equilibrium" is established only for a limited amount of time, perhaps a few hours, and is easily shifted by changes in pressure, temperature, rainfall and wind.

According to Eisenbud (1977, p. 167), "Without doubt, the most common of the radioactive anomalies are the thousands of mineral springs that are known to have extraordinarily high levels of radioactivity." In a typical aquifer, groundwater movement is slow, the amount of ^{222}Rn withdrawal by springs and wells is small (compared with the amount left in the aquifer), and the distance that ^{222}Rn migrates is short (see p. 300 for the decay products of ^{238}U). Under such conditions, it is possible that (approximate) radioactive equilibrium exists among ^{226}Ra, ^{222}Rn and other decay products, so that the source of the radioactivity may be several isotopes (but chiefly Ra and Rn). Water drawn from wells and springs in the St. Peters sandstone in various parts of Illinois and Iowa is elevated in radium (and radon) which may possibly be attributed to the fact that the aquifer is in contact with non-economic black shales (hence the anomalies are false or non-significant).

Numerous radioactive springs are known in which radon is the main radioactive element (e.g., the springs at the Badgastein spa, Austria, are estimated to discharge about 200 mCi of ^{222}Rn each day into the atmosphere). One of the more remarkable recent discoveries, according to Eisenbud (1977), is that some groundwater in the Helsinki, Finland area contains substantial quantities of ^{226}Ra, ^{228}Ra, and ^{222}Rn, and natural uranium (a mixture of ^{238}U and ^{235}U). The radon concentrations have been found to be as high as 309 nCi/l. Other values are: ^{226}Ra, as high as 256 pCi/l; ^{228}Ra, as high as 5 - 6 pCi/l; and uranium as high as 15 ppm in some of the Helsinki wells. (These are all very high values for drinking water, and are substantially in excess of accepted permissible limits; the normal ranges for ^{226}Ra and

^{222}Rn in groundwater are generally from 1 to 10 pCi/1). The source of these anomalous values does not appear to be known.

With respect to the distance of radon migration, and its significance in prospecting for uranium, reference to the work of Tanner (1975) is particularly timely and appropriate. Knowing that ^{222}Rn has a short half-life of 3.8 days and is only fairly soluble (p. 300), he recognized that the distance it can migrate in the ground before it decays is limited, even though the ground may be fractured (see pp. 431-432). Even by assuming very favorable conditions, such as a sizeable uranium body containing 1 % uranium, and a reasonable diffusion coefficient for radon in air, Tanner (1975) calculated that the maximum mobility of radon in soil would be about 42 m (a figure he really believed to be a gross over-estimation). Accordingly, he pointed out the need for great caution in the interpretation of radon surveys. Rather than postulate upward movement of radon from a uranium ore body, he suggested that it is much more reasonable to assume that fluctuations in the water table over periods of thousands of years result in the migration of ^{238}U, ^{234}U, and ^{226}Ra (which are precursors of ^{222}Rn; see p. 300) and to interpret radon anomalies as radium anomalies. There are, however, some geochemists who insist that groundwater can move much faster than commonly believed, especially in good aquifers, and they may reject the preceding arguments.

Considering the relative ease of mobility of uranium and radium in some groundwaters, and the adsorption capacity of organic matter and Fe-Mn oxides for these elements, it is possible to explain many radon anomalies as false anomalies. In other words, radon anomalies may merely represent uranium and/or radium, which have been adsorbed by organic matter or Fe-Mn oxides, and which are unrelated to mineralization.

Additional aspects of radon geochemistry, including false anomalies, are discussed by Smith et al. (1976), Levinson and Coetzee (1978), Severne (1978), and Hambleton-Jones (1978).

INTERPRETATION OF GEOCHEMICAL DATA
(pp. 210-222)

Introduction   (pp. 210-211)

In the earlier edition of this book the statement was made (p. 210), "The decision as to the true meaning of geochemical data is probably the most difficult part of exploration geochemistry." That statement is as true at this time (5 years later) as it was when it was made. It is even possible to argue that the interpretation of geochemical data is more difficult now because of the sheer volume of data which confronts the exploration geochemist, in part generated by the ever increasing use of multi-element analytical methods. In general, however, the interpretive techniques described earlier are used more or less in the same way as they have been in the recent past. Maps and diagrams are plotted, background and threshold are determined in order to recognize anomalies, and various graphical plots, specifically histograms and frequency distribution curves, are usually prepared. Computers continue to be used in the same capacities as discussed in Chapter 12 (pp. 469-508). It is important to realize, however, that there is no substitute for maps which present the *spatial distribution* of the geochemical data; in fact, in many cases, maps of this type are far more valuable than histograms, cumulative frequency curves, or other methods of data presentation.

Maps and Diagrams   (pp. 211-213)

No additional comments.

Background   (pp. 213-214)

Background was defined (p. 213) "as the normal range of concentrations for an element, or elements, in an area (excluding mineralized samples)." Contained within the definition are two concepts which require emphasis and repetition.

The first is *range*. It is a common failing of many geochemists (the present author included), when discussing background, to think in terms of a single value rather than a *range;* this practice should be avoided. Another proper way to consider the concept of background may be seen on p. 470 where reference is made to " . . . the normal or background *variation* (italics added) of an element . . ."; this is an alternative method of referring to "range".

The second concept requiring emphasis and repetition is *"(excluding mineralized samples)"*. If mineralized samples are included in the consideration of background, the result will be that the background will have a long range and it will be too high. This problem was recognized by Hawkes and Webb (1962, p. 31) who, when suggesting that threshold for single-population data may be conveniently considered to be the mean plus two standard deviations, specifically stipulated "excluding markedly high erratic values"; such values would include mineralized samples.

Threshold   (pp. 214-215)

Being able to determine the upper limit on background values, that is, the threshold, is of utmost importance in the interpretation of geochemical data. It is not surprising, therefore, that much attention has been given, and continues to be given, to this topic.

As pointed out by Sinclair (1974) thresholds in geochemical data are chosen in a variety of ways. One of these, the way originally recommended by Hawkes and Webb (1962) and used by Lepeltier (1969) and a great many practicing geochemists today, involves the arbitrary choice of a threshold at a value corresponding to the mean plus two standard deviations *of the background population*. However, Sinclair (1974, p. 130) points out:

"In some cases this procedure might be adequate but it ignores the fact that no *a priori* reason exists for exactly the upper $2^{1}/_{2}\%$ of every data set being anomalous. Furthermore, the method does not take into account adequately, the fact that anomalous and background populations have fairly extensive ranges of overlap in some cases, and as they are two populations, the mean and standard deviation derived from the whole data set really have no statistical validity and are just numbers."

A second approach, used by many field practitioners, is to rely on subjective visual examination of histograms to choose threshold values; in practice, this procedure may well be the most commonly used method. A third approach is to define thresholds at points of maximum curvature in cumulative probability (frequency) diagrams (e.g., Fig. 4-12, right) but, according to Sinclair (1974) this method can have drawbacks (e.g., the method is approximate; a high proportion of anomalous values may go unrecognized).

To correct some of the problems and limitations inherent in the above three general methods, Sinclair (1974) has described a method of choosing threshold values between anomalous and background geochemical data based on partitioning a cumulative probability plot. The procedure, although somewhat arbitrary, can be applied to any polymodal distribution containing adequate values and populations with appropriate density distribution. Along the same line, Parslow (1974) has also proposed an improved graphical method for the determination of both background and threshold (the method is a cumulative plot on log-probability paper with the two component linear distributions being asymptotic to the sigmoid tails and having a symmetry with each other about the inflection point). The technique optimizes, in the sense of accentuating and defining more precisely, the weak anomalous distributions rather than obvious anomalous distributions. The methods discussed by both Sinclair (1974; see also Sinclair, 1976) and Parslow (1974) are, however, beyond the scope of this chapter. For the sake of completeness, it is worthwhile noting that some very highly regarded exploration geochemists have found it unnecessary to use the concept of threshold. Bolviken (1971), for example, found the concept of threshold meaningless, based on the statistical approach he uses for interpreting geochemical data; only "background" and "anomaly" are necessary. Others will look in the literature for compilations, (if available) of, for example, anomalous zinc values in granites for a particular terrain and use these as a guide.

From the above discussion it should be clear that there is no universally accepted method for selecting background and threshold, although most methods do have some points in common. Each geochemist must make his own selection in a way that best fits the geology, explains multi-element distributions, and locates mineralization. In addition it should be noted that some geochemists believe that the concepts of threshold and background are far too overworked, and that they are just guides for an interpretation, to be changed as need be; these individuals often will rely heavily on "experience" and subjective judgments. Interpretations should be based primarily on: (1) the spatial distribution of the elements; (2) the correlation between geochemically related elements, such as Ni-Co, Pb-Zn and Cu-Mo; and (3) the relationship between element distributions and geology.

Anomaly (pp. 216-219)

No additional comments.

Multiple Populations (pp. 219-220)

The entire concept of cumulative probability diagrams, with particular emphasis on their use in exploration, is discussed in a book by Sinclair (1976) which exploration geochemists will undoubtedly find very useful.

Computers (pp. 221-222)

No additional comments.

RELATIONSHIP BETWEEN GEOPHYSICS AND GEOCHEMISTRY
(A new topic)

It is essential in exploration to be aware of the limitations of a particular technique, as well as its advantages, and to know what level of confidence can be placed in it. Seldom can one method, be it geology, geophysics or geochemistry, be optimal for all stages of exploration. The question which should be asked is, "What method or combination of methods constitute the optimum to be employed at any particular stage of any exploration project?" Specifically, the main concern is to select the procedure capable of detecting a particular anomaly, which is conspicuous and sufficiently unique to justify follow-up work, most of which is far more costly than the preceding survey.

One inherent weakness in geochemical methods, and all techniques have comparable problems, is that they are often unable to discriminate between a weak anomaly caused by insignificant mineralization, often at shallow depth, and an anomaly from economic mineralization at greater depth (see Fig. 8A-3). For example, a weak geochemical anomaly in or near a fault zone is caused by non-economic mineralization in the vast majority of cases but, rarely, it may represent a leakage halo (p. 52) emanating from an ore body at depth. Geophysics, on the other hand, can often indicate where and how large a target may be, but it can not determine with certainty whether the target contains any of the sulfide minerals of economic interest. Furthermore, in some areas such as in the Canadian and South African Precambrian shields, geophysics is also plagued by detecting an over-abundance of non-economic anomalies. In many instances, however, geophysics can provide negative information; specifically, it can fairly reliably eliminate ground from further consideration, at least in the case of possible massive sulfide and porphyry copper deposits (depth limitations not considered). A case in point was illustrated in Fig. 9-5 (pp. 377) in which it was shown, and stated in the text, that mineralization did not occur in areas of low magnetic relief. Obviously, geophysics and geochemistry compliment and supplement each other in many exploration programs and coincidence of anomalies by these two fundamentally different techniques greatly enhances the confidence placed in such anomalies, e.g., such as encouraging deep overburden drilling on EM targets.

Although integration of geophysics, geochemistry and geology has been standard practice with many mining companies for many years, few case histories have been published. In recent years, however, a few reports have appeared. By way of example, Curtin and King (1974) have reported on the association of geochemical anomalies with a negative gravity anomaly in Colorado. Specifically, anomalously high amounts of Au, Ag, Pb, Zn, Cd and Bi found in mull ash, as well as Cu and Hg in soils, correspond to a negative gravity anomaly in the Front Range mineral belt (Fig. 4A-1). The correspondence of the geochemical anomalies and the negative gravity anomaly suggested to them the presence of an altered and mineralized (Tertiary) pluton — the negative gravity anomaly indicating the presence of rock less dense than the Precambrian rocks exposed at the surface. By analogy to other mineralized districts known to the authors, they suggested " . . . that the anomalies reflect leakage from a concealed Tertiary pluton."

A considerable number of studies which integrate geophysical and geochemical results in glaciated areas have been published in recent years. For example, during extensive uranium prospecting in Sweden, Lundberg (1973) used results from various geophysical methods (airborne and ground magnetic, and radiometric surveys), radon determinations, boulder tracing procedures, and geochemical analyses of water and organic samples taken from streams. Several interesting prospects were discovered which, at the time of publication, had not been evaluated. Other examples in which both geophysical and geochemical methods were used concurrently in exploration for base metals deposits in glaciated areas may be found in Jones (1975).

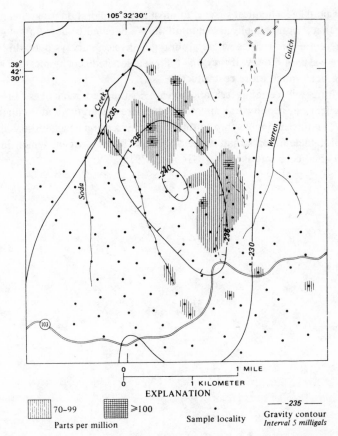

Fig. 4A-1. Copper in the B-soil horizon (shaded areas) in relation to a negative gravity anomaly. From Curtin and King (1974).

In a discussion of the status of exploration geochemistry in southern Africa, well-documented with examples, Buhlmann et al. (1975) concluded that geological as well as geophysical programs become more effective if geochemical techniques are incorporated. One of the main advantages of the geochemical methods is that they permit unmineralized targets detected by geophysical methods to be eliminated from consideration and, at the same time, interest in those prospects which yield positive anomalies by both geochemical and geophysical methods is enhanced.

Finally, it is worth mentioning certain differences between airborne and ground geophysical surveys, and their relationship to geochemical surveys. Airborne geophysical surveys have much the same sort of aims as regional geochemical surveys and are equally as useful. Ground geophysical surveys, however, while more like detailed geochemical surveys, are unlikely to detect mineralization; instead they indicate the size and shape of conductors, magnetic anomalies, density differences, and so forth. An airborne geophysical survey could well be carried out at the same time as a regional geochemical survey. In most large exploration programs it is probably better that ground geophysics should follow a detailed (follow-up) geochemical survey. There are many cases where ground geophysical surveys operating "blind" have been a waste of money. If they had followed detailed geochemical surveys they would have been much more cost-effective.

The future will certainly see an increase in the number of case histories in which geochemical, geophysical and other data are integrated and compared. *Exploration 77*, an international symposium on geophysics and geochemistry applied to the search for metallic ores, had at least six case history papers which integrated geophysical and geochemical results (Hood, 1979).

$5A$ / *Field Methods*

A discussion of field methods is essentially a discussion of sample collection. In a simplified form, sample collection can be considered to consist of four basic concepts: *what, when, where* and *how* to sample — topics all considered in the first edition. However, this simplified approach can be misleading because it doesn't force the sampler to ask some of the more critical questions, such as how many anomalous samples should be collected before moving to the next exploration stage. Some geochemists require three anomalous samples whereas others will investigate every individual anomaly. Another critical question is how much sample (weight) must be collected. These are among the topics to be discussed below.

Following the procedure used in Chapter 5, the discussion of field methods which follows is based on the type of material being sampled. Most attention is given to refinements and recent advances in the past several years. With respect to refinements in sample collection, it is appropriate to note that the use of helicopters has revolutionized reconnaissance surveys in many types of terrains. These range from stream sediment (drainage basin) surveys in tropical Brazil, to lake surveys in northern Canada.

STREAM SEDIMENTS (pp. 223-227)

The general principles of stream sediment collection remain unchanged and one method applicable in those areas with an adequate supply of water is illustrated in Fig. 5A-1. Grab sampling (as described on pp. 223-224) is used in areas with inadequate water. These principles are usually based on the use of the minus 80 mesh fraction, but this procedure must be modified for local environmental or climatic conditions (e.g., humid vs. desert) which affect or control the availability of various size fractions, the type of weathering, etc.

It was pointed out on p. 226 that "Occasionally it may be desirable to separate the heavy minerals (e.g., detrital sulfide grains) from drainage basin sediments to detect or enhance anomalies . . . ". Recently it has become standard practice in many areas to collect "pan concentrates", along with the usual stream sediments. These concentrates consist of sulfide grains, heavy resistate minerals such as cassiterite, unweathered silicates such as garnet, rock fragments with admixed heavy constituents, and anything else which is collected by 'gold panning' methods (Fig. 5A-2). With the concentration of such heavy materials prior to analysis, the likelihood of detecting and recognizing anomalous samples is greatly enhanced although, on occasion, special problems will be encountered, such as the presence of iron-manganese oxides

711

with adsorbed metals which coat the concentrated grains. Also, it is often very difficult to compare analytical results from pan concentrates due to differences in the "degree of panning" from place to place; that is, a little extra panning immediately enhances an anomaly. This panning procedure is essentially analogous to the collection of "heavy clasts" (p. 456) from till samples in glaciated areas.

Fig. 5A-1. Conventional wet screening of stream sediments to obtain clay and silt sized material during a reconnaissance survey in Scotland. In this case, the samples are taken from about 0.2 m below the stream bed. From Plant and Rhind (1974). Courtesy of Institute of Geological Sciences, London.

It must be emphasized that stream sediment sampling at the reconnaissance stage must be done very carefully. For follow-up work this may not be so critical because there should be many more samples covering the anomaly.

It is logical to suppose that the collection of pan concentrates would be most useful in arid or desert environments, or in areas with rugged topography, where mechanical erosion is predominant and clastic grains would be expected to occur. This has been confirmed by Bugrov and Shalaby (1975) who collected pan concentrates from wadis in the Eastern Desert of Egypt. Although such is the case, there is also undeniable evidence that valuable information can be obtained from pan concentrates collected in areas where chemical weathering is known to be predominant.

For example, in Scotland and England, Leake and Aucott (1973) found that with the light felsic phase removed, analysis of the pan concentrates was particularly useful in assisting in geological mapping in areas showing little contrast between major rock types. Similar techniques have been used in Czechoslovakia (Pokorny, 1975), Canada, and elsewhere, as integral parts of geochemical programs. Many potentially favorable areas (for example, in the mountainous areas of western Canada) which were originally explored the basis of −80 mesh stream sediments with negative results are now being reconsidered for exploration by means of pan concentrates, because it is realized that mineralization may be detectable by analysis of these materials. It should be noted, however, that sampling reproducibility using pan concentrates is much lower than for stream sediments; problems caused by "degree of panning" were mentioned above. Some studies have shown that sampling reproducibility is much better if the samples are sieved to −30 mesh before analysis.

Fig. 5A-2. Panning to obtain a heavy mineral concentrate during a reconnaissance survey in Scotland. From Plant and Rhind (1974). Courtesy of Institute of Geological Sciences, London.

The collection of pan concentrates from alluvium and other media presents difficulties in deserts and in any area with limited supplies of water although Bugrov (1974) has described a water panning technique applicable for arid environments in which water is conserved. Previously the separation had to be made in streams well away from the sample site, or even in the laboratory, which necessitated the transportation of large volumes of material for long distances (an example of the results from such an effort is illustrated in Fig. 8-5). However, in such environments, it is possible to obtain the equivalent of pan concentrates by means of an air separator. Truck-mounted air separators have been successfully used in certain areas of Africa by the Institute of Geological Sciences, London. Where only the magnetic portion of the heavy mineral content of stream sediments is of interest, it is possible to collect this fraction in the field by means of a strong magnet. Callahan (1975) has described a procedure whereby such samples can be obtained directly from wet sediments with an automagnet.

Finally, it is worth noting that Bradshaw (1975a, p. 203) recommends that stream sediment sample analyses should be done on five mesh sizes, plus the heavy-mineral (pan concentrate) fraction in the ideal case. The size fractions (in mesh) are: (a) $-38, +80$; (b) -80; (c) $-80, +160$; (d) $-160, +296$; (e) -296. Clearly, this is an expensive and time-consuming operation which can only be justified in special circumstances.

SOILS (pp. 227-231)

A major problem facing geochemists conducting soil (and rock) surveys is determining the optimum sample spacing. This subject was briefly introduced earlier on pp. 227-228, and the statistical aspects of "survey sampling" were discussed on pp. 492-496. The problems of sampling in the geosciences, of which geochemical sampling is just one very small part, have produced a voluminous literature indicative of their importance.

Sinclair (1975) has considered the problem anew, and has shown that simple geometric considerations are enough to indicate the importance of the shape and size of an anomaly in determining the optimum grid orientation and sample spacing for a geochemical soil survey. Several simple geometric examples will illustrate the point (all are taken from Sinclair, 1975).

Case 1. A rectangular anomaly $2S \times S$ would definitely be found by using a rectangular sampling grid with the same dimensions ($2S \times S$) *provided* the sampling grid was oriented in the same direction as the anomalous zone (Fig. 5A-3, a). This would be true no matter where the origin of the sampling grid was located. However, if the sampling grid were rotated 90° (Fig. 5A-3, b), the probability of locating a random anomaly would be only 50%. Although anomalies are not generally rectangular, the example is not completely unrealistic and it does emphasize the importance of grid orientation.

Case 2. A circular anomaly is illustrated in Fig. 5A-4. A square grid will intersect the anomaly with certainty provided grid (sample) spacing is $\sqrt{2}r$ or less.

From this figure it can be seen that grid orientation is unimportant, and the probability of detecting the anomaly depends only on the grid dimensions (sample spacing) in relation to the diameter of the target.

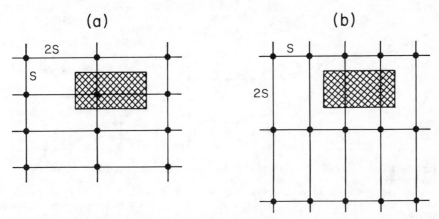

Fig. 5A-3. A simple example which illustrates the importance of grid orientation. The rectangular target (hatchured) and the sampling grid are equivalent in shape and size. In (a), the probability of locating the target is 100%; in (b), it is 50%. After Sinclair (1975).

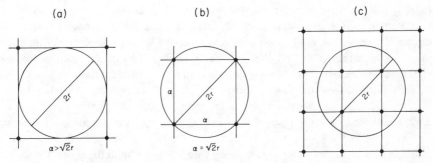

Fig. 5A-4. Circular anomalies. In (a), the probability of success in locating the target with the grid shown is < 1.0. In (b), the probability of success is 1.0, and the grid spacing, $\alpha = \sqrt{2}r$, is optimal. In (c), the probability of success is also 1.0, but the grid spacing is not optimal as more than one anomalous sample per target would be collected (however, this is often desirable). After Sinclair (1975).

Case 3. Most anomalous areas can be approximated by ellipses of which the circle is a special case. Elliptical anomalies are illustrated in Fig. 5A-5 in which the semi-major and semi-minor axes are labelled α and β, respectively; in the cases shown, one axis (α) is twice as long as the other (β). For individual elongated targets of this type, that is, where one axis is twice as long as the other, to ensure that at least one sampling point will intersect the anomaly: (1) the sampling grid must be oriented parallel to axes of the elliptical target, and (2) the sample spacing parallel to the major axis (2α) must be $\sqrt{2}\alpha$, and the sample spacing parallel to the minor axis (2β) must be $\sqrt{2}\beta$; this is illustrated in Fig. 5A-5(b). (Under these optimum conditions there is a 73% probability of two sample sites falling within the anomalous zone.)

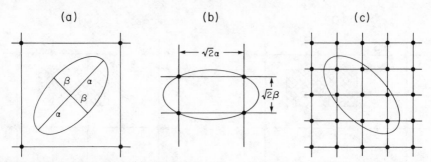

Fig. 5A-5. Elliptical anomalies. In (a), the probability of success in locating the target with the grid and orientation shown is < 1.0. In (b), the probability of success is 1.0; this is also the optimal sample spacing and anomaly orientation. In (c), the probability of success is also 1.0 but the grid spacing is not optimal as more than one anomalous sample per target would be collected (however, this is often desirable). After Sinclair (1975).

Simple and obvious as the above discussion may appear, on a practical basis it is often a major problem to predict the size, shape and orientation of an anomaly. An orientation survey run in the same area may be the best method, but it must be borne in mind that the mineralization may be only partially exposed, or even buried, and it is entirely possible (even likely) that such a survey will have to cover an uneconomic deposit. Nevertheless, having decided on the size of a target, a good rule of thumb is to choose a spacing for the original grid that enables two samples on each of two adjacent lines to be included within the *smallest* anomaly desired. Otherwise, one must be prepared to follow up single anomalous values. Alternatively, one could collect two to four times as many samples as are required and then, in the first instance, analyze only one-half or one-quarter of them thus eliminating the costs of re-visiting distant or difficult access areas. Another possibility is to make composites of two or four samples, a procedure used extensively in the Soviet Union (Beus and Grigorian, 1977).

In soils, of course, the target (anomaly) can be much larger than the actual ore body (see Fig. 1-2, p. 12) and many factors affecting metal dispersion can influence the choice of sample density. A case in point was mentioned by Scott (1975), based on experiences in the arid regions of South West Africa. Because of the alkaline soils in this area, copper dispersion is low (the pH of hydrolysis for copper is exceeded; Table 3-6), and soil reconnaissance samples were collected on 10 m intervals along traverses spaced 200-500 m apart, although for analysis they were combined into 20 m composites. In other parts of Africa where rainfall is more abundant and the soils more acid, sample intervals are as large as 50-60 m, presumably when searching for similar targets.

ROCKS (pp. 231-233)

In general, the concepts discussed above concerning the selection of the proper sample spacing and grid orientation for soil surveys are applicable to the collection of rock samples. Under certain conditions, however, special problems may be encountered. For example, nowhere in the preceding discussion on soils has the matter of sample size (weight) been considered. In the case of rocks, the varying size of its mineral constituents must be considered, and suggested sample weights to be collected are given on p. 232 based on the average grain size of the rock. There is a special problem in the case of the precious metals, specifically gold and platinum in the native state, because: (1) extremely small concentrations are significant (the ore at the world's richest platinum mine is reported to average less than 10 ppm of this metal); (2) they are usually in discrete particles of significant size, non-uniformly distributed (non-uniform distribution was discussed on p. 298); and (3) because of their high specific gravity relative to the other mineral fragments, they tend to segregate easily (this is especially troublesome following the grinding stage, and during the analytical phase).

Brown and Hilchey (1975) reviewed one aspect of the problem of sampling and analyzing precious metals. They are concerned with the effect of sample size on the precision of analyses (see also Clifton et al., 1969), and suggested that the problem could be visualized by means of the following example (Brown and Hilchey, 1975, p. 683):

"Assume that a sample contains 5 g Au per ton. A 200-g sample will then contain 1 mg or 1 ppm. However, a 1-mg piece of gold is a rather common size found in placer deposits [but not in rocks; author]. Thus if the gold is all in a single 1-mg piece, there is no way to obtain a representative 5-g sample for chemical extraction or a 30-g sample for fire assay. Even if the gold were in forty 0.025-mg pieces, more than one-third of the 5-g samples would contain no gold (Clifton et al., 1969, Fig. 5)."

There are three generalized ways to solve the problem of sample size: (1) increase the size of the portion analyzed; (2) reduce the size of the gold by grinding or pulverizing; and (3) concentrate the precious metal and analyze it separately. All have their cost limitations and drawbacks, the details of which are of limited interest here (recognition of the problem is the main concern). For the particular project with which they were concerned, Brown and Hilchey (1975) found that about 300 g of material were sufficient for their analytical purposes, and pulverizing the entire sample (as opposed to obtaining a specific size fraction) was preferable, even though gold is ductile and malleable making grinding and pulverizing difficult. Pre-ashing of a 10-g sample followed by an aqua regia digestion and analysis by atomic absorption, was the preferred analytical procedure. However, their particular emphasis appeared to be the detection of gold in B-horizon soil samples (although rocks were investigated) and so their conclusions can be considered applicable only to that specific case (in British Columbia). Each case of rock sampling and subsequent analysis (including sample preparation), should be considered as a separate, and unique problem.

Additional aspects of rock sampling are discussed on p. 761.

WATERS AND LAKE SEDIMENTS   (pp. 233-238)

Water   (p. 233-237)

Recently interest has been shown in the sampling of interstitial water of stream sediments and bog soils (the latter in northern latitudes; see pp. 440-451). This is because in these environments metals may be in interstitial solutions although absent from the waters above them. Interstitial waters in marine sediments, for example, have been found to be very different chemically from associated sea water.

Nowlan and Carollo (1974) described a probe for sampling interstitial waters of stream sediments and bog soils. Samples can be obtained within a specific interval of 2-3 cm, to a depth of 60-80 cm. Little contamination, if any at all, is claimed for samples collected with their probe.

Lake Sediments   (pp. 237-238)

The use of lake sediments for geochemical purposes, particularly in Canada and Scandinavia, continues to increase as methods of sample collection, and especially methods for the interpretation of the resultant data, continue to improve. The general subject of lake sediments was discussed earlier (pp. 187-191, 237-238, 438-440) and further discussion will be found in Chapter 11A.

There are now at least a dozen different types of lake samplers designed to take bottom samples, to obtain cores, or for other special purposes. Sly (1969) described and illustrated ten, and discussed their specific advantages (e.g., whether a sampler creates little disturbance of the bottom sediments); they are shown in Fig. 5A-6. Davenport et al. (1975) have described a simple sampler which has been found to be very effective in the lakes of western Newfoundland for geochemical purposes, and Coker et al. (1979) have shown a similar lake sediment sampler ("1976 model") used by the Geological Survey of Canada. Many variables (discussed earlier) must be considered not only in the collection and interpretation of lake sediment samples, but also in the type of sampler or corer to be used. Such decisions can be properly made only with experience. However, lake center samples are very fine grained and homogeneous and, consequently, the problem is a practical one of costs and efficiency of collection of the required material. Lake margin samples, on the other hand, are more similar to stream sediments in size and homogeneity.

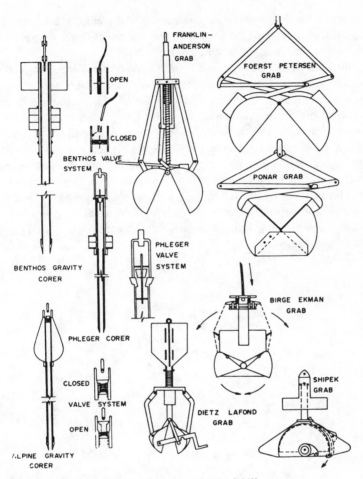

FRANKLIN – ANDERSON GRAB

OPEN

CLOSED

BENTHOS VALVE SYSTEM

FOERST PETERSEN GRAB

BENTHOS GRAVITY CORER

PHLEGER VALVE SYSTEM

PONAR GRAB

PHLEGER CORER

BIRGE EKMAN GRAB

CLOSED

VALVE SYSTEM

OPEN

DIETZ LAFOND GRAB

SHIPEK GRAB

ALPINE GRAVITY CORER

Fig. 5A-6. Bottom samplers and corers used for lakes. From Sly (1969).

DUTIES AND RESPONSIBILITIES OF SAMPLE COLLECTORS
(pp. 238-240)

Field data cards (Fig. 5-4) are being used with increasing frequency by certain government exploration parties. The advantages of such systems include: (1) the field data can be readily stored; (2) the data is collected and recorded in a uniform manner by different collectors; and (3) the survey becomes very easy to computerize. Clearly, the development of the cards is closely linked to the development of computer methods as an aid to interpretation, and the requirements for standardization, in data acquisition. The history of the development of field data cards as used for geochemical work by the Geological Survey of Canada is described by Garrett (1974). The five cards in use by that organization are illustrated and explained. These include: field analytical card; rock sample card; glacial till and soil sample card; stream water and sediment sample card; and lake sediment sample card.

One difficulty which has emerged from the use of data cards is that subjective categories recorded by collectors can appear unduly respectable. For example, in one experiment involving student collectors in the United Kingdom, the categorization of certain types of drift in an area varied from 10 to 90 percent. Clearly, data of this quality are useless and, with the use of data cards, may be difficult to recognize or even suspect. If these cards are to be useful, all sample collectors, whether they be professional geochemists, students or professors, must use reliable and reproducible methods of measuring grain size, water flow rates, percent of organic matter, etc., in the field. This requires careful training. In view of the foregoing, as well as other problems, it is not surprising that some individuals who were originally very impressed with the potential of data cards now believe that there is a need for simpler cards with fewer categories.

6A / Analytical Methods

INTRODUCTION   (p. 241)

In the Preface to the first edition the statement was made that (p. viii) "At the present time, discussions of the analytical methods most commonly used in exploration geochemistry are appropriate as developments in analytical techniques seem to have reached a temporary plateau". The analytical techniques referred to in that Preface, particularly atomic absorption, colorimetry, emission spectrography and X-ray fluorescence, are still the most widely used methods in exploration today. However, as always, new methods of analysis continue to be developed (and older ones improved), some of which clearly indicate that the analytical techniques in most common use a decade from now will not necessarily be those mentioned above.

In addition to such well-established objectives as selectivity (specificity) and sensitivity (limit of detection), the choice of methods to be used in exploration, and also in such fields as environmental and quality control, is now motivated by three other factors: (1) multiple element capabilities of the technique; (2) the simplification (and preferably elimination) of sample preparation required before a sample can be analyzed; and (3) cost. The new techniques which presently satisfy the above objectives, at least to some degree, and thus are beginning to attract serious attention for exploration purposes, are plasma emission spectroscopy, and neutron activation analysis. The latter is one of several nuclear-based techniques which, although known and used for over a decade, has now reached the application stage in exploration; others include fission (etch) track methods for uranium, and nuclear bore-hole logging. Progress in the improvement of some of the more established techniques such as atomic absorption, X-ray fluorescence, and the use of specific ion electrodes, has also been made. Advancements have usually been concerned with increasing sensitivity and productivity, as well as extending the list of elements which can be conveniently determined by the particular method. New techniques, and refinement of older procedures, continue to be reported in the field of colorimetry. Ward and Bondar (1979) have prepared a general review of the various analytical techniques used in the search for metallic ores and have stressed recent advances. Reeves and Brooks (1978) have written a book devoted to the determination of trace elements in geological materials.

Valiant efforts are being made by some individuals and organizations to stress the importance of: (1) the recognition and systematic determination of accuracy and precision in geochemical data; (2) the value of partial, as well as total, chemical extractions; and (3) the importance of analytical standards, in part because recent results of inter-laboratory analyses of some standards have been quite disturbing.

GENERAL PRINCIPLES   (pp. 242-254)

Accuracy versus Precision   (pp. 242-243)

Whereas in the previous discussion stress was placed on the *distinction* between accuracy and precision, it is appropriate now to extend the discussion stressing the *need* for accuracy and precision in exploration geochemistry. In other words, it is necessary to emphasize that accuracy and precision must be carefully monitored because errors (which are the cause of poor accuracy and precision) can appear from many unsuspected sources. Furthermore, they are likely to go unnoticed and unsuspected especially in the modern automated laboratory designed to process and analyze up to several thousand samples per day. Some errors are *sporadic* and, although they may affect many samples over a variable period of time, once identified are not likely to occur again. Sporadic errors, according to Hill (1975), include calculation errors, decimal errors, failure to add a necessary reagent to a sample, and so forth. Although difficult to identify, once found, the individual analyses can usually be corrected with relative ease. Allcott and Lakin (1975) reported similar examples of sporadic errors and, in the case of misplaced decimal points, they note that this is particularly serious when data are being transcribed; in this situation misplaced decimal points may affect as much as 1% of the results.

It is the *systematic* errors which are the most troublesome, often resulting from continuing problems in the laboratory. Although precision and accuracy can be monitored by running a great many standards (p. 312) this is not always practical and, further, will not detect such things as errors caused by sample preparation. Plant (1973) and Plant et al. (1975) have given this matter the serious consideration it deserves because such errors may invalidate any attempt at interpretaton. Table 6A-1 lists some sources of systematic errors which may occur in the laboratory.

Systematic errors, such as those listed in Table 6A-1, are particularly onerous because, among other things, they can produce false anomalies and genuine anomalies can be falsely extended. Such could happen if, for example, instrument drift (p. 313), or any one of many temporary instrumental pitfalls, occurs over a short interval between standards designed to detect such problems.

Limit of Detection   (pp. 243-244)

No additional comments.

Table 6A-1.   Some sources of systematic error.
(Modified from Plant, 1973)

Sample Preparation	Possible Cause
1. Sieving	Contamination of background samples following preparation of anomalous samples.
2. Grinding	Same as above.
3. Weighing	analyst error
4. Digestion	"fall out" from fume hood; contamination following bubbling

Analytical Methods	Possible Cause
1. Instrumental methods	Changes in conditions between determination of standards (such as instrument drift).
2. Methods depending on visual comparison (for example, photographic emission spectrography, or comparison of intensity of colors in colorimetry).	Tendency for analyst to read "high" or "low" depending on previous determination.

Wet Analyses   (p. 244)

No additional comments.

Instrumental Analysis   (pp. 244-245)

No additional comments.

Partial and Total Analyses   (pp. 245-247)

The previous discussion of this topic ended with the statement (p. 247), "Some geochemists recommend that at least one partial and one total extraction be used on samples collected in any new area." The vagueness of that statement was unavoidable because, at the time, there were no published definitive suggestions or recommendations as to what specific extractions, or group of extractions, should be used. As a result, a wide variety of partial extraction methods involving acids, alkaline solutions, buffer solutions, complexing agents, reducing agents and fluxes have been used to improve the contrast of anomalies.

Bradshaw (1975) must be given credit for proposing his tentative minimum list (Table 6A-2) of recommended analytical extractions (the base metal extractions were briefly mentioned on p. 361). Readers should particularly note the words *tentative* and *minimum* when considering the application of these extractions to their particular projects. They are based on many years of practical experience by Dr. Bradshaw and associates and certainly seem reasonable. If, as suggested by Bradshaw (1975), these extractions were to be used by more geochemists there would be some hope of standardizing geochemical data. (The equivalent of the total extractions listed in Table 6A-2 may, of course, be obtained by emission spectroscopy, X-ray fluorescence, and certain other analytical techniques.)

Table 6A-2.   Tentative *minimum* list of recommended analytical extractions that should be available.

Class of Attack	Suggested Details
	For Base Metal Samples (from Bradshaw, 1975)
1. Total	hot $HClO_4$ or $HClO_4$-HNO_3 mixture at reflux temperature for a minimum of 4 hours.
2. Hot Extraction	0.5N HCl boil for 20 minutes.
3. Cold Extraction	0.25% EDTA shaken cold for 2 minutes.
	For Uranium Samples (from Bradshaw, personal communication)
1. Total	HF-$HClO_4$-HNO_3 mixture (or equivalent) at reflux temperatures for a minimum of 4 hours.
2. Partial	5N HNO_3 (this is the extraction most commonly used by mining companies).
3. Weak	5% Na_2CO_3 solution (with H_2O_2) at 90°C for 2 hours; alternatively oxalic or acetic acid digestions.

A very practical and important weak partial extraction procedure which permits the selective extraction of uranium from rocks is listed in Table 6A-2. It was mentioned by Beus and Grigorian (1977), and is apparently commonly used in the U.S.S.R. When uranium-bearing samples are digested at about 90°C for several hours in a 5% Na_2CO_3 solution (5 gm in 100 ml of water) containing 15 ml of 30% H_2O_2, tetravalent uranium, such as that in uraninite, is oxidized by the hydrogen peroxide to the hexavalent state. This results in the formation of a soluble uranyl-carbonate complex which can be analyzed for uranium fluorometrically. In many cases, satisfactory results may be obtained without the added hydrogen peroxide according to Thomson and Read (1979) who have confirmed the value of this method. This weak partial extraction is particularly valuable because it enables halos from uraninite, as well as any secondary hexavalent uranium minerals, or adsorbed uranium, to be detected in the ppb range in rocks; such halos may represent buried uranium mineralization.[*] At the same time, uranium within rock-forming minerals (e.g., micas, amphiboles) or resistate minerals (e.g., zircon, sphene) are not detected. Rose and Keith (1976) reported a similar partial extraction technique for uranium in stream sediments in which a mixture of 5 ml of 30% H_2O_2 and 15 ml of 50% glacial acetic acid is digested with the samples for 1 hour in a sonic cleaner. They report this leach is intended to dissolve uranium in organic matter and in exchange

---

[*]According to Dr. P. M. D. Bradshaw (personal communication) the Na_2CO_3-H_2O_2 extraction procedure is particularly valuable in samples from the primary environment, such as granites and pegmatites, because uranium mineralization is likely to occur as uraninite. However, in the secondary environment, such as stream sediments and till, it may be difficult to distinguish between mechanical (e.g., uraninite) and hydromorphic dispersion unless the extractions are performed on the heavy and light fractions, respectively.

sites, as well as in uranium minerals, but without dissolving zircon, feldspars, apatite and other minerals not associated with local uranium ore deposits. Because the procedure used by Rose and Keith (1976) may dissolve Fe-Mn oxides from the sediments, analyses by delayed neutron activation would be preferred to fluorometric determinations of uranium because of possible "quenching" problems (see p. 300).

Two other valuable studies which recognize the importance of partial extractions are worthy of note. The first, by Gatehouse et al. (1977), involved the sequential leaching of soil samples from Tasmania with various extractants so that from one 2-gram soil sample the elemental concentrations in the following phases may be determined: water-extractable, ion-exchangeable, reducible Mn, reducible Fe, organic material, clay and residue. The second, by Hoffman and Fletcher (1979), used different sequential partial extraction procedures to determine certain elemental concentrations in organic matter, sulfides, carbonate minerals, Mn oxides, Fe oxides and silicate minerals from soils and stream and lake sediments from central British Columbia. Both studies stress the practical importance of knowing the concentrations of elements in the Fe and Mn oxides. Partial and total extractions may also be accomplished by heating, such as in the case of mercury (pp. 332-334), sulfur and some volatile elements (e.g., Te) but no new definitive studies in this area are available. Clearly, *partial analysis involves the isolation of a particular phase which can be more specific to mineralization.*

Sample Attack (Decompositon) Commonly Used (pp. 247-252)

Some minerals in silicate rocks can be quantitatively recovered from rock fragments by prolonged (from 3 - 4 weeks) digestion in cold, concentrated hydrofluoric acid. (*Caution:* all work with this acid must be done under a hood with good ventilation; it is also dangerous because of reactions with the skin; see p. 250). The procedure, described in great detail by Neuerburg (1975), yields clean, unharmed crystals, mostly of minerals which occur largely in trace amounts in rocks (such as, many sulfides, cassiterite, rutile, zircon, gold). The procedure must be followed exactly as described by Neuerburg (1975) if good results are to be obtained.

Reporting Units Used In Exploration Geochemistry (pp. 252-254)

No additional comments.

PREPARATION OF SAMPLES FOR ANALYSIS (pp. 254-257)

No additional comments.

ANALYTICAL METHODS — GENERAL COMMENTS
(pp. 257-258)

No additional comments.

ATOMIC ABSORPTION SPECTROMETRY
(pp. 258-263)

Atomic absorption (A.A.) continues to be the main analytical technique used in the laboratory analysis of geochemical samples. Refinements and improvements in established procedures have been reported from many laboratories. Ward (1975), for example, has published new and refined methods used by the U.S. Geological Survey for 14 elements of interest in geochemical prospecting, most of which are based on atomic absorption. Abbey et al. (1974) have described an improved analytical scheme, based on A. A., used by the Geological Survey of Canada.

Research in atomic absorption instrumentation has not yet produced the long-sought, and long-awaited, truly simultaneous (as opposed to sequential) multi-element instrument, although one company manufactures an instrument with which two elements can be determined simultaneously. However, highly automated, computer controlled systems do result in even greater productivity than has been the case in the past (Bristow, 1975) but such systems are not recommended in remote areas or in under-developed countries where service facilities are limited. Research has been quite successful along other lines, such as perfecting A.A. techniques, or combining wet chemical-A.A. methods, to obtain lower detection limits for elements of particular geochemical importance. For example, Hubert and Lakin (1973) described a method of obtaining Tl and In down to 0.2 ppm in geologic materials. Another case in point is the work of Watterson and Neuerberg (1975) who used a wet chemical-A.A. method to determine Te as low as 5 ppb in rocks. As discussed previously (p. 326), this element, at Coeur d'Alene for example, forms primary geochemical aureoles at considerable distances from mineralization and being able to detect it in the 5 ppb range is very desirable. Chao and Sanzolone (1973) developed a special A.A. procedure for determining low quantities of Co, Ni, Cu, Pb and Zn in soils and sediments containing large amounts of hydrous manganese and iron oxides. And finally, in the way of examples, Jonasson et al. (1973) have described the methods used by the Geological Survey of Canada for the determination of mercury in ores, rocks, soils and water by flameless atomic absorption (see p. 263 for an explanation of the "flameless" variation).

By far the most important development with respect to atomic absorption has been the adaptation of the heated graphite furnace (atomizer) to the point where it is now used routinely for many analyses previously beyond the scope of the conventional flame A.A. The heated graphite furnace (which rated only one sentence in the first edition; p. 263) is one of the "flameless" techniques and has the following advantages over the flame models discussed on pp. 258-263: (1) the ability to detect extremely low concentrations of some metallic elements (in some cases sensitivities are increased from 100 - 1000 times); (2) the ability to analyze extremely small samples; (3) the ability to analyze certain types of solids directly; and (4) a reduction in the likelihood of contamination during sample preparation, as sample preparation is often easier (especially with the solid samples). The heated graphite furnace may be used in conjunction with a basic flame instrument such as shown in Fig. 6-5 (p. 260) as these intruments can operate in either the flame or flameless mode. Essentially, the flame excitation section is replaced by the heated graphite crucible and the remainder of the instrumentation remains the same. It is important to realize that there are still many analyses which are done better with flame techniques (for example, high concentration and high precision determinations), and for many exploration purposes the advantages listed above are not necessary.

Fig. 6A-1 illustrates a simplified cross section of a heated graphite furnace (atomizer) which replaces the burner-flame sample chamber (Fig. 6-5; p. 260). The ends of the graphite tube (which is open at both ends) are supported by two graphite rings which are connected to an electrical power supply enabling rapid heating to about 2500°C. The graphite tube is heated by passing a high current at low voltage (up to 500A at 10V) directly through the tube. As drying, ashing and atomizing of the sample have to be carried out at different temperatures, the power supply includes three different temperatures and several selectable temperature programs. An inert gas (nitrogen) is commonly used to prevent burning and oxidation of the sample.

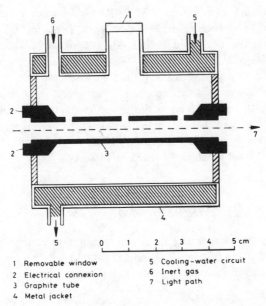

1	Removable window	5	Cooling-water circuit
2	Electrical connexion	6	Inert gas
3	Graphite tube	7	Light path
4	Metal jacket		

Fig. 6A-1. Simplified cross-section of a graphite furnace (atomizer). From Friedrich et al. (1973).

COLORIMETRY   (pp. 264-283)

General Comments   (pp. 264-267)

Meaningful figures on the current use of colorimetry are not available. However, colorimetric tests, particularly for field determinations, continue to be used from places such as the Arctic Island of Canada to the mountains of Ecuador. In several developing African countries, colorimetry has been replaced by some extremely well-equipped laboratories in major centers in which atomic absorption is the basic analytical method. However, colorimetric methods, particularly when used in remote, active exploration areas, have advantages which make them indispensable (see p. 265). Further, even in some of the more advanced countries, colorimetric procedures still play an important role. For example, Shapiro (1975) has described the "rapid" analysis methods used by the U.S. Geological Survey for the analysis of the major constituents of silicate, carbonate and phosphate rocks. Six determinations (for SiO_2, Al_2O_3, Fe_2O_3, TiO_2, P_2O_5 and MnO) are accomplished by spectro-photometric (colorimetric) procedures, whereas four (CaO, MgO, Na_2O and K_2O) are determined by atomic absorption. Other techniques are used for H_2O, FeO, CO_2, F and S.

Apparatus   (pp. 267-269)

Apparatus used for colorimetry in the laboratory has not changed but, several new innovations have been reported in field work. These include: (1) pre-calibrated scoops, similar to that illustrated in Fig. 6-8, but designed for several sample weights (0.1, 0.2 and 0.4g); (2) paper tissues impregnated with exact amounts of dithizone to obviate the need for weighing this reagent in the field*; and (3) "kits" (p. 268), once the mainstay of the exploration geochemist, are once again readily available (and highly advertised), but this time by the environmentalists. Several companies offer field kits to test for trace elements in the biosphere. Kits are available for Cu, Pb, Zn, As, Hg, Cd and so forth, and generally are based on the same colorimetric reagents (e.g., dithizone) used for the tests listed in Table 6-2.

General Comments On Reagents   (pp. 269-270)

The quality of the reducing agent hydroxylamine hydrochloride referred to in the copper biquinoline test (pp. 279 and 553) and in the total heavy metal (Bloom) test (p. 554) seems to vary considerably. Accordingly, it is recommended by Prof. H.

---

*These papers are prepared by taking a solution of dithizone in an appropriate solvent (e.g., toluene) and evaporating a specific volume on to filter paper. In the field, the dithizone is re-dissolved in fresh solvent.

Bloom (personal communication) that ascorbic acid be substituted weight for weight for hydroxylamine hydrochloride as was originally suggested by Stanton in 1966 (see p. 279).

Colorimetric Reagents   (pp. 270-272)

No additional comments.

Standard Solutions and Matching Standards   (pp. 272-273)

No additional comments.

Calculation of Trace Element Concentrations   (pp. 273-274)

No additional comments.

Erroneous Analytical Data   (pp. 274-275)

No additional comments.

Selected Colorimetric Analyses   (pp. 275-283)

Several new colorimetric procedures have been reported which have application to exploration geochemistry, including some which can be done in the field (e.g., Pt; Kothny, 1974). They are listed below (including those with a spectrophotometric final stage):

As: Boyle and Jonasson (1973, p. 279); Marshall (1978).

Mo: (a) Hoffman and Waskett-Myers (1974); this is a slightly modified zinc dithiol procedure (see pp. 282-283; 556-557).

   (b) Quin and Brooks (1975) have developed a rapid procedure for the determination of molybdenum, also based on zinc dithiol, which is applicable to geochemical, biological and steel samples, and which involves changing only the preliminary sample treatment.

Nb: Greenland and Campbell (1974).

Pt: Kothny (1974).

Re: Mahaffey (1974).

S: Shapiro (1973).

V: Marinenko and Mei (1974).

Zr: Meyowitz (1973).

A valuable qualitative colorimetric "spray-type" field test for *secondary* zinc minerals, such as smithsonite and cerussite (but not sphalerite), has been successfully used in the Yukon, South Africa and elsewhere. The procedure involves mixing one solution (a 3% solution of potassium ferricyanide in water) and a second solution (a solution of 3% oxalic acid and 0.5% diethylanaline in water) in equal amounts and spraying (or dropping) the combined solutions on a rock outcrop or on a mineral specimen. The appearance of a red color indicates the presence of zinc. (Note: *caution* is advised for the handling and spraying of the ferricyanide solution).

Stanton (1976) has summarized important colorimetric methods for B, Bi, Mo, Pd, Pt, Th, V and W in a new, small book on analytical methods used in geochemical exploration (some of the procedures were included in an earlier book by the same author). Also included are cold extraction techniques for As, Cu, Ni, Pb, Zn, and total heavy metals.

EMISSION SPECTROGRAPHY   (pp. 283-289)

Few new advances or innovations have been reported in the theory, techniques, or interpretation of results of emission spectrography as related to exploration geochemistry. Use of this method continues to be about the same as reported earlier for North America (Table 6-1), and in the Soviet Union the technique still remains the analytical basis of most exploration programs (Beus and Grigorian, 1977). One noteworthy achievement by the U.S. Geological Survey has been the adaptation of computer analysis to the interpretation of emission spectrographic results obtained on instruments which record data photographically (Walthall, 1974). The computer system of spectral analysis performs the determination of 68 elements in geologic materials. Very carefully controlled experimental conditions are required to obtain the necessary precision and accuracy.

The installation of new direct reading emission spectrometers for the analysis of geochemical samples continues to be reported from widely scattered locations such as South Africa (Watson and Russell, 1974) and Mexico (de Pablo Galan, 1975). However, some direct reading emission spectrometers are being converted to plasma emission spectrometers simply by changing the excitation source (for a discussion of plasma emission spectroscopy, see p. 740). So the use of emission spectography is likely to decline in the future.

X-RAY FLUORESCENCE   (pp. 289-295)

Laboratory Analysis   (pp. 289-293)

X-ray fluorescence (XRF) analysis continues to be a standard analytical technique employed primarily for: (1) elements which cannot be economically, or practically, determined by any other method; and (2) whole rock major element analyses, often in conjunction with rock geochemical surveys.

Levinson and de Pablo (1975) published a paper which illustrates the use of XRF in the first instance. They showed how the method can be made rapid, economical and sufficiently accurate and precise for those elements which are normally difficult to determine (e.g., Th, Nb, Ta, W), and especially those heavier than iron. Their procedure is based upon the measurement of the ratio of a characteristic peak and nearby background, while primary scattered radiation is used to correct or compensate for matrix variations. In the second instance, Fabbi et al. (1976) described their methods for the quantitative analysis of selected major, minor and trace elements in rocks by means of a computerized automatic x-ray spectrometer.

Leake and Aucott (1973) were able to show that XRF analysis of panned concentrates for 21 elements was particularly useful in assisting with geological mapping in areas of northern England and the Scottish Highlands where there is little contrast between the major rock types. Their interpretations were based mainly on elemental ratios, and interrelationships, rather than on absolute quantities. Plant et al. (1975) also used XRF to analyze for essentially the same elements as Leake and Aucott (1973), also in heavy mineral concentrates, but from other areas in Great Britain.

Field Analysis (Portable Equipment) (pp. 293-294)

Friedrich et al. (1974) have reported what can be considered an advance in field analysis based on the principles of X-ray fluorescence. They have been successful in analyzing manganese nodules from the Pacific Ocean on board a ship (hence "field" but not "portable") for six elements: Mn, Fe, Co, Ni, Cu and Zn. Their technique consisted of exciting the nodules with a radio-isotope source (^{238}Pu) and analyzing the resultant X-ray radiation by means of an energy-dispersive system (the details of which are beyond the scope of this book). The important point is that the future will likely see more of this type of field-portable equipment designed for on-site analysis.

OTHER METHODS (pp. 295-304)

Paper Chromatography (pp. 295-296)

In exploration geochemistry paper chromatography continues to be used either for special situations or where the analyst has particular expertise. An example of the latter would be the field test for extractable copper which reacts with rubeanic acid to form a "spot" on filter paper (Delavault, 1977). A very early paper chromatographic analysis technique for uranium was described by Thompson and Lakin (1957) and it is still used to some extent by the U.S. Geological Survey (Allcott and Lakin, 1975, p. 663).

One very interesting example of how paper chromatography is used to separate uranium from associated elements in sediments, soils and water, is described in detail by Grimbert (1972) based on successful experiences with the method in France. In the procedure the uranium is first separated by paper chromatography from other elements. The part of the paper in which the uranium is concentrated is then calcined, after which the uranium-containing residue is mixed with a flux of sodium fluoride and sodium carbonate, as is described on p. 299 for fluorometry. The final determination is made fluorometrically. One advantage of Grimbert's (1972) combined paper chromatography-fluorometry method is that possible "quenching" elements (p. 300) are not present to interfere with the final fluorometric analysis.

Specific Ion Electrodes   (pp. 296-297)

Specific ion electrodes continue to be used mainly for the determination of fluorine in water and in all solid exploration media (e.g., soils), and to a lesser extent for chlorine (e.g., Van Loon et al., 1973). The work of Pluger and Friedrich (1973) showed that fluoride adsorption in soils was dependent chiefly on time, nature of the extracting solution (e.g., HCl, NaCl), and the nature of the clay mineral. Hence, partial extraction techniques for fluorine can, and should, be used on soils and sediments. The fluorine from such extractions, which can be determined by a fluoride ion electrode, are necessary in order to obtain better contrasts between background, threshold and the anomaly.

Other specific ion electrodes have been used for the determination of iodine in rocks and soils (Ficklin, 1975). The use of copper specific ion electrodes has been reported by Friedrich et al. (1973) who, on the basis of their experiences, concluded that they can be recommended in water surveys if only rough estimates of the copper content are necessary; however, the application is restricted to waters with low levels of interfering ions (such as chloride, bromide, ferric, mercury and phosphate). Schuller et al. (1975) have used a copper electrode for on-site soil determinations.

Gravimetric, Volumetric and Fire Assay   (pp. 297-299)

Haffty et al. (1977) have published a manual which describes in detail fire assay techniques and the methods of analysis used for the determination of the noble metals (Ag, Au, Ir, Os, Pd, Pt, Rh, Ru) in geological materials. Also included are photographs and detailed discussions of the specialized equipment used.

Fluorometry   (pp. 299-300)

Fluorometry still appears to be the most widely used method for the determination of uranium in all kinds of exploration materials, although other techniques, such as delayed neutron activation and fission track analysis, are coming into wider use.

Recently, Barringer Research Ltd. (Rexdale, Ontario) has offered a field kit for the analysis of uranium which is based on the fluorometric technique. Samples are fused with a fluoride-carbonate flux on platinum dishes by means of a flame from a hand-held propane cylinder. The intensity of the green fluorescence of the fused disc illuminated by an ultra-violet light, is a measure of the amount of uranium, as it is with the laboratorty model (Fig. 6-12, p. 299). This field instrument, however, is not intended to be a substitute for accurate and precise analysis in a laboratory.

Fluorescence from the uranyl ion in solution may be induced, under certain conditions, when the solution is irradiated with a very short pulse of ultraviolet light from a nitrogen laser. Scintrex Ltd. (Concord, Ontario) have used this phenomenon as the basis of their laser-induced fluorescence analyzer (Model UA-3; described by Robbins, 1978) which can be used in either the field or laboratory (Fig. 6A-2). Although natural waters contain varying amounts of dissolved organic matter which also fluoresce under the same operating conditions, the fluorescence of the organic matter ceases almost immediately upon termination of the laser pulse (whereas the uranyl fluorescence persists). By electronically rejecting any short-lived (from organic matter) component of the fluorescent signal, the instrument is capable of making interference-free measurements of uranium to the sub-ppb level in most natural waters (Robbins, 1978). This rapid, sensitive, field-adaptable technique has

Fig. 6A-2. The UA-3 laser-induced fluorescence analyzer for uranium. Courtesy of Scintrex Ltd., Concord, Ontario.

great potential in those areas with abundant waters, such as in Canada, Scandinavia and the U.S.S.R.

Two similar combined wet chemical-fluorometric methods for the determination of selenium described by Crenshaw and Lakin (1974) and Schnepfe (1974) give very good sensitivities (at least as low as 0.1 ppm Se). A fluorometer different from that illustrated in Fig. 6-12 is required for this procedure as the fluorometric determination is made on aqueous solutions.

Radon Counter   (pp. 300-301)

Interest in the detection of radon continues with several portable devices similar to that shown in Fig. 6-13 (p. 301) now commercially available. All, of course, are based on the same principle of detecting and measuring the α-decay of radon. Problems concerned with the detection and interpretation of radon data in water have been discussed earlier (pp. 430-432), however, equally serious problems have been encountered with the use of radon in soils (a topic not previously discussed herein).

The instrument shown in Fig. 6-13 employs a "pump" (the bulb in the foregound) which extracts air from the hole and the surrounding overburden material. Another portable model which has been illustrated by both Michie et al. (1973, Fig. 6) and Bowie (1974, Fig. 4) measures radon in ground air in the immediate vicinity of a shallow hole in which the alpha detector ("probe") is inserted. According to Bowie (1974), the first instrument is more satisfactory in the search for more deeply buried uranium deposits of large volume whereas the second instrument tends to reflect radon concentrations from a limited volume of overburden, and is hence suitable for defining relatively narrow mineralized structures or tracing faults along which radon may escape. The "pump" method (Fig. 6-13) has the advantage of being able to measure radon in waters and, therefore, is particularly useful in the areas of abundant lakes in Canada and Scandinavia. Excellent results have been obtained by both methods through sand, peat, glacial drift, and other relatively unconsolidated types of cover, although the physical nature of the soil cover is an important factor to be considered.

It is well established that diurnal variations in the determination of gases, such as radon and mercury, particularly from porous soils, can be quite large (Fig. 6A-3). These variations are attributed chiefly to the effect of changes in atmospheric pressure. In the case of high clay or organic soils, changes in humidity, temperature, and groundwater conditions are most important. It is for reasons such as these that many geochemists experienced in uranium exploration consider radon determinations as a qualitative technique. However, during the extended period of a radon survey in an area, say three weeks, the variations can be compensated for by measuring the fluctuations in one hole and applying appropriate corrections to others in the area which can be assumed to be similarly affected. Smith et al. (1976) and Severne (1978) have

discussed other aspects of radon methods for uranium exploration which are of value
to those interested in further details on this subject.

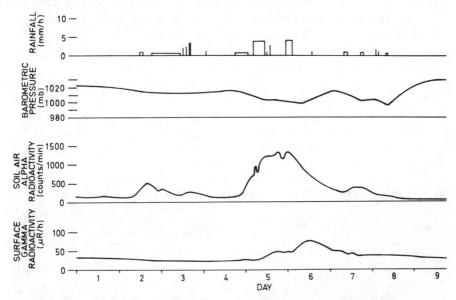

Fig. 6A-3. The effects of variations in atmospheric conditions on radon concentrations (designated "soil air
alpha radioactivity") in soil air. When saturated by rainfall, humic top soil seals off the escape of radon to the
atmosphere and causes a build-up of soil air radon concentration. If this is sufficiently prolonged, radon decays
into gamma-emitting nuclides, producing a rise in surface gamma activity. The effect disappears as the rain
ceases and the ground dries. From Miller and Ostle (1973).

To this point, "electronic" monitoring of the alpha particles emitted as ^{222}Rn
decays has been considered. Instruments such as that illustrated in Fig. 6-13 (p. 301)
are typical. There are also scintillation counters (e.g., "alphanuclear") commercially
available which are left in the ground to integrate the radon flux over a period of
hours or days thus eliminating short-term fluctuations related to atmospheric variations. In addition, several various types of alpha-sensitive film have been employed
in radon measurements, the most widely used being cellulose nitrate emulsions. The
best known are called TRACK ETCH (covered by patents owned by Terradex Corporation, Walnut Creek, California) and Kodak LR 115 (see Severne, 1978, for a discussion of, and further references to, these materials). One major advantage of the
alpha-sensitive films is that they are placed in the ground for about 30 days and,
therefore, it is possible to "iron out" the effect of diurnal variations and thus obtain
"quantitative" results from soil radon surveys.

The method consists of placing small pieces of the alpha-sensitive film attached
to the insides of small plastic cups in shallow (about 2 feet) holes over the area being
explored (Fig. 6A-4). The cups are inserted on a grid pattern of 50 - 1500 feet,
depending on whether a reconnaissance or detailed survey is being run, and on other

factors. After the sampling cups containing the alpha-sensitive film are in place, the holes are covered and left undisturbed for several weeks during which time radon gas migrates into each cup, and produces tracks on the film, the number of which is dependent on the abundance of radon. At the end of the period, the films are retrieved and processed by an etching procedure using NaOH after which the tracks are visible under a microscope. The density of the tracks is related to the radon concentration. The film is sensitive only to α-particles, and is unaffected by light or beta or gamma particles. However, these films are unable to discriminate between ^{222}Rn (half-life 3.8 days) and ^{220}Rn (half-life 56 seconds; generated during the decay of ^{232}Th). To alleviate this problem a porous polymer membrane (described by Ward et al. 1977) may be placed over the opening of the cup creating a "diffusional barrier". The movement of the ^{220}Rn (thoron) radio-isotope is slowed sufficiently for it to decay before reaching the alpha-sensitive film.

Fig. 6A-4. TRACK ETCH film cups being placed upside down in shallow holes to trap radon in soils. Courtesy of Terradex Corporation.

Radon detection instruments, particularly the electronic devices such as that illustrated in Fig. 6-13, are sometimes called emanometers because the detecton of radon is one aspect of emanometry. *Emanometry* is defined as the detection of gaseous emanations characteristic of radioactive deposits. Radon and helium, both gaseous products in the U (and Th) decay series, are the detectable signatures of such deposits.

Other important practical aspects of radon surveys (e.g., instrumentation, disequilibrium, advantages and disadvantages of specific techniques) are discussed by Hambleton-Jones (1978).

Radiometric Techniques (pp. 301-302)

Surface and airborne radiometric techniques continue to be used in more or less the conventional manner with improvements in instrumentation constantly taking place. The most important radiometric techniques are still those concerned with detecting and interpreting gamma radiations emanating from uranium, thorium and potassium in order to locate radioactive mineral deposits and also to differentiate between common types of rocks and sediments (i.e., geologic mapping).

One of the newer innovations has been the perfection of radiometric instrumentation for *in situ* seabed gamma ray spectrometry. Miller and Symons (1973) towed a seabed gamma ray spectrometer to produce continuous profiles of the natural radioactivity on the sea floor from which the total count, potassium, uranium and thorium measurements were made. Similar instrumentation is also being used on large lakes in Canada and elsewhere.

NUCLEAR TECHNIQUES
(A new topic)

Among the many techniques of analysis based on the detection and interpretation of induced nuclear reactions, only a few have been applied to exploration for mineral deposits. These include neutron activation analysis, nuclear borehole logging, and the fission track technique.

Neutron Activation Analysis (pp. 302-303)

Until recently, neutron activation analysis (NAA), and the essentially equivalent instrumental neutron activation analysis (INAA; the subtle distinction being that the former may involve some radiochemical separation whereas the latter is entirely instrumental), have been used principally in fundamental geochemical investigations in which accuracy, precision and sensitivity were of utmost concern. Simply stated, the technique requires that a sample be irradiated with a neutron source (for example, in a nuclear reactor) after which the resultant gamma spectra is then analyzed. Gijbels (1973) has presented a most useful account of the application of NAA to the analysis of ores and minerals. Lunar and certain environmental studies have been particular benefactors of this powerful technique. However, in the last few years, several significant advances have been made which now permit the technique to be seriously considered for certain exploration purposes. These advances are principally: (1) the development of high resolution Ge (Li) detectors for gamma spectrometry; and (2) the development of computer programs which permit the efficient processing of the complex spectra which result from the neutron irradiation. Inherent in the technique is multi-element capabilities because all elements are irradiated simultaneously. However, the sensitivity for different elements varies widely, including some which can be analyzed at very low levels, and some which cannot be measured at all. Where sensitivity is good, the accuracy is also good. Furthermore, neutron activation analysis is only rarely influenced by matrix variations (a case in which it would be influenced is a rock which has a high boron content). Sample preparation is simple although, for high accuracy, the samples and standards must have the same geometric configurations. To accomplish this, for example, some investigators employ powdered samples which are pressed into pellets of standard size.

Adaptation of NAA techniques to exploration geochemistry have largely been pioneered and made practical by members of the Institute of Geological Sciences, London. Ostle et al. (1972) determined uranium for a reconnaissance program in the United Kingdom by the delayed neutron method (a form of NAA) in a wide range of materials (e.g., stream sediments, rock, soil, water) at low cost. Plant and Coleman (1972) applied NAA analysis to the evaluation of placer gold concentrates from Scotland and were able to use large samples (up to 500 g) of unground alluvium and fluvioglacial detritus; the use of large samples for the evaluation of gold prospects is important because more representative results can be obtained by comparison with other techniques which usually involve the analysis of 50 g or less of material. In a more recent paper, Plant et al. (1976) have been able to expand the previous individual analyses of U and Au to multi-element capabilities and, in fact, they can now determine 32 elements at good sensitivities (e.g., U, 1 ppm; As, 2 ppm; W, 2 ppm) and with very satisfactory precision and accuracy. Their productivity is 40 samples per man week, although the time taken for analysis is dependent upon the decay time necessary before counting which, in some cases, may be one month. The approximately 40 samples per man week yield about 1,200 determinations and, as a result, the costs are considered competitive. The particular procedure was specifically designed for elements difficult to determine by standard analytical methods and, as such, the costs are especially reasonable in that context. (Others may not find the costs satisfactory depending upon the financial arrangements they must make for amortization of equipment, irradiation, counting, etc.). Plant et al. (1976) concluded that the method is considered to have the greatest potential for orientation studies, error control and follow-up investigations. The technique is not designed, at least at present, for routine analyses which can be more economically and rapidly obtained by atomic absorption or colorimetry.

In connection with NAA it is important to recognize that "multi-element capabilities" does not mean "all elements". In fact, the 32 elements mentioned by Plant et al. (1976) are probably the main ones which NAA techniques can determine in geological samples at realistically useful sensitivities, at least at present. These include (with the average ppm detection limits obtained by Plant et al., 1976, in parentheses): Na (20), Al (500), K (1000), Sc (0.05), Ti (1000), Cr (5), V (20), Mn (5), Fe (500), Co (0.05), Zn (50), Se (5), As (2), Br (1), Rb (20), Mo (2), Ag (5), Cd (3), Sb (0.5), Ba (200), Cs (1), La (0.1), Ce (3), Sm (4), Eu (1), Hf (0.3), Ta (2), W (2), Ir (0.01), Au (0.005), Th (0.5), and U (1).

There are several extremely important elements in exploration geochemistry which cannot be determined at all by NAA, for example, Pb, Li, Be and Nb, because they have no gamma ray emitting isotope. Other elements are very insensitive at levels of interest for prospecting (for example, Pt; and even Zn at 50 ppm). Finally, it is important to keep in mind two significant limitations of NAA: (1) necessary access to an irradiation facility; and (2) relative slowness of analyses.

Nuclear Borehole Logging
(A new topic)

Closely associated with NAA is the technique of nuclear borehole logging; both techniques depend upon neutron irradiation and the measurement of gamma rays. In borehole logging a portable neutron source (usually ^{252}Cf) is lowered down the borehole along with a gamma ray counter, and the gamma ray spectrum is measured *in situ*. Such small neutron sources are too weak, in general, to produce radioactive isotopes in sufficient quantities for NAA. Instead one measures the "prompt" gamma rays which are emitted at the moment that a neutron is captured by an element (isotope). One commercial instrument marketed under the name "Metalog" (Scintrex Ltd., Concord, Ontario) has been tested in lateritic nickel deposits where it has been demonstrated to give good statistical accuracy at the 1% nickel level, and on a porphyry copper deposit at the 0.5% copper level (Annonymous, 1975a). Similar nuclear borehole logging devices able to detect uranium are known to be in the development stage (Dodd and Eschliman, 1972; Steinman et al., 1975).

Fission Track Analysis   (FTA)
(A new topic)

This technique (also called "Charged Particle Track Analysis" or PTA) received brief mention on page 302 in which it was stated that it had been applied to the search for uranium in Scotland (Bowie et al. 1973), and by Wollenberg (1973) who showed how the method could be used to determine the location and abundance of radioactive elements and minerals in rocks. More recently, the method has received further attention from some exploration geochemists, particularly with respect to uranium exploration and, therefore, further discussion is in order. The technique has extremely wide geologic application, ranging from the study of the distribution of the light elements Li and B, to the heaviest element U, "in every conceivable type of matrix, from blood to ultramafic rocks, in every conceivable locale, from the bottom of the seas to the surface of the moon" (Fisher, 1975, p. 291). The entire subject is discussed in detail in the book by Fleischer et al. (1975).

The technique depends on the "damage" produced when a charged particle moves through a non-conducting solid. As applied to uranium analysis, the charged particles are the high-energy fission fragments produced when [235]U is irradiated with neutrons. The specimen to be analyzed is placed in juxtaposition to a polycarbonate plastic such as Lexan. Fission products ejected from the surface of the rock specimen produce damage tracks in the plastic. These tracks must be etched (usually with NaOH) in order to make them visible under the microscope. The number of tracks per unit area gives the uranium content when compared with a uranium-containing standard similarly irradiated. A very significant advantage of the method is that it can indicate the exact location of uranium in a mineral sample; fission track analysis is superior to all other techniques when it is desirable to obtain information on the spatial distribution of uranium content. The method is quite sensitive with detection limits of 0.05 ppm U reported in rocks, and as little as 1 ppb U in water (Reimer, 1975). Access to a nuclear reactor is necessary, and the accurate counting of the fission tracks is required. The technique is considered to be inexpensive, rapid, sensitive, and sufficiently accurate for exploration purposes by those who use the method.

Mercury Detector (pp. 303-304)

Several mercury detectors, both for laboratory and field use, are now available which are based on the principle that elemental mercury released from geological materials by heating will be adsorbed by gold (amalgamation). In one version, the mercury is subsequently removed from the gold by heating and determined by atomic absorption (Joensuu, 1971). In another version, the increase in electrical resistance of the mercury-containing gold is quantitatively related to the mercury concentration (McNerney et al., 1972; McNerney and Buseck, 1973).

PLASMA EMISSION SPECTROSCOPY (PES)
(A new topic)

The new technique of plasma emission spectroscopy (PES) is the most promising of the multi-element analytical methods and has several valuable advantages over atomic absorption and emission spectroscopy. These advantages include: (1) improved sensitivity particularly for the volatile elements (detection limits are at least comparable to atomic absorption and they are generally significantly better than those attainable by emission spectrography); (2) fewer inter-element and matrix problems, such as those which occur in A.A. (p. 261), are encountered although the method does suffer from some of the problems of interference (such as with iron; p. 284) which plague emission spectrography; (3) no graphite electrodes are required as in emission spectroscopy (p. 284); (4) sample preparation is essentially the same as for A.A., but significantly easier than for emission spectrography; (5) because of the multi-element capabilities and easy sample preparation, the cost of PES analyses are relatively low. At least one company (Barringer Research Ltd.) was offering a 20-element quantitative analysis for about $12.00 in 1979. Further, although A.A. does have multi-element capabilities, at present, it is a *sequential* determination in that the solution must be analyzed with successive lamps; PES has true *simultaneous* multi-element capabilities comparable to emission spectrography.

For the exploration geochemist it suffices to know that plasma emission spectroscopy is another, more efficient, method of sample excitation. Meyer and Lam Shang Leen (1973) define a plasma (p. 325) "as a mass of ionized gas in which the concentration of electrons and positive ions are in equilibrium". The plasma state may be produced in several ways. Most analytical methods are based on "induced coupled plasma" (ICP) — a special type of plasma that derives its sustaining power by induction from high-frequency electromagnetic fields. Commercial instruments appear to be based mostly on radio-frequency magnetic fields (e.g., Fassel and Kniseley, 1974). In actual practice, energy from a radio-frequency generator is transmitted to an induction coil, surrounding a "plasma torch" creating a magnetic field. A plasma is formed when argon gas (an inert carrier gas) is ionized after passing through the field (Fig. 6A-5). Sample in a liquid form (solids may also be used under certain conditions) is carried by the stream of gas into the core of the plasma. Very high temperatures (5,000 - 10,000°K) are developed in the plasma and free atoms of the metallic ions to be analyzed are formed. Characteristic radiations from the plasma, including those of the elements being determined, pass into a detection device which, for multi-element determinations, is identical to the detection system of a direct reader emission spectrograph (pp. 287-288), where the intensity of the radiations from all elements are measured simultaneously.

Fig. 6A-6 shows a schematic diagram for a typical direct reading spectrometer which can be used for emission spectroscopy, plasma emission spectroscopy, and other techniques (compare with Fig. 6-9, p. 286, which is not a direct reader). The plasma torch, in essence, replaces the graphite electrodes of a typical emission spectrograph. This is why some direct reader emission spectrographs can, and are, being converted to PES at relatively low cost with the resultant analytical advantages listed above. At present, published reports show that good detection limits are achievable for over 60 elements, including some such as U, Zn and Te for which sensitivities are poor by emission spectroscopy (see Table 6-3). However, detection limits, as well as reproducibility, will vary considerably with the type of plasma generator. Specifically, an inductively coupled radio-frequency plasma is superior in all ways (sensitivity, reproducibility) to a capacitively coupled microwave plasma (Boumans et al. 1975). Multi-channel spectrometers as shown in Fig. 6A-6 achieve spectral isolation with precisely located photomultiplier tubes. At present, only about 30 of these tubes can be included in a spectrometer so that all the theoretically obtainable elements cannot be determined once an instrument has been programmed.

Finally, it is essential to realize that because sample preparation for PES is similar to that employed for atomic absorption (in other words, the sample must be put into solution with acids or other solvents), PES is not automatically a "total" method, as is emission spectrography, unless the entire sample is digested. Looking at it in another way, PES has partial extraction capabilities.

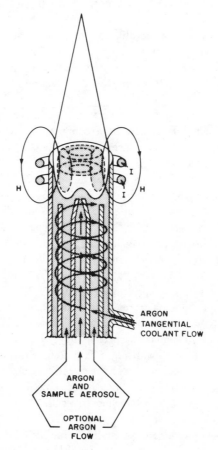

Fig. 6A-5. Complete plasma configuration. Magnetic fields (and "eddy currents") are generated by the induction coil. High-frequency currents flowing in the induction coil (I) generate oscillating magnetic fields (H). After Fassel and Kniseley (1974). Reprinted with permission from *Analytical Chemistry.*

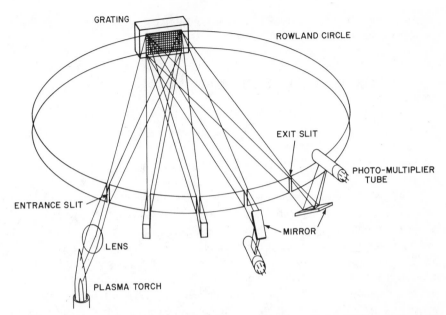

Fig. 6A-6. Schematic diagram of a typical direct reading spectrometer for multi-element determinations, with a plasma emission (plasma torch) excitation source. (The Rowland Circle is the arc along which all radiations are focused for the particular instrument). After Fassel and Kniseley (1974). Reprinted with permission from *Analytical Chemistry*.

SUMMARY OF ANALYTICAL TECHNIQUES (pp. 304-309)

On the basis of the size of this chapter, which reflects changing trends in analytical geochemistry, one might assume that modifications would be forthcoming for Figs. 6-15 and 6-16 (pp. 308-309) which are graphical summaries of suggested analytical techniques. Actually, such is not the case because: (1) much of the preceding discussion is only an up-date of existing methods (such as techniques to increase sensitivities); (2) some of the discussion concerns instrumentation, such as neutron activation analysis, which, although potentially of great significance is still not readily available to most exploration geochemists; and (3) the exact capabilities of some of the instruments are not yet known (for example there are conflicting reports in the literature on the detection limits obtainable by plasma emission spectroscopy, possibly because of variations in the excitation sources used).

It is noteworthy, however, that new methods have been devised for elements which have been particularly difficult to determine in low concentrations previously. These include As, Nb, Pt, Re, V and Zr which are colorimetrically determined, and Se which is a fluorometric method. These new methods, which have been discussed above, may be considered as additions to Fig. 6-15. At the present time, analytical methods used in exploration geochemistry, and specifically multi-element methods, appear to be in the initial stage of a transition period (particularly towards plasma emission spectroscopy).

COMMERCIAL LABORATORIES   (pp. 309-311)

No additional comments.

STANDARDS   (pp. 312-314)

The use of standards is still the most common method of controlling the accuracy and precision of geochemical analyses (for example, see Abbey, 1977). New standards useful in the various areas of the earth sciences continue to proliferate. Flanagan (1974) has compiled another comprehensive list (the previous one was in 1970) of more than 150 standard reference samples of use to geoscientists, but the more recent compilation also has entries of interest to agronomists, archaeologists, environmentalists and others. Some new standards of special interest to exploration geochemists are listed in Table 6A-3 which may be considered as a supplement to Table 6-5 (p. 314).

Of great interest to exploration geochemists are the results of interlaboratory analyses of the six samples specifically prepared for exploration purposes by the U.S. Geological Survey (the last entry in Table 6-5, p. 314). Allcott and Lakin (1975) have presented the results obtained by the U.S. Geological Survey Laboratories on 35 elements in these samples based on more than 46,000 determinations by several methods. In addition, they have been able to conclude, based on statistical (coefficient of variation) studies of the data, that the samples are sufficiently uniform (homogeneous) to be geochemical reference samples. In an "open-file" report of the U.S. Geological Survey, Allcott and Lakin (1974) have made available a statistical summary of the analytical results for 15 elements furnished by 85 unidentified laboratories (representing industry, university and government in 26 countries) on the above-mentioned six samples; a more recent compilation has also been issued (Allcott and Lakin, 1978). The results for Zn, probably the most commonly determined element in geochemical exploration, on a sample analyzed by atomic absorption, are presented in Fig. 6A-7. Clearly, the results obtained by the most commonly used method in exploration, are shocking in some cases. The element and samples selected for the figure are not atypical; in fact, worse examples could easily have been selected.

Although the results may not surprise some analytical chemists, they will certainly startle many geologists who are accustomed to accepting values from the laboratory as gospel. In a most tactful manner, Allcott and Lakin (1974, p. 4) state, "The results of this study suggest that we should critically evaluate our methods." From this one might be correct in concluding that a trace element analysis today is as

Table 6A-3. Selected additional analytical standards of particular importance in exploration geochemistry
(supplement to Table 6-5, p. 314)

Type	Designation	Source	Comments
Rocks (eight new rock standards)	miscellaneous	1	see Flanagan (1976)
Copper-Molybdenum ore	HV-1	2	Recommended values: 0.522% Cu and 0.058% Mo (Faye, 1978). $50 for 200 grams.
Nickel-copper-cobalt ore	SU-1	2	Recommended values: 1.51% Ni, 0.87% Cu and .063% Co (Faye, 1978). $25 for 100 grams.
Nickel-copper-cobalt ore	UM-1	2	Recommended values: 0.88% Ni, 0.43% Cu, and 0.035% Co (Faye, 1978). $25 for 100 grams.
Zinc-lead-tin-silver ore	KC-1	2	Recommended values: 20.37% Zn, 6.87% Pb, 0.67% Sn, 0.112% Cu and 0.112% Ag (Faye, 1978). $50 for 200 grams.
Gold ore	MA-1	2	Recommended value: 0.519 oz/ton (or 17.8 ppm) (Faye, 1978). $50 for 200 grams.
Radioactive ores	DH-1, DL-1, BL-1, BL-2, BL-3, BL-4	2	Recommended values for uranium range from 0.0041-1.02% (as U); Thorium values (on 3 samples only) are 15 and 83 ppm and 0.104% (as Th). (Faye, 1978). DH-1, DL-1, and BL-4 cost $50 for 200 grams; the others are $25 for 100 grams.
Tungsten ores	CT-1, BH-1, TLG-1	2	Recommended values for tungsten range from 0.083-1.04% (as W). (Faye, 1978). $50 for 200 grams.
Noble-metals-bearing sulfide concentrate	PTC-1	2	Recommended values; 3.0 ppm Pt, 12.7 ppm Pd, 0.62 ppm Rh, 0.65 ppm Au and 5.8 ppm Ag (Faye, 1978). $50 for 200 grams.
Copper and molybdenum	Copper standards (29001-29010) and molybdenum standards (42001-42005)	3	10 copper standards ranging from 0.078-0.94% total Cu and 5 molybdenum standards ranging from 0.051-0.30% total Mo. Acid soluble Cu and total S values also included. Catalogue and price list supplied by source.
Glasses with trace elements added	GSA thru GSE	4	Powdered glasses with traces of many elements added at 0, 0.5, 5, 50 and 500 ppm. (See Myers et al., 1976).

Sources:
1. Mr. F. J. Flanagan, U.S. Geological Survey, Reston, Virginia 22092
2. Chairman, Canadian Certified Reference Materials Project, c/o Mineral Sciences Laboratories, Canada Centre for Mineral and Energy Technology (CANMET), 555 Booth St., Ottawa, Ontario K1A 0G1.
3. Hazen Research Inc., 4601 Indiana Street, Golden, Colorado 80401.
4. R. G. Havens, U.S. Geological Survey, Federal Center, Denver, Colorado 80225.

difficult as it was 30 years ago, especially since the improvements in sensitivity and ease of operation are balanced by demands for accurate analyses in the low ppm, and ppb, range. Apparently, analytical problems in these ranges are easily concealed today by the elegant appearance of sophisticated and automated instruments! Again, it is worth emphasizing the point originally made on p. 311. It is, that important samples should be confirmed for accuracy by at least two different laboratories and preferably by two different methods; the two methods should differ from each other in principle and, ideally, should include different methods of sample preparation, digestion and final determination.

Finally, attesting to the importance of geochemical standards, a new journal, *Geostandards Newsletter,* devoted to the study and promotion of geochemical reference samples, has recently (1978) begun publications in France. (It is published

by the International Working Group of the Association Nationale de la Recherche
Technique, 101 Ave. Raymond Poincare, 75116 Paris).

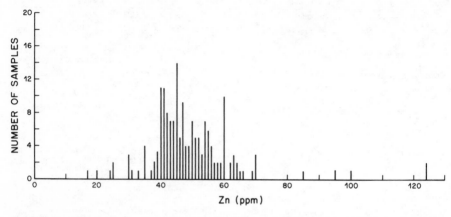

Fig. 6A-7. Variations in the zinc content of sample GXR5 by atomic absorption based on 170 reported values.
Range: 17 - 124 ppm. Mean: 49.19 ppm. Median: 47.00 ppm. Standard deviation: 13.89 ppm. Drawn from data
in Allcott and Lakin (1974).

MOBILE LABORATORIES   (p. 315)

No additional comments.

$7A$ / *Primary Distribution**
(Selected Topics)

INTRODUCTION (p. 317)

In this chapter the theoretical concepts presented in Chapters 2 and 2A are developed into practical criteria applicable to exploration and they are illustrated with examples from many parts of the world. A significant proportion of the discussion is based on results obtained in the Soviet Union, due mainly to the recent availability of the English translation of a book by Beus and Grigorian (1977) which emphasized the importance of lithogeochemical methods in the primary environment. By use of lithogeochemical methods in this environment, Soviet geochemists and geologists have successfully discovered blind** ore bodies as much as several hundred meters below the surface. In fact, it is the possibility of detecting blind ore deposits by lithogeochemical methods that has made these procedures, which are based on the concepts of primary distribution, so attractive. Further, by means of the geochemical analysis of rock specimens or their mineral separates, Soviet geochemists have shown how, in certain cases, it is possible to distinguish mineralized from barren host rocks (a topic introduced earlier on pp. 318-319). Under favorable conditions, from the analysis of rock samples they are able to determine whether anomalous lithogeochemical samples indicate that the main mass of ore lies at depth below the observer, or whether the sampling site is below ore which has already been eroded (the "supra-ore" and "sub-ore" halos discussed below).

The theoretical basis of lithogeochemistry in mineral exploration is dependent upon the existence of a genetic relationship between the primary distribution patterns and mineralization. Following the procedure of Govett and Nichol (1979) in their state-of-the-art review of lithogeochemistry, regional patterns will be discussed in terms of their geochemical response in: (1) intrusive rocks; and (2) extrusive and sedimentary rocks (specifically exhalatives). Recently, sufficient information and positive results have been obtained from primary distribution patterns in rocks of the second category to justify their being considered separately. Exhalative deposits (sedimentary deposits of volcanic origin; also called volcanogenic) are generally regarded as having originated from metal-bearing fumarolic exhalations on the sea

---

*For reasons discussed in the Introduction to Chapter 2A the word *distribution* is used in place of *dispersion*.

**"Blind" ore deposits are defined as those which are covered by rock; "buried" ore deposits are covered by overburden. Both can be considered as "concealed".

floor that are contemporary with volcanic or sedimentary activity. Clearly, their genesis and post-mineralization history differ from those characteristic of ore deposits of intrusive origin, resulting in different geochemical responses in the wall rock, as well as in the characteristics of their primary distribution patterns.

PRIMARY DISTRIBUTION ON A REGIONAL SCALE: GEOCHEMICAL PROVINCES   (pp. 318-324)

Total Analysis Techniques   (pp. 318-323)

1. **Geochemical Responses in Intrusive Rocks.**

Probably the first successful attempts to discriminate between productive and barren intrusives on a regional scale involved the analysis of granitic rocks for tin (see pp. 319-321). The use of any one element, or group of elements, to determine the mineralization potential of a pluton is the basis of geochemical specialization (characterization) explained in Chapter 2A (p. 638). In connection with that topic, Table 2A-8 and Table 2A-9 based on the tin and lithium contents of granites, respectively, were presented in the way of illustration.

Although the data in Table 2A-8 suggest that the discrimination between economic tin-bearing and barren granites is straightforward, such is not universally the case, as will be shown with examples below. Nevertheless, confirmation of the value of tin determinations of granites may be found in the study by Sheraton and Black (1973). They showed that the cassiterite-bearing granites of northeastern Queensland, Australia, which range in age from Precambrian to Permian, have significantly higher tin contents (average from 5-16 ppm) than granites without economic tin mineralization (average less than 4 ppm) in the area. However, it is worth noting that the anomalous granites have lower tin contents than those reported for productive tin-bearing granites elsewhere (see Table 2A-8). The economic stanniferous granites are also enriched in volatile elements such as B, Be, Li and F, and are characterized by a low Mg/Li ratio in agreement with data in Table 7-2 (p. 320). None of the economic tin-bearing granites are uniformly high in tin and, in fact, a very variable tin content is characteristic. This means that the sampling must be detailed if the tin-bearing potential is to be realistically assessed (the authors collected about 25 samples per intrusion).

Hesp and Rigby (1975), in a continuation of the earlier studies mentioned on pages 319 and 321, made an extensive study of granitic rocks from the Tasman geosyncline, Eastern Australia (ranging from northern Queensland to Tasmania) and found that granites associated with cassiterite mineralization have a higher average tin content (26 ppm) compared to similar, but barren, granitic rocks (average 3.4 ppm). Although there are some exceptions, particularly in the low range of tin concentration, generally all the granitic samples which have more than 20 ppm tin have associated tin mineralization. This is consistent with findings in the Soviet Union (Table 2A-8). Others have observed that Sn deposits may occur within or near highly fractionated leucogranites (see p. 318) characterized by high SiO_2 and K_2O, and low FeO, MgO and CaO. Thus, detailed exploration is focused on these particular types of intrusions.

Apart from the tin discussed above, lithium shown in Table 2A-9, and tungsten discussed below, most workers have found no correlation between anomalous abundances of other elements and mineralization in granites. This applies particularly to Pb, Zn and Cu (Sheraton and Black, 1973; Beus and Grigorian, 1977; Govett and Nichol, 1979). The lack of correlation can be ascribed to the fact that when mineralization involving these chalcophile elements does occur, it results from processes other than those related to magmatic differentiation.

There are situations in which intense hydrothermal alteration of granitic rocks over a large area can produce significant differences in the average element content between mineralized and barren intrusions. A case in point is the intrusion that hosts the Coed-y-Brenin porphyry-type copper deposit in Wales in which there is enrichment of K_2O, Cu and Mo, and depletion of MnO, Na_2O, CaO and Zn relative to similar but barren intrusions in the same region (Allen et al., 1976). Analogous trends can be seen at Bingham, Utah (John, 1978; Moore, 1978). The enhancement and depletion effects are the result of mineralization and associated hydrothermal alteration. Specifically the K_2O, Na_2O and CaO variations are associated with phyllitic

Table 7A-1. Data on selected trace elements in ore-bearing and barren (for Ni and Cu) basic and ultrabasic rocks. Modified from Beus and Grigorian (1977).

Region	Type of rock	Character of mineralization	Ni(%)	Co(%)	Cu(%)	Cr(%)	Ti(%)	S(%)	$\frac{Ni}{S}$
Kola Peninsula	Basic	Absent	0.8	0.6	0.8	1.2	0.05	0.05	16.0
		Sulfide	2.1	1.3	0.8	1.2	0.3	1.1	1.1
	Ultrabasic	Absent	1.0	0.6	1.7	1.3	0.05	0.06	16.0
		Sulfide	2.3	1.0	5.1	0.9	0.3	0.8	2.9
Northern Baikal	Ultrabasic	Absent	0.6	0.5	1.1	1.4	0.15	0.01	60.0
		Sulfide	1.4	0.7	2.1	0.6	0.08	0.06	23.4
Norilsk	Basic	Absent	1.2	1.2	0.9	1.3	1.4	0.3	4.0
		Sulfide	1.9	1.1	2.7	3.9	0.8	1.0	1.9
Central Russia	Strongly metamorphosed ultrabasic rocks	Absent	0.12	0.57	2.31	0.84	0.02	2.30	0.05
		Sulfide	0.43	0.50	1.89	0.80	0.06	0.76	0.56
Northern Baikal	Strongly metamorphosed ultrabasic rocks	Absent	0.40	0.10	0.07	0.30	0.007	0.02	20
		Sulfide	0.95	0.46	1.60	0.60	0.064	0.06	16

Note:
At this time, only deposits in the Kola Peninsula and Norilsk regions are economic. The examples from the Northern Baikal region are not typical, nor are they likely to become economic deposits, because they only contain small quantities of sulfides (S = 0.06%) in the "mineralized" rocks.

alteration, whereas Zn depletion may reflect migration to the propylitic or argillic zone (see Fig. 7-6, p. 330). These points are discussed again below in connection with patterns on a more local scale (Fig. 7A-4).

Soviet geochemists have been able to discriminate between ore-bearing (Ni and Cu) and barren basic and ultrabasic rocks by means of total analysis for certain trace and minor elements. This is illustrated in Table 7A-1. Of particular importance in this table is the Ni/S ratio. An ideal ratio would be 1.0 which implies that all the Ni is in sulfide minerals; a high ratio (low S) implies that most of the Ni is dispersed in silicates. It is also important to recognize that the significance of this ratio must be determined for each area. Thus, for example, in the Kola Peninsula a low Ni/S ratio (e.g., 1.1) in basic rocks will suggest mineralization, but in the Norilsk region this ratio will be higher (1.9).

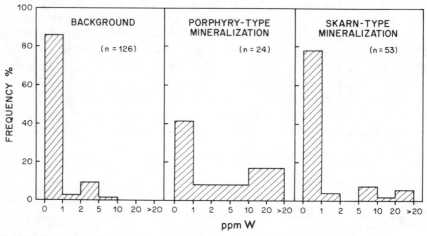

Fig. 7A-1. Fequency distribution of tungsten in barren (background) and mineralized granitic intrusions, and skarn-type mineralization, northern Canadian Cordillera. From Govett and Nichol (1979) but based on data from Garrett (1971).

Govett and Nichol (1979) pointed out that, in certain cases, the form of the frequency distribution of an element in a region or a particular rock unit is more diagnostic of mineralization than the mean value of that element. They selected data from Garrett (1971) to illustrate their point (Fig. 7A-1) and this shows that the form of the frequency distribution for tungsten differs between barren and mineralized granitic plutons in northwestern Canada. The distribution of tungsten in barren (background) plutons is only slightly skewed, whereas it has a marked positive skewness in mineralized (porphyry-type) plutons. This is interpreted as reflecting the presence of tungsten mineralization superimposed on the original tungsten distribution. The plutons that have porphyry-type tungsten mineralization are significantly easier to distinguish from barren plutons on the basis of the distribution of their tungsten contents than are those which have skarn-type tungsten mineralization.

Garrett (1971) concluded that by collecting duplicate samples from 15 sites, an unzoned pluton of "average" size could be characterized geochemically.

The use of major components of rocks for exploration purposes continues to be employed with increasing frequency. Of the numerous possible examples which could be used to illustrate the point, only two will be cited. The first, by Kesler et al. (1975b), showed that intrusive rocks associated with porphyry copper mineralization in island arc areas are lower in potassium than their counterparts on cratons. In the second, Beus (1979) showed that the sodium content of black shales which are host to the emerald mineralization in Colombia, can be used for exploration purposes. This is possible because the mineralization occurs only at the intersection of certain fault patterns in the vicinity of which sodium metasomatism has been intense. The content of sodium in black shales from the emerald-bearing zones averages about 3.5%, whereas in the non-faulted zones it approximates 1.0%. The K_2O contents in the respective zones are approximately 0.30% and 1.10%. The sharp differences in the contents of sodium and potassium, as well as the ratio of these elements, between those zones which have been affected by the sodium metasomatism and those which have not, are also clearly reflected in stream sediments.

2. Geochemical Responses in Extrusive and Sedimentary Rocks (specifically exhalatives).

Since the appearance of the first edition of this book significant progress has been made on the recognition and interpretation of the chemical composition of exhalative horizons for regional exploration. Exhalative deposits (sedimentary deposits of volcanic origin) within a particular stratigraphic horizon commonly cover wide areas. In the Red Sea area, where metalliferous sediments are presently being deposited, and where it is possible to study initial dispersion patterns before they have been disturbed by later geological events, anomalous contents of Cu, Zn, Mn and Hg in the exhalative sequences extend as far as 10 km from the Atlantis II Deep, and as far as 8 km from the Nereus Deep (Bignell et al., 1976). In both cases the anomalous elements show an increase in the gradient of the contrast toward the center of the Deeps, the source of the exhalatives. There is a 6-fold contrast in metal contents between samples from these locations and those from distant sites.

The practical use of such halos for exploration purposes may be illustrated by examples from Germany and Ireland. In the first case, Gwosdz and Krebs (1977) showed that at the Meggen Pb-Zn-Ba stratiform deposit which occurs in Devonian pelagic limestones, manganese enrichments (halos) of exhalative origin extend more than 5 km beyond the ore body (Fig. 7A-2). This is particularly important and valuable for exploration purposes because in the sediments above or below the exhalative formation only very local anomalous values of the indicator or pathfinder elements are found. As can be seen from Fig. 7A-2, there is a 3 to 4-fold contrast in manganese concentration near the mineralization, although the peak manganese content is not directly over the deposit. This lateral shift can be explained by aqueous dispersion of certain elements around, or away from, an active brine source. Similar hydrothermal (exhalative) manganese halos can be found associated with the Tynagh Pb-Zn-Cu-barite deposit in Lower Carboniferous limestones in Ireland (Russell, 1974; 1975). At this locality the manganese halo has a radius of about 7 km around the deposit, and the contrast can reach 20-fold. However, many samples are required to ensure that such high contrast is obtained (note the great range of Mn values at each sampling location in Fig. 7A-2). As in the Meggen deposit, the maximum enrichment of manganese is displaced from the ore body, 0.5 km to the north in this case.

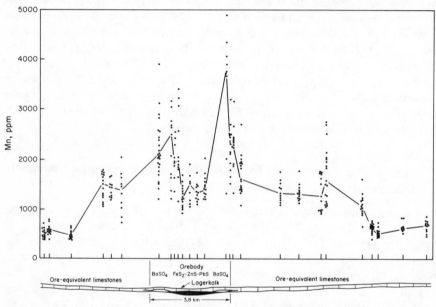

Fig. 7A-2. Manganese distribution in the Lagerkalk and ore-equivalent limestones (which are 2-4 metres thick) in the area of the Meggen ore deposit, Germany, based on the analysis of 342 samples. Dots indicate the Mn contents at numerous outcrops, underground workings and drill holes at each location. The solid line is the average value of samples taken from the many different sampling sites within the limestone sequence. After Gwosdz and Krebs (1977).

Govett and Nichol (1979) give many more examples, and details, of extensive regional-scale exhalative lithogeochemical anomalies from many parts of the world. These include the McArthur River, Australia, Zn-Pb-Ag deposit in which anomalous values of Zn, Pb, As and Hg in shales extend as much as 20 km from the deposit. They also discuss: (1) the Cyprus cupiferous pyrite massive sulfide deposits which occur in a sequence primarily of basaltic pillow lavas with only minor sedimentary rocks, and in which sulfide mineralization occurs in those regions where the Cu/Zn ratio of the basalts is less than 1.2; and (2) several of the volcanogenic massive sulfide deposits of the Canadian Shield in which there are geochemically distinctive productive and non-productive cycles of volcanism. The productive cycles are generally characterized by slightly higher contents of Fe, Mg and Zn, and by lower contents of Na and Ca, although there are aspects other than the bulk chemical composition of the rock which must also be taken into account during exploration (Nichol et al., 1977).

In considering stratiform exhalative or volcanogenic deposits it is especially important to appreciate the fact that the form of their primary distribution pattern differs from those typical of mineralization in intrusive rocks where the primary halo will often take the form of an enveloping or overlying aureole. In the exhalative or volcanogenic examples, the primary distribution patterns are extensive along the lateral equivalents of the mineralized horizons (up to 20 km in the case of the McArthur River deposit discussed above), where they grade more or less evenly from ore abundances to background values (Fig. 7A-2). In contrast, in the vertical direction there is an abrupt change from mineralized to completely barren horizons, as a direct result of the stratigraphic control of the mineralization.

Partial Extraction Techniques   (pp. 323-324)

There are very few new reports of lithogeochemical surveys based on partial extraction techniques being used either successfully or unsuccessfully in regional exploration. In their review of lithogeochemistry in mineral exploration, Govett and Nichol (1979) only discuss one such case and this involved an attempt to discriminate between mineralized and barren ore bodies. This is the work of Cameron et al. (1971), briefly mentioned on p. 323, which employed a sulfide selective digestion (ascorbic acid-hydrogen peroxide mixture) to determine the Cu, Ni and Co contents present in sulfide minerals from ultramafic rocks in the Canadian Shield. (In Table 7A-1 a total analysis procedure, based on the same concept, was employed).

The procedure of Cameron et al. (1971) is based on the assumption that in order for significant Cu-Ni mineralization to form from an ultramafic magma, the magma must be enriched to such an extent that the solubility products of the metal sulfides will be exceeded, causing metal sulfides to separate. Hence, an indication of possible

Table 7A-2. Average contents of sulfur and sulfide-held Cu, Ni and Co in mineralized and barren ultramafic intrusives of the Canadian Shield; based on data from 1079 samples. After Govett and Nichol (1979) but based on data from Cameron et al. (1971).

Group	No. of Intrusions	Cu (ppm)	Ni (ppm)	Co (ppm)	$S(^0/_0)$
"Ore Group" (deposits with more than 5000 tons Ni-Cu)	16	67.8	715	57.4	0.166
"Minore Group" (deposits with less than 5000 tons Ni-Cu)	5	6.8	560	25.2	0.036
"Barren Group"	40	6.9	354	31.3	0.031

mineralization may be obtained by determining the amount of sulfur and sulfide-containing metals in the lithogeochemical samples. The results obtained by Cameron et al. (1971) are presented in Table 7A-2. Samples from the ore-bearing localities ("Ore Group") are clearly high in the sulfide-held metals and total sulfur, relative to mineralized localities containing deposits with less than 5000 tons of Cu-Ni ("Minore Group"), and to barren localities ("Barren Group"). The last two groups can be distinguished from each other only on the basis of their Ni contents.

In a similar manner, Olade and Fletcher (1974) evaluated a potassium chlorate-hydrochloric acid leach on granodiorite samples containing copper sulfides for its possible use in bedrock geochemistry studies. For their particular samples, which were from the Highland Valley porphyry copper district in British Columbia, this partial extraction procedure has certain advantages over the ascorbic acid-hydrogen peroxide leach mentioned above. These include: (1) the method appears to be more sulfide selective for their particular samples; and (2) the procedure is extremely rapid and simple.

PRIMARY DISTRIBUTION ON A LOCAL SCALE: DETAILED PATTERNS   (pp. 325-332)

In the past few years there have been many studies concerned with primary geochemical distribution patterns which can be classified as local or detailed in scale. These include patterns associated with individual deposits or districts in which either total or partial extraction procedures have been employed. Accordingly, only a selection of six examples will be presented in the next section. The reader is referred to Beus and Grigorian (1977) for many more examples from the Soviet Union, and to Govett and Nichol (1979) for examples from other parts of the world.

Patterns Due To Chemical Zoning   (pp. 325-328)

1. Patterns Based On One Element; Total Analysis.

Zonal patterns which involve only one element are, obviously, the simplest example of primary distribution to be found in lithogeochemical samples. Tellurium is an example of one trace element which forms distinctive zones around certain ore deposits, characteristically as well-defined halos at considerable distances (up to several kilometers) from the center of mineralization. This is shown in Fig. 7-3 (p. 326) at one deposit in the Coeur d'Alene district, Idaho. For this reason, Te has had special appeal as a pathfinder in the primary environment.

By using a new analytical procedure which enabled them to analyze Te in rocks and, in some cases, soils to as low as 5 ppb (Watterson and Neuerburg, 1975), it was possible for Watterson et al. (1977) to show that tellurium is, indeed, a valuable guide to various types of mineral deposits. They showed that tellurium distribution patterns; (1) closely define Au-Te fissure veins in the Cripple Creek district, Colorado; (2) define a concentric zone around monzonite stocks in the Coeur d'Alene district (Fig. 7-3); (3) form a nearly continuous halo peripheral to porphyry copper ores near Ely, Nevada; and (4) reflect, and extend beyond, the alteration patterns associated with ores in the Montezuma and Crater Creek areas in Colorado. It does not appear likely, however, that Te will yield significant halos in deposits associated with mafic igneous rocks.

2. Patterns Based On One Element; Partial Analysis.

On a smaller scale, primary halos with a radius of 600-800 meters from mineralization have been shown by Goodfellow and Wahl (1976) to exist for single elements and combinations of elements in volcanic rocks of the Bathurst District, New Brunswick, containing two massive sulfide deposits (Brunswick No. 12 and Heath Steele). The interesting point here is that halos were delimited for Ca, Mg, Na, K, F and Cl by analyzing 1 gram samples of rock powder which were crushed to minus 200-mesh and merely placed in 10 ml of water and stirred for one minute. For example, the halo zone in the mafic volcanic rocks overlying the Brunswick No. 12 deposit is outlined by high Cl and low K. The intensity and size of the anomalous halos associated with these New Brunswick deposits, as well as the ease and speed of analysis of the water-extractable elements, led Goodfellow and Wahl (1976) to suggest that this procedure has potential in the search for blind mineralization in the Bathurst district, and probably elsewhere. In general, the concentrations of the water-soluble elements mentioned above are low, usually less than 150 ppm, but they are easily determined by modern analytical methods, specifically atomic absorption (Ca, Mg, Na and K) and specific ion electrodes (F and Cl).

3. Patterns In Carbonate Rocks; Unusual Method For Detection.

Primary halos in carbonate rocks are typically of limited width, for the sake of discussion generally less than 50 meters, as shown in Fig. 2-3 (p. 50). The small size is generally due to the limited mobility of metals in solution in the alkaline environment characteristic of carbonate rocks (see pH of hydrolysis; p. 128). The problem is especially difficult when veins as little as 1 meter wide are being sought. This was the problem which faced Barnes and Lavery (1977) when they attempted to explore for Mississippi Valley-type zinc mineralization in the carbonate rocks of Wisconsin. The problem was further complicated by the fact that the Zn content of the carbonates, which is less than 4.7 ppm, varies with the clay content.

In order to detect and measure the primary distribution patterns in the district it was necessary for them to be able to distinguish between concentrations of metals added by mineralization, and those initially present in the carbonate host rocks (i.e., background values). Barnes and Lavery (1977) accomplished this by subtracting an amount of Zn proportional to the clay content (effectively the insoluble residue), and the remainder was assumed to have been added hydrothermally. These adjusted Zn values permitted primary halos to be detected as much as 53 meters away from one major ore body. Of course, a close sampling pattern is essential to detect ore bodies with a minimum width of 1 meter; in this case the optimum sampling interval was found to be 9 meters with a threshold concentration, adjusted for the value of Zn in the clay minerals, of only 5 ppm. (Although apparently not attempted, it is possible that Barnes and Lavery (1977) might have obtained similar results by means of a partial extraction procedure which released the hydrothermal Zn characteristic of the primary distribution pattern.)

4.   Patterns Based On Metal Ratios To Outline Paths Of Solutions.

The use of certain metal ratios can result in a quantitative characterization of chemical zoning which, in turn, may be used to define precisely the paths followed by hydrothermal solution. This was convincingly demonstrated by Goodell and Petersen (1974) within the Julcani mining district, Peru, where polymetallic mineralization occurs as fracture fillings in Tertiary volcanics. Of particular value in this case are Pb/Cu ratios obtained from rocks (Fig. 7A-3); variations in the contents

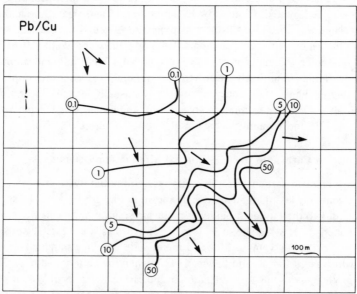

Fig. 7A-3. Contours of Pb/Cu at the 360 level, Herminia mine, Julcani district, Peru. Arrows indicate inferred direction of solution movement. After Goodell and Petersen (1974).

of the individual elements alone do not give a clear picture of the zoning. Metal ratio variations result from mineralogic changes, as well as from variations in the relative amounts of each phase present. These changes may be in response to local lithologic controls, or to the overall physicochemical variations experienced by a solution as it travels through the rock.

Although the Pb/Cu ratio is particularly effective in the Julcani district, Ag/Pb and Ag/Cu are also useful in providing insights into similarities and differences among mineralizing solutions and depositional environments in this area. The use of other metal ratios, such as Zn/Pb, Fe/Pb and Fe/Zn, had been proposed earlier (see Goodell and Petersen, 1974, for details).

In the Mimosa mine (Fig. 7A-3), one of several in the Julcani district, the solution movement was primarily lateral; however, in other mines it was primarily vertical. Clearly, knowing the direction of movement of mineralizing solutions in a mining district has great value in exploration. There may, however, be problems in interpreting changes in metal ratio if, for example, the deposit contains metals from different sources or depositional periods. In the case of the Pb/Cu ratio, it cannot be used in areas where either metal is impoverished.

5. Patterns Associated With Porphyry Copper Mineralization; An Example From British Columbia.

It is frequently necessary to distinguish zones of mineralization within porphyry-type copper deposits from barren zones, and this can be accomplished by means of the primary distribution patterns of certain elements. Govett and Nichol (1979) have reviewed and summarized available lithogeochemical data on the distribution of major and trace elements in the host rocks of this type of deposit. They note that, at the present time, the most comprehensive investigations have been carried out on the various deposits in the Highland Valley, British Columbia. Noteworthy among these studies are those of Olade and Fletcher (1976) and Olade (1977) in which it has been shown that the geochemical halos around the deposits are related to parent lithology, hydrothermal alteration, and mineralization.

The distributions of Mn, Zn, Sr, Ba, Rb, Mo, S and Cu at the 3600 level of the Valley Copper deposit are shown in Fig. 7A-4 in relation to the ore zone and the alteration zones (phyllic and argillic). Clearly, halos of Cu, S and Rb extend beyond the alteration zones and, therefore, constitute larger exploration targets. The depletion of some of the elements, specifically Ba, Sr and Mn in this case, is closely related to the phyllic alteration zone, and this can be useful as a guide to zoning in the deposit.

It must be stressed that although the concept of primary halos presented in Fig. 7A-4 is correct, the details of the element distributions vary from one deposit to another. Three other large deposits within 10 km of the Valley Copper deposit show significantly different distribution patterns for the same elements (see Olade and Fletcher, 1976, Figs. 8-11). Not only must geological factors (for example, differences in the composition of the mineralizing fluids) be taken into account in the interpretation of these primary halos, but so also must the erosion level (Fig. 7-6, p. 330). Different erosion levels result in different alteration zones being exposed, each of which has its own characteristic trace element associations. Elements other than those shown in Fig. 7A-4 which are sometimes determined in detailed studies of

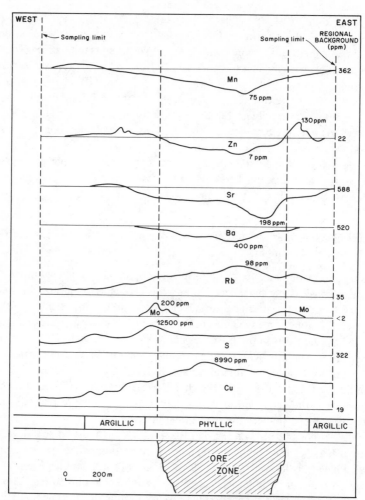

Fig. 7A-4. Distribution of selected elements in granitic rocks along a traverse at the 3600 level at the Valley Copper porphyry copper deposit, British Columbia. After Olade and Fletcher (1976).

halos in the vicinity of porphyry copper deposits include Ag, Au, B, Cl, Hg, Pb, Sn and W. Each of these elements, like others discussed previously, tends to be concentrated at specific positions lateral to the mineralization, and at different levels below and above the mineralization. For example, Pb and Zn are characteristic of the upper levels, whereas Sn and W are enriched at lower levels. These concepts are discussed in greater detail below in connection with Figs. 7A-6 to 7A-10.

6. **Patterns Based on Major Elements.**

To this point, only trace element primary distribution patterns have been stressed in this discussion of lithogeochemical techniques. However, Boyle (1974) has emphasized that the utilization of major element geochemistry in defining and quantifying wall rock anomalies may be easier, more rapid and less expensive. Boyle (1974) pointed out that the ratio K_2O/Na_2O is the most suitable for estimating proximity to ore under certain conditions. His suggestion is based on studies in at least a half-dozen different Canadian localities with different types of mineralization (gold-quartz, polymetallic, massive sulfide) which showed that potash metasomatism is especially common in the vicinity of many types of epigenetic deposits. This metasomatic process results in the formation of K-bearing minerals such as sericite, potash feldspar, and K-containing clays; sometimes these minerals can be detected in thin section, or by staining methods. During the process there is usually an exchange between the potassium and sodium so that the former increases uniformly toward the ore, and the latter shows a concomitant decrease. Therefore, the ratio K_2O/Na_2O usually shows an increase as the ore is approached. Although Boyle (1974) suggests that most (perhaps 95%) of the wall rock alteration sequences reported in the literature show an increase in the K_2O/Na_2O ratio toward the ore, exceptions do occur. These are usually related to sodium metasomatism, as was discussed earlier in this chapter with respect to the search for emerald mineralization in Colombia (Beus, 1979); however, sodium metasomatism can often be detected in thin section by the recognition of albitization. Tin mineralization also is often associated with an increase in Na_2O.

Other ratios involving major elements which Boyle (1974) considered to have potential under certain geological conditions include: (1) SiO_2/CO_2; this ratio shows a consistent decrease as ore is approached, particularly in ultrabasic to intermediate igneous rocks which have veins with a quartz gangue; and (2) $SiO_2/H_2O +$ $CO_2 + S$, particularly in granitic rocks and siliceous sediments; this ratio also exhibits a fairly consistent decrease in the direction of ore bodies (vein deposits).

Boyle (1974) pointed out that there are some difficulties in using the above-mentioned ratios. In the case of very narrow veins, for example, major element ratios have no advantages over trace element distribution patterns. The best results are to be expected where there are wide, well-developed alteration zones, as in the case of wide shear zones or in alteration zones associated with porphyry copper deposits (Fig. 7A-4). In all cases, however, it must be recognized that these major element ratios are only indicative of mineralization, and not necessarily of ore bodies.

That the K_2O/Na_2O ratio does, indeed, show a consistent increase from the outer toward the inner margins of two porphyry copper deposits emplaced in granitic rocks in the Highland Valley district, British Columbia, has been confirmed by Olade (1977). Furthermore, he concluded that the use of major element halos is more advantageous, and less time-consuming, than quantitative mineralogical techniques in delineating alteration zones and proximity to mineralization. Gunton and Nichol (1975), on the other hand, found the reverse relationship, that is, Na_2O enriched and K_2O depleted, as mineralization is approached from certain directions in the porphyry copper ore bodies they studied in another part of British Columbia. These deposits are localized in andesitic wall rocks where albitization is commonly a dominant process. Therefore, as with all exploration methods, the K_2O/Na_2O ratio must be used with geological acumen.

Wallrock Alteration (pp. 329-331)

Wallrock alteration has been primarily concerned with mineralogical changes which accompany the introduction of hydrothermal solutions. Chemical variations in altered wallrock (i.e., primary geochemical halos) are intimately related to the mineralogical changes and so there has been an increasing emphasis on wallrock studies among many of those concerned with exploration.

In Fig. 7-6 (p. 330) the various alteration zones (propylitic, phyllic, argillic, potassic) associated with porphyry copper mineralization are diagrammatically illustrated. Within each of these zones there are characteristic element associations, such as those shown in Fig. 7A-4. Thus, not only may trace and minor elements be used in determining the approximate distance from mineralization, but so also may be the mineralogical phases as well as the major elements (e.g., potassium, sodium) comprising these phases. Although, on occasion, visual (e.g., bleaching) or microscopic observations are sufficient to characterize the alteration zones at a particular deposit, usually semi-quantitative estimation of the clay minerals, secondary feldspar, as well as other secondary minerals, by means of X-ray diffraction techniques are also required.

Hausen (1979) has reviewed the use of quantitative measurement of wallrock alteration in exploration, particularly for blind mineral deposits. He observed that, in the past, most "monomineralic contouring" has been applied to porphyry copper-type mineralization in, for example, the Highland Valley, British Columbia (Jambor and Delabio, 1978), and Bougainville, Papua, New Guinea (Ford, 1978). In the latter case, it is interesting to note that Ford (1978) recognized three broad alteration events overprinted upon each other and, obviously, the resulting chemical-mineralogical relationships are very complex.

Recently, wallrock alteration studies have been applied to other types of deposits, such as the volcanogenic (exhalative) sulfide deposits in the Canadian Shield (Franklin et al., 1975; Hausen, 1979) and in Turkey (Cagatay and Boyle, 1977). In addition to monomineralic contouring of alteration zones, ratios of minerals (for example, clay mineral/feldspar ratios) have been used in: (1) determining the presence of alteration zones; (2) detecting the presence of hydrothermal activity related to sulfide mineralization; (3) estimating the proximity to ore; and (4) assisting in the direction of drilling and mining operations. A limiting factor in the use of mineralogical halos is that, generally, chemical halos are more extensive and, hence, detectable over greater distances.

Discussion (pp. 331-332).

No additional comments.

LITHOGEOCHEMISTRY IN THE SOVIET UNION
(A new topic)

Lithogeochemistry has received considerable attention in the Soviet Union and certain procedures used there are innovative and unique by comparison with those used in the Western World. Possibly of greatest significance is the fact that, on an almost routine basis, Soviet geochemists are able to detect certain types of deposits far below the surface (up to 850 meters is claimed) by patterns of different element halos.

In this section selected examples of some of the simpler Soviet concepts and methods which are not routinely used in the West are presented. All are taken from Beus and Grigorian (1977).

Sampling Procedures

It is essential that simple, yet effective, sampling procedures be used for lithogeochemical surveys, including those concerned with geochemical specialization (p. 638). Surface samples for the study of primary geochemical halos are collected along cross sections or profiles, usually oriented across the strike, and include material from both the country rock and the ore body. The sampling grid is selected in such a way that each potential geochemical anomaly will be crossed by at least two traverses (see Figs. 5A-3 to 5A-5). The profiles may be of different lengths, but they should extend beyond the areas of any altered rock. Geochemical samples may also be taken along the walls of a mine, or from cores obtained during exploratory drilling, at an interval of between 0.5 and 10 meters. Geochemical sampling of bedrock, whether it be at or below the surface, must be accompanied by a detailed geological description of all the intervals sampled. Later, detailed petrographic and mineralogical studies must be integrated with the geochemical results.

Exploration for ore deposits using primary geochemical halos or the concept of geochemical specialization requires the selection and analysis of a great number of samples. In order to reduce the expense of sampling and analysis, and yet obtain all necessary information, Soviet geochemists rely heavily on the *chip-channel* method of sampling (mentioned on p. 231). It consists of selecting, over a specific, interval, 5 or 6 small broken fragments which are combined into one sample for subsequent analysis. The chip-channel method is much faster than other methods, such as obtaining continuous channel samples.

Garrett (1979) discussed various aspects of sample design for regional lithogeochemical studies with particular attention to the optimal sample spacing and sample size. For his statistical procedures to be applicable, prior orientation surveys must have been carried out so that plutons in the particular area may be classified as "barren" or "productive".

Methods of Enhancing Primary Halos

Whereas the primary halos from some polymetallic vein deposits may be several hundred meters in vertical or horizontal extent, halos from other types of deposits, such as gold ores, are generally very small and weak with the result that it is difficult to detect them during geochemical exploration. In view of this problem, Soviet geochemists have developed special procedures for enhancing weak halos and also extending the size (area of recognition) of those which they traditionally classified as large. The four methods employed are discussed below. All, in effect, attempt to reduce the interfering effect of the background content, as this is what limits detection of a halo to its full extent.

1.   Method based on quantitative analysis.

Variations in the contents of the chemical elements in rocks, when a large number of samples are being compared, are due to: (1) the non-uniform distribution of the elements in the rocks being studied, and; (2) errors introduced by sampling, processing and analysis of the geochemical samples. Thus:

$$S^2 = S^2_{ntr} + S^2_{smpl} + S^2_{prc} + S^2_{an}$$

Where S^2 is the total variance of the contents:

S^2_{ntr} is the natural variance caused by the non-uniform distribution of the element content (i.e., lithological and weathering variance);

S^2_{smpl} is the variance due to sampling;

S^2_{prc} is the variance due to the processing of the samples;

S^2_{an} is the analytical variance.

Because much of analytical data in the Soviet Union is based on semi-quantitative emission spectrographic methods, analytical errors can sometimes reach 60%. Therefore, the use of quantitative analytical methods can result in lower analytical variance and, consequently, larger primary halos of the elements. In the Western World, where most analytical results are considered quantitative, this may have no practical significance (but see Fig. 6A-7).

2.   Techniques using partial chemical analysis.

The value of partial analyses in exploration geochemistry has been discussed at length in various places in this book (e.g., Chapters 6 and 6A) in connection with the analysis of stream sediments and other sampling media. Partial extractions can also be used in lithogeochemical surveys to enhance the size of primary halos as the example below will illustrate.

Fig. 7A-5 shows primary uranium halos around a blind uranium ore body approximately 200 meters below the surface. The halos around the ore body are outlined in two ways: (A) by determining the total uranium in the samples; and (B) by determining the "mobile" uranium by partial extraction techniques, specifically the Na_2CO_3-H_2O_2 method described on p. 724. Comparison of the results from the two types of uranium determinations show that the "mobile" uranium halo is wider and longer than that which would be detected from the total uranium content. It is especially important to note that in all the samples collected at the surface, the contents of the "mobile" uranium are anomalous. Such is not the case with the total uranium results and the deposit would have gone undetected if only total uranium analyses were employed in the exploration sequence. To be sure, the partial extraction procedure yielded threshold uranium values in the rocks on the surface above

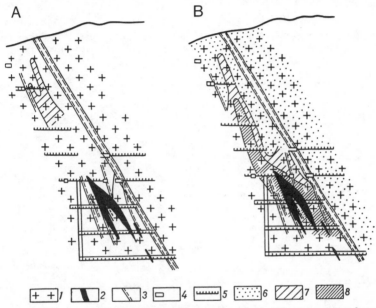

Fig. 7A-5. Primary uranium halos around a blind ore body approximately 200 m below the surface which is revealed by partial analysis (but not total analysis) methods on surface samples. *A:* halo from total analysis. *B:* halo from partial analysis. From Beus and Grigorian (1977).
1. granite; *2.* ore body; *3.* fracture; *4.* underground workings and boreholes; *5.* sampling interval; *6.* U contents of 2-5 ppb; *7.* U contents of 5-50 ppm for A, and 5-50 ppb for B; *8.* U contents of more than 50 ppb.

the deposit of only 2-5 ppb but, nevertheless, these values show significant contrast over background. Threshold for the total uranium content in rocks in this area is 16 ppm. (Note: In Fig. 7A-5 (A) the total uranium content was determined only on samples from the surface and the uppermost sampling horizon. Even though total uranium data are missing from the lower horizons, this illustration was selected for inclusion here because it clearly shows how partial extraction of surface samples can be used to detect blind mineralization).

3.   Analysis of the heavy fractions from rocks.

This method of enhancement is based on the selective concentration and analysis of the heavy fractions from crushed rock samples which, in primary halos, consist of sulfides, oxides (such as, cassiterite or uraninite), etc., depending on the type of deposit. Usually about 2 kg (or a representative split) of each sample is used and this is divided into heavy and light fractions by means of bromoform(S.G. = 2.8). Then each fraction is sieved into concentrates of various particle sizes. A selective enrichment of the indicator or pathfinder elements (e.g., Cu, Pb, Zn, U) will usually be found in the heavy fractions, particularly in those samples collected from the geochemical halos, and generally (but not always) in the sample class with the smallest sized particles (Table 7A-3).

Table 7A-3. Distribution of elements and enrichment indexes in different size fractions of geochemical samples obtained from within and outside the halo of a uranium deposit in granite. After Beus and Grigorian (1977).

Frac-tion	Size group mm	Weight g	Content, %				Enrichment index, $n \cdot 10^3$			
			Pb	Mo	Cu	Zn	Pb	Mo	Cu	Zn
			General sample from within the halo							
Light	+0.6	98	0.02	0.001	0.003	0.01	–	–	–	–
Light	−0.6+0.4	408	0.02	0.0006	0.001	0.01	–	–	–	–
Light	−0.4+0.25	320	0.01	Traces	0.0006	0.01	–	–	–	–
Light	−0.25+0.1	382	0.03	0.0006	0.002	0.01	–	–	–	–
Light	−0.1	125	0.03	0.002	0.003	0.02	–	–	–	–
Heavy	+0.6	19.1	0.5	0.005	0.1	0.15	696	7	139	209
Heavy	−0.6+0.4	7.5	0.3	0.003	0.03	0.1	164	1.64	16	55
Heavy	−0.4+0.25	7.45	0.3	0.008	0.1	0.1	163	4.34	54	54
Heavy	−0.25+0.1	3.67	1.0	0.008	0.2	0.3	268	2.14	54	80
Heavy	−0.1	0.45	1.0	0.03	0.2	0.3	33	0.98	7	10
			General sample from outside the halo							
Light	+0.6	187	0.006	0.0002	0.0006	0.007	–	–	–	–
Light	−0.6+0.4	349	0.002	0.0002	0.0002	0.007	–	–	–	–
Light	−0.4+0.25	250	0.006	Traces	0.0003	0.007	–	–	–	–
Light	−0.25+0.1	325	0.006	Traces	0.0008	0.007	–	–	–	–
Light	−0.1	130	0.01	Traces	0.003	Traces	–	–	–	–
Heavy	+0.6	3.88	0.09	0.0008	0.0115	0.028	28	0.24	3.6	8.7
Heavy	−0.6+0.4	1.0	0.1	0.001	0.003	0.03	8	0.08	0.24	2.4
Heavy	−0.4+0.25	0.98	0.06	0.002	0.003	0.02	4.7	0.15	0.23	1.5
Heavy	−0.25+0.1	1.85	0.1	0.0003	0.02	0.03	14.8	0.04	2.98	4.5
Heavy	−0.1	0.05	0.1	0.0003	0.03	0.03	0.4	0.001	0.12	0.1

In addition to the content of the individual elements, the *enrichment index (Q)* is another quantitative index of the accumulation of an element in each specific size fraction. It is expressed by the equation:

$$Q = C \cdot b$$

where: C is the content of the element in the fraction (in %);

            b is the quantity of this fraction in the whole sample (in %).

Calculations show that the parameter Q is much larger for the elements from anomalous samples than it is for samples collected outside an anomaly. This is illustrated in Table 7A-3 which is based on two samples obtained from a granite, one of which was collected within the halo of a uranium deposit, and the other outside. The first differs from the second in having: (1) higher contents of the indicator elements; (2) a considerably larger percentage of the heavy mineral fractions. As expected, the maximum content of the indicator elements from the sample from within the mineralized halo are found in the smallest size fraction (-0.1 mm).

Table 7A-3 also illustrates the point that the highest enrichment indexes are found in the heavy fractions within the halo. Further, it is possible to recognize anomalies with greater ease by means of the enrichment index. For example, the total lead content of the anomalous sample is only 0.02%, and with a background value of 0.006%, the contrast is 3.3. However, the contrast for lead from the +0.6 mm fraction is 25 (696 divided by 28).

Once orientation studies have shown which fraction is best for specific deposits in an area, it may be possible to use only this fraction on a routine basis.

4. Methods based on composite halos.

In many cases, the primary halos around ore deposits can be more easily detected, that is, enhanced, if a group of indicator elements are combined. The reason for this is that, by doing so, there is a reduction both in the effects of analytical errors and of local reversals in any zoning sequence.

There are two types of these multi-element (composite) halos, *additive* and *multiplicative* (Fig. 7A-6). Additive halos are constructed by the simple addition of the elemental contents, whereas multiplicative halos are constructed by multiplication of the element values; values in background areas are similarly determined. From Fig. 7A-6 it can be seen that both additive and multiplicative halos are larger and more distinct compared to the monoelement halos, adding to the reliability of any interpretation. (For additive halos, it is necessary that all the elements being added have the same concentration range).

In selecting the elements to be used in composing the additive or multiplicative indexes, one must consider elements which have definite geochemical significance for the type of mineralization sought, for their mobility characteristics in hydrothermal or other types of solutions, and so forth. In particular, it is essential that the elements being combined have "unidirectional correlations" (Beus and Grigorian, 1977, p. 63). Thus, Mg and Li for example, which have opposing concentration tendencies in magmas (see Table 7-2, p. 320) would not be combined. It would appear that elements known to be generally in close association, i.e., in the normal zoning patterns for steeply dipping or horizontal bodies, should be considered for combination. These general zoning patterns are discussed below under "standard zonation sequence."

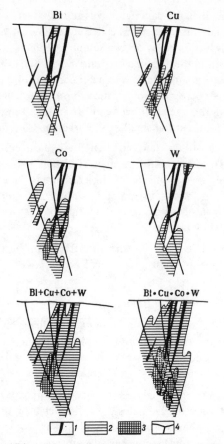

Fig. 7A-6. Monoelement, additive, and multiplicative halos around gold ore deposits, U.S.S.R. From Beus and Grigorian (1977).
1. gold-quartz veins; *2.* Contents of the individual elements (in ppm): Bi, 1-10; Cu, 100-1000; Co, 5-10; W, 3-10; *3.* Contents of the individual elements (in ppm); Cu, 1000-5000; Co, 10-80; *4.* cross section sampled; curved vertical lines represent drill holes. *(Note:* Contents are *not* given for the additive and multiplicative halos as implied by patterns *2* and *3*).

Methods of Interpreting Primary Zoning Patterns

Soviet geochemists have devised numerous ways of studying or interpreting primary halos, most of which are based on the concept of *zonation.* Zonation, as used in exploration geochemistry, is the term given to natural variations in elemental contents which are broken down into divisions (zones) of any shape, usually distributed vertically or horizontally. Zonation is the result of a regular alteration of the geological environment and of the content of different elements due to the processes involved in the formation of ore deposits. Particular emphasis is placed on vertical zonation because this can be used to assess the exposed level of erosion of geochemical anomalies relative to an ore body or, stated in another way, where the ore body lies with respect to a primary halo.

Fig. 7-5 (p. 328), Fig. 7A-5 and Fig. 7A-6 illustrate vertical zonation based on the halos of individual elements, as well as composite halos in Fig. 7A-6. Thus, if one were to have available the As or Ba vertical zonations (primary distribution patterns) in an unknown area represented by Fig. 7-5 (p. 328), and also evidence (from orientation studies or the literature) of the characteristic size, shape and relative position of As and Ba halos associated with skarn-type ore bodies, one would infer that there was mineralization at depth, as these elements characterize the upper horizons of such deposits. Conversely, if one obtained a prominent W anomaly in an area of potential gold-quartz veins (Fig. 7A-6), one would likely be dealing with a case in which the ore deposit was eroded away as tungsten is characteristic of the roots of such deposits. The former example involving As and Ba would be called a *supra (or above) - ore halo,* whereas the latter, involving W, would be called a *sub (or below) - ore halo,* the distinction resting on whether the primary halo produced by an element is characteristically above or below the main ore zones.

The concept of supra-ore and sub-ore halos is further illustrated in Fig. 7A-7 where Ba and W are used as hypothetical examples. The bottom portion of the figure shows that the relative contents of these elements will vary at the two different chosen erosion levels. Knowledge of the primary halo characteristics of these and

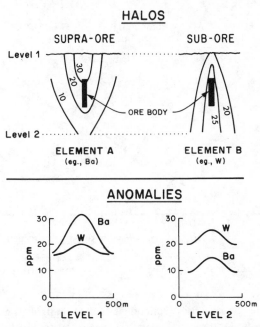

Fig. 7A-7. *Top.* Diagrammatic representation of supra-ore and sub-ore halos. *Bottom.* Relative intensities of anomalies obtained from rock samples containing Ba and W at erosion levels above and below the ore body.

other elements will greatly assist the geochemist in determining the position of the halos (anomalous values) relative to mineralization.

Soviet geochemists (e.g. Ovchinnikov and Grigoryan, 1971) have added another parameter to the concept of supra-ore and sub-ore halos, i.e., the *dimension* of the halo. This is used to calculate the *linear productivity* of an anomaly by multiplying the average element content (in percent) by the width of the anomaly (in metres). An example of the use of linear productivities is given in Fig. 7A-8. In this illustration the ratios of linear productivities of the elements which form supra-ore halos (Pb and Zn) and the elements which form sub-ore halos (Cu and Bi) show that there are more than four orders of magnitude in the values of these ratios between erosion levels 1 and 4; this should be compared with a maximum difference in simple ratios of about 2. Clearly, this approach can yield very valuable information in attempting to determine whether a halo reflects a significant ore deposit or only the root zone of a deposit that has been eroded.

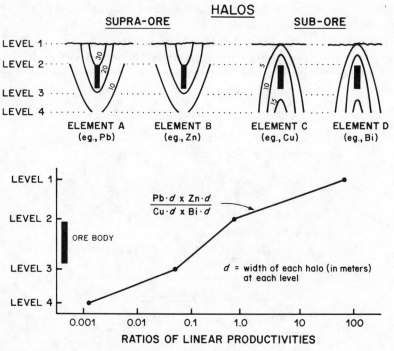

Fig. 7A-8. *Top.* Profile of elements which form supra-ore and sub-ore halos along levels 1 to 4. *Bottom.* Ratios of the linear productivities (see text for definition) at the various levels. After Ovchinnikov and Grigoryan (1971).

Fig. 7A-9 illustrates vertical zonation based on individual elements around a uranium deposit about 80 meters below the surface in the U.S.S.R. Pb, Zn and Ag halos are characteristically associated with, or are above, the ore deposit; they produce supra-ore halos in this type of mineralization. From the surface it is possible to speculate on the likely position of the ore body by means of certain element *ratios*, specifically Pb/U and Mo/U, as shown in Fig. 7A-10 (based on data from Fig. 7A-9). The Pb/U ratio is particularly useful in this case because at the surface this ratio is high and it decreases regularly with depth, reflecting the zonal distribution, and also indicating the approach to ore.

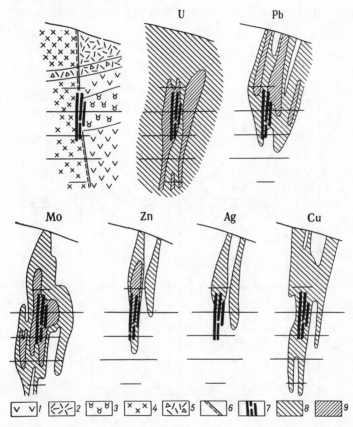

Fig. 7A-9. Primary halos around uranium ore bodies in granitic rock about 80 m below the surface. From Beus and Grigorian (1977).
1. quartz porphyry; *2.* tuffs of quartz porphyry composition; *3.* spherulite-porphyry; *4.* felsite-porphyry; *5.* tuffaceous breccia of quartz porphyry composition; *6.* fractures; *7.* ore bodies; *8* and *9:* element contents in primary halos (compared to geochemical background); (*8.* 2 to 10 times background; *9.* more than 10 times background).

Whereas the Pb/U ratio in Fig. 7A-10 reaches a maximum value of 4.25 above the uranium deposit, Shipulin et al. (1973) report, based on experiences in the Soviet Union, that the ratio of other elements in other types of uranium, base metal, and rare element deposits can reach much higher values. For example, the Mo/Cu ratio increases 5-7 times, and the Mo/W ratio increases 8-10 times, with depth in the vicinity of metasomatic molybdenum mineralization. The Mo/U ratio sometimes increases 10-15 times with depth in certain types of uranium deposits. The Mo/Be ratio increases 2-4 times with depth in wolframite-beryl-molybdenite deposits associated with greisen formation. Even though Shipulin et al. (1973) did not indicate the distance such ratios could be detected above or below the respective ore deposits, it is clear that the use of such selected element ratios associated with supra-ore and sub-ore halos are likely to be of great value in locating blind mineralization.

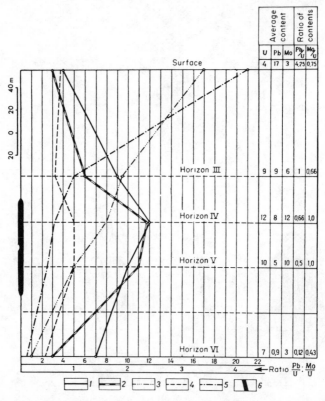

Fig. 7A-10. Graphs showing variations in the average contents of indicator elements, and ratios of these elements, with depth. These graphs are based upon the data from Fig. 7A-9.
1. uranium; *2.* molybdenum; *3.* lead; *4.* Mo/U; *5.* Pb/U; *6.* ore body. (*Note:* Units 2, 4 . . . 20, 22 along abscissa represent contrast over background for the individual elements U, Pb and Mo. Units 1, 2, 3 and 4 represent values for the ratios Pb/U and Mo/U). From Beus and Grigorian (1977).

Zonation in Primary Halos

In the preceding section some examples of vertical (sometimes called axial) zonation, which are characteristic of steeply dipping ore bodies, were given. Other types of zonations (called transverse and longitudinal) yield characteristic horizontal halos, the result of ore-bearing fluids travelling in this plane. From a practical point of view, vertical zonality is the most important type because by understanding and using the concept it is possible to suspect blind mineralization, and to distinguish between supra-ore and sub-ore halos.

The technique of detecting blind mineralization, and deducing the relationship of a halo to mineralization, are based on the interpretation of the zoning patterns of the elements, i.e., zonation. Beus and Grigorian (1977, Table 30) have tabulated Soviet experiences with the sequences of elements which characterize many different types of endogenic deposits (e.g., gold-ore, porphyry copper, uranium) formed by hydrothermal processes. Despite variations in the type of ore, local geological conditions, and other factors, the halos generally conform to a *standard zonation sequence* for steeply dipping bodies which is essentially independent of geological environment and ore type. From the top (supra-ore halos) to the bottom (sub-ore halos) the sequence is:

Ba-(Sb, As, Hg)-Cd-Ag-Pb-Zn-Au-Cu-Bi-Ni-Co-Mo-U-Sn-Be-W.

This sequence is only generalized and variations will, and do, occur but the validity of the generalization can be seen by perusal of the halos in Fig. 7-5 (p. 328), Fig. 7A-6 and Fig. 7A-9. According to Beus and Grigorian (1977), the sequence can be explained by the stability of compounds (e.g. chloride complexes) of the elements in the ore-bearing solutions. When the zonations are out of sequence at a particular deposit, the cause may be found in variations in the mineralogical form in which an element occurs (e.g. As in arsenopyrite and tennantite might be found at different depths), or in the fact that the deposit formed during several stages of mineralization. In general, the above standard zonation sequence does not apply to deposits exhibiting horizontal zonations; deposits in this category seem to vary so much that a general sequence cannot be proposed.

Discussion

The preceding discussions on lithogeochemistry in the Soviet Union are only examples of how Soviet geochemists use data obtained from rock surveys. Other uses of primary halos also described by Beus and Grigorian (1977): (1) show how it is possible to differentiate anomalies representative of ore from those which are simply an expression of disseminated, non-economic mineralization; and (2) illustrate how chemical studies of ores and their wall rocks may indicate whether more than one type of ore is present in a given mine. However, these topics are beyond the scope of this book.

There can be no doubt that Soviet geochemists have an impressive understanding of primary halos and their uses, and these provide a new, interesting and stimulating viewpoint of mineral exploration for the Western geochemist. At this particular time, however, Western geochemists lack the experience in interpreting the type of lithogeochemical data routinely used in the Soviet Union. For example, it is not immediately clear as to which halos are most important, which additive or multiplicative halos should be constructed, or when and where partial extractions should be used. Further, some Western geochemists have doubts about the practicality of using lithogeochemistry for general mineral exploration and, in particular, for regional surveys. They would suggest that there would be great difficulty in obtaining representative and fresh rock samples, especially: (1) in deeply weathered tropical environments; and (2) of coarse pegmatitic material (probably requiring a 5 kg sample). Sample preparation (crushing, grinding, sieving, etc.) would be time-consuming and expensive. In short, as far as regional exploration is concerned, they would prefer stream sediment or soil samples, although they would certainly agree that lithogeochemical methods are very desirable for more detailed follow-up work. However, sooner or later, Soviet exploration methods based on the use of lithogeochemistry will make an impact on the future of mineral exploration in the Western World.

PRIMARY DISTRIBUTION OF MERCURY, HELIUM AND RELATED ELEMENTS   (pp. 332-336)

Mercury. The basic geochemical facts necessary to conduct and interpret mercury surveys are reasonably well known, and a portable highly sensitive gold-film mercury detector is commercially available (p. 740). Yet, the use of the mercury content of rocks and soils for exploration purposes appears to have fallen into disfavor in the past few years, except possibly in the search for certain types of gold ores. The reason is probably because the results obtained by some investigators have been inconsistent, equivocal or negative. For example, Cagatay and Boyle (1977, p. 49) state, "the dispersion of Hg is restricted, and the element is not a good indicator", based on their study of volcanogenic sulfide deposits in Turkey. Similarly, less-than-satisfactory results with mercury surveys have been reported by Olade and Fletcher (1976) from porpyry copper deposits in British Columbia. Even for mercury deposits themselves, Beus and Grigorian (1977, p. 115) note that there are difficulties involved in the study of the distribution patterns because the great mobility of mercury results in broad dispersion patterns and detached halos, and, in their case, there were also analytical problems. Others have suggested that "old" deposits may have lost the bulk of any Hg initially present.

On the other hand, satisfactory or excellent results have been reported by McNerney and Buseck (1973) from soils over a variety of base and precious metal deposits. Bignell et al. (1976) found that Hg anomalies in the metalliferous sediments being deposited in the Red Sea may be detected up to 10 km from the deposits. Brooks and Berger (1978) were able to detect fault zones in Nevada, some of which may contain disseminated gold mineralization, by means of mercury surveys in soils (analogous to the helium surveys illustrated in Fig. 7-10, p. 335). Wu and Mahaffey

(1979) found anomalous mercury areas up to about 600 meters wide in soils over one massive sulfide deposit in Arizona, but another deposit 90-150 meters below the surface gave no measurable soil mercury response.

From the few examples given above, one can only conclude that the value of mercury surveys in bedrock, or soils above ore deposits, is not predictable, and each survey must be considered as a separate case. Because most Hg anomalies are in the ppb range, the sensitivity, accuracy, precision, and reliability of the instrumentation is extremely important, as is the experience of the operator or analyst.

Helium. Helium is a by-product of radioactive decay (the alpha particle; see p. 300) and may eventually prove to be a valuable indicator of the presence of uranium and thorium deposits, in addition to its use in locating faults and porous zones (p. 335), geothermal zones, and hydrocarbons. The reports by Clarke and Kugler (1973), Clarke et al. (1977), and Torgersen and Clarke (1978) are examples of some of the Canadian studies in which high helium concentrations were found in ground and lake waters in uraniferous areas. They indicate the potential of helium detection as an exploration tool for uranium. The entire subject of the use of helium in mineral exploration has been reviewed by Dyck (1976).

One major advantage of the use of helium in exploration is that it is a stable isotope and, therefore, there is no time restriction on its migration (^{222}Rn, for example, has a half-life of 3.8 days). Another is that it is a very light and small atom and, as a result, it has a much greater ability to diffuse through rocks than does radon and other gases. Radon, however, is used much more extensively than helium because it is much easier to sample and measure (see p. 300 and p. 734 for a discussion of methods of radon detection). Measurement of helium abundances requires the use of a mass spectrometer. Semi-portable (for example, truck mounted) instruments designed or modified specifically for uranium exploration have recently been described by members of the U.S. Geological Survey (Reimer, 1976; Reimer et al., 1979) and the Geological Survey of Canada (Dyck and Pelchat, 1977). A few commercial laboratories in North America offer helium surveys and determinations but no private exploration company is known to have the instrumentation and facilities.

In some respects, there are similarities between helium and radon surveys: both are conducted primarily in soils and waters, and both are affected by similar environmental problems (such as, their concentrations being influenced by soil moisture and diurnal variations in temperature and pressure). Dyck and Jonasson (1977) have discussed some aspects of the nature and behavior of helium and other gases in natural waters with particular emphasis on exploration. They concluded (p. 711), "Rn and He are clearly related to the presence of U-enriched rocks. Rn can be used as a tracer to such mineralization but the extreme mobility manifest in He limits its usefulness in this regard."

Notwithstanding some of the apparent difficulties involved with helium surveys, results to date with this relatively new technique have been encouraging. It has all the potential for being a very valuable adjunct to present methods used in uranium (and thorium) exploration, and especially in locating deposits which are deep, or associated with no detectable radon.

Iodine. Very small amounts of iodine are associated with some mineral deposits. Because it migrates upward as a vapor, it forms primary halos analogous to those of Hg and He. Iodine has significant potential for the detection of blind mineralization based on numerous studies in the Soviet Union. By way of example, Krylova (1978) has shown that iodine halos can be used to detect ore bodies 100 metres or more below the surface and, in the particular case of the Pb-Zn-Cu hydrothermal ore body she described, the maximum vertical extent of associated Pb-Zn-Cu halos was 40 meters.

Relatively little has been written in English on the use of iodine in mineral exploration. At this time, the difficulty in detecting iodine at the low levels required (in the range 0.1 - 1.0 ppm in rocks and soils) discourages its use on a routine basis.

TRACE ELEMENT CONTENT OF MINERALS   (pp. 336-341)

Mica   (pp. 336-338)

On p. 337 of the first edition the statement was made ". . . the evidence seems to indicate that the chlorine and fluorine content of micas will become useful in exploration". This was based on studies available at that time on the differing Cl and F contents of biotite from porphyry copper and barren plutons in the Basin and Range Province of the western United States (Fig. 7-11, p. 338). However, subsequent investigations of this province by Parry and Jacobs (1975) have shown that, although the Cl and F contents of biotites differ significantly between individual plutons, no differences which might prove useful for exploration purposes could be recognized. In a subsequent study, extended to include Fe, Mg, Ti and Ba in biotites from this same area, Jacobs and Parry (1976) reached the same conclusion.

The conclusions with respect to the halogens in biotite seem to be confirmed by the work of Kesler et al. (1975a) on porphyry copper and barren plutons in the Caribbean and Central America. Although they performed whole rock analyses only, their general conclusion is that it seems unlikely that halogen contents can be used to distinguish between mineralized and barren intrusive rocks. Nevertheless, the trace elements in biotite, such as tin (p. 320), still have significant exploration potential. From Table 2A-6, which lists mineral concentrators and accumulators of the trace elements, it can be seen that the potential importance of biotite and other micas in exploration for numerous types of mineralization remains significant.

All porphyry copper deposits contain hydrothermal (as distinct from magmatic) biotite which is characteristic of the potassic zone (Fig. 7-6, p. 330). This type of biotite has long been recognized as an important indicator of the lateral and vertical extent of better grade mineralization and, as such, was noted in mapping and core logging. Recent studies, such as those by Ford (1978) at the Panguna deposit on Bougainville, and by Parry et al. (1978) and Moore (1978) at Bingham, Utah, show that there are definite interesting and potentially useful variations in some of the major elements, particularly Fe and Mg, in the hydrothermal biotites. Ford (1978) noted that the ratio Mg/Mg+Fe in this mineral increases toward the ore zone, the same trend being found in hydrothermal amphibole. Mason (1978, 1979) also observed that the Mg/Mg+Fe ratio in biotites and amphiboles may be useful for exploration in porphyry copper areas; however, he suggests that variations in these elements within individual crystals be used. Specifically, by means of electron probe studies he found that there is Mg enrichment toward the rims of biotite and amphibole crystals obtained from mineralized plutons.

At porphyry copper deposits in the Highland Valley, British Columbia, Jambor and Beauline (1978) observed that the distribution of hydrothermal biotite has a well-defined correlation with copper zones, and that the intensity of biotite alteration (both chemical and physical) is correlative with copper grades. Moreover, the character of the biotite changes systematically from the ore zone outward, and reflects progressive decline in intensity of potassic alteration and dilution of copper grades. They consider that these characteristics constitute valid criteria for assessing exploration possibilities in the Highland Valley.

Finally, Mohsen and Brownlow (1971) observed that biotites from the Philipsburg batholith show a two-fold increase in Mn content toward manganese replacement deposits in adjacent country rock over a distance of 1,500 meters.

Feldspar   (p. 338)

Studies by Lanier et al. (1978) at Bingham, Utah, have shown that in the mineralized (porphyry copper), monzonite pluton the plagioclase becomes more sodic, and the orthoclase more potassic, toward the center of the deposit.

Magnetite   (p. 339)

Magnetite at the stratiform Gamsberg zinc ore body in the arid region of South Africa contains 2400-4400 ppm Zn. By sampling the magnetite fraction from stream sediments, McLaurin (1978) was able to detect an anomalous Zn dispersion train 14 km from the deposit.

Tourmaline   (p. 339)

No additional comments.

Apatite   (pp. 339-340)

Williams and Cesbron (1977) have shown that accessory apatite and rutile are useful prospecting guides for porphyry copper deposits. Apatites from such deposits are enriched in chlorine, whereas the rutile has an unusually high ratio of Cr + V/Nb + Ta. Apatites from porphyry copper deposits also exhibit characteristic crystallographic properties, specifically the relationship between their *a* and *c* axes.

Chalcopyrite, Sphalerite and Pyrite   (p. 340)

Ryall (1977) studied 76 pyrite samples from the vicinity of a volcanogenic massive sulfide deposit at Woodlawn, N.S.W., Australia, and was able to divide them into four groups based on the distribution of 12 analyzed metals. The group characterized by enrichment in Zn, Pb, Cu and Ag, and a low (0.8) Co/Ni ratio, reflected the base metal mineralization.

Tan and Yu (1970) reported on the separation of pyrite from stream sediments in an area of Taiwan and their analysis for Cu. Orientation studies showed that pyrite from copper mineralization in the drainage area contained 1100-1700 ppm Cu, pyrite from gold deposits contained 40-480 ppm Cu, and pyrite from coal seams contained 100-120 ppm Cu. This information was used successfully in detecting new anomalies and in delineating known deposits.

Discussion   (pp. 340-341)

As can be gleaned from the topics discussed in this section, the trace element content of individual mineral phases continues to be used, with varying degrees of success, to discriminate between barren and productive plutons, to define regional geochemical provinces and, to a lesser extent, to identify specific exploration targets. Most of the work has been done on biotite but additional minerals continue to be proposed for this purpose. A case in point is the recent interest in the uranium content of accessory minerals, such as zircon and sphene (Adler, 1977), as indicators of uraniferous plutons or provinces.

FLUID INCLUSIONS   (pp. 341-343)

Roedder (1977) published what time will probably prove to be a classic paper on the *potential* use of fluid inclusions in exploration for many types of ore deposits. The paper considers seven major topics:

1.   the general nature of inclusions;
2.   types of data available from inclusions;

3. applications of inclusion data to clarify regional or local geology;
4. applications of inclusion data in the search for blind ore bodies;
5. applications of inclusion data to the evaluation of altered or weathered material in outcrop or sediments;
6. applications of inclusion data toward an understanding of the environment of ore deposition; and
7. practical problems.

Many of these topics are explained in purely geochemical terms, such as the K/Na ratios of the inclusion fluids, and geochemical-type maps of the distribution of high-salinity fluid inclusions. On the other hand, physical studies of the fluids, such as decrepitation temperatures, geothermometry, and geobarometry, are also discussed. This paper by Roedder (1977) is required reading for all those interested in a detailed explanation, or review, of fluid inclusions and their possible use in mineral exploration.

ISOTOPE METHODS (pp. 343-346)

The use of isotopes in mineral exploration is still limited by the expense involved in obtaining the necessary data. Accordingly, only brief comments will be given on some of the topics covered previously. The uses of isotopic abundances and variations in the uranium decay series are discussed in Chapter 3A.

Lead Isotopes (p. 344-346)

Doe and Stacey (1974) and Loveless (1975) have reviewed the use of lead isotopes in mineral exploration. Both discuss how lead isotope data on ores may indicate the potential for a major deposit, as opposed to minor mineral occurrences. The $^{206}Pb/^{204}Pb$ ratio is particularly valuable (although $^{206}Pb/^{207}Pb$ had previously been proposed for this purpose; see p. 345). For example, based on the evaluation of certain Cretaceous - Tertiary "magmatothermal" deposits in the western United States, Doe (1979) has recently shown that the lead isotopic composition of the ores in all the large mining districts is characterized by $^{206}Pb/^{204}Pb$ ratios of less than 18; where this ratio exceeds 20, the potential is minor. One advantage of these types of studies is that only one or a very few samples need be analyzed because there is very little variation in the lead isotope ratios of specimens from the same district.

Sulfur Isotopes (p. 346)

Seccombe (1977) showed how sulfur isotope data, in combination with the trace element content of stratiform (volcanogenic) sulfide deposits, may be used to discriminate between economic and non-economic mineralization in one part of the Canadian Shield. Specifically, economic ores exhibit a narrow range (2‰) with the $^{34}S/^{32}S$ ratio near zero per mil, whereas non-economic mineralization has a broad range (more than 6‰) in the isotopic ratio which is generally lighter than zero per mil.

CONCLUDING REMARKS
(A new topic)

From the discussions and examples presented in this chapter it is clear that, under favorable circumstances, the various types of lithogeochemical surveys have the potential to assist in mineral exploration in the following ways:

1. Regional scale exploration to identify potentially productive plutons, such as large low-grade deposits of Cu or Sn, the recognition of which may not be otherwise possible because of the low grades involved or the inconspicuous nature of the mineral sought;
2. Regional scale exploration to identify sedimentary horizons containing exhalative (volcanogenic) deposits which produce extensive bedrock dispersion patterns outside the limits of recognizable mineralization;
3. Local scale exploration to locate blind ore deposits;
4. Mine-scale exploration.

It would appear that in the Western countries it is imperative that immediate emphasis be placed on the third point, that is, locating blind ore deposits by lithogeochemical methods using the Soviet experience as a guide, and their success as encouragement.

In this connection, it is appropriate to quote the admonition of Boyle (1974, p. 346):

"It is obvious that any further progress in locating deeply buried mineral deposits will require extensive drilling of favourable structures and beds outlined by good geological field work and intuition, surface geochemical surveys, and geophysical surveys. In drilling programs geochemistry has largely been neglected, although it has received attention by some companies in a more or less desultory fashion. Today, considering the rapidity and low cost of analyses, it is ludicrous not to have the drill core from exploratory or development programs analyzed for suites of both major and trace elements that may indicate the location of orebodies. How many orebodies have been missed by this neglect, and how many times have hopeful ore hunters failed to snatch fortunes from adversity by failing to have their drill core analyzed for the elements which they sought in economic bodies? I suspect quite a few. One large orebody found by such analytical methods would certainly pay for all of the chemical analyses ever done throughout the world on geochemical prospecting work."

8A / Secondary Dispersion

(Selected Topics)

INTRODUCTION (pp. 347-348)

One fundamental difference between primary distribution patterns (discussed in the previous chapter) and secondary dispersion patterns is that the former are developed at the same time as the ore accumulation whereas the latter form later; exceptions include: (a) some leakage halos; and (b) where the ore itself is secondary, e.g., eluvial monazite or pyrochlore deposits. From a practical point of view, this means that primary distribution patterns occur only in one sampling medium — rocks. On the other hand, secondary dispersion patterns can be found in a multitude of unconsolidated materials (e.g., soils, stream sediments, glacial till), waters, vegetation and air, just to mention the major types. Stated another way, in general, primary patterns are the result of processes operative during ore formation and before weathering whereas secondary dispersion halos form during, or as a result of, weathering. Thus, secondary dispersion halos, which result from the erosion of ore deposits, form within a broad range of natural environments. They are affected by chemical, biological, mechanical and environmental factors operative in the secondary environment and, as a result, there is a plethora of variables (a selection of 18 are listed on p. 348) affecting the size, shape, extent and contrast of secondary dispersion halos. For example, factors affecting dispersion patterns in stream sediments include: variability in the composition, grain size and sorting of the sediments; geology; topography (relief); climate; density and orientation of the drainage network; nature of weathering (chemical and/or mechanical); contamination; water chemistry; and vegetation. Clearly it is not possible to discuss comprehensively, for example, stream sediment surveys in environments ranging from arid to tropical rain forest, and climates ranging from tropical to polar, in the confines of the space available here (although this has been achieved by Meyer et al., 1979, in their review of stream sediment geochemistry). Fortunately, this is not necessary as the basic techniques applicable in any one sampling medium of the secondary environment can be altered in detail, at least as a first approximation, to suit the local environmental, geological, and other conditions. Thus, in this chapter only "selected topics" concerned with secondary dispersion will be considered.

DISPLACED ANOMALIES (pp. 348-352)

For the purposes of exploration, we may assume that stream sediments are produced when soils (or glacial drift) are reworked by water. If the sediments contain a

779

component which represents mineralization, any anomalies which result and can be detected are considered to be displaced from the original site of mineralization. However, anomalies are often rapidly diluted with barren material making recognition of the anomaly difficult.

Hawkes (1976c) has attempted to quantify the effects of dilution of an anomalous stream sediment dispersion train and thereby provide a basis for planning sediment sample spacings and locations, and for interpreting stream sediment data obtained from reconnaissance surveys. Toward this end he developed the following empirical formula:

$$Me_m A_m = A_a (Me_a - Me_b) + A_m Me_b$$

where: Me_m is the total metal content (ppm) of a mineralized area (and not any selectively extractable fraction).

A_m is the area (km^2) of mineralization.

A_a is the area (km^2) of the drainage basin above the sample site.

Me_a is the metal content (ppm) of the anomalous sample.

Me_b is the metal content (ppm) in the background area.

Fig. 8A-1 illustrates the various terms contained in the formula.

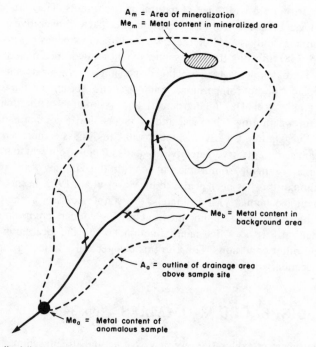

Fig. 8A-1. Idealized diagram illustrating the various terms used to calculate the downstream dilution of stream sediment anomalies (i.e., the stream sediment anomaly dilution formula). After Hawkes (1976c).

Components in the formula relate to: (1) the metal content of the sediment sample; (2) the size of the catchment area; and (3) the size and grade of the surface expression of the mineralization causing the anomaly. One important requirement is that the background metal content, Me_b, be known and relatively constant. Based on actual field checks at four porphyry copper deposits, Hawkes (1976c) confirmed the validity of the formula within reasonable limits. There is no reason why it should not apply equally well to any type of deposit, although difficulty is likely to arise with smaller deposits, such as high-grade veins, where the area of the source of the anomaly is much more difficult to measure. Theoretically, the mineralized source can be anywhere within the drainage basin.

If it were necessary to establish a suitable stream sediment sample spacing during a reconnaissance survey, the formula could be used to predict the maximum size of the drainage basin below which sample collection will result in the detection of an anomaly, i.e., the quantity A_a. To achieve this, an assumption must be made as to minimum size and grade of an economically acceptable deposit in the specific locality, as well as the smallest anomaly contrast that can be recognized above background levels.

Hawkes (1976c) presented the problem in this manner:

"For example, suppose we want to know how far downstream we can go below a porphyry copper deposit of 1 km² area (A_m) and grade (Me_m) of 0.4% or 4000 ppm and still have an anomaly of, say, twice background. If local background (Me_b) is found to be 15 ppm Cu, then we are looking for a limiting value of 30 ppm (Me_a) in our anomalous stream. We now can compute A_a, the area above our anomalous field site, as follows:

$$A_a = \frac{A_m (Me_m - Me_b)}{Me_a - Me_b} = \frac{1 (4000 - 15)}{30 - 15} = 266 \text{ km}^2$$

This area of 266 km² would be equivalent to a linear stream distance of 15 to 20 km, depending on the shape of the drainage basin. Thus, this figure could be used as a limiting distance between samples to assure detecting the porphyry copper deposit under the specified conditions.

There are, of course, limitations to the procedure since certain assumptions have to be made, and certain conditions satisfied. The matter of a uniform background metal content (Me_b) was mentioned earlier, but other assumptions and conditions include: a constant erosion rate (not the case in mountainous areas undergoing rapid erosion or in very low-lying areas); stable sediment-water chemistry; absence of contamination; minimal sampling and analytical error; and, of course, the deposit must be exposed. This procedure, and other, would have no value (except in finding a false anomaly) in the case where a sub-ore halo is exposed but the ore deposit itself has been eroded away (see Fig. 8A-3). Nevertheless, the successful field tests conducted by Hawkes (1976c) do show that the method has potential in important situations. Similar formulae have been proposed by Soviet geochemists, for example Polikarpochkin (1971), in which other parameters, such as the shape of its drainage basin, have been introduced.

Along the same lines, Mackenzie (1977) has been able to demonstrate empirically that in the terrain of Papua-New Guinea, and nearby areas such as the Solomon Islands, the metal values in the sediments of any catchment area approximate the average soil value of that catchment. From this he has been able to show that, for his particular exploration projects, the most important point to consider in reconnaissance stream sediment surveys is the size of the catchment area in relation to the size of the target. Further, in agreement with Hawkes (1976c), he showed that the position of the soil anomaly within the drainage basin is immaterial, as is the length of the drainage train. Mackenzie's (1977) conclusions are based on exploration for porphyry copper deposits using the -80 mesh fraction in a tropical environment.

In the way of documentation, Mackenzie (1977) presented results obtained in the catchment area containing the Panguna porphyry copper deposit (see Baldwin et al., 1978 for a description of the deposit). Fig. 8A-2 shows three drainage basins in the general area, as well as the average copper contents of the stream sediment samples at strategic locations. The stream sediment value of 150 ppm below the Jaba River confluence is still outstanding at 14 km below the ore body where the catchment area is at least 130 km². A weak anomaly and mineralized float have been recorded in the Jaba River 28 km downstream from the ore body.

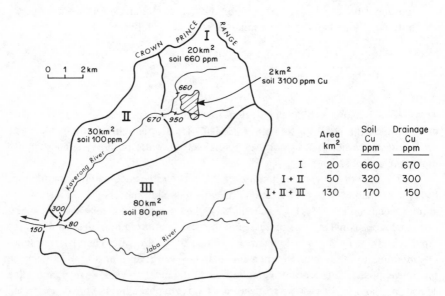

	Area km²	Soil Cu ppm	Drainage Cu ppm
I	20	660	670
I + II	50	320	300
I + II + III	130	170	150

Fig. 8A-2. The catchment areas, and the copper contents of the soils and stream sediments within the catchment areas, in the vicinity of the Panguna porphyry copper deposit, Bougainville, Papua New Guinea. After Mackenzie (1977).

The relationship between the copper content of the soils and that of the stream sediments in the catchment area is most important. The similarity of these values is shown on Fig. 8A-2, for example, 660 compared to 670 ppm Cu in the smallest catchment area. As a consequence, once the size and type of target being sought have been defined, Mackenzie (1977) found it unnecessary to be concerned with threshold; only the "excess value" of the stream sediment drainage samples over the background is important.

For example, if a target can be defined by the local geological parameters as 1 km² with a copper content of 3000 ppm (similar to the conditions at Panguna), and the background copper content of the soils is 100 ppm (see Fig. 8A-2 for background soil values in areas II and III), an "excess value" of 145 ppm in the stream sediments will indicate an important catchment area.

The calculation for the above statement, based on the catchment area containing the Panguna porphyry copper deposit, is as follows:

Calculation of "excess value" of a drainage anomaly:

1. Target sought: 1 km² x 3000 ppm = 3000

2. Background: 19 km² x 100 ppm   = 1900
     Total   4900

3. Anomaly required: 4900 ÷ 20 = 245 ppm, or
    $$\frac{(3000 \times 1) + (19 \times 100)}{20} = 245$$

4. Anomaly required − Background = "Excess value" over background
    (245 ppm)(100 ppm)(145 ppm for a 20 km² catchment area).

Thus, by means of a straightforward dilution, it is possible to predict a stream sediment value at the exit of a given drainage basin which includes a target of a specific size. This predicted value assumes ideal dilution conditions. Lower values could not only indicate smaller or more obscure targets or a higher background, but also some non-ideal erosion-dilution conditions. Mackenzie (1977), again similar to Hawkes (1976c), has presented examples based only on porphyry copper deposits with large surface expressions, but he believes there is no reason why this principle should not be applicable to targets of any size under similar tropical weathering conditions as found at Panguna. In other situations, such as in the search for small high grade veins in Alaska discussed on p. 359, these methods are not likely to be as successful.

PHYSICAL FORM AND CLASSIFICATION OF SECONDARY DISPERSION PATTERNS   (pp. 352-356)

One important point to consider in the interpretation of secondary halos is the erosion level of the mineral deposit as it affects the size and extent of anomalies in soils. The point is illustrated in Fig. 8A-3 which shows four variants of the erosion surface of an ore body which is: (1) blind with only its primary halo exposed; (2) slightly eroded; (3) half eroded; and (4) entirely eroded. Soil anomalies associated with the second and third variants would be stronger than the first, and these (Nos. 2 and 3) may be erroneously assumed to be more promising than the first unless the erosion levels are taken into account. The fourth soil anomaly might well be similar in intensity to that associated with the blind mineralization (No. 1) and, if not properly interpreted, this false anomaly could lead to fruitless exploration. Root zones (No. 4) of some types of ore deposits typically have a different metal association from the ore zone and leakage (upper) zones, and these associations may be helpful in identifying the relationship of a soil anomaly to mineralization. At the present time, however, geochemists outside of the Soviet Union do not appear to have the expertise necessary to make such interpretations (see Beus and Grigorian, 1977, pp. 186-189 for an example and details of how Soviet geochemists tackle this problem. The concept is similar to that employed in distinguishing between above -ore and below-ore primary halos as discussed in the previous chapter (see Figs. 7A-6 to 7A-9 and accompanying discussions).

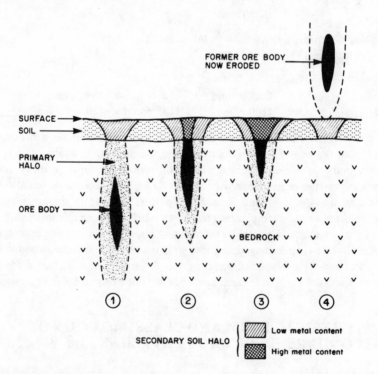

Fig. 8A-3. Variations in the amount of metal in dispersion halos in residual soils depending on the level of the erosion surface. After Beus and Grigorian (1977).

MECHANICAL DISPERSION (pp. 356-359)

The general principles of mechanical dispersion in the secondary environment were firmly established at the time of the first edition and no new concepts or procedures have appeared in recent years. However, there have been many published reports which show refinements in the practical use of mechanical dispersion in various media, particularly stream sediments. In their state-of-the-art review of stream sediment geochemistry, Meyer et al. (1979) have presented and discussed numerous useful examples on this subject.

With respect to soils, Barbier and Wilhelm (1978) have reported some very interesting and important results based on many years of detailed study of surficial dispersion around sulfide deposits in France. Their work provides quantitative data to support the commonly held view that dispersion of anomalies in residual soils over mineralization is more predominantly mechanical than chemical (hydromorphic); hydromorphic dispersion only becomes significant, and identifiable as such, some distance away (for example, see Fig. 9-7, p. 380). Their evidence for this is found in the constant association of certain element pairs, such as Pb-As and Cu-Ag, which they believe would become separated if dispersion were chemical because of the different mobilities of the paired elements in an aqueous medium (see pp. 142-150). They find that mechanical dispersion predominates even in the very fine fractions (less than 100 μm). They explain their observations by the presence in the residual soils of supergene minerals, gossan microfragments, and some mineralized fragments. Although their studies were based on deposits in France, they suggest that the same mechanisms and processes will be operative in all regions which experienced the same type of paleoclimate during the Pleistocene. These would include Canada and the northern United States, and central and northern Europe. In these areas the mechanical migration is the result of solifluction, the formation of frost boils, and other dispersion processes operative in regions which have experienced permafrost conditions (Fig. 11-8, p. 421).

Verification of the findings of Barbier and Wilhelm (1978) can be found in the Russian literature. Specifically Beus and Grigorian (1977, p. 171) state, "Residual dispersion halos always contain clastic disintegration products of the mineral deposit and its primary geochemical halos". Although such would be expected in arid environments, they further state (pp. 171-172), "In temperate humid regions, the residual halos are usually the combined mechanical-hydromorphic type". However, Beus and Grigorian (1977) never do explicitly state that the mechanical component predominates.

DISPERSION PATTERNS DETECTABLE BY
PARTIAL EXTRACTION TECHNIQUES (pp. 359-363)

The importance and use of partial extraction procedures for the detection and interpretation of secondary dispersion patterns have been discussed earlier in Chapter

3A (e.g., in connection with iron-manganese oxides) and in Chapter 6A (e.g., in connection with analytical methods for the analysis of uranium; Table 6A-2).

GEOTHERMAL PATTERNS (pp. 363-365)

In addition to geological and geophysical studies, geochemistry is now considered essential in exploration for geothermal resources. The analysis of waters in potential geothermal areas is the most widely used geochemical method and the procedures were discussed earlier (pp. 364-365).

Numerous recent studies, such as those of Matlick and Buseck (1976), Phelps and Buseck (1978, 1979), and Klusman and Landress (1978), have established that anomalous contents of mercury (and arsenic) are found in the soils above nearly all geothermal areas. The mercury contents of the waters in these areas are generally too low to be of value in this type of exploration. Although the magnitude and shape of the anomalies differ significantly from one geothermal area to another, careful measurements of the content of Hg in the soils can outline the geothermal areas. For example, all Hg anomalies detected in the Yellowstone area are narrow, extending outwards 0.5 km or less from the thermal feature (Phelps and Buseck, 1979). Soils in the Yellowstone geothermal area contain more than 500 ppb Hg, whereas those in non-geothermal areas have less than 30 ppb. Hg minerals (but not necessarily ore deposits) are found in the active geothermal areas of New Zealand, Iceland, California, Italy and the Soviet Union (Matlick and Buseck, 1979).

9A / *Regional and Detailed Surveys*

INTRODUCTION (pp. 367-370)

In the first edition of this book the objective of Chapter 9 was to explain *how* regional and detailed surveys were conducted. In this edition more emphasis will be on the *philosophy* of regional and detailed surveys.

At the outset it is important to realize that there are numerous reasons why a geochemist finds himself exploring in a particular area. These range from such extremes as: (1) search of a particular area for all possible economic mineralization; and (2) a company is interested in a particular mineral commodity and the area was selected on the basis of sound geological and geophysical evidence. In each of these examples the exploration programs will differ because the objective in the first case is to find anything, whereas in the second case, a specific mineral is being sought. Nevertheless, there are certain fundamental factors which should be carefully considered in both cases, usually before the project is begun, and which should be constantly reviewed during the course of exploration. These points have been enumerated by Joyce (1976) under the title of "Design of Geochemical Prospecting Programs", although some may not be applicable in all situations.

1. *Decision on the type of ore deposit sought.*

   An exploration geochemical program designed to find a large, low-grade porphyry copper deposit will be carried out in an entirely different manner to one designed to find small high-grade silver veins. For example, the sample spacing will vary with the level of exploration.

   Once the type of mineral deposit being sought is decided upon, using data from known similar mineralization in the area or, if such is not available, from accepted theories on ore genesis, the following features of the potential ore deposits should be considered:

   a. **The primary ore mineral(s).** This is important because it will influence the size and nature of the secondary dispersion halo. For example, chalcopyrite and wolframite will produce entirely different types of halos in the secondary environment.

   b. **Associated minerals and elements.** Information of this type will assist in the selection of pathfinder elements. Also, the association of certain

minerals, such as pyrite and carbonates, with an ore deposit occurring in a
humid environment will have a major influence on the pH of waters in the
vicinity of the deposit and, as a result, on the mobility of the elements.

c. **Probable localizing structures or rock types.** A knowledge of such factors
as the relative importance of stratigraphic and structural control of the particular type of mineralization in the locality being studied, enables greater
emphasis to be placed on the most favorable geological areas, and will influence the sampling program and the sampling density.

d. **Predicted grade and size of the deposit.** These factors affect such variables
as the probable contrast of an anomaly, as well as the sampling density
which will be required to locate the deposit. Grade prediction generally is
based on the metal content of the known deposits in the area.

2. *Decision on what type of dispersion pattern to utilize.*

Taking into account the nature of the ore, the environmental (climatic) conditions in the area, probable depth to the ore, and the size of the target, it must
be decided what type of dispersion pattern or patterns should be utilized. The
selection may change from one stage of exploration to another. For example, at
the reconnaissance stage, dispersion patterns of mobile elements in stream sediments may be used in those areas with adequate relief and precipitation that are
covered by a suitable drainage system, but at the detailed stage, primary distribution patterns in rocks may be more appropriate.

With respect to the stream sediments, the geochemist must recall that there
are at least five basic types of stream sediment surveys each of which may
delimit dispersion patterns of different length and contrast: (1) analysis of the
fine (-80 mesh) fraction; (2) analysis of some other (usually coarser) fraction;
(3) analysis of the entire sample; (4) analysis of the heavy mineral fraction; and
(5) use of the heavy minerals without resort to chemical analysis (e.g., panning
for gold, scheelite, cassiterite or chromite).

These variations result from the fact that trace elements in stream sediments occur in five basic modes (Rose, 1975): (1) as major elements in trace
minerals (e.g., Pb in galena); (2) as trace constituents of primary rock-forming
minerals (e.g., Cu in biotite; Pb in potash feldspar); (3) as trace constituents in
minerals formed during weathering (e.g., Ni, Co or As in Fe-Mn oxides); (4) as
ions adsorbed on colloidal particles or in the exchange position on clays (e.g., Zn
in the exchange site in clays); and (5) adsorbed by organic matter (e.g., Cu and
Mo; see Fig. 3-18, p. 139).

3. *Choice of the element or elements to be analyzed.*

Usually the target elements are determined, but in certain cases a pathfinder
may be used. However in recent years multi-element determinations have been
encouraged, particularly at the reconnaissance level, since the cost of analyses is
generally a minor part of an exploration budget, and the potential benefit of the
data is great.

4. *Choice of sample material and method of collection.*

The choice of material sampled and how much is collected is governed by what is available in addition to the type of dispersion pattern sought. Methods of collecting various types of samples are discussed in chapters 5 and 5A under "Field Methods".

It is advisable to list the attributes of the ideal sampling medium for geochemical exploration (from Lovering and McCarthy, 1978, p. 253): (1) it should accumulate and concentrate ore elements, or other elements and compounds, that are intimately and uniquely associated with ore bodies; (2) it should be abundant in the exploration area; (3) it should be easily and quickly sampled in the field; (4) it should delineate a large primary or secondary halo, or a long dispersion train, which increases or decreases in a uniform or predictable manner toward the ore target; (5) it should be capable of revealing the presence of blind ore deposits concealed beneath a considerable depth of overburden or barren rock; and (6) it should be easy to analyze. However, as Lovering and McCarthy (1978, p. 254) point out "No single sample medium has yet been investigated that fulfills all these requirements, with respect to any type of ore deposit, under all combinations of geologic and geographic variables; nor does any single medium work for all types of deposits, even in areas where geographic and geologic factors are relatively constant."

An important consideration in soil sampling is the nature of the overburden, specifically if it is residual soil, glacial till, or material of remote origin (e.g., alluvium, glacio-lacustrine deposits). In the case of residual soils, thickness is of no serious consequence, but if the overburden is composed of 2-3 meters of material of remote origin, this may result in insurmountable difficulties for geochemical methods. If the deposit is blind, i.e. covered by rock, then rock materials which may reveal leakage halos, primary distribution halos which reach to the surface or which enable the detection of other types of lithogeochemical halos discussed in Chapters 7 and 7A, should be enployed.

5. *Choice of suitable sample spacing.*

Any decision with respect to sample spacing is difficult, being based upon the predicted size and shape of an anomaly, as well as economic factors. This topic has been discussed in several places in this book, for example in chapters 5 and 5A (Figs. 5A-3 to 5A-5), and in chapters 12 and 12A under the heading of "Survey Sampling".

6. *Choice of an appropriate analytical technique.*

This problem has been discussed under "Summary of Analytical Methods" (pp. 304-309).

7. *Integration of theoretical and practical considerations.*

In some cases, it simply is not possible, for practical reasons, to accomplish what is theoretically best. For example, in an area of very rugged terrain it may be impossible for logistical reasons to collect soil samples on a grid. In that case, one may use a ridge-and-spur soil sampling pattern or take stream sediment samples with extreme caution being exercised when these data are integrated into the interpretation of the results.

8. *Choice of data to be recorded.*
This topic has been discussed under the title "Duties and Responsibilities of Sample Collectors" in chapters 5 and 5A.

9. *Orientation Surveys.*
See "Orientation Surveys" in chapters 4 and 4A.

10. *Choice of interpretation methods.*
Thought should be given at an early stage to the method to be used for the interpretation of the data which will eventually be obtained. Further, the geochemist must always be prepared to change his procedures if meaningful and satisfactory results are not forthcoming. "Interpretation of Geochemical Data" is discussed in chapters 4 and 4A. Statistical techniques in data interpretation are discussed in various parts of chapters 12 and 12A.

11. *Recommendation on follow-up studies to assess anomalies.*
The geochemist must ascertain whether an anomaly is true or false; whether it is displaced; and what follow-up work, if any, should be recommended. All good exploration programs will at least attempt to integrate geology and geophysics with geochemistry, and it is possible that geophysics or geology will be employed in follow-up work on a geochemical anomaly. Each stage of an exploration program must be redesigned (or at least reconsidered) after taking into account all the factors listed in this discussion.

An example of how some of these concepts may be profitably applied early in an exploration project can be appreciated from the conceptual models of Lovering and McCarthy (1978). They prepared three dimensional models (similar to those illustrated in Figs. 9-6 to 9-9, pp. 380-381) in order to relate the type of concealed ore to the best geochemical exploration method for deposits in the Basin and Range Province of the western United States and northern Mexico. In this important mineralized belt, potential ore deposits are concealed by volcanics, alluvium (Fig. 9-1, p. 369), post-ore sedimentary rocks and soils. In such a situation, one of the first items a geochemist should consider is the types of geological materials which are likely to conceal an ore deposit. This should be followed by thoughts about which exploration techniques will have the best chances of success in detecting deposits buried by these materials. Perusal of Figs. 9A-1, 9A-2 and 9A-3 will show that the possible (or preferred) sampling media for each situation will vary.

It should be clear to the reader that the results obtained, and conclusions reached, by Lovering and McCarthy (1978) are designed specifically for the Basin and Range Province. Geological, environmental, and all other exploration parameters will, of course, differ from place to place and so each project must be considered separately although the basic concepts are the same.

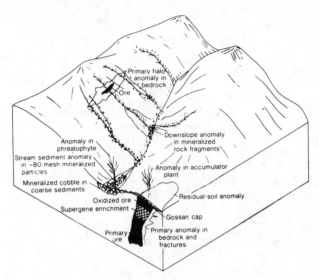

Fig. 9A-1. Sketch diagram of geochemical anomalies related to ore deposits that crop out at the surface or are concealed only by residual soil or colluvium. From Lovering and McCarthy (1978).

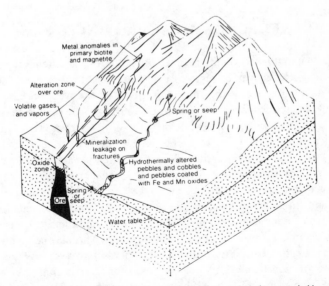

Fig. 9A-2. Sketch diagram of geochemical anomalies related to blind ore deposits concealed beneath host rock. From Lovering and McCarthy (1978).

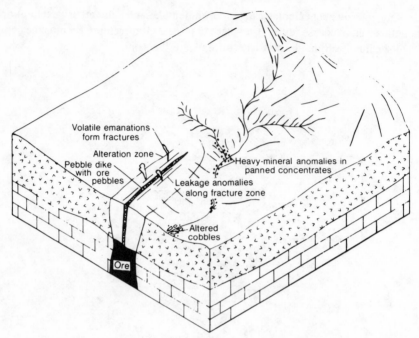

Fig. 9A-3. Sketch diagram of geochemical anomalies related to blind deposits concealed by pre-mineralization rocks. From Lovering and McCarthy (1978).

LOCATION OF GEOCHEMICAL PROVINCES   (pp. 370-374)

As indicated in this section of the first edition, locating geochemical provinces is one of the major objectives of regional exploration, particularly when the region under consideration is poorly known geologically, e.g., the Amazon basin of Brazil and eastern Bolivia. Low-density reconnaissance surveys (one sample per 20-100 sq. mi.) at this scale attempt to detect and delineate areas to be covered at a later stage of more detailed exploration (this, of course, is the objective at almost all scales of exploration) but there is no intention of identifying specific areas of mineralization. Where the drainage systems are sufficiently well developed, concurrent stream sediment and geological surveys are likely to be the first exploration techniques employed, and these will be followed by more detailed surveys, including ground geophysical work. Remote sensing methods, such as Landsat imagery, may be properly used as the initial source of geological information. This may be followed by airborne geophysics after which the ground survey work is undertaken. Whether stream sediments or lithogeochemical methods are used for the recognition of geochemical provinces, the geochemist should be concerned with recognizing features of the regional geochemical specialization (discussed on p. 638). However, lithogeochemical surveys run at this reconnaissance scale are usually very difficult to interpret. In some circumstances, hydrogeochemical surveys may be effectively used at this stage of exploration.

For the most effective use of samples collected at this scale of exploration, multi-element analyses are recommended (p. 373). However, particularly in the case of stream sediments, analyses should be obtained from several size fractions, as well as on the total sample, and the use of various partial extractions should also be considered.

SEDIMENT COLLECTION AT ONE SAMPLE PER 5-20 SQUARE MILES   (pp. 374-375)

In actual practice, there are few organizations, either governmental or commercial, which attempt just to detect geochemical provinces. Surveys conducted at this density (one sample per 5-20 sq. mi.) should be able to delimit both geochemical provinces and large areas favorable for mineralization. In regions where bedrock is exposed, lithogeochemical methods may be considered, and large potentially ore-bearing plutons and formations may be sampled. Lithogeochemical techniques may attain particular importance in arid regions where stream sediment sampling is likely to be less effective.

SEDIMENT COLLECTION AT ONE SAMPLE PER 1-5 SQUARE MILES   (pp. 375-377)

At this stage of exploration an effort is made to locate both primary and secondary geochemical halos around ore deposits and to delineate areas worthy of more detailed study. The geochemical procedures most effective at this stage depend on the landscape-environmental conditions in the region and, in particular, on the relative amounts of exposed and hidden bedrock. Where the area has a developed drainage system, stream sediments are recommended. Lake sediment surveys are particularly effective in certain parts of Canada. Lithogeochemical methods may be used as an adjunct to the stream sediment methods, particularly for sampling individual plutons.

When none of the above methods are likely to be effective, hydrogeochemical surveys using surface or subsurface water may be employed. Such circumstances might arise in arid regions overlain by sands, or in certain areas covered by glacial deposits, where a suitable density of agricultural or domestic wells is found. In flat arid and semi-arid regions overlain by transported unconsolidated deposits greater than 10 to 15 meters thick, hydrogeochemical sampling from springs and wells may be the only source of geochemical information. In some respects this technique is to be recommended wherever it can be used because, as discussed earlier (Fig. 3-23), groundwater halos can be very informative, although they may be very difficult to interpret.

In those areas where anomalous values have been detected, the sampling density should be increased by, for example, collecting more stream sediments or, if these are not locally available, a soil survey at an appropriate smaller grid capable of detecting primary or secondary halos. The objective, of course, is to outline as closely as possible those areas worthy of detailed surveys. In order to ensure that the maximum information is obtained from the collected samples, which is still at a *relatively* low sampling density, analytical and interpretative procedures which enhance weak anomalies should be employed. These procedures have been discussed in Chapter 7A in connection with the enhancement of primary halos, but they are also applicable, at least in part, to stream sediments and soils. Enhancement procedures include: (1) obtaining quantitative analyses; (2) using selective and specific partial extractions; (3) analyzing the heavy fractions; and (4) using composite (additive and multiplicative) halos in the interpretative phase. At this stage of exploration, geophysical ground surveys are commonly used in conjunction with geochemical and geological techniques, as they are at other stages. Maps prepared at the conclusion of this stage of exploration should indicate all geochemical and geophysical anomalies, as well as all relevant geological information.

DETAILED SURVEYS   (pp. 378-382)

Detailed surveys are conducted in promising areas delineated during earlier stages of exploration for the purpose of evaluating anomalies and outlining the location of mineralization as closely as possible prior to drilling or initiating surface/subsurface excavations (e.g., trenches, adits). The sample spacing can vary considerably from as dense as a 2 or 3 meter soil grid, to one of perhaps 100 meters depending on the size and shape (see Figs. 5A-3 to 5A-5) of the possible deposit and its halos, and many other interrelated factors. In any event, the systematic sampling is done in such a way, and using such material, so that an understanding of the nature and extent of anomalies in the area will be obtained. Any, and all, sampling media may be used in the detailed stage of exploration to achieve the above objectives; soils, stream sediments, bedrock, hydrogeochemistry and biogeochemistry are the most common (in that order).

Among the principal problems which should be considered are:
1. The type of deposit giving rise to the anomaly (e.g., Pb-Zn in a vein with Ag, or in a stratiform deposit; U in a vein or in a roll front deposit). This can often be determined by analogy with anomalies and halos from other deposits in the area, by means of element associations, or by geological studies.
2. Evaluation of the erosion level of any geochemical anomaly. This problem was discussed in connection with Fig. 7A-7 and Fig. 8A-3. An understanding of vertical zonations in the primary distribution patterns (Chapter 7A) can assist in

determining whether an anomaly is of the supra-ore or sub-ore type. However, in those areas with limited exposure of bedrock, sampling of the limited number of outcrops will, obviously, result in misleading conclusions with respect to the size and other characteristics of primary halos.

When interpreting secondary dispersion halos, the landscape-environmental conditions in the region must be thoroughly evaluated. Under favorable circumstances, usually involving the interpretation of multi-element data, it may be possible to determine the geochemical origin of the anomaly. However, the only way to prove unequivocally that an anomaly covered by soils, glacial deposits, and unconsolidated materials, is true, rather than false, is to expose the source of mineralization by trenching or drilling. Nevertheless, it is possible to deduce that an anomaly is true if it: (1) is in a favorable geological location; (2) has a realistic size in relation to the type of mineralization which could produce such an anomaly, taking into account the erosion level of the ore body (Fig. 8A-3), and the possibility of small leakage halos being developed over deeply buried deposits; (3) has a geologically realistic shape and orientation; (4) has anomalous element abundances and contrasts comparable to those obtained during orientation surveys on similar deposits in the area; and (5) yields realistic results from soil profile sampling. All of the above-mentioned methods must be used and interpreted with caution as they are not infallible.

It is at this stage of exploration, particularly, that ground geophysics and detailed geology are integrated with geochemistry. Surface geophysical methods, primarily based on differences between the resistivity (conductivity), magnetic susceptibility and density of the ores and host rocks, can often confirm the presence of host rocks which may contain mineralization, as well as assist in the selection of locations for drill holes. Certain geophysical methods can generally assist in the location of geological structures, for example faults.

The detailed stage of exploration is usually considered to be completed with the preparation of a report which includes maps incorporating all the geochemical, geological and geophysical data available, as well as any mineralogical information on the source of the anomaly. Cross sections and profiles showing the relationship between secondary dispersion halos and the primary mineralization and halos, should also be prepared. The report should also contain recommendations with respect to further exploration in the area, specifically drilling.

An important aspect which is sometimes overlooked is the necessity to determine the mineral(s) source of anomalous values as early as possible, as this will influence any decision on additional work. For example, it is important to know that Ni is in sulfides and not primarily in silicates, that uranium is in an economically recoverable mineral rather than within the structure of zircon, or that tin occurs as millable cassiterite as opposed to micron-sized "needle tin". Information of this type may require various chemical extraction procedures, mineralogical studies (e.g., X-ray diffraction, thin section, polished section, electron probe), and beneficiation tests. These types of studies can only be performed when proper material is available, such as during the detailed stage of exploration if a mineralized outcrop is located.

However, such material may not be available until later when trenching or drilling reveal the unweathered mineralization.

DRILLING STAGE OF EXPLORATION
(A new topic)

If drilling of an anomaly is approved on the basis of results obtained from a detailed survey, geochemistry can also be of great value during this phase of exploration. Drilling is usually the most expensive, but also the most revealing, part of any exploration program. Therefore, if any exploration technique can increase its effectiveness (e.g., reduce costs, assist in the evaluation of data obtained), such a technique should be used.

Geochemistry can assist in the drilling phase primarily by interpreting data on the primary halos as they are revealed in the third dimension by chemical analysis of the drill core. According to Beus and Grigorian, (1977) such lithogeochemical methods can be used advantageously in solving several problems which are likely to appear at this stage. These include:

1. **Determining the erosion level of the ore-bearing zones.** In essence, this is the same problem that was discussed above in connection with detailed surveys. However, during the drilling stage, new information may be forthcoming from the analysis of the drill core or cuttings which may permit, or even require, revision of the earlier conclusions.

    Beus and Grigorian (1977) have pointed out that, in general, in multi-deposit ore-bearing regions which are highly eroded, the number of anomalies originating from sub-ore halos is much greater than those originating from supra-ore halos. This is because halos formed below ore bodies have greater vertical extents by comparison with halos formed above ore bodies. Halos below ore bodies reflect the paths of movement for the ore-bearing solutions and can, therefore, be traced to considerable depths. This concept is diagrammatically illustrated in Fig. 9A-4.

    From Fig. 9A-4 it can be seen that at erosion level 3 (relatively uneroded surface) there is only one sub-ore halo (plus one supra-ore halo and one exposed ore body). At erosion level 7 there are five sub-ore halos (plus one supra-ore halo and one exposed ore body). At the deepest erosion surface, level 11, there are nine sub-ore halos (and no supra-ore halos or ore bodies). The sub-ore and supra-ore halos are identified by: (a) constructing composite (additive and multiplicative) halos: (b) using ratios of the various elements in the halos; and (c) interpreting the "standard zonation sequence" for steeply dipping ore bodies. These points have been discussed in Chapter 7A. In any particular circumstance, it is clear that the geochemist should always be aware that it is essential to differentiate between sub-ore halos of no economic significance, and supra-ore halos indicative of blind mineralization, even if this evaluation can only be made during the drilling stage.

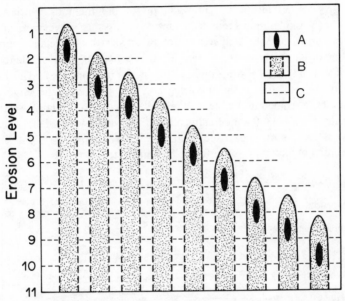

Fig. 9A-4. An idealized diagram showing possible multi-ore occurrences within a mineralized district, and the relative vertical extents of halos above and below the ore bodies. The number of sub-ore halos increases as the level of the erosion surface increases (level 11 represents the most extensive erosion surface). After Beus and Grigorian (1977). A. ore body. B. halos (both supra-ore and sub-ore). C. level of the erosion surface.

2. **Adjustment in the direction of exploratory work.** During drilling, lithogeochemical studies may reveal that the original direction of drilling, or the original drilling plan, should be altered. This may result from many causes, such as a revised interpretation of the shape and dimensions of primary halos, or the suspected occurrence of more valuable mineralization in another direction.

CONCLUDING REMARKS
(A new topic)

Geochemistry is just one exploration method whose use increases the probability of success in the search for mineral deposits. Geology, geophysics (see pp. 707-710) and remote sensing are the others of particular importance at present. All have the same basic objective, i.e., to detect and record any feature of the Earth's crust which, when properly interpreted, will guide the explorationist to an ore deposit. However, each method has advantages and disadvantages in the various stages of an exploration program, and in different geologic-environmental situations. Similarly, each geochemical method has specific advantages and disadvantages and it is the responsibility of the geochemist to use the one (or those) that will yield the most useful information, in the most cost-effective manner, at each stage of exploration.

Fig. 9A-5 is an idealized representation of the methods which might be employed in the exploration sequence from reconnaissance surveys (recognition of geochemical provinces) through several follow-up stages, to a detailed survey, and finally to the drilling stage. During and following each stage, unfavorable regions are eliminated and additional effort is put into the more promising areas. This figure was constructed on the basis of the following assumed landscape and environmental parameters: (1) a temperate climate with rainfall sufficient for chemical weathering to predominate; (2) moderate topographic relief; (3) a well-developed drainage system; (4) a reasonable amount of exposed bedrock; (5) predominantly residual soils, with lesser amounts of alluvial or other types of transported materials.

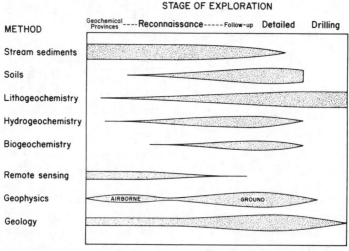

Fig. 9A-5. Idealized diagram illustrating the sequence in which various exploration techniques might be used in an area with a well-developed drainage system and a temperate climate (see text for additional landscape-climatic conditions).

From Fig. 9A-5 it is clear that each geochemical sampling medium, and each exploration technique (remote sensing, geophysics and geology), has a preferred position in the exploration sequence under the conditions specified. In a different environment, such as a desert or tropical rain forest, some of the above methods might not be used at all, but, if they are, their relative positions in the exploration sequence may well be different. Further, the selection of geochemical sampling media in a specific exploration area is governed and limited by the media which are available and also by the objectives of the particular phase of the exploration. Usually, many more types of sampling media are available in any particular area than can be effectively used. Certain media are more effective in the detailed phase of exploration for ore bodies of small area, whereas others are best suited for the rapid reconnaissance evaluation of a large region. In some cases, one sampling medium may be ideal in the search for a specific type of mineralization under one set of local environmental conditions but, because of changes in these conditions, this medium may be ineffective relatively short distances away. This is particularly true of those sampling media which vary not only from place to place, but also from time to time; such media include certain types of water and vegetation samples, and (almost) all vapor samples.

In view of the difficulties involved in selecting the appropriate sampling media that will result in a meaningful and successful geochemical program, it is not surprising that some experienced exploration geochemists can conclude that ". . . despite recent technological advances in analytical techniques, the choice of sample media for geochemical exploration programs remains today far more of an art than a science" (Lovering and McCarthy, 1978, p. 254). Thus, the often-heard statement "geochemical methods of exploration don't work" can usually be explained by one or more of the following: (1) inappropriate materials being sampled; (2) appropriate materials being sampled at the wrong time; (3) analyses being conducted for the wrong group of elements; or (4) incorrect extraction procedures used for the correct group of elements. As difficult as it may be to make the correct decisions, it is clear than *an important aspect of exploration is matching a particular exploration method to the objectives of a survey and the environment in which the survey will be conducted.*

10A / *Vegetation Surveys*

INTRODUCTION   (pp. 383-391)

It was stressed earlier (pp. 398 and 406) that both geobotanical and bio-geochemical surveys require, or are aided by, the services of an expert versed in botany, ecology and plant physiology. As a result, the number of professional geobotanists and biogeochemists is relatively low and this, in part, explains why the number of vegetation samples analyzed is also relatively low (Table 1-1). It is not surprising, therefore, that there will be lean periods with respect to new developments in these specialties such as the past few years in which relatively little has been published on vegetation surveys, exceptions being the excellent state-of-the-art review papers by Cannon (1979) and Brooks (1979). Nevertheless, some progress continues to be reported such as the orientation studies by Cole et al. (1974) in western Queensland, Australia, which attempted to integrate geobotanical and biogeochemical results with remote sensing techniques. Furthermore, there is definite evidence that geobotany and biogeochemistry are being utilized, presumably successfully, by certain mining companies, for example in Australia and South Africa, but the results have not been published.

In recent years, some of the most important work on vegetation surveys has been concerned with nickel. This is probably fortuitous and a result of the interest in the element by the few workers in the field, as well as by exploration companies in the late 1960's. But because of the time lag between research and publication, the results have only appeared relatively recently. By way of example, Cole (1973) working in the nickel regions of Western Australia was able to find regional and local relationships between vegetation and local bedrock geology. Geobotanical studies showed that shrub communities containing *Hybanthus floribundus* were associated with high nickel concentrations in soils at some localities, but not at others. The relationships were complex, with low soil pH (which increases the availability of individual elements) playing an important role. Cole (1973) also found that geobotany was of limited assistance where there is deep cover (transported overburden, laterite, etc.) or where there is abundant calcium carbonate in the soil. Biogeochemical studies showed that, apart from *Hybanthus floribundus* which accumulates large quantities of nickel (as much as 23% Ni in the ash of leaves, equivalent to about 1.4% on a dry-weight basis; Severne and Brooks, 1972) and thrives on nickeliferous soils, all other shrub species restrict nickel uptake (values of 50 ppm on a dry-weight basis are high for the other shrubs growing over nickel deposits). None of the plants contained large quantities of other elements, such as chromium, which are present in very small amounts even in samples collected from species growing in enriched soil. Consequently, soil sampling is often more useful than geobotany and biogeochemistry except (1) for remote sensing applications in the case of geobotany, and (2) in arid environments in the case of biogeochemistry because here the decay products of vegetation are blown away and soils do not form.

Brooks et al. (1977a) have recently discovered five previously unknown hyperaccumulators of nickel from analysis of about 2,000 herbarium specimens of the genera *Homalium* and *Hybanthus,* collected from many parts of the world. *(Hyperaccumulators* are plants which have unusual concentrations of an element; in the case of nickel, values which exceed 1,000 ppm on a dry-weight basis are defined as unusual). Previously, all examples of hyperaccumulators of the above genera such as *Hybanthus floribundus,* were known only from New Caledonia and Australia. However, the study of Brooks et al. (1977a) showed that specimens of the genera *Homalium* and *Hybanthus* delineated many of the world's ultramafic areas (e.g., Cuba, Puerto Rico, Philippines, Western Australia and New Caledonia) in which nickel is mined. An unusual and important adjunct of this particular study that the authors pointed out, is the usefulness of analyzing herbarium specimens. This could be a simple, rapid and inexpensive method of carrying out wide-ranging biogeochemical surveys. Credit for first suggesting the use of herbarium specimens in mineral exploration probably belongs to Persson (1948) who studied collection localities of "Copper Mosses" in the Stockholm herbarium and as a result was able to delineate copper anomalies. Following work on *Homalium* and *Hybanthus,* Brooks and Wither (1977) discovered large quantities of nickel (up to 1.72 % on a dry-weight basis) in *Rinorea bengalensis* and used the collection localities of herbarium specimens to delineate previously unknown ultrabasic areas in Indonesia. A similar technique was used by Wither and Brooks (1977) using herbarium specimens of *Trichospermum kjellbergii* and *Planchonella oxyedra.* Both of these plants are hyperaccumulators of nickel and it was possible thereby to identify a previously unknown ultrabasic region on Ambon Island, South Moluccas (Indonesia). It should be emphasized that whereas these herbarium surveys were used to identify nickeliferous rocks, this did not necessarily indicate economic-grade nickel mineralization.

Mention should also be made of a herbarium survey of African species of the genera *Crotalaria* (Brooks et al. 1977b) and *Haumaniastrum* (Brooks, 1977). The former survey identified new accumulators of copper and cobalt in the copperbelt of Zaire. The latter showed that the "copper flower" of Shaba Province, Zaire (Duvigneau and Denaeyer-de-Smet, 1963) contains much more cobalt (up to 1.02 % dryweight) than copper (0.11 %) and its distribution may, therefore, be controlled by cobalt rather than by copper. Hence it may more properly be called a "cobalt flower."

Other elements recently found to be useful in biogeochemical surveys under the proper conditions include:

a.   **Lithium.** Cannon et al. (1975) found that vegetation in the Basin and

Range province of California and Nevada may contain 100 times as much Li as average plants; analysis of vegetation is suggested as a means of prospecting for deep-seated lithium brines in that area.

b.   **Tungsten.** Quin et al. (1974) found that biogeochemical prospecting by analyzing the leaves of certain shallow-rooted tree ferns can be useful in prospecting for scheelite in New Zealand. Surprisingly, larger trees with deep root systems were not effective for reasons which are not clear. The environmental conditions, specifically very high rainfall with resultant leaching and low pH, are an important factor for the success reported in this situation.

Finally, it is worth noting that Peterson (1971) reviewed and tabulated unusual accumulations of elements by plants and animals primarily from a biological point of view; however, their importance in exploration geochemistry has yet to be demonstrated.

Mobilization Of Trace Elements   (pp. 383-385)

No additional comments.

Soil-Plant Transfer Of Elements; Plant Nutrition   (pp. 385-387)

No additional comments.

Biogeochemical Classification Of Elements   (pp. 387-389)

No additional comments.

Metal Uptake By Plants   (pp. 389-391)

Chemical analysis of plants or plant parts can be used as a geochemical prospecting technique only if the chemical content of the vegetation sampled has some predictable relationship to the chemical content of the nutrient soil, bedrock, or to metal-containing groundwater moving through the soil or bedrock. The chemical content of the vegetation may be the specific element sought, or a pathfinder.

Based on work in Siberia, Kovalavskiy (1975) has concluded that there are two main reasons why biogeochemical anomalies (aureoles) may be absent in plants whose roots are in contact with ores, or their dispersion halos, in soils, rocks or groundwater. First, are the physiological barriers that plants present to absorption from areas of high concentration of ore-forming elements (these have been discussed previously under various headings on pp. 387-390). Second, the ore-forming elements may be present in a form unavailable for absorption by plant roots (usually this results from the fact that the minerals are stable in the weathering zone or, when weathered, they form insoluble compounds). In other words, elements must be in a mobile (available or soluble) form to be taken up by plant roots. Thus tungsten in the form of scheelite, $CaWO_4$, can be sufficiently chemically mobilized to form biogeochemical halos such as those mentioned above (Quin et al., 1974), but when this element is present as wolframite, (Fe, Mn) WO_4, it is resistant to chemical weathering, and biogeochemical anomalies will not be produced. Analogous variations in mobility can be shown for any element which forms both soluble and insoluble minerals (e.g., boron, as soluble borates and insoluble tourmaline). A third explanation, not mentioned by Kovalavskiy (1975), is that vegetation may release metals into the atmosphere by transpiration and, under favorable conditions, the release may be rapid (P. M. D. Bradshaw, personal communication).

Kovalavskiy (1975) was apparently the first to attempt to estimate the likelihood of plants having absorption (physiological) barriers which prevent or inhibit the formation of biogeochemical anomalies by chemically mobilized elements in soils and related materials; information of this type is of value in predicting which elements are likely to be good candidates for biogeochemical surveys. Based on the analysis and classification of 800 different "biological objects" (various plant species, plant organs, etc.) he produced data which are the basis of Table 10A-1. Twelve elements were considered, and Kovalavskiy (1975) was able to characterize them on the basis of the responses they evoked on plants, in other words, whether or not a plant (or part) had a physiological barrier to the particular element (such as the partial exclusion of copper in black spruce shown in Fig. 10-3, p. 389). The "concentration factor" as used in Table 10A-1, is the content of the element found in the vegetation sampled (over ore) divided by the content of the same element in background vegetation samples. It is a particularly useful concept because four broad groupings can be recognized. Only elements in the first two categories (barrier-free and low-barrier), based on the numerical value of the concentration factor being greater than 30, are recommended for biogeochemical prospecting. Therefore, only Ra and Mo can be reliably recommended among the group of 12, and even here Mo will likely give poor results in 19 % of cases because its concentration factor, on occasion, can be as low as 1. Elements which are likely to belong in the "medium-barrier" group should be considered only when no better option is available, but the risks are obvious (i.e., Pb, W, Ag, Ba). Little chance for biogeochemical success can be expected for "high-barrier" elements, specifically U, F, Mn, and Fe. However, some accumulators of uranium, for example, are definitely known as should be suspected by the fact that this element does have some representation in the low-barrier and medium-barrier categories. By way of example, Bowie (1973) states that it seems that trees growing near uranium deposits contain 2 to 200 times as much U in their leaves compared with ones growing in barren ground.

Although Kovalavskiy's (1975) study does not include certain elements of geochemical interest (e.g., Cu, Zn, Ni) and is based mainly on Siberian examples, it does generally conform with experiences reported in the literature. The last column of Table 10A-1 is particularly revealing and, at the same time, disturbing.

Table 10A-1. Grouping of twelve selected elements for biogeochemical exploration purposes in terms of the responses (physiological barriers) they invoke in biological objects. Modified from data published by Kovalavskiy (1975) based on Siberian experiences.

Grouping	Concentration Factor	Percentage of Biological Objects (which fall within each group delimited by the Concentration Factor)												
		Ra	Mo	Au	Pb	W	Ag	Ba	Be	U	F	Mn	Fe	Average
1. Barrier-free (most favorable)	300 or more	100	16	10	10	3	2	0	0	0	0	0	0	12
2. Low-barrier (favorable)	30 - 300	0	65	26	22	20	7	0	4	3	3	0	0	13
3. Medium-barrier (least favorable)	1 - 30	0	19	44	35	32	49	65	45	17	7	6	0	25
4. High-barrier (not recommended)	less than 1	0	0	20	33	45	42	35	51	80	90	94	100	50
Total		all sum to 100%												

Notes:

1. Data based on 800 "biological objects" which include typical materials used for biogeochemical surveys such as needles, bark, branches (the latter is generally a poor accumulator of elements).

2. "Concentration Factor" is defined here as the maximum content of the element found in the vegetation sampled, divided by the content of the same element in background vegetation samples. Kovalavskiy (1975) does not use this term but refers to "maximum, background and minimum anomalous contents of elements in biological objects" from which a concentration factor, as used here, can be deduced.

3. Kovalavskiy's (1975) original table, upon which this table is based, related the groupings of biological objects in terms of concentration of the 12 elements (that is, the reverse of this table heading). For the purposes of this book the emphasis has been modified somewhat. Although the results possibly are not directly comparable, they certainly convey the same intent on the probable success, or lack of it, in biogeochemical prospecting. Specifically, if Ra is barrier-free in 100% of the biological objects sampled and forms high biogeochemical anomalies, it is logical to conclude, at least as a first approximation, that Ra will produce a biogeochemical anomaly 100% of the time.

Specifically, Kovalavskiy (1975) found that fully 50 % of all biological objects are in the "high-barrier" category. As he pointed out for the 12 elements considered (p. 185), "This means that in a collection of various species of plants whose roots are in contact with ores of the above elements or the ore haloes, one-half of the samples will show no anomalous concentrations of such elements." Thus, the theoretical basis for biogeochemical prospecting, that is, the direct correlation between the concentration of an element and the medium in which it grows (or is nourished, in the case of groundwater), is only likely to be met no more than 25 % of the time, at least on the basis of Kovalavskiy's (1975) study. Clearly he has refuted the simplified view held by some that biogeochemical halos are formed in all species by showing that physiological ("anti-concentrational") barriers exist which impede the incorporation of elements into vegetation.

In view of the 100 % rating obtained by radium in Table 10A-1, it is worthwhile to discuss, and illustrate, Ra biogeochemical anomalies in vegetation. By way of review, there are two relatively long-lived radium isotopes (none are stable): ^{226}Ra (in the ^{238}U series; half-life, 1,602 years); and ^{228}Ra (in the ^{232}Th series; half-life, 6.7 years). Two other isotopes, ^{223}Ra (^{235}U series) and ^{224}Ra (^{232}Th series), have short half-lives of 11 and 3 days, respectively, and are not important contributors to the radium content in vegetation. Working with specimens from the uraniferous areas of New Zealand, Whitehead et al. (1971) found that the α-activity in plants in that area was caused by absorption of several isotopes: ^{210}Pb, ^{227}Ac, ^{238}U, ^{235}U, and ^{226}Ra. The ratio of the abundance of these isotopes in the plants compared with their abundance in associated soils indicated that radium is the most mobile of all, and that ^{226}Ra can migrate into the plants. (This may appear inconsistent with the discussion on p. 79, but there does not appear to be sufficient sulfate in the environment to cause the precipitation of insoluble radium sulfate, either in the New Zealand locality, or the Brazilian locality described immediately below).

The source of radiation originating from plants growing over thorium mineralization in Brazil has been thoroughly investigated over the past decade by Professor E. Penna-Franca, Dr. M. Emmerich, and other scientists at the Institute of Biophysics at the Federal University, Rio de Janeiro (for example, see Penna-Franca and Gomes de Freitas, 1963; Penna-Franca et al., 1972). Their interests are concerned mainly with health and the environment (i.e., the effects on man and animals of long term exposure to natural radiations) but their work has application to transportation of elements and their eventual uptake by plants. Autoradiographs (Fig. 10A-1), made by exposing X-ray films (the type used for chest X-rays) to vegetation for periods of one month, were obtained from leaves of several plants growing in Morro do Ferro (Hill of Iron), in the state of Minas Gerais. Professor Penna-Franca and associates found that the source of radioacitivity is mostly ^{228}Ra and daughters, a result of the fact that the plant grew on soil over a rich thorium deposit (specifically a thorium-containing monazite); only a small amount of activity was from ^{226}Ra and daughters as the deposit contains only traces of uranium. Thorium itself is not taken up by the plants because it is insoluble and immobile (see Fig. 3-19); however, radium, one of the products of the decay of thorium, is soluble and mobile in that particular environment (there is no sulfate) and it is accepted into the plants. Professor Penna-Franca (personal communication) states that the radium uptake among different species growing in the locality is quite variable. *Mellas tomatacea* shows the highest radium content (in the range of 104 pCi per gram of ash). Some species incorporate very little radium, not dissimilar to the exclusion encountered by many other metals (pp. 389-390). Of great significance, is the fact that *no recognizable gross abnormalities were found in plants growing in the radioactive soils* (compare with statements on p. 398). However, high rates of chromosome aberrations were found in scorpions living in (on) the radioactive soils by comparison with control specimens (however, sampling scorpions for exploration purposes does not appear likely to be productive!). From Fig. 10A-1 it is possible to see that ^{228}Ra is concentrated in the stems, apparently uniformly, but also that it is preferentially concentrated in the vein system of the leaves (this point will be discussed further under "Element Distribution In Vegetation", below).

MICONIA THEAEZANS
MORRO DO FERRO - POÇOS DE CALDAS
BRAZIL - 1975
M. Emmerich and E. Penna Franca

Fig. 10A-1. Autoradiograph produced by ^{228}Ra (which is a decay product in the ^{232}Th series). Leaves are from the species *Miconia theaezans* growing over a thorium-bearing monazite deposit, Morro do Ferro, Pocos de Caldas Brazil. (The holes in the leaves have no significance and are probably caused by insects). Courtesy of Professor E. Penna-Franca and Dr. M. Emmerick.

In addition to several studies concerned with biogeochemical methods applicable to nickel exploration (Severne and Brooks, 1972; Cole, 1973; Brooks et al., 1974; Lee, 1977a), two other studies concerned with the uptake of copper and/or molybdenum merit brief mention here:

1. Wolfe (1974) confirmed previous work (Fig. 10-3, p. 389) which showed that copper and molybdenum typically display diverse metal uptake behavior in coniferous species. Contrast (defined as the comparison of values in the ash of vegetation and in supporting soils of background and anomalous areas) can range as high as 20:1 for molybdenum, but is limited to 3:1 for copper even in soils containing as much as 30 times background levels of copper. From this he concluded that the copper content of vegetation is determined mainly by the specific requirements of the plant for this metal; in other words, the copper content of vegetation tends to be controlled *internally* by characteristics of the vegetation rather than *externally* by the copper content of soils (Timperley et al., 1970, reached similar conclusions).

2. Yates et al. (1974), in agreement with the conclusion of Wolfe (1974) above, recognized that copper is a "difficult" element (p. 407) for biogeochemical prospecting and suggested the use of the Cu/Zn ratio (a suggestion originally made earlier by others).

GEOBOTANY   (pp. 392-399)

Introduction   (p. 392)

No additional comments.

Indicator Communities   (pp. 393-394)

No additional comments.

Indicator Plants   (pp. 394-396)

No additional comments.

Morphological, Physiological and Mutational Changes   (pp. 396-398)

The only point which will be mentioned here is that the vegetation, specifically leaves (Fig. 10A-1), growing in highly radioactive soils in Brazil in which radium has definitely been determined, shows no gross abnormalities. (The same applies to the vegetation illustrated in Fig. 10A-5, below.)

Discussion (pp. 398-399)

In the earlier discussion (p. 399) it was noted that geochemists and geologists who have little training in botany (the vast majority) should keep in mind the fact that many ore deposits occur in areas that are (relatively) bare of vegetation. The use of this form of geochemical observation, together with the recognition of a "copper flower", resulted in the discovery of a copper deposit at Kalengwa, Zambia (Ellis and McGregor, 1967).

Cole (1973) noted, however, that the absence of vegetation may, or may not, reflect mineralization owing to many interacting factors. Specifically, she discussed the fact that at the main Kambalda, Western Australia nickel geochemical anomaly, there was a complete absence of trees and most of the shrubs over the eastern half, but vegetation was present over the western half. This unusual circumstance was attributed to the presence of calcium carbonate in the soils of the western half which reduced mobility of the elements Ni and Cr. The eastern half lacked calcium carbonate thus permitting greater mobility and uptake of Ni and Cr with the result that plant growth was inhibited.

Lag and Bolviken (1974) reported occurrences of naturally lead-poisoned soils and vegetation in five different regions of Norway. Samples from lead-affected patches showed lead contents of as much as 2.5% in soils and 400 ppm in vegetable dry matter, corresponding to 400 and 70 times background, respectively. In the case of lead poisoning, Lag and Bolviken (1974) noted there are different stages during which the characteristics of the vegetation (i.e., plant communities) change. Other elements, such as copper, can also be toxic to vegetation under the proper conditions (Bolviken and Lag, 1977; Bolviken et al., 1977). A naturally poisoned area over copper and zinc mineralization is illustrated in Fig. 10A-2. The advanced stage of all poisoning is characterized by: abnormal, dying or deficient vegetation; an apparently high stone content at the soil surface; and a poorly developed or deficient bleached layer in podzol areas. As a consequence of their observations, natural heavy metal poisoning of soil and vegetation in connection with sulfide mineralization in Norway has been found to be more common than was formerly believed. Fig. 10A-3 shows a bare area in northern Montana. However, one must *not* assume that all barren areas (many of which appear at seepages) are the result of metal poisoning; in fact most are probably not.

BIOGEOCHEMISTRY (pp. 399-408)

Sample Collection (pp. 399-403)

No additional comments.

Fig. 10A-2. Naturally lead-poisoned area at Tverrfjellet, Hjerkinn, Norway. Light patches in front are stones and boulders in barren soil; the absence of vegetation is typical of advanced poisoning (see text). Mineralization consists of pyrite, chalcopyrite and sphalerite. From Lag and Bolviken (1974).

Fig. 10A-3. Bare areas in northern Montana in which copper mineralization has been found. Courtesy of Dr. David Grimes.

Element Distribution In Vegetation   (pp. 403-406)

On page 403, the statement was made that "although some plants can concentrate many elements to a striking degree, the accumulations are not uniformly distributed throughout the plant". The concept in mind at the time was that different plant organs (e.g., stems, needles, leaves) have different metal contents, a fact illustrated by variations in the arsenic contents of different organs in Douglas Fir (Table 10-4). However, from Fig. 10A-1 it can be readily seen that Ra can be concentrated in the veins of leaves. From Fig. 10A-4 it can also be shown that certain elements are likely to be concentrated in exudates (the significance of which has been discussed previously on p. 621). Finally, again using autoradiographs of radium-containing Brazilian specimens, it is possible to show that concentrations of elements can also be found in other specific parts of leaves; in Fig. 10A-5 they are highly concentrated at the outer edges. Thus, using evidence of Figs. 10A-1 and 10A-5, element variations *within* individual organs can now be added to the list of variables.

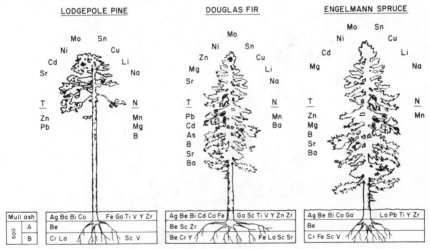

Fig. 10A-4. Location of highest concentrations of elements in a group of vegetation samples collected in Colorado. Especially note that the "element halos" above the trees denote elements concentrated in the ash of exudate residue. N denotes highest concentration for elements in the ash of needles. T denotes highest concentration of elements in the ash of twigs. After Curtin et al. (1974).

It is well known that vegetation is composed of many different organic compounds and phases (wood, for example, is made of cellulose, ligins, lipids, resins and many other compounds) and it is reasonable, therefore, to assume that each compound (organic phase) may have its own tendency to concentrate specific elements. Such tendencies, whether they be by absorption or other mechanisms, may result in some sort of "biological fractionation" within the plant organ. (Analogous fractionations have been reported by hydrogen and carbon isotopes and deuterium in wood.)

If such be the case, and Figs. 10A-1, 10A-4 and 10A-5 tend to support it, analyses of selective phases or components of a specific organ (e.g., the cellulose) may yield more reliable biogeochemical anomalies than bulk plant organ analyses (e.g., whole leaves). Selective extractions, analogous to those now well-established for soils, sediments and rocks, should be attempted and evaluated (for example, a benzene-alcohol mixture is a selective extractant for lipids in plants). Such extractions might overcome the problems associated with different ratios of cellulose, lipids, resins, etc., each of which may have different trace element incorporation characteristics. Only a very few studies of this type appear to have been made specifically for exploration purposes. Whitehead et al. (1971a) investigated the distribution and chemical form of uranium in leaves of New Zealand plants and found that it occurred in several different forms (e.g., uranium-RNA complex; uranium-protein complex). Kelly et al. (1975) found that nickel in plants from New Caledonia occurred in several forms, including one which is water soluble. This humic compound was later isolated and identified by Lee et al. (1977b) who found it to be a citrato-complex of nickel.

ADIANTUM LORENTZII HIERN
MORRO DO FERRO - POÇOS DE CALDAS
BRAZIL - 1975
M. Emmerich and E. Penna Franca

Fig. 10A-5. Autoradiograph produced by ^{228}Ra present in the leaves of *Adiantum lorentzii hiern*. Same locality as that in Fig. 10A-1. Courtesy of Professor E. Penna-Franca and Dr. M. Emmerick.

In the field of botany, there is now a vast literature concerned with the distribution of elements in plants, much of which is based on studies with various radioactive isotopes (e.g., Kamen, 1957). Among the many findings of interest to exploration geochemists is the fact that the element distribution depends on the isotope introduced into the root system. In addition, time is an important factor as the rate of movement of elements into the intercostal (between veins) fields varies.

Finally, it is worth pointing out that there is a lack of agreement with respect to the trace element content of the ash of plants when results obtained from different parts of the world by different investigators are compared. A compilation by Beus and Grigorian (1977) based on Soviet experiences is shown in Table 10A-2. When this is compared with the compilation of metal contents in ashes from unmineralized ground (background values) in the United States presented in Table 10-5 (p. 406), it is clear that there are significant variations. The cause of these discrepancies is not known, but they certainly do stress one of the many potential difficulties likely to be encountered when using biogeochemical surveys.

Table 10A-2.   Accumulation of ore trace elements in the ash of plants, ppm.
From Beus and Grigorian (1977).

	Over barren areas (local biogeochemical background)	*Over mineral deposits*
Cu	15-30	From 50-60 to 100-200 (depending on the concentration of copper in ores); in individual cases more than 1000
Pb	0.2 to 30; seldom to 50	40 to 500, in places up to 1000 or higher
Zn	20 to 100 (in grasses); 100 to 1000 (in wood species)	More than 300 in grasses, 500 to 3000 in wood species
Ag	0.01 to 3	0.04 to 10 (depending on the background), maximum contents 30 to 60
Au	Less than 0.01	More than 0.01; rarely up to 10
Mo	0.1 to 20	20 to 350 (depending on the background); in individual cases up to 1000
Sn	1 to 6	More than 10, usually 40 to 150 sometimes approaching 550
Ni	5 to 30	More than 100, sometimes up to 2000
Co	Up to 30	More than 50
Bi	Less than 5	10 or more (up to 100)
Li	From less than 1 to 15	More than 10 to 15, usually 40 to 200, in individual plant species more than 1000
Be	0.5 to 4, rarely up to 7 or 8	5 to 20, in individual cases reaching 30 to 40
Nb	Less than 10	10 to 50
Zr	Up to 10	More than 10, sometimes approaching 100

Discussion   (pp. 406-408)

It is perhaps appropriate to quote a paragraph from Beus and Grigorian (1977, p. 217) which sums up Soviet experience:

"The most important advantage of the biogeochemical (phytogeochemical) method of mineral exploration is the increased depth range which can be attained as a result of the development of the plant root system. However, the biogeochemical method is more labor-consuming and more expensive than other methods of geochemical exploration. This is because of the specific requirement for the sampling and processing of phytogeochemical samples. (It is also desirable to have a geobotanist or botanist on the exploration team during biogeochemical exploration.) Therefore, the biogeochemical method may be used most effectively in cases where the less expensive lithogeochemical [which includes soils and stream sediments] methods cannot be used for some reason or other."

11A / *Geochemical Exploration in Canada*

INTRODUCTION (pp. 409-410)

Geochemical exploration activity in Canada and in other areas in the northern latitudes continues at a high rate, and significant discoveries in which geochemistry has played an important role continue to be made. Part of this success can be attributed to the fact that geochemists concerned with exploration in these areas are now generally much more aware of the special problems posed by each of the four characteristic features or environments of the northern latitudes: permafrost; water (including lake sediments); muskeg and bogs; and glaciated areas. Glaciation and poor drainage are the two dominant factors complicating the use of geochemical exploration in Canada at present.

Geochemists have made a major effort at advancing our understanding of the specific techniques to be used in glaciated areas, and this effort has been reasonably successful. On the other hand, although much exploration activity has been based on the sampling of water and lake sediments, our understanding of the fundamental aspects of this exploration environment does not appear to have advanced significantly by comparison with glaciated areas. Progress in the other two environments (permafrost; muskeg and bogs) has been even more limited. Advances in each of these topics are presented below.

In the past five years there has been a significant increase in the use of surface water and groundwater, mainly for uranium exploration, and also in gas sampling (principally radon and helium), again for uranium exploration.

PERMAFROST (pp. 410-426)

Introduction (pp. 410-412)

Washburn (1973) has recently written probably the most definitive book on periglacial processes and environments which includes valuable information on permafrost and permafrost terminology.

Nature, Distribution And Occurrence Of Permafrost (pp. 412-417)

No additional comments.

Weathering And Mobility Of Elements In Permafrost (pp. 417-423)

Evidence continues to accumulate indicating that chemical weathering does take place in permafrost, and that permafrost does not inhibit the use of soil or sediment geochemistry. Cameron (1975) has reviewed the subject and, based on Soviet experiences, stated (p. 40), "The critical element to an understanding of how chemical weathering takes place was the discovery that in frozen ground thin films of water are present along mineral/ice interfaces and along ice/air interfaces." Anderson and Morgenstern (1973) found that ionic mobility within these films is only slightly less than in normal aqueous solutions. The water and the ions contained in these films move in response to several gradients, of which thermal gradients are apparently the most important. The movement is from warmer to colder regions which, in permafrost areas during the long, cold winter, is in an upward direction; some of the water and ions may even move into the overlying snow (see Fig. 11-7). Should chemical weathering take place in areas with sulfide mineralization, deep thawing and the development of taliks (Fig. 11-5) can result because of the exothermic nature of sulfide oxidation. In addition, the resulting solutions will have a high ionic concentration.

Geochemical Methods In Permafrost Terrain (pp. 423-425)

On page 424 it was stated, "In conclusion it can be said that, in general, as was pointed out by Boyle in 1958, the occurrence of permafrost appears to present no insurmountable problems for exploration geochemistry except in obtaining soil samples and possibly in using vegetation." Since the above statement was made, much work has been done in northern Canada by exploration companies on the assumption that the statement was true, but there has been little published in the way of verification.

One of the few exceptions is the study of Allan (1974), which was concerned with trace metal dispersions and mechanisms of concentration of metals in various surficial materials, from a Mississippi Valley type Pb-Zn deposit located in a cold (Arctic) desert landscape in permafrost terrain. The Arctic desert landscape is typical of most of the Arctic Islands (Baffin Island is the major exception). The area studied is on Little Cornwallis Island, with over 1,000 feet of permafrost and a typical high Arctic vegetation, but no ice cap cover. (This small island is off the northwest coast of the island on which the settlement of Resolute is shown in Fig. 11-1; Little Cornwallis itself is not shown.) Allan (1974) studied distributions of Zn, Pb, Ag, Cu, Ni,

Co and Hg in stream and lake sediments, stream and lake waters, frost boils, and gossan soils.

The basic conclusions were that weathering and dispersion of trace metals from anomalous sources do occur in the Arctic desert landscape underlain by permafrost, and that exploration geochemistry is a valid tool for mineral prospecting in this somewhat unique area of Canada. Allan's (1974) work indicated: (1) for reconnaissance purposes, analyses of waters for Zn and Hg, and drainage sediments for Zn and Ag, may locate anomalous zones several square miles in area; (2) for more detailed surveys, Zn and Pb in stream sediments were more diagnostic; and (3) gossan samples analyzed for Pb and Zn reflected mineralization very close to the source.

More recent confirmation of the geochemical dispersion of elements in permafrost environments was presented by Cameron (1977) based on studies over the Agricola Lake massive sulfide deposit in the northwestern Canadian Shield. He concluded, in agreement with others previously, that: (1) permafrost is no deterrent to active oxidation of sulfide bodies (in fact, oxygen is more soluble in cold water than in warm water); (2) because of the exothermic nature of many oxidation processes, a continuing energy source is provided; and (3) in frozen ground, thin, intergranular water films allow chemical processes to be active even in winter. Further, he was able to establish the order of mobility in this particular acid (carbonate-free) weathering environment as: $Zn > Cu > Fe > As > Ag > Pb > Hg > Au$. This information allows the selection of indicator elements appropriate to reconnaissance, intermediate and detailed levels of geochemical exploration in such environments.

Dyck et al. (1976) studied the effect of frozen terrain on the migration of the gas ^{222}Rn. However, this gas is only moderately soluble in water which is believed to restrict its migration. In addition, the 3.8 day half-life of ^{222}Rn will further restrict migration. Their results, obtained in several parts of Canada, indicate that permafrost, snow, and ice do restrict the escape of radon. As a result, the sampling of air and waters below these frozen barriers, where increased contents of radon are found, can be an effective exploration technique for uranium in frozen terrain during the winter period. Similarly, Dyck and Tan (1978) found Rn values obtained from lake water under the ice during the winter in the Key Lake, Saskatchewan area were three times higher than those obtained in the summer.

Snow (pp. 425-426)

The use of snow in the search for uranium mineralization by analyzing it for radon has been discussed above (Dyck et al., 1976). This possibility had been suggested earlier (e.g., see Ketola and Sarikkola, 1973), along with the use of the bottom layers of snow for the measurement of gamma activity from ^{214}Bi.

WATER   (pp. 426-440)

Introduction   (p. 426)

At present, hydrogeochemical methods in Canada are mainly centered on lake water and lake sediments for both base metal and uranium exploration. Lake sediment geochemistry, in particular, has attracted much attention at the reconnaissance level of exploration. In those areas where there is a high density of lakes, such as the Canadian Shield, lake sediment sampling has superseded stream sediment sampling. Lake sediments represent a readily available, easily obtainable, low cost, source of information on general lithology and the occurrence of mineralization at the reconnaissance stage. In the Cordillera and in the Maritimes, on the other hand, relatively little attention has been paid to lake sediment methods because lakes are generally sparse and streams are very common. Where lakes are present, however, they do form a good sampling medium.

Groundwater   (pp. 426-428)

In those areas in which the density of wells is sufficient to obtain a representative coverage, groundwater can be a useful geochemical sampling medium. A case in point is the study by Dyck et al. (1976a) who collected and analyzed about 2,000 well water samples from an area in eastern maritime Canada of approximately 25,000 km^2 (1 sample per 13 km^2) for 20 constituents (elements, pH, suspended matter, conductivity and alkalinity). They were attempting to relate well water geochemical anomalies, on a regional basis, with known mineral occurrences and/or rock formations. The highlights of their findings were: (1) the elements U, Rn, He, F, along with conductivity and alkalinity, showed systematic regional patterns indicating broad regional belts of element enrichment which are presently being leached by groundwaters; and (2) the metals Zn, Cu, Pb, Mn and Fe showed positive correlation with each other but their spatial distribution is more spotty than that of the uranium-associated elements. Dyck et al. (1976a) observed that the former distribution of elements suggests a similarity to those found in the waters of some roll-front (Fig. 3-24) uranium mining districts in Colorado and Wyoming, whereas the latter group of elements could be explained in terms of known mineral occurrences. They also pointed out that the contrast between background and anomalous trace elements is greater in well waters, and that variations in the depth of the sampled wells (difference in aquifers) must be taken into consideration in evaluating well water (groundwater) data.

That the interpretation of groundwater data is difficult, yet possible, has been shown by Hoag and Webber (1976). Working with chemical analyses from surface and drill-hole waters, together with geomorphic, hydrologic and geologic evidence, they were able to locate the source of anomalous waters in the vicinity of an old mine in the Eastern Townships of Quebec. They were also able to show that the travel distance and location of the approximate source area of the water can be estimated from a knowledge of hydraulic head, sodium concentration of the water, and a constant which must be derived for the specific area.

River Water   (pp. 428-432)

No additional comments.

Lake Water   (pp. 432-438)

Lake water is collected and analyzed primarily in connection with exploration for uranium (although it is used on a limited basis for base metal exploration). A case in point is the study by Dyck and Cameron (1975) in which both uranium and radon were determined in the Lineament Lake area, District of Mackenzie, N.W.T. The concept is similar to that illustrated in Fig. 11-9, and discussed on pp. 430-431.

A few new studies have been reported in which vertical profiles have been obtained in lakes for important parameters (dissolved oxygen, temperature, pH, trace metal content, etc.) in order to determine lake characteristics such as stratification. One such study was made by Coker and Nichol (1975) as part of a larger survey of five lakes in western Ontario. Although the lakes were all in the same general area, they were found to be limnologically different. Some were stratified, others were not; some were oxygen-deficient on the bottom, the bottom waters of others contained oxygen; and so forth. Such variations would result in different environments which would not only affect contents of metals in the waters, but also in the bottom sediments. (These topics are discussed on pages 179-182 and 436-437.) Limnological instruments ("probes") capable of measuring these and other parameters simultaneously, are now available and are being used by some exploration geochemists.

Small lakes (2 km² or less) above the treeline are not stratified (because they are mixed by wind) and can be used for exploration (Cameron, 1978).

Lake Sediments   (pp. 438-440)

Although lake sediments are now widely used for reconnaissance exploration in the Canadian Shield and in northern Canada, there is no uniformity of opinion as to where and how to sample a lake for sediments, and how to interpret the results. This is, perhaps, proper considering the great number of variables involved (discussed on pp. 187-191 and 438-440), and each case must be considered individually. For example, in a recent study by Davenport et al. (1975) concerned with the detection of zinc mineralization in western Newfoundland on a reconnaissance basis (about 1 sample per 3 sq. km), it was found that the collection of organic-rich bottom sediments from the center of lakes was preferable. The zinc distribution strongly correlated with the distributions of iron and manganese which suggested co-precipitation, and zinc correlation with loss on ignition suggested chelation by organic matter. With respect to organic matter, Garrett and Hornbrook (1976) concluded that center-lake bottom sediments whose organic content is dominantly in the range of 12 - 50% LOI (loss on ignition) form the most effective sample media for regional lake sediment surveys (based only on studies with zinc).

Coker and Nichol (1975) reported similar, but by no means identical, results to those of Davenport et al. (1975). In western Ontario lakes Coker and Nichol (1975) found that Zn/Mn and Ni/Mn ratios, because of adsorption or co-precipitation of Zn and Ni by Mn, were essential in dealing with sediments from the center of lakes to discriminate between lakes adjacent to sulfide mineralization and those in barren terrain.

In view of the above differences, the comments and advice of Bradshaw (1975) are particularly appropriate and worthwhile. With regards to lake sediments he said (p. 204), "the collection of lake samples is more difficult to standardize at this time because the controlling factors for their use in exploration are not yet fully documented. However, as a minimum, sampling should include collection of mineral matter from the margins of the lake in shallow water, and also from the deeper parts of the lake. If the mineralization occurs under the lake then a regular grid of samples from the lake bottom is recommended. Both anomalous and background lakes must be included." Unless an orientation survey indicates otherwise, it would appear that Bradshaw's (1975) conclusions and advice should be followed, for the present at least, to ensure that an anomaly is not overlooked.

Coker et al. (1979) have prepared the most extensive review to date on the subject of lake sediment geochemistry applied to mineral exploration. Careful study of this paper, which expands upon many aspects of lake water and lake sediments covered in this book, is essential for those seriously engaged in mineral exploration using these media. Equally important is the review by Dyck and Miller (1979) which is concerned with the more general topic of the application of hydrogeochemistry to the search for uranium and base metals.

MUSKEG AND BOGS   (pp. 440-451)

Definition And Occurrence   (pp. 440-442)

No additional comments.

Bog Water Chemistry   (pp. 442-443)

No additional comments.

Origin Of Bogs   (pp. 443-444)

No additional comments.

Geochemical Studies   (pp. 444-449)

Bradshaw (1975) published three idealized models depicting the geochemical dispersion of mobile elements in the vicinity of bogs, and the characteristics of the anomalies which would result in the bogs themselves and in the underlying materials (e.g., till). The three models are: (1) a bog in the Canadian Cordillera overlying till, with mineralization to one side; (2) the same as (1), but with mineralization underneath the bog; and (3) the same as (1), but in the Canadian Shield. Fig. 11A-1, which can be considered representative of the others, illustrates the first of these idealized models.

From Fig. 11A-1 it can be seen that in the bog itself the magnitude of the geochemical response tends to be much higher than in the well drained soils to one side (designated SL (R and M)), using both total and partial extractions, giving values which can reach 1,000 ppm or even the percent range (see Fig. 5-2, p. 229). When the source lies directly underneath the bog (case 2 above), values tend to be exceptionally high in the overlying bog, up to the percent range for the mobile elements (e.g., copper), as Gunton and Nichol (1974) showed in their studies in the Whipshaw Creek area in southern British Columbia. Because of the strength of these bog anomalies, Bradshaw (1975) stated that it is mandatory to treat bogs and well-drained soils separately when calculating thresholds, otherwise genuine anomalies in well-drained ground may be regarded as background fluctuations because their magnitude is so small by comparison with values in the bogs. A case in point is the Valley Copper deposit in the Highland Valley, British Columbia, where an anomaly in well-drained ground over mineralization is only 75 ppm against a background of 20 ppm, compared with anomalies in the bog downslope as high as 4,000 ppm. This example is the basis for Fig. 5-2 (although it was not identified as such originally).

When the source of the anomaly lies to one side of the bog, anomalies in soil profiles decrease sharply downward. This is seen in the till in Fig. 11A-1 where only background concentrations are detected, except very close to the top. However, when the source lies directly under the bog (case 2, above), the values tend to remain the same in both the bog and the till, and may possibly increase in the latter. This was also shown by Gunton and Nichol (1974). Clearly, there is an advantage to sampling till under an anomalous bog to determine whether the source of the anomaly lies to one side or underneath.

To this point, it has been assumed that bogs are relatively uniform in the nature of their organic constituents (e.g., see Fig. 11-15), and that their peat sections will adsorb elements uniformly. However, Eriksson (1973) has shown that the above assumptions are not necessarily true. Working in bogs formed on till in central Sweden, she found that the types of peat within one bog can vary greatly within a

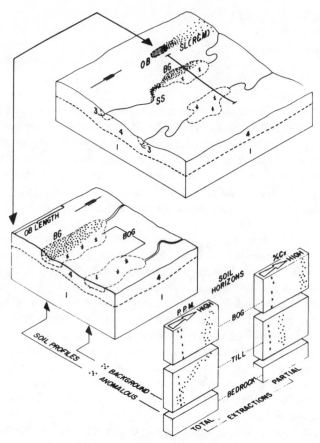

Fig. 11A-1. Idealized model for geochemical dispersion of mobile elements in bogs overlying till, with mineralization to one side of the bog. OB = ore body. BG = bog anomaly. SL (R & M) = soil anomaly, residual (R) and mechanically smeared (M) (by glacial action). SS = stream sediment anomaly. Arrow = direction of glacial movement. 1 = bedrock. 2 = residual soil (not present). 3 = recent alluvium. 4 = till. Intensity of the dots (stippling) indicates the strength of the anomaly. From Bradshaw (1975).

relatively short distance (certainly within 100 m). This is shown in Fig. 11A-2 where three different kinds of peat are found in varying amounts in bog profiles taken 150 m apart from the same bog. (The bog is about 100 m from Zn, Cu and Pb mineralization and, therefore, is an actual example of the idealized situation illustrated in Fig. 11A-1.) Further, the element distribution will vary greatly as shown in Fig. 11A-3. Zinc anomalies have their maximum values in the southern (left) parts of the N-S bog section, but there are also two isolated areas with high contents of this element. Similar anomalies were found for copper (not shown), but no significant anomalies of this element reached the surface even in the southern end of the bog. Lead values are comparatively low throughout the bog profiles with the highest concentrations found on the surface. In addition, comparison of the distribution of metals in the peat with the distribution of different kinds of peat, revealed no correlation between the kind of peat and the Zn or Pb distributions (compare Figs. 11A-2, 3). On the other hand, Cu and Fe (not shown) appeared to be low in areas of *Sphagnum* peat, contrary to what would have been expected on the basis of the discussion on p. 443.

Several recent studies involving the use of bog materials (i.e., peat) in Scandinavia are worthy of note. Eriksson (1976) pointed out that in central Sweden, where the relief is low and the surface is covered with ablation till with many hummocks and depressions in which there are bogs, stream sediment sampling is neither useful nor practical. Regional prospecting by use of peat sampling is a proven alternative to stream sediment sampling (10 prospects were found using peat). About 175 samples are collected at an average density of 7 per km². Based on studies of the distribution of elements (mainly Zn and Cu) carried into bogs by groundwater, samples are collected: (a) as close to the border of the bog as possible at a depth of 2 m; and (b) at the bottom of the organic-inorganic interface in the pure organic matter (core samplers or equivalent would be required). The samples are analyzed after drying, and a correction is made for the inorganic content of the peat.

Tanskanen (1976) studied the distribution of seven elements (V, Ni, Cr, Cu, Co, Zn and Pb) in 103 peat profiles from central Lappland, Finland. He found that each of four characteristic variables in the peat profiles has a different effect on the metal

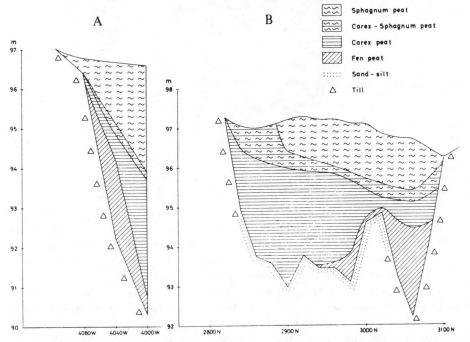

Fig. 11A-2. Distributions of different kinds of peat in two profiles (A and B) taken about 150 m apart within the same bog (formed on till) in central Sweden. Zn, Cu and Pb mineralization occurs about 100 m from the nearest point in either bog. From Eriksson (1973).

abundances. These variables are: (1) pH; (2) degree of humification; (3) percentage of
ash; (4) relative depth. Pb-Zn, Cr-V, and Co-Ni constitute analogous pairs as far as
their behavior in peat is concerned, but each pair has a characteristic behavior in peat
bogs.

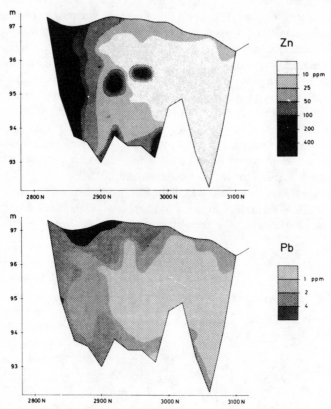

Fig. 11A-3. Distribution of zinc and lead in bog profile B shown in Fig. 11A-2. From Eriksson (1973).

    Larsson (1976) pointed out that bog sampling can be easily carried out during
the winter months, thus helping to maintain the continuity of work on exploration
projects, and employment of sampling crews and laboratory personnel. Fig. 11A-4
shows results obtained by Larsson (1976) during a follow-up survey in the winter in
the Pajala district in northern Sweden, when the ground was covered with up to a
meter of snow. From this idealized figure the sharp contrast in copper content between background and anomalous peat samples can be seen at all depths (50 - 70
ppm Cu in the background bogs, and more than 2,000 ppm in the anomalous bogs).
    And, recently, Boyle (1977) described in detail two cupiferous bogs in the
Sackville area of New Brunswick in which the organic matter contains from 2 to 6 %
copper. The copper, which occurs as copper humate(s), is derived from springs carry-

ing copper ranging from 5 to 1,000 ppb. The metal probably originates from the leaching of cupiferous (chalcocitic) deposits in grits and conglomerates. Boyle (1977) reviewed the occurrences of copper bogs in many parts of the world and pointed out that such bogs are good geochemical indicators of cupiferous deposits.

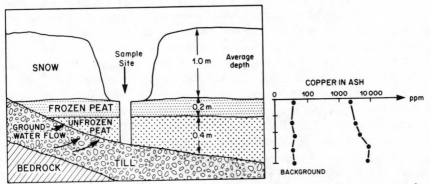

Fig. 11A-4. Idealized model illustrating the results of sampling bog margins in winter. Based on example from Sweden. After Larsson (1976).

Finally, it is important to note some of the limitations of bog sampling. In this connection, the report by Gunton and Nichol (1974) who conducted profile sampling for Cu, Fe, Mn, organic matter and pH in several organic-rich samples overlying till in southern British Columbia which were in the vicinity of copper mineralization, can be cited. The results of their study showed that, although strong and extensive anomalous copper values were obtained in swamps near the mineralization, basal till sampling gave more significant and useful information with respect to outlining the mineralization. In other words, the sampling of swamps in this area appears good for reconnaissance purposes (that is, to indicate generally favorable areas), but leaves something to be desired during the detailed (follow-up) stages.

Bog Iron And Manganese   (pp. 449-451)

No additional comments.

GLACIATED AREAS   (pp. 451-467)

Introduction   (pp. 451-452)

In the first edition there is the statement (p. 452) that "At the present time great strides continue to be made in understanding the optimum procedures which must be employed, and in developing methods of interpretation of geochemical data from glaciated areas." That statement is as true today as it was when it was written. Fortunately, very favorable results have been achieved in recent years from geochemical surveys conducted in glaciated areas which are covered with drift. As a result, there is now a tendency to consider glacial deposits, particularly till, in terms of indicators of mineralization rather than obscurers of mineral deposits (Shilts, 1976, Szabo et al., 1975). Indicative of this change of attitude is the large number of conferences and symposia dedicated to furthering the understanding of glacial sediments and prospecting in glaciated terrain. Several of these have been published as symposia proceeding or special volumes (Jones, 1973 and 1975; Bradshaw, 1975; Kauranne, 1976; IMM, 1977; Govett, 1977). Much of the discussion which follows in this section is based upon information presented by Shilts (1976) at one such conference in Canada because of the new, interesting and useful concepts he has proposed. However, as in the earlier edition, where advisable, new information is drawn from Scandinavian, Soviet, and United Kingdom experiences. The serious student of glacial processes will be particularly interested in the volumes edited by Jopling and McDonald (1975) and Legget (1976) which contain many papers on fundamental aspects of glacial geology.

Definitions And Basic Principles (pp. 452-455)

The implication of Fig. 11-19 and accompanying statements is that ablation till must be avoided because it is composed of material derived from a distant source(s). Specifically, with respect to overburden drilling, the statement was made (p. 460), " . . . the samples must be taken in lodgment till". Although this may be generally true, Shilts (1976) has pointed out two cases in which ablation till may be of local origin. Such cases are: (1) in mountainous areas, where ice tongues flow through valleys or the glacier is pierced by nunataks, debris may slide or roll onto the ice from adjacent rock slopes (see Fig. 11-23); and (2) where ice flows regionally over mountainous terrain, topping the highest ridges by a considerable amount, debris may be supplied throughout a thickness of ice corresponding to the maximum relief of the bedrock surface. (Apparently, this mode of entrainment and transport is particularly common in the Appalachian region of eastern Canada and the United States, where the amount of englacial debris that is ultimately deposited on the lodgment till is generally proportional to the ruggedness of the terrain.) Depending upon the amount of the above-mentioned relief, distance of glacial advance before the process of ablation begins, and many other factors, it is possible for ablation till to be composed of material of very local origin. Thus, both lodgment and ablation till may be locally derived and, in such cases, the ablation facies is likely to preferentially reflect the resistant rock types that often form hills, and the lodgment facies will likely represent locally less resistant rock types that form valleys. Finally, it is important to note that in areas of strong relief (any type of glaciation), it is possible for debris carried in the lodgment position (along the base) to be carried long distances along the valleys.

Glacial Dispersion And Provenance   (pp. 455-458)

Shilts (1976) has observed that glaciers appear to disperse material in the form of a negative exponential curve (Fig. 11A-5); that is, there is a concentration of elements (minerals, heavy clasts, etc.) reaching a peak in the till at or close to the source, and this is followed by an exponential decline in the direction of transport. The slope and dimensions of the curve appear to be determined by the physical characteristics of the components and the mode of transport. That portion of the curve which exhibits a rapid increase and decrease in geochemical concentration at and immediately down ice from the source, is called the "head", whereas the "tail" exhibits a gradual decrease, in some cases for great distances (a tail of about 30 miles is shown in Fig. 11A-5). Fig. 11A-6 shows a dispersal tail and head in perspective, but recognizable tails as long as 60 km, as shown in this figure, are unusual. More likely, the dispersal

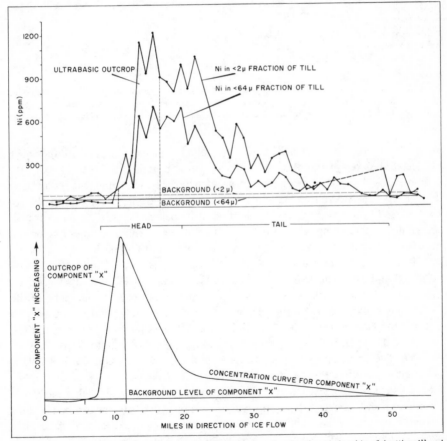

Fig. 11A-5. Actual (top) and idealized (bottom) dispersal curves showing the relationship of the "head" and "tail" of the negative exponential (glacial) curve. From Shilts (1976).

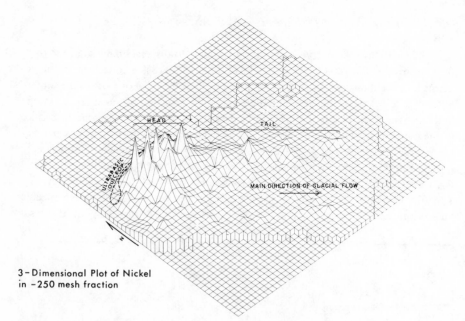

3-Dimensional Plot of Nickel
in -250 mesh fraction

Fig. 11A-6. Perspective plot of nickel values in the −250 mesh (64μ) fraction of surface till in the Thetford Mines, Quebec, ultrabasic indicator train, showing graphically the head and tail regions of dispersal. The tail is easily traceable for about 60 km in the main direction of glacial flow (SE). From Shilts (1976).

tail is difficult to detect at great distances. Detection of the tail is dependent upon such factors as the area of the dispersal train (which is proportional to the area of the source exposed to glacial erosion), and how distinctive the component sought is in the barren, non-mineralized material of the dispersal area. Thus, for example, Cr or Ni dispersed from ultrabasic rocks might be easily detectable great distances from a source if the main component of the associated till were dolomite, limestone or granite, because such rocks have extremely low contents of Cr and Ni.

In two more recent papers, Shilts (1977) and Klassen and Shilts (1977) have amplified and extended the head and tail concept, and applied it to exploration for uranium and base metals. The details of these papers are beyond the scope of this review other than to say that the purpose of introducing the tail and head concept is largely to explain the rationale of describing a sampling program as "reconnaissance" or "detailed" in glaciated terrain. The scales of both types of sampling programs are variable. The sampling spacing at the reconnaissance stage is such that it is able to "guarantee" detecting the tail of a size typical for a particular type of mineralization, whereas the sampling spacing at the detailed stage is close enough to find the source within a head. Ultrabasic rocks and porphyry copper deposits, for example, would probably have large tails and the reconnaissance sample spacings could be miles apart. Stratabound sulfide deposits, on the other hand, may have small tails requiring samples at very much closer spacings (e.g., at a fraction of a mile). Thus, "reconnaissance" means finding parts of tails; heads are found only accidentally. "Detailed" is sampling where a head should be, based on a target indicated by the trend of the tail (or by some other means).

Although mapping the distribution of ore-bearing boulders is probably still the most common direct use of till in mineral prospecting, it is certainly not the panacea for Canada. This procedure has been very successful in Finland where non-geologists send about 10,000 samples annually to the Geological Survey and mining companies, on the basis of which, ten mines have been found (Hyvarinen et al., 1973). To be generally effective, a high rural population density (as in Finland and Scandinavia) is desirable so that non-geologists may be trained to spot and report ore-boulders uncovered during construction and working of the land. "The chances of the relative handful of geologists engaged in exploration ever constituting an effective boulder or mineral tracing force using visual methods, or even dogs, are slight" (Shilts, 1976, p. 208). However, once mineralized boulders are found, then boulder tracing is valuable. This technique has been an important factor in locating several ore deposits in Canada, including at least two uranium deposits (Rabbit Lake and Key Lake) in Saskatchewan.

Modern methods of boulder tracing in Finland, where the techniques have been particularly successful, have been described by Hyvarinen et al. (1973). Among the points they mention is the fact that a train will be narrow if the boulders were transported by only one glaciation, but it may be very broad if it was affected by several glacial advances (Fig. 11A-7). Further, the length of the train depends on the amount of material transported, the area of outcrop exposure, and the abrasion resistance of the material. Other factors being equal, easily weathered rock will yield a shorter boulder train than mechanically and chemically resistant rock. Hyvarinen et al. (1973) noted that the most difficult task in boulder tracing is finding the source of what might be called the "lone boulder" — ore boulders which occur in an area where there are no other boulders of the same type. Although at least one such case resulted in the eventual discovery of a mine (the Hallinmaki copper mine) with the assistance of additional information over a 10-year span, the sources of many lone boulders still remain obscure.

Exploration Using Till   (pp. 459-463)

Till at any given site can exhibit considerable vertical variation in its components. One obvious case is when ablation and lodgment tills are sampled at one location (Fig. 11-19). Another cause of such variations may be due to an included zone of sulfide grains within a dispersal train (Fig. 11-20). Variations such as shown in Fig. 11-20 certainly do occur, but the direction and thickness of such sulfide-containing zones in till are likely to vary in complex ways, sometimes corresponding to bedrock irregularities, and other times to rising relic shear planes in the till. These features are illustrated in Fig. 11A-8, and are probably particularly common in the head zone

(Fig. 11A-5). In the tail area of a dispersal curve, the indicator components of mineralization (or their oxidized products) are probably more evenly distributed vertically through any one facies of a till sheet than elsewhere.

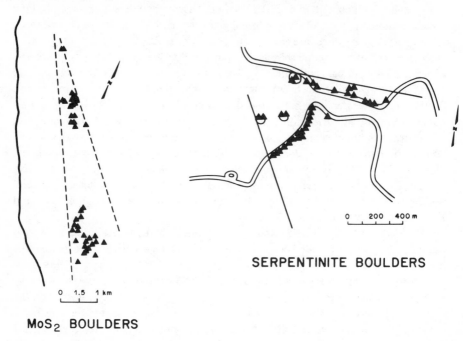

SERPENTINITE BOULDERS

MoS$_2$ BOULDERS

Fig. 11A-7. Narrow boulder train produced by one glacial movement (left), and broad boulder train produced by at least two separate glacial movements (right). After Hyvarinen et al. (1973).

Vertical chemical variation which may be important in any single till sheet can result from the oxidation and subsequent leaching of certain components, specifically sulfides. This possibility was mentioned in connection with the removal of sulfides from eskers and tills (p. 458), and as the obvious source of hydromorphic and cold extractable anomalies in till (pp. 461-463). However, Shilts (1976) has stressed the importance of the likelihood of the partial or complete destruction, by weathering, of transported sulfide grains and mineralized pebbles or boulders in till (assuming the till is not overlain by peat, peaty alluvium, or by water, all of which will inhibit weathering by limiting the oxygen supply). This is well illustrated in Fig. 11A-9. Post-glacial destruction of glacially transported sulfide grains and mineralized fragments generally occurs in the upper 2 - 5 meters of till, and within the active layer in permafrost regions. *In till that has been weathered, the clay-sized fraction should be separated and analyzed for elements of interest, as the clays are likely to have adsorbed the mobilized cations (metals) in a manner analogous to the mechanisms operative in the B horizon of soils.* Fig. 11A-9 convincingly shows that the metal contents of clays are higher than the metal content of the coaser (heavy) fractions in the zone of oxidation

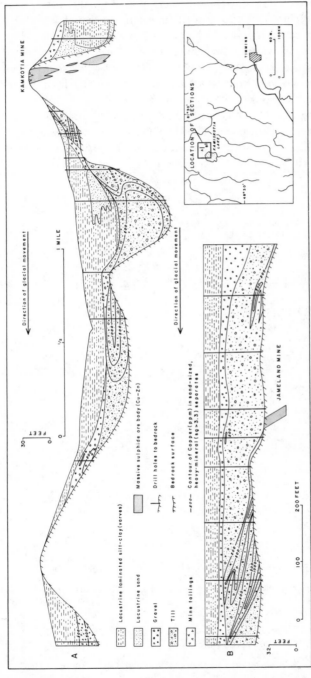

Fig. 11A-8. Cross-sections of glacial deposits in the Timmins, Ontario area of the Canadian Shield showing sheet-like zones of high copper concentration in the down-ice direction from copper sulfide-bearing ore bodies. Samples were collected and analyzed at four-foot vertical intervals. From Shilts (1976).

of the till, whereas in the unoxidized (lower) zone of till, the reverse situation prevails. In the oxidized zone, partial extraction techniques should be useful in enhancing the contrast between anomalous and background samples. It is important to
note, however, that oxidation and other types of chemical weathering have not
everywhere advanced to the stage where all evidence of sulfide mineralization is lost;
if they had, there would be no basis for sulfide ore boulder tracing, "dog sniffing" (p.
30), and other techniques which have been used successfully in areas of glacial drift.
In a recent study around Mt. Pleasant, New Brunswick, Szabo et al. (1975) found
that useful information could be obtained from *both* the fine and coarse fractions of
till. At this locality, cold extraction analyses of the fine fraction (−80 mesh) for Cu,

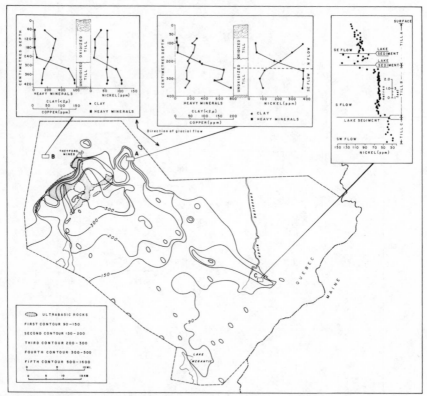

Fig. 11A-9. Dispersal from ultrasbasic outcrops in the Thetford Mines area, Quebec.

Section A illustrates the postglacial destruction of Cu and Ni sulfides in oxidized till. It also illustrates a reverse
in the Ni content in clay in the upper part of the till caused by a late glacial shift in ice flow from SE to N; compare Ni and clay in Sections A and B.

Section B illustrates the normal weathering profiles for Ni and Cu outside the dispersal train. Cu in the clay is
high whereas Cu in the heavy minerals is low in the oxidized layer.

Section C illustrates the Ni content of three tills from a section about 50 km down-ice from the Ni source. Only
the upper till was deposited by ice flowing consistently from the Thetford Mines area. From Shilts (1976).

Pb, and Zn (elements adsorbed following weathering of sulfides) outlined dispersal trains 2 - 5 km from the mineralized source, whereas analyses of the sulfide-containing coarse fraction resulted in the recognition of the dispersal train for more than 16 km. Although the analysis of the coarse fraction in this area is clearly a superior reconnaissance-scale technique, such may not be the case elsewhere. Accordingly, unless an orientation survey or other information indicates otherwise, it seems advisable to analyze both the fine and coarse fractions of till for the best results. In an analogous manner, Brundin and Bergstrom (1977) found that certain heavy minerals resistant to chemical weathering (e.g., scheelite, wolframite, cassiterite and chromite) separated from till could be used for regional prospecting. This procedure may also be applicable for the detection of sulfide mineralization in favorable situations (see "heavy clasts" p. 456).

The final cause of vertical variations in till at a given site to be discussed here, can be explained by variations in the direction of ice flow, with time, during the deposition of a single till sheet. Such a shift is illustrated by the sharp increase in Ni in the clay near the top of Fig. 11A-9 (Section A). At this site, ultrabasic debris which was transported toward the SE was re-directed northward by a late glacial shift in flow (caused by downdraw as the sea flooded up the St. Lawrence valley). Somewhat analogous would be the case in which several distinct till beds (or alternating till and stratified drift, such as glaciolacustrine sediments) were laid down at one specific sampling site. Three such tills are illustrated in Fig. 11A-9 (Section C). Kokkola (1975) and Kujansuu (1976) have reported six different till sheets in central and northern Finland and the ice direction varied for each. Hirvas (1977) determined that there were at least five different stages of glacial transport in the area of northern Finland he studied, and the till from each differ in age, composition, and direction and distance of flow.

Even though glacial processes are reasonably well understood, and the fact that certain aspects of the dispersal of ore materials (i.e., boulders, weathering products) in till have been known for a long time, Shilts (1976) believes that drift prospecting using till is still in its infancy. Much more research, such as that by Hirvas (1977), is needed into the modes of entrainment, transportation, and deposition of glacial sediments. Dispersal curves for various rock and mineral types need further elucidation. Methods of sample and data processing and interpretation, for all major glacial and climatic landscapes, need further study. As information of this type is collected (such as the many case histories assembled by Bradshaw, 1975), geochemical exploration in areas of glacial cover will become ever more successful. And, finally, it is important to mention that Bolviken and Gleeson (1979) have recently produced an excellent review of the principles, application and techniques of "soil" (here defined as any type of glacial overburden except recent stream sediments and other drainage channel material) geochemistry as they apply to glacial terrain. All of the concepts mentioned here, as well as others, are discussed in a very thorough manner.

In closing, it is worth mentioning that because of the complexity of glacial deposits and dispersion in many areas, the interpretation of geochemical data is difficult. Accordingly, geologically favorable areas should not be eliminated from consideration solely on the basis of geochemical data obtained from glacial overburden.

Comparison Of Continental And Valley Glaciers　(pp. 463-465)

Levinson and Carter (1979) have published one of the few recent studies on the use of geochemical methods in areas of alpine (valley) glaciation. They employed profile sampling of the glacial overburden from 14 locations in the Babine Lake area of British Columbia in which there are several porphyry copper mines. The area experienced several directions of ice advance during the Pleistocene, and the glacial deposits range from coarse, gravelly till to glaciolacustrine deposits (varved clays). Because of the complexity of both the dispersion and glacial deposits, interpretation of the data was difficult. Nevertheless, for the area they studied they concluded: (1) dispersion is primarily (about 80%) mechanical and a total extraction analytical technique should be used; (2) the overburden material with the highest partial extractable metal content (hydromorphic transport) is the silty-clay glaciolacustrine sediments; (3) secondary dispersion of metals is primarily by means of groundwater beneath the tills; (4) mechanical dispersion of sulfides can be detected in the tills up to about 1 mile down ice from the deposits; (5) the same concepts applied to exploration in areas of continental glaciation are applicable in regions with alpine glaciation; and (6) because of the difficulties in using geochemical methods in the glacial overburden of central British Columbia, areas with mineral potential should not be eliminated from further consideration solely on the basis of negative or inconclusive geochemical data obtained by sampling these materials.

Masking　(pp. 465-466)

No additional comments.

Soil In Glaciated Areas　(pp. 466-467)

No additional comments.

$12A/$ The Statistical Treatment of Geochemical Data

R. B. McCammon
U.S. Geological Survey
Reston, Virginia 22092

INTRODUCTION (pp. 469-470)

There is little doubt that the statistical treatment of geochemical data in the past five years continues to have an important role in the interpretation of data collected in geochemical surveys. Its importance can be measured by the increased difficulty of finding the "easy" targets and the need for greater objectivity in devising sampling programs. Moreover, advances in the methods of acquiring geochemical data have produced an amount of data which has far surpassed the amount collected in earlier years. Coupled with greater access to time-share computing and the rapid growth in computer graphics, the result has been to produce among exploration geochemists a greater working knowledge of the basic statistics required to interpret geochemical data. An indication of the progress which has been made in the last five years is the routine incorporation of the more advanced techniques such as factor and regression analysis to data interpretation. In large part, this has resulted from the recognition that such advanced techniques can, in fact, uncover subtle though significant features in multi-element data that are not apparent in the untreated data (Closs and Nichol, 1975). Clearly, as orebodies become harder to find, such subtle features take on significance.

Developments in statistical techniques in the past five years are discussed in the following sections.

BACKGROUND VALUES (pp. 470-473)

Perhaps the main reason for establishing background values of trace element contents in a region is so that anomalous contents of the element can be recognized. In many instances, what is observed in studying the frequency distribution of the content of a given element for samples collected in an area is the presence of one or more subpopulations. Cumulative frequency plots are especially useful for identifying such mixed populations. To assist in this problem, Sinclair (1976) has described graphical techniques for separating mixtures of populations observed in geochemical exploration data. The generalized graphical approach to dissection is shown in Fig. 12A-1.

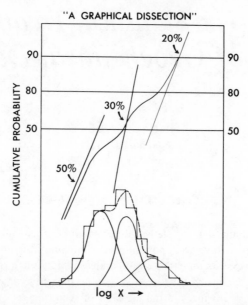

Fig. 12A-1.   A generalized graphical dissection of a mixture of three lognormal distributions. From McCammon et al. (1976).

As most element distributions are lognormal, such data can be plotted on arithmetic normal probability graph paper. Fig. 12A-1 represents a mixture of three lognormal populations. Slopes of different parts of the distribution curve can be used to estimate standard deviations for each of the subpopulations. Inflection points of the curve can be used to estimate the proportion of each subpopulation in the mixture. In the example, inflection points occur at the 50 and 80th percentiles which indicate that the three subpopulations contribute 50, 30, and 20 percent, respectively. The mean of each subpopulation can be estimated by taking the percentile midway between the adjacent inflection points. Thus, by purely graphical methods, it is possible to dissect mixtures of lognormal distributions. More recently, McCammon (1976) devised a combined graphic-analytic approach for dissecting mixtures which takes advantage of the more accurate analytical methods while preserving the greater flexibility of the graphical approach. In a study of geochemical populations in rocks at Ely, Nevada, McCammon et al. (1979) identified mixtures composed of two subpopulations in distributions for copper, zinc, antimony, lead and silver in mineralized samples. These results suggested two major stages of mineralization. The ability to dissect such mixtures of distributions easily and accurately offers a powerful tool for exploration.

REGIONAL VARIATION   (pp. 474-480)

The most important purpose of low-density regional geochemical reconnaissance is to delineate broad-scale patterns of element distributions. These patterns provide the basic framework for interpreting anomalous concentrations observed in more detailed follow-up investigations. To more effectively display regional patterns, Lowenstein and Howarth (1973), as mentioned earlier (p. 502), introduced automated color mapping for multi-element data. Applying the approach, Webb et al. (1973) have produced a geochemical atlas of northern Ireland using a series of grey-scale maps based on concentration intervals for twenty different elements. An example for one of these elements, lead, is shown in Fig. 12A-2. It has subsequently been

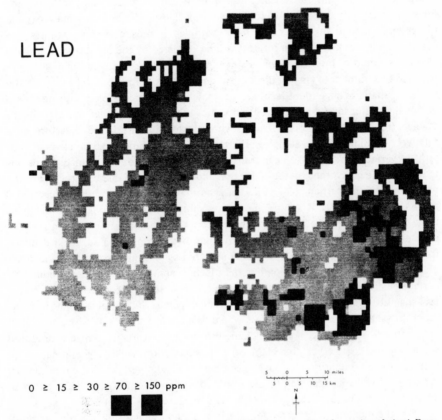

LEAD

0 ≥ 15 ≥ 30 ≥ 70 ≥ 150 ppm

5   0   5   10 miles
5   0   5   10 15 km
N

Fig. 12A-2.   Grey-scale map based on logarithmic concentration intervals of lead for northern Ireland. From Webb et al. (1973).

determined that the anomalous areas of high concentration generally are related to mineralization. More recently, Mancey and Howarth (1978) have produced factor score maps of regional geochemical data from England and Wales (Webb et al., 1978). These maps, based on moving-averaged smoothed regional patterns for 21 elements, reflect the overall geochemical trends. In the color-combined maps, a relatively simple picture of the regional geochemistry emerges with patterns reflecting the principal rock types. Regions of contamination from mining or industrial activity have also been identified. Such information is of invaluable assistance in evaluating current programs and for planning future national geochemical surveys.

CORRELATION BETWEEN ELEMENTS   (pp. 480-487)

In many instances, the observed correlation between any two elements based on a large number of samples can be misleading. It can be misleading because of the inherent variability and often heterogeneous background versus anomaly populations represented in any large collection of samples. In most correlation studies, it is assumed that areas of greatest interest will be reflected by the highest values of the element content of interest. Thus, a single homogeneous population is assumed as opposed to a number of smaller distinct subpopulations. Botbol et al. (1978) have described a spatially dependent technique in which importance is placed on measured values that differ locally from measured values at neighboring locations. Geochemical occurrence models derived from such consideration are based therefore on a relatively small number of samples. The technique is designed to treat element contents as variables which have been transformed to ternary form, in which +1 means that a variable is favorable in the sense that a location is favorable with respect to the occurrence of the particular model, -1 means that a variable is unfavorable, and 0 means that a variable is of indeterminate value for determining favorability or unfavorability with respect to a model. The ternary transformation is performed for each element by taking the second derivative of the contoured surface calculated from a set of data. Positive anomalies considered as favorable are values which are greater than the values corresponding to inflection points defined by a mathematical surface whereas negative anomalies considered as unfavorable are values which are less than the values corresponding to inflection points. Indeterminate values correspond to values corresponding to areas where data are lacking or else where the values are relatively nonvarying. For each element treated as a variable, a map is produced which identifies whether the element is positively or negatively anomalous or indeterminate for appropriately sized geographic cells defined inside the area. Spatial patterns derived with this method are amplitude independent in that the height of an anomaly has no influence on an area being considered anomalous.

Geochemical occurrence models are formulated by choosing a particular set of elements in a given geographic area. These models usually are chosen in mineralized areas so that the model becomes a "fingerprint" of a particular type of deposit. The weight assigned to each element which comprises the model reflects the degree of mutual overlap of anomalies with other model elements in the mineralized area. Thus, local as opposed to regional correlation is the controlling factor. Once a model is formulated, region cells outside the model area can be evaluated in terms of similarity to the model. Using this method, Botbol et al. (1978) found that an aggregated mineralized model comprised of the major mining areas of the Coeur d'Alene district, Idaho led to the result that nonmodel cells located between model cells exhibited high degrees of association following structural restoration. Not only did this confirm continuity of mineralization for the structural interpretation of the region but also it resulted in the delineation of areas of high potential for future exploration.

CLASSIFICATION OF SAMPLES (pp. 487-491)

It is now recognized that misleading results can be obtained if the effects of differing geochemical behaviour of trace elements within a survey area are ignored. For instance, a regional survey area may be chosen independent of geologic considerations. In such situations, when considering a sample analyzed for p elements from n localities, it is of interest to know if the n localities fall within some natural geochemical grouping and, if so, how such groups are distributed geographically. To answer such a question, Sinding-Larsen (1975) has devised a computer method for dividing a regional geochemical survey area into geochemically homogeneous subareas to facilitate statistical interpretation. The method is based on a statistical measure of partitioning a p-variate population into k-group subpopulations (k<p) based on n samples. The k-group subpopulations are chosen such that the p-variates within each subpopulation are clustered as tightly as possible to the relative multivariate distance between the various k-groupings. The method was applied to 482 samples of humus soils collected from an area in central Norway. Each sample was analyzed for nineteen chemical elements. The resulting classification of the 482 samples into 4 geochemical subgroups led to the recognition of subareas with homogeneous geochemical subpopulations. The difference in readily soluble potassium and phosphorus, pH, ash content and chromium, cobalt, iron, molybdenum and nickel served to identify four geographic areas with respect to threshold values and environmental effects. Thus, each of the four areas could be considered separately in evaluating local versus regional geochemical heterogeneity.

SURVEY SAMPLING (pp. 492-496)

Although the design of geochemical surveys continues to be based largely on rule-of-thumb practices, there is considerable evidence which suggests that exploration geochemists are turning toward more quantitative design criteria. In large part, this is due to greater recognition among exploration geochemists of the importance before commencing a sampling program to have some prior knowledge of the size and shape of the targets being sought. As noted earlier in Figs. 5A-3 to 5A-5, Sinclair (1975) has shown how geometric considerations can aid in determining optimal sample spacings for assumed sizes and shapes of anomalies. More recently, Garrett (1979) and Garrett and Goss (1979) have described sampling procedures for conducting regional geochemical surveys in which particular attention is paid to optimal sample spacing and collection. Their objective is to reduce the cost of surveys without affecting their quality.

One of the techniques proposed is an improved analysis of variance method based on an inverted-nested sampling design. To appreciate the improvement, it is necessary to recall that in more conventional sampling progressively greater numbers of samples are required at the lower levels of sampling — that is, at individual localities. For the proposed technique, an inverted design has been implemented in which the degrees of freedom are concentrated at the top of the design. This results in the determination of the sampling variance precisely at the regional level. The resulting estimated variance components provide useful information for interpreting the geochemical trends simultaneously allowing evaluation of laboratory analyses. Using the results obtained in a regional reconnaissance survey for uranium in Canada, Garrett and Goss (1979) demonstrated that an earlier two-stage analysis of variance could be replaced by a single analysis using the inverted-nested design. The latter, it is argued, is relatively easy to administer, is reasonably statistically efficient, and useful for supplementary analyses.

Finally, special mention must be made of a syllabus for a short course on sampling designs for geochemical surveys prepared by Miesch (1976). This is perhaps the most complete offering of the general principles that can be applied to the development of sampling designs suited to particular needs. The main point to be made, according to Miesch, is that proper emphasis on requirements that the data must satisfy comes before any meaningful statistical analysis. To rephrase this point, one cannot collect data and then think about the statistics. The main consequence of this reasoning is that randomization is an essential element of any sampling design. In particular, any applications associated with analysis of variance methods are based on this assumption. Although the notion of random sampling disturbs many geochemists, it offers the only path if the results of statistical methods applied to geochemical data are to be taken seriously. It should be noted, however, that the more formal statistical approaches have proved most effective in establishing regional geochemical trends from the data whereas the less formal but more robust statistical approaches have proved most effective in the recognition of anomalies.

ANOMALY DETECTION   (pp. 496-503)

The main objective in exploration geochemistry continues to be the detection of anomalous patterns related to mineralization. Current efforts have focused on statistical techniques for anomaly recognition. The key has been the recognition of local anomalies which in general are superposed on regional trends. What complicates the issue is the observance of statistical noise in geochemical data which only contributes "false" positives in the search for geochemical anomalies. It is for this reason that robust methods for locating anomalies have been developed (Bement et al., 1977; Koch and Link, 1977; Conover et al., 1977). The general problem is one of multivariate statistics. The values for one or more chemical species contribute to finding the sought-for anomalies but rarely do values for all species. Instead, the values for one or more of the species add "noise" to the statistical analysis and obscure or conceal the anomalies that could be found if the excess data were deleted. The problems that arise, therefore, are these: How can an anomalous value best be defined? How many anomalous values are needed to define an anomaly? Are the anomalous points clustered? These are some of the questions which the following models have attemped to answer.

As part of the National Uranium Resource Evaluation (NURE) program sponsored by the U.S. Department of Energy, Bement et al. (1977) have described a procedure for locating outliers or anomalous areas of uranium concentration based on aeroradiometric data. Andrews' robust sine function M-estimate (Andrews et al., 1972) was used to locate areas of anomalous contents of uranium in the Lubbock quadrangle, Texas. The technique assigns weights to observations based on their proximity to the center of the main body of data. The advantage of the technique is that it is not dependent on the type of distribution assumed for the observations. Within the Lubbock quadrangle, the anomalies were associated with the geologic formations being investigated.

Using a different approach, Conover et al. (1977), using the same data, developed a method for calculating the exact probability for obtaining no clusters of the data for prescribed "local" windows. With this probability, it was possible to define subareas within the Lubbock quadrangle which contained anomalies which could not be ascribed to chance. The anomalies defined by this method were similar but not identical to the anomalies defined by Bement et al. (1977).

Finally, a method described by Koch and Link (1977) has been devised to recognize anomalies with compact shapes. A simple situation in the case of a single variable is shown in Fig. 12A-3 for a 3 by 3 square grid of 9 anomalous points. Table 12A-1 presents the results of trying to identify this anomaly by quadratic regression as the contrast between the mean of the anomalous points and the mean of the background

Table 12A-1. Recognition of anomalies defined by points in a 3 by 3 square.
From Koch and Link (1977)

Number of runs	μ_{1A}	σ_{1A}	μ_{1B}	σ_{1B}	Percent of F-values > 2.08
10	50	10	20	10	20
10	60	10	20	10	40
10	70	10	20	10	50
10	80	10	20	10	70
10	90	10	20	10	80
10	100	10	20	10	90
5	70	10	20	10	40
5	70	10	20	7	60
5	70	10	20	6	80

The Statistical Treatment of Geochemical Data

points. The criteria for successful identification is a quadratic regression that yields a calculated F-value larger than the F-value of 2.08 corresponding to the 10 percent significance level. For instance, in line one of the table, the mean of the anomalous observations is 50, the mean of the background observations is 20, and the standard deviation of all observations is 10. Because 20 percent of the calculated F-values exceed 2.08, we conclude that the anomaly is successfully recognized by quadratic regression 20 percent of the time. The last three lines of the table show that the recognition rate rises from 40 to 80 as the standard deviation of the background observations falls from 10 to 6. This type of approach leads to the generalization that for anomalies formed by compact groups of contiguous points where values are "large" compared to those of the background values, the chance of detection is proportional to the number of points multiplied by the magnitude of the point values measured in standardized units.

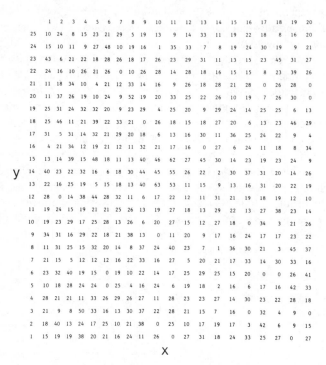

Fig. 12A-3. Anomalous (A) and background (B) observations corresponding to pseudo-random numbers drawn from distributions $\mu_{1A} = 50$, $\sigma_{1A} = 10$ and $\mu_{1B} = 20$, $\sigma_{1B} = 10$. The nine anomalous points are in a square centered on x = 9, y = 15. From Koch and Link (1977).

COMPUTERIZED DATA SYSTEMS (pp. 503-508)

It would be difficult, if not impossible, in a few words to describe the rapid change and development that has occurred in computerized data systems in the past five years. The most notable changes, however, have occurred in time share computing and computer graphics. This has given geochemists easy access to the most advanced computer programs for processing geochemical data and, with the aid of computer graphics, to display the results in a variety of forms. There is hardly a geochemist today who does not enjoy some form of access to a computer.

The U.S. Geological Survey's RASS-STATPAC system for management and statistical reduction of geochemical data is now operating in a time share computing environment and affords the user flexibility both in data searches and retrievals and in the manner of statistical treatment of data (Van Trump and Miesch, 1976). The STATPAC programs are general purpose, but contain features designed specifically for the treatment of geochemical data. In particular, the programs were written to accept qualified values, that is, values accompanied by such qualifiers as less than, greater than, not detected, no analysis performed, and so forth. The principal guideline followed in the development of the STATPAC programs was to make each program sufficiently flexible to treat the data in a manner one chooses. Within a time share environment, a user of STATPAC can also bridge to other program systems for statistical processing and automatic plotting. The latter is especially desirable in view of the growth of readily available automatic contouring programs.

Apart from computing, a major effort is underway in many countries to develop national geochemical data banks (Horder, 1978). The primary objective is to provide for the storage and exchange of analytical, locational and geologic data on the large number of geochemical samples that have been collected over the past several years. It is anticipated such data banking systems will serve to act as the link between organizations or individual geochemists wishing to deposit or retrieve data and a computer facility. Such activity should encourage higher standards of data collection and recording among geochemists.

13A/ Application Of Geochemistry To The Search For Crude Oil And Natural Gas*

BRIAN HITCHON
Special Assistant to the President
Alberta Research Council, Edmonton, Alberta, Canada

INTRODUCTION (pp. 509-511)

The Preface to the first edition of *Introduction to Exploration Geochemistry* stated that the purpose of that book was to present an up-to-date introduction to exploration geochemistry sufficient to enable the reader to understand the literature and to discuss the subject intelligently with experienced individuals or consultants; it was also recognized that the search for crude oil and natural gas was a very specialized aspect of exploration geochemistry. That purpose and recognition remain true for this second edition. In the intervening years there has been a world-scale change in the economics of the petroleum industry, accompanied by an acceleration of trends that were in evidence in the late 1960's and early 1970's with respect to alternate energy resources, energy conservation, and the manner in which mankind utilizes the minerals of the Earth. The corporate and government entities that previously had essentially explored for petroleum are now becoming energy exploration entities, and are increasingly turning their attention to exploration for non-energy minerals. Now, and in the next few decades, liquid fuels will likely be the dominant energy source for the world. There will thus be continued and increased effort put into both exploration for petroleum and coal conversion — particularly coal liquefaction. Thus the business of manufacturing liquid fuels (from oil sands, coal and oil shales) will accelerate. Concurrent with the need to explore for and manufacture more liquid fuels is the recognition of potential shortages of some non-energy minerals. It is within this scenario that the reader must consider recent geochemical research that may be applied to the search for crude oil and natural gas. The author has selected what he considers are the most important publications appearing in the period 1973 to mid-1979, primarily in the English language, in easily accessible journals and publications, and which appear to him to be most directly applicable to suc-

---

*Contribution No. 975 from Alberta Research Council.

cessful geochemical exploration. In addition, and bearing in mind the cited scenario, consideration will also be given to selected aspects of water-rock interaction.

Without doubt, the most important, comprehensive, single publication in the field of petroleum exploration geochemistry to appear in the past decade is the book *Petroleum Formation and Occurrence* by Professors Tissot and Welte (1978). It covers topics ranging from the production and accumulation of organic matter in modern environments, through the fate of this organic matter after its incorporation in sedimentary basins and subsequent transformation to crude oil and natural gas, to such difficult geological and geochemical problems as the migration and accumulation of these two fluids, their classification and their alteration in the reservoir. Finally, Tissot and Welte apply this knowledge to the identification of source rocks, oil-oil and oil-source rock correlations, and the search for petroleum prospects by geochemical and mathematical model techniques. This single volume amply covers the sections on regional evaluation of petroleum potential, search for high quality crude oil and classification into crude oil families in Chapter 13 of the first edition of *Introduction to Exploration Geochemistry*. Accordingly, emphasis in these three sections in this revised chapter will be placed on selected details not covered by Tissot and Welte, and additional references not cited by these authors. The remaining sections of this chapter will receive a more general review of the literature appearing during the past few years.

REGIONAL EVALUATION OF PETROLEUM POTENTIAL (pp. 512-519)

The concept that crude oil and natural gas originate via a kerogen pathway from organic matter deposited with the sediments continues to form the keystone to regional evaluation of petroleum potential. Studies carried out to better our understanding of this concept range from laboratory simulations of the thermal alteration of organic matter found in modern sediments to sophisticated chemical and physical examinations of kerogen in sedimentary rocks. Tissot and Welte (1978, Part II, chapters 1 to 5) have covered this concept in considerable detail and the interested reader is referred to these chapters for a modern review of this subject, and to the important paper of Dow (1977).

In thermal alteration experiments on fractionated organic matter from a young marine sediment from the Tanner Basin, offshore southern California, Ishiwatari et al. (1977) found that the lipid and kerogen fractions (with the former fraction quantitatively an order of magnitude less abundant than the latter) produced comparable amounts of the more than 99 per cent of the total identified n-alkanes that these two fractions yielded; the balance originated from the humic acid fraction. Further studies on kerogen from this same offshore area by these authors (Ishiwatari et al., 1978) indicate that in nearshore sediments, the kerogen apparently is sufficiently different from marine-derived kerogen to represent a poor source for liquid hydrocarbons. Hence, deltaic environments which derive most of their organic matter from land will produce source rocks possibly high in organic carbon but low in oil potential. This difference in kerogen types resulting from differences in source of the original organic matter has been studied by Philippi (1974) and can be observed also in detailed studies of sedimentary basins (Huc, 1977). The importance of adjacent offshore marine shales as potential source rocks when exploring in deltaic environments is obvious.

The potential of phosphorites as source rocks for petroleum was investigated by Powell et al. (1975). Kerogen isolated from phosphorites is rich in nitrogen, sulphur and oxygen compared to that from other sedimentary rocks, and reflects the highly euxinic nature of the phosphorite-forming environment. The composition of the kerogen supports the contention that crude oils with high nitrogen contents are derived from phosphatic source rocks.

Important so-called biological markers present in crude oils, such as steranes and triterpanes, or their immediate percursors, have been generated by thermal alteration experiments from both recently deposited algal mats and oozes (Philp et al., 1978) and kerogen from shales (Seifert, 1978), indicating that the individual chemical components of crude oils may be generated, admittedly in variable quantities, at all stages from deposition of the organic matter in the original sedimentary environment up to the stage of thermal destruction of the kerogen. Similar experiments have been carried out with respect to the normal paraffins (Harrison, 1978; Young and Yen, 1977).

Pyrolysis techniques are now sufficiently developed that they may be used on quite small samples, e.g. drill cuttings, to evaluate the petroleum generating capacity of the rock (Barker, 1974; Claypool and Reed, 1976; Leventhal, 1976; Harwood, 1977). Claypool and Reed (1976) have used the fact that heating of sedimentary rocks under controlled conditions (28°C/minute) causes a sequential two-stage release of organic compounds to produce a rapid (½ hour) method of determining organic richness. The material released during the 30 to 400°C heating interval corresponds to volatile compounds pre-existing as such in the rock. At higher temperatures (400 to 800°C) non-volatile organic matter thermally decomposes to yield volatile compounds. For a given sample weight, the logarithm of the integrated response of the hydrogen-flame-ionization detector during the 30 to 400°C heating period is directly proportional to the logarithm of the concentration of extractable heavy hydrocarbons in that sample, as independently determined by conventional methods (Fig. 13A-1a). Similarly, the logarithm of the detector response during the 400 to 800°C heating period is approximately proportional to the logarithm of the organic-carbon content, as determined by a separate combustion analysis (Fig. 13A-1b). Combining these trends (Fig. 13A-1c) produces a plot which quantitatively evaluates organic richness using parameters that are comparable to those given by the conventional solvent-extraction and combustion methods. Similar, simple, approximate methods for evaluating source rock potential have been developed by Saxby (1977), in which oil, methane, carbon dioxide, water and hydrogen yields can be

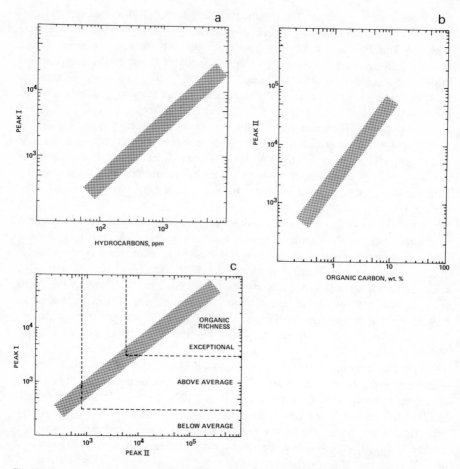

Fig. 13A-1. a. Trend of data points (shaded area) in log-log plot of integrated response of hydrogen-flame-ionization detector (peak I), during heating of sedimentary rocks from 30 to 400°C, versus content of independently determined solvent extractable heavy hydrocarbons (after Claypool and Reed, 1976).

b. Trend of data points (shaded area) in log-log plot of integrated response of hydrogen-flame-ionization detector (peak II), during heating of sedimentary rocks from 400 to 800°C, versus independently determined organic carbon content (after Claypool and Reed, 1976).

c. Trend of data points (shaded area) in combined log-log plot of peak I versus peak II (after Claypool and Reed, 1976).

deduced from elemental analyses of kerogen; and by Waples (1979), who used commonly determined measures of kerogen quantity (total organic carbon content of the rock), quality (H/C ratio of the kerogen) and thermal maturity (vitrinite reflectance of the kerogen) to produce a scaled value for both "total oil source potential" and "oil already generated".

The ultimate thermal destruction of a modern estuarine kerogen through conversion to a graphite-like atomic arrangement has been demonstrated by means of X-ray diffraction techniques by Harrison (1976). It will be interesting to see if this study, and knowledge gained on the thermal history of both kerogen (Marchand, 1976; Yen and Sprang, 1977) and asphaltenes (Elofson et al., 1977) through the use of electron spin resonance methods, can be usefully applied to understanding reactions taking place during *in situ* combustion of oil sands and during underground gasification of coal.

The kerogen pathway, by which crude oil and natural gas are generated, is influenced by variations in the amount and type of organic matter, the geothermal gradient and the time available for petroleum generation. These variables explain most of the differences in petroleum potential among sedimentary basins. Connan (1974) has applied chemical kinetic laws for first-order reactions to the threshold of intense oil generation, which is defined by the age (t in millions of years) of the formation in which it occurs, as well as by the corresponding existing temperature at that depth (T°K). There is a good correlation between $\ln t$ and $1/T$ (Fig. 13A-2), suggesting that kerogen degradation obeys the laws of chemical kinetics. An interesting paper

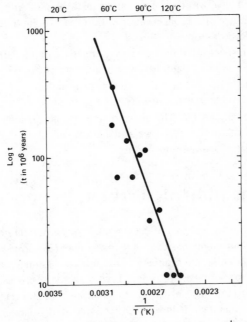

Fig. 13A-2   Plot of log t, age of the reservoir, (in millions of years) versus $\dfrac{1}{T(°K)}$, reservoir temperature, showing first order kinetic relationship (after Connan, 1974).

(Young et al., 1977) which bears on the matter of chemical kinetics, uses the changes in the composition of crude oils as they mature to calculate the ages of the hydrocarbons in the oils, based on the premise that the hydrocarbons in crude oils are not in kinetic equilibrium but are approaching equilibrium with time and maturation (temperature). The hydrocarbon ages calculated by this method are usually in good agreement with geologically interpreted ages of the crude oils.

Selected examples of recent publications that have evaluated the petroleum potential of regions based on the kerogen pathway concept range from those dealing with the Canadian Arctic (Snowdon and Roy, 1975; Henao-Londoño, 1977) and Canadian east coast offshore (Cassou et al., 1977; Rashid, 1979), to the United States (Nixon, 1973; Laplante, 1974; Claypool et al., 1978; Tissot et al., 1978), Australia (Powell and McKirdy, 1973; Powell, 1975; Shibaoka, et al., 1978), Africa (Albrecht et al., 1976), Borneo (Combaz and Matharel, 1978) and the North Sea (Oudin, 1976; Brooks and Thusu, 1977). Dow (1978) has discussed this concept with specific reference to petroleum source beds on continental slopes and rises.

The close interrelation between organic and inorganic geochemical indicators of basin maturity is exemplified by the work of Foscolos et al. (1976) and Powell et al. (1978) in which comparative evaluations were made in suites of shales of such maturation parameters as vitrinite reflectance, extractable organic matter and mineralogical characteristics of the discrete illites and mixed layer clays. These authors found that the first clay dehydration occurs prior to hydrocarbon generation and is accompanied by adsorption of K^+ and substitution of Al^{3+} for Si^{4+} in the clay lattice. The significance of this observation with respect to the time of primary migration remains for further study.

Finally, with respect to kerogen, we may conclude that information obtained in the past few years on the kerogen pathway demonstrates not only the very complex nature of this biopolymer but also the capability of modern geochemical techniques to elucidate the multiple sources of kerogen and to indicate through pyrolysis and other methods the physics and chemistry of the transformations that take place in nature. Based on this knowledge, we can better seek data that will indicate the petroleum potential of sedimentary basins.

SEARCH FOR HIGH QUALITY CRUDE OIL   (pp. 519-528)

The model for processes that affect the quality of crude oil in the reservoir, described in the first edition of *Introduction to Exploration Geochemistry,* continues to be the guide to the search for high quality crude oil. Tissot and Welte (1978, Part IV, chapters 4 and 5) present an up-to-date review of these processes, together with some useful background information on the composition and classification of crude oils (*ibid.,* Part IV, chapters 1 and 2). Other important reviews of this model are those of Philippi (1975, 1977), Bailey et al. (1974) and particularly that of Milner et al. (1977), which includes some previously unpublished examples of the practical use of the model as an exploration tool.

With respect to biodegradation of crude oils, no mention was made in the first edition of this book of the type of bacteria causing the chemical degradation. It was assumed (p. 524) that the bacteria were either introduced into the reservoir, presumably primarily with fresh water recharging from the surface, or that the biodegradation resulted from reactivation of dormant pre-existing bacteria, speculatively through reduction of salinity and addition of nutrients. An important paper by Jobson et al. (1979) has presented evidence to show that sulphate-reducing bacteria cannot initiate degradation of crude oil but rather they grow on the residues resulting from the aerobic degradation of the oil. This observation supports the concept of the initial introduction of aerobic bacteria through fresh water recharging from the surface, and when there is sufficient degraded crude oil and the reservoir conditions become anaerobic, the sulphate-reducing bacteria complete the degradation.

In an examination of some twenty crude oils from ten wells drilled in the Mackenzie Delta of the Beaufort Basin, Northwest Territories, Canada, Burns et al. (1975) distinguished two groups of crude oils. From geochemical data they interpreted that the generally deeper group, with higher API gravities and pour points and lower viscosities than the second group, has been bacterially degraded in the shallower reservoirs with resultant loss of n-paraffins, including the wax components. However, the over-all biodegradation effects have not appreciably altered the viscosities or sulphur content but have drastically lowered the pour point, which has a highly beneficial effect on the flow properties of the oils in the arctic climate. In this sense, the loss of n-paraffins by bacterial action has actually enhanced the pipeline characteristics of the shallow oils, and therefore they should not be equated to the biodegraded "heavy" crude oils of the Western Canada Basin, which have undesirable physical properties. Their study shows the importance of a detailed geochemical evaluation of all crude oils, including the heavy ones, if the explorationist is not to miss potentially pipelinable crude oils. From the work of Jobson et al. (1979) cited previously, it might be speculated that these Beaufort Basin crude oils have only been degraded by aerobic bacteria, hence the appreciable lack of change in the sulphur content between the two groups of oils.

With the general acceptance of biodegradation as the dominant process in the degradation of crude oils, there is now a need for more detailed study of the chemical effects produced. A start has been made (Rubinstein et al., 1977; Seifert and Moldowan, 1979) but progress will probably be slow because of the complex biochemical processes involved. A by-product of bacterial degradation is carbon dioxide, and in a study on the Barrow Island field, Northwest Shelf, Western Australia by Gould and Smith (1978) it was shown, through the use of carbon isotopes, that the extensive deposits of secondary carbonates resulted from reaction between bacterially produced carbon dioxide and metal ions in the formation waters. This type of reaction may be more widespread than realized heretofore.

Carbonate rocks commonly contain solid bituminous materials which may originate either from thermal alteration of the crude oils to dry gas and solid bitumen, or deasphalting of crude oil following solution of large amounts of gas. Rogers et al. (1974) examined a suite of these bitumens from the Western Canada Basin and developed criteria based on the H/C ratio and solubility in carbon disulphide to distinguish these two processes. With respect to their study area, they found gas deasphalting more important than thermal alteration.

CLASSIFICATION INTO CRUDE OIL FAMILIES
(pp. 528-531)

Once crude oil has been found in a sedimentary basin, it is important to identify the source rock from which it orginated and to determine if nearby crude oils have the same or a different source. This information, together with knowledge of the geology and the hydrodynamics of the fluid system, enable the explorationist to formulate a series of source-migration-accumulation-alteration scenarios for each suite of crude oils. It is the development of these scenarios that forms one of the most important tools in locating petroleum prospects. Tissot and Welte (1978, Part V, chapters 2 and 3) have reviewed the subject of source rock-oil and oil-oil correlations but like most authors, unfortunately, have not linked this geochemical information to the hydrodynamic situation.

A detailed chemical examination of the components in a crude oil, together with stable isotope studies, continue to be the main geochemical tools used for source rock-oil and oil-oil correlations. In addition to the cited book by Tissot and Welte (1978), Erdman and Morris (1974) have summarized the most commonly used techniques, and practical examples of their use have been published for the Williston Basin (Dow, 1974; Williams, 1974), Gulf Coast of the United States (Koons et al., 1974), Australasia (Powell and McKirdy, 1975), and the Maranon Basin, Peru (Illich et al., 1977). All these practical studies used comparatively simple analysis and evaluation of the normal and isoprenoid alkanes, and stable carbon isotopes in the total crude oil and the main fractions (saturates, aromatics, eluted NSO compounds, asphaltenes). The degree of chemical complexity that may be obtained through an examination of the steranes, terpanes and monoaromatics is well exemplified in a paper by Seifert and Moldowan (1978) who studied these components in crude oils from the McKittrick field, California. Although the results obtained by these authors at the molecular level provided far more detailed information than that available from studying the 'bulk' parameters, such as yields of the main fractions and the stable carbon isotope ratios, both suites of data were nevertheless mutually supporting. This observation suggests that examination of the 'bulk' parameters is sufficient at the exploration level but that if uncertainties remain they may perhaps be resolved through detailed study at the molecular level. In this respect, Deroo (1976) found the aromatic and thiophenic compounds to be of value in the more mature basins, such as the Western Canada Basin, where catagenesis had strongly affected the crude oils. From a detailed study of 78 crude oils from many of the world's major fields, Ho et al. (1974) were able to group the sulphur compounds according to the various processes which had affected the crude oils.

In the past few years much attention has been paid to the use of stable isotopes in source rock-oil and oil-oil correlations. Excellent reviews of the practical applications of stable isotope studies to petroleum exploration have been provided by Fuex (1977) for carbon isotopes and Krouse (1977) for sulphur isotopes. Orr (1974) has used sulphur isotopes as a means of evaluating thermal maturation processes in

Paleozoic crude oils from the Big Horn Basin, Wyoming. Further pertinent papers are those of Stahl (1977, 1978). Figure 13A-3 shows a typical source rock-oil correlation based on stable isotope ratios. The importance of stable isotope analysis is based on

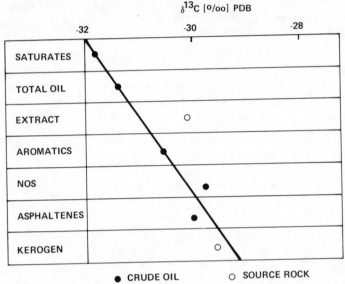

Fig. 13A-3.   Source rock — crude oil correlation based on stable carbon isotope ratios (after Stahl, 1977).

the fact that it links the kerogen of the source rock directly with the crude oil; the extract from the source rock represents the portion of the crude oil that remains at its place of origin. Therefore, the carbon isotope ratios of a crude oil found in a reservoir should be similar to that of the extract in the source rock. Because of the kinetics of the stable isotope separation process and of the manner in which crude oil is generated, the stable isotope ratios of the crude oil and extract from the source rock are similar, whereas both are lighter than that found in the parent kerogen.

ASPECTS OF EXPLORATION FOR NATURAL GAS
(A new topic)

In the first edition of this chapter, minimal attention was paid to exploration for natural gas. It was recognized (pp. 517-520) that natural gas may be immature and dominantly bacterially formed, or mature (with or without associated crude oil) and resulting from thermal alteration processes. The stable carbon isotope ratios of the high-molecular-weight hydrocarbon components of crude oils and source rocks are relatively free of both primary and secondary isotopic fractionations, hence their value for correlation purposes. In contrast, the stable carbon isotope ratios of components of natural gas are greatly affected by primary fractionations and, in some cases,

secondary fractionations. Thus stable carbon isotope data primarily provide information on gas type (i.e. bacterial versus thermal origin), maturity and nature of the
parent organic material. Because of this, the stable carbon isotopic composition of
natural gas components provides information with respect to the association or non-
association of the natural gas with crude oil. In this sense, study of the stable carbon
isotopes in natural gas is useful in both oil and gas exploration. Fuex (1977) has
reviewed this entire aspect of exploration for gaseous hydrocarbons and has cited an
example from the Western Canada Basin (Fig. 13A-4) in which both the stable carbon isotope ratio of methane from natural gas and the ratio of methane to methane-
plus-ethane can be used to show that natural gas in Cretaceous reservoirs of southern

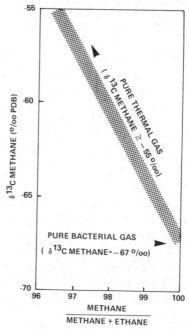

Fig. 13A-4.   Plot of $\delta^{13}C$ in methane versus ratio methane to methane-plus-ethane for natural gases from Cretaceous reservoirs in southern Alberta, Canada (after Fuex, 1977).

Alberta has a mixed bacterial-maturation origin. The deep dry-gas fields of the Val
Verde and Delaware basins of west Texas have been studied by Stahl and Carey
(1975) who found $\delta^{13}C$ values of methane in the range -35 to -47, and high levels
of maturity of the kerogen (high R_0 values), both consistent with their being very
mature gases without associated crude oil.

One aspect of natural gas that was not considered at all in the first edition of this
chapter is the occurrence of natural gas in the form of natural gas hydrates. These are
solid compounds, resembling ice or wet snow in appearance, which form below and
above the freezing point of water under specific PT conditions. Under such conditions the water molecules form pentagonal dodecahedra that can be arranged into two

different structures, leaving interstitial space into which can be accommodated some of the components found in natural gas, e.g. methane, ethane, hydrogen sulphide, carbon dioxide. The most likely way to produce natural gas hydrates in sedimentary basins is through a reduction in temperature, rather than an approach to lithostatic pressures, and the most pertinent situation is that found in regions with relatively thick permafrost sections. Hitchon (1974) has reviewed the pressure-temperature-composition limitations to the occurrence of natural gas hydrates and provided descriptions of permafrost regions with proven or potential occurrences of natural gas hydrates. Makogon (1978) has described all aspects of natural gas hydrates, including the Russian occurrences, and Bily and Dick (1974) the natural gas hydrates found in the Mackenzie Delta, NWT, Canada. Although the detection of natural gas hydrates *in situ* and their subsequent production are both subjects requiring considerably more research, it is important, in view of increased exploration effort in the Arctic, that geochemists at least be aware of their existence and the problems they present during drilling.

PETROLEUM OCCURRENCE INDICATORS IN FORMATION WATERS (pp. 531-537)

The study of formation waters continues to present only a low priority in the overall application of geochemistry to the search for crude oil and natural gas; nevertheless some important advances in our knowledge have been made since the first edition of this chapter was published. Collins (1975) has provided a broad review of the geochemistry of oilfield waters. Two aspects deserve special attention in this chapter, namely, the part played by formation waters in primary migration, and the significance of the occurrence of aliphatic acid anions to the generation of natural gas. In addition, water-rock interaction, a topic not covered in the first edition, will be considered.

Geochemists continue to be concerned with the mechanisms of primary migration, and Cordell (1973) and Price (1976) have each extensively reviewed different aspects of this subject. Both have recognized the difficulties in determining the precise form in which the hydrocarbons from crude oils exist during the primary migration phase, but it now appears, at least to this author, that micellar solution (colloidal soaps) is the most likely form. High temperatures promote micellar solution, and high salinities, especially those over 150,000 ppm NaCl, depress micellar solution. The relative importance of these two factors assists in determining the amount of hydrocarbons accommodated in the formation water. In addition to the new hydrocarbon solubility data provided by Price (1976), Eganhouse and Calder (1976) report information on the solubility of medium molecular weight aromatic hydrocarbons and the effects of hydrocarbon co-solutes and salinity. A very recent paper by McAuliffe (1979) suggests that the hydrocarbons generated in the source rock flow in a three-dimensional, hydrophobic kerogen network where they are not subjected to interfacial forces until they enter the much larger pores of the reservoir rock. He proposes that secondary migration is by separate-phase buoyant flow which does not require the flow of water, suggesting that water flow probably disperses water-soluble constituents instead of concentrating them in reservoir traps. This latter observation on secondary migration is curious in view of the fact that hydrocarbon accumulations all occur in an aqueous environment. The mechanism of primary migration is far from solved and undoubtedly further studies can be anticipated.

The presence of aliphatic acid anions in formation waters has been known, largely from Russian work, since the early 1960's, when the possibility of their use as petroleum occurrence indicators was suggested. Willey et al. (1975) were probably the first to use modern analytical techniques to determine the content of short chain aliphatic acid anions (C_2 to C_5, namely, the acetate, propionate, *n*-butyrate, *iso*-butyrate, *n*-valerate and *iso*-valerate species) in several formation water samples from the Kettleman North Dome oil field, California. They showed that the measured alkalinity values of these formation waters are due mainly (75 to 100 per cent) to these acid anions. Their study demonstrates the danger of interpreting alkalinity in terms of the carbonate species only. In a further study, Carothers and Kharaka (1978) determined these same aliphatic acid anions in 95 formation waters from 15 oil and gas fields in the San Joaquin Valley, California and in the Houston and Corpus Christi area of Texas. Their results are summarized in Table 13A-1.

Table 13 A-1.   Aliphatic acid anions in formation waters. After Carothers and Kharaka (1978)

Zone	Formation temperature (°C)	Concentration of total aliphatic acid anions (mg/l)	Predominant. type
1	< 80	< 60	Propionate
2	80 - 200	up to 4,900, but decreasing with increasing temperature	Acetate (more than 90 per cent)
3	> 200	none (by extrapolation of data in zone 2)	—

Microbiological degradation of acetate and dilution by mixing with meteoric waters most probably explain the composition and concentration of aliphatic acid anions in zone 1. The trends in zone 2, and the absence of acid anions in zone 3, are explained by thermal decarboxylation of these acid anions, such as in the reaction:

$$CH_3COO^- + H_2O \rightarrow CH_4 + HCO_3^-$$

Iodide concentrations increase with increasing concentrations of aliphatic acid anions, and this observation strengthens the general belief that marine organisms are the source of iodide in formation waters and of petroleum. This common source would explain why high iodide concentrations are a good indicator of petroleum (Hitchon and Horn, 1974).

The foregoing observations have significance to the search for, and origin of, natural gas, Carothers and Kharaka (1978) suggest that the aliphatic acid anions mainly result from the thermocatalytic degradation of kerogen. They propose that these highly soluble anions are produced and dissolved in the pore waters of the source rocks, and are expelled to the reservoir rocks during dehydration of clays. Decarboxylation of these acid anions to the components of natural gas is believed to occur mainly in the reservoir rocks, thus providing an alternative explanation for natural gas occurrence to that of primary migration. Evidence for the formation of natural gas from decarboxylation of acid anions is provided by the stable carbon isotope ratio values of total bicarbonate and methane, and the good correlation between the proportions of these anions in formation waters (acetate 94 per cent; propionate 5 per cent; butyrate 2 per cent) and their decarboxylated gases in the natural gas produced (methane 90 per cent; ethane 5 per cent; propane 2 per cent). Their proposition appears to present a viable variation to the bacterial-maturation origin of natural gas described in the previous section of this chapter. However, it should be clear to the perceptive reader that the whole complex relationship among crude oil, natural gas, aliphatic acid anions, iodine and phosphate (see comments in Hitchon et al., 1977, pp. 8-13) require considerably more research. The advent of techniques such as ion chromatography, which enable rapid detection and determination of organic anions, will undoubtedly enhance progress in this field. There will likely be no simple solution, and in any one situation more than one process may have produced the observed relations.

Water-rock interaction, although not at this time of direct application to the search for crude oil and natural gas, is nevertheless becoming an increasingly important aspect of geochemistry which deserves the attention of the petroleum exploration geochemist. Economically important mineral deposits in sedimentary rocks that are the result of *natural* water-rock interaction include petroleum and Mississippi - type Pb-Zn deposits. Understanding of the origin of these deposits through water-rock interaction requires knowledge of the relations between hydrochemistry and hydrodynamics. Hitchon (1976, 1977a) has outlined some of the principles involved; however, the detailed discussion of these are beyond the scope of this chapter. In the latter paper the author explored the genetic link between oil fields and ore deposits in sedimentary rocks, which takes physical form through moving formation waters. An analogy is developed between the generation and leaching of hydrocarbons from shale source rocks, and their subsequent migration and accumulation as oil fields, and the leaching of shales by hot saline formation waters, with their subsequent migration and accumulation as ore deposits. Thus the source-migration-accumulation scenario for oil fields and ore deposits in sedimentary rocks is controlled by the temperature, pressure, hydrodynamic and fluid-rock-interaction continuum that exists in all the sedimentary basins of the world. The most important element for migration is the presence of a conduit for channel flow, and the best candidates for future channel-flow conduits for hydrocarbon- and metal-laden formation waters are reefal or shelf carbonate fronts, close to both evaporites and organic-rich muds. The length and attitude of the carbonate front with respect to the future sedimentary basin are crucial. Accumulation of dissolved hydrocarbons and metals results from their removal from solution (accommodation) in the saline formation waters through decrease in temperature, pressure or salinity; changes in pH, P_{H2S}, P_{CO2}; or adverse fluid-rock interaction. Subsequently, the oil fields and ore deposits may be subjected to waterwashing, biodegradation and weathering in the zone of local groundwater flow. For both oil fields and ore deposits in sedimentary rocks, the five-stage conceptual model is therefore deposition-burial and leaching - migration - accumulation - uplift and dissipation.

The recovery of some of these mineral deposits, specifically hydrocarbon deposits, involves man-imposed water-rock interactions, for example, during water flooding of petroleum reservoirs and *in situ* steam injection into oil sands. In addition, man-imposed water-rock interactions are also important in underground coal gasification. In all these situations man-imposed water-rock interactions may result in (1) subsurface reactions which can reduce permeability, (2) the production of toxic or deleterious substances which require removal before reuse of the produced water, (3) contamination of local potable groundwater, or (4) cause problems in waste injection wells because of subsequent water-rock reactions. Hitchon (1977b) has evaluated some of these effects with respect to *in situ* recovery from oil sands and in a subsequent paper (Hitchon, 1979) has pointed out some of the economic aspects of water-rock interaction as they relate to both *in situ* recovery from oil sands and underground coal gasification. In any situation involving moving water, and more particularly petroleum with which this chapter is concerned, the geochemist must be cognizant of potential water-rock reactions. These may occur at any stage between the generation of the petroleum and the many facets of recovery processes. As long as the geochemist is aware of this potential, he may be of value in assisting utilization of this resource.

SURFACE PROSPECTING   (pp. 537-543)

Surface prospecting for petroleum on land remains equivocal. A detailed study of the Cement field, Oklahoma, by Donovan (1974) presents some interesting observations but fails to provide the key hydrodynamic information on the near-surface flow regime as required and outlined in the first edition of this chapter. With respect to offshore areas, there is still considerable merit to the technique if cognizance is taken of the various factors that may modify anomalies. Sackett (1977) has presented an important review of the use of "hydrocarbon sniffing" in offshore exploration and has indicated that because the methane from the seep gases may be of biological origin, based on studies of the stable carbon isotopes, sniffing for the C_3 to C_6 hydrocarbons in near-bottom waters maximizes the chances for success in offshore petroleum exploration. A subsequent paper by Reitsema et al. (1978) confirms the multiple origin of methane and the necessity of sampling from near-bottom waters, as well as noting that petroleum production activities, at least in the Gulf of Mexico, did not increase the hydrocarbon content of the deeper waters beyond that often found above petroliferous structures. In the same geographic area, even cores from the deep sea deposits show evidence of migrating liquid hydrocarbons (McIver, 1974). Data published by Reed and Kaplan (1977) on the chemistry of marine petroleum seeps demonstrate that analyses of the water column for enrichment in high-molecular-weight hydrocarbons or trace metals adjacent to seep areas may produce equivocal results. Sediment analysis appears to give a more reliable indication of petroleum contribution from non-anthropogenic sources in this case.

Understanding the flow of fluids in sedimentary basins is important to several aspects of petroleum exploration. In basins still undergoing considerable compaction the explorationist should be concerned with the vicissitudes of primary migration. By contrast, in geologically mature basins (i.e. tectonically stable and non-compacting) attention should be directed to appreciating the concept of cross-formational, topography-induced gravity flow. Clearly, because the completion of compaction is ill-defined, and the deeper strata in a sedimentary basin may be fully compacted compared to the shallower strata, both phenomena require attention. Together with knowledge of the kerogen pathway for the generation of the components of crude oil and natural gas from organic matter in the sediments, understanding primary migration will assist in identifying broad regional source-reservoir relationships, and application of the concept of cross-formational topography-induced gravity flow will help in indicating areas in which accumulation may have taken place. Neither the phenomenon of primary migration, cross-formational flow, nor kerogen evolution can be considered in isolation when evaluating the hydrocarbon potential for a given region.

Primary migration, as its name implies, is concerned with the initial movement of the components of crude oil and natural gas from their place of generation (source rocks) into the mainstream of the regional flow system of the basin. The precise physical form these components take and the exact exit mechanisms remain equivocal. Nevertheless, understanding of the process of primary migration has improved because of studies by Magara (1974, 1975, 1976a, b, 1977, 1978a, b) and Neglia (1979). There is now increased awareness of the importance of temperature (Magara, 1974), montmorillonite dehydration (Magara, 1975) and other dehydration reactions (Fyfe, 1973) to the process. Further, Magara (1976b) has developed work on shale compaction, which can be used to determine the direction of primary migration, into a method for calculating the thickness of sedimentary rocks removed by erosion in the geological past. By integrating the shale compaction data with homogenization temperatures of fluid inclusions in secondary quartz from fractures in the sedimentary rocks, it is now possible to estimate the paleotemperature and paleogeothermal gradients in a basin. This has been done for the southwestern part of the Western Canada Basin (Magara, 1976b), where calculations show that as much as 4,600 feet (1,400 meters) of strata have been removed by erosion, and although the geothermal gradient probably has not changed significantly in the geologic past, the fractured petroliferous Upper Cretaceous Cardium sandstone which presently has a temperature range in the study area of 140 to 160°F (60 to

71°C) had much higher temperatures of about 300°F (149°C) at the time of maximum burial, prior to extensive erosion of the previously overlying strata. This type of study is clearly of considerable significance to the explorationist, not only from the point of view of primary migration but also with respect to the time and generation of crude oil and natural gas components from the shales surrounding this specific sandstone reservoir.

The concept of cross-formational, topography-induced gravity flow was outlined briefly in the first edition of this chapter. Subsequently, a number of regional hydrodynamic studies have been published which to a greater (Toth, 1978) or lesser (Stone and Hoeger, 1973; Chiarelli, 1978) extent recognize the nature of the cross-formational flow network in sedimentary basins. Jacquin and Poulet (1973) have modelled the hydrodynamic history of a sedimentary basin, and Coustau (1977) has presented an important paper on the application of hydrochemistry and hydrodynamics to the search for crude oil and natural gas in the Aquitaine Basin of southwestern France. However, the most significant contribution to our understanding of the concept is found in a recent paper by Toth (1979), which could well become a classic in its field. The message presented by Toth is that topography-induced cross-formational flow of formation water has a significant genetic effect on the distribution of hydrocarbon deposits in geologically mature sedimentary basins. There are three fundamental, sequential premises to the concept:

1. On a regional scale the rock framework of geologically mature (i.e. tectonically stable and non-compacting) areas is hydraulically continuous.
2. In a hydraulically continuous continental area formation water flow is generated by the relief of the water table and modified by permeability differences; flow systems of different orders develop; and the natural unit of subsurface fluid-hydraulics is the drainage basin.
3. Formation water is able to, and does, mobilize, transport and deposit hydrocarbons as it moves along its flow paths from regions of high energy to regions of low energy.

The cornerstone of the concept is, therefore, regional hydraulic continuity. Or, as put in popular terms by Hitchon (1979) "a sedimentary basin can be considered as analogous to a giant sponge comprising layers of sedimentary rocks with varying degrees of porosity and permeability through which fluids are more or less free to move from regions of high energy to regions of lower energy, with the degree of movement being related to the degree of difference in porosity and permeability between the high energy and low energy regions. While this analogy may not be strictly correct or commonly acceptable, it does at least have the merit of being simple and easy to understand."

Using hydraulic theory, together with field examples from many basins throughout the world, Toth has shown that in geologically mature basins, gravity-induced cross-formational flow is the principal agent in the transport and accumulation of hydrocarbons. The mechanism becomes operative after compaction of the sediments and the concomitant primary migration have ceased, and subaerial topographic relief has developed. Hydrocarbons from source sediments are then moved along well-defined migration paths toward discharge foci of converging flow systems, and may accumulate en route in hydraulic or hydrodynamic traps. Accordingly, hydrocarbon deposits are expected and observed to be associated preferentially with the ascending limbs and stagnant zones of flow systems (which are characterized by relative potentiometric minima), downward increase in hydraulic heads (possibly reaching artesian conditions), and reduced or zero lateral hydraulic gradients with accompanying high formation water salinity. Continuous flow of formation waters imports hydrocarbons into such traps until the trap capacity is reached, and then the excess hydrocarbons become source material for new accumulations in the downflow direction. Temporal changes in surface topography result in proportionate but delayed readjustment of the flow pattern and redistribution of petroleum, although some hydrocarbons may remain in place, constituting residual deposits in discharge and stagnant regions of relict flow systems.

Although this concept may not be accepted by all explorationists of this time, there is, in this author's opinion, effectively irrefutable evidence that it is applicable in mature sedimentary basins. Variances from theoretical predictions may result more from the fact that in many so-called mature basins some compaction of the younger sediments is still operative, than from major deficiencies in the concept. Cognizance of the concept should assist in developing petroleum plays and make the explorationist more aware of the difficulties of surface prospecting.

IDEAL APPLICATION — A PERSONAL VIEW
(pp. 544-545)

In the first edition of this chapter the author indulged himself and presented a personal view of the ideal application of geochemistry to the search for crude oil and natural gas on the premise that the prodigious problems of preparing a review paper carried with them the privilege of presenting such a personal viewpoint. These views still remain an ideal.

Political entities with abundant natural resources (especially energy and mineral resources), strong economic (financial) resources and adequate, talented human resources are fortunate, but few. The exploration concepts expressed in this book may, hopefully, assist in developing the energy and mineral resources of the political entities to which the individual reader owes allegiance. Perhaps financial resources will flow from that development. What is more certain, in my opinion, is that human development, while it can be encouraged by the political milieu in which the individual is located, is mainly dependent on personal initiative. Over the past century there has been a gradual change from the broad-based "natural historians" of the early Victorian era to the high degree of specialization found today. With respect to geochemistry, itself previously regarded as a specialization, there is a tendency for the individual to confine attention to a narrow field, such as mineral exploration or organic geochemistry. Yet, with the trend for petroleum exploration entities to become energy exploration entities with currently increasing emphasis on exploration for non-energy minerals, it is clear that, while specialist geochemists will always be in demand, there is a particular need for geochemists with a broad background to span the entire field of inorganic-organic-isotope geochemistry as well as ancilliary aspects such as hydrodynamics. In addition, the modern concept of the project-team approach to problems means that the geochemist is constantly called upon to interact with individuals in different scientific and engineering disciplines. If the student geochemist will bear these thoughts in mind and use initiative, it will soon become apparent that one of the most satisfying careers is that of a geochemist.

Note added in press:

The recently published book *Petroleum Geochemistry and Geology* by John M. Hunt (W. H. Freeman and Company, San Francisco, 1979) provides an excellent background to the applications of geochemistry to the search for crude oil and natural gas with attention directed to students who have had the basic courses in geology and chemistry.

Appendix

E

GEOCHEMICAL CHARACTERISTICS
OF SELECTED ELEMENTS

In this appendix, modelled after a similar appendix originally presented by Hawkes and Webb (1962), important geochemical characteristics are compiled for the trace elements which are of most importance in exploration geochemistry. Emphasis has been given to those characteristics which affect the development of secondary dispersion halos (anomalies).

The main sources of the data used in this compilation are:

IGNEOUS ROCKS: Table 2-1 (pp. 43-44).

SEDIMENTARY ROCKS: Table 2-1; sandstone, Beus and Grigorian (1977); black shale (for a limited number of elements), Vine and Tourtelot (1970); and Wedepohl (1969-1974) in some cases.

SOILS: Table 2-1. Also, Aubert and Pinta (1977) for soils in temperate, arid, and tropical humid regions; data are only available for a limited number of elements. In the case of the Aubert and Pinta (1977) data, ranges are generally given which exclude exceptionally high and low values.

SURFACE WATERS: Table 2-1.

VEGETATION ASH: Beus and Grigorian (1977; Table 46), and Table 10A-2 (this volume). For biological response, Table 10A-1, Boyle (1974a), and other miscellaneous sources.

GEOCHEMICAL ASSOCIATIONS: Table 2-4, Fig. 2-5 and Table 2A-7 of this volume; Boyle (1974a); and Wedepohl (1969-1974).

SUBSTITUTIONS IN: Tables 2-2, 2-5 and 2A-6 of this volume; Boyle (1974a); and Wedepohl (1969-1974).

MOBILITY − SECONDARY ENVIRONMENT: Fig. 3-19, Tables 3A-3 and 3A-4 of this volume; Boyle (1974a); and Perel'man (1977).

GEOCHEMICAL BARRIERS: Fig. 3-19, Tables 2A-1 and 3A-3 of this volume; Boyle (1974a); and Perel'man (1977).

PATHFINDERS: Tables 2-3 and 3A-4 of this volume; and Boyle (1974a).

Data for all other categories were obtained from various scattered sources. Where there are no data, entries for the particular category (e.g., aqueous phase) have been omitted.

Boyle (1974a) is highly recommended to those who are interested in further information on the geochemical characteristics of the elements. This is clearly the most comprehensive compilation of elemental associations and indicator (pathfinder) elements available which is designed primarily for exploration geochemistry.

Abbreviations: AV = average; INTER = intermediate (granodiorite); SH = shale; BLSH = black shale; SS = sandstone; LS = limestone; TEMP = temperate; TROP (HUM) = tropical humid.

ANTIMONY (Sb)

IGNEOUS ROCKS (ppm): Av 0.2; MAFIC 0.2; INTER 0.2; FELSIC 0.2.
SEDIMENTARY ROCKS (ppm): SH 1; Ss < 0.1; Ls 0.2.
SOILS (ppm): Av 5.
SURFACE WATERS (ppb): Av 1; up to 1 ppm in hot springs.
GEOCHEMICAL ASSOCIATIONS: Chalcophile; As, Bi, Pb, Ag, Cu and other chalcophile
   elements; particularly enriched in Pb-Zn-Ag deposits.
ORE MINERALS: stibnite, native antimony, sulphosalts (e.g., tetrahedrite).
SUBSTITUTIONS IN: galena, sphalerite, pyrite and other sulfides (Table 2-5); ilmenite,
   and magnesian olivine.
SECONDARY MINERALS: kermesite, stibiconite.
MOBILITY – PRIMARY ENVIRONMENT: high; concentrates in late differentiates and
   veins; forms supra-ore halos.
MOBILITY – SECONDARY ENVIRONMENT: low (Table 3A-4).
GEOCHEMICAL BARRIERS: sulfide; adsorption (by Fe-Mn oxides at spring orifices).
SB IS A PATHFINDER FOR: certain types of gold and silver deposits.
PATHFINDERS FOR SB ARE: As, Hg, Au, base metals.
COMMENTS: Boyle (1974a) notes that stibnite and other Sb minerals weather slowly
   and suggests heavy mineral surveys of soils and sediments using Sb as an indicator
   of deposits containing this element. See Boyle (1974a) for other aspects of Sb
   geochemistry.

ARSENIC (As)

IGNEOUS ROCKS (ppm): Av 1.8; MAFIC 2; INTER 2; FELSIC 1.5.
SEDIMENTARY ROCKS (ppm): SH 15; Ss 1; Ls 2.5.
SOILS (ppm): Av 5. RANGE 1-50.
SURFACE WATERS (ppb): Av 2; up to 300 ppb in waters from hot springs.
VEGETATION ASH (ppm): 0.3; up to 6000 ppm in some coal seams in mineralized
   areas; highly concentrated in some vegetation (Fig. 10-3).
GEOCHEMICAL ASSOCIATIONS: Chalcophile; in Au hydrothermal veins; in Ag veins;
   with Cu, Ni, Co, Fe, Ag, etc., in Ni-Cu massive sulfide ores; in some U deposits
   (veins); with Cu in copper shales (values up to 5000 ppm As); with Cu, V, U, Ag
   in copper sandstone deposits; up to about 175 ppm in some phosphate rocks.
ORE MINERALS: arsenopyrite, realgar, orpiment.
SUBSTITUTIONS IN: pyrite, galena, sphalerite (Table 2-5), and other sulfides; apatite.
SOILS; OCCURRENCE IN: limonite, pyrite.
SECONDARY MINERALS: iron arsenates (e.g., scorodite, $FeAsO_4 \cdot 2H_2O$); in limonite.
AQUEOUS PHASE: forms complexes (e.g., arsenites) but exact nature poorly known.

MOBILITY – PRIMARY ENVIRONMENT: high; concentrates in late differentiates and veins; forms supra-ore halos (Fig. 7-5).

MOBILITY – SECONDARY ENVIRONMENT: fair in oxidizing and gley environments (Table 3A-3; Fig. 3-19).

GEOCHEMICAL BARRIERS: sulfide; adsorption (e.g., by limonite or clay, Table 3-3).

AS IS A PATHFINDER FOR: gold, silver (Cobalt type), copper, lead, and zinc deposits using soils, waters, or sediments.

COMMENTS: Arsenic forms volatile inorganic and organic compounds, such as arsine (see also Table 1-2), which may have application in soil gas and atmogeochemical surveys. See Boyle (1974a) for additional information.

BARIUM (Ba)

IGNEOUS ROCKS (ppm): AV 425; MAFIC 250; INTER 500; FELSIC 600.

SEDIMENTARY ROCKS (ppm): SH 700; BLSH 300-700; SS 10-100; LS 100.

SOILS (ppm): AV 500. RANGE 100-3000; TEMP 150-3000; ARID 10-1500; TROP (HUM) 10-3000.

SURFACE WATERS (ppb): AV 10.

VEGETATION ASH (ppm): 600; fair biological response (Table 10A-1).

GEOCHEMICAL ASSOCIATIONS: Lithophile; K, Rb, and Ba in K-feldspars and micas of granites and syenites; as barite gangue in many Pb-Zn sulfide ores and in carbonatities; high in some brines; high in some carbonates, Mn-rich sediments, and clays.

ORE MINERALS: barite, witherite.

SUBSTITUTIONS IN: K-feldspar, micas, apatite, calcite.

SOILS; OCCURRENCE IN: clays; as barite.

SECONDARY MINERALS: in Fe-Mn and carbonate precipitates at springs; clays.

AQUEOUS PHASE: as ionic Ba^{2+}.

MOBILITY – PRIMARY ENVIRONMENT: high; concentrates in acid igneous rocks; forms supra-ore halos (Figs. 7-5, 7A-7).

MOBILITY – SECONDARY ENVIRONMENT: low, because of precipitation by sulfate, and adsorption by clays (p. 97).

GEOCHEMICAL BARRIERS: sulfate; carbonate; adsorption.

BA IS A PATHFINDER FOR: Pb-Zn-Ag veins; carbonatites; recognizing mineralized granites (Table 7-1).

PATHFINDERS FOR BA ARE: Sr, F, and base metals.

COMMENTS: Barite and witherite (both moderately stable) have high specific gravities and tend to accumulate in heavy fractions of soils and sediments. See Boyle (1974a) for additional information.

BERYLLIUM (Be)

IGNEOUS ROCKS (ppm): AV 2.8; MAFIC 0.5; INTER 2; FELSIC 5.

SEDIMENTARY ROCKS (ppm): SH 3; SS 0.1-1.0; LS 1.

SOILS (ppm): AV 6.

SURFACE WATER (ppb): AV ~0.2.

VEGETATION ASH (ppm): 3; poor biological response (Table 10A-1); see also Table 10A-2.

GEOCHEMICAL ASSOCIATIONS: Lithophile; Li, Be, B, Rb, Cs and other pegmatite elements; Be-F associations in greisens and non-pegmatite metasomatic occurrences (Be often as bertrandite, as at Spor Mtn., Utah).

ORE MINERALS: beryl, bertrandite.

SUBSTITUTIONS IN: mainly in plagioclases and micas in granites.

SOILS; OCCURRENCE IN: clays, bauxites (follows aluminum).

AQUEOUS PHASE: probably as fluoroberyllate complex.

MOBILITY – PRIMARY ENVIRONMENT: high; concentrates in granites and pegmatites; forms sub-ore halos.

MOBILITY – SECONDARY ENVIRONMENT: low; of some potential value in waters (Table 3A-4).

GEOCHEMICAL BARRIERS: adsorption (by clays, organic matter, Fe-Mn oxides).

BE IS A PATHFINDER FOR: pegmatite mineralization; carbonatites; some fluorite mineralization; Sn-W mineralization.

PATHFINDERS FOR BE ARE: Li, B, Nb, Ta, F in pegmatites; F (from fluorite) in non-pegmatite occurrences.

COMMENTS: Beryl and bertrandite (both stable) have low specific gravities and will appear in the light fractions of soils and sediments. See Boyle (1974a) for additional information.

BISMUTH (Bi)

IGNEOUS ROCKS (ppm): AV 0.17; MAFIC 0.15; FELSIC 0.1.

VEGETATION ASH (ppm): < 5 (Table 10A-2).

GEOCHEMICAL ASSOCIATIONS: Chalcophile; Mo, Sn, W, Cu, Pb, Ag, Au, in polymetallic deposits.

ORE MINERALS: native bismuth, bismuthinite.

SUBSTITUTIONS IN: galena, sphalerite, chalcopyrite and other sulfides; also possibly in apatite.

SECONDARY MINERALS: bismutite, bismite.

MOBILITY – PRIMARY ENVIRONMENT: high; concentrates in pegmatites and hydrothermal Sn-Mo-W deposits.

MOBILITY – SECONDARY ENVIRONMENT: very low; that Bi which is mobilized is rapidly precipitated by Fe oxides. Also found in spring precipitates.

Bi is a pathfinder for: sulfide, polymetallic, and other deposits containing Bi; because of its low mobility, its presence indicates close proximity to deposits; soils, stream sediments and panned concentrates have been used successfully.

Comments: Relatively little is known about Bi geochemistry in soils, waters, etc. See Boyle (1974a) for additional information.

BORON (B)

Igneous rocks (ppm): Av 10; Mafic 5; Inter 20; Felsic 15.

Sedimentary rocks (ppm): Sh 100; Ss 35; Ls 10.

Soils (ppm): Av 12. Range 2-100; Temp 2-145; Arid 25-190; Trop (Hum) 1-3.

Surface waters (ppb): Av 10.

Vegetation ash (ppm): 400; toxic to plants (p. 388) in excessive amounts; toxic symptoms can be used for exploration.

Geochemical associations: Lithophile; Li, Be, B, Nb, Sn and other pegmatite elements; B, Be, Cu, Zn, Pb, Mo, W in skarns; B, Be, Sn, F and W in greisens; B, Na, K, Li, Mg, Ca in evaporates (borates).

Ore minerals: borates (e.g., borax, colemanite); tourmaline (very stable).

Substitutions in: many rock-forming silicates (e.g., micas, feldspars).

Soils; occurrence in: tourmaline, clays, soluble borates.

Secondary minerals: in clays; in Fe-oxides.

Aqueous phase: very soluble as $B(OH)_4^-$, also as undissociated boric acid; high in certain brines and hydrothermal waters.

Mobility – primary environment: high; accumulates in late magmatic stages.

Mobility – secondary environment: high as borate in solution in all environments; tourmaline is stable in all environments and moves mechanically.

Geochemical barriers: adsorption (by clays) of soluble borates; mechanical in case of tourmaline.

B is a pathfinder for: borate (evaporate) deposits; all types of deposits which contain tourmaline, e.g., skarns, pegmatites, Sn-W veins, Au-quartz veins.

Pathfinders for B are: boron itself in sedimentary deposits.

Comments: Boron may form either hydromorphic halos (in water; adsorbed on to sediments) or heavy mineral dispersion patterns (from tourmaline). See Boyle (1974a) for additional information. Also see "Comments" under Fluorine in this Appendix.

CADMIUM (Cd)

Igneous rocks (ppm): Av 0.2; Mafic 0.2; Inter 0.2; Felsic 0.2.

Sedimentary rocks (ppm): Sh 0.2; Ss < .1; Ls 0.1.

Soils (ppm): Av 1.

VEGETATION ASH (ppm): .01.

COMMENTS: The geochemistry of Cd is essentially identical to that of Zn, for reasons explained on p. 66, and sphalerite always contains Cd generally in the range of 1000-5000 ppm (Table 2-5). In some situations Cd may become enriched relative to Zn (e.g., hydrothermal zoning; see Fig. 7-3). During the weathering of Cd-containing zinc minerals, secondary Cd minerals, such as greenockite (CdS) and otavite (CdCO$_3$), may form. See Zinc in this appendix.

CERIUM (Ce)

IGNEOUS ROCKS (ppm): Av 60; MAFIC 35; INTER 40; FELSIC 46.

SEDIMENTARY ROCKS (ppm): SH 50; SS 90; LS 10.

SOILS (ppm): Av~5.

SURFACE WATERS (ppb): Av 0.06.

VEGETATION ASH (ppm): 30; although Ce has been found in numerous plants, its value in exploration is poorly documented.

GEOCHEMICAL ASSOCIATIONS: Lithophile; RE, Li, Rb, Cs, Be, Nb, Ta, Zr, B, Th, U, and F in pegmatites; RE, Th, P, Zr, Fe, and Cu in monazite veins; RE, Th, Ba, Sr, P, F, and C in carbonatites; RE, U, P, and F in phosphorites; RE, Au, Ti, Sn, Zr, and Th in placers.

ORE MINERALS: monazite, bastnaesite, cerite, xenotime.

SUBSTITUTIONS IN: apatite, allanite, sphene, fluorite, niobate-tantalates (e.g., euxenite), feldspar (Eu concentrated).

SOILS; OCCURRENCE IN: limonite, clays; possibly as insoluble phosphates (e.g., rhabdophane, florencite).

AQUEOUS PHASE: soluble only under acid conditions (pH of hydrolysis is 2.7; Table 3-6); also forms complexes with organics and carbonates (Wedepohl, 1969-1974).

MOBILITY – PRIMARY ENVIRONMENT: high; concentrates in late stage granites and pegmatites; highly concentrated in some alkalic rocks, including carbonatites.

MOBILITY – SECONDARY ENVIRONMENT: low in solution even though some Ce minerals (e.g., the fluorcarbonates such as bastnaesite) are unstable; primarily as detrital monazite, xenotime, etc.

GEOCHEMICAL BARRIERS: pH; mechanical.

CE IS A PATHFINDER FOR: other rare earths and Th in carbonatites; other elements in associated deposits (e.g., Ti or Sn in placers).

PATHFINDERS FOR CE ARE: The rare-earth elements, P, F, and other elements in the specific geochemical association (e.g., Th, Nb, U); however, orientation studies should precede actual surveys to determine nature of dispersion (chemical or mechanical) and general suitability of proposed survey.

COMMENTS: Cerium is the most abundant (about 30% of the total) of the 16 rare earth (RE) elements which consist of Y and La-Lu (atomic numbers 39 and 57-71). As a rule, all the rare earth (RE) elements accompany each other in specific relative abundances, and they all have essentially identical geochemical characteristics (e.g., similar mobility in all environments) but, in fact, separations are frequent. For example, Ce, La and the other "light" RE elements concentrate in monazite, allanite, and other minerals in carbonatites, whereas Y, Er, Yb and Lu and the other "heavy" RE elements concentrate in xenotime, and fluorite. Likewise, Gd, Tb and Dy concentrate in certain minerals (e.g., gadolinite).

CHROMIUM (Cr)

IGNEOUS ROCKS (ppm): AV 100; MAFIC 200; INTER 20; FELSIC 4.

SEDIMENTARY ROCKS (ppm): SH 100; SS 35; LS 10.

SOILS (ppm): AV 50. RANGE 5-1000; TEMP 7-300; ARID 200-500; TROP (HUM) 150-300.

SURFACE WATERS (ppb): AV 1.

VEGETATION ASH (ppm): 90; poor biological response.

GEOCHEMICAL ASSOCIATIONS: Siderophile; Cr, Co, Ni, Cu, Pt-group.

ORE MINERALS: chromite.

SUBSTITUTIONS IN: micas (fuchsite, mariposite, Cr-micas) and garnets; in absence of chromite, Cr substitutes in pyroxene and magnetite in basic rocks.

SOILS; OCCURRENCE IN: chromite, limonite.

SECONDARY MINERALS: none; but serpentine may contain minute chromite grains.

AQUEOUS PHASE: effectively insoluble except under very acid-oxidizing conditions as chromate ion.

MOBILITY – PRIMARY ENVIRONMENT: low; concentrates in ultrabasic rocks.

MOBILITY – SECONDARY ENVIRONMENT: only as detrital chromite (very stable); concentrates in heavy fraction of soils and sediments.

GEOCHEMICAL BARRIERS: mechanical.

CR IS A PATHFINDER FOR: chromite, platinum, and other ultramafic deposits.

PATHFINDERS FOR CR ARE: Ni, Co, Cu.

COMMENTS: Cr has been successfully used in soil, sediment and heavy mineral surveys (but not hydrogeochemical) based primarily on the stability of chromite.

COBALT (Co)

IGNEOUS ROCKS (ppm): AV 25; MAFIC 50; INTER 10; FELSIC 1.

SEDIMENTARY ROCKS (ppm): SH 20; BLSH 10-20; SS 0.3; LS 4.

SOILS (ppm): AV 10. RANGE 1-40; TEMP 1-45; ARID 10-100; TROP (HUM) 1-50.

SURFACE WATERS (ppb): AV 0.2.

VEGETATION ASH (ppm): 15; biological response satisfactory (Table 10A-2 but see p. 384).

GEOCHEMICAL ASSOCIATIONS: Siderophile; also chalcophile. Ni, Co, Pt, Fe, Cu, Ag, Au, Se, Te, and S in Ni-Cu massive sulfide deposits; Ni, Co, Ag, Fe, Cu, Pb, Zn, As, Sb, Bi, and U in Cu-Co sulfide ores; with certain Au and Ag ores; Ni, Co, Fe, Mn, and Cr in laterites; Mn, Ni, Cu, Zn, and Co in deep-sea Mn nodules (Table 3-10); see Uranium in this appendix for U associations.

ORE MINERALS: cobaltite, smaltite.

SUBSTITUTIONS IN: Ni minerals; pyrite, sphalerite and other sulfides (Table 2-5); in some Fe-Mg silicates.

SOILS; OCCURRENCE: strongly adsorbed by Mn oxides (pp. 135, 384); also adsorbed by limonite and clay (Table 3-3).

SECONDARY MINERALS: erythrite (cobalt bloom).

AQUEOUS PHASE: as ionic Co^{2+}.

MOBILITY – PRIMARY ENVIRONMENT: high; concentrates in hydrothermal veins where it forms sub-ore halos (Fig. 7A-6); also low, as it occurs in some ultrabasic occurrences.

MOBILITY – SECONDARY ENVIRONMENT: good; pH of hydrolysis is 6.8 (see also Fig. 6-3).

GEOCHEMICAL BARRIERS: sulfide; adsorption; pH.

CO IS A PATHFINDER FOR: all deposits in which it occurs, e.g., Cu (in shales), Ni-Cu (massive sulfide), U (vein type), and certain Ag and Au veins; see Table 3A-4.

COMMENTS: Co can be used in practically all types of surveys: soil, sediment, water, vegetation, and till. See Boyle (1974a) for additional information.

COPPER (Cu)

IGNEOUS ROCKS (ppm): AV 55; MAFIC 100; INTER 30; FELSIC 10.

SEDIMENTARY ROCKS (ppm): SH 50; BLSH 70-150; SS 10; LS 15.

SOILS (ppm): AV 20. RANGE 2-100; TEMP 25; ARID 15-100; TROP (HUM) 10-150.

SURFACE WATERS (ppb): AV 7.

VEGETATION ASH (ppm): 20; biological response fair (Table 10A-2) although Cu is often considered a "difficult" element (p. 407); Cu can cause barren areas (Fig. 10A-3).

GEOCHEMICAL ASSOCIATIONS: Chalcophile; Cu, Pb, Zn, Cd, Ag, Fe, As, Sb in massive sulfide (volcanogenic) type deposits; Cu, Mo, Re, Fe in porphyry Cu type deposits; Ni, Cu, Pt, Cr in ultrabasic Pt deposits; Ag, Zn, Pb, Mo, Co in copper shale deposits (e.g., Kupferscheifer).

ORE MINERALS: chalcopyrite, bornite, other sulfides, native Cu.

SUBSTITUTIONS IN: pyroxene, amphiboles, magnetite, biotite.

SOILS; OCCURRENCE: adsorbed by clays, Fe-Mn oxides, and organic matter.

SECONDARY MINERALS: many carbonates, sulfides, oxides, sulfates, and silicates.

AQUEOUS PHASE: ionic (Cu^{2+}, Cu^+); organic complexes.

MOBILITY – PRIMARY ENVIRONMENT: high, as in the case of porphyry copper deposits in felsic rocks; low, as in case of Pt-Cr occurrences in ultrabasics.

MOBILITY – SECONDARY ENVIRONMENT: high in oxidizing, acidic waters; low in alkaline and reducing waters.

GEOCHEMICAL BARRIERS: sulfide; adsorption; pH (pH of hydrolysis = 5.5).

CU IS A PATHFINDER FOR: all types of deposits in which it occurs.

PATHFINDERS FOR CU ARE: Mo in porphyry copper deposits; other chalcophile elements (e.g., Zn, Co) are used in hydrogeochemical surveys for Cu deposits when waters are alkaline.

COMMENTS: Cu is one of the most commonly determined elements in exploration geochemistry using all types of sampling media (soils, sediments, waters, vegetation, etc.). See Boyle (1974a) for additional information.

FLUORINE (F)

IGNEOUS ROCKS (ppm): AV 625; MAFIC 400; INTER 500; FELSIC 735.

SEDIMENTARY ROCKS (ppm): SH 740; SS 270; LS 330.

SOILS (ppm): AV 200.

SURFACE WATERS (ppb): AV 100.

VEGETATION ASH (ppm): 10; poor biological response (Table 10A-1).

GEOCHEMICAL ASSOCIATIONS: Lithophile; F, Ca, Fe, S, Si, Ba, Sr, Pb, Zn, and Cu in veins and stockworks; F, Al, Ca, Sn, Mo, and W in greisen zones; Nb, Ta, P, F, Ti, and rare-earths in carbonatites; F, U, V, Se, As, and rare-earths in phosphorites; Pb, Zn, Ba, and F in carbonates (Mississippi Valley type).

ORE MINERALS: fluorite, topaz.

SUBSTITUTIONS IN: apatite, mica, amphibole, tourmaline.

SOILS; OCCURRENCE IN: secondary phosphates, apatite, fluorite, adsorbed by Al oxides.

AQUEOUS PHASE: as fluoride ion (F^-).

MOBILITY – PRIMARY ENVIRONMENT: high; concentrates in granites, pegmatites, late stage hydrothermal veins, and carbonatites.

MOBILITY – SECONDARY ENVIRONMENT: high; soluble under all environmental conditions.

GEOCHEMICAL BARRIERS: calcium barrier (causes precipitation of fluorite) limits solubility of F; adsorption (by Al oxides).

F IS A PATHFINDER FOR: many types of deposits (see below).

PATHFINDERS FOR F ARE: F, and associated base metals.

COMMENTS: Boyle (1974a, p. 36) states, "Fluorine is a persistent constituent of almost all types of mineral deposits, occurring in amounts from traces to minor and major. The element is, therefore, almost the universal indicator of mineralization. The other elements of almost universal occurrence in mineral deposits are boron and sulphur. These three elements used on conjunction with one another provide a formidable array of indicators in geochemical prospecting for practically all types of deposits. Where one element fails to serve another will in most cases." (Note: phosphate fertilizer may contain F which can result in high values for streams).

GOLD (Au)

IGNEOUS ROCKS (ppm): Av 0.004; MAFIC 0.004; INTER 0.004; FELSIC 0.004.

SEDIMENTARY ROCKS (ppm): SH 0.004; LS 0.005.

SOILS (ppm): Av 0.001.

SURFACE WATERS (ppb): Av 0.002.

VEGETATION ASH (ppm): 0.005; the value of biogeochemical methods in gold exploration are dubious at present.

GEOCHEMICAL ASSOCIATIONS: Siderophile (also noble); in quartz veins generally with SiO_2, Ag, As, Sb, S, and Fe; Fe, Zn, Pb, Cu, and Mo in sulfide deposits; in placers with U (South Africa), or other elements in heavy minerals.

ORE MINERALS: native gold, tellurides.

SUBSTITUTIONS IN: native silver, in sulfides it probably occurs as minute inclusions and not in isomorphous substitution.

SOILS; OCCURRENCE IN: dissolved and re-precipitated native gold.

AQUEOUS PHASE: soluble organic complexes, chloride complexes, and other mechanisms have been proposed.

MOBILITY − PRIMARY ENVIRONMENT: concentrates in hydrothermal veins; also concentrated in some porphyry copper deposits.

MOBILITY − SECONDARY ENVIRONMENT: for all practical purposes only mechanical mobility as native gold.

GEOCHEMICAL BARRIERS: mechanical (Table 2A-1).

AU IS A PATHFINDER FOR: has been proposed for porphyry copper (p. 54).

PATHFINDERS FOR AU ARE: Ag, As, Sb, Bi, and W in placers or rocks (Te and Hg also in some cases); the same elements plus Cu, Se and Te in plant ash may be useful indicators.

COMMENTS: See Boyle (1974a) for further information.

LEAD (Pb)

IGNEOUS ROCKS (ppm): Av 13; MAFIC 5; INTER 15; FELSIC 20.

SEDIMENTARY ROCKS (ppm): SH 20; BLSH 20-70; SS 7; LS 8.

SOILS (ppm): Av 20. RANGE 2-200; TEMP 40; ARID 20; TROP (HUM) 20.

SURFACE WATERS (ppb): Av 3.

VEGETATION ASH (ppm): 10; generally fair to good biological response; can cause barren areas (Fig. 10A-2).

GEOCHEMICAL ASSOCIATIONS: Chalcophile; most lead deposits contain Ag, Zn, Cd and Cu; sulfide deposits with Pb contain Ag, Zn, Cd, Cu, Ba, Sr, V, Cr, Mn, Fe, Ga, In, Tl, Ge, Sn, As, Sb, Bi, Se, Hg, and Te. In carbonates (Mississippi Valley type) with Zn and Cd.

ORE MINERALS: galena.

SUBSTITUTIONS IN: K-feldspar, plagioclase, mica, zircon, magnetite (Table 2A-4).

SOILS; OCCURRENCE IN: relatively stable galena, anglesite, and cerussite in heavy fractions; ionic Pb^{2+} accumulates by adsorption in humic and clay horizons.

SECONDARY MINERALS: cerussite, anglesite, pyromorphite.

AQUEOUS PHASE: ionic Pb^{2+} in most waters; complexing with chloride in brines.

MOBILITY — PRIMARY ENVIRONMENT: high; accumulates in granitic rocks and hydrothermal veins; forms supra-ore halos (Figs. 7-5, 7A-10, 7A-11).

MOBILITY — SECONDARY ENVIRONMENT: generally low in most oxidizing and gley environments but immobile in reducing sulfide environment; forms very insoluble carbonate, sulfate, and phosphate minerals; galena, cerussite, and anglesite move mechanically in the heavy fractions of sediments and soils.

GEOCHEMICAL BARRIERS: sulfide; carbonate; sulfate; adsorption; pH (pH of hydrolysis = 6.0).

PB IS A PATHFINDER FOR: Pb deposits; U deposits; polymetallic vein deposits.

PATHFINDERS FOR PB ARE: Zn, Cd, Ag, Cu, Ba, As, and Sb.

COMMENTS: The use of isotopic analysis of lead minerals (discussed in Chapters 7 and 7A) has significant exploration potential.

LITHIUM (Li)

IGNEOUS ROCKS (ppm): AV 20; MAFIC 10; INTER 25; FELSIC 30.

SEDIMENTARY ROCKS (ppm): SH 60; Ss 15; Ls 20.

SOILS (ppm): AV 30. RANGE 5-200; TEMP 30; ARID 30; TROP (HUM) 30.

SURFACE WATERS (ppb): AV 3.

VEGETATION ASH (ppm): 6; biological response good.

GEOCHEMICAL ASSOCIATIONS: Lithophile; Li, Be, B, K, Rb, Cs, Nb, Ta, F, P, Sn, W and rare earths in granite pegmatites; Li, B, F, Be, Sn, W, and Mo in greisens; Li, Na, K, B, P, W, F, Br, Cl, I, SO_4 and CO_3 in brines and saline evaporites (e.g., Searles Lake, California).

ORE MINERALS: spodumene, lepidolite, amblygonite; also economic deposits in brines (Searles Lake; Silver Peak, Nevada; Great Salt Lake, Utah).

SUBSTITUTIONS: for Mg in micas (especially biotite) and hornblende.

SOILS; OCCURRENCE IN: Mn oxides (p. 135) and clays.

AQUEOUS PHASE: ionic Li^+.

MOBILITY — PRIMARY ENVIRONMENT: high; concentrates in granites and pegmatites.

MOBILITY — SECONDARY ENVIRONMENT: relatively mobile in all environments (although it is the least mobile of the alkalis).

GEOCHEMICAL BARRIERS: adsorption (by Mn oxides and clays).

LI IS A PATHFINDER FOR: pegmatite deposits; hydrothermally altered tuffs containing Be and fluorite (Spor Mtn., Utah); greisens; deep-seated brines containing Li by analyzing vegetation (Cannon et al., 1975); recognizing mineralized granites (Tables 7-1, 7-2, 2A-9).

PATHFINDERS FOR LI ARE: Li, Sn, Nb, W, etc., in heavy resistate minerals of soils and sediments in pegmatite and greisen areas; B and W for Li-containing brines.

COMMENTS: See Boyle (1974a) for additional information.

MERCURY (Hg)

IGNEOUS ROCKS (ppm): AV 0.08; MAFIC 0.08; INTER 0.08; FELSIC 0.08.

SEDIMENTARY ROCKS (ppm): SH 0.5; SS 0.07; LS 0.05.

SOILS (ppm): AV 0.03.

SURFACE WATERS (ppb): AV 0.007; up to 200 ppb in waters from hot springs.

VEGETATION ASH (ppm): 0.001; biological response for exploration purposes unknown.

GEOCHEMICAL ASSOCIATIONS: chalcophile; Hg, Sb, As and sometimes Sn and W; in some gold-quartz and silver (e.g., Cobalt type) veins; in some Zn vein deposits.

ORE MINERALS: cinnabar, native mercury.

SUBSTITUTIONS IN: sphalerite (Table 2-5), tetrahedrite, tennatite, and other sulfides.

SOILS; OCCURRENCE IN: cinnabar, native mercury, Hg vapor, and organic complexes.

AQUEOUS PHASE: probably as soluble sulfide complex (especially in thermal waters), and as organic complexes.

MOBILITY – PRIMARY ENVIRONMENT: highly mobile; concentrates in late stage, low temperature veins and stockworks; forms supra-ore halos.

MOBILITY – SECONDARY ENVIRONMENT: low in surface waters; high in thermal waters; high in vapor phase; also mechanical transport (in the heavy fraction) because cinnabar is very stable.

GEOCHEMICAL BARRIERS: sulfide, adsorption (by organic matter).

HG IS A PATHFINDER FOR: base metal (particularly Zn-Pb-Ag) deposits; certain types of Au deposit (e.g., Carlin-type).

PATHFINDERS FOR HG ARE: Sb and As; high content of Hg in placer gold suggests proximity to Hg deposits.

COMMENTS: Discussions on the use of Hg in geochemical exploration will be found in Chapters 7 and 7A. Instrumentation is discussed in Chapters 6 and 6A.

MOLYBDENUM (Mo)

IGNEOUS ROCKS (ppm): AV 1.5; MAFIC 1; INTER 1; FELSIC 2.

SEDIMENTARY ROCKS (ppm): SH 3; BLSH 10; SS 0.2; LS 1.

SOILS (ppm): AV 2. TEMP 1-5; ARID 2-5; TROP (HUM) 1-5.

SURFACE WATERS (ppb): AV 1.

VEGETATION ASH (ppm): 9; good biological response (Tables 10A-1, 10A-2).

GEOCHEMICAL ASSOCIATIONS: Chalcophile and siderophile; Mo, W, Re, Cu, Sn, Be, B, F, P, Zn, Bi, and Fe in pegmatites; Mo, Bi, W, F, and Be in griesens; Mo, Cu, Re, Ag, Au, and Zn in prophyry copper deposits; Mo, U, Se, V, and Cu in sandstone-type U deposits.

ORE MINERALS: molybdenite; minor powellite and wulfenite.

SUBSTITUTIONS: for W in tungsten minerals (e.g., scheelite and wolframite); does not substitute in silicates or sulfides to any significant degree.

SOILS; OCCURRENCE IN: molybdenite, ferrimolybdite, wulfenite, and powellite; adsorbed on organic matter (Fig. 3-18) and limonite.

SECONDARY MINERALS: ferrimolybdite, powellite, wulfenite; secondary molybdenite (in hydrogen sulfide environment); also as ilsemannite ($Mo_3O_8 \cdot nH_2O$), which occurs as a deep blue stain, and is the result of Mo in molybdate-containing solutions being reduced from Mo^{6+} to Mo^{5+} (as in the sandstone U deposits in the U.S.).

AQUEOUS PHASE: molybdate (MoO_4^{2-}; Fig. 3-20).

MOBILITY – PRIMARY ENVIRONMENT: high; enriched in granites, pegmatites, high temperature hydrothermal veins, and porphyry copper deposits.

MOBILITY – SECONDARY ENVIRONMENT: high in oxidizing, alkaline environments but low in acidic (the reverse of copper); low in reducing environments; mobility restricted by presence of Pb (forms wulfenite), Fe (forms ferrimolybdite), and carbonate (forms powellite); molybdenite weathers slowly in some climates and may be found as detrital grains in the heavy fractions of soils and sediments.

GEOCHEMICAL BARRIERS: sulfide; reducing; adsorption; special ions (e.g., Pb, Fe, carbonate).

MO IS A PATHFINDER FOR: molybdenum, porphyry copper, and sandstone-type U deposits.

PATHFINDERS FOR MO ARE: Cu, W, and Bi; F and some of the other associated elements in some types of occurrences.

COMMENTS: Mo is an excellent indicator of deposits containing this element and can be determined in all types of surveys utilizing soils, sediments, spring precipitates, vegetation, bogs, etc. Boyle (1974a) notes that during the weathering of deposits that contain both quartz and molybdenite, the highly resistant quartz in eluvium, soils and sediments should be examined closely for the presence of molybdenite; this applies particularly to cobbles and boulders which are not normally collected for analysis. See Boyle (1974a) for additional information on the geochemistry of Mo.

NICKEL (Ni)

IGNEOUS ROCKS (ppm): AV 75; MAFIC 150; INTER 20; FELSIC 0.5.

SEDIMENTARY ROCKS (ppm): SH 70; BLSH 50-200; SS 2; LS 12.

SOILS (ppm): AV 30. RANGE 5-500; TEMP 25; ARID 50; TROP (HUM) 40.

SURFACE WATERS (ppb): AV 0.3.

VEGETATION ASH (ppm): 20; biological response good (see discussion of Ni hyperaccumulators in Chapter 10A).

GEOCHEMICAL ASSOCIATIONS: Siderophile (also chacophile); Ni, Co, Fe, Cu, Au, Ag, Pt metals, Se, Te, As and S in massive sulfide deposits (e.g., Sudbury); Ni, Co, Fe, Cu and S in veins and sulfide lenses; U, Cu, Ag, Co, Ni, As, V, Se, Au and Mo in unconformity U deposits (e.g., Key Lake, Sask.); Ni, Co, Fe, Mn and Cr in residual laterite deposits (Fig. 3-12); Mn, Ni, Cu, and Co in deep sea Mn nodules (Table 3-10).

ORE MINERALS: pentlandite, nickeliferous pyrrhotite, niccolite, nickeliferous laterites, Mn nodules.

SUBSTITUTIONS IN: olivine (up to 5000 ppm), Mg-pyroxenes, amphiboles, and micas; pyrite, chalcopyrite, and other sulfides.

SOILS; OCCURRENCE: adsorbed on Fe oxides (limonite and laterites) and Mn oxides; adsorption by organic matter in soils and bogs; hydrous Ni silicates.

SECONDARY MINERALS: garnierite and other hydrated Ni silicates (including Ni-bearing serpentine).

AQUEOUS PHASE: ionic Ni^{2+}.

MOBILITY – PRIMARY ENVIRONMENT: low; concentrates in ultramafic and basic rocks (average content 2000 ppm).

MOBILITY – SECONDARY ENVIRONMENT: good in oxidizing acidic or reducing gley environments; immobile in alkaline (pH of hydrolysis = 6.7) or reducing hydrogen sulfide environments; often enriched in spring precipitates.

GEOCHEMICAL BARRIERS: sulfide, adsorption, pH.

NI IS A PATHFINDER FOR: all types of deposits in which it occurs (e.g., massive sulfide, Pt metal, certain uranium deposits).

PATHFINDERS FOR NI ARE: nickel itself; Cu, Co, As, and Cr (heavy mineral surveys of soils and glacial tills using these elements are often very effective); Pt and Pd in deeply weathered lateritic terrain as in Western Australia (Table 3A-1).

COMMENTS: Ni is a good indicator in practically all types of surveys using most sampling media.

NIOBIUM (Nb)

IGNEOUS ROCKS (ppm): AV 20; MAFIC 20; INTER 20; FELSIC 20.

SEDIMENTARY ROCKS (ppm): SH 20; SS < 1; LS 0.3.

SOILS (ppm): AV 15.

VEGETATION ASH (ppm): 0.3; biological response possible, but not well documented (see Table 10A-2).

GEOCHEMICAL ASSOCIATIONS: Lithophile; Nb, Ta, Sn, W, Li, Be, Ti, Rb, Cs, U, Th, B, Zr, Hf, P, F, and rare-earths in granite and syenitic pegmatites; Nb, Ta, Na, K, Ba, Sr, Ti, Zr, U, Th, Cu, Zn, P, S, F, and rare-earths in carbonatites; Nb, Ti, Ga, Be, and Al in bauxite developed on alkalic rocks.

ORE MINERALS: pyrochlore, columbite-tantalite.

SUBSTITUTIONS IN: rutile, ilmenite, sphene, cassiterite, zircon, and biotite.

SOILS; OCCURRENCE: as heavy resistate minerals.

AQUEOUS PHASE: Nb is insoluble in most natural waters, however, precipitates from springs in carbonatite areas may have high contents of Nb.

MOBILITY – PRIMARY ENVIRONMENT: high; concentrates in albitic granites, pegmatites, and carbonatites.

MOBILITY – SECONDARY ENVIRONMENT: as detrital heavy minerals.

GEOCHEMICAL BARRIERS: mechanical.

NB IS A PATHFINDER FOR: carbonatites; placer deposits which may also contain Au, Ti, Sn, W, rare-earths, and other elements.

PATHFINDERS FOR NB ARE: F, P, Ba, Sr, rare-earths, etc. for carbonatite occurrences; Li, Be, W, Zr, B, Rb, Cs, Mo, etc., for pegmatite occurrences; depending on the geochemical characteristics of the individual pathfinder elements, they may be either in the light or heavy fraction of soils or sediments, and some may be in solution.

COMMENTS: Tantalum always accompanies Nb in all Nb minerals (solid solution series Nb-Ta) and mineral deposits. Usually Nb predominates over Ta by a factor of 10; however, in some minerals and in some carbonatites, Ta > Nb. Ta has essentially the same geochemical characteristics as those listed above for Nb, and the same geochemical techniques used to search for Nb apply to Ta.

PLATINUM (Pt)

IGNEOUS ROCKS (ppm): AV 0.002; MAFIC 0.02; FELSIC 0.008.

GEOCHEMICAL ASSOCIATIONS: Noble and siderophile; Pt, Ni, Cu, Co, As, Ag, and Au, with minor Te, Bi, Sn and Sb, where Pt is a minor constituent (< 0.5 ppm Pt) in massive Ni-Cu sulfide ores containing sperrylite (Sudbury type); Pt, Ag, Au, Cr, Fe, Cu, Ni, Co and S in sulfides (Merensky Reef, South Africa; see below).

ORE MINERALS: various Pt arsenides (e.g., sperrylite), sulfides (e.g., braggite), Bi-tellurides (e.g., moncheite), and compounds with Sb, Sn, Hg and Fe; also native Pt, Os-Ir, etc.

SUBSTITUTIONS IN: chromite; reportedly in solid solution in pyrite, pyrrhotite, and pentlandite at Merensky Reef (Boyle 1974a).

SOILS; OCCURRENCE IN: native platinum; chromite; intergrown with heavy resistate minerals (e.g., chrome-spinel).

MOBILITY − PRIMARY ENVIRONMENT: low; concentrates in dunitic ultrabasic rocks containing Cu-Ni sulfides (typically pyrrhotite, pentlandite, chalcopyrite); associated minerals are olivine, chrome-spinel, chromite, magnetite.

MOBILITY − SECONDARY ENVIRONMENT: only as detrital grains of native platinum which form placers.

GEOCHEMICAL BARRIERS: mechanical.

PT IS A PATHFINDER FOR: Ni sulfide deposits in Western Australia by analyzing gossans (see Table 3A-1).

PATHFINDERS FOR PT ARE: Ni, Cu, Cr, Co, Se, and Te; mineralogical identification of Cr-spinel, chrome-magnetite, olivine, and chromite in geologically favorable areas with ultrabasic rocks (including serpentinites) using heavy mineral concentrates.

COMMENTS: The term "platinum metals" or "platinoids" applies to six elements: Ru, Rh, Pd, Os, Ir and Pt which are geochemically closely associated. Minerals of the Pt group may contain all six elements in varying proportions (e.g., in Alaskan placers, usually Pd > Pt). Platinum element mines in South Africa (Merensky Reef), which are among the richest in the world, contain only approximately 8 ppm (0.25 oz/ton) total platinoids.

Data are lacking for the behaviour of Pt during weathering, transport mechanisms in aqueous media, biogeochemistry, and for many other geochemical parameters. Excellent reviews of Pt geochemistry have been prepared by J. H. Crocket (in Wedepohl, 1969) and Boyle (1974a). The latter lists 14 types of deposits in which Pt group metals are concentrated; most are not economic at this time.

RADIUM (Ra)

ROCKS (ALL TYPES): The nuclei ratio Ra/U at equilibrium is 3.42×10^{-7}; half-life of ^{226}Ra is about 1600 years.

SURFACE WATERS: Av 0.1 − 1.0 picocuries/liter (pCi/l).

VEGETATION ASH (ppm): average values not available; biological response excellent (see Table 10A-1; Figs. 10A-1 and 10A-5).

GEOCHEMICAL ASSOCIATIONS: Lithophile; ^{238}U, ^{230}Th, ^{226}Ra, ^{222}Rn, ^{214}Bi, and ^{206}Pb in uranium deposits which are in equilibrium (Fig. 3A-12); Ra and Ba in weathering products (e.g., gossans) generally as radiobarite (Ba, Ra) SO_4; Ra, Mn, Fe in precipitates from springs; Ra and organic matter in bogs, or as thucolite. (Several other Ra isotopes are known of which ^{228}Ra, with a half-life of 6.7 years, has potential in exploration for Th.)

ORE MINERALS: none; if recovered (which today is for environmental as opposed to economic reasons) it is as a by-product of U or Th processing.

SUBSTITUTIONS IN: radiobarite; also in radio-calcite, radio-opal, and Mn-Fe oxides at spring deposits.

Soils; occurrence in: radiobarite, adsorbed on Mn-Fe oxides, or in U or Th minerals.

Secondary minerals: radiobarite, Mn-Fe oxides, and other minerals mentioned above.

Aqueous phase: ionic Ra^{2+}; very soluble.

Mobility – primary environment: high; similar to Ba.

Mobility – secondary environment: high, but mobility reduced by the geochemical barriers listed below; soluble in all types of waters (surface, ground, and lake).

Geochemical barriers: cocrystallization (with Ba in radiobarite or Ca in radiocalcite); coprecipitation (with Fe-Mn oxides); adsorption (by Fe-Mn oxides, clays and organic matter).

Ra is a pathfinder for: U and Th deposits of all types; other types of deposits containing U or Th as accessories (e.g., rare metal pegmatites, carbonatites).

Comments: Modern exploration geochemists concerned with search for uranium must understand: the concepts of disequilibrium in the U series; the relationships between Ra and U, and between Ra, Rn and ^{214}Bi; false anomalies which may result from the fractionation of Ra from its parent and daughter nuclides; geochemical barriers for Ra; limitations and pitfalls in the use of radon techniques; and other aspects of the mobility of U and its daughter nuclides in the secondary environment. These are discussed in Chapter 3A and in other parts of this book (see index).

RADON (Rn)

Rocks (all types): The nuclei ratio Rn/U at equilibrium is 2.3×10^{-12}; half-life of ^{222}Rn is about 3.8 days.

Surface waters: 0-5 pCi/1 in lakes; 5-100 pCi/1 in streams; 100-1000 pCi/1 in wells and springs.

Vegetation ash (ppm): Average value not available; dependent upon amount of Ra in the vegetation.

Geochemical associations: Lithophile; ^{222}Rn is a decay product of ^{226}Ra in the ^{238}U series, hence, the association ^{222}Rn and ^{226}Ra, and also ^{222}Rn and ^{214}Bi (see Fig. 3A-12). Several other Rn isotopes are found in nature (^{220}Rn in the ^{232}Th series and ^{219}Rn in the ^{235}U series) but because of their very short half-lives (in seconds) they are not important in exploration.

Soils; occurrence in: all uranium minerals; radiobarite; soil gas; dissolved in soil waters and groundwater.

Aqueous phase: Rn gas is soluble in water.

Mobility – primary environment: potentially high, but limited by its short half-life (3.8 days) and the emanation coefficient of radioactive minerals.

Mobility – secondary environment: potentially high; limited by its short half-life, emanation coefficient, and physical characteristics of the host formation.

GEOCHEMICAL BARRIERS: short half-life; emanation coefficient of parent Ra-containing mineral; physical characteristics (e.g., permeability) of host rock or other formations through which Rn attempts to transgress.

RN IS A PATHFINDER FOR: uranium and thorium deposits; other types of deposits containing U or Th as accessories (e.g., rare metal pegmatites, carbonatites); the value of Rn is based upon the fact that it is a gas which, in theory at least, can migrate upwards so that concealed deposits can be detected.

COMMENTS: Comments under "Radium" are applicable to radon.

RARE EARTH ELEMENTS (RE or REE) (see Cerium)

SELENIUM (Se)

IGNEOUS ROCKS (ppm): AV 0.05; MAFIC 0.05; FELSIC 0.05.

SEDIMENTARY ROCKS (ppm): SH 0.6; LS 0.08.

SOILS (ppm): AV 0.2.

SURFACE WATERS (ppb): AV 0.2.

VEGETATION ASH (ppm): 1; biological response good particularly as a geobotanical indicator and as a biogeochemical sampling medium for Se-containing U ores in the western U.S.

GEOCHEMICAL ASSOCIATIONS: Chalcophile; Se, Hg, As, Sb, Ag, Cu, Zn, Cd and Pb in polymetallic sulfide ores; Cu, Ni, Se, Ag, Co, etc., in copper-pyrite ores (e.g., Sudbury); U, V, Se, Cu, Mo in sandstone U ore; Au and Ag selenide ores.

ORE MINERALS: none; by-product from refining of sulfide (primarily Cu) ores; also recovered as native Se from some sandstone-type U ores.

SUBSTITUTIONS IN: pyrite, chalcopyrite, pyrrhotite, and sphalerite (Table 2-5), and in other sulfide minerals, substituting for S.

SOILS; OCCURRENCE: adsorbed on iron oxides (gossans) and on clay minerals; native Se.

SECONDARY MINERALS: native Se; secondary selenides.

AQUEOUS PHASE: as ionic selenite (possibly as selenate); similar to sulfate in its mobility.

MOBILITY – PRIMARY ENVIRONMENT: accompanies S separated from magmas; hence may be found in sulfides from basic to hydrothermal deposits with some changes in S:Se ratio; also forms selenides at hydrothermal stages.

MOBILITY – SECONDARY ENVIRONMENT: good in oxidizing, acid, and alkaline environments; immobile in reducing environments (similar to S); Se-bearing sulfides and selenides weather easily resulting in selenite ions.

GEOCHEMICAL BARRIERS: reducing, adsorption.

SE IS A PATHFINDER FOR: sandstone, classical vein, and unconformity types of uranium deposits; polymetallic sulfide deposits; some Au-Ag deposits; Ni sulfide deposits in Western Australia by analyzing gossans for Pt, Pd, and Se (see Table 3A-1).

COMMENTS: Se is related to S in its biological behavior and certain plants are Se accumulators whereas it is toxic to other forms of vegetation (also to cattle and other animals). Most of the successes with vegetation surveys (both geobotanical and biogeochemical) based on Se are limited, at least to this point, to the U-bearing regions of the western United States (pp. 393-395). However, Se in plants has potential in exploration for gold and other ores in which Se is present.

SILVER (Ag)

IGNEOUS ROCKS (ppm): AV 0.07; MAFIC 0.1; INTER 0.07; FELSIC 0.04.

SEDIMENTARY ROCKS (ppm): SH 0.05; BLSH < 1-5; SS < 0.01; LS 1.

SOILS (ppm): AV 0.1.

SURFACE WATERS (ppb): AV 0.3.

VEGETATION ASH (ppm): 1; biological response of vegetation for Ag is poor (Table 10A-1) but anomalies can be detected (Tables 10-2, 10A-2).

GEOCHEMICAL ASSOCIATIONS: Chalcophile; Pb, Zn, Cd, Ag, Hg, As, Sb, Se in complex sulfides; Ag, Ni, Co, Fe, S, As, Sb, Bi (and U) in native Ag deposits containing Ni-Co arsenides (Cobalt type); U, V, Se, As, Mo, Pb, Cu, Ag, etc., in Cu-U-V "red bed" sandstone deposits; U, Cu, Ag, Co, Ni, As, V, Se, Au and Mo in unconformity vein uranium deposits (e.g., Key Lake, Sask.); Au, Ag, Te and Hg in veins; Cu, Mo, Ag (and Au) in some porphyry copper deposits.

ORE MINERALS: native silver, argentite, tetrahedrite, Ag-bearing galena.

SUBSTITUTIONS IN: galena, sphalerite, chalcopyrite (Table 2-5); other sulfides; arsenides and antimonides (particularly tetrahedrite); native copper; also in nearly all silicates but seldom exceeding 500 ppb (0.5 ppm), possibly replacing Na.

SOILS; OCCURRENCE: enriched in A (humic) horizons (Fig. 3-3); adsorbed by Mn and Fe oxides.

SECONDARY MINERALS: cerargyrite, secondary native silver.

AQUEOUS PHASE: various chloride complexes with Na and K; complex sulfide, polysulfide and hydrosulfide ions; soluble organic complexes.

MOBILITY − PRIMARY ENVIRONMENT: high; concentrates in late hypogene veins and stockworks (e.g., some porphyry copper deposits); in veins with carbonates or barite (Au is usually found with quartz); forms supra-ore halos (Figs. 7-5, 7A-9).

MOBILITY − SECONDARY ENVIRONMENT: slightly mobile in oxidizing, acid and gley environments; immobile in reducing and alkaline environments; concentrated in supergene enrichment zones (p. 84); mobility reduced by precipitates (e.g., Pb) and adsorption; mobility increased by certain complexes (e.g., thiosulfate); Ag only rarely forms placers.

GEOCHEMICAL BARRIERS: sulfide; pH; adsorption (by humic materials, Fe and Mn oxides); ionic precipitates (precipitated by Pb, Cl, chromate, arsenate).

AG IS A PATHFINDER FOR: Au-Ag veins; some types of U deposits; some types of porphyry copper deposits; other types of deposits in which it occurs.

PATHFINDERS FOR AG ARE: Ni, Co, As, Sb and Bi in native Ag deposits (Cobalt type); where U occurs in native Ag veins, it (U) may be used; Hg in some types of deposits; Ag is a good indicator of its own deposits.

COMMENTS: Boyle (1974a) notes that ground and surface waters in the vicinity of Ag deposits are frequently enriched in Ag and other elements, particularly Ni, Co, U and As (Table 3-2; see also Table 3A-4 which confirms the use of Ag in certain hydrogeochemical surveys).

TANTALUM (Ta) (see Niobium)

TELLURIUM (Te)

IGNEOUS ROCKS (ppm): AV 0.001; MAFIC 0.001; INTER 0.001; FELSIC 0.001.

SEDIMENTARY ROCKS (ppm): SH 0.01.

GEOCHEMICAL ASSOCIATIONS: Chalcophile; Ni, Cu, Co, Te, etc., in pyrrhotite-pentlandite mafic sulfide ores with chalcopyrite (e.g., Sudbury); Au-Te in sulfide-bearing veins (e.g., Cripple Creek, Colorado); Au, Ag, Te, Hg in high temperature veins; Cu, Mo, Te, S in some porphyry copper deposits (e.g., western U.S.); polymetallic sulfide deposits with Pb may contain Te.

ORE MINERALS: none; Te is a by-product of the smelting of Ni-Cu massive sulfide ores, and polymetallic ores (particularly those containing Pb).

SUBSTITUTIONS IN: pyrite (particularly of high temperature sedimentary origin); molybdenite; native gold and silver.

SOILS; OCCURRENCE: as stable tellurides in some environments; in highly oxidizing environments, Te may occur as native tellurium or be oxidized to TeO_2; also as tellurites and tellurates.

SECONDARY MINERALS: various rare tellurites (e.g., emmonsite) and tellurates (e.g., ferrotellurite), TeO_2 (tellurite or paratellurite), and native tellurium.

AQUEOUS PHASE: mostly as tellurite (TeO_3^{2-}).

MOBILITY – PRIMARY ENVIRONMENT: accompanies sulfur in sulfides separated from magmas ranging from mafic to high temperature hydrothermal; forms extensive primary halos (see Fig. 7-3).

MOBILITY – SECONDARY ENVIRONMENT: very low mobility in solution; primary tellurides are very stable in most environments and may accumulate in placers with gold and other heavy minerals.

GEOCHEMICAL BARRIERS: mechanical.

TE IS A PATHFINDER FOR: Au-Te veins; polymetallic sulfide deposits (e.g., Coeur d'Alene, Idaho, Fig. 7-3); some porphyry ores (e.g., Ely, Nevada).

COMMENTS: Many aspects of the geochemistry of Te (e.g., abundances in sedimentary rocks, soils, vegetation, waters) are poorly known. Interest in Te is based on its potential in lithogeochemical surveys as it forms extensive primary halos which may permit the detection of blind ore bodies. Te may also be detected in soils developed on primary halos (see Watterson et. al., 1977, for additional information).

THORIUM (Th)

IGNEOUS ROCKS (ppm): AV 10; MAFIC 2; INTER 10; FELSIC 17.

SEDIMENTARY ROCKS (ppm): SH 12; SS 2; LS 2.

SOILS (ppm): AV 13.

SURFACE WATERS (ppb): AV 0.1.

VEGETATION ASH (ppm): insufficient data; biological response to Th is negligible; however, associated elements may be good (particularly Ra; Figs. 10A-1 and 10A-5).

GEOCHEMICAL ASSOCIATIONS: Lithophile; in general the association K, Th and U is found throughout the magmatic crystallization sequence; the deposits and associations of Th are essentially the same as those mentioned for cerium (as representative of the rare-earths) and uranium; see entries for Cerium and Uranium.

ORE MINERALS: Chiefly thorium-rich (up to 18% Th) monazite; the polymorphs thorite and huttonite ($ThSiO_4$), and thorianite (ThO_2).

SUBSTITUTIONS IN: monazite, zircon, sphene, allanite, xenotime, uraninite.

SOILS; OCCURRENCE IN: primarily resistate, clastic and transported minerals and rock fragments; Th (up to 50 ppm) in laterites is due in part to Th in zircon and in part to possible adsorption by aluminum hydroxides.

SECONDARY MINERALS: none.

AQUEOUS PHASE: none.

MOBILITY – PRIMARY ENVIRONMENT: high; concentrates in late stage granites and pegmatites; highly concentrated in some alkalic rocks, including carbonatites.

MOBILITY – SECONDARY ENVIRONMENT: only as clastic and detrital mineral phases in the heavy fractions of soils, sediments, glacial debris, etc.; forms placers (e.g., Th-rich monazite).

GEOCHEMICAL BARRIERS: mechanical (placers); (adsorption by clays and aluminum hydroxides is minor).

TH IS A PATHFINDER FOR: deposits containing the rare-earths, Nb and Ta (e.g., monazite and other minerals in carbonatites or pegmatites); some U deposits in hydrothermal veins (e.g., Bokan Mtn., Alaska); U, Au, Th, RE, etc., in placers.

PATHFINDERS FOR TH ARE: rare-earth elements, Nb, Ta, and F for carbonatite occurrences; U, RE, Nb, Ta, Zr, Ti, etc., for placer occurrences.

COMMENTS: Th may be detected by gamma radioactivity from the isotope ^{208}Tl (which has a different energy level than that of ^{214}Bi); caution must be exercised to ensure that the source of the gamma radiation (whether it is from a decay product of U or Th, or both) is properly interpreted.

TIN (Sn)

IGNEOUS ROCKS (ppm): AV 2; MAFIC 1; INTER 2; FELSIC 3.

SEDIMENTARY ROCKS (ppm): SH 4; SS < 1; LS 4.

SOILS (ppm): AV 10. TEMP 2-10; ARID 3-50; TROP (HUM) 3-20.

SURFACE WATERS (ppb): insufficient data.

VEGETATION ASH (ppm): 5; biological response to Sn is apparently satisfactory based on data in Table 10A-2 but specific details are lacking.

GEOCHEMICAL ASSOCIATIONS: Lithophile; Sn, W, Nb, Ta, Be, B, Li, Rb, Cs, and rare-earths in pegmatites; Sn, W, B, F, Be, etc., in veins and greisens (gangue minerals include quartz, fluorite, Li-micas, topaz and tourmaline); Sn, B, F (and As) in cassiterite pipes (e.g., Australia and South Africa).

ORE MINERALS: cassiterite, stannite.

SUBSTITUTIONS IN: biotite, muscovite, sphene, rutile, tourmaline, magnetite, amphiboles.

SOILS; OCCURRENCE IN: residual cassiterite; possibly some Sn in aluminum oxides.

SECONDARY MINERALS: none.

AQUEOUS PHASE: possibly as ionic Sn^{2+} in very acidic waters (pH of hydrolysis = 2.0).

MOBILITY − PRIMARY ENVIRONMENT: high; enriched in high temperature pneunatolytic and hydrothermal deposits, granites, pegmatites, and greisens.

MOBILITY − SECONDARY ENVIRONMENT: effectively immobile in solution; mainly as detrital cassiterite which forms placers.

GEOCHEMICAL BARRIERS: mechanical for cassiterite; pH for Sn^{2+}.

SN IS A PATHFINDER FOR: rare metal pegmatite and greisen deposits by determining cassiterite in the heavy fractions of soils, sediments (in streams, lakes or oceans), or in glacial materials.

PATHFINDERS FOR SN ARE: Sn is a good indicator of its own deposits. High Sn contents in pegmatitic muscovites indicate favorable pegmatites. High Sn in granites indicate potential Sn-bearing plutons (see Tables 7-2 and 2A-8). Cassiterite in heavy fractions of soils, sediments or glacial material is particularly effective in locating primary tin lodes and their derived placers. Fe-Mn precipitates at spring orifices may contain Sn, W, Li or other elements associated with Sn deposits.

COMMENTS: See Boyle (1974a) for additional information.

TUNGSTEN (W)

IGNEOUS ROCKS (ppm): Av 1.5; MAFIC 1; INTER 2; FELSIC 2.

SEDIMENTARY ROCKS (ppm): SH 2; Ss 1.6; Ls 0.5.

SOILS (ppm): insufficient data.

SURFACE WATERS (ppb): Av 0.03.

VEGETATION ASH (ppm): average concentration value not available; biological response reasonably good (Table 10A-1); probably best for scheelite ores in soils with low pH in order to mobilize the W (see Quin et al., 1974).

GEOCHEMICAL ASSOCIATIONS: Lithophile; W, Mo, Sn, Cu, As, Nb, Ta, Bi, Li, B, F and rare earths in pegmatites and aplites with principal W mineral being wolframite; W, Mo, Bi, Cu, Pb, Zn, S, As, Au, Ag, B and F in skarn deposits with scheelite as the principal W mineral.

ORE MINERALS: scheelite, wolframite.

SUBSTITUTIONS IN: muscovite, Nb-Ta minerals (up to 1% W), Mn oxides (particularly near hot springs); in powellite and wulfenite substituting for Mo.

SOILS; OCCURRENCE IN: resistate scheelite and wolframite, and possibly the secondary minerals listed below.

SECONDARY MINERALS: tungstite ($WO_3 \cdot H_2O$) and ferritungstite under conditions of acid weathering.

AQUEOUS PHASE: as tungstate in alkaline waters (see below), polytungstates and complexes (but not ionic).

MOBILITY – PRIMARY ENVIRONMENT: high; concentrates in late crystallizing phases such as granites, pegmatites, and high temperature hydrothermal veins; forms sub-ore halos (Fig. 7A-6).

MOBILITY – SECONDARY ENVIRONMENT: slightly mobile in alkaline solutions; primarily as detrital grains of wolframite and scheelite in heavy fractions.

GEOCHEMICAL BARRIERS: pH (slightly soluble only in alkaline waters); adsorption by Mn in sediments, soils, spring precipitates and some bogs (which contain Mn); mechanical for placers.

W IS A PATHFINDER FOR: non-pegmatitic beryllium, and beryllium-fluorite deposits (Table 3A-4).

PATHFINDERS FOR W ARE: W is an excellent indicator of its own deposits particularly by obtaining scheelite and wolframite in the heavy fractions of eluvium near W deposits, in soils or stream sediments (scheelite fluoresces in ultraviolet light whereas wolframite does not). B, F, As, Li, and Cu in some districts.

COMMENTS: Boyle (1974a) list 11 different types of W occurrences, including the alkaline brines at Searles Lake, California, which contain 70 ppm WO_3.

URANIUM (U)

IGNEOUS ROCKS (ppm): Av 2.7; MAFIC 0.6; INTER 3; FELSIC 4.8.

SEDIMENTARY ROCKS (ppm): SH 4; BLSH 3-1250; Ss 0.45-3.2; Ls 2.

SOILS (ppm): Av 1.

SURFACE WATERS (ppb): Av 0.4.

VEGETATION ASH (ppm): 0.5; poor biological response (Table 10A-1) but associated elements such as Ra, Mo and Se may be useful.

GEOCHEMICAL ASSOCIATIONS: Lithophile; U, Th, rare-earths, P, F, Zr, Ti, Mo, Bi, Cu, Ag, Zn, etc., depending on the type of igneous association (e.g., pegmatite, carbonatite); U, Cu, V, Se, Mo, C in sandstone type; U-Au in placers; see below under "Pathfinders" for additional associations.

ORE MINERALS: uraninite, brannerite, carnotite.

SUBSTITUTIONS IN: zircon, apatite (including phosphorites), allanite, niobate-tanta-lates (e.g., euxenite), monazite (Table 2A-5).

SOILS; OCCURRENCE IN: resistate, clastic and transported minerals and rock fragments; adsorbed on organic matter, clays, and iron oxides.

SECONDARY MINERALS: phosphates, vanadates, carbonates; also uraninite.

AQUEOUS PHASE: uranyl carbonate and phosphate complexes (Figs. 3A-10, 3A-11).

MOBILITY – PRIMARY ENVIRONMENT: highly mobile; concentrates in late phases (granites, hydrothermal veins).

MOBILITY – SECONDARY ENVIRONMENT: highly mobile in the oxidizing environment, especially alkaline (Figs. 3A-10, 3A-11).

GEOCHEMICAL BARRIERS: reduction (Eh); adsorption; special ion precipitates (e.g., vanadates such as carnotite).

U IS A PATHFINDER FOR: uranium deposits; it has potential for Au-U placers, certain Ag-Au veins, and carbonatites.

PATHFINDERS FOR U ARE: Depending upon type of deposit: (a) sandstone or roll-front type: Mo, Se, V, Cu, C; (b) classical vein (e.g., Beaverlodge, Sask.): Cu, Ag, Co, V, Ni, As, Au, Mo, Bi, Se; (c) unconformity vein (e.g., Key Lake, Sask.): Cu, Ag, Co, Ni, As, V, Se, Mo, Au; (d) pegmatite: Th, Mo, Nb, Ti, rare-earths; (e) car-bonatite: Nb, Th, Cu, F, P, Ti, Zr, rare-earths; (f) placer (e.g., Elliot Lake, Ont.): Th, Ti, Au, Zr, rare-earths. In addition, Rn, He and Ra for all types of uranium deposits.

COMMENTS: All modern exploration geochemists must understand the factors governing the mobility of uranium and its daughter products (e.g., Rn, Ra) in the secondary environment, complexing, disequilibrium, and other topics discussed throughout this book, particularly in Chapter 3A (but see additional entries in the Index). Also, see entries under Radium and Radon in this Appendix.

VANADIUM (V)

IGNEOUS ROCKS (ppm): Av 135; MAFIC 250; INTER 100; FELSIC 20.

SEDIMENTARY ROCKS (ppm): SH 130; BLSH 150-700; Ss 20; Ls 15.

SOILS (ppm): Av 80. RANGE 20-500; TEMP 10-400; ARID 10-300; TROP (HUM) 10-300.

SURFACE WATERS (ppb): AV 0.9.

VEGETATION ASH (ppm): 60; geobotanical indicators of V are known (Table 10-2) but the overall biological response for exploration purposes are poorly documented.

GEOCHEMICAL ASSOCIATIONS: Siderophile and lithophile; V, Ti, Fe and P in vanadiferous magnetite; V, Cu, Pb, Zn, Mo, Ag, Au and As in polymetallic sulfide deposits; U, V, Se, Mo, Cu, K, Ca and C in sandstone-type U ores; P, U, V, F, Se, As, etc., in phosphorites and black shales; V, Fe, Mn and P in certain V-rich sedimentary iron ores; V, S, C, Ni, Fe and Ca in deposits of asphalt or other solid hydrocarbons (e.g., Minas Ragra, Peru); V is a common constituent of petroleum (generally from 5-50 ppm) from which it may be economically recovered; V may be high in the ash of some coals; and V and other elements in Mn and Fe oxides.

ORE MINERALS: principally vanadiferous (up to 0.5% V) magnetite; V-containing uranium minerals (e.g., carnotite); patronite (VS_4) is one of only two known vanadium sulfides and is mined in Peru.

SUBSTITUTIONS IN: magnetite, sphene, rutile, muscovite (var. roscoelite), biotite, apatite (in iron ores), amphiboles; the silicates release V upon weathering.

SOILS; OCCURRENCE IN: within resistates (e.g., magnetite, sphene), or in decomposing mafic minerals; adsorbed or coprecipitated with clays (e.g., laterites); in Fe-Mn oxides.

SECONDARY MINERALS: many secondary vanadates (e.g., vanadinite, descloizite, carnotite, tyuyamunite); also V oxides (e.g., montroseite) and other mineralogical types.

AQUEOUS PHASE: as various soluble vanadates, e.g. $(H_2VO_4)^-$.

MOBILITY — PRIMARY ENVIRONMENT: low; concentrates in the early-formed rocks and minerals (e.g., magnetite, mafics) as a trace-minor constituent replacing Fe; vanadium forms no primary magmatic minerals.

MOBILITY — SECONDARY ENVIRONMENT: highly mobile in oxidizing, acid-alkaline waters; immobile in reducing environments (Fig. 3-19, Table 3A-3); spring precipitates often enriched in V; resistates (e.g., V-magnetites) form placers.

GEOCHEMICAL BARRIERS: reducing; adsorption; mechanical (placers).

V IS A PATHFINDER FOR: sandstone, classical vein, and unconformity vein types of uranium deposits.

PATHFINDERS FOR V ARE: V is a good indicator of its own deposits and most types of surveys (e.g., soils, sediments) work well; analysis of magnetites in heavy mineral surveys for V may lead to vanadiferous magnetite deposits.

COMMENTS: Vanadium is a widely dispersed, relatively abundant element (more abundant than Cu, Pb, Zn or Ni; Table 2A-1). In magmas V occurs as V^{3+} and replaces Fe^{3+} in magnetite, mafic minerals, and micas (siderophile association); in the secondary environment (vanadates; lithophile association) it occurs as V^{5+}. V may also occur in the tetravalent state, such as VO^{2+} complexes in soil organic matter. Along with Fe, Mn, S and U (and to a lesser extent, Mo and Se), vanadium is one of the few elements whose variable valence state is of importance in exploration geochemistry.

ZINC (Zn)

IGNEOUS ROCKS: (ppm): AV 70; MAFIC 100; INTER 60; FELSIC 40.

SEDIMENTARY ROCKS (ppm): SH 100; BLSH < 300-1000; SS 16; LS 25.

SOILS (ppm): AV 50. RANGE 10-300; TEMP 10-600; ARID 10-900; TROP (HUM) 10-400.

SURFACE WATERS (ppb): AV 20.

VEGETATION ASH (ppm): 900; biological response fair to poor (Zn is a "difficult" element; see p. 407); high Zn can produce barren areas (Fig. 10A-2) as a result of its toxicity to vegetation even though it is an essential nutrient in small amounts.

GEOCHEMICAL ASSOCIATIONS: Chalcophile; Zn, Cd in essentially all occurrences; Zn, Cd, Pb, Ba, F in Mississippi Valley type deposits; Zn, Pb, Mn, Ba, Fe in stratiform (volcanogenic) deposits; Zn, Pb, Fe, Cu, Ag, Ba, Te, etc., in veins and massive sulfide deposits; Zn, Pb, Cu, Ag, B, Mo, W, Be in skarns; Mn, Ni, Cu, Co, Zn in deep-sea nodules; Cu, Mo, Re, Fe, Au, Ag, Zn in some porphyry copper deposits (Fig. 7A-4); Cu, Pb, Zn in copper shales.

ORE MINERALS: sphalerite (which contains numerous trace elements such as Cd, Se, Mn, Ag, Cu, Ga, Hg, etc.; see Table 2-5); sphalerite is easily weathered.

SUBSTITUTIONS IN: biotite and amphibole (in acid and intermediate rocks replacing Fe^{2+} and Mg^{2+}; these minerals are easily weathered); magnetite (in mafic rocks; stable).

SOILS; OCCURRENCE IN: Fe-Mn oxides (adsorbed and coprecipitated); clays and organic matter (adsorbed); in some secondary Zn minerals.

SECONDARY MINERALS: smithsonite ($ZnCO_3$); hemimorphite ($Zn_4Si_2O_7(OH)_2 \cdot H_2O$); hydrozincite ($Zn_5(CO_3)_2(OH)_6$); willemite ($ZnSiO_4$).

AQUEOUS PHASE: ionic Zn^{2+}; soluble organo-metallic complexes; some Zn may be "temporarily" incorporated in floating organisms (e.g., algae) or adsorbed on suspended matter.

MOBILITY – PRIMARY ENVIRONMENT: in most igneous rocks Zn substitutes in mineral structures (e.g., magnetite, biotite, amphibole); Zn ore deposits in igneous rocks are usually found associated with felsic rocks and hydrothermal veins; transport probably as chloride and bisulfide complexes.

MOBILITY – SECONDARY ENVIRONMENT: high in oxidizing acidic and neutral waters (pH of hydrolysis = 7.0; Table 3-6), hence low mobility in carbonate-rock terrain; high in reducing gley environment (Table 3A-3); immobile in sulfide environment; mobility in waters may be greatly reduced by adsorption or coprecipitation on Fe-Mn oxides (in soils, stream sediments, spring precipitates), and by adsorption on organic matter.

GEOCHEMICAL BARRIERS: pH; reducing hydrogen sulfide; adsorption (on clays, Fe-Mn oxides, organic matter); coprecipitation (on Fe-Mn oxides); precipitated by high abundances of carbonate and phosphate in waters.

Zɴ ɪꜱ ᴀ ᴘᴀᴛʜꜰɪɴᴅᴇʀ ꜰᴏʀ: Pb deposits; Pb-Ag deposits; some fluorite deposits; most polymetallic sulfide deposits; skarn deposits; porphyry copper deposits; numerous other deposits (see Table 3A-4).

Pᴀᴛʜꜰɪɴᴅᴇʀꜱ ꜰᴏʀ Zɴ ᴀʀᴇ: Hg in some vein and massive sulfide deposits; F in some carbonate-hosted Zn deposits; F in some skarn deposits; Pb in soils, glacial materials, and other sampling media (usually except waters) where Zn and Cd have been leached from sphalerite-galena deposits; Mn in stratiform (exhalative) deposits (Fig. 7A-2); Zn is an excellent indicator of its own deposits.

Cᴏᴍᴍᴇɴᴛꜱ: Zn is the most commonly determined element in geochemical exploration because: (1) it is easily determined by several analytical methods to sufficiently low abundances; (2) it is mobile in most geochemical environments (important exceptions include alkaline, and reducing hydrogen sulfide environments); (3) it forms extensive halos in soils, sediments, glacial materials, most types of waters, bog materials, etc.; (4) it is a common constituent of many different types of deposits and, therefore, it is a good pathfinder (Table 3A-4).

Zn (up to 0.1%) may be found in some phosphate fertilizers (see pp. 196 and 204 for other examples of Zn contamination and false anomalies). Wedepohl (1969-1974; Chapter 30) prepared an outstanding compilation on the geochemical characteristics of Zn.

GRAVIMETRIC CONVERSION FACTORS

Element	Found (from)	Sought (to)	Factor (multiply by)	Element	Found (from)	Sought (to)	Factor (multiply by)
Aluminum	Al	Al_2O_3	1.889	Holmium	Ho	Ho_2O_3	1.145
	Al_2O_3	Al	0.529		Ho_2O_3	Ho	0.872
Antimony	Sb	Sb_2O_5	1.328	Hydrogen	H	H_2O	8.936
	Sb_2O_5	Sb	0.752		H_2O	H	0.111
Arsenic	As	As_2O_5	1.533	Indium	In	In_2O_3	1.209
	As_2O_5	As	0.651		In_2O_3	In	0.827
Barium	Ba	BaO	1.116	Iridium	Ir	IrO_2	1.165
	BaO	Ba	0.895		IrO_2	Ir	0.857
Beryllium	Be	BeO	2.775	Iron	Fe	FeO	1.286
	BeO	Be	0.360		Fe	Fe_2O_3	1.429
Bismuth	Bi	Bi_2O_3	1.114		FeO	Fe	0.777
	Bi_2O_3	Bi	0.896		Fe_2O_3	Fe	0.699
Boron	B	B_2O_3	3.219	Lanthanum	La	La_2O_3	1.172
	B_2O_3	B	0.310		La_2O_3	La	0.852
Cadmium	Cd	CdO	1.142	Lead	Pb	PbO	1.077
	CdO	Cd	0.875		PbO	Pb	0.928
Calcium	Ca	CaO	1.399	Lithium	Li	Li_2O	2.152
	CaO	Ca	0.714		Li_2O	Li	0.464
Carbon	C	CO_2	3.664	Lutecium	Lu	Lu_2O_3	1.137
	CO_2	C	0.272		Lu_2O_3	Lu	0.879
Cerium	Ce	Ce_2O_3	1.171	Magnesium	Mg	MgO	1.658
	Ce_2O_3	Ce	0.853		MgO	Mg	0.603
Cesium	Cs	Cs_2O	1.060	Manganese	Mn	MnO	1.291
	Cs_2O	Cs	0.943		Mn	MnO_2	1.582
Chromium	Cr	Cr_2O_3	1.461		MnO	Mn	0.774
	Cr_2O_3	Cr	0.684		MnO_2	Mn	0.631
Cobalt	Co	CoO	1.271	Mercury	Hg	HgO	1.079
	CoO	Co	0.786		HgO	Hg	0.926
Copper	Cu	CuO	1.251	Molybdenum	Mo	MoO_3	1.500
	CuO	Cu	0.798		MoO_3	Mo	0.666
Dysprosium	Dy	Dy_2O_3	1.147	Neodymium	Nd	Nd_2O_3	1.166
	Dy_2O_3	Dy	0.871		Nd_2O_3	Nd	0.857
Erbium	Er	Er_2O_3	1.143	Nickel	Ni	NiO	1.272
	Er_2O_3	Er	0.874		NiO	Ni	0.785
Europium	Eu	Eu_2O_3	1.157	Niobium	Nb	Nb_2O_5	1.430
	Eu_2O_3	Eu	0.863		Nb_2O_5	Nb	0.699
Gadolinium	Gd	Gd_2O_3	1.152	Osmium	Os	OsO_4	1.336
	Gd_2O_3	Gd	0.867		OsO_4	Os	0.748
Gallium	Ga	Ga_2O_3	1.344	Oxygen	O	H_2O	1.126
	Ga_2O_3	Ga	0.743		H_2O	O	0.888
Germanium	Ge	GeO_2	1.440	Palladium	Pd	PdO	1.149
	GeO_2	Ge	0.694		PdO	Pd	0.869
Gold	Au	Au_2O	1.040	Phosphorus	P	P_2O_5	2.291
	Au_2O	Au	0.961		P_2O_5	P	0.436
Hafnium	Hf	HfO_2	1.179	Platinum	Pt	PtO_2	1.163
	HfO_2	Hf	0.847		PtO_2	Pt	0.859

Element	Found (from)	Sought (to)	Factor (multiply by)	Element	Found (from)	Sought (to)	Factor (multiply by)
Potassium...........	K	K_2O	1.204	Tellurium	Te	TeO_2	1.250
	K_2O	K	0.830		TeO_2	Te	0.799
Praseodymium.......	Pr	Pr_2O_3	1.170	Terbium	Tb	Tb_2O_3	1.150
	Pr_2O_3	Pr	0.854		Tb_2O_3	Tb	0.869
Rhenium............	Re	Re_2O_7	1.300	Thallium............	Tl	Tl_2O	1.039
	Re_2O_7	Re	0.768		Tl_2O	Tl	0.962
Rhodium............	Rh	Rh_2O_3	1.233	Thorium	Th	ThO_2	1.137
	Rh_2O_3	Rh	0.810		ThO_2	Th	0.878
Rubidium	Rb	Rb_2O	1.093	Thulium	Tm	Tm_2O_3	1.141
	Rb_2O	Rb	0.914		Tm_2O_3	Tm	0.875
Ruthenium..........	Ru	RuO_2	1.314	Tin	Sn	SnO_2	1.269
	RuO_2	Ru	0.760		SnO_2	Sn	0.787
Samarium	Sm	Sm_2O_3	1.159	Titanium............	Ti	TiO_2	1.668
	Sm_2O_3	Sm	0.862		TiO_2	Ti	0.599
Scandium	Sc	Sc_2O_3	1.532	Tungsten............	W	WO_3	1.261
	Sc_2O_3	Sc	0.651		WO_3	W	0.792
Selenium............	Se	SeO_2	1.405	Uranium	U	UO_2	1.134
	SeO_2	Se	0.711		U	U_3O_8	1.179
Silicon	Si	SiO_2	2.139		UO_2	U	0.881
	SiO_2	Si	0.467		U_3O_8	U	0.847
Silver...............	Ag	Ag_2O	1.074	Vanadium...........	V	V_2O_5	1.785
	Ag_2O	Ag	0.930		V_2O_5	V	0.560
Sodium	Na	Na_2O	1.347	Ytterbium...........	Yb	Yb_2O_3	1.138
	Na_2O	Na	0.741		Yb_2O_3	Yb	0.878
Strontium	Sr	SrO	1.182	Yttrium.............	Y	Y_2O_3	1.269
	SrO	Sr	0.845		Y_2O_3	Y	0.787
Sulfur...............	S	SO_2	1.997	Zinc	Zn	ZnO	1.244
	SO_2	S	0.500		ZnO	Zn	0.803
Tantalum	Ta	Ta_2O_5	1.221	Zirconium...........	Zr	ZrO_2	1.350
	Ta_2O_5	Ta	0.818		ZrO_2	Zr	0.740

References

Note: This list of references only contains entries for *"The 1980 Supplement."*

ABBEY S. (1977) Studies in "standard samples" for use in the general analysis of silicate rocks and minerals. Part 5: 1977 edition of "usable" values. *Geol. Surv. Canada Paper* 77-34.

ABBEY S., LEE N. J. and BOUVIER J. L. (1974) Analysis of rocks and minerals using an atomic absorption spectrophotometer. Part 5. An improved lithium - fluoborate scheme for fourteen elements. *Geol. Surv. Can. Paper* 74-19.

ADLER H. H. (1977) Geochemical factors contributing to uranium concentration in alkalic igneous rocks. In, *Recognition and Evaluation of Uraniferous Areas*, pp. 35-45. STI/PUB/450. Inter. Atomic Energy Agency.

ALBRECHT P., VANDENBROUCKE M. and MANDENGUE M. (1976) Geochemical studies on the organic matter from the Douala Basin (Cameroon) — 1. Evolution of the extractable organic matter and the formation of petroleum. *Geochim. Cosmochim. Acta* 40, 791-799.

ALLAN R. J. (1974) Trace metal dispersion in an Arctic desert landscape: A Pb-Zn deposit on Little Cornwallis Island, District of Franklin. *Geol. Surv. Canada, Paper* 74-1, Part B, 51-56.

ALLAN R. J. (1974a) Metal contents of lake sediment cores from established mining areas: An interface of exploration and environmental geochemistry. *Geol. Surv. Can. Paper* 74-1, Part B, 43-49.

ALLAN R. J. and TIMPERLY M. H. (1975) Prospecting by use of lake sediments in areas of industrial heavy metal contamination. In, *Prospecting In Areas of Glaciated Terrain 1975* (editor M. J. Jones), pp. 87-111, IMM London.

ALLCOTT G. H. and LAKIN H. W. (1974) Statistical summary of geochemical data furnished by 85 laboratories for six geochemical exploration reference samples. *U.S. Geol. Surv.* Open-file report (unnumbered).

ALLCOTT G. H. and LAKIN H. W. (1975) The homogeneity of six geochemical exploration reference samples. In, *Geochemical Exploration 1974* (editors I. L. Elliott and W. K. Fletcher), pp. 659-681. Elsevier.

ALLCOTT G. H. and LAKIN H. W. (1978) Tabulation of geochemical data furnished by 109 laboratories for six geochemical exploration reference samples. *U.S. Geol. Surv.* Open File Report 78-163.

ALLEN P. M., COOPER D. C., FUGE R. and REA W. J. (1976) Geochemistry and relationships to mineralization of some igneous rocks from Harlech Dome, Wales. *Trans. IMM* (Sect. B, Appl. Earth Sci.) 85, B100-B108.

ANDERSON D. M. and MORGENSTERN N. R. (1973) Physics, chemistry, and mechanics of frozen ground: A review. In, *Permafrost,* North American Contrib., Second Inter. Conf., pp. 257-288. National Acad. Sciences. Washington.

ANDREWS D. F., BICKEL P. J., HAMPEL F. R., HUBER P. J., ROGERS W. H. and TUKEY J. W. (1972) *Robust Estimates Of Location.* Princeton Univ. Press.

ANDRIANOVA G. A. (1971) Trace-element contents in soil-forming rocks, soils, and plants of central Yakutia: Characterization of the biogeochemical province. *Geochem. Intern.* 8, 287-299.

ANNONYMOUS (1975) Project Radam maps the unknown in Brazil. *Eng. Mining J.* 176 (Nov.; No. 11), 165-168.

ANNONYMOUS (1975a) Nuclear borehole logging. *Mining Mag.* 133, 478.

AUBERT H. and PINTA M. (1977) *Trace Elements In Soils.* (Developments in Soil Science 7). Elsevier.

BAILEY N. J. L., EVANS C. R. and MILNER C. W. D. (1974) Applying petroleum geochemistry to search for oil: examples from Western Canada basin. *Amer. Assoc. Petroleum Geologists Bull.* 58, 2284-2294.

BAKER W. E. (1978) The role of humic acid in the transport of gold. *Geochim. Cosmochim. Acta* 42, 645-649.

BALDWIN J. T., SWAIN H. D. and CLARK G. H. (1978) Geology and grade distribution of the Panguna porphyry copper deposit, Bougainville, Papua New Guinea. *Econ. Geol.* 73, 690-702.

BAMPTON K. F., COLLINS A. R., GLASSON K. R. and GUY B. B. (1977) Geochemical indications of concealed copper mineralization in an area northwest of Mount Isa, Queensland, Australia. *J. Geochem. Explor.* 8, 169-188.

BARBIER J. and WILHELM E. (1978) Superficial geochemical dispersion around sulphide deposits: Some examples in France. *J. Geochem. Explor.* 10, 1-39.

BARKER C. (1974) Pyrolysis techniques for source-rock evaluation. *Amer. Assoc. Petroleum Geologists Bull.* 58, 2349-2361.

BARNES H. L. and LAVERY N. G. (1977) Use of primary dispersion for exploration of Mississippi Valley-type deposits. *J. Geochem. Explor.* 8, 105-115.

BARRINGER A. R. (1977) AIRTRACE — An airborne geochemical exploration technique. *U.S. Geol. Surv., Prof. Paper* 1015, 231-251.

BARRINGER A. R. (1979) The application of atmospheric particulate geochemistry in mineral exploration. In, *Geophysics and Geochemistry in the Search for Metallic Ores* (editor P. J. Hood). *Geol. Surv. Canada, Econ. Geol. Report* 31.

BARRINGER A. R. and BRADSHAW P. M. D. (1979) SURTRACE — An airborne and ground exploration technique based on surface micro-layer geochemistry (Abstract). In, *Geochemical Exploration 1978* (editors J. R. Watterson and P. K. Theobald), p. 497. Assoc. of Explor. Geochemists.

BEAMISH R. J. and VAN LOON J. C. (1977) Precipitation loading of acid and heavy metals to a small acid lake near Sudbury, Ontario. *J. Fish. Res. Board Canada* 34, 649-658.

BEMENT T. R., SUSCO D. V., WHITEMAN D. E. and ZEIGLER R. K. (1977) Geostatistics project of the National Uranium Resource Evaluation Program. *Los Alamos Sci. Lab. Prog. Rept. LA-6804-PR.*

BEUS A. A. (1979) Sodium — A geochemical indicator of emerald mineralization in the Cordillera Oriental, Colombia. *J. Geochem. Explor.* 11, (in press).

BEUS A. A. and GRIGORIAN S. V. (1977) *Geochemical Exploration Methods For Mineral Deposits.* Applied Publishing Ltd.

BIGNELL R. D., CRONAN D. S. and TOOMS J. S. (1976) Metal dispersion in the Red Sea as an aid to marine geochemical exploration. *Trans. IMM* (Sect. B, Appl. Earth Sci.) 85, B274-B278.

BILY C. and DICK J. W. L. (1974) Naturally occurring gas hydrates in the Mackenzie Delta, N.W.T. *Bull. Canadian Petroleum Geol.* 22, 340-352.

BIRKELAND P. W. (1974) *Pedology, Weathering, and Geomorphological Research.* Oxford Univ. Press.

BLAIN C. F. and ANDREW R. L. (1977) Sulphide weathering and the evaluation of gossans in mineral exploration. *Minerals Sci. Engng.* 9, 119-150.

BOLLINGBERG H. J. (1975) Geochemical prospecting using seaweed, shellfish and fish. *Geochim. Cosmochim.* 39, 1567-1570.

BOLVIKEN B. (1971) A statistical approach to the problem of interpretation in geochemical prospecting. *Geochemical Exploration.* CIM Spec. Vol. 11, 564-567.

BOLVIKEN B. and GLEESON C. J. (1979) Focus on the use of glacial soils in geochemical exploration. In, *Geophysics and Geochemistry in the Search for Metallic Ores* (editor P. J. Hood). *Geol. Surv. Canada, Econ. Geol. Report* 31.

BOLVIKEN B., HONEY F., LEVINE S. R., LYON R. J. P. and PRELAT A. (1977) Detection of naturally heavy-metal-poisoned areas by LANDSAT-1 digital data. *J. Geochem. Explor.* 8, 457-471.

BOLVIKEN B. and LAG J. (1977) Natural heavy-metal poisoning of soils and vegetation: an exploration tool in glaciated terrain. *Trans. IMM* (Sect. B., Appl. Earth Sci) 86, B173-B180.

BOLVIKEN B. and LOGN O. (1975) An electrochemical model for element distribution around sulphide bodies. In, *Geochemical Exploration 1974* (editors I. L. Elliott and W. K. Fletcher), pp. 631-648. Elsevier.

BOSTROM K. (1978) Book review of *Geochemical Exploration Methods for Mineral Deposits* (by Beus and Grigorian, 1977). *Earth-Science Rev.* 14, 67-69.

BOTBOL J. M., SINDING-LARSEN R., McCAMMON R. B. and GOTT G. B. (1978) A regionalized multivariate approach to target selection in geochemical exploration. *Econ. Geol.* 73, 534-546.

BOLT G. H. and BRUGGENWERT M. G. M. (1976) (editors) *Soil Chemistry. Part A. Basic Elements.* Elsevier.

BOUMANS P. W. J. M., DEBOER F. J., DAHMEN F. J., HOELZEL H. and MEIER A. (1975) A comparative investigation of some analytical performance characteristics of an inductively-coupled radio frequency plasma and a capacitively-coupled microwave plasma for solution analysis by emission spectrometry. *Spectrochim. Acta* 30B, 449-469.

BOWIE S. H. U. (1973) Methods, trends and requirements in uranium exploration. In, *Uranium Exploration Methods*, pp. 57-65. Inter. Atomic Energy Agency, Vienna, (STI/PUB/334).

BOWIE S. H. U. (1974) Modern methods in the search for metalliferous ores in Britain. *Proc. R. Soc. London*, A, 339, 299-311.

BOWIE S. H. U., SIMPSON P. R. and RICE C. M. (1973) Application of fission-track and neutron activation methods to geochemical exploration. In, *Geochemical Exploration 1972.* (editor, M. J. Jones), pp. 359-372. IMM.

BOYER S. J. (1975) Chemical weathering of rocks on the Lassiter Coast, Antarctic Peninsula, Antarctica. *New Zeal. J. Geol. Geophys.* 18, 623-628.

BOYLE R. W. (1974) The use of major elemental ratios in detailed geochemical prospecting utilizing primary halos. *J. Geochem. Explor.* 3, 345-369.

BOYLE R. W. (1974a) Elemental associations in mineral deposits and indicator elements of interest in geochemical prospecting (revised). *Geol. Surv. Canada Paper* 74-75.

BOYLE R. W. (1977) Cupiferous bogs in the Sackville area, New Brunswick, Canada. *J. Geochem. Explor.* 8, 495-527.

BOYLE R. W. (1979) Geochemistry overview. In, *Geophysics and Geochemistry in the Search for Metallic Ores* (editor P. J. Hood). *Geol. Surv. Canada, Econ. Geol. Report No. 31.*

BOYLE R. W. and JONASSON I. R. (1973) The geochemistry of arsenic and its use as an indicator element in geochemical prospecting. *J. Geochem Explor.* 2, 251-296.

BRADSHAW P. M. D. (1975) (editor) Conceptual models in exploration geochemistry. The Canadian Cordillera and Canadian Shield. *J. Geochem. Explor.* 4, 1-213.

BRADSHAW P. M. D. (1975a) Orientation sampling and standardization of data collection and presentation. *J. Geochem. Explor.* 4, 201-205.

BRADSHAW P. M. D. and THOMSON I. (1979) The use of surface soil geochemistry in the search for buried and blind ore deposits (Abstract). In, *Geochemical Exploration 1978* (editors J. R. Watterson and P. K. Theobald), p. 498. Assoc. of Explor. Geochemists.

BRADSHAW P. M. D., THOMSON I., SMEE B. W. and LARSSON J. O. (1974) The application of different analytical extractions and soil profile sampling in exploration geochemistry. *J. Geochem. Explor.* 3, 209-225.

BRADY N. C. (1974) *The Nature and Properties of Soils* (8th edition). MacMillan.

BRISTOW Q. (1975) A computer-controlled geochemical analysis system for use with existing Perkin Elmer double-beam atomic absorption spectrophotometers. *J. Geochem. Explor.* 4, 371-383.

BROOKS J. and THUSU B. (1977) Oil-source rock identification and characterization of the Jurassic sediments in the northern North Sea. *Chem. Geol.* 20, 283-294.

BROOKS R. A. and BERGER B. R. (1978) Relationship of soil-mercury values to soil type and dissentimated gold mineralization, Getchell mine area, Humboldt County, Arizona. *J. Geochem. Explor.* 9, 186-194.

BROOKS R. R. (1977) Copper and cobalt uptake by *Haumaniastrum* species. *Pl. Soil.* 48, 541-544.
BROOKS R. R. (1979) Advances in botanical methods of prospecting for minerals. Part II. Advances in biogeochemical methods of prospecting. In, *Geophysics and Geochemistry in the Search for Metallic Ores* (editor P. J. Hood). *Geol. Surv. Canada, Econ. Geol. Report* 31.
BROOKS R. R., LEE J. and JAFFRE T. (1974) Some New Zealand and New Caledonian plant accumulators of nickel. *J. Ecol.* 62, 493-499.
BROOKS R. R., LEE J., REEVES R. D. and JAFFRE T. (1977a) Detection of nickeliferous rocks by analysis of herbarium specimens of indicator plants. *J. Geochem. Explor.* 7, 49-57.
BROOKS R. R., MCCLEAVE J. A. and MALAISSE F. (1977b) Copper and cobalt in African species of *Crotalaria L. Proc. Roy. Soc. Lond.* B. 197, 231-236.
BROOKS R. R. and WITHER E. D. (1977) Nickel accumulation by *Rinorea bengalensis* (Wall.) O.K. *J. Geochem. Explor.* 7, 295-300.
BROWN B. W. and HILCHEY G. R. (1975) Sampling and analysis of geochemical materials for gold. In, *Geochemical Exploration 1974* (editors I. L. Elliott and W. K. Fletcher), pp. 683-690.
BROWNLOW A. H. (1979) *Geochemistry.* Prentice Hall.
BRUNDIN N. H. and BERGSTROM J. (1977) Regional prospecting for ores based on heavy minerals in glacial till. *J. Geochem. Explor.* 7, 1-19.
BUGROV V. (1974) Geochemical sampling techniques in the Eastern Desert of Egypt. *J. Geochem. Explor.* 3, 67-75.
BUGROV V. A. and SHALABY I. M. (1975) Geochemical prospecting in the Eastern Desert of Egypt. In, *Geochemical Exploration 1974* (editors I. L. Elliott and W. K. Fletcher), pp. 523-530. Elsevier.
BUHLMANN E., PHILPOTT D. E., SCOTT M. J. and SANDERS R. N. (1975) The status of exploration geochemistry in southern Africa. In, *Geochemical Exploration 1974* (editors I. L. Elliott and W. K. Fletcher), pp. 51-64. Elsevier.
BUOL S. W., HOLE F. D. and MCCRACKEN R. J. (1973) *Soil Genesis and Classification.* Iowa State Univ. Press.
BURNS B. J., HOGARTH J. T. C. and MILNER C. W. D. (1975) Properties of Beaufort Basin liquid hydrocarbons. *Bull. Canadian Petroleum Geol.* 23, 295-303.
BUTT C. R. M. and NICHOL I. (1979) The identification of various types of geochemical stream sediment anomalies in Northern Ireland. *J. Geochem. Explor.* 11, 13-32.
BUTT C. R. M. and WILDING I. G. P. (1977) (editors) *Geochemical Exploration 1976.* Elsevier. Also issued as *J. Geochem. Explor.* 8, 1-494.
CADIGAN R. A. and FELMLEE J. K. (1977) Radioactive springs geochemical data related to uranium exploration. *J. Geochem. Explor.* 8, 381-395.
CAGATAY M. N. and BOYLE D. R. (1977) Geochemical prospecting for volcanogenic sulphide deposits in the Eastern Black Sea ore province, Turkey. *J. Geochem. Explor.* 8, 49-71.
CALLAHAN J. E. (1975) A rapid field method for extracting the magnetic fraction from stream sediments. *J. Geochem. Explor.* 4, 265-267.
CAMERON E. M. (1975) Geochemical methods of exploration for massive sulphide mineralization in the Canadian Shield. *Geochemical Exploration 1974* (editors I. L. Elliott and W. K. Fletcher), pp. 21-49. Elsevier.
CAMERON E. M. (1977) Geochemical dispersion in mineralized soils of a permafrost environment. *J. Geochem. Explor.* 7, 301-326.
CAMERON E. M. (1977a) Geochemical dispersion in lake waters and sediments from massive sulphide mineralization, Agricola Lake area, Northwest Territories. *J. Geochem. Explor.* 7, 327-348.
CAMERON E. M. (1978) Hydrogeochemical methods for base metal exploration in the northern Canadian Shield. *J. Geochem. Explor.* 10, 219-243.
CAMERON E. M., SIDDELEY G. and DURHAM C. C. (1971) Distribution of ore elements in rocks for evaluating ore potential: Nickel, copper, cobalt and sulphur in ultramafic rocks of the Canadian Shield. *Geochemical Exploration.* CIM Spec. Vol. 11, 298-313.

CANADA DEPARTMENT of AGRICULTURE (1978) The Canadian system of soil classification. *Canada Dept. Agric. Publ.* 1646.

CANNON H. L. (1979) Advances in botanical methods of prospecting for minerals. Part I. Advances in geobotanical methods of prospecting. In, *Geophysics and Geochemistry in the Search for Metallic Ores* (editor P. J. Hood). *Geol. Surv. Canada, Econ. Geol. Report* 31.

CANNON H. L., HARMS T. F. and HAMILTON J. C. (1975) Lithium in unconsolidated sediments and plants of the Basin and Range province, southern California and Nevada. *U.S. Geol. Surv. Prof. Paper* 918.

CAROTHERS W. W. and KHARAKA Y. K. (1978) Aliphatic acid anions in oil-field waters-implications for origin of natural gas. *Amer. Assoc. Petroleum Geologists Bull.* 62, 2441-2453.

CARPENTER R. H. and HAYES W. B. (1978) Precipitation of iron, manganese, zinc and copper on clean, ceramic surfaces in a stream draining a polymetallic sulfide deposit. *J. Geochem. Explor.* 9, 31-37.

CARPENTER R. H., POPE T. A. and SMITH R. L. (1975) Fe-Mn oxide coatings in stream sediment geochemical surveys. *J. Geochem. Explor.* 4, 349-363.

CARROLL D. (1970) *Rock Weathering.* Plenum.

CASSOU A. M., CONNAN J. and PORTHAULT B. (1977) Relations between maturation of organic matter and geothermal effect, as exemplified in Canadian east coast offshore wells. *Bull. Canadian Petroleum Geol.* 25, 174-194.

CHAFFEE M. A. (1975) Geochemical exploration techniques applicable in the search for copper deposits. *U.S. Geol. Surv. Prof. Paper* 907-B.

CHAMP D. R., GULENS J. and JACKSON R. E. (1979) Oxidation-reduction sequences in ground water flow systems. *Can. J. Earth Sci.* 16, 12-23.

CHAO T. T. and ANDERSON B. J. (1974) The scavenging of silver by manganese and iron oxides in stream sediments collected from two drainage areas of Colorado. *Chem. Geol.* 14, 159-166.

CHAO T. T. and SANZOLONE R. F. (1973) Atomic absorption spectrophotometric determination of microgram levels of Co, Ni, Cu, Pb, and Zn in soil and sediment extracts containing large amounts of Mn and Fe. *J. Res. U.S. Geol. Surv.* 1, 681-685.

CHAO T. T. and THEOBALD P. K. (1976) The significance of secondary iron and manganese oxides in geochemical exploration. *Econ. Geol.* 71, 1560-1569.

CHERDYNTSEV V. V. (1971) *Uranium-234.* Israel Program for Scientific Translations (Keter). Jerusalem.

CHIARELLI A. (1978) Hydrodynamic framework of eastern Algerian Sahara — influence on hydrocarbon occurrence. *Amer. Assoc. Petroleum Geologists Bull.* 62, 667-685.

CLARKE W. B. and KUGLER G. (1973) Dissolved helium in groundwater. A possible method for uranium and thorium prospecting. *Econ. Geol.* 68, 243-251.

CLARKE W. B., TOP Z., BEAVAN A. P. and GANDHI S. S. (1977) Dissolved helium in lakes: Uranium prospecting in Precambrian terrain of central Labrador. *Econ. Geol.* 72, 233-242.

CLAYPOOL G. E., LOVE A. H. and MAUGHAN E. K. (1978) Organic geochemistry, incipient metamorphism, and oil generation in black shale members of Phosphoria Formation, western interior United States. *Amer. Assoc. Petroleum Geologists Bull.* 62, 98-120.

CLAYPOOL G. E. and REED P. R. (1976) Thermal-analysis technique for source rock evaluation: quantitative estimate of organic richness and effects of lithologic variation. *Amer. Assoc. Petroleum Geologists Bull.* 60, 608-626.

CLAYTON J. S., EHRLICH W. A., CANN D. B., DAY J. H. and MARSHALL I. B. (1977) *Soils of Canada* (in two vols.). Canada Dept. of Agriculture.

CLIFTON H. E., HUNTER R. E., SWANSON F. J. and PHILLIPS R. L. (1969) Sample size and meaningful gold analyses. *U.S. Geol. Surv. Prof. Paper* 625-C.

CLOSS L. G. and NICHOL I. (1975) The role of factor and regression analysis in the interpretation of geochemical reconnaissance data. *Can. J. Earth Sci.* 12, 1316-1330.

COETZEE G. L. (1974) Trends in exploration geochemistry in Africa. Summary Progress Report. *Newsletter No. 12* (January, 1974). Assoc. Explor. Geochem.

COKER W. B., HORNBROOK E. H. W. and CAMERON E. M. (1979) Lake sediment geochemistry applied to mineral exploration. In, *Geophysics and Geochemistry in the Search for Metallic Ores* (editor P. J. Hood). *Geol. Surv. Canada, Econ. Geol. Report* 31.

COKER W. B. and NICHOL I. (1975) The relation of lake sediment geochemistry to mineralization in the northwest Ontario region of the Canadian Shield. *Econ. Geol.* 70, 202-218. Discussion: 71, 952-963 (1976).

COLE M. M. (1973) Geobotanical and biogeochemical investigations in the sclerophyllous woodland and shrub associations of the Eastern Goldfields area of Western Australia, with particular reference to the role of *Hybanthus floribundus* (Lindl.) F. Muell. as a nickel indicator and accumulator plant. *J. Appl. Ecol.* 10, 269-320.

COLE M. M., OWEN-JONES E. S., CUSTANCE N. D. E. and BEAUMONT T. E. (1974) Recognition and interpretation of spectral signatures of vegetation from aircraft and satellite imagery in western Queensland, Australia. *Proc. Sympos. at Frascati, Italy, Jan. 28-Feb. 1, 1974*. European Space Res. Organ. (European Space Agency) SP-100, pp. 243-287.

COLLINS A. G. (1975) *Geochemistry of Oilfield Waters,* Elsevier.

COMBAZ A. and MATHAREL M. DE (1978) Organic sedimentation and genesis of petroleum in Mahakam Delta, Borneo. *Amer. Assoc. Petroleum Geologists Bull.* 62, 1684-1695.

CONNAN J. (1974) Time-temperature relation in oil genesis *Amer. Assoc. Petroleum Geologists Bull.* 58, 2516-2521.

CONNOR J. J. and SHACKLETTE H. T. (1975) Background geochemistry of some rocks, soils, plants, and vegetables in the conterminous United States. *U.S. Geol. Surv. Prof. Paper* 574-F.

CONOVER W. H., BEMENT T. R., and IMAN R. L. (1977) On a method for detecting clusters of possible uranium deposits. *Proc. Dept. Ener. Stat. Symp.,* pp. 33-37, Oct. 26-28th, Richland, Wash.

COOKE R. U. and WARREN A. (1973) *Geomorphology In Deserts.* Batsford.

CORDELL R. J. (1973) Colloidal soap as proposed primary migration medium for hydrocarbons. *Amer. Assoc. Petroleum Geologists Bull.* 57, 1618-1643.

COUSTAU H. (1977) Formation waters and hydrodynamics. *J. Geochem. Explor.* (Special Issue: Application of Geochemistry to the Search for the Crude Oil and Natural Gas; B. Hitchon, Compiler and Editor) 7, 213-241.

COWAN G. A. and ADLER H. H. (1976) The variability of the natural abundance of ^{235}U. *Geochim. Cosmochim. Acta* 40, 1487-1490.

COWART J. B. and OSMOND J. K. (1977) Uranium isotopes in groundwater: Their use in prospecting for sandstone-type uranium deposits. *J. Geochem. Explor.* 8, 365-379.

COX R. (1975) Geochemical soil surveys in exploration for nickel-copper sulphides at Pioneer, near Norseman, Western Australia. In, *Geochemical Exploration 1974* (editors I. L. Elliott and W. K. Fletcher), pp. 437-460. Elsevier.

CRENSHAW G. L. and LAKIN H. W. (1974) A sensitive and rapid method for the determination of trace amounts of selenium in geologic materials. *J. Res. U.S. Geol. Surv.* 2, 483-487.

CULBERT R. R. and LEIGHTON D. G. (1978) Uranium in alkaline waters — Okanagan area, British Columbia. *CIM Bull.* 71 (No. 793; May), 103-110.

CURTIN G. C. and KING H. D. (1974) The association of geochemical anomalies with a negative gravity anomaly in the Chief Mountain-Soda Creek area, Clear Creek County, Colorado. *J. Res. U.S. Geol. Surv.* 2, 581-592.

CURTIN G. C., KING H. D. and MOSIER E. L. (1974) Movement of elements into the atmosphere from coniferous trees in subalpine forests of Colorado and Idaho. *J. Geochem. Explor.* 3, 245-263.

CUSTODIO E. and LLAMAS M. R. (1976) *Hidrologiá subterránea* (in Spanish). 2 vols. Ediciones Omega. Barcelona.

CZEHURA S. J. (1977) A lichen indicator of copper mineralization, Lights Creek District, Plumas County, California. *Econ. Geol.* 72, 796-803.

DAVENPORT P. H., HORNBROOK E. H. W., and BUTLER A. J. (1975) Regional lake sediment geochemical survey for zinc mineralization in western Newfoundland. In, *Geochemical Exploration 1974* (editors I. L. Elliott and W. K. Fletcher), pp. 555-578. Elsevier.

DELAVAULT R. (1977) Simplified field test for copper. *J. Geochem. Explor.* 8, 537-540.

DEROO G. (1976) Corrélations huiles brutes — roches mères a l'échelle des bassins sédimentaires. *Bull. Centre Rech. Pau-SNPA* 10, 317-335.

DIJKSTRA S., VAN DEN HUL H. J. and BILL E. (1979) Experiments on the usefulness of some selected chemical quantities in geochemical exploration in a former mining district. In, *Geochemical Exploration 1978* (editors J. R. Watterson and P. K. Theobald), pp. 283-288. Assoc. of Explor. Geochemists.

DIXON J. B. and WEED S. B. (1977) (editors) *Minerals In Soil Environments.* Soil. Sci. Soc. Amer.

DODD P. H. and ESCHLIMAN D. H. (1972) Borehole logging techniques for uranium exploration and evaluation. In, *Uranium Prospecting Handbook* (editors S. H. U. Bowie, M. Davis and D. Ostle), pp. 244-276. IMM.

DOE B. R. (1979) The application of lead isotopes to mineral prospect evaluation of Cretaceous-Tertiary magmatothermal ore deposits in the Western United States. In, *Geochemical Exploration 1978* (editors J. R. Watterson and P. K. Theobald), pp. 227-232. Assoc. of Explor. Geochemists.

DOE B. R. and STACEY J. S. (1974) The application of lead isotopes to the problems of ore genesis and ore prospect evaluation: A review. *Econ. Geol.* 69, 757-776.

DOMENICO P. A. (1972) *Concepts and Models in Groundwater Hydrology.* McGraw-Hill.

DONOVAN T. J. (1974) Petroleum microseepage at Cement, Oklahoma: evidence and mechanism. *Amer. Assoc. Petroleum Geologists Bull.* 58, 429-446.

DOW W. G. (1974) Application of oil-correlation and source-rock data to exploration in Williston Basin. *Amer. Assoc. Petroleum Geologists Bull.* 58, 1253-1262.

DOW W. G. (1977) Kerogen studies and geological interpretations. *J. Geochem. Explor.* (Special Issue: Application of Geochemistry to the Search for Crude Oil and Natural Gas; B. Hitchon, Compiler and Editor) 7, 79-99.

DOW W. G. (1978) Petroleum source beds on continental slopes and rises. *Amer. Assoc. Petroleum Geologists Bull.* 62, 1584-1606.

DREGNE H. E. (1976) *Soils Of Arid Regions.* Elsevier.

DUTRIZAC J. E. and MACDONALD R. J. C. (1974) Ferric ion as a leaching medium. *Minerals Sci. Engng.* 6, 59-100.

DUVIGNEAUD P. and DENAEYER-DE-SMET S. (1963) Copper and vegetation in Katanga (in Fr.). *Bull. Soc. Roy. Bot. Belg.* 96, 93-231.

DYCK W. (1976) The use of helium in mineral exploration. *J. Geochem. Explor.* 5, 3-20.

DYCK W. (1978) The mobility and concentration of uranium and its decay products in temperate surficial environments. In, *Short Course in Uranium Deposits: Their Mineralogy and Origin,* pp. 57-100. Mineral. Assoc. of Canada.

DYCK W. and CAMERON E. M. (1975) Surface lake water uranium-radon survey of the Lineament Lake area, District of Mackenzie. *Geol. Surv. Canada* 75-1A, pp. 209-212.

DYCK W., JONASSON I. R. and LIARD R. F. (1976) Uranium prospecting with [222]Rn in frozen terrain. *J. Geochem. Explor.* 5, 115-127.

DYCK W., CHATTERJEE A. K., GEMMELL D. E. and MURRICANE K. (1976a) Well water trace element reconnaissance, eastern maritime, Canada. *J. Geochem. Explor.* 6, 139-162.

DYCK W. and JONASSON I. R. (1977) The nature and behavior of gases in natural waters. *Water Research* 11, 705-711.

DYCK W. and MILLER W. R. (1979) Application of hydrogeochemistry to the search for uranium and base metals. In, *Geophysics and Geochemistry in the Search for Metallic Ores* (editor P. J. Hood). *Geol. Surv. Canada, Econ. Geol. Report* 31.

DYCK W. and PELCHAT J. C. (1977) A semiportable helium analysis facility. *Geol. Surv. Canada Paper* 77-1C, 85-87.

DYCK W. and TAN B. (1978) Seasonal variations of helium, radon, and uranium in lake waters near the Key Lake uranium deposit, Saskatchewan. *J. Geochem. Explor.* 10, 153-167.

EATON A. (1979) The impact of anoxia on Mn fluxes in the Chesapeake Bay. *Geochim. Cosmochim. Acta* 43, 429-432.

EGANHOUSE R. P. and CALDER J. A. (1976) The solubility of medium molecular weight aromatic hydrocarbons and the effects of hydrocarbon co-solutes and salinity. *Geochim. Cosmochim. Acta* 40, 555-561.

EISENBUD M. (1977) Summary Report. In, *Inter. Sympos. On Areas of High Natural Radioactivity.* (Proc. of Conf. held in Pocos de Caldas, Brazil, June 16-20, 1975), pp. 167-179. Brazil Acad. Sciences. Rio de Janeiro.

ELLIOTT I. L. and FLETCHER W. K. (1975) (editors) *Geochemical Exploration 1974.* Elsevier.

ELLIS M. W. and McGREGOR J. A. (1967) The Kalengwa copper deposit, northwestern Zambia. *Econ. Geol.* 62, 781-797.

ELOFSON R. M., SCHULZ K. F. and HITCHON B. (1977) Geochemical significance of chemical composition and ESR properties of asphaltenes in crude oils from Alberta, Canada. *Geochim. Cosmochim. Acta* 41, 567-580.

ERDMAN J. G. and MORRIS D. A. (1974) Geochemical correlation of petroleum. *Amer. Assoc. Petroleum Geologists Bull.* 58, 2326-2337.

ERIKSSON K. (1973) Prospecting in an area of central Sweden. *Prospecting In Areas Of Glacial Terrain* (editor M. J. Jones), pp. 83-86. IMM.

ERIKSSON K. (1976) Regional prospecting by the use of peat sampling. *J. Geochem. Explor.* 5, 387-388.

FABBI B. P., ELSHEIMER H. N. and ESPOS L. F. (1976) Quantitative analysis of selected minor and trace elements through use of a computerized automatic X-ray spectrograph. *Adv. in X-ray Anal.,* 19, 273-292.

FASSEL V. A. and KNISELEY R. N. (1974) Inductively coupled plasma-optical emission spectroscopy. *Anal. Chem.* 46, 1110A-1120A; and 1155A-1164A.

FAYE G. H. (1978) Certified and provisional reference materials available from Canada Centre for mineral and energy technology, 1978. *Can. Centre for Mineral and Energy Technology (CANMET), Can. Energy Mines and Resources, CANMET Report* 78-3.

FICKLIN W. H. (1975) Ion-selective electrode measurements of iodine in rocks and soils. *J. Res. U.S. Geol. Surv.* 3, 753-755.

FISHER D. E. (1975) Geoanalytic applications of particle tracks. *Earth-Sci. Rev.* 11, 291-335.

FLANAGAN F. J. (1974) Reference samples for the earth sciences. *Geochim. Cosmochim. Acta* 38, 1731-1744.

FLANAGAN F. J. (1976) Descriptions and analyses of eight new USGS rock standards. *U.S. Geol. Surv. Prof. Paper* 840.

FLEISCHER R. L., PRICE P. B. and WALKER R. M. (1975). *Nuclear Tracks In Solids.* University of California (Los Angeles) Press.

FORD J. H. (1978) A chemical study of alteration at the Panguna porphyry copper deposit, Bougainville, Papua New Guinea. *Econ. Geol.* 73, 703-720.

FOSCOLOS A. E., POWELL T. G. and GUNTHER P. R. (1976) The use of clay minerals and inorganic and organic geochemical indicators for evaluating the degree of diagenesis and oil generating potential of shales. *Geochim. Cosmochim. Acta* 40, 953-966.

FOTH H. D. (1978) *Fundamentals of Soil Science* (6th edition). Wiley.

FRANKLIN J. M., KASARDA J. and POULSEN K. H. (1975) Petrology and chemistry of the alteration zone of the Mattabi massive sulfide deposit. *Econ. Geol.* 70, 63-79.

FREEZE R. A. and CHERRY J. A. (1979) *Groundwater.* Prentice-Hall.

FRIEDRICH G. H. W. and CHRISTENSEN S. M. (1977) Geochemical dispersion patterns associated with the Lake Yindarlgooda sulphide mineralization, Western Australia. *J. Geochem. Explor.* 8, 219-234.

FRIEDRICH G. H. W., KUNZENDORF H. and PLUGER W. L. (1974) Ship-borne geochemical investigations of deep-sea manganese-nodule deposits in the Pacific using a radio-isotope energy-dispersive X-ray system. *J. Geochem. Explor.* 3, 303-317.

FRIEDRICH G. H., PLUGER W. L., HILMER E. F. and ABU-ABED I. (1973) Flameless atomic absorption and ion-sensitive electrodes as analytical tools in copper exploration. In, *Geochemical Exploration 1972* (editor M. J. Jones), pp. 435-443.

FUEX A. N. (1977) The use of stable carbon isotopes in hydrocarbon exploration. *J. Geochem. Explor.* (Special Issue: Application of Geochemistry to the Search for Crude Oil and Natural Gas; B. Hitchon, Compiler and Editor) 7, 155-188.

FYFE W. S. (1973) Dehydration reactions. *Amer. Assoc. Petroleum Geologists Bull.* 57, 190-197.

GABELMAN J. W. (1977) *Migration of Uranium and Thorium — Exploration Significance.* Amer. Assoc. Petrol. Geol., Studies In Geology No. 3.

GARRETT R. G. (1971) Molybdenum, tungsten and uranium in acid plutonic rocks as a guide to regional exploration, S. E. Yukon. *Can. Mining J.* 92 (No. 4), 37-40.

GARRETT R. G. (1974) Field data acquisition methods for applied geochemical surveys at the Geological Survey of Canada. *Geol. Surv. Canada Paper* 74-52.

GARRETT R. G. (1979) Sampling considerations for regional geochemical surveys. *Geol. Surv. Canada Paper* 79-1A, 197-205.

GARRETT R. G. and GOSS T. I. (1979) The evaluation of sampling and analytical variation in regional geochemical surveys. *Geochemical Exploration 1978* (editors J. R. Watterson and P. K. Theobald), pp. 371-383. Assoc. of Explor. Geochemists.

GARRETT R. G. and HORNBROOK E. H. W. (1976) The relationship between zinc and organic content in centre-lake bottom sediments. *J. Geochem. Explor.* 5, 31-38.

GATEHOUSE S., RUSSELL D. W. and VAN MOORT J. C. (1977) Sequential soil analyses in exploration geochemistry. *J. Geochem. Explor.* 8, 483-494.

GEDEON A. Z., BUTT C. R. M., GARDNER K. A. and HART M. K. (1977) The applicability of some geochemical analytical techniques in determining "total" compositions of some lateralized rocks. *J. Geochem. Explor.* 8, 283-303.

GIJBELS R. (1973) Neutron-activation analysis of ores and minerals. *Miner. Sci. Eng.* 5, 304-348.

GLENNIE K. W. (1970) *Desert Sedimentary Environments.* Elsevier.

GOODELL P. C. and PETERSEN U. (1974) Julcani mining district, Peru: A study of metal ratios. *Econ. Geol.* 69, 347-361.

GOODFELLOW W. D. and WAHL J. L. (1976) Water extracts of volcanic rocks — Detection of anomalous halos at Brunswick No. 12 and Heath Steel B-zone massive sulfide deposits. *J. Geochem. Explor.* 6, 35-59.

GOULD K. W. and SMITH J. W. (1978) Isotopic evidence for microbiologic role in genesis of crude oil from Barrow Island, Western Australia. *Amer. Assoc. Petroleum Geologists Bull.* 62, 455-462.

GOVETT G. J. S. (1975) Soil conductivities: assessment of an electrogeochemical exploration technique. In, *Geochemical Exploration 1974* (editors I. L. Elliott and W. K. Fletcher), pp. 101-118. Elsevier.

GOVETT G. J. S. (1976) Detection of deeply buried and blind sulphide deposits by measurement of H^+ and conductivity of closely spaced surface soil samples. *J. Geochem. Explor.* 6, 359-382.

GOVETT G. J. S. (1977) (editor) Exploration geochemistry in the Appalachians. *J. Geochem. Explor.* 6, 1-298.

GOVETT G. J. S., GOODFELLOW W. D. and WHITEHEAD R. E. S. (1976) Experimental aqueous dispersion of elements around sulfides. *Econ. Geol.* 71, 925-940.

GOVETT G. J. S. and NICHOL I. (1979) Lithogeochemistry in mineral exploration. In, *Geophysics and Geochemistry in the Search for Metallic Ores* (editor P. J. Hood). *Geol. Surv. Canada Econ. Geol.* Report 31.

GREENLAND L. P. and CAMPBELL E. Y. (1974) Spectrophotometric determination of niobium in rocks. *J. Res. U.S. Geol. Surv.* 2, 353-355.

GRIMBERT A. (1972) Use of geochemical techniques in uranium prospecting. In, *Uranium Prospecting Handbook* (editors S. H. U. Bowie, M. Davis and D. Ostle), pp. 110-120. IMM.

GUNTON J. E. and NICHOL I. (1974) Delineation and interpretation of metal dispersion patterns related to mineralization in the Whipsaw Creek area. *CIM Bull.* 67, 66-75.

GUNTON J. E. and NICHOL I. (1975) Chemical zoning associated with the Ingerbelle-Copper Mountain mineralization, Princeton, British Columbia. In, *Geochemical Exploration 1974* (editors I. L. Elliott and W. K. Fletcher), pp. 297-312. Elsevier.

GWOSDZ W. and KREBS W. (1977) Manganese halo surrounding Meggen ore deposit, Germany. *Trans. IMM* (Sect. B., Appl. Earth Sci.) 86, B73-B77.

HAFFTY J., RILEY L. B. and GOSS W. G. (1977) A manual on fire assaying and determination of the noble metals in geological materials. *U.S. Geol. Surv. Bull.* 1445.

HALCROW W., MACKAY D. W. and THORNTON I. (1973) The distribution of trace metals and fauna in the Firth of Clyde in relation to the disposal of sewage sludge. *J. Mar. Biol. U.K.* 53, 721-739.

HAMBLETON-JONES B. B. (1978) Theory and practice of geochemical prospecting for uranium. *Minerals Sci. Engng.* 10, 182-197.

HARRISON W. E. (1976) Laboratory graphitization of a modern estuarine kerogen. *Geochim. Cosmochim. Acta* 40, 247-248.

HARRISON W. E. (1978) Experimental diagenetic study of a modern lipid-rich sediment. *Chem. Geol.* 21, 315-334.

HARWOOD R. J. (1977) Oil and gas generation by laboratory pyrolysis of kerogen. *Amer. Assoc. Petroleum Geologists Bull.* 61, 2082-2102.

HAUSEN D. M. (1979) Quantitative measurement of wallrock alteration in the exploration of buried mineral deposits. *Trans. AIME*, 266 (in press).

HAWKES H. E. (1976a) The early days of exploration geochemistry. *J. Geochem. Explor.* 6, 1-11.

HAWKES H. E. (1976b) *Exploration Geochemistry Bibliography* (period January 1972 to December 1975). Spec. Vol. No. 5, Assoc. of Explor. Geochemists.

HAWKES H. E. (1976c) The downstream dilution of stream sediment anomalies. *J. Geochem. Explor.* 6, 345-358.

HAWKES H. E. (1977) *Exploration Geochemistry Bibliography* (period January 1976 to June 1977). 1977 Bibliography Supplement. Assoc. of Explor. Geochemists.

HAWKES H. E. and WEBB J. S. (1962) *Geochemistry In Mineral Exploration*. Harper and Row.

HEM J. D. (1977) Reactions of metal ions at surfaces of hydrous iron oxides. *Geochim. Cosmochim. Acta* 41, 527-538.

HEM J. D. (1978) Redox processes at surfaces of manganese oxide and their effects on aqueous metal ions. *Chem. Geol.* 21, 199-218.

HENAO-LONDONO D. (1977) A preliminary geochemical evaluation of the Arctic Islands. *Bull. Canadian Petroleum Geol.* 25, 1059-1084.

HESP W. R. and RIGBY D. (1975) Aspects of tin metallogenesis in the Tasman Geosyncline, eastern Australia, as reflected by cluster and factor analyses. *J. Geochem. Explor.* 4, 331-347.

HILL W. E. (1975) The use of analytical standards to control assaying projects. In, *Geochemical Exploration 1974* (editors I. L. Elliott and W. K. Fletcher), pp. 651-657. Elsevier.

HIRVAS H. (1977) Glacial transport in Finnish Lapland. In, *Prospecting In Areas Of Glaciated Terrain 1977* (editor G. R. Davis), pp. 128-137. IMM.

HITCHON B. (1974) Occurrence of natural gas hydrates in sedimentary basins. *Natural Gases in Marine Sediments* (editor I. R. Kaplan), 195-225, Plenum Press.

HITCHON B. (1976) Hydrogeochemical aspects of mineral deposits in sedimentary rocks. *Handbook of Strata-bound and Stratiform Ore Deposits* (editor K. H. Wolf), 2, 54-66, Elsevier, Amsterdam.

HITCHON B. (1977a) Geochemical links between oil fields and ore deposits in sedimentary rocks. *Proceedings of the Forum on Oil and Ore in Sediments* (editor P. Garrard), 1-37, Imperial College, London, England.

HITCHON B. (1977b) Geochemical aspects of in-situ recovery. *The Oil Sands of Canada-Venezuela 1977* (editors D. A. Redford and A. G. Winestock), 80-86, The Canadian Inst. Mining and Metall., Toronto, Canada.

HITCHON B. (1979) Some economic aspects of water-rock interaction. *Physical and Chemical Constraints of Migration* (editors W. H. Roberts and R. J. Cordell), Amer. Assoc. Petroleum Geologists (in press).

HITCHON B. and HORN M. K. (1974) Petroleum occurrence indicators in formation waters from Alberta, Canada. *Amer. Assoc. Petroleum Geologists Bull.* 58, 464-473.

HITCHON B., LEVINSON A. A. and HORN M. K. (1977) Bromide, iodide, and boron in Alberta formation waters. *Alberta Res. Council, Econ. Geol. Rept. 5.*

HO T. Y., ROGERS M. A., DRUSHEL H. V. and KOONS C. B. (1974) Evaluation of sulfur compounds in crude oils. *Amer. Assoc. Petroleum Geologists Bull.* 58, 2338-2348.

HOAG R. B. and WEBBER G. R. (1976) Hydrogeochemical exploration and sources of anomalous waters. *J. Geochem. Explor.* 5, 39-57.

HOAG R. B. and WEBBER G. R. (1976a) Hydrogeochemical exploration and sources of anomalous waters. *J. Geochem. Explor.* 5, 39-57.

HOAG R. B. and WEBBER G. R. (1976b) Significance for mineral exploration of sulphate concentrations in groundwaters. *Can. Inst. Min. Met. (CIM) Bull.* 69, No. 776 (Dec.), 86-91.

HOEVE J. and SIBBALD T. I. I. (1978) On the genesis of Rabbit Lake and other unconformity-type uranium deposits in northern Saskatchewan, Canada. *Econ. Geol.* 73, 1450-1473.

HOFFMAN S. J. (1977) Talus fine sampling as a regional geochemical exploration technique in mountainous regions. *J. Geochem. Explor.* 7, 349-360.

HOFFMAN S. J. and FLETCHER W. K. (1979) Extraction of Cu, Zn, Mo, Fe, and Mn from soils and sediments using a sequential procedure. In, *Geochemical Exploration 1978* (editors J. R. Watterson and P. K. Theobald), 289-299. Assoc. of Explor. Geochemists.

HOFFMAN S. J. and WASKETT-MYERS M. J. (1974) Determination of molybdenum in soils and sediments with a modified zinc dithiol procedure. *J. Geochem. Explor.* 3, 61-66.

HOLMES A. (1978) *Principles of Physical Geology* (3rd edition). Nelson.

HOOD P. J. (1979) (editor) *Geophysics and Geochemistry in the Search for Metallic Ores. Geol. Surv. Canada, Econ. Geol. Report* No. 31.

HORDER M. (1978) An introduction to the national geochemical data bank. *Comput. Unit. Inst. Geol. Sci., London.*

HUBERT A. E. and LAKIN H. W. (1973) Atomic absorption determination of thallium and indium in geologic materials. In, *Geochemical Exploration 1972* (editor M. J. Jones), pp. 383-387. IMM.

HUC A. Y. (1977) Contribution de la géochimie organique a une esquisse paléoécologique des schists bitumineux du Toarcien de l'est du Bassin de Paris: Étude de la matière organique insoluble (kérogènes). *Rev. l'Inst. Français du Petrole.* 32, 703-718.

HYVARINEN L., KAURANNE K. and YLETYINEN V. (1973) Modern boulder tracing in prospecting. *Prospecting In Areas Of Glacial Terrain* (editor M. J. Jones), pp. 87-95. IMM.

ILLICH H. A., HANEY F. R. and JACKSON T. J. (1977) Hydrocarbon geochemistry of oils from Maranon Basin, Peru. *Amer. Assoc. Petroleum Geologists Bull.* 61, 2103-2114.

IMM (1977) *Prospecting in Areas of Glaciated Terrain 1977.* Institution of Mining and Metallurgy, London.

ISHIWATARI R., ISHIWATARI M., ROHRBACK B. G. and KAPLAN I. R. (1977) Thermal alteration experiments on organic matter from recent marine sediments in relation to petroleum genesis. *Geochim. Cosmochim. Acta* 41, 815-828.

ISHIWATARI R., ROHRBACK B. G. and KAPLAN I. R. (1978) Hydrocarbon generation by thermal alteration of kerogen from different sediments. *Amer. Assoc. Petroleum Geologists Bull.* 62, 687-692.

JACKSON K. S., JONASSON I. R. and SKIPPEN G. B. (1978) The nature of metals-sediment-water interactions in freshwater bodies, with emphasis on the role of organic matter. *Earth-Sci. Rev.* 14, 97-146.

JACKSON K. S. and SKIPPEN G. B. (1978) Geochemical dispersion of heavy metals via organic complexing: A laboratory study of copper, lead, zinc, and nickel behaviour at a simulated sediment-water boundary. *J. Geochem. Explor.* 10, 117-138.

JACOBS D. C. and PARRY E. T. (1976) A comparison of the geochemistry of biotite from some Basin and Range stocks. *Econ. Geol.* 71, 1029-1035.

JACQUIN C. and POULET M. (1973) Essai de restitution des conditions hydrodynamiques régnant dans un bassin sédimentaire au cours de son évolution. *Rev. l'Inst. Français du Petrole* 28, 269-297.

JAMBOR J. L. and BEAULNE J. M. (1978) Sulphide zones and hydrothermal biotite alteration in porphyry copper-molybdenum deposits, Highland Valley, British Columbia. *Geol. Surv. Canada Paper* 77-12.

JAMBOR J. L. and DELABIO R. N. (1978) Distribution of hydrothermal clay minerals in the Valley Copper porphyry deposit, Highland Valley, British Columbia. *Geol. Surv. Canada Paper* 77-9.

JOBSON A. M., COOK F. D. and WESTLAKE D. W. S. (1979) Interaction of aerobic and anaerobic bacteria in petroleum biodegradation. *Chem. Geol.* 24, 355-365.

JOENSUU O. I. (1971) Mercury-vapor detector. *Appl. Spectrosc.* 25, 526-528.

JOHN E. C. (1978) Mineral zones in the Utah Copper orebody. *Econ. Geol.* 73, 1250-1259.

JONASSON I. R., LYNCH J. J. and TRIP L. J. (1973) Field and laboratory methods used by the Geological Survey of Canada in geochemical surveys. No. 12. Mercury in ores, rocks, soils sediments and water. *Geol. Surv. Can. Paper* 73-21.

JONES M. J. (1973) (editor) *Prospecting In Areas of Glacial Terrain.* IMM. London.

JONES M. J. (1975) (editor) *Prospecting In Areas of Glaciated Terrain 1975.* Institution of Mining and Metallurgy. London.

JOPLING A. V. and McDONALD B. C. (1975) (editors) Glaciofluvial and glaciolacustrine sedimentation. *Soc. Econ. Paleon. and Mineral.,* Spec. Public. No. 23.

JOYCE A. S. (1976) *Exploration Geochemistry* (second edition). Techsearch Inc., and The Australian Mineral Foundation. Adelaide and Glenside, South Australia.

KAMEN M. (1957) *Isotope Tracers In Biology* (Third Edition). Academic Press.

KAURANNE L. K. (1976) (editor) Conceptual models in exploration geochemistry. Norden, 1975. *J. Geochem. Explor.* 5, 173-420.

KELLY P. C., BROOKS R. R. , DILLI S. and JAFFRE T. (1975) Preliminary observations on the ecology and plant chemistry of some nickel-accumulating plants from New Caledonia. *Proc. R. Soc. Lond.* B., 189, 69-80.

KESLER S. E., ISSIGONIS M. J. and VANLOON J. C. (1975a) An evaluation of the use of halogen and water abundances in efforts to distinguish mineralized and barren intrusive rocks. *J. Geochem. Explor.* 4, 235-245.

KESLER S. E., JONES L. M. and WALKER R. L. (1975b) Intrusive rocks associated with porphyry copper mineralization in island arc areas. *Econ. Geol.* 70, 515-526.

KETOLA M. and SARIKKOLA R. (1973) Some aspects concerning the feasibility of radiometric methods for uranium exploration in Finland. In, *Uranium Exploration Methods,* pp. 31-43. Intern. Atomic Energy Agency. Vienna.

KLASSEN R. A. and SHILTS W. W. (1977) Glacial dispersion of uranium in the District of Keewatin, Canada. In, *Prospecting In Areas Of Glaciated Terrain 1977* (editor G. R. Davis), pp. 80-88. IMM.

KLEIBER P. and ERLEBACH W. E. (1977) Limitations of single water samples in representing mean water quality. III. Effect of variability in concentration measurements on estimates of nutrient loadings in the Squamish River, B.C. *Tech. Bull No. 103,* Inland Water Directorate, Pacific and Yukon Region, Water Quality Branch. Fisheries and Environment Canada.

KLUSMAN R. W. and LANDRESS R. A. (1978) Secondary controls on mercury in soils of geothermal areas. *J. Geochem. Explor.* 9, 75-91.

KOCH G. S. and LINK R. F. (1977) Anomaly recognition in exploration geochemistry through a statistical analysis of multivariate data. *Fourteenth Appl. Comput. Math. Min. Ind.,* 312-321.

KOKKOLA M. (1975) Stratigraphy of till at Hitura open-pit, Nivala, western Finland, and its bearing on geochemical prospecting. In, *Prospecting In Areas Of Glaciated Terrain 1975* (editor M. J. Jones), pp. 149-154. IMM.

KOLOTOV B. A., KISELEVA Y. A. and RUBEYKIN V. Z. (1965) On the secondary dispersion aureoles in the vicinity of ore deposits. *Geochem. Inter.* 2, 675-677.

KOONS C. B., BOND J. G. and PEIRCE F. L. (1974) Effects of depositional environment and postdepositional history on chemical composition of Lower Tuscaloosa oils. *Amer. Assoc. Petroleum Geologists Bull.* 58, 1272-1280.

KOTHNY E. L. (1974) Simple trace determinations of platinum in geological materials. *J. Geochem. Explor.* 3, 291-299.

KOVALAVSKIY A. L. (1975) Conditions for successful use of the biogeochemical method of prospecting for ore deposits. *Doklady, Earth Sci. Sect.,* 218, Sept.-Oct. 1974. Translation by Amer. Geol. Inst. Nov. 1975, 183-186.

KRAUSKOPF K. (1979) *Introduction to Geochemistry* (second edition). McGraw Hill.

KRONBERG B. I., FYFE W. S., LEONARDOS O. H. and SANTOS A. M. (1979) The chemistry of some Brazilian soils: Element mobility during intense weathering. *Chem. Geol.* 24, 211-229.

KROUSE H. R. (1977) Sulfur isotope studies and their role in petroleum exploration. *J. Geochem. Explor.* (Special Issue: Application of Geochemistry to the Search for Crude Oil and Natural Gas; B. Hitchon, Compiler and Editor) 7, 189-211.

KRYLOVA L. Y. (1978) Exploration significance of iodine distribution in the bedrock and weathering crust of the Maykain deposit. *Intern. Geol. Rev.* 20, 357-361.

KUJANSUU R. (1976) Glaciogeological surveys for ore prospecting purposes in northern Finland. In, *Glacial Till — An Inter-disciplinary Study* (R. F. Legget, Editor), pp. 225-239. Royal Soc. Canada, Spec. Public. No. 12.

LAG J. and BOLVIKEN B. (1974) Some naturally heavy-metal poisoned areas of interest in prospecting, soil chemistry, and geomedicine. *Norges Geol. Unders.* 304, 73-96.

LANGMUIR D. (1978a) Uranium solution-mineral equilibria at low temperatures with applications to sedimentary ore deposits. In, *Short Course in Uranium Deposits: Their Mineralogy and Origin.* pp. 17-55. Mineral. Assoc. of Canada.

LANGMUIR D. (1978b) Uranium solution-mineral equilibria at low temperatures with applications to sedimentary ore deposits. *Geochim. Cosmochim. Acta* 42, 547-569.

LANIER G., RAAB W. J., FOLSOM R. B. and CONE S. (1978) Alteration of equigranular monzonite, Bingham mining district, Utah. *Econ. Geol.* 73, 1270-1286.

LAPLANTE R. E. (1974) Hydrocarbon generation in Gulf Coast Tertiary sediments. *Amer. Assoc. Petroleum Geologists Bull.* 58, 1281-1289.

LARSEN E. S. and GOTTFRIED D. (1961) Distribution of uranium in rocks and minerals of Mesozoic batholiths in western United States. *U.S. Geol. Surv. Bull.* 1070-C, pp. 63-103.

LARSEN E. S., PHAIR G., GOTTFRIED D. and SMITH W. L. (1956) Uranium in magmatic differentiation. *U.S. Geol. Surv. Prof. Paper* 300, pp. 65-74.

LARSSON J. O. (1976) Organic stream sediments in regional geochemical prospecting, Precambrian Pajala district, Sweden. *J. Geochem. Explor.* 6, 233-249.

LEAKE R. C. and AUCOTT J. W. (1973) Geochemical mapping and prospecting by use of rapid automatic X-ray fluorescence analysis of panned concentrates. In, *Geochemical Exploration 1972* (editor M. J. Jones), pp. 389-400. IMM.

LEE J., BROOKS R. R., REEVES R. D. and JAFFRÉ T. (1977a) Plant-soil relationships in a New Caledonian serpentine flora. *Pl. Soil,* 46, 675-680.

LEE J., REEVES R. D., BROOKS R. R. and JAFFRE T. (1977b) Isolation and identification of a citrato-complex of nickel from nickel-accumulating plants. *Phytochemistry* 16, 1503-1505.

LEGGET R. F. (1976) (editor) *Glacial Till — An Inter-Disciplinary Study*. Royal Soc. Canada Spec. Public. No. 12.

LEPELTIER C. (1969) A simplified statistical treatment of geochemical data by graphical representation. *Econ. Geol.* 64, 538-550.

LERMAN A. (1978) (editor) *Lakes — Chemistry, Geology, Physics*. Springer-Verlag.

LEVENTHAL J. S. (1976) Stepwise pyrolysis-gas chromatography of kerogen in sedimentary rocks. *Chem. Geol.* 18, 5-20.

LEVINSON A. A. (1977) Hydrogen — a reducing agent in some uranium deposits. *Can. J. Earth Sci.* 14, 2679-2681.

LEVINSON A. A. and BLAND C. J. (1978) Examples of the variability of disequilibrium and the emanation factor in some uraniferous materials. *Can. J. Earth Sci.* 15, 1867-1871.

LEVINSON A. A., BLAND C. J. and PARSLOW G. R. (1978) Possible pitfalls in the search for uranium deposits using lake sediments and lake waters. *Can. Inst. Mining Met. (CIM) Bull.* 71, 59-62.

LEVINSON A. A. and CARTER N. C. (1979) Glacial overburden profile sampling for porphyry copper exploration: Babine Lake area, British Columbia. *Western Miner* 52, No. 5 (May), 19-31.

LEVINSON A. A. and COETZEE G. L. (1978) Implications of disequilibrium in exploration for uranium ores in the surficial environment using radiometric techniques — A review. *Minerals. Sci. Engng.* 10, 19-27.

LEVINSON A. A. and DE PABLO L. (1975) A rapid X-ray fluorescence procedure applicable to exploration geochemistry. *J. Geochem. Explor.* 4, 399-408.

LOGN O. and BOLVIKEN B. (1974) Self potentials at the Joma pyrite deposit, Norway. *Geoexploration* 12, 11-28.

LOPATKINA A. P. (1964) Characteristics of migration of uranium in the natural waters of humid regions and their use in the determination of the geochemical background for uranium. *Geochem. Inter. 1964,* 788-795.

LOVELESS A. J. (1975) Lead isotopes — A guide to major mineral deposits. *Geoexploration* 13, 13-27.

LOVERING T. G. and MCCARTHY J. H. (1978) Conceptual models in exploration geochemistry. The Basin and Range Province of the western United States and northern Mexico. *J. Geochem. Explor.* 9, 113-276.

LOWENSTEIN P. L. and HOWARTH R. J. (1973) Automated colour mapping of three component systems and its application to regional geochemical reconnaissance. *Geochemical Exploration 1972.* IMM. London.

LUNDBERG B. (1973) Exploration for uranium through glacial drift in the Arjeplog district, northern Sweden. In, *Prospecting In Areas of Glacial Terrain* (editor M. J. Jones), pp. 31-43.

MABBUTT J. A. (1977) *Desert Landforms*. MIT Press.

MACKENZIE D. H. (1977) Empirical assessment of anomalies in tropical terrain. *Newsletter No. 21*, pp. 6-10. Assoc. of Explor. Geochemists.

MAGARA K. (1974) Aquathermal fluid migration. *Amer. Assoc. Petroleum Geologists Bull.* 58, 2513-2521.

MAGARA K. (1975) Reevaluation of montmorillonite dehydration as cause of abnormal pressure and hydrocarbon migration. *Amer. Assoc. Petroleum Geologists Bull.* 59, 292-302.

MAGARA K. (1976a) Water expulsion from elastic sediments during compaction — directions and volumes. *Amer. Assoc. Petroleum Geologists Bull.* 60, 543-553.

MAGARA K. (1976b) Thickness of removed sedimentary rocks, paleopore pressure, and paleotemperature, southwestern part of Western Canada basin. *Amer. Assoc. Petroleum Geologists Bull.* 60, 554-565.

MAGARA K. (1977) A theory relating isopachs to paleocompaction water-movement in a sedimentary basin. *Bull. Canadian Petroleum Geol.* 25, 195-207.

MAGARA K. (1978a) The significance of the expulsion of water in oil-phase primary migration. *Bull. Canadian Petroleum Geol.* 26, 123-131.

MAGARA K. (1978b) Time of oil generation and migration and oil window. *Bull. Canadian Petroleum Geol.* 26, 152-155.

MAHAFFEY E. J. (1974) A spectrophotometric method for the determination of rhenium in geologic materials. *J. Geochem. Explor.* 3, 53-59.

MAKOGON YU. F. (1978) *Hydrates of natural gas.* Translated by W. J. Cieslewicz, Geoexplorers Associates, Inc., Denver, Colo.

MANCEY S. J. and HOWARTH R. J. (1978) Factor score maps of regional geochemical data from England and Wales. *Appl. Geochem. Res. Grp., Imp. Coll. Sci. and Tech., London.*

MARANZANA F. (1972) Application of talus sampling to geochemical exploration in arid areas: Los Pelambres alteration area, Chile. *Trans IMM* (Sect. B, Appl. Earth Sci.) 81, B26-B33.

MARCHAND A. (1976) La résonance paramagnétique électronique (R.P.E.) sa mise en oeuvre pour l'étude des kérogènes. *Bull. Centre Rech. Pau-SNPA* 10, 253-266.

MARCHANT J. W. (1974) Effect of fallout of windborne zinc, lead, copper and cadmium from a fuming kiln on soil geochemical prospecting at Berg Aukas, South West Africa. *J. Geochem. Explor.* 3, 191-198.

MARINENKO J. and MEI L. (1974) Spectrophotometric determination of vanadium in rutile and in mafic igneous rocks. *J. Res. U.S. Geol. Surv.* 2, 701-703.

MARSHALL C. E. (1977) *The Physical Chemistry and Mineralogy of Soils. Vol. II: Soils in Place.* Wiley.

MARSHALL N. J. (1978) Colorimetric determination of arsenic in geochemical samples. *J. Geochem. Explor.* 10, 307-313.

MASON D. R. (1978) Compositional variations in ferromagnesian minerals from porphyry copper-generating and barren intrusions of the Western Highlands, Papua New Guinea. *Econ. Geol.* 73, 878-890.

MASON D. R. (1979) Ferromagnesian mineral chemical variations: A new exploration tool to distinguish porphyry copper provinces. In, *Geochemical Exploration 1978.* (editors J. R. Watterson and P. K. Theobald), pp. 243-249. Assoc. of Explor. Geochemists.

MATLICK J. S. and BUSECK P. R. (1976) Exploration for geothermal areas using mercury: a new geochemical technique. In, *Proc. Second U.N. Symp. on the Development and Use of Geothermal Resources* (editor C. Pezzotti), vol. 1, 785-792.

MCAULIFFE C. D. (1979) Oil and gas migration — chemical and physical constraints. *Amer. Assoc. Petroleum Geologists Bull.* 63, 761-781.

MCCAMMON R. B. (1976) An interactive computer graphics program for dissecting a mixture of normal (or lognormal) distributions. *Proc. 9th Interface Symp. Comput. Sci. and Stat.,* pp. 36-43. Harvard Univ.

MCCAMMON R. B., BRIDGES N. J., MCCARTHY J. H., and GOTT G. B. (1979) Estimate of mixed geochemical populations in rocks at Ely, Nevada. *Geochemical Exploration 1978* (editors J. R. Watterson and P. K. Theobald), pp. 385-390. Assoc. of Explor. Geochemists.

MCCARTHY J. H., VAUGH W. W., LEARNED R. E. and MEUSCHKE J. L. (1969) Mercury in soil gas and air — a potential tool in mineral exploration. *U. S. Geol. Surv. Circ.* 609.

MCGINNIES W. G., GOLDMAN B. J. and PAYLORE P. (1968) (editors) *Deserts Of The World.* Univ. of Arizona Press.

MCIVER R. D. (1974) Evidence of migrating liquid hydrocarbons in deep sea drilling project cores. *Amer. Assoc. Petroleum Geologists Bull.* 58, 1263-1271.

MCLAURIN A. N. (1978) Geochemical dispersion from the Gamsberg orebody, northwestern Cape, South Africa. *J. Geochem. Explor.* 10, 295-306.

MCNERNEY J. J. and BUSECK P. R. (1973) Geochemical exploration using mercury vapor. *Econ. Geol.* 68, 1313-1320.

MCNERNEY J. J., BUSECK P. R. and HANSON R. C. (1972) Mercury detection by means of thin gold films. *Science* 178, 611-612.

MEANS J. L., CRERAR D. A., BORCSIK M. P. and DUGUID J. O. (1978) Adsorption of Co and selected actinides by Mn and Fe oxides in soils and sediments. *Geochim. Cosmochim. Acta* 42, 1763-1773.

MEIGS P. (1953) World distribution of arid and semi-arid homoclimates. *Reviews of Research on Arid Zone Hydrology,* pp. 203-209. UNESCO, Paris.

MEYER T., THEOBALD P. K. and BLOOM H. (1979) Stream sediment geochemistry. In, *Geophysics and Geochemistry in the Search for Metallic Ores* (editor P. J. Hood). *Geol. Surv. Canada, Econ. Geol. Report* 31.

MEYER W. T. and LAM SHANG LEEN K. C. Y. (1973) Microwave-induced argon plasma emission system for geochemical trace analysis. In, *Geochemical Exploration 1972* (editor M. J. Jones), pp. 325-335. IMM.

MEYROWITZ R. (1973) Chemical analysis of rutile — a pyrocatechol violet spectrophotometric procedure for the direct microdetermination of zirconium. *J. Res. U.S. Geol. Surv.* 1, 549-554.

MICHIE U. M., GALLAGHER M. J. and SIMPSON A. (1973) Detection of concealed mineralization in northern Scotland. In, *Geochemical Exploration 1972* (editor M. J. Jones), pp. 117-130. IMM.

MIESCH A. T. (1976) Sampling designs for geochemical surveys — syllabus for a short course. *U.S. Geol. Surv. Open File Rept.* 76-772.

MILLER J. M. and OSTLE D. (1973) Radon measurement in uranium prospecting. In, *Uranium Exploration Methods* (Proc. of a Panel 10-14 April, 1972), pp. 237-247. Inter. Atomic Energy Agency. STI/Pub/334.

MILLER J. M. and SYMONS G. D. (1973). Radiometric traverse of the seabed off the Yorkshire coast. *Nature* 242 (No. 5394), 184-186.

MILNER C. W. D., ROGERS M. A. and EVANS C. R. (1977) Petroleum transformations in reservoirs. *J. Geochem. Explor.* (Special Issue: Application of Geochemistry to the Search for Crude Oil and Natural Gas; B. Hitchon, Compiler and Editor) 7, 101-153.

MOESKOPS P. G. (1977) Yilgarn nickel gossan geochemistry — a review, with new data. *J. Geochem. Explor.* 8, 247-258.

MOHSEN L. A. and BROWNLOW A. H. (1971) Abundance and distribution of manganese in the western part of the Philipsburg batholith, Montana. *Econ. Geol.* 66, 611-617.

MOORE W. J. (1978) Chemical characteristics of hydrothermal alteration at Bingham, Utah. *Econ. Geol.* 73, 1260-1269.

MURRAY J. W. (1975a) The interaction of metal ions at the manganese dioxide-solution interface. *Geochim. Cosmoshim. Acta.* 39, 505-519.

MURRAY J. W. (1975b) The interaction of cobalt with hydrous manganese dioxide. *Geochim. Cosmochim. Acta* 39, 635-647.

MYERS A. T., HAVENS R. G., CONNER J. J., CONKLIN N. M. and ROSE H. J. (1976) Glass reference standards for the trace-element analysis of geological materials — Compilation of interlaboratory data. *U. S. Geol. Surv. Prof. Paper* 1013.

NEGLIA S. (1979) Migration of fluids in sedimentary basins. *Amer. Assoc. Petroleum Geologists Bull.* 63, 573-597.

NEUERBERG G. J. (1975) A procedure, using hydrofluoric acid, for quantitative mineral separations from silicate rocks. *J. Res. U.S. Geol. Surv.* 3, 377-378.

NICKEL E. H., ALLCHURCH P. D., MASON M. G. and WILMSHURST J. R. (1977) Supergene alteration at the Perseverance nickel deposit, Agnew, Western Australia. *Econ. Geol.* 72, 184-203.

NICKEL E. H. and THORNBER M. R. (1977) Chemical constraints on the weathering of serpentinites containing nickel-iron sulphides. *J. Geochem. Explor.* 8, 235-245.

NICHOL I., BOGLE E. W., LAVIN O. P., McCONNELL J. W. and SOPUCK V. J. (1977) Lithogeochemistry as an aid in massive sulphide exploration. In, *Prospecting In Areas Of Glaciated Terrain 1977,* pp. 63-71. IMM.

NICHOL I., COKER W. B., JACKSON R. G. and KLASSEN R. A. (1975) Relation of lake-sediment composition to mineralization in different limnological environments in Canada. In, *Prospecting In Areas of Glaciated Terrain 1975* (editor M. J. Jones), pp. 112-125. IMM.

NIXON R. P. (1973) Oil source beds in Cretaceous Mowry shale of northwestern interior United States. *Amer. Assoc. Petroleum Geologists Bull. 57,* 136-161.

NOWLAN G. A. (1976) Concretionary manganese-iron oxides in streams and their usefulness as a sample medium for geochemical prospecting. *J. Geochem. Explor.* 6, 193-210.

NOWLAN G. A. and CAROLLO C. (1974) A probe for sampling interstitial waters of stream sediments and bog soils. *J. Geochem. Explor.* 3, 199-205.

OLADE M. A. (1977) Major element halos in granitic wall rocks of porphyry copper deposits, Guichon Creek batholith, British Columbia. *J. Geochem. Explor.* 7, 59-71.

OLADE M. and FLETCHER K. (1974) Potassium chlorate-hydrochloric acid: A sulphide selective leach for bedrock geochemistry. *J. Geochem. Explor.* 3, 337-344.

OLADE M. A. and FLETCHER W. K. (1976) Trace element geochemistry of the Highland Valley and Guichon Creek batholith in relation to porphyry copper mineralization. *Econ. Geol.* 71, 733-748.

d'OREY F. L. C. (1975) Contribution of termite mounds to locating hidden copper deposits. *Trans. IMM* (Sect. B, Appl. Earth Sci.) 84, B150-B151.

ORR W. L. (1974) Changes in sulfur content and isotopic ratios of sulfur during petroleum maturation — study of Big Horn Basin Paleozoic oils. *Amer. Assoc. Petroleum Geologists Bull.* 58, 2295-2318.

OSTLE D., COLEMAN R. F. and BALL T. K. (1972) Neutron activation analyses as an aid to geochemical prospecting for uranium. In, *Uranium Prospecting Handbook* (editors S. H. U. Bowie, M. Davis and D. Ostle), pp. 95-109. IMM.

OUDIN J. L. (1976) Étude géochimique du Bassin de le Mer du Nord. *Bull. Centre Rech. Pau-SNPA* 10, 339-358.

OVCHINNIKOV L. N. and GRIGORYAN S. V. (1971) Primary halos in prospecting for sulphide ore deposits. *Geochemical Exploration.* CIM Spec. Vol. 11, 375-380.

de PABLO-GALAN L. (1975) Direct reading emission spectroscopy analysis of geochemical samples. In, *Geochemical Exploration 1974* (editors I. L. Elliott and W. K. Fletcher), pp. 707-720. Elsevier.

PALACHE C., BERMAN H. and FRONDEL C. (1944) *The System of Mineralogy* (7th Edition), Vol. 1. Wiley.

PARRY W. T., BALLANTYNE G. H. and WILSON J. C. (1978) Chemistry of biotite and apatite from a vesicular quartz latite porphyry plug at Bingham, Utah. *Econ. Geol.* 73, 1308-1314.

PARRY W. T. and JACOBS D. C. (1975) Fluorine and chlorine in biotite from Basin and Range plutons. *Econ. Geol.* 70, 554-558.

PARSLOW G. R. (1974) Determination of background and threshold in exploration geochemistry. *J. Geochem. Explor.* 3, 319-336.

PEACHEY D. and ALLEN B. P. (1977) An investigation into selective dissolution of sulphide phases from stream sediments and soils. *J. Geochem. Explor.* 8, 571-577.

PENNA-FRANCA E., COSTA-RIBEIRO C., CULLEN P., BARCINSKI M. and GONZALEZ D. E. (1972) Natural radioactivity in Brazil: A comprehensive review with a model for dose-effect studies. In, *Natural Radiation Environment.* Vol. 2, (Editors: J. A. S. Adams, W. M. Lowder and T. F. Gesell); (Proc. Sec. Int. Symp. on Nat. Rad. Environ., Houston, Texas). U. S. Energy Res. Dev. Admin., CONF-720805-P2, pp. 929-940.

PENNA-FRANCA E. and GOMES DE FREITAS O. (1963) Radioactivity of biological materials from Brazilian areas rich in thorium compounds. *Nature* 197, 1062-1063.

PEREL'MAN A. I. (1967) *Geochemistry of Epigenesis.* Plenum Press.

PEREL'MAN A. I. (1977) *Geochemistry of Elements in the Supergene Zone.* Israel Program for Scientific Translations, Keter Publishing, Jerusalem. John Wiley.

PERSSON H. (1948) On the discovery of *Merceya ligulata* in the Azores, with a discussion of the so-called "copper mosses". *Revue Bryol. Lichen* 17, 76-78.

PETERSON P. J. (1971) Unusual accumulations of elements by plants and animals. *Sci. Progress* (Oxford) 59, 505-526.

PHELPS D. W. and BUSECK P. R. (1978) Natural concentrations of Hg in the Yellowstone and Coso geothermal fields. *Geothermal Resources Council Trans.* 2, 521-522.

PHELPS D. W. and BUSECK P. R. (1979) Mercury in soils as an indicator of geothermal activity: Yellowstone National Park, Wyoming. In, *Geochemical Exploration 1978* (editors J. R. Watterson and P. K. Theobald), pp. 153-160. Assoc. of Explor. Geochemists.

PHILIPPI G. T. (1974) The influence of marine and terrestrial source material on the composition of petroleum. *Geochim. Cosmochim. Acta* 38, 947-966.

PHILIPPI G. T. (1975) The deep subsurface temperature controlled origin of the gaseous and gasoline-range hydrocarbons of petroleum. *Geochim. Cosmochim. Acta* 39, 1353-1373.

PHILIPPI G. T. (1977) On the depth, time and mechanism of origin of the heavy to medium-gravity naphthenic crude oils. *Geochim. Cosmochim. Acta* 41, 33-52.

PHILP R. P., CALVIN M., BROWN S. and YANG E. (1978) Organic geochemical studies on kerogen precursors in recently deposited algal mats and oozes. *Chem. Geol.* 22, 207-231.

PLANT J. (1973) Random numbering system for geochemical samples. *Trans. IMM* (Sect. B., Appl. Earth Sci.) 82, B64-B65.

PLANT J. and COLEMAN R. F. (1972) Application of neutron activation analysis to the evaluation of placer gold concentrations. In, *Geochemical Exploration 1972* (editor M. J. Jones), pp. 373-381. IMM.

PLANT J., GOODE G. C. and HERRINGTON J. (1976) An instrumental neutron activation method for multi-element geochemical mapping. *J. Geochem. Explor.* 6, 299-319.

PLANT J., JEFFERY K., GILL E., and FAGE C. (1975) The systematic determination of accuracy and precision in geochemical exploration data. *J. Geochim. Explor.* 4, 467-486.

PLANT J. and RHIND D. (1974) Mapping minerals. *Geograph. Mag.* 47, 123-126.

PLUGER W. L. and FRIEDRICH G. H. (1973) Determination of total and cold-extractable fluoride in soils and stream sediments with an ion-sensitive fluoride electrode. In, *Geochemical Exploration 1972* (editor M. J. Jones), pp. 421-427. IMM.

POKORNY J. (1975) Geochemical prospecting for ores in the Bohemian Massif, Czechoslovakia. In, *Geochemical Exploration 1974* (editors I. L. Elliott and W. K. Fletcher), pp. 77-83.

POLIKARPOCHKIN V. V. (1971) The quantitative estimation of ore-bearing areas from sample data on drainage systems. *Geochemical Exploration.* CIM Spec. Vol. 11, 585-586.

POWELL T. G. (1975) Geochemical studies related to the occurrence of oil and gas in the Damper sub-basin, Western Australia. *J. Geochem. Explor.* 4, 441-466.

POWELL T. G., COOK P. J. and McKIRDY D. M. (1975) Organic geochemistry of phosphorites: relevance to petroleum genesis. *Amer. Assoc. Petroleum Geologists Bull.* 59, 618-632.

POWELL T. G., FOSCOLOS A. E., GUNTHER P. R. and SNOWDON L. R. (1978) Diagenesis of organic matter and fine clay minerals: a comparative study. *Geochim. Cosmochim. Acta* 42, 1181-1197.

POWELL T. G. and McKIRDY D. M. (1973) The effect of source material, rock type and diagenesis on the *n*-alkane content of sediments. *Geochim. Cosmochim. Acta* 37, 623-633.

POWELL T. G. and McKIRDY D. M. (1975) Geologic factors controlling crude oil composition in Australia and Papua, New Guinea. *Amer. Assoc. Petroleum Geologists Bull.* 59, 1176-1197.

PRICE L. C. (1976) Aqueous solubility of petroleum as applied to its origin and primary migration. *Amer. Assoc. Petroleum Geologists Bull.* 60, 213-244.

QUIN B. F. and BROOKS R. R. (1975) The rapid colorimetric determination of molybdenum with dithiol in biological, geochemical and steel samples. *Anal. Chim. Acta* 74, 75-84.

QUIN B. F., BROOKS R. R., BOSWELL C. R. and PAINTER J. A. C. (1974) Biogeochemical exploration for tungsten at Barrytown, New Zealand. *J. Geochem. Explor.* 3, 43-51.

RASHID M. A. (1979) Pristane-phytane ratios in relation to source and diagenesis of ancient sediments from the Labrador Shelf. *Chem. Geol.* 25, 109-122.

RATTIGAN J. H., GERSTELING R. W. and TONKIN D. G. (1977) Exploration geochemistry of the Stuart Shelf, South Australia. *J. Geochem. Explor.* 8, 203-217.

REED W. E. and KAPLAN I. R. (1977) The chemistry of marine petroleum seeps. *J. Geochem. Explor.* (Special Issue: Application of Geochemistry to the Search for Crude Oil and Natural Gas; B. Hitchon, Compiler and Editor) 7, 255-293.

REEVES R. D. and BROOKS R. R. (1978) *Trace Element Analysis of Geological Materials.* Wiley.

REIMER G. M. (1975) Uranium determination in natural water by fission-track technique. *J. Geochem. Explor.* 4, 425-431.

REIMER, G. M. (1976) Design and assembly of a portable helium detector for evaluation as a uranium exploration instrument. *U. S. Geol. Surv.* Open-File Rept. 76-398.

REIMER G. M., DENTON E. H., FRIEDMAN I. and OTTON J. K. (1979) Recent developments in uranium exploration using the U.S. Geological Survey's mobile helium detector. *J. Geochem. Explor.* 11, 1-12.

REITSEMA R. H., LINDBERG F. A. and KALTENBACK A. J. (1978) Light hydrocarbons in Gulf of Mexico water: sources and relation to structural highs. *J. Geochem. Explor.* 10, 139-151.

RICH R. A., HOLLAND H. D. and PETERSEN U. (1977) *Hydrothermal Uranium Deposits.* Elsevier.

ROBBINS J. C. (1978) Field technique for the measurement of uranium in natural waters. *CIM Bull.* 71 (No. 793; May), pp. 61-67.

ROBERTS T. M., HUTCHINSON T. C., PACIGA J., CHATTOPADHYAY A., JEVIS R. E., VANLOON J. and PARKINSON D. K. (1974) Lead contamination around secondary smelters: Estimation of dispersal and accumulation by humans. *Science* 186, 1120-1123.

ROEDDER E. (1977) Fluid inclusions as tools in mineral exploration. *Econ. Geol.* 72, 503-525.

ROGERS J. J. W. and ADAMS J. A. S. (1969) Uranium. In, *Handbook of Geochemistry,* II-1 (editor K. H. Wedepohl). Springer-Verlag.

ROGERS M. A., MCALARY J. D. and BAILEY N. J. L. (1974) Significance of reservoir bitumens to thermal-maturation studies, Western Canada basin. *Amer. Assoc. Petroleum Geologists Bull.* 58, 1806-1824.

ROMBERGER S. B. (1979) personal communication.

ROSE A. W. (1975) The mode of occurrence of trace elements in soils and stream sediments applied to geochemical exploration. In, *Geochemical Exploration 1974* (editors I. L. Elliott and W. K. Fletcher), pp. 691-705. Elsevier.

ROSE A. W., HAWKES H. E. and WEBB J. S. (1979) *Geochemistry In Mineral Exploration* (second edition). Academic Press.

ROSE A. W. and KEITH M. L. (1976) Reconnaissance geochemical techniques for detecting uranium deposits in sandstones of northeastern Pennsylvania. *J. Geochem. Explor.* 6, 119-137.

RUBINSTEIN I., STRAUSZ O. P., SPYCKERELLE C., CRAWFORD R. J. and WESTLAKE D. W. S. (1977) The origin of oil sand bitumens of Alberta: a chemical and microbiological simulation study. *Geochim. Cosmochim. Acta* 41, 1341-1353.

RUSSELL M. J. (1974) Manganese halo surrounding the Tynagh ore deposit, Ireland: a preliminary note. *Trans. IMM* (Sect. B, Appl. Earth Sci.) 83, B65-B66.

RUSSELL M. J. (1975) Lithogeochemical environment of the Tynagh base-metal deposit, Ireland, and its bearing on ore deposition. *Trans. IMM* (Sect. B, Appl. Earth Sci) 84, B128-B133.

RYALL W. R. (1977) Anomalous trace elements in pyrite in the vicinity of mineralized zones at Woodlawn, N.S.W., Australia. *J. Geochem. Explor.* 8, 73-83.

SACKETT W. M. (1977) Use of hydrocarbon sniffing in offshore exploration. *J. Geochem. Explor.* (Special Issue: Application of Geochemistry to the Search for Crude Oil and Natural Gas; B. Hitchon, Compiler and Editor) 7, 243-254.

SATO M. and MOONEY H. M. (1960) The electrochemical mechanism of sulfide self-potentials. *Geophysics* 25, 226-249.

SAXBY J. D. (1977) Oil-generating potential of organic matter in sediments under natural conditions. *J. Geochem. Explor.* 7, 373-382.

SCHNEPFE M. M. (1974) Spectrofluorimetric procedure using 2,3-napthalenediamine for determining selenium in rocks. *J. Res. U.S. Geol. Surv.* 2, 631-636.

SCHOELLER H. (1962) *Les Eaux Souterraines* (in French). Masson. Paris.

SCHULLER R. M., HUNTSMAN B. E., WARNER B. J. and MALONE P. G. (1975) A rapid on-site electrode technique for determining copper in soil for geochemical exploration (abstract). *Program, Geol. Soc. Amer. Ann. Meeting,* Salt Lake City, Utah.

SCOTT M. J. (1975) Case histories from a geochemical exploration programme — Windhoek district, South-West Africa. In *Geochemical Exploration 1974* (editors I. L. Elliott and W. K. Fletcher), pp. 481-492.

SECCOMBE P. K. (1977) Sulphur isotope and trace metal composition of stratiform sulphides as an ore guide in the Canadian Shield. *J. Geochem. Explor.* 8, 117-137.

SEIFERT W. K. (1978) Steranes and terpanes in kerogen pyrolysis for correlation of oils and source rocks. *Geochim. Cosmochim. Acta* 42, 473-484.

SEIFERT W. K. and MOLDOWAN J. M. (1978) Applications of steranes, terpanes and monoaromatics to the maturation, migration and source of crude oils. *Geochim. Cosmochim. Acta* 42, 77-95.

SEIFERT W. K. and MOLDOWAN J. M. (1979) The effect of biodegradation on steranes and terpanes in crude oils. *Geochim. Cosmochim. Acta* 43, 111-126.

SEVERNE B. C. (1978) Evaluation of radon systems at Yeelirrie, Western Australia. *J. Geochem. Explor.* 9, 1-22.

SEVERNE B. C. and BROOKS R. R. (1972) A nickel-accumulating plant from Western Australia. *Planta* 103, 91-94.

SHACKLETTE H. T. and CONNOR J. J. (1973) Airborne chemical elements in Spanish moss. *U.S. Geol. Surv. Prof. Paper* 574-E.

SHAPIRO L. (1973) Rapid determination of sulfur in rocks. *J. Res. U. S. Geol. Surv.* 1, 81-84.

SHAPIRO L. (1975) Rapid analysis of silicate, carbonate and phosphate rocks — revised edition. *U.S. Geol. Surv. Bull.* 1401.

SHERATON J. W. and BLACK L. P. (1973) Geochemistry of mineralized granitic rocks of northeast Queensland. *J. Geochem. Explor.* 2, 331-348.

SHIBAOKA M., SAXBY J. D. and TAYLOR G. H. (1978) Hydrocarbon generation in Gippsland Basin, Australia — comparison with Cooper Basin, Australia. *Amer. Assoc. Petroleum Geologists Bull.* 62, 1151-1158.

SHILTS W. W. (1976) Glacial till and mineral exploration. In, *Glacial Till — An Inter-Disciplinary Study* (editor, R. F. Legget), pp. 205-224. Royal Soc. Canada, Spec. Public. No. 12.

SHILTS W. W. (1977) Geochemistry of till in perennially frozen terrain of the Canadian Shield — application to prospecting. *Boreas* 6, 203-212.

SHIPULIN F. K. and eleven others (1973) Some aspects of the problem of geochemical methods of prospecting for concealed mineralization. *J. Geochem. Explor.* 2, 193-235.

SHVARTSEV S. L. (1972) Hydrogeochemical prospecting for blind ores in permafrost. *Internat. Geol. Rev.* 14, 1037-1043.

SINCLAIR A. J. (1974) Selection of threshold values in geochemical data using probability graphs. *J. Geochem. Explor.* 3, 129-149.

SINCLAIR A. J. (1975) Some considerations regarding grid orientation and sample spacing. In, *Geochemical Exploration 1974* (editors I. L. Elliott and W. K. Fletcher), pp. 133-140.

SINCLAIR A. J. (1976) *Applications of Probability Graphs In Mineral Exploration.* Spec. Vol. 4 Assoc. of Explor. Geochemists.

SINDING-LARSEN R. (1975) A computer method for dividing a regional geochemical survey area into homogeneous subareas prior to statistical interpretation. *Geochemical Exploration 1974* (editors I. L. Elliott and W. K. Fletcher), pp. 191-217. Elsevier.

SLY P. G. (1969) Bottom sediment sampling. In, *Proc. Twelfth Conf. Great Lakes Res.,* pp. 883-898. Intern. Assoc. Great Lakes Research.

SMITH A. Y., BARRETTO P. M. C. and POURNIS S. (1976) Radon methods in uranium exploration. In, *Exploration for Uranium Ore Deposits*, pp. 185-211. STI/PUB/434. Intern. Atomic Agency. Vienna.

SMITH B. H. (1977) Some aspects of the use of geochemistry in the search for nickel sulphides in lateritic terrain in Western Australia. *J. Geochem. Explor.* 8, 259-281.

SMITH W. L. (1977) (editor) *Remote-Sensing Applications for Mineral Exploration*, Halstead Press.

SNOWDON L. R. and ROY K. J. (1975) Regional organic metamorphism in the Mesozoic strata of the Sverdrup basin. *Bull. Canadian Petroleum Geol.* 23, 131-148.

STAHL W. J. (1977) Carbon and nitrogen isotopes in hydrocarbon research and exploration. *Chem. Geol.* 20, 121-149.

STAHL W. J. (1978) Source rock - crude oil correlation by isotopic type-curves. *Geochim. Cosmochim. Acta* 42, 1573-1577.

STAHL W. J. and CAREY B. D. (1975) Source-rock identification by isotope analyses of natural gases from fields in the Val Verde and Delaware Basins, west Texas. *Chem. Geol.* 16, 257-267.

STANTON R. E. (1976) *Analytical Methods For Use In Geochemical Exploration*. Edward Arnold.

STEINMAN D. K., COSTELLO D. G., PEPPER C. S., GALLOWAY R. N. and JOHN J. (1975) ^{252}Cf based borehole logging system for *in-situ* assaying of uranium ore. Abstracts of the 1975 Uranium and Thorium Research and Resources Conf., *U. S. Geol. Surv. Open-file Report* 75-595 pp. 48-49.

STONE D. S. and HOEGER R. L. (1973) Importance of hydrodynamic factor for formation of Lower Cretaceous combination traps, Big Muddy - South Glenrock area, Wyoming. *Amer. Assoc. Petroleum Geologists Bull.* 57, 1714-1733.

STUMM W. and MORGAN J. J. (1970) *Aquatic Chemistry*. Wiley-Interscience.

SZABO N. L., GOVETT G. J. S. and LAJTAI E. Z. (1975) Dispersion trends of elements and indicator pebbles in glacial till around Mt. Pleasant, New Brunswick, Canada. *Can. J. Earth Sci.* 12, 1534-1556.

TALVITIE J. and PAARMA H. (1973) Reconnaissance prospecting by photogeology in northern Finland. *Prospecting In Areas of Glacial Terrain* (editor M. J. Jones), pp. 73-81. IMM. London.

TAN L. P. and YU F. S. (1970) Heavy-mineral reconnaissance for gold and copper deposits of the Chinkuashih area, Taiwan. *United Nations Mineral Resources Dev. Series* No. 38, 162.

TANNER A. B. (1975) Radon migration as applied to prospecting for uranium. In, *Abstracts of the 1975 Uranium and Thorium Research and Resources Conference* (editors L. C. Craig, R. A. Brooks and P. C. Patton). *U.S. Geol. Surv. Open-file Report* 75-595.

TANSKANEN H. (1976) Factors affecting the metal contents in peat profiles. *J. Geochem. Explor.* 5, 412-414.

THOMPSON C. E. and LAKIN H. W. (1957) A field chromatographic method for the determination of uranium in soils and rocks. *U.S. Geol. Surv. Bull.* 1036-L, 209-220.

THOMSON I. and READ D. (1979) The character and application of various chemical analytical techniques in uranium exploration. In, *Geochemical Exploration 1978* (editors J. R. Watterson and P. K. Theobald), pp. 309-316. Assoc. of Explor. Geochemists.

THORNBER M. R. and WILDMAN J. E. (1979) Supergene alteration of sulphides, IV. Laboratory study of the weathering of nickel ores. *Chem. Geol.* 24, 97-110.

TIMPERLEY M. H., BROOKS R. R. and PETERSON P. J. (1970) The significance of essential and non-essential trace elements in plants in relation to biogeochemical prospecting. *J. Appl. Ecol.* 7, 429-439.

TISSOT B., DEROO G. and HOOD A. (1978) Geochemical study of the Uinta Basin: formation of petroleum from the Green River Formation. *Geochim. Cosmochim. Acta* 42, 1469-1485.

TISSOT B. P. and WELTE D. H. (1978) *Petroleum Formation and Occurrence*. Springer-Verlag.

TORGERSEN T. and CLARKE W. B. (1978) Excess helium-4 in Taggau Lake: Possibilities for a uranium ore body. *Science* 199, 769-771.

TOTH J. (1978) Gravity-induced cross-formational flow of formation fluids, Red Earth region, Alberta, Canada: analysis, patterns, and evolution. *Water Resour. Res.* 14, 805-843.

Toth J. (1979) Cross-formational gravity-flow of groundwater: a mechanism of the transport and accumulation of petroleum − or − the generalized hydraulic theory of petroleum migration. *Physical and Chemical Constraints of Migration* (editors W. H. Roberts and R. J. Cordell), Amer. Assoc. Petroleum Geologists (in press).

UNESCO (1971) *Soils And Tropical Weathering.* Proc. of the Bandung Symposium 16 to 23 November, 1969. UNESCO Natural Resources Research XI.

United States Department of Agriculture (1975) *Soil Taxonomy:* A basic system of soil classification for making and interpreting soil surveys. Agriculture Handbook No. 436.

Van Loon J. C., Kesler S. E. and Moore C. M. (1973) Analysis of water-extractable chloride in rocks by use of a selective ion electrode. In, *Geochemical Exploration 1972* (editor M. J. Jones), pp. 429-434. IMM.

Van Trump G. and Miesch A. T. (1976) The U.S. Geological Survey RASS-STATPAC system for management and statistical reduction of geochemical data. *Comput. and Geosci.* 3, 475-488.

Venkatavaradan V. S. (1973) Geochemical reconnaissance with human hair. *Curr. Sci.* (India), 42, 477.

Viljoen R. P., Viljoen M.J., Grootenboer J. and Longshaw T. G. (1975) ERTS-1 imagery: An appraisal of applications in geology and mineral exploration. *Minerals Sci. Eng.* 7, 132-168.

Vine J. D. and Tourtelot E. B. (1970) Geochemistry of black shale deposits − A summary report. *Econ. Geol.* 65, 253-272.

Wagner G. H., König R. H. and Steele K. F. (1978) Stream sediment geochemical investigations − comparisons of manganese, zinc and lead-zinc districts with an unmineralized area. *J. Geochem. Explor.* 9, 63-74.

Walthall F. G. (1974) Spectrochemical computer analysis − program description. *J. Res. U.S. Geol. Surv.* 2, 61-71.

Waples D. W. (1979) Simple method for oil source bed evaluation. *Amer. Assoc. Petroleum Geologists Bull.* 63, 239-248.

Ward F. N. (1975) New and refined methods of trace analysis useful in geochemical exploration. *U.S. Geol. Surv. Bull.* 1408.

Ward F. H. and Bondar W. F. (1979) Analytical methodology in the search for metallic ores. In, *Geophysics and Geochemistry in the Search for Metallic Ores* (editor P. J. Hood), *Geol. Surv. Canada, Econ. Geol. Report* 31.

Ward W. J., Fleischer R. L. and Mogro-Campero A. (1977) Barrier technique for separate measurement of radon isotopes. *Rev. Sci. Instrum.* 48, 1440-1441.

Washburn A. L. (1973) *Periglacial Processes And Environments.* Edward Arnold.

Watson A. E. and Russell G. M. (1974) The analysis of geological samples for trace elements by direct-reading emission spectrometry. *Dept. Nat. Inst. Metall., South Africa,* No. 1656.

Watterson J. R., Gott G. B. Neuerberg G. J., Lakin H. W. and Cathrall J. B. (1977) Tellurium, a guide to mineral deposits. *J. Geochem. Explor.* 8, 31-48.

Watterson J. R. and Neuerburg G. J. (1975) Analysis for tellurium in rocks to 5 parts per billion. *J. Res. U.S. Geol. Surv.* 3, 191-195.

Watterson J. R. and Theobald P. K. (1979) (editors) *Geochemical Exploration 1978.* Association of Exploration Geochemists.

Webb J. S. (1973) Applied geochemistry and the community. *Trans. IMM* (Sect. B) 82, 23-28.

Webb J. S. (1975) Environmental problems and the exploration geochemist. In, *Geochemical Exploration 1974* (editors I. L. Elliott and W. K. Fletcher), pp. 5-17.

Webb J. S., Nichol I., Foster R., Lowenstein P. L. and Howarth R. J. (1973) Provisional geochemical atlas of northern Ireland. *Appl. Geochem. Res. Grp., Imp. Coll. Sci. and Tech.,* London.

Webb J. S., Thornton I., Thompson M., Howarth R. J. and Lowenstein P. L. (1978) *The Wolfson Geochemical Atlas of England and Wales.* Oxford Univ. Press.

WEBBER G. R. (1975) Efficacity of electrochemical mechanisms for ion transport in the formation of geochemical anomalies. *J. Geochem. Explor.* 4, 231-233.

WEDEPOHL K. H. (1969-1974) (editor) *Handbook of Geochemistry* (vols. I and II). Springer-Verlag.

WEISS O. (1971) Airborne geochemical prospecting. *Geochemical Exploration.* CIM Spec. Vol. 11, 502-514.

WHITEHEAD N. E., BROOKS R. R. and COOTE G. E. (1971) Gamma radiation of some plants and soils from a uraniferous area in New Zealand. *New Zealand J. Sci.* 14, 66-76.

WHITEHEAD N. E., BROOKS R. R. and PETERSON P. J. (1971a) The nature of uranium occurrences in the leaves of *Coprosma australis* (A. Rich.) Robinson. *Austral. J. Biol. Sci.* 24, 67-73.

WHITNEY P. R. (1975) Relationship of manganese-iron oxides and associated heavy metals to grain size in stream sediments. *J. Geochem. Explor.* 4, 251-263.

WILLEY L. M., KHARAKA Y. K., PRESSER T. S., RAPP J. B. and BARNES I. (1975) Short chain aliphatic acid anions in oil field waters and their contribution to the measured alkalinity. *Geochim. Cosmochim. Acta* 39, 1707-1711.

WILLIAMS J. A. (1974) Characterization of oil types in Williston Basin. *Amer. Assoc. Petroleum Geologists Bull.* 58, 1243-1252.

WILLIAMS R. S. and CARTER W. D. (1976) (editors) ERTS-1. A new window on our planet. *U. S. Geol. Surv. Prof. Paper* 929.

WILLIAMS S. A. and CESBRON F. P. (1977) Rutile and apatite: Useful prospecting guides for porphyry copper deposits. *Mineral Mag.* 41, 288-292.

WILMSHURST J. R. (1975) The weathering products of nickeliferous sulfides and their associated rocks in Western Australia. *Geochemical Exploration 1974* (editors I. L. Elliott and W. K. Fletcher), pp. 417-436. Elsevier.

WITHER E. D. and BROOKS R. R. (1977) Hyperaccumulation of nickel by some plants of Southeast Asia. *J. Geochem. Explor.* 8, 579-583.

WOGMAN N. A., BRODZINSKI R. L. and MIDDLESWORTH L. V. (1976) Radium accumulation in animal thyroid glands — a possible method of uranium and thorium prospecting. *Rept. BNWL-SA-5560.* Available from National Technical Information Service, U.S. Dept. of Commerce, Springfield, VA. 22161. Cited in Hawkes (1977).

WOLFE W. J. (1974) Geochemical and biogeochemical exploration research near Early Precambrian porphyry-type molybdenum-copper mineralization, northwestern Ontario, Canada. *J. Geochem. Explor.* 3, 25-41.

WOLL P. W. and FISCHER W. A. (1977) (editors) Proceedings of the First Annual William T. Pecora Memorial Symposium, October 1975, Sioux Falls, South Dakota. *U.S. Geol. Surv. Prof. Paper* 1015.

WOLLENBERG H. A. (1973) Fission-track radiography of uranium and thorium in radioactive minerals. In, *Geochemical Exploration 1972* (editor M. J. Jones), pp. 347-358. IMM.

WU I. J. and MAHAFFEY E. J. (1979) Mercury in soils geochemistry over massive sulfide deposits in Arizona. In, *Geochemical Exploration 1978* (editors J. R. Watterson and P. K. Theobald), pp. 201-208. Assoc. of Explor. Geochemists.

YARIV S. and CROSS H. (1979) *Geochemistry of Colloid Systems* (for Earth Scientists). Springer-Verlag.

YATES T. E., BROOKS R. R. and BOSWELL C. R. (1974) Biogeochemical exploration at Coppermine Island, New Zealand. *New Zealand J. Sci.* 17, 151-159.

YEN T. F. and SPRANG S. R. (1977) Contribution of E. S. R. analysis toward diagenetic mechanisms in bituminous deposits. *Geochim. Cosmochim. Acta* 41, 1007-1018.

YOUNG A., MONAGHAN P. H. and SCHWEISBERGER R. T. (1977) Calculation of ages of hydrocarbons in oils — physical chemistry applied to petroleum geochemistry. *Amer. Assoc. Petroleum Geologists Bull.* 61, 573-600.

YOUNG D. K. and YEN T. F. (1977) The nature of straight-chain aliphatic structures in Green River kerogen. *Geochim. Cosmochim. Acta* 41, 1411-1417.

ZAJIC J. E. (1969) *Microbial Biogeochemistry.* Academic Press.

ZUMBERGE J. H. and NELSON C. A. (1976) *Elements of Physical Geology.* Wiley.

Index

Note: This index only contains entries for *"The 1980 Supplement"* starting on page 615 (except for the inclusion of a few important entries inadvertently omitted from the index to the First Edition)